David Blume's
ALCOHOL CAN BE A GAS!

Fueling an Ethanol Revolution for the 21st Century

With a Foreword by **R. Buckminster Fuller**

Edited by **Michael Winks**

 iiEA The International Institute for Ecological Agriculture

Santa Cruz, California • permaculture.com

First printing 4500 October 2007
Second printing 5000 December 2007

ISBN 9780979043789 (hardbound);
9780979043772 (paperback);
LCCN 2007926369

To sign up for our mailing list, visit <www.permaculture.com>.
To contact us by email, please send to <info@permaculture.com>.

ATTENTION ORGANIZATIONS: Quantity discounts may be available on bulk purchases of this book for educational or gift purposes, or as premiums. Special books or book excerpts can also be created to fit specific needs. For information, please contact the International Institute for Ecological Agriculture, 309 Cedar Street #127, Santa Cruz, CA 95060 USA.

 Printed on 30% post-consumer recycled paper.

COVER BY RON HARPER

PRAISE FOR *ALCOHOL CAN BE A GAS!...*

"David Blume's Alcohol Can Be a Gas! *is the most comprehensive and understandable book on renewable fuels ever compiled. Over a quarter century in the making, the book explains the history, technology, and even the sociology of renewable fuels in a fashion that can be appreciated by the most accomplished in the ethanol and biodiesel fields, as well as the novice and young students of the issues. You will laugh out loud at his sharp wit and the dozens of cartoons. When you finish reading Dave's book, you will have a much better understanding of how our nation's energy policy evolved, why it is what it is today, and what needs to be done for the future. I have worked in the renewable energy sector in one form or another for close to four decades, and I can recommend* Alcohol Can Be a Gas! *as the best book I have ever read on the subject."*

—LARRY MITCHELL, CHIEF EXECUTIVE OFFICER, AMERICAN CORN GROWERS ASSOCIATION

"Humanity has used up roughly half of the world's oil and topsoil. Just in time, David Blume has given us Alcohol Can be a Gas! *It's a practical road map for supplying all of our energy needs without drilling, strip-mining, and/or depleting the soil. In fact, following Blume's model, soil fertility would actually increase worldwide; energy production would be not only sustainable, but democratic— and highly profitable on the small scale. This is a brilliant visionary work. And, with Mr. Blume's witty personality, reading it is certainly a gas."*

—LARRY KORN, SOIL SCIENTIST, TRANSLATOR, AND EDITOR OF *THE ONE-STRAW REVOLUTION: AN INTRODUCTION TO NATURAL FARMING*

"As intersections of the food-energy-climate matrix form in Iowa cornfields, Amazonian rain forests, and Canadian gene-splicing labs, and as endgame battles for their control pit theocratic flat-worlders against biologists, climatologists, and tree-huggers over the very survival of life on Earth, David Blume emerges like a wizard on a misty pinnacle, backlit by the full moon, revealing a gemstone in his extended palm."

—ALBERT BATES, AUTHOR OF *THE POST-PETROLEUM SURVIVAL GUIDE AND COOKBOOK: RECIPES FOR CHANGING TIMES*

"Brilliant! This book should be on the reading list of every American!!"

—THOM HARTMANN, *NEW YORK TIMES* BESTSELLING AUTHOR AND NATIONALLY SYNDICATED HOST OF *THE THOM HARTMANN PROGRAM* ON AIR AMERICA

...MORE PRAISE FOR *ALCOHOL CAN BE A GAS!*

"Dave Blume has written the definitive opus on alcohol as a fuel. From the 30,000-foot view to the most minute technical detail, Alcohol Can be a Gas! makes a strong case for the practical, ecological, political, and economic sense in converting to ethanol. It's heartening to see the world's original 'alcohol pioneer' stay abreast of the times with a book that has the promise to knock some sense into our insidious fossil-fueled economy. This book is much needed in this era of Peak Oil and fast-accelerating climate change."

—JOHN SCHAEFFER, PRESIDENT AND FOUNDER OF REAL GOODS, AND EXECUTIVE DIRECTOR OF THE INSTITUTE FOR SOLAR LIVING

"What a tour de force! This is the most comprehensive and authoritative guide through all the controversy about ethanol as transportation fuel, showing it as a clear winner in the quest for solutions to our environmental and geopolitical problems. Engagingly written, full of important and amazing information and resources, this book meets every challenge to the vision for a clean, democratic path to a prosperous future for all."

—JOE JORDAN, ATMOSPHERIC RESEARCHER, NASA/AMES RESEARCH CENTER, SETI INSTITUTE, AND CABRILLO COLLEGE

"Finally, an alcohol book for the layman and backyard enthusiast. In our culture's collective, industrialized love affair with mega-everything, Blume cuts across the government-subsidized factories with ecologically practical models. Here is a viable energy system that can be embedded in a region, linking rural producers to urban users of energy and food. Self-reliance and resiliency follow community-based alcohol production, and we all owe a debt of gratitude to Blume for codifying his life's passion in what is a veritable compendium of information."

—JOEL SALATIN, FARMER, AND AUTHOR OF *YOU CAN FARM* AND *EVERYTHING I WANT TO DO IS ILLEGAL*

"Ethanol champion David Blume has completed his opus, Alcohol Can Be a Gas! It is a great read. The history of petroleum, history of alcohol, technical coverage of production process, vehicle development (conversion), and feedstocks—it's all in the text, complete with charts and pictures. David's wit, wisdom, and hardcore experience illuminate this biofuel's potential. We have eagerly awaited this publication and will use it in our Sustainable Transportation and Biofuels courses."

—DR. JACK MARTIN, APPROPRIATE TECHNOLOGY PROGRAM, APPALACHIAN STATE UNIVERSITY; VICE-CHAIR OF RENEWABLE FUELS AND TRANSPORTATION DIVISION, AMERICAN SOLAR ENERGY SOCIETY

... MORE PRAISE FOR *ALCOHOL CAN BE A GAS!*

Books for Wiser Living from *MOTHER EARTH NEWS*

Today, more than ever before, our society is looking for ways to live more conscientiously. To help bring you the very best inspiration and information about greener, more sustainable lifestyles, we've joined forces with *MOTHER EARTH NEWS*. For more than 35 years, *MOTHER EARTH NEWS* has been North America's original guide to living wisely, creating books and magazines for people with a passion for self-reliance and a desire to live in harmony with nature. Across the countryside and in our cities, *MOTHER EARTH NEWS* is leading the way to a wiser, more sustainable world.

ALCOHOL CAN BE A GAS!
THE TWO-MINUTE SUMMARY

1. With alcohol fuel, almost every country can become energy-independent. Anywhere that has sunlight and land can produce alcohol from plants. Brazil, the fifth largest country in the world, imports no oil, since half their cars run on alcohol fuel made from sugarcane, grown on 1% of its land.

2. We can reverse global warming. Since alcohol is made from plants, its production takes carbon dioxide out of the air, sequestering it, with the result that it reverses the greenhouse effect (while potentially vastly improving the soil). Recent studies show that in a permaculturally designed mixed-crop alcohol fuel production system, the amount of greenhouse gases removed from the atmosphere by plants—and then exuded by plant roots into the soil as sugar—can be 13 times what is emitted by processing the crops and burning the alcohol in our cars.

3. We can revitalize the economy instead of suffering through Peak Oil. Oil is running out, and what we replace it with will make a big difference in our environment and economy. Alcohol fuel production and use is clean and environmentally sustainable, and will revitalize families, farms, towns, cities, industries, as well as the environment. A national switch to alcohol fuel would provide many millions of new permanent jobs.

4. No new technological breakthroughs are needed. We can make alcohol fuel out of what we have, where we are. Alcohol fuel can efficiently be made out of many things, from waste products like stale donuts, grass clippings, food processing waste—even ocean kelp. Many crops produce many times more alcohol per acre than corn, using arid, marshy, or even marginal land in addition to farmland. Just our lawn clippings could replace a third of the auto fuel we get from the Mideast.

WITH ALCOHOL FUEL, YOU CAN BECOME ENERGY-INDEPENDENT, REVERSE GLOBAL WARMING, AND SURVIVE PEAK OIL IN STYLE. ALCOHOL FUEL IS "LIQUID SUNSHINE" AND CAN'T BE CONTROLLED BY TRANSNATIONAL CORPORATIONS. YOU CAN PRODUCE ALCOHOL FROM A WIDE VARIETY OF PLANTS AND WASTE PRODUCTS, FROM ALGAE TO STALE DONUTS. IT'S A MUCH BETTER FUEL THAN GASOLINE, AND YOU CAN USE IT IN YOUR CAR, RIGHT NOW. YOU CAN EVEN USE ALCOHOL TO GENERATE YOUR ELECTRICITY. ALCOHOL FUEL PRODUCTION IS ECOLOGICALLY SUSTAINABLE, REVITALIZES FARMS AND COMMUNITIES, AND CREATES HUGE NEW OPPORTUNITIES FOR SMALL-SCALE BUSINESSES. ITS BYPRODUCTS ARE CLEAN AND VALUABLE. ALCOHOL HAS A PROUD HISTORY AND A VITAL FUTURE.

5. Unlike hydrogen fuel cells, we can easily use alcohol fuel in the vehicles we already own. Unmodified cars can run on 50% alcohol, and converting to 100% alcohol or flexible-fueling (both alcohol and gas) costs only a few hundred dollars. Most auto companies already sell new dual-fuel vehicles.

6. Alcohol is a superior fuel to gasoline! It's 105 octane, burns much cooler with less vibration, is less flammable in case of accident, is 98% pollution-free, has lower evaporative emissions, and deposits no carbon in the engine or oil, resulting in a tripling of engine life. Specialized alcohol engines can get at least 22% better mileage than gasoline or diesel.

7. It's not just for gasoline cars. We can easily use alcohol fuel to power diesel engines, trains, aircraft, small utility engines, generators to make electricity, heaters for our homes—and it can even be used to cook our food.

8. Alcohol has a proud, solid history. Gasoline is a refinery's toxic waste; alcohol fuel is liquid sunshine. Henry Ford's early cars were all flex-fuel. It wasn't until gasoline magnate John D. Rockefeller funded Prohibition that alcohol fuel companies were driven out of business.

9. The byproducts of alcohol production, instead of being oil refinery waste, are clean and are worth more than the alcohol itself. In fact, they can make petrochemical fertilizers and herbicides obsolete. The alcohol production process concentrates and makes more digestible all protein and non-starch nutrients in the crop. This "byproduct" is so nutritious that when used as animal feed, it produces more meat or milk than the corn it comes from. That's right, fermentation of corn increases the food supply *and* lowers the cost of food.

10. Locally produced ethanol supercharges regional economies. Instead of fuel expenditures draining capital away to foreign bank accounts, each gallon of alcohol produces local income that gets recirculated many times. Every dollar of tax credit for alcohol generates up to $6 in new tax revenues from the increased local business.

11. Alcohol production brings many new small-scale business opportunities. There is huge potential for profitable local, integrated, small-scale businesses that produce alcohol and related byproducts, whereas when gas was cheap, alcohol plants had to be huge to make a profit.

12. Scale matters—most of the widely publicized potential problems with ethanol are a function of scale. Once production plants get beyond a certain size and are too far away from the crops that supply them, closing the ecological loop becomes problematic. Smaller-scale operations can more efficiently use a wide variety of crops than huge specialized one-crop plants, and diversification of crops would largely eliminate the problems of monoculture.

13. The byproducts of small-scale alcohol plants can be used in profitable, energy-efficient, and environmentally positive ways. For instance, spent mash (the liquid left over after distillation) contains all the nutrients the next fuel crop needs and can return it back to the soil if the fields are close to the operation. Big-scale plants, because they bring in crops from up to 45 miles away, can't do this, so they have to evaporate all the water (and sell the resulting byproduct as low-price animal feed), which accounts for half the energy used in the plant.

14. The design and implementation of a revolution based on small-scale ethanol production is simple common sense, and at the same time a radical departure from the way corporations and the government currently do things. So it's truly up to us citizens to make the change. You will see early on in this book that MegaOilron historically stops at nothing to make sure that the public perception of ethanol is tightly managed. But once you know the truth, you can't be swayed by their propaganda. It's time we share what we know, organize to bring it about, and win.

DEDICATION

First of all, I want to recognize my parents, Gerry and Louise Blume, who taught me three things: Always question what the government tells you; never bend your knee to anyone or anything; and never run away from or ignore a problem (or someone in trouble.) In return for their efforts on behalf of democracy and civil rights, the IRS audited them five years in a row, paying them a refund each time.

There are many people I consider heroes for their bravery. Some of my inspiration comes from Karen Silkwood, a union organizer at a Kerr-McGee nuclear facility, and Ben Linder, an engineer who brought hydroelectricity to El Cua, Nicaragua—two ordinary people who did remarkably brave work and were murdered for it.

To the Sandinistas of Nicaragua, who taught me that a small band of people willing to do whatever was necessary could free a nation from a dictatorship sponsored by the most powerful government on Earth—and then, in the aftermath of victory, pardon the very same men who had tortured them for decades.

I draw special inspiration from two women who speak powerfully for the Earth and for those who work it: Arundhati Roy and Vandana Shiva.

As a farmer, I am justifiably proud of my brethren in India who, when confronted with a Cargill warehouse poised to sell proprietary hybrid seed, descended en masse, damaged the building, shut down the site, and went to jail to fight against the

I AM MOVED BY MY FELLOW PERMACULTURISTS ALL OVER THE PLANET, WHO CONSTANTLY DEMONSTRATE THAT THERE IS NO NEED FOR DESPAIR WHEN IT COMES TO THE WORK OF HEALING THE WORLD. IN FACT, TASKS THAT SEEM INTIMIDATING TO OTHERS ARE MERELY A FINE EXCUSE TO DESIGN AND EXECUTE PARTICULARLY DELIGHTFUL SOLUTIONS.

subtle slavery of patented seed. A more eloquent statement of truth to power, I cannot imagine.

I am also moved by my fellow permaculturists all over the planet, who constantly demonstrate that there is no need for despair when it comes to the work of healing the world. In fact, tasks that seem intimidating to others are merely a fine excuse to design and execute particularly delightful solutions.

I would also like to recognize the sacrifices made by my fellow alcohol fuel pioneers in the informal group known (only to themselves) as the OFECKERS.

And, most of all, I wish to dedicate this book to R. Buckminster Fuller, who was denied the satisfaction of seeing it in print, but who continued to encourage me in spirit for decades.

WARNING! DISCLAIMER

AND NOW A WORD FROM THE LAWYERS...

This book is designed to provide the reader with a wide range of information about the history, production, and use of alcohol fuel. It must be understood that in the publishing of this book, the publisher and author are not engaged in rendering legal, accounting, engineering, or other professional services or advice. In pursuing any number of interests or business ventures that may be stimulated by reading this volume, you should always consult the appropriate experts to advise you and not rely solely on what you read here.

Every effort has been made to ensure that the information herein is as thorough and accurate as possible. But in traversing such a breadth of subject matter, there may be mistakes in either content or typography. So use this manual as a starting point in your investigation of any of the subjects discussed within, and educate yourself fully from multiple sources.

Information and technology are always in a constant state of change, and what may have been up-to-date when this book was written may have changed by the time you buy it. Always seek to verify prices quoted herein, as they change according to the market and the area in which you live.

While some of the things described in this book may not be technically legal where you live, or some suggested actions may have uncertain legal ramifications, neither the author nor publisher advocates your breaking any laws.

DANGER

Making alcohol, although it has been done for thousands of years, is essentially an industrial process and has potential risks for injury or even

MAKING ALCOHOL, ALTHOUGH IT HAS BEEN DONE FOR THOUSANDS OF YEARS, IS ESSENTIALLY AN INDUSTRIAL PROCESS AND HAS POTENTIAL RISKS FOR INJURY OR EVEN DEATH—EVEN IF YOU DO EVERYTHING CORRECTLY.

death—even if you do everything correctly. You may be dealing with hand tools, power tools, welding equipment, farm equipment, earth-moving equipment, material-handling equipment, logging equipment, motorized vehicles, hot surfaces, laboratory equipment, and other devices that have the potential to harm or kill the operator or others. You will also be handling potentially hazardous acidic or alkaline materials, flammable or hot liquids, and possibly other chemicals. If you work on your vehicle, there are countless ways to cause property damage, injuries to people, or damage to your vehicle.

So, if you are not qualified to perform any of the tasks in this book, you should either become qualified or get help from someone who is. *You* need to be responsible for your safety in undertaking anything to do with the subjects discussed herein.

The author and publisher of this book shall have no liability nor responsibility to any person or entity with respect to any loss or damage caused, or alleged to have been caused, directly or indirectly, by the information contained herein. The purpose of the book is to educate and entertain the reader.

By purchasing this book, you agree to be responsible for yourself as described above. *If you do not agree to accept this responsibility, you may return this book to the publisher for a full refund.*

ACKNOWLEDGMENTS

Alcohol Can Be a Gas! would not be in your hands if not for the contributions, help, cooperation, and peer review of scores of colleagues from around the globe. In covering this wide breadth of material, I have relied on the work of so many others.

It isn't possible to provide space for full citations of all the references I've used in research. I have chosen to cite, in endnotes, the references that I thought would be most important to the reader. Every effort has been made to have peers review and check all the facts in this manuscript; any errors are my responsibility alone.

In the four years I spent assembling this book, I was supported by the financial contributions of nearly three dozen primary funders and the hundreds of members of Alcoholics Unanimous. Since not a penny was provided by corporations or other institutions, I had the luxury few authors enjoy: the ability to explore and say what I thought should be said, without fear of loss of a patron or publisher. For that, I am deeply grateful.

I can't say enough about Michael Winks and Gayla Groom, who made invaluable contributions to my work. Mike was in on the book from the get-go, supplying numerous research materials and support. I owe him a lot for his unflagging support and enthusiasm. As the manuscript's first editor, Mike winnowed my mammoth work to a more manageable size (yes, this book used to be a lot longer!). Gayla then did a great job of wrestling the book into shape, with both editing and layout.

The book owes much of its crispness to Laurie Masters, who runs Precision Revision in the San Francisco Bay Area. Her motto is "I turn what you wrote into what you meant," and the truth of that is evident throughout these pages. She not only made many insightful copyediting changes, she also proofread every word.

Wendy Stegall did a super job indexing the book, so that you have a detailed map at your disposal to find darned near anything.

SINCE NOT A PENNY WAS PROVIDED BY CORPORATIONS OR OTHER INSTITUTIONS, I HAD THE LUXURY TO EXPLORE AND SAY WHAT I THOUGHT SHOULD BE SAID.

Graphic artist Ron Harper, who had laid out the original *Alcohol Can Be a Gas!* in 1982, was on board again with this new book. He did the cover artwork and the lovely drawings between the sections. With his ability to see things sideways, he was also able to look at what I had written and create humorous and instructive illustrations and cartoons. His whimsical, illuminating work speaks for itself.

A special thanks to the cartoonists: Ken Alexander, Khalil Bendib, Phil Frank, Mike Keefe, Guemsey Le Pelley, Dan O'Neill, Mike Peters, Bill Schorr, Andy Singer, and Matt Wuerker. All of the cartoons scattered throughout the book were donated by the artists, as were all the photographs, tables, and illustrations not created specifically for this book.

Photographers Matt Farruggio and Bob Fitch deserve special thanks for doing such a fine job of illustrating concepts, tools, processes, and people with their cameras, bringing an extra dimension to the book.

The following friends and allies provided critical moral support when things looked dark: Joe Chernick, Paul Gaylon, Lea Karlssen, Margaret Koster, Eric Lang, Paul Robbins, Gray Shaw, Loralyn Shepard, Jay Trudeau, and Steve Zeifman. Many, many thanks; I needed the TLC.

Although many allies have been there for this project right when I needed them, special thanks need to go to the Second Foundation in Summertown, Tennessee, Eugene Scott, Bernd Schaefers, Rich Martini, Scott Fitzmorris, Press Maycock, Vivek Chandra, and Molly Armour, for coming though with loans in the final stages of the book, and especially right when it was time to go to press.

I also want to thank all the people who provided aid and assistance in the research and financing, and in so many other ways. If I've left anyone off this list, I apologize in advance.

Salvado Aceves, Lawrence Livermore National Laboratory; Airborne Studios; Mary Andrew Montgomery, the Botanical Center; David Ange; Joyce Anne; Meredith Appy, Factor Mutual Engineering; Richard Archer; Deborah Baker; Shirley Ball, W.I.F.E.; Albert Bates; Buck Blalock, Nebraska Energy Commission; Dr. J.P Barford, University of Sydney, Australia; Dr. Bemis, University of Arizona; Tom Bender; Khalil Bendib; Hal Bernton; Sarah Bianco; Morris Bitzer, National Sweet Sorghum Producers; Carson Blanton; Francis Blanton; Harriet Blue; Paul Blue; Gerry Blume; Louise Blume; Peter Blume; Gertrude Bock, Jim Bock; Michael Bock; Oliver Bock; Peter Bock; Roger Boulton, University of California Davis, Department of Viticulture and Enology; Briarpatch Network; Pete Britz, Rhodes University; Floyd Butterfield; Cactus Pete Graphics; Guillermo Camacho; Michael Caravatta; Gerald D. Carr; Judith Champagne; Ben Chesley; Frank Cieciorka; Carmen Clar, Lightnin Mixers; Virginia Clarke-Laskin; Clifford Research; Barry Commoner, Center for the Biology of Natural Systems; Paula J. Corso; Keira Costic, Friends of the Earth; Jeff Cox; Ian Crawford; Vicky Crowe, Tomco; Gene Culver; Howard A. Dahl, Amity Technology, LLC; Lemon DeGeorge; Iris DeMent; Lane deMoll; Bob Derby, Nuvera Fuel Cells; Philip Donaldson; Susan Donaldson; Allen Dong; Akara Draper; Joanne Dufiho, *Mother Earth News*; Benjamin Eric, Grimmy, Inc.; Mark Evanoff, Friends of the Earth; Rod Ferronato, Santa Rosa Stainless Steel; Keith Field, Capstone Electricity; Steve Fitzpatrick, Biocon, Inc.; E.P. Flemyng; Benjamin Froke, Bug Music; R. Buckminster Fuller; Linda Fultz, Archer Daniels Midland Greenhouse; Michael Funk; Dave Gancher, *Sierra Magazine*; Marilyn Garrett; John A. Gollin, *San Francisco Examiner*; Paul Goodman; Maribeth Grant; Valerie Greene; Noah Griffin; Joe Gruber, The Wittemann Company, LLC; Mike Guiry, National University of Ireland; Jim Hall; William Hard; Hollie Hartley; Guine Harwood-Shaw, *Christian Science Monitor*; Royce Hays; Kent Heintz; Thurly Heintz; The Henry Ford Museum; Richard Highland, U.S. Bureau of Standards; Nancy Ho, Purdue University; Mark Hoffman; Jenny Hoffman, Morrison Brothers Co.; Kerstin Hoffmann, Robert Bosch GmbH; John Holliday, Aloha Medicinals; Gerald J. Holmes, North Carolina State University; Roy Howell, Red Line Synthetic Oil Corporation; Bob Hughes; David Hull, Humboldt Bay Harbor Recreation & Conservation District; Judy Hummel; IPD Corporation; David Jennings; Delene Jennings; Yancy Jergenson; Al Kasperson; Karl Kessler, *The Furrow*, John Deere Co.; Ann Kinsey; Gene Kinsey; John Knabb; N. Kosaric, University of Western Ontario; Bill Kovarik; Kevin Kraus, Red Star Yeast; Mike Ladisch, Purdue University; Barbara Lawrence; Perrie Layton, *Organic Gardening* Magazine; Steve Lemas; Robert Leonard, University of Arizona; Patrick Lofthouse; Ken Luboff, John Muir Publications; Dr. Vicky Macarian, University of Arizona; Carissa Maloney; Shawn Maloney; David Martin; Diana Matthews; Michael McCormick; James McMurtry; Merit Academy; Jill Merritt; Gene Meseck; Frank Michael; Lewis E. Middleton, Indiana Farm Bureau; Alok Mitra; Charlie Moran; Ruth Morgan, Battelle Institute; Gary Mule Deer; U.S. National Park Service; Ray Newkirk, Pacific Biofuel; Staffan Nillson, Syngenta; Don O'Brian; Doris Ober; Roger D. Oglesby, *Seattle Post Intelligencer*; Miriam Ornstein; Dr. Josmar Pagliuso; Dr. David Palzkill, University of Arizona; William Paynter; C. Jay Paynter, Union Flights; Jessica Paynter, Union Flights; Michael Pekosh; Peter & Wolf Co.; John Petersen, Viking Pump, Inc.; Claire Peterson; Utah Phillips; Lorina Poland; John Powell, Caneharvesters; R & D Research; Bo Rader, *Wichita Eagle*; Laura Rainey; Phil Randall; Kimberly Ranick, Goodnature Products, Inc.; Gerow Reece; Anthony Reed, *Wichita Beacon*; Bob Reinfield, American Chain Association; Rob Reynoldson; Fernando Augusto Moreira, Ribeiro, UNICA; Jim Richter; Kathleen Robbins, John Wiley & Sons Permissions; Roy L. Robinson; Simon Robinson, Nottingham University Press; James Ruebsamen, *LA Herald Examiner*; Barry Rugg, New York University; Lisa Schaller, Martin Williams Advertising; Diane Schatz; Karl Schleicher; Paul Schultz; Tom Schultz, Seaward Stoves; Dr. C. Senthil, Rajshree Sugars; Carolyn Shefler; Les Shook; Scott Sklar; Patricia Slingerland; Christina Smith, *Science News*; Jon Spar; Michael Spencer; Harry Stokes, Dometic/Origo; Joann Stout, MacKissic, Inc.; Robert Sturdivant; Orrie Swayze; Bob Theis; Rick Tilby, Tilby Systems, Ltd.; Brett Tinling; Easter Trabulse; Tom Turner, Friends of the Earth; Heather Tuttle, Union of Concerned Scientists; United States Department of Energy; Phyllis Yoshida, U.S. Department of Energy; U.S. Geological Survey; David Umberger, *Purdue News*; Tom Valens; Philip J. Valvo, Lesaffre Yeast; Martine Van Bogaert, Ministry of the Flemish Government; Corey Waggoner; Waterstar Motors; Doris Webb; Earl Webb; Jerry C. Weigel; Charles Welch; Bryan Welch, Ogden Publications, Inc.; Walter Wendt; Don Wenger, Wenger Manufacturing; Wetlands Engineering; Rick White, California Solar; Tish Wilson, Moyno, Inc.; Alinda Worley; Mark Zabel; Steve Zenofsky, FM Global; and Mary Zimmerman.

ABOUT THE AUTHOR

David Blume started his ecological training young. He and his father Jerry grew almost all the food their family ate, organically—on a city lot in San Francisco in the mid-'60s!

Dave taught his first ecology class in 1970. After majoring in Ecological Biology and Biosystematics at San Francisco State University, he worked on experimental projects, first for NASA, and then as a member of the *Mother Earth News* Eco Village alternative building and alternative energy teams.

When the energy crisis of 1978–79 struck, Dave started the American Homegrown Fuel Co., an educational organization that taught upwards of 7000 people how to produce and use low-cost alcohol fuel at home or on the farm.

KQED, San Francisco's Public Broadcasting System station, asked Dave to put his alcohol workshop on television, and together they spent two years making the ten-part series, *Alcohol as Fuel.* To accompany the series, Dave wrote the comprehensive manual on the subject, the original *Alcohol Can Be A Gas!* Shortly after the first show aired, in 1983, oil companies threatened to pull out their funding if the series was continued. KQED halted the distribution of the series and book (see this current book's Introduction for the whole story).

In 1984, Dave founded Planetary Movers, an award-winning social experiment and commercial venture, well known for productive activism (e.g., on behalf of Nicaragua's Sandinistas), as well as for pioneering practices of progressive employment, green marketing, and the sharing of a percentage of profits for peace and the environment.

In 1994, he started Our Farm. This community-supported agriculture (CSA) farm was also a teaching farm, based on sustainable practices, that hosted over 200 interns and apprentices from all over the world, and held regular tours for thousands of people. Our Farm grew as much as 100,000 pounds of food per acre, without a tractor, using only hand tools, on a terraced, 35-degree slope.

DAVID BLUME IS THE EXECUTIVE DIRECTOR OF THE INTERNATIONAL INSTITUTE FOR ECOLOGICAL AGRICULTURE. HE HAS BEEN AN ALCOHOL FUEL PIONEER SINCE THE SEVENTIES, AND HAS CONSULTED FOR A WIDE ARRAY OF CLIENTS, INCLUDING GOVERNMENTS, FARMERS, AND COMPANIES INTERESTED IN TURNING WASTE INTO VALUABLE AND PROFITABLE PRODUCTS.

BOB FITCH PHOTO, <WWW.BOBFITCHPHOTO.COM>

The International Institute for Ecological Agriculture (IIEA), founded by Dave in 1993, is dedicated to healing the planet while providing for the human community with research, education, and the implementation of socially just, ecologically sound, resource-conserving forms of agriculture—the basis of all sustainable societies. The IIEA teaches permaculture, an ethical system of ecological land design, which incorporates the disciplines of agriculture, hydrology, energy, architecture, economics, social science, animal husbandry, forestry, and others.

Dave and his IIEA associates are establishing a biofuels station in Santa Cruz, California, that will offer alcohol fuel in a driver-owned cooperative, as detailed in this book. Dave is currently Executive Director of the IIEA.

He has consulted for a wide array of clients, including governments, farmers, and companies interested in turning waste into valuable and profitable products. Recent work includes a feasibility study for a macadamia growers' cooperative in Mexico, and a water harvesting/reforestation project in Antigua, West Indies. He is working with a farming college connected to the government of Ghana to develop alternative fuels, to train agricultural extension agents in organic farming, and to design an ecological strategy to stop the Sahara Desert from advancing. He also recently inspired the city of Urbana, Illinois, to hold a conference between builders, lenders, developers, municipalities, building inspectors, architects, and engineers, to coordinate the mainstreaming of natural building technologies. He has helped the Ford Motor Company demonstrate alcohol-fuel-powered vehicles at a series of U.S. events.

"Farmer Dave" is often called upon to testify before agencies on issues related to the land and democracy. He is a frequent speaker at ecological, sustainability, Peak Oil, and agricultural conferences in the Americas, and has appeared in interviews over 1000 times in print, radio, and television. Dave firmly believes in Emma Goldman's view of, "If I can't dance, I don't want to be in your revolution," and he can frequently be found on the dance floor when he isn't flagrantly inciting democracy.

TABLE OF CONTENTS

LIST OF FIGURES

PREFACE

I am often accosted by people whom I would normally consider my colleagues. They are typically environmental activists informed by what they read on the Internet, people who watched *An Inconvenient Truth*, people who are aware of Peak Oil, sustainable agriculture, and climate change.

They say "Dave, don't you know that fossil fuels, (and fossil-fuel-based fertilizers) are beginning to run out and 'There Is No Alternative'? [My acronym for this is "TINA".] The only thing we can do is stop driving, stop using energy, walk to our green jobs in the new localized economy, and go back to farming by hand. Power down."

They also say, "The reason we are in this mess—polluted air, lung cancer, melting ice caps, drowning polar bears, food traveling 10,000 miles from farm to eater, horrible wars for oil, MTBE-poisoned groundwater, massive monoculture farms growing animal feed to be shipped to other countries to feed their rich people's cattle, requiring billions of gallons of pesticides—is all because of the internal combustion engine and cheap fossil fuels. Why are you writing a book which is all about making it possible for American soccer moms to drive their massive SUVs while the rest of the world starves for basic energy and food?"

They say, "Don't you get it?!!"

Believe me, I get it. I agree that we, particularly in the United States, are using and wasting a disgustingly huge quantity of energy. My fellow ecologists have a deep knowledge of natural systems, and we find ourselves every day walking through a world of horrible environmental wounds that we cannot help seeing.

Conservation and, more importantly, good design are the basic foundations on which to plan our energy future.

This book is not about providing unlimited clean fuels for SUVs. It's about shaping energy policy now with our own individual and group actions, to make sure the energy future we get is the one we

THIS BOOK CHARTS A CLEAR, ATTAINABLE PATH THAT WILL WORK—AND THAT IS, IN FACT, ALREADY IN MOTION. THIS PATH, IN THE BEST TRADITIONS OF MARTIAL ARTS, TURNS THE FORCES DESTROYING OUR PLANET BACK AGAINST THE PLUNDERERS, AND PUTS BOTH THE POWER AND THE RESPONSIBILITY FOR IMPLEMENTING THE SOLUTION IN THE HANDS OF ORDINARY PEOPLE, WORKING TOGETHER AT THE LOCAL LEVEL.

want and not the one the Oilygarchy is planning for us. This book is about maintaining your power and hope in the face of "this mess." It charts a clear, attainable path that will work—and that is, in fact, already in motion. This path, in the best traditions of martial arts, turns the forces destroying our planet back against the plunderers, and puts both the power and the responsibility for implementing the solution in the hands of ordinary people, working together at the local level.

This is why I refuse to give in to a philosophy of despair, why I refuse to surrender to those who plunder the planet, although so many of us have given them our permission to destroy the Earth under the banner of TINA. TINA allows the powerful to decide which energy we will use to pollute our planet.

And while people in developed countries may practice a little well-meaning conservation, they will not be willing to return to a world where the basic unit of energy is their human labor. The multinational energy corporations know this, and they are planning their transition to fuels that make petroleum look like Mr. Clean (see Chapter 4). A dying planet is of secondary importance to these people.

In many ways, my strident colleagues are correct when they say that the central cause of our planet's woes can be characterized by our use of engines and fossil fuels. But they are wrong about TINA. The central solution, which ripples out to every corner of the planet, is to replace those fuels with available solar-based fuels all over the planet. As you will see, even in a cursory examination of this book, this alternative is powerful, inexpensive, fast, and effective—and will regenerate ecological systems, if done properly.

As I point out in this book, it is possible to make ethanol using a badly designed industrialized model, one that corporate agribusiness currently employs. In this system, biofuels production would amplify the abuses to our environment.

But this book shows how a permaculturally designed ethanol system provides us with surpluses of local food, energy, community, and power—all while deepening our topsoil, eliminating the use of toxic agrichemicals, and reversing global warming.

It's about doing things on a human scale and as you will see, the human scale has virtually all the advantages in this struggle. Our energy/food production system can either affirm our living environment or treat the Earth as one big strip mine to exploit.

Yes, there is an alternative to a chaotic post-petroleum world. But it will take your help, blood, and sweat to make it happen. In fact, it's going to take a revolution.

Share. Organize. Win.

1983 FOREWORD

BY R. BUCKMINSTER FULLER

What follows was originally the foreword for the 1983 version of this book. I was greatly honored that my colleague R. Buckminster Fuller felt that my project was important enough to take time out of his busy schedule to write something to my readers to communicate just what it might mean to have a solar-powered future.

Unlike a lot of solar visionaries of the day, Bucky had been around long enough to witness and understand the interlocking corporate/governmental structure, and in his last book, The G.R.U.N.C.H. of Giants, *he talked about how it would all end. The acronym stood for "Gross Universal Cash Heist." He prophesied that not only would corporations grow bigger than giants, but that predatory transnational capitalism was leading us to a very bad end indeed. Bucky always inspired me and millions of others with his emphasis on direct positive action as a strategy for rapid change. This perspective is a source of what he refers to as the change from a world based on weaponry to one based on livingry.*

During the press conference marking the premiere of my television series, Alcohol As Fuel, *Bucky said that he had been in charge of America's alternative energy research for the military during World War II. He himself had run an engine on alcohol for two years during his war research. In a very thinly veiled comment, he noted that his voluminous notes and research done within the military-industrial establishment are now missing, apparently stolen from military archives in 1970.*

Shortly after the series went on the air in 1983, Bucky's wife passed away. They were so joined at the heart that he, too, passed away a few days later. I still miss him and wish somehow he could see this book finally published—finally out doing its intended work.

PHYSICISTS, ASTROPHYSICISTS, ASTRONOMERS, CHEMISTS, AND ENGINEERS EMPLOY … STANDARDS OF MEASUREMENT TO DISCOVER HOW THE PHYSICAL UNIVERSE IN WHICH WE LIVE OPERATES. SUCH METHODICAL MEASURING HAS DISCOVERED A COMPLETE INTEGRITY OF ENERGY ACCOUNTING IN THE UNIVERSE.

Fig. f-1 R. Buckminster Fuller and David Blume. *At the KQED San Francisco press conference for the launching of Dave's PBS alcohol how-to series.*

The area of the surface of a sphere is exactly four times the area of the sphere's great-circle disk, as produced by a plane cutting through the center of the sphere. The surface of a hemisphere is, then, twice the area of the sphere's great-circle plane. When we look at the "full" Moon, we are looking at a hemisphere's surface which is twice the area of the seemingly flat, bright, circular disk in the sky. The total surface of the Moon's invisible other side plus the visible hemisphere is exactly four times that of the flat disk area of the full Moon as our optical illusion views it.

All of Earth's operational energy comes from the stars, but primarily the star Sun. It comes either as radiation or as inter-astrogravitational pull. One half of Earth's 200-million-square-mile surface is always sunlit. Twenty-four hours a day the Sun radiation is drenching the outside of the aurora and cloud-islanded biosphere's 100-million-square-mile hemispherical surface. The planet Earth's total biospheric mantle, which travels with it through space, has a depth of 400 miles. Forty billion cubic miles of Earth's Sun-radiation-impounding biosphere is always exposed to Sun, while the other 40 billion cubic miles of Earth's biosphere is always in Earth's night shade.

All of humanity's world-around energy uses are so inefficient that 95% of it is wasted. Despite that wasting, the total amount of energy consumed by humanity as of 1982 amounts to less than one five-hundred-thousandth of 1% of our daily energy income from the Sun....

While its thermosphere mantle is 400 miles deep, its mesosphere 50 miles deep, its stratosphere 30, Earth's propeller-flyable atmosphere is only 10 miles deep. This gives us one billion cubic miles of wind turbulent atmosphere on the sunny (day) side, and one billion cubic miles on the shadow (night) side. The one billion cubic miles of atmospheric molecules in the hemisphere which is constantly saturated by Sun are kinetically accelerated, while simultaneously the one billion cubic miles of atmospheric molecules in the shaded night hemisphere are kinetically decelerated.

The shadow side of Earth consists of one billion cubic miles of contracting atmosphere, while the one billion cubic miles on the sunny side is sum-totally expanding. All around Earth, both yesterday's and today's Sun radiation impound-ments countered by gravity pulsatingly perturbate the atmosphere by anti-gravity thermal columns and Sun-evaporated Earth surface water risings, here and there outwardly against gravity, from oceans and lands. The rotation of Earth brings about a series of high-low relative atmospheric pressure differentials and their world-around semi-vacuumized drafts, which altogether produce the "high" and "low" atmospheric pressures and the complex turbulence which we speak of as "the weather."

The combined two billion cubic miles of atmospheric kinetics continually and anew convert the solar energy into weather, which in turn differentiates into wind power and gravity-accelerated "rain" and highlands-landed water power. A fraction of the wind power again interacts with gravity to produce the great waves, rolling power-laden, onward across the 150 million square miles of surface of the oceans, lakes, rivers, and ponds, to pound thunderingly and grind ceaselessly with fury upon the rocks, producing the gravels, sands, and dusts of Earth's million miles of coastlines. Wind power is Sun power in its most abundant, day and night, anywhere and everywhere, by a relative energy abundance factor of better than 99 to 1 in contradistinction to any of the other known means of terrestrial impoundments of Sun energy.

Three-quarters of Earth's total surface is covered with water. Employing photosynthesis, the water-borne algae convert Sun radiation energy into hydrocarbon molecules which, in a complex succession of swallowings by progressively larger marine organisms, metabolically accomplish a vast impoundment of Sun-emanated energy within Earth's waters.

Since human lungs and other organisms are designed for initial success only on dry land, and since the development of boats capable of mastery of major life support through offshore fishing means required millions of years of development, we must first look to the dry land for comprehension of humanity's initial life support.

The water-free one-quarter of Earth's surface consists of land which is largely covered by deserts, ice and snow fields, and rugged mountains. Only about 10% of our planet's surface is blessed with a total complex of natural conditions suitable for humans' outdoor cultivation of animal and vegetable food products. This suitability occurs only where the properly watered and temperatured topsoil vegetation can impound Sun's radiation by photosynthesis.

Among the solar-energy-impounding dry land vegetation species, none can now match sweet corn's performance efficiency. Sweet corn converts and stores as recoverable energy 25% of the received ultraviolet radiation, whereas wheat and rice average only 18 to 20%. Sustained by the initial botanical impounders of solar energy, animals and other creatures automatically proliferate the production of orderly hydrocarbon molecules in their own DNA-RNA-programmed cell growth.

Humans in turn consume the energy-rich hydrocarbons proliferated by creature and vegetation cell growth to sustain their own subconsciously (DNA-RNA) accomplished corporeal cell multiplication, as well as for sustaining their conscious expenditures of physical work or play effort. Objectively employing the cosmic principles discovered by human mind, human work can design and produce the extracorporeal tools which in turn produce the commercial alcohol, methane, gas, etc. Or humans can leave untampered-with the slow process of nature's energy production of solid and liquid fossil fuels produced through heat and pressure conditions accomplished by nature within Earth's crust.

The requisite pressures and heat that must be maintained continuously and steadily over the multimillenia involved in the natural production of liquid fossil fuels, when accounted at the kilowatt-per-hour prices charged to retail customers by the public utilities for that much energy for that vast span of time, amount to well over a million dollars per each gallon of petroleum. Kilowatts are energy units employed by scientists and engineers which are mathematically convertible into (and predicated upon) humans' prime measure of energy as work, i.e., the amount of energy expended to oppose gravity by lifting a given weight a given vertical distance in a given amount of time, which is expressed as foot-pounds-per-minute, or meter-kilograms-per-minute, or as centimeter-gram-seconds (cgs), etc., ergo kilowatt-hours. Physicists, astrophysicists, astronomers, chemists, and engineers employ these standards of measurement to discover how the physical Universe in which we live operates. Such methodical measuring has discovered a complete integrity of energy accounting in the Universe.

Among the solar-energy-impounding dry land vegetation species, none can now match sweet corn's performance efficiency. Sweet corn converts and stores as recoverable energy 25% of the received ultraviolet radiation, whereas wheat and rice average only 18 to 20%.

Newton's discovery of the geometrical rate of interattractiveness variance, in respect to arithmetical change of celestial bodies' interdistancing, from which we derive the gravitational constant, plus the measurement of the linear speed of radiation in a vacuum (186,000+ miles per second) is compounded with quantation of photons radiated per second to provide the radiation constants. Nonsimultaneous-scenario Universe's eternally regenerative, 100%-efficient integrity is predicated upon the ever-energetic intertransformings of gravity vs. radiation—from matter to radiation and vice versa. Einstein's equation of $E=mc^2$ (as eventually proven by fission) provided humanity with the competence to participate in the integrity of cosmic accounting.

All of humanity's world-around energy uses are so inefficient that 95% of it is wasted. Despite that wasting, the total amount of energy consumed by humanity as of 1982 amounts to less than one five-hundred-thousandth of 1% of our daily energy income from the Sun and only one 12-millionth of a percent of our daily energy income from both the Sun and our geothermal sources.

It is incontestably clear that it is now technologically feasible to harvest enough of our daily income of extraterrestrial energy … generated at an inexorable, nature-sustained rate, to provide all humanity and all their generations to come with a higher standard of living and greater freedoms than ever have been experienced by any humans and to do so within ten years, while completely phasing out all further use or development of fossil fuels, atomic and fusion energies.

It is incontestably clear that it is now technologically feasible to harvest enough of our daily income of extraterrestrial energy as well as of the surface eruptive steams of internal Earthian infernos, all generated at an inexorable, nature-sustained rate, to provide all humanity and all their generations to come with a higher standard of living and greater freedoms than ever have been experienced by any humans and to do so within ten years, while completely phasing out all further use or development of fossil fuels, atomic and fusion energies.

This means it is possible for Earthian humanity to live on its daily energy income as generated by star radiation and cosmic gravity—primarily that of the star Sun—rather than: (a) by exhaustion of the millions of years of celestial energy photosynthetically impounded by the terrestrial vegetation and deposited into Earth's crust as a cosmic savings account possibly to be used many billions of years hence to convert planet Earth into a star; or (b) by burning up the atoms of which Spaceship Earth is structured; or (c) by fusion's disruption of the biosphere's delicate hydro and thermal balancing, the incisive integration of which governs the comprehensive metabolic chemistries of terrestrially regenerative ecology.

Three-quarters of our planet Earth's surface is covered by water. Water constitutes about 60% of the physical substance of planet Earth's biological organisms. Sixty-five percent of the human body consists of water. The Earth's oceans contain 97% of all our planet's water. The surface of the Earth's waters is being continually vaporized into clouds to be redistributed around Earth as rain or snow. Sum-totally, Earth's waters are being continually recirculated throughout its combined ecological and geological biospheric system. Water is the "blood" of Earthian life. The average depth of its oceans is less than one four-thousandth of the Earth sphere's diameter. It is a gossamer film so thin that it is proportionately less than the depth of the blue-ink printing of the oceans on a 24-inch Earth globe. This almost ethereal film is kept from instant evaporation by the Sun only through the energy-reshunting properties of the plurality of additional concentrically enshrouding chemical and electromagnetic spherical mantles of Earth's biosphere.

The entropic energy losses to Universe occasioned by our emergency-urged fortuitous exploitation of Spaceship Earth's inventory of integral atoms either by fission or fusion will probably violate the integrity of the complex cosmic design for successful maintenance of human life aboard Earth.

This means it is possible for Earthian humanity to live on its daily energy income as generated by star radiation and cosmic gravity ... rather than ... by exhaustion of the millions of years of celestial energy photosynthetically impounded by the terrestrial vegetation and deposited into Earth's crust as a cosmic savings account....

INTRODUCTION

FUELING A REVOLUTION

This book is the distilled essence of the most pertinent information ever assembled in one place on alcohol fuel—the technology that can help us finally become producers of almost limitless energy, instead of extractors of finite resources. How we produce our energy from here on out will determine how we govern ourselves and how we relate to nature and the environment; it will also create a sea change in where wealth concentrates. It will determine if the future is ruled by a small number of armed dictatorships backed by military and industrial interests (a cabal I like to refer to as MegaOilron or the Oilygarchy), or if energy, and therefore power, is held by a diffusion of democratic entities, based on their ingenuity and ability to gather a portion of their daily solar income.

How did I first learn about alcohol's ability to run vehicles? I can still remember, as clear as a bell, talking about brewing beer (which was still illegal in 1974) with Doc Sweeney, one of my ecological biology professors at San Francisco State. He was infamous for telling students outrageous tall tales with a straight face just to see their reaction, or better yet, to see if he could get away with it.

He said to me, "That beer you're brewing could even run your car."

His deadpan expression was daring me. "You're lying," I said.

And, as any really excellent teacher would, he said, "Prove it."

So I went to the library, figuring this was going to be the needle in the haystack search of all time. Much to my surprise, I found more than 30 books from the early part of the 20th century. There was a whole hidden history of alcohol as a fuel that my friends and I had never known existed. Damn that Sweeney; he hooked me good.

A few years later, during the mid-1970s gas crises, the knowledge that I gained would hold me in good stead as I started making fuel alcohol,

IT TOOK OVER 25 YEARS TO FINALLY GET THIS BOOK TO YOU. IT REPRESENTS THE CONFIDENCE OF ALMOST 30 PEOPLE WHO COLLECTIVELY LOANED MORE THAN $250,000 TO SEE THIS PROJECT THROUGH. IT'S THE MOST COMPREHENSIVE BOOK EVER WRITTEN ABOUT ALCOHOL FUEL. ITS PRODUCTION HAS BEEN A MASSIVE EFFORT THAT HAS DEPENDED ON THE COOPERATION OF HUNDREDS OF PEOPLE WHO CONTRIBUTED BOTH THEIR KNOWLEDGE AND, MORE IMPORTANTLY, THEIR EXPERIENCES. ALMOST EVERY SECTION OF THIS BOOK WAS REVIEWED BY COLLEAGUES AROUND THE WORLD....

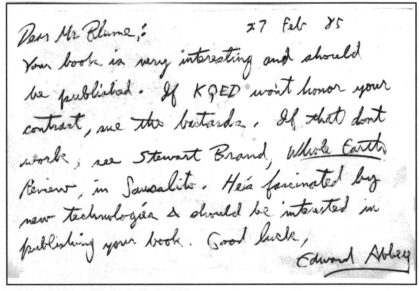

Fig. 0-1 Postcard from Edward Abbey. *The iconoclastic, irascible curmudgeon and author of* The Monkey Wrench Gang *suggested I sue KQED. Well, Ed, I tried.*

and started working at the *Mother Earth News* Ecovillage research facility. *Mother Earth News* was a pioneering publication that was the guiding voice of the back-to-the-land movement, and its ecovillage was a seething hotbed of part-time and full-time inventors building all sorts of equipment, tools, and simple machinery that really defined appropriate technology in the U.S.

The MegaOilron full-court press was like a black cloud blotting out the sun. The reign of George Bush the Elder, first with Ronald Reagan as nominal president and then with himself as president, would eventually crush virtually all alternative energy programs everywhere.

In 1979, I started American Homegrown Fuel, an educational organization that would spread the word. Over the next few years, I taught nearly 180 workshops to over 7500 people on how to produce alcohol fuel and how to convert their vehicles and equipment to use it. At least 97 fuel-making cooperatives formed out of my workshops. I made over 700 press appearances and worked as a consultant all over the U.S., as well as in many small foreign countries that were reeling from the impact of spiking oil prices.

The late 1970s were an exhilarating time for those of us in the alternative energy movement. The Carter administration passed a windfall profits tax on the obscene oil industry profits that were made when OPEC embargoed oil to the U.S. and jacked up the price. Carter took the money from the oil companies and plowed it into alternative energy. He generated tax credits for alternative energy projects such as cogeneration, alcohol fuel, solar panel installation, and biomass energy equipment. Many states got into the act by matching or exceeding the federal credits with state tax credits. The state of California, under Governor Jerry Brown, established an alcohol fuel design competition and also put together a revolving loan fund for small farm-based alcohol plants. The federal government forced utilities to buy alternatively produced electricity from small producers at the "avoided cost" of what power would have cost from a new nuclear plant.

Those of us working in those heady days knew with an unfounded certainty that the solar revolution was happening and that we were going to take MegaOilron down to its foundations.

It was in this climate that, in 1980, I was approached by PBS station KQED in San Francisco to take my alcohol fuel seminar and put it into the format of a ten-part how-to television series. The

GUESS WHAT JUST BLEW IN FROM WASHINGTON...

SOLAR DEVELOPMENT

Fig. 0-2

MICHAEL KEEFE

THE DENVER POST 1981

original version of this present book was to be produced as the companion volume to the series.

At the time, I was concerned that since several corporations sponsored various PBS offerings, KQED could be pressured about the content of the show. I was assured that I had nothing to worry about, since most of KQED's funding came from the Corporation for Public Broadcasting, funded by the public's tax dollars; corporate grants were only a small part of the station's budget. In fact, PBS had recently aired something called *Death of a Princess*, the secret filming of the execution of a Saudi princess who was stoned to death for adultery. Oil companies had tried to pressure PBS not to air the show, but PBS had refused. I was reassured, and we went to work on the project, which took two years.

We filmed the series at my distillery in Napa and in the studio. We also took a 1000-mile road trip all over the West Coast and filmed my students who had built various distilleries and done car conversions, and got a lot of great footage telling folks about how it felt to be independent of big oil.

At that time, dark clouds were forming. MegaOilron had organized and funded political campaigns in 1980 to overtake the White House and sweep the progressives who were spearheading the solar revolution out of office. Its presidential candidate was a former actor and spokesperson for General Electric; the choice for vice president was an oilman born into wealth derived in part from fascist collaboration with Hitler.

According to the best evidence, MegaOilron colluded with the government of Iran to delay releasing American hostages held in that country until the day after the U.S. election, in exchange for weapons Iran could use in its conflicts with Iraq. It also manipulated oil prices and supplies to embarrass President Carter. The MegaOilron full-court press was like a black cloud blotting out the sun. The reign of George Bush the Elder, first with Ronald Reagan as nominal president and then with himself as president, would eventually crush virtually all alternative energy programs everywhere.

This was the climate in which, finally, in 1983, the PBS alcohol fuel series premiered on KQEC, KQED's smaller UHF station in San Francisco. The premiere press conference featured visionary R. Buckminster Fuller, author of the original

book's foreword; Ernest Callenbach, author of *Ecotopia;* and Bill Paynter, Ronald Reagan's private pilot and partner of astronaut Gordon Cooper in an alcohol aviation company.

KQED had already successfully marketed the series at the PBS convention, and approximately 120 stations nationwide had agreed to air it. During the first three segments, which aired in the San Francisco Bay Area, thousands of people called to order the book. I traveled to Portland, Oregon, with KQED's check and the layout boards in hand, to print the first 10,000 books. I stayed in town to be on hand in case any last-minute line changes or layout changes needed to be approved.

I got a call from the printer the day after the airing of the fourth segment. He told me that KQED had called him and told him to stop work and to send back the check. I raced down to the printer's, barely in time to retrieve my book before he boxed it up to mail to KQED. The station refused to take my calls from Portland.

"Cowardice asks the question, Is it safe? Expediency asks the question, Is it politic? Vanity asks the question, Is it popular? But conscience asks the question, Is it right? And there comes a time when one must take a position that is neither safe, nor politic, nor popular, but he must take it because his conscience tells him that it is right."

—DR. MARTIN LUTHER KING, JR.

I returned to San Francisco, and much to my dismay saw the last five segments of the show air without the book advertisement. KQED cancelled the distribution of the series, and when I tried to ask why, they stonewalled me for weeks, which finally resulted in my invoking the arbitration clause in our contract. Despite the arbitration clause, KQED was provided with a three-lawyer team of attorneys paid for by Chevron. To obfuscate the case, using a nifty legal maneuver, they claimed I breached the contract and countersued me. While they knew they had no case, they used the legal discovery process to put me through weeks of depositions and document requests, which they couldn't legally do in arbitration.

When we finally got to arbitration, they used the reams of material they generated in the lawsuit to turn what should have been a three-day arbitration process into weeks of hearings, which wore the volunteer arbitrators down and ran me into substantial debt. In an understandable fit of frustration, the arbitrators called an end to the proceedings. KQED retained the rights to the video, with no obligation to air it, but I received the rights to my book back. (I estimate KQED outspent me ten to one.)

Just after Bush/Reagan were elected, they cut the funding for the Corporation for Public Broadcasting by two-thirds. Then MegaOilron went to the handful of PBS stations that did original productions at that time and stepped in to make up the lost government funding. This made the PBS core stations fully dependent on corporate funding. So when someone at Chevron heard about my series, it took a simple phone call. I was caught in the pacification of PBS without knowing it was going on.

I firmly believe that if the series had gone on in 1983, we would not have had the first Gulf War, and we would not have found ourselves in wars with Afghanistan and Iraq—because we wouldn't have been dependent on Middle East oil.

Remember ... ecologically, Nature favors creatures that cooperate; often, in situations of competition for the same ecological niche, both competitors become extinct. (Charles Darwin did not say "survival of the fittest"—that comes from right-wing economist Herbert Spencer, who created the crackpot theory of Social Darwinism.)

In designing this new version of the book, I had to make some choices early on. Various prospective publishers argued that putting all of this material into one large volume might scare off readers who just want a recipe book of how to make alcohol. They said, "All this history and politics is fascinating, but aren't you afraid that including it in your how-to book would scare away some buyers?" "Put it in a separate publication," their marketing experts said. But in the final analysis, I decided that this book should be a complete tool kit to revolutionize our transportation energy system, combining a broad, sweeping vision with intricate detail.

I spent four years working on this book with a small team of researchers. I traveled all over the United States in search of the most up-to-date information. In frozen South Dakota, I talked to Orrie Swayze and his farmer and VFW buddies who are taking on the oil companies, and to alcohol combustion engineer and alcohol aviation expert, Jim Behnken. I went to Decatur, Illinois, to see the largest alcohol plant in the U.S., Archer Daniels Midland's 200-million-gallon-per-year plant. My travels also took me to Brazil to document the world's largest alcohol fuel program.

It took over 25 years to finally get this book to you. It represents the confidence of almost 30 people who collectively loaned more than $250,000 to see this project through. It's the most comprehensive book ever written about alcohol fuel. Its production has been a massive effort that has depended on the cooperation of hundreds of people who contributed both their knowledge and, more importantly, their experiences. Almost every section of this book was reviewed by colleagues around the world to make sure that its contents meet the high standards of peer review.

You probably will not personally need all the parts of this book. If you don't, then that's good, because it means you recognize that you need other people with their particular interests and skills to work with you. Interdependence, not self-sufficiency, is the focus of this book, although those who like to do things on their own will relish the details I provide. Remember, though, that ecologically, Nature favors creatures that cooperate; often, in situations of competition for the same ecological niche, both competitors become extinct. (Charles Darwin did not say "survival of the fittest"—that comes from right-wing economist Herbert Spencer, who created the crackpot theory of Social Darwinism.)

So what might that cooperation look like? Farmers among your circle might produce an energy crop, and provide you a place to operate a distillery to produce the fuel. Or maybe you're a lawyer and can write up the limited liability corporation paperwork of this cooperative effort so the members can get the hefty tax credits usually lapped up by big corporations. You may be a mechanic, and your bowling buddy is showing this book to you so you can convert his vehicle, generator, lawn mower, or heater to alcohol fuel. You might be a truck driver who decides to get her own fuel truck to deliver alcohol to tanks people have at their homes, just like propane or fuel oil. Maybe you are a patriotic soccer mom/anti-oil-war activist who has decided that it's your role to organize the folks together because a barrel of oil is not worth a bucket of blood. Maybe you are someone who feels really sick every time you fill your tank with gas, knowing that you are part of the problem but working two jobs just to stay even, and you realize that you need to be a member of a fuel co-op.

I want to say to all of you, it's okay that you yourself might not understand every part of this book. Relax, you don't need to. Share. Organize. Win.

There's a lot that goes on in the world of energy that you never see on the 11 o'clock news—things people really ought to know if they are to be informed participants in a democracy, or even savvy taxpayers. There's also a lot going on with our environment that is inextricably connected to energy, underreported news that affects our health as individuals and as a species. So you will find that I've liberally peppered facts, quotes, figures, and personal stories in bite-sized pieces throughout the book. I hope you find them informative, useful to start heated conversations with at parties, and maybe the source of motivation to write a few letters to Congress. I hope they inspire you if the going gets a little tough.

So after all that, what is this book about? Well, as I remind you in Chapter 2, it's not just about getting unlimited cheap, clean fuel for your SUV. It's also not just about having your own fuel when declining oil supplies mean that rationing becomes necessary. In raw terms, it's about power. After all, democracy is the quaint idea that the people ought to have the last say in how they are governed.

The energy corporations are the biggest in the world. They are larger in economic size than most countries. For years now, they and their ilk have acted as the government of the world—because the control of a country's energy is the ultimate control of its people.

As we reach a point where there isn't enough petroleum to go around, the battle for who gets the last barrels is already engaged. MegaOilron—that relatively small clan of government, military, and energy industry elites—is ready to govern at any cost—war without end, terrified and battered people, a world without civil rights. It has shown it is willing to kill large numbers of people in the pursuit of money and power. It is willing to lie, cheat, and steal from our treasury; willing to tax us and send our children to die in the fight for oil.

Nature, through ecological collapse, will fight back against the oil culture. Farmers are fighting back by refusing to sell their products for nothing and joining together to produce fuel. Whole nations, knowing that the Oilygarchy won't let them near the last oil, are gearing up to harvest solar energy for biofuels.

But the most important resistance comes from you and me. Do we let MegaOilron take the world down into the hell it is creating, or do we withdraw our support in the way that capitalism knows best? If I don't give transnational corporations my capital and if you don't give them yours, then this corrupt Oilygarchy will stagger and fall. We can and we must do it together; there is no one else to do it but ourselves. We are more than strong enough to do this without our leaders if they won't follow us.

Share. Organize. Win.

"I say to you, this morning, that if you have never found something so dear and precious to you that you will die for it, then you aren't fit to live.

"You may be 38 years old, as I happen to be, and one day, some great opportunity stands before you and calls upon you to stand up for some great principle, some great issue, some great cause. And you refuse to do it because you are afraid.

"You refuse to do it because you want to live longer. You're afraid that you will lose your job, or you are afraid that you will be criticized or that you will lose your popularity, or you're afraid that somebody will stab you or shoot at you or bomb your house. So you refuse to take the stand.

"Well, you may go on and live until you are 90, but you are just as dead at 38 as you would be at 90.

"And the cessation of breathing in your life is but the belated announcement of an earlier death of the spirit.

"You died when you refused to stand up for right.

"You died when you refused to stand up for truth.

"You died when you refused to stand up for justice."

—DR. MARTIN LUTHER KING, JR., *BUT IF NOT* (SERMON), NOVEMBER 5, 1967

Fig. 0-3

BOOK 1

UNDERSTANDING ALCOHOL: VISIONS AND SOLUTIONS

Climate change and the proliferation of greenhouse gases mean that continued use of fossil fuels is suicidal for us all. So you are about to participate in a turning point in history.

Aside from learning how to make and use alcohol fuel and its related products, it helps to know what has come before.

You may want to think that producing alcohol is a politically neutral act, but, as you'll see, it's anything but neutral. What you are and are not permitted to hear is largely determined by who has the power. History is almost always written by the victors—and for the past 100 years, that means the Oilygarchy has been doing the writing.

So now it's time you heard about "the forbidden fuel."

Fermentation, the making of **alcohol**, has often been referred to as humankind's second-oldest profession. **Yeast**, the organism that turns sugars into alcohol, is found on every piece of fruit and on most plants all over the planet. Given the least encouragement, yeast will ferment sugar into alcohol.

All through history, alcohol has figured into the affairs of humans. Each culture decides for itself which drugs are acceptable. At times, alcohol has been considered a blessing and at other times a curse, but always an item of trade. And whenever there is trade, it seems that some entity, be it a government or the lord of a serfdom, levies a tax on it.

FROM THE WHISKEY REBELLION TO THE CIVIL WAR

Alcohol has long been a focus of class warfare in the United States. Consider the area known around the time of the Revolutionary War as Monongahela. With Pittsburgh at its center, and with both the Monongahela and Ohio Rivers flowing through it, this rich, forested region could reach the Mississippi with exports of coal, whiskey, and all sorts of goods for settlers going west.

Prior to the Revolutionary War, though, rum from the Caribbean was so cheap that farmers couldn't compete with the slave-produced import, causing hard times for the Monongahela region. When the war came, the imports stopped, and distilleries sprouted in Monongahela like mushrooms. A 100-gallon **distillery** was worth a 200-acre farm within ten miles of Pittsburgh.[1] And, although there were supposedly taxes in effect on liquor between 1771 and 1775, none were collected. (President George Washington himself ran the largest distillery in the country.)

Many of the men who fought the Revolutionary War returned to their farms afterwards to try to pick up where they left off in their various enterprises.

THE MODEL T AND A WERE DUAL-FUEL VEHICLES. THE DISTRIBUTOR, WHICH SENT THE SPARK TO EACH CYLINDER, WAS "DIGITAL": YOU USED THE FIVE DIGITS OF YOUR LEFT HAND TO PULL THE LEVER ON THE STEERING COLUMN TO ADVANCE THE SPARK TIMING AS YOU DROVE. WHEN RUNNING ON ALCOHOL, YOU WOULD ADVANCE THE SPARK TIMING A GREAT DEAL MORE TO ACCOMMODATE ALCOHOL'S COOLER-BURNING, HIGH-OCTANE QUALITIES. THE MODEL T GOT 34 MILES PER GALLON OF GAS.

INDIANA FARM BUREAU

Fig. 1-1 Farm alcohol tractors. These early alcohol-powered tractors with bulldozer-style traction were designed to run on alcohol right from the factory.

But they found that the merchants who controlled transport and trade to the East Coast would pay very little for the products of the land. According to Leland Baldwin, author of *Whiskey Rebels: The Story of a Frontier Uprising*, "[t]he Revolution was over, and a federal government was already consolidating the fruits of the victory in the hands of the Eastern moneyed classes. The people of Monongahela, perfectly aware of this fact, complained bitterly that they had been induced to pour out the blood of its men, women, and children simply to enrich speculators and manufacturers."[2]

The main items of value that the region could trade for on the Eastern Seaboard were whiskey, Scotch (whiskey made from rye), or furs. The government was the main local buyer for whiskey (for soldier rations), but it paid only 50 cents a gallon. In the East, prices were more than double, so many farmers took their grain to the Eastern cities. A horse could carry two eight-gallon kegs of 114.2-**proof** whiskey to market; this was the equivalent of over five bushels of grain (280 pounds), but fetched a much higher price.[3]

"For my part, whatever anguish of spirit it may cost, I am willing to know the whole truth; to know the worst, and to provide for it."

—PATRICK HENRY, AMERICAN PATRIOT

When it came time to pay for Revolutionary War bills, a property tax was proposed. This solution seemed fair to the Monongahelans, since those who would profit the most—by getting title to property that formerly had been owned by Britain—would pay the proportionate amount of tax. But the big landholders in the East quickly shot down that proposal and replaced it with an excise tax on alcohol, taxing citizens of Pennsylvania at rates even higher than those on the East Coast. The tax was a crushing blow; turning their grain into alcohol was essential to the Monongahelans' economy, since they had little money and poor means to get the grain to faraway markets on bad roads.

Monongahelans revolted, putting up liberty poles—flags showing the snake-in-a-ring symbol from the Revolution and the words, "An equal tax—no excise tax." Excisemen from the new nation's capital were chased out of Monongahela, terrified, returning to Philadelphia with stories of farmers in revolt.

The press called the dissidents "White Indians" and traitors. Alexander Hamilton, one of the moneyed elite, denounced the insurgents. Using the pen name "Tully," he wrote in the press that anyone who criticized the excise tax or spoke of corruption in President Washington's administration was an enemy not only to the Constitution, but to all orderly government.

Led by President Washington, the propertied elites tried to get citizens to volunteer for an army to put down the freedom fighters/distillers in Pennsylvania. They got virtually no response, and Washington quickly resorted to drafting soldiers from several Eastern states. After a great deal of resistance, an army of 12,950 men (11,000 infantry, 1500 cavalry, and 450 artillery) was sent by two routes to converge on the dissident farmers. It was a miserable long march that included an entire week of unbroken rain.

As the conscripted army approached, over 2000 Monongahelan men successfully hid. General Henry Lee, who was in charge of the troops, berated the citizens of Pittsburgh (the area's population center), accusing them of harboring more insurgents, and ordered all citizens to swear new loyalty oaths to the government (for which there was a fee).

Lee then sent lists of suspects to his officers. The listed men were dragged out of their beds at two o'clock in the morning—most were not allowed to dress completely—and were herded through the mud by cavalry for several miles out of Pittsburgh, then herded back. They were corralled in an open pen, where they were held in rain and snow while soldiers jeered at them. A fire was built, but when the prisoners would approach to get warm, they were driven away with bayonets. The next day they were moved to a foul "waste house," where they were held for five days, after which they were lodged in an unheated jail.

After ten days of this inhumane treatment, they were brought before a judge, who ordered them all released for lack of evidence. It was only then that the authorities realized that almost all of the "traitors" were actually witnesses or distillers who had plea-bargained by signing up for an amnesty offered before the arrival of the army.

Over the next few months, many distillers registered their **stills**, ready to pay the excise tax. A group of 20 alleged leaders of the rebellion were taken into custody to be tried in Philadelphia. After a month of trudging through snow, rain, and mud, they arrived in Philadelphia on Christmas Day and were paraded and humiliated before 20,000 people, then put in jail for six months while awaiting trial. Not a single one of them was found guilty.

Once the troops left the Monongahela Valley, the farmers who had fled returned and immediately resumed making whiskey without paying taxes. But now they had to do it in stealth, by the light of the moon, and so became known as "moonshiners."

In the U.S. in the 1800s, alcohol and blends of alcohol were competitive with other heating and lighting fuels. Alcohol/camphor blends cost half the price of lard oil and a third the price of whale oil,[4] the most common illuminants at the time. By mid-century, alcohol was being produced on farms and in distilleries in almost every U.S. county. Much of the alcohol came from apple orchards. Although the Bible warned against drinking the beverage of the grape, it said nothing about apples. So applejack, or apple whiskey, was distilled from "spitters." These were bitter, tart fruits (*ptui!*) that made great hard cider, distilled spirits, and vinegar.

In 1861, Congress once again passed an excise tax on alcohol. This $2.08 per gallon, which applied regardless of how the alcohol was used, helped foot the bill for the Civil War.[5] The thriving alcohol industry, which had been producing over 25 million gallons per year for lighting and industrial purposes, was effectively demolished by the end of the war.[6]

GERMANY'S FARMER COOPERATIVES

In 1877, Nicholas Otto of Germany was credited with patenting the first **internal combustion engine (ICE)** vehicle designed to operate on alcohol. The earliest ICEs had run on coal gas and were used as stationary engines in mines.

Germany had little native petroleum, but lots of agriculture. Its economy was heavily invested in agriculture, with the capacity to produce much more than could be marketed. Germany had a surplus of sorghum and potatoes, and the government had subsidized its potato farmers for years—during which time bumper crop after bumper crop depressed the market price, until the entire farm economy was on the verge of collapse.

The government hoped that by encouraging distillers, it could eliminate the subsidy program, and regain its loans.[7] So, in 1887, the government enacted financial incentives (tax exemptions, loans, and guarantees) for distillers and farmer cooperatives. The distilleries and farmer cooperatives were not government-owned.

A farmer would bring his excess potato crop into a cooperative distillery. No money changed hands. The distillery marketed its share of the fuel and some of the fermentation byproduct (spent **mash**). The farmer received back a portion of his alcohol, and most of the spent mash. The farmer used his mash (especially the liquid portion) to fertilize his land, and his mash-fed animals became meat he could sell at a profit.[8] His alcohol provided heat, light, and power for the new machinery, or he sold it to others in his area.

These cooperative distilleries were also a great deal for the government. Within a few years, Germany's farm economy was stabilized; the country had a healthy export business in potatoes, alcohol, alcohol engines, and meat; and it didn't have to buy foreign oil. Employment increased; commodity surpluses were checked. And all this was accomplished without depleting the soil, which increased in depth during this period from one to five inches a year.[9]

"Gasoline is not a substance characterized by any definite physical or chemical limits but is really only a name applied to reasonably volatile hydrocarbon mixtures which are chiefly for motor fuel. These hydrocarbon mixtures are exceedingly complex and the possibilities of variation in physical and chemical properties are many."[10]
—BUREAU OF MINES, OCTOBER 19, 1916

FORD AND ROCKEFELLER

Back in the United States, President Teddy Roosevelt spearheaded legislation in 1906 that lifted alcohol taxes for industrial use and simplified paperwork for farmers and others who might build their own stills. Farmers now had a way to convert their grain surpluses to a useful product.

One enterprising automaker of the era preferred alcohol as a fuel. In fact, until 1931, Henry Ford's automobiles, including the Model T and later the Model A, were designed to run on either alcohol or **gasoline**.[11] Mass production, led by Ford and copied by others, brought the price of cars—and especially tractors—within reach of the majority of people involved either directly or indirectly in farming. Ford envisioned a day when the Industrial

Revolution would benefit the rural landowner, and technological innovations would make work easier and life more ideal.

Ford's contemporary, John D. Rockefeller, had his own ideas. His primary industry was kerosene, which he distilled from oil, for use in boilers and for lighting. At night, the toxic wastes from processing—the **volatile**, explosive components of oil—were flushed into rivers. Eventually, Rockefeller was able to use some of this toxic waste to run internal combustion engines. The fuel was dirty and dangerous, but he sold it cheap, since it would otherwise simply be toxic waste. This was gasoline. It soon became the prevalent fuel in the cities, backed by oil company money and organization.

By 1906, because of the rising use of gasoline for fuel and electricity for lighting, alcohol production in the U.S. was down to ten million gallons a year.

Fig. 1-2 Spark advance lever. *The lever just under the wheel is the spark advance lever, used for changing* **ignition timing** *when you go from one fuel to another fuel or when you're accelerating.*

DAVID BLUME

"*The Rockefeller interests have been busy strangling the interest in denatured alcohol, not only in this country, but in Germany, France, England, and Canada, but they were not as successful there as they were here. They know that promulgation and introduction of alcohol sounds their death-knell, for it will enter into every phase of rural and urban life, unifying the two and bringing about improvements not dreamt of by the present generation.*"

—CORRESPONDENCE FROM MR. PAUL PFEDHNER TO GEORGE HAMPTON OF *FARMERS OPEN FORUM* MAGAZINE; PFEDHNER TRAVELED TO GERMANY AND EXAMINED THE ALCOHOL PROGRAM THERE FROM 1913–14.

Nevertheless, in rural areas—and that was most of America—alcohol was the main fuel, and often the only fuel there was. Motorists out for a weekend drive could either carry all their gasoline from town, a dangerous practice, or stop at almost any farm to fill up on alcohol.

The Model T and Model A were dual-fuel vehicles. The **distributor**, which sent the spark to each **cylinder**, was "*digital*": You used the five digits of your left hand to pull the lever on the steering column to advance the spark timing as you drove. When running on alcohol, you would advance the spark timing a great deal more than for gasoline, to accommodate alcohol's cooler-burning, high-**octane** qualities. The dual-fuel vehicles also had a **carburetor** that could be adjusted from the inside of the automobile by turning a knob (incorrectly termed the **choke**) on the dashboard.

By the way, the Model T got 34 miles per gallon, compared to our **corporate average fuel economy (CAFE)** of 35 mpg today.

The Henry Ford Museum

In 2004, I did something I had wanted to do for 20 years: I made a pilgrimage to The Henry Ford Museum and Library in Dearborn, Michigan. Most of the exhibits I examined dated from about 1915 to 1921. World War I was prominently featured in news clippings, as were Prohibition and extremely high gasoline prices. Although the press claimed the prices were due to a shortage of gasoline, they appear in hindsight to have been one of the first incidents of oil companies manipulating supply in order to hike prices during a wartime economy.

Major oil corporations or their associates had already quietly taken over ownership of the major distilleries. These distilleries put up almost no significant fight when Rockefeller funded Prohibition initiatives—first in states and finally at the federal level—that threatened their existence. Scores of pre-Prohibition letters from brewery owners beseeched Ford to buy their plants—or, more commonly, begged in as dignified a way possible for the information the breweries desperately needed to make their **beer**, which was about to become illegal, into distilled fuel (essentially whiskey). But that **distillation** information was carefully guarded expertise that now resided largely in the oil industry, which had hired all the top distillery engineers to work in its **refineries**.

It was clear from his secretary's correspondence files that Ford certainly saw the hand of Rockefeller in the whole Prohibition movement but was nevertheless conflicted. A teetotaler, Ford was in favor of doing something about alcohol as a beverage, but thought that a Constitutional amendment was "too radical."[12]

With the perfidy of a bought-off Congress, thousands of small brewing companies faced ruin. Just as in America's first months, when George Washington's support for taxation and regulation of distilling had eliminated small-farmer participation in the beverage industry, corporations had again used the government regulatory burden to stifle competition, this time on behalf of oil. Prohibition represented the destruction of hundreds of thousands, perhaps millions, of jobs and the reduction to scrap of hundreds of millions of dollars of capital equipment.

In 1916, Ford was quoted in the *Detroit News:* "The 60 brewing plants in the state need not be abandoned. Millions of dollars are invested in these plants. Economically it would be a shameful waste to have them become idle. But there is no reason why they should become so. Every standard brewing plant can be transformed from a brewery into a distillery for manufacturing **denatured** alcohol for use in automobiles or other internal combustion engines."[13]

Ford received probably thousands of letters from many countries in the two years following the *Detroit News* article, which is remarkable when you realize there was no *USA Today* or Internet. Hundreds of the letters still exist in his archives today. They reminded me of the outpouring of sentiment when I first taught my workshops in the 1980s.

Between 1915 and 1921, engineer Jack Dailey served as Henry Ford's alcohol fuel project director. Dailey performed more than 100 trials on **feedstocks**, looking for the best crops for alcohol fuel production. The correspondence in the museum showed that he was working on engine modifications, various fuel mixtures of kerosene and alcohol, power output of various dilutions of water and alcohol, and much more.

Dailey was a main engineer on the revolutionary Fordson Tractor, which, right from the start, was able to run on alcohol, and which was exported to many countries in that configuration. Before World War II, there was still no extensive system of tanker transportation for crude oil. New Zealand

INDIANA FARM BUREAU

Fig. 1-3 Seagram alcohol tractor. Seagram, the beverage company, kept this tractor running on alcohol all the way into the early 1970s.

made alcohol fuel from beets, the Philippines used nipa palms, and much of South America made fuel from sugarcane. Henry Ford produced and exported Dailey's alcohol-powered tractors, trucks, and cars all over the world.

Jack Dailey also handled the bidding to build a commercial distillery at a Ford farm. He tangled with the taxing authorities, who held up legalizing Ford's plans to build his alcohol plant for years, arguing over technicalities—for instance, whether the permit should be rejected and redone because of such details as whether the name of the applicant was "Henry Ford" or "Henry Ford and Son."

The amazing thing is that the six years of work done by Dailey and almost all the records of his actual trials, experiments, and results have been

FROM THE COLLECTIONS OF THE HENRY FORD MUSEUM

Fig. 1-4 Jack Dailey with a Henry Ford tractor. This tractor, bound for England, was designed to run on alcohol. Jack Dailey, Ford's Chief Alcohol Engineer and Research Director, is seen leaning on the large rear wheel.

completely expunged from the archives. No records of the still or experiments at the farm exist. None of the mileage and horsepower trials liberally referenced in correspondence appear. No draft of the booklet that Ford promised to hundreds of correspondents about alcohol fuel production and use—which, according to the various notes, was nearing completion—exists today. No photographs of the still, no legal paperwork in application to the government for his permit, survive.

Jack Dailey does not even appear in the library's card catalog index. I found one of his notebooks (which apparently survived because his name wasn't on it, although his distinctive handwriting, which I had seen in correspondence, was unmistakable). In finding that notebook, I experienced both pleasure and sadness in holding something tangible from a man with whom I came to feel kinship through his papers. I felt a lot of connection to Jack, whose passion for alcohol fuel was wiped from history. I am pleased to honor him and his work today.

THE BEGINNINGS OF BIODIESEL

Dr. Rudolf Diesel was the inventor of the **compression-ignition engine**. His design came along at the end of the 19th century, well after the **spark-ignition engine** that works with gasoline or alcohol. He originally was hoping for an engine that would run on finely ground coal, but due to technical difficulties he gave that up for liquid fuels, namely vegetable oils. This was the beginning of biodiesel.

After having success with peanut oil and animal fat, Diesel focused on African nut palm oil, due to its having the highest oil yield per acre of all land plants. With industry and the military enthusiastically embracing these engines, the U.S. and European governments promoted the growing of palms in underdeveloped tropical countries.

Nut palms take about ten years to reach maturity to produce a good yield, so Rockefeller had to work fast. While the vast acreages of palms were growing, his chemists were busily working to create synthetic diesel. They succeeded. By the time the African palms were producing, very cheap petroleum-based synthetic diesel had already captured the market.

So what was a tropical nut palm farmer to do? Market a food oil that had the ability to withstand very high heat. For decades, people ate the stuff, until medical proof showed that consuming fully saturated palm oil was disastrous for human health. That's what happens when you eat diesel fuel.

WORLD WAR I

Alcohol experienced new life during World War I—war has a way of creating huge energy demands. Alcohol was used in the production of gunpowder, mustard gas, explosives, and so forth. Over the four years of the war, alcohol production bounced back up to an average of 50 million gallons a year.[14]

At the same time, well-known scientists and automakers were beginning to warn that the incredibly fast changeover from steam to internal combustion power for industry and transportation was depleting oil as a natural resource. The weapons of war were rapidly becoming petroleum-dependent, and the U.S. was using up a large part of existing oil supplies. Rockefeller, and the chemical companies that relied on petroleum products, became very rich during World War I.

Just before the war, the Grange—an activist, progressive, national farmers' "union"—had begun to investigate implementing the German cooperative alcohol system in the U.S.; they obtained good working designs for farm-scale distilleries. But this was potential financial disaster for the early-20th-century equivalent of **MegaOilron**.

When World War I broke out, with Germans as the enemy, the oil industry set to work branding the cooperative alcohol fuel movement as a "dangerous German idea," fueling public suspicion. Then in the hubbub at the end of the war, Rockefeller, flush with cash, decided to deal his competition a killing blow, doing it the same way it had been done during the Whiskey Rebellion: with federal power.

Under the guise of being a devout Christian, Rockefeller supported a fringe movement of holy rollers who decreed that "demon rum" was to blame for the ills of society. This fringe movement was composed of small-time operations that should have had no more effect on a community's drinking habits than an irate grandmother raising hell in the local pub.

But those small clusters of uptight citizens found a powerful patron and benefactor in Rockefeller, who donated from $1.5 to $4 million (the figure varies according to whose report you read) to the movement. Even the $4 million figure may be low, since Rockefeller was conducting similar campaigns in many countries around the globe. (Four million 1918 dollars equal 50 million of today's dollars.[15]) Members of the Rockefeller family sat on the boards of several

World Map of Developed and Potential Petroleum Reserves, 1919

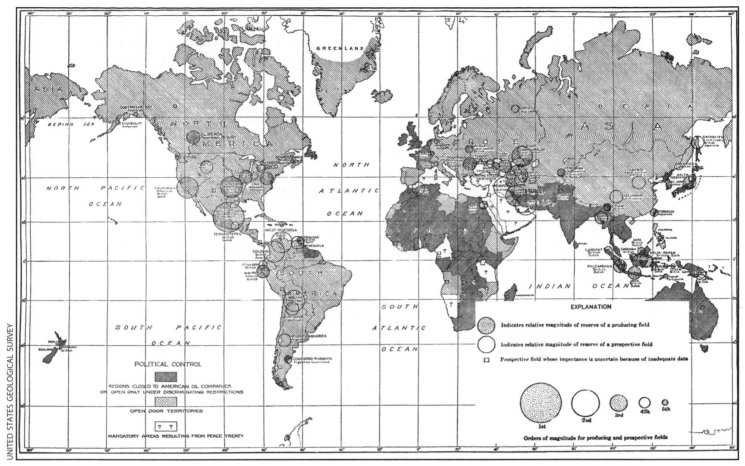

Fig. 1-5 *World map of developed and potential petroleum reserves, 1919.* *What's striking about this very early map is how much of the world's large oil reserves were already discovered. A few deposits, such as in the North Sea, had not been located, but most of the Middle East had been well explored by this time. Note the super-giant field in Mexico. This reserve was reported newly "discovered" to the public during the 1980 fuel crisis.*[16] *A similar field was "discovered" yet again in September 2006 amidst high gasoline prices and public demands for ending dependence on* **fossil fuels***.*

temperance organizations, including the Anti-Saloon League.

In 1919, Congress passed the 18th Amendment to the Constitution, known as the Volstead Act, or Prohibition, which made the production or transportation of alcohol illegal. Imagine, if you will, an all-male (probably hard-drinking) Congress drafting and voting for a constitutional amendment—not simply a piece of legislation—to ban the production of alcohol! It had little to do with demon rum and everything to do with the money, power, and connections the oil industry developed in supplying the U.S. military with the fuel it devoured in World War I.

It wasn't until 1933 that Congress finally passed the 21st Amendment to the Constitution, revoking Prohibition. The restriction had had almost no effect on drinking habits. But during the 14 years that alcohol was illegal, gasoline had become

entrenched as the national fuel, and alcohol was driven way underground. Even Henry Ford had given up making dual-fuel vehicles.

THE GREAT DEPRESSION TO WORLD WAR II

Even before the Great Depression, the 1920s were a dismal time for farmers. During World War I, the growth of agribusiness in America had been encouraged by war-torn Europe's need for food. Farmers in the U.S. had increased production by investing in machinery and some of the first chemical fertilizers, and the tractor had replaced the horse and mule for farm power. When the war was over, so was demand, but farmers couldn't afford to cut production—they had to pay for their expensive equipment. Something had to give, and it was price: Grains fell twelvefold to a low of 25 cents a bushel.

For the American farmer, Prohibition couldn't have been in effect at a worse time. Converting their grain into alcohol would have saved a huge number of farmers from ruin. Instead, Prohibition gave oil companies nearly 15 years of competition-free marketing of their product.

When Prohibition was finally repealed in 1933, oil companies weren't thrilled about sharing their market with a bunch of farmers. The age of mass communication had begun, and the **American Petroleum Institute (API)** coordinated a nation-wide campaign to discredit alcohol as a fuel. Most damaging were the published reports of a test conducted by the API and the American Automobile Association, which indicated that alcohol blends measured significantly less mileage and less drivability, and underwent **phase separation** in humid air. The rigged test was less a testament to the power of the written word (although the press had a field day with it) than to the power of the API, the public relations arm of U.S. oil companies.[17] (Recently, in cahoots with the California Air Resources Board, the API tested **permeation emissions**. Strategically, this is a carbon copy of this amicable 1930s collaboration.)

The oil industry's campaign did not stop the farmers, **chemurgists**, scientists, or organizations dedicated to using crops and natural materials instead of oil to make industrial materials. In a 1936 fuel symposium hosted by Henry Ford, the opposition was silenced when Francis Garvan, a leader of the alcohol fuel movement in the '20s and '30s, rose and said, "We have been fed volumes [by the oil industry] to the effect that power alcohol is not a practical fuel. Were they quite sincere? I think you can judge."

Garvan then distributed copies of an English fuel advertisement published by Standard Oil of New Jersey's subsidiary Cleveland Discol, and read from it.[18] "If you take your little pamphlet, you will find that all these worries have been settled for us. All this chemical research has been done for us, and all the testing. The Standard Oil Company of New Jersey has gone over to England, and in its delightful international aspect of life has joined hands with the English Distillers Company and they together have produced, in their own words, 'The most perfect motor fuel the world has ever known.'"

The product was Discol,[19] an alcohol blend about which the pamphlet boasted, "It is possible to pour almost a pint of water into a car tank containing ten gallons of Discol without the slightest trouble—in fact in some circumstances with better running."

All along, oil companies (including Texaco, Esso, and Ethyl) were doing a lively alcohol blend business in Europe (Europeans could buy Koolmotor Alcohol Blend, comparable to the best racing fuels of the day)—at the same time that they were insisting in the U.S. that alcohol production was impractical and produced an inferior fuel that destroyed engines.

COURTESY OF HAL BERNTON

ALCOHOL
FOR HIGH SPIRITED HORSE-POWER FROM A THOROUGHBRED CAR

COOL Alcohol adds volatility for quick-starting, supercharges the cylinder by cooling and contracting the mixture, and adds power by lowering exhaust valve temperature. EXTRA POWER

CLEAN Alcohol saves overhaul costs by burning with a carbon-free flame and eliminates any existing carbon deposits. EXTRA ECONOMY

CONTROLLED Alcohol smoothes vibration by timing power production. EXTRA TUNE

COMBUSTION Alcohol, by burning coolly, cleanly and completely, adds to all-round engine efficiency. EXTRA EFFICIENCY

Write for a copy of informative booklet full of interest to keen motorists
CLEVELAND PETROLEUM PRODUCTS CO., Central House, Upper Woburn Place, London, W.C.1

CLEVELAND DISCOL
BRITISH MOTOR ALCOHOL SPIRIT

Fig. 1-6 1936 alcohol advertisement in England. While saying just the opposite to American motorists, oil companies were claiming alcohol was a superior fuel in horsepower, carbon deposits, and improved engine life. Recently, a replay of this embarrassing disconnect happened, with oil-company propaganda in Australia against alcohol as a fuel, while other Commonwealth countries such as Canada were praising its virtues.

imported molasses. Not willing to support foreign agriculture when agriculture at home was at its low point, American farmers, the backbone consumers of the fuel, abandoned it.[20] It was a PR coup that the sophisticated oil company spin doctors accomplished with their money and access to the media. Today, oil companies are agitating to import alcohol from abroad in a transparent attempt to duplicate their success in stifling alcohol in the 1930s.

Between 1933 and 1939, more than 40 countries legislated assistance for their alcohol fuel movements. But by 1940, U.S. alcohol fuel distilling was almost nonexistent. World War II was beginning; oil company executives held major offices during the war (under Democrat Franklin D. Roosevelt), dictating many of America's energy policies. Unbeknownst to the government, many so-called U.S. oil companies had already evolved to think of themselves as transnational corporations beholden to no country. They secretly had deals with the Nazis and even provided them with fuel at sea for their submarines.

The Japanese used alcohol in their aircraft, the Zero. The Germans used alcohol made from potatoes; since fuel production was decentralized in distilleries, it took the Allies several years to bomb enough distilleries to dry up Germany's fuel supply.

Despite oil company efforts, many alcohol ventures grew and flourished in the Midwestern U.S. for several years. These included the Alcogas Company, the Vegehol Company, and, perhaps the largest, the Agrol Company, of Atchison, Kansas. At the height of the alcohol movement in the 1930s, several thousand stations carried alcohol blends in eight Midwestern and several Northwestern states. At its peak, Agrol sold its blend of alcohol and gasoline at over 2000 stations.

Oil companies wouldn't prevent their distributors from selling the alcohol blends, but they did demand removal of all signs and markings showing the oil company's affiliation with stations that chose to sell the blends. The federal government had to pass legislation prohibiting such activities.

What finally broke the Agrol Company, when propaganda and manipulation couldn't, and what broke the Midwest alcohol fuel movement as well, was a rumor spawned by the oil industry that Agrol's alcohol was being made from

[WHO WOULD PAY-

COST OF ALCOHOL GASOLINE

60% OF IOWA MOTOR FUEL-BILL

IOWAN'S POCKETBOOK

40% OF IOWA MOTOR FUEL BILL

IOWA MOTORIST

IOWA FARMER

FOR CORN-ALCOHOL?]

COURTESY OF HAL BERNTON

Fig. 1-7 Illustration in pamphlet circulated in 1933 by the Iowa Petroleum Council. The constant refrain that alcohol is a subsidized fuel, while supposedly gasoline is not, continues in American Petroleum Institute propaganda today. Oil companies attack any subsidy or tax benefit for alcohol fuel, while they reap billions in tax breaks.

The U.S. used alcohol fuel, too, made from wood (**cellulose**!) and other **biomass**.

Synthetic Rubber

One of the United States' problems in World War II was that the enemy had managed to cut off the country's supply of natural rubber by bombing Pearl Harbor. Six hundred million dollars were entrusted, mostly to Standard Oil and its subsidiaries, to develop synthetic rubber from oil, in order to provide the U.S. war effort the necessary feedstock for boots, tires, and other essential items.

At the same time, Standard Oil had an agreement with the largest chemical manufacturer in the world, Nazi-controlled I.G. Farben Company of Germany (supplier of chemicals to concentration camps), to restrain trade. As part of that agreement, Standard secretly dragged its feet and held up rubber production (and other forms of chemical production from oil); I.G. Farben meanwhile was to restrain its participation in the production of synthetic fuel from coal (which it had no intention of producing).

Annual Value of Imported Oil from Selected Countries

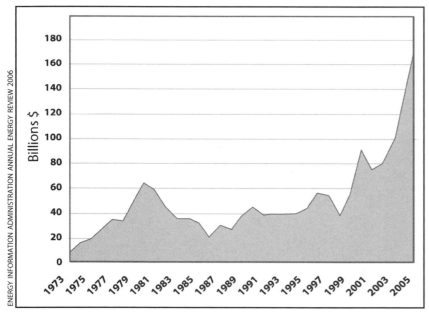

Fig. 1-8 Annual cost of petroleum imports. *Once the U.S. peaked in oil production in 1970, it could no longer control world pricing of oil by increasing domestic production. The U.S. repeated its control strategy eventually by making Saudi Arabia an ally that could increase production to prevent price spikes through the 1990s. With Saudi production peaking and the worldwide functional peak of oil production, which started in late 2003, there is no longer any ability to manipulate the price of oil downward. Insufficient refining capacity worldwide inflames the issue. 2006 costs will take the graphed line far off the chart.*

The oil-to-rubber process started losing credibility. It was obvious the oil companies couldn't (or wouldn't) have enough rubber available quickly enough to do any good. Around the middle of World War II, Senator Guy Gillette (D–Iowa) presented Congress with evidence that a synthetic rubber process using alcohol would cost one-tenth as much to set up as the oil process, would use one-fifth as much rationed steel, and could be operational within six to eight months, as compared to an oil-to-rubber plant requiring 18 months or more.

Two Polish scientists who testified at that congressional hearing had been smuggled into the U.S., protected by a British spy ring. The British Security Coordination had declared Standard Oil a "hostile and dangerous element of the enemy," and took these precautions to avoid a run-in with any corporation affiliated with the synthetic rubber program of the "enemy" before the hearing.

Oil-based-rubber production plants rose only toward war's end; they were still not operational as late as the Normandy invasion in 1944. Wartime alcohol plants were a smashing success, however,[21] providing not only rubber, but explosives and fuel for aircraft—which provided reliable torpedo power, gave the U.S. a power edge, and made possible higher-elevation and extended operations using blends. (Many GIs remember "torpedo juice" hangovers.)

The nation's alcohol production capacity rebounded and grew fourfold between 1942 and 1944, to 600 million gallons. But at the end of the war, those late-starting petroleum-based synthetic rubber plants stood there, finally completed, built with our tax dollars, fueled by an expanded cheap petroleum supply from taxpayer-subsidized wells, delivered by taxpayer-subsidized pipelines and tankers—and supposedly able to produce synthetic rubber at 11 cents per pound compared to alcohol's 21 cents.[22]

In 1942, Assistant U.S. Attorney General Thurman Arnold exposed Standard Oil in addressing Congress: "Not only was the production of synthetic rubber in this country absolutely stifled by Standard's adherence to the restrictions imposed on them by I.G. Farben—which they always loyally preserved—but after 1939, when Standard received permission from I.G. Farben to enter into negotiations with rubber companies, Standard proceeded to further retard the development of synthetic rubber because of its natural monopolistic desire to keep complete domination over this industry.[23]

There it was, the bottom line, as various Defense and Energy Department boards (often composed of oil company executives) pointed out in their recommendations to the government: a "natural monopolistic desire" to dominate. They weren't even subtle about it. Oil company lobbyists pushed legislators to abandon grain and cellulose alcohol plants. Contracts to dismantle the wartime alcohol plants were in many cases awarded to subsidiaries of oil corporations.

Immediately, oil-based synthetic rubber jumped in price—first to 15 cents per pound; then, with no alcohol-based competition to hold the price down, abruptly to 60 cents per pound.[24] Over the next few decades, the alcohol fuel movement went underground again, while the nation wallowed in surplus grain and commodities.

POST-WORLD WAR II TO NOW

In the 1950s through early '70s, most Americans believed most everything Uncle Sam had to say. A country that could defeat Hitler, invent the atomic bomb, and provide a standard of living hitherto unknown in the history of the planet was a government to believe in. Things seemed pretty rosy.

But this period marked the peak and subsequent decline of American crude oil production, as had been foreseen years earlier. The decline in oil production at home didn't make big news. The oil industry and the few government officials who were monitoring the situation didn't see it as advantageous to advertise the fact. Besides, Vietnam was a lot more interesting than oil. In fact, it's been suggested that a principal reason for U.S. involvement in Vietnam was to protect vital tungsten mines and rich offshore oil leases, which the United States felt would be lost if South Vietnam fell to the North.[25,26,27]

In 1970, not long after oil production peaked in the U.S., a freak accident occurred in Libya: A bulldozer backed into a pipeline, causing it to be shut down for repairs. Suddenly there was a slight shortage of oil, and a bidding war erupted, causing a temporary spike in the price of oil worldwide. Up until that time, very few people realized how tight supply was with respect to demand.

But this incident sure woke up a sleeping giant. In October 1973, after the U.S. supported Israel in the Yom Kippur War (when Arab military forces

*Fig. 1-9*

GUEMSEY LE PELLEY, COPYRIGHT 1979 THE CHRISTIAN SCIENCE MONITOR

invaded Israel), a group called the Organization of Petroleum Exporting Countries (OPEC)—certainly not a household word then—raised the price of oil $3 a **barrel**, finally reaching $11 over the next three months. It has been argued that the Saudis initiated this in order to scale back production so they could install water-injection equipment in their biggest oil field.[28]

U.S. leaders appeared to be caught completely by surprise. There was nothing to do, said the oil companies; U.S. companies were all pumping at maximum capacity. The new high price of oil fanned inflation, closed factories, reduced employment, and gave Americans a hint of what was to come.

MegaOilron has always been deft in the exploitation of national crises such as war. The first oil crisis of 1973–74, in which this country's imported Middle Eastern supplies were cut off, is a case in

---

The technologies for alcohol production are proven. Although additional production capacity is needed to make ethanol cost-competitive with petroleum-based gasoline, ethanol technology has advanced to a greater degree of commercialization than that for any other proposed synthetic fuel. Yet, we do not have the national commitment, we have not drafted the clear strategy, to let the private sector know that the developing alcohol fuels industry is a legitimate and important contributor to the Nation's energy needs.

—FORMER INDIANA SENATOR BIRCH BAYH, MAY 1, 1980

point. The Alaska pipeline, the oil industry's big project in those days, had been dying under the public's opposition; it had no chance of being approved. But during the oil crisis, the oil industry held out its proposed pipeline as some salvation. Congress fell for the ploy, and the project went through, even though the production pipeline took many years to complete and never had any effect on resolving the crisis.

In the last couple of years, during our current price rise, MegaOilron pushed to open the Arctic National Wildlife Refuge. Some longtime Democrats had seen this scam before and were not swayed by the emotional appeal of being promised relief from oil shortages now, knowing that any project of this nature takes a decade to result in any oil. Republicans had to resort to a back-door dirty trick to get the scam approved this time.

The oil crisis years of the 1970s turned everyone's attention to energy and its uses and costs. In 1978, it was calculated (by General Motors, no less) that **Fig. 1-10** given the amount of energy needed to maintain the minimal healthy annual growth rate in the economy (3%), the world supply of oil would be exhausted in 25 to 30 years, assuming all known oil supplies and those expected to be verified were developed.[29] Due to extended depressions and a couple of unexpected finds like the North Sea oil field, the point at which production of **conventional oil** will dwindle each year has been pushed back a little. Persuasive arguments have been made by many petroleum geologists, such as Colin Campbell, that we may have already hit world peak production.

By the late '70s and early '80s, Congress was asking hard questions of the oil industry, and looking at the prospect of world oil production peaking, as U.S. production had peaked in 1970. The hard questioning was prompted in part by price hikes brought about by the revolt against the Shah in Iran, Iraq's war with Iran, and OPEC oil embargoes.

Senators such as George McGovern, Frank Church, Birch Bayh, and William Proxmire (and, on the Republican side, Charles Percy), and even

ALTERNATIVE FUELS BAIT AND SWITCH

In January 2007, Democrats in Congress introduced the Biofuels Security Act of 2007, which would raise the **Renewable Fuel Standard** to 60 billion gallons per year, force oil companies to carry alcohol at the pump, and repeal tax breaks for oil companies in order to pay for it all. This, of course, is reminiscent of Carter's windfall profits tax. This time around, however, MegaOilron isn't going to let the law pass and then subvert it; they are going in for the preemptive strike.

Also in January 2007, George Bush's State of the Union Address seemed to support alcohol, calling for America to replace 30% of its imported fuel. But hidden in the Republican proposal is a conversion of the Renewable Fuel Standard to an Alternative Fuel Standard—which would reward oil companies for converting coal, which destroys the climate and environment, to liquid fuel and other fossil-based alternatives that would elbow renewable fuels out of the way.

President Jimmy Carter, recognized the need for alternative fuels. They spearheaded the return of the chemurgists' vision of producing their own fuel. Jerry Brown, an early supporter of alcohol fuel as governor of California, established a revolving loan fund to finance small-farm-based distilleries.

When MegaOilron raised its prices to match OPEC's—even while receiving massive U.S. public subsidies for decades—Congress passed the Crude Oil Windfall Profit Tax Act of 1980 to take back part of those obscene, unpatriotic profits. The proceeds of the tax were dedicated to alternative energy: **tax credits** for alcohol fuel plant construction, for **cogeneration**, solar panels, and more.

By 1979, it appeared that America was heading for a revolution in energy, toward a future that was farmer- and solar-based. For those of us involved in alternative energy, the future seemed sunlit and bright indeed. Research in alternatives was a priority of the **Department of Energy (DOE)**. The White House had solar water-heating panels installed on its roof as symbolic proof of the intent of the government. Brazil, with a long history of alcohol fuel production, instituted a national program to return to it—dragging auto companies into line by requiring them to convert assembly lines to make dedicated alcohol vehicles within six months. It was a heady time, and revolution was in the air. Little did we know.

MegaOilron engineered a massive retaliation. Carter's windfall profits tax earned him enemy status, and massive campaign funding flowed into Republican party coffers to run MegaOilron insiders. Senators who had spearheaded the tax were all targeted and driven out of Congress. In came Reagan (a longtime darling of General Electric) and George Bush (an established oilman), heavily backed by MegaOilron. The solar panels were ripped off the White House, windfall profits taxes were redirected to be paid out to MegaOilron for such "alternative fuels" as spraying diesel oil on coal; federal loan guarantees for alcohol plants were reneged upon; my television series and book were shut down; and oil became relatively inexpensive for a time.

With Saudi cooperation, OPEC price controls fell apart, as managed overproduction kept surplus oil in the market, driving down the price. By the mid-'90s, with a massive increase in subsidies, oil briefly dropped to less than $10 per barrel, economically making alternatives impossible to justify. All the predictions of peak oil production were ignored, and those who dared to raise the subject were ridiculed. Discussion of alternative energy became dangerous to the professional lives of scientists; government funding for research into alternatives pretty much dried up and was not replaced in the emerging era of corporate funding of research. The '80s and '90s were a party for the Oilygarchy, built on the dance floor of a flood of cheap oil.

"This is a government of the people, by the people, and for the people no longer. It is a government of corporations, by corporations, and for corporations."

—RUTHERFORD B. HAYES, U.S. PRESIDENT, 1877–1881

But now the party's over. History is colliding with current events. With oil production peaking all over the planet, not just in the U.S., it's no longer possible to use the production of a single country to manage supply, and therefore price. World refinery capacity is now inadequate for seasonal peaks in world oil use. In fact, the dual effects of skyrocketing demand and peaking production are combining for a perfect economic storm. OPEC once again is in command of the price of oil. Compounding the problem is global warming, so even if there were a way to produce more oil, using it could very well deliver the final blow to climate stability.

As George Santayana said, "Those who cannot learn from history are doomed to repeat it." We are now at a critical point where we must learn from history and chart a new course. Everything depends on the choices we make next.

Endnotes

1. Leland D. Baldwin, *Whiskey Rebels: The Story of a Frontier Uprising* (Pittsburgh: University of Pittsburgh Press, 1939), 26.

2. Baldwin, 3.

3. Baldwin, 3.

4. Rufus Frost Herrick, *Denatured or Industrial Alcohol: A Treatise on the History, Manufacture, Composition, Uses, and Possibilities of Industrial Alcohol in the Various Countries Permitting Its Use and the Laws and Regulations Governing the Same, Including the United States* (New York: J. Wiley & Sons, 1907), 287. Alcohol was 54 cents a gallon—half the price of lard oil and one-third that of whale oil.

5. Herrick.

6. Herrick.

7. Charles Edward Lucke and S.M. Woodward, *The Use of Alcohol and Gasoline in Farm Engines* (Washington, DC: U.S. Dept. of Agriculture, 1907).

8. Harvey Washington Wiley, *Industrial Alcohol: Uses and Statistics*, Farmer's Bulletin (Washington, DC: U.S. Dept. of Agriculture, 1906).

9. Wiley. Soil depth based on author's calculations.

10. *Proposed Specifications and Methods of Analysis for Gasoline to Be Used [as] a Motor Fuel*, U.S. Bureau of Mines (October 19, 1916). Unpublished draft of regulations sent to Henry Ford for feedback.

11. Melvin Kranzberg, *Technology and Culture: An Anthology* (New York: Schocken Books, 1972).

12. *The Reminiscences of Mr. Charles Voorhees, Volume II*, from Ford Motor Company Archives Oral History Section (November 1952).

13. Ford R. Bryan, *Friends, Families & Forays: Scenes from the Life and Times of Henry Ford* (Detroit: Wayne State University Press, 2002).

14. *Industrial Alcohol*, Report No. 2, War Changes in Industry Series (Washington, DC: U.S. Tariff Commission, January 1944).

15. Robert Sahr, *Inflation Conversion Facts for Dollars 1665 to Estimated 2016*, http://oregonstate.edu/politsci/faculty/sahr.htm (2006).

16. George Raine, "Big Oil Find by Chevron Team Deep in Gulf," *San Francisco Chronicle*, September 6, 2006, Sec. A.

17. Hal Bernton, Bill Kovarik, and Scott Sklar, *The Forbidden Fuel: Power Alcohol in the Twentieth Century* (New York: Griffin, 1982), 19.

18. Bernton, Kovarik, and Sklar, 21.

19. Bernton, Kovarik, and Sklar, 23.

20. Bernton, Kovarik, and Sklar, 26.

21. Bernton, Kovarik, and Sklar, 30–32.

22. Bernton, Kovarik, and Sklar, 33.

23. U.S. Congress, Senate, Special Committee Investigating the National Defense Program, *Statement of Assistant Attorney General Thurmond Arnold, Hearings on Senate Resolution 71*, 77th Cong., 2nd Sess., March 26, 1942, p. 4313, cited in Bernton, Kovarik, and Sklar.

24. Bernton, Kovarik, and Sklar, 33.

25. Archimedes L.A. Patti, *Why Viet Nam?: Prelude to America's Albatross* (Berkeley: University of California Press, 1980).

26. Harrison E. Salisbury, *Vietnam Reconsidered: Lessons from a War* (New York: Harper & Row, 1984).

27. U.S. National Security Council, *United States Objectives and Courses of Action with Respect to Southeast Asia, Statement of Policy*, Document 20, NSC 5405 (June 25, 1952).

28. Edward Jay Epstein, "OPEC: The Cartel That Never Was," *The Atlantic*, March 1983.

29. Council on Environmental Quality and Gerald O. Barney, *The Global 2000 Report to the President: Entering the Twenty-First Century* (Penguin, 1982), 161.

CHAPTER 2

BUSTING THE MYTHS

For more than 30 years, I have known of the potential of alcohol fuel (**ethanol**)—an enthusiasm I shared with my late colleague, master designer R. Buckminster Fuller, who had been involved in ethanol research way back in the '40s. In 1983, in the waiting room prior to our press conference launching the *Alcohol as Fuel* television series, Bucky and I discussed the sheer elegance of the alcohol solution.

In design terms, what we were doing was producing a powerfully compact, convenient, non-toxic, liquid form of solar energy that was able to run nearly every device we ever created. Bucky's delight was palpable. He relished the potential that my book and television series might have on the G.R.U.N.C.H (Gross Universal Cash Heist). Bucky had recently written a book of that title that foretold today's domination of the world by corporate power.

Since then, I have watched in helpless despair as our country pulled back in its commitment to alternative energy in the 1980s. I have cheered as tremendous advances in ethanol production were made with little help from government, and groaned at the misconceptions that continue to be promoted about ethanol in our popular media.

Most significantly, I have had to feel the burden of responsibility that comes with being a Cassandra, who knew the oil wars were coming, who knew they could be prevented, and who knew the Oilygarchy would get its way and blunder us into ruin. When I became an organic farmer in the early '90s, I told myself that maybe battles were best fought locally, that the world would come to an alternative energy future in its own time, that the environmental and economic catastrophes would make it imperative. That I should turn away from alcohol fuel and let my garden grow.

But once awakened, one can never again sleep so soundly. Arundhati Roy wrote in her 2001 book, *Power Politics*, "In the midst of putative peace, you

THE VITUPERATIVE BILE AROUND FUEL ALCOHOL IS TOTALLY MISPLACED. USED WITH A VISION THAT INCORPORATES ORGANIC FARMING—WHICH MEANS SHIFTING TOTALLY AWAY FROM INDUSTRIAL FARMING METHODS AND IMPLEMENTING SUSTAINABLE PRACTICES— ETHANOL IS AN EXCELLENT OPTION TO SOLVE OUR ENERGY PROBLEMS. ALL OF THEM, IF WE WISH.

Fig. 2-1

could, like me, be unfortunate enough to stumble on a silent war. And once you've seen it, keeping quiet, saying nothing, becomes as political an act as speaking out. There's no innocence. Either way, you're accountable." Maybe I had a destiny to fulfill.

"This is a time for a loud voice, open speech, and fearless thinking."

—HELEN KELLER

It's become a sort of heresy to talk about alcohol fuel or any form of alternative energy as a viable way out of our energy dilemma. Debate rages around available technologies and the readiness of our economic system to absorb massive change, but primarily the concern is with practicality. Ethanol, despite its promise, has been trashed in just about every publication and weblog in America.

Rest assured, there is enough land to produce solar energy in many forms, including alcohol, for a world that makes energy-efficient design a priority. We can have a large cooperative cellulose distillery operation in each county, producing ethanol and biomass electricity to keep our essential services running. We can have small integrated farms that produce fuel, food, and building materials. We can eat well on locally produced food and locally processed products. We can even cogenerate our electricity and hot water at our homes using our cars running on alcohol in a pinch, if we are clever enough (see Chapter 24).

NEW GEORGE BUSH NATIONAL PARKS

OLD FAITHFUL

COPYRIGHT GRIMMY INC., DISTRIBUTED BY KING FEATURES SYNDICATE

Fig. 2-2

The vituperative bile around fuel alcohol is totally misplaced. Used with a vision that incorporates organic farming—which means shifting totally away from industrial farming methods and implementing sustainable practices—ethanol is an excellent option to solve our energy problems. All of them, if we wish.

Now that you have picked yourself up off the floor, I'll explain.

I spent nine years as an organic farmer 30 miles south of San Francisco. I had a little over an acre on a 35-degree slope that I terraced and a little over an acre of flat valley bottom. From those bits of land, I produced enough vegetables to provide food for as many as 450 people. The USDA says this isn't possible. Over time, my organic matter content went from 2% to nearly 22%, the biological equivalent of converting desert sand to deep forest soil. My adobe clay soil went from one inch of topsoil to 16 inches. My loss to insect pests dropped more and more, so that by the fourth year I stopped spending any time worrying about it. I had a very nicely functioning, self-regulating, and self-maintaining ecological system that permitted me to produce huge surpluses of a great diversity of crops and make a decent living.

The key to the success of that long-term experiment was adherence to basic tenets of **permaculture**. Work with Nature, not against it. Everything is a yield; it's up to you to realize its value and find what to use it for. Be allergic to any extra work. Put things in the right place in relation to other things. Never fight gravity; it wins. The problem is the solution. Biology is constantly responding to stimulus, "learning" in response, and optimizing itself.

I'll talk more about the synthesis of permaculture (including organic farming) and ethanol production in our next chapter, but keep it in mind as we address the myths about alcohol fuel.

HERE'S THE BIGGEST MYTH:

If you ask anyone what they know about alcohol fuel, you will find they almost always dredge up from their vague memory the certainty that it takes more energy to make alcohol than is contained in the fuel. No matter that the person can't remember where they heard it from, why they know that, or why it's important. Such is the mark of truly excellent propaganda.

One thing I have learned about corporate propaganda, however, is that it is rarely imaginative.

Once something works, it is repeated in another form in a successive campaign.

Take the upcoming discussion about **energy returned on energy invested**. **EROEI** propaganda has its origins in the 1970s in crude, obviously phony studies done by oil companies, using the energetics of beverage distilleries. Such distilleries made so much profit on alcohol as a beverage that no effort whatsoever went into energy efficiency, since it made far less than a 1% difference in the retail price. But for the short time before these studies were debunked, the energy lie was obviously a powerful sound bite that affected the public. This spawned the more sophisticated attack that follows.

Myth #1: It Takes More Energy to Produce Alcohol than You Get from It!

Most research done on ethanol over the past 25 years has been on the topic of energy returned on energy invested (EROEI), or energy balance. In Appendix A, we detail how public discussion of this issue has been dominated by the American Petroleum Institute's aggressive distribution of the work of Cornell professor David Pimentel and his numerous studies. We cite his distortion of key calculations, his unfamiliarity with farming in general,

BOB FITCH PHOTO, <WWW.BOBFITCHPHOTO.COM>

BOB FITCH PHOTO, <WWW.BOBFITCHPHOTO.COM>

his ignoring of studies from Brazil that disagree with him, and his poor understanding of the value of **co-products** and their contribution to an accurate portrayal of energy accounting in the ethanol manufacturing process. In fact, he stands virtually alone in portraying alcohol as having an EROEI that is negative—producing less energy than is used in its production (see Appendix A, Figure A-2).

In fact, it's oil that has a negative EROEI. Because oil is both the raw material and the energy source for production of gasoline, it comes out to about 20% negative. That's just common sense; some of the oil is itself used up in the process of refining and delivering it (from the Persian Gulf, a distance of 11,000 miles in tanker travel).

Fig. 2-3 Sugar plant generator. Built with government loans, this steam-powered generator puts three-quarters of its electricity into the national **grid**, since the alcohol plant only uses about one-quarter of the power.

"If they can get you asking the wrong questions, they don't have to worry about answers."

—THOMAS PYNCHON

As Dr. Barry Commoner of the Center for the Biology of Natural Systems once said, "It's always possible to do a good thing stupidly,"[1] and some existing scenarios for making alcohol on a grand scale prove just that. However, the most exhaustive (and least-cited) study on the energy balance, by Isaias de Carvalho Macedo of Brazil, shows an alcohol energy return of more than eight units of output for every unit of input—and this study accounts for everything right down to smelting the ore to make the steel for tractors.[2]

But perhaps there's a more important measurement to consider than EROEI. What is the energy return for fossil fuel energy input? Using this

LEFT: Fig. 2-4 Boilers burning bagasse. After crushing to remove the sugary juice, surplus sugarcane fiber, **bagasse**, is burned to produce all the heat and electricity used in the alcohol plant, without using fossil fuels.

This book provides you with the means to make alcohol fuel using no nitrogen fertilizer, pesticides, or herbicides; using machinery powered only by clean-burning fuels; and using almost no nonrenewable energy sources to power the fuel plant. It's going on right now....

criterion, the energy returned from alcohol fuel per fossil energy input is much higher. Since the Brazilian system supplies almost all of its energy from biomass, the ratio of return could be positive by hundreds to one.[3]

Even with massive subsidies, the price of natural gas has now risen high enough that U.S. alcohol plants will be fiscally irresponsible to their shareholders if they don't start taking some of the spent liquid mash to self-produce all their own natural gas (**methane**). I predict that this system will sweep alcohol plants in this country, and that by 2012 every alcohol plant will be providing its own energy this way. New U.S. plants are already being built that feed the alcohol byproduct grain to animals on-site, and then turn their manure into methane to run the alcohol plant.

This book provides you with the means to make alcohol fuel using no nitrogen fertilizer, **pesticides**, or **herbicides**; using machinery powered only by clean-burning fuels; and using almost no nonrenewable energy sources to power the fuel plant. It's going on right now: India runs its plants using self-produced methane boilers/generators, while Brazilian alcohol plants actually generate large surpluses of electricity from their biomass-

fueled boilers. That's the bar we need to set for ourselves in producing fuel.

And since permaculture should be an integral part of the alcohol fuel revolution, no easy-to-dismiss studies of **annual**, monocultural crops such as corn are acceptable here. Those arguments belong to another era, and we will be showing farmers how to make **monoculture** obsolete by switching to permaculturally based organic farming.

The bottom line is that it's oil that is energy-negative, nonrenewable, and running out. Alcohol in America is already energy-positive, even when using coal or natural gas for process heat, and will become dramatically positive in the immediate future, running on its own renewable **process energy**.

THE OTHER MEGAOILRON MYTHS ABOUT ALCOHOL:

There are other common themes in the American Petroleum Institute's lexicon of propaganda. These are targeted to deny alcohol a constituency from among environmentalists, policy makers, labor, or anyone who cares about the people or planet. They are designed to frame production of alcohol as somehow immoral, wasteful, destructive, or foolish. After all, who wants to be thought of as an immoral fool?

These appeals can be even more powerful than the energy return fable, and, because they reach out to us more on the emotional level, they frequently are not backed up with any real science or meaningful statistics. What's appalling is watching this propaganda being absorbed and then regurgitated as science. It's like the historical reference to "survival of the fittest" being thought of as a cornerstone of evolutionary theory and quoted in scientific papers. In reality, as I mentioned earlier, the term was nothing more than economic quackery by Herbert Spencer in his absurd theory of Social Darwinism.

When you hear these issues come up in the future, think "How am I being manipulated against my own best interests by the Oilygarchy?"

Myth #2: There Isn't Enough Land to Grow Crops for Both Food and Fuel!

According to the U.S. Department of Agriculture (USDA), the United States has 434,164,946 acres of "cropland."[4] This is a very conservative number, describing land that is able to be worked in an industrial fashion (monoculture), primarily for

Fig. 2-5 Blessed by agriculture; percent of U.S. land in agriculture. The United States, one of the largest countries in the world, has an enormous amount of agriculturally productive land, as narrowly defined by the USDA. An equally huge area could be considered agriculturally productive when nontraditional crops, lawns, arid land energy crops, cellulose-producing crops, and urban/suburban green waste are included.

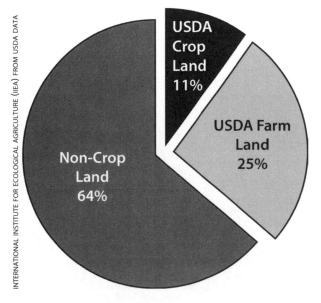

Percent of U.S. Land in Agriculture

INTERNATIONAL INSTITUTE FOR ECOLOGICAL AGRICULTURE (IIEA) FROM USDA DATA

USDA Crop Land 11%

USDA Farm Land 25%

Non-Crop Land 64%

Percent of U.S. Agricultural Land in Corn

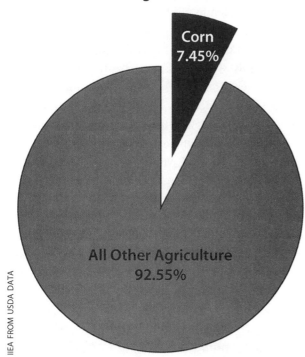

Corn
7.45%

All Other Agriculture
92.55%

IIEA FROM USDA DATA

Fig. 2-6 Percent of U.S. agricultural land in corn. This graph shows the percentage of U.S. agricultural acreage devoted to corn, not the percentage of all U.S. land. Although corn and soybeans are the two largest single crops, you can see that corn uses very little of the U.S. agricultural potential. Pundits who claim that to fuel our cars we would have to plant every square foot of the U.S. with corn are clearly prevaricating.

annual crops. This cropland is the prime, level, and generally deep agricultural soil.

Of this nearly half a billion acres of prime cropland, the U.S. uses only 72.1 million acres for corn in an average year.[5] The land used for corn takes up only 16.6% of our prime cropland! And corn takes up only 7.45% of our total agricultural land. When statements are made saying that the U.S. can replace only 10–15% of its gasoline by using agriculture, only the corn **starch** portion of the grain, produced on this small fraction of prime cropland, is used in the calculation!

Even if, for alcohol production, we used only what the USDA considers prime flat cropland, we would still have to produce only 368.5 gallons of alcohol per acre to meet 100% of the demand for transportation fuel at today's levels.[6] Although I am not proposing it, corn starch alone, at the modern average of 140 bushels of grain per acre, and not even counting use of corn's cellulosic stalks, could technically meet all of this goal, while actually increasing the meat supply (see Myth #4 below)—

and corn isn't a particularly stellar energy crop. A wide variety of standard crops yield up to triple this level (see Chapter 8). Dr. Barry Commoner did substantial research in the 1980s that showed that a simple shift away from starch crops to sugar crops, such as beets, would dramatically increase yields of both alcohol and animal feed per acre compared to corn.[7]

In addition to cropland, the U.S. has 939,279,056 acres of "farmland."[8] This land is also good for agriculture, but it's not as level and the soil not as deep as "cropland." Much of this farmland could support **perennial** crops that don't require the soil to be plowed every year, or that allow annual crops to be cultivated as long as the soil is plowed on contour (where the rows follow the land's contours, like the lines on a topographic map, to minimize soil erosion).

Many people argue that a substantial portion of this land is arid. As we'll discuss in Chapter 8, there are already 70 million acres of producing mesquite trees, essentially the same amount of acreage as cropland planted to corn. Considered a weed by farmers, mesquite grows partially on farmland but mostly on land that is too arid to even be considered farmland. Mesquite's harvested seedpods would generate 33 billion gallons of alcohol, without irrigation, fertilizer, or annual planting. That's another 21% of our annual gasoline needs from only 7.45% of our "farmland" (if we generously credit the land where the mesquite grows as farmland).

There is a vast amount of additional land that the USDA doesn't count as either cropland or farmland, but which is still suitable for growing

Alcohol Production from Selected Feedstocks

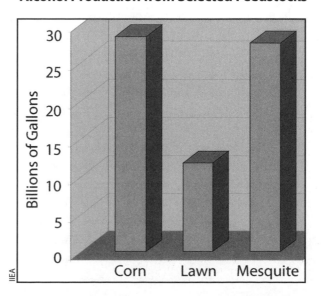

Billions of Gallons — Corn, Lawn, Mesquite

Fig. 2-7 Who needs OPEC?; alcohol production from selected feedstocks. Simply making alcohol from these three existing crops would replace 34% of our total transportation fuel needs—that's pretty close to what we import from OPEC. We already have lawn clippings and mesquite as unused perennial (permanent) crops. Combining these with the annual corn, we can be free of Middle East oil almost overnight without having to change anything in agriculture. As we'll see in Chapter 13, almost every engine in the U.S. can run on 34% alcohol right now—so what's stopping us?

specialized energy crops. This includes places like the Texas Panhandle and arid Western areas that currently require over 100 acres of low-quality grazing land to raise one beef steer. The same land can be used far more profitably to grow crops adapted to arid climates (such as pimelon, buffalo gourd, and prickly pear) to produce alcohol, biodiesel, and animal feed.

Some of the land that the USDA doesn't classify as cropland or farmland has plenty of water but is more highly sloped. Tree and bush crops, such as hybrid chestnut and hazelnut, would make good use of sloped land; they halt soil erosion, and they can produce a substantial amount of alcohol and biodiesel (rivaling energy production from corn on prime cropland—but with essentially no inputs).

Lowlands that are considered swamps or wetlands (not counted as farmland or cropland) could be restored to allow cultivation of high-energy crops like cattails (see Chapter 8), while dramatically enhancing wildlife habitat value. And these crops can also be grown in artificial marshes that are economical to build.

For example, cattails are now used in constructed marshes to inexpensively treat sewage. Yields of starch and cellulose from cattails easily top 10,000 gallons per acre in such a nutrient-rich environment. If all the sewage in the U.S. were sent to such constructed marshes, the 3141 U.S. counties would

---

CELLULOSE

The whole issue of "food versus fuel," as it is often put, becomes moot when a new alcohol feedstock is considered: cellulose. Cellulose is a **carbohydrate** produced by plants as their basic structural building material for stalks, roots, and leaves. Even though it is composed of countless sugar **molecules**, cellulose is the tough basic fiber that makes up the great majority of plant matter, other than the fruit, grain, or storage tuber of the crop.

Cellulose as a feedstock makes the amount of alcohol possible from sugar and starch sources seem insignificant by comparison. For instance, for every pound of corn grain, there are two and a half pounds of cellulosic corn stalk and more than four pounds of cellulosic root matter.

Biologist Jeffrey Dukes estimates that almost five times the plant matter needed to replace fossil fuels is produced each year on Earth.[9] Bucky Fuller, always ahead of his time, told me back in 1982 that he thought worldwide cellulose would provide six times the energy needed by humanity. Other estimates for annual production of biomass (organic matter used as a fuel) range up to 15 times all human energy use.

All of these estimates are based on current natural cellulose production worldwide. By focusing on designing cellulose production, we could increase available cellulose by a factor of ten, using practices that would optimize conversion of solar energy to carbohydrates and simultaneously build soil fertility. So there is an unlimited potential to design and grow polyculture cellulose crops to be **biorefined** into alcohol and multiple byproducts.

Now that cellulose-to-alcohol production is being commercialized, potential total alcohol yields from several fibrous crops could routinely top 1500 to 5000 gallons/acre per year.[10] Crops such as hemp, sudan grass, switchgrass, and many fast-growing trees can be grown for a high yield of alcohol per acre, even on land that isn't considered cropland or farmland. So, at 5000+ gallons/acre, the U.S. might even use less than 15% of its prime cropland to serve all of its transportation fuel needs.

In addition, paper waste from cities, timber waste, and sawdust are examples of cellulose feedstocks that could produce alcohol at a high rate per ton. If cellulose becomes the overwhelmingly single most important carbohydrate for alcohol production, as many feel is inevitable, just the yeast recovered from cellulose fermentation after distillation would be sufficient to provide livestock with all the **protein** and fat they currently get from grain.

A CELLULOSE/STARCHY STEW

What might be a potential polyculture yield from all carbohydrates, including cellulose, over a year's time with return of spent mash and **carbon dioxide (CO$_2$)**? Here's one example: In southern areas with moderate winter temperatures (above 17 degrees **Fahrenheit (°F)**) and moderate rainfall, a **coppiced overstory** of the **leguminous** tree Tipuana tipu would yield 24 tons (4320 gallons/acre). Add to that a mixed summer **understory** of large-cane sorghum (1200–2500 gallons/acre), with a vine such as pumpkins (600 gallons), and turnips (500 gallons); a winter understory of fodder beets (1500+ gallons); and, as a permanent perennial crop in the row with the trees, thornless Opuntia cactus (600 gallons).

A marshy or water-based example good to 0°F would include bands of fast-growing willows or bamboo on **berms** separating long canals of cattails, with spent mash and fermentation carbon dioxide returned to the canals. Such a setup would routinely top 10,000 gallons/acre.

need only 6360 acres each to fulfill all of our foreseeable transportation fuel needs, both gasoline and diesel, at 200 billion gallons per year. That equals 1.46% of our agricultural land. And we'd be doing it with no chemical fertilizer or irrigation water, since an average county would generate 15 million pounds of liquefied "humanure" per year.

And crops aren't limited to land, either. When we look at the huge potential of kelp (marine algae) grown in coastal river mouths rich in nitrates and sewage (oops, I mean "surplus nutrients"!), there is no doubt that we can reap bumper crops of alcohol, replacing all petroleum fuel without ever planting a single seed on land (see Chapter 8).

And, as we'll see in Chapter 8, the potential alcohol production from cellulose could dwarf all other crops. When it comes to corn, the cellulose in the stalks, cobs, and grain itself is two to nearly three times the weight of the corn starch from the grain.

Production from cellulose requires us to think quite differently about yield. For instance, the United States has nearly 30 million acres of lawns. That's 41% of the total acreage we use for corn, and it isn't counted as either farmland or cropland! Grass clippings are the number one irrigated crop in the U.S., and would generate over 11 billion gallons of fuel per year. In addition to grass clippings, there is a huge amount of green waste from landscapes, adding to the cellulose total in every county.

Cellulose makes energy crops out of unlikely plants. For instance, turnips and rutabagas are mostly cellulose and would generate tremendous yields in comparison to corn starch. Fast-growing trees are usually about 75% cellulose and can yield several thousand gallons per acre on a sustained basis through planned pruning (coppicing). **Polycultures** of mixed cellulose/starch/sugar crops can yield more than 10,000 gallons per acre, compared to the hundreds of gallons per acre yielded by most starch and sugar crops grown singly.

So can we produce enough alcohol for both food and fuel? As you'll see, the question should be: After we replace all the gasoline, diesel, and heating oil, do we sell our surplus alcohol to the rest of the world—or do we use it to replace all the electricity coming from nuclear and coal plants? Let's do both.

Myth #3: Ethanol's an Ecological Nightmare!

History shows us that in order to mechanize farming—in order to remove as much labor as possible, for the benefit of corporate elites with capital—planting, cultivation, and harvesting of the land had to be simplified from husbandry to brainless machines. And so monoculture—one crop grown in one area—was born.

Industrial agriculture is not a means of feeding more people, but a way to produce food with the least amount of human labor for maximum corporate profit. As I point out in Myth #4 below, food is a commodity, not a human necessity, in this system.

The major crop used for alcohol fuel in the world is sugarcane. Unlike corn, which is an annual crop that must be planted each year, sugarcane is perennial, planted only once every five to ten years, and it can be harvested continually. Since Brazilian and Indian alcohol plants return most of the byproducts of alcohol production to the fields, little fertilizer is

---

"The potential impact of biomass ethanol on our available fuel supply and economy is enormous. According to the National Renewable Energy Lab study, an average of 2.45 billion metric tons of cellulosic biomass could sustainably be available on an annual basis for ethanol production in the United States. This is enough biomass to produce over 270 billion gallons of ethanol, approximately two times the level of current U.S. gasoline supply".[11]

—STEVEN GATTO, JUNE 2000

needed, and soil builds in fertility rather than loses vigor. In Brazil, sugarcane is planted on carefully laid-out contours to prevent soil loss. (The same can be done with corn and is in some areas, but where corn is not contour-planted, there is soil loss.)

If, over time, alcohol production from cellulose (see Myth #4 below and Chapter 8) takes the lead, then perennial grass, shrubs, and trees—either singly or mixed in polyculture—will dominate energy agriculture. These practices are soil-building, not soil-degrading.

The effect that estrogenic pesticides and herbicides, particularly those used in industrial farming, are having on our species is perhaps the most devastating ecological disaster. Use of these chemicals is made necessary by growing crops in monocultures for agribusiness. Sometimes the most horrendous effects are when the dose is so small that the body doesn't recognize the substance as dangerous, and does not mount a response to eliminate it. Instead, the body reacts to the chemical as if it is the hormone estrogen.

Estrogen is, of course, the female hormone. One of the main messages estrogen shouts is, "Be a girl!"—which is an alarming message if you are male. Since the advent of chemical agriculture, sperm counts have dropped to below 50% viability. In the mid-1990s, the National University Hospital in Copenhagen, Denmark, published studies that showed a precipitous drop in sperm count and rising infertility of men in that country.[14] This study caused most other countries to examine their sperm banks, only to find the same drop developing since the 1940s. At 50% sperm viability, mammals are considered functionally sterile.[15]

"I have always maintained that if we were to mix orange fluorescent dye with all pesticides, the public would demand all-organic food overnight."

—DAVID BLUME, AUTHOR AND PERMACULTURIST

The estrogenic effects of pesticides and herbicides have been known since the 1940s, when scientists "chemically castrated" roosters by including pesticides with their feed. Today's pesticides are 3000 times more toxic than these original **chlorinated hydrocarbons** and **organophosphates**, all of which were made from crude oil.

In the U.S., we spray well over one billion pounds of pesticides, made from oil, on our food every year. The amount of herbicides may be in excess of four billion pounds, with use quadrupling since the creation of **GMO (genetically modified organism)** crops engineered to withstand direct herbicide exposure. That's about 15 pounds per person of these chemicals put directly on our food every year.

Estrogenic effects that can permanently affect the development of fetuses can occur at the almost unbelievably small doses of parts per quadrillion in the bloodstream. At higher doses, these chemicals can either directly or indirectly cause cancer and can stimulate the growth of existing cancer. Prostate cancer and breast cancer rates have both closely followed the rate of exposure to agricultural chemicals in our food and water. When I farmed, I had dozens of customers who had been advised by their cancer doctors to eat only organic food.

These chemicals are also toxic to the microlife that is responsible for soil fertility, and, of course, are toxic to all the predators of pest insects. So here we are, stewing in a mess of our own making, much of which is caused by combining fossil fuel products and agriculture.

JUST HOW INEFFICIENT IS OIL PRODUCTION ANYWAY?

University of Massachusetts biologist Jeffrey Dukes took a good hard look at what goes into making oil. Unlike other researchers, who appear to assume coal or crude oil are almost free forms of energy, Dukes studied just what Nature does to produce, and what man does to retrieve, crude oil. He found that about 90 metric tons (198,000 pounds) of ancient plant matter are required to make one gallon of gasoline. He calculated that the fossil fuels burned in 1997 were created from ancient plant matter and energy equal to 400 times today's annual **net primary productivity (NPP)**.[12] NPP represents the photosynthesis of both land plants and waterborne algae and plankton.[13]

Processing of oil and natural gas were found to be less than 0.01% efficient. If we assume gasoline contains 120,000 **British thermal units (Btu)** of **heating value** per gallon, that would mean it took 1.2 billion Btu to produce the fuel that might propel an SUV 15 miles. Having 10,000 Btu of energy input for one Btu of output gives fossil fuels an incredibly negative energy return.

In stark contrast, Dukes estimated that it would take only about 22% of today's land-based NPP to replace all fossil fuel energy, not just transportation fuel. That's right, plants could replace all the energy we use. And permaculturally designed energy polycultures can be more than ten times as productive as the current NPP.

Switching to organic-style crop rotation will cut energy use on farms by a third or more:[16] no more petroleum-based herbicides, pesticides, or chemical fertilizers. Fertilizer needs can be served either by applying the byproducts left over from the alcohol manufacturing process directly to the soil, or by first running the byproducts through animals as feed. In the latter case, manure is run through a **methane digester**; the resulting liquid is then sprayed back on the land that the crop came from—via an inexpensive pipeline, or by using trucks, or by making **compost** for later application.

The pipeline system is standard in India. The same pipeline can be used to deliver the liquid fertilizer at night, and deliver carbon dioxide (an alcohol fermentation co-product) during the day, possibly tripling yield (see Chapter 11). This system is much more practical with smaller alcohol plants (making 500,000 gallons or less a year) than with the very large plants, where the farms and production are geographically divorced from one another.

You'd be hard-pressed to find another route that so elegantly ties the solutions to the problems as does growing our own energy. Far from destroying the land and ecology, a permaculture ethanol solution will vastly improve soil fertility each year.

Myth #4: It's Food Versus Fuel—Crops for Food Will Compete with Crops for Fuel! We Should Be Growing Our Crops for Starving Masses, Not Cars!

The first thing to realize is that there is no food shortage and none impending at this point. Anyone can look at world crop production and see that we produce around twice the **calories** we need to feed everyone. What we do have is a money shortage, since food is a commodity and not a right: Whoever can pay for it gets to buy it.

Also, there is a big imbalance in the amount of starch produced in relation to protein (we've already mentioned the preponderance of corn in the U.S. agricultural system). In malnourishment, the problem is usually lack of protein. The four biggest crops worldwide are rice, wheat, corn, and potatoes. These grains and tubers can be as much as 75% starch. Beans, a significant source of protein for the poor of the world, don't even make it into the top four crops.

When I studied mushroom cultivation back in the 1980s, I concluded that we could provide the

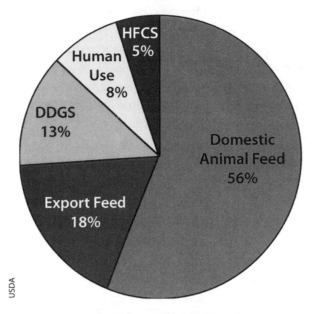

United States Corn Uses 2005

HFCS 5%
Human Use 8%
DDGS 13%
Domestic Animal Feed 56%
Export Feed 18%

USDA

ANIMAL FEED	MILLION BUSHELS
Domestic Animal Feed	6075
Export Feed	1950
DDGS	1425
HUMAN USES	
(Food, Seed, Industrial Products)	
High-Fructose Corn Sweetener	529
Starch	280
Sweeteners	219
Cereal/Other	189
Alcohol (Potable)	133
Seed	20

entire world's balanced protein needs just by using 25% of the grain straw that was annually burned off fields, to produce **oyster mushrooms**. The fungi efficiently extract protein from the **lignified** dry straw. So if we really wanted to feed everyone, even without using a single animal as a food source, it would not be difficult.

Humankind has barely begun to work on designing farming as a method of harvesting solar energy for multiple uses. Given the massive potential for polyculture yields, monoculture studies of ethanol production seem almost silly in comparison when viewed from economic, energetic, or ecological perspectives.

Because the United States grows a lot of it, corn has become the primary crop used in making ethanol in the U.S. This is supposedly controversial, since corn is identified as a staple food in poverty-stricken parts of the world. What most of us don't realize is that 87% of the *United States'* corn crop

Fig. 2-8 Uses of U.S. corn. *It is abundantly clear that 87% of our corn ends up as animal feed. Since cattle cannot efficiently digest the starch in the corn, we can actually increase the meat supply by first processing the corn into alcohol and its byproduct DDGS. Energy use and* **greenhouse gas** *emissions resulting from corn exports and domestic shipping would be reduced by two-thirds, since DDGS is a much more compact source of protein and fat. All human uses of U.S. corn total only a single-digit percentage, other than high-fructose corn sweetener (HFCS),[17] which I would argue is not a human food but an industrial product. Now, when it comes to the less than 1% of corn that goes for whiskey, well, that might be considered food.*

is fed to animals.[18] In most years, the U.S. sends close to 20% of its corn to other countries. While it is assumed that these exports could feed most of the hungry in the world, the corn is actually sold to wealthy nations to fatten their livestock. Plus, virtually no impoverished nation will accept our corn, even when it is offered as charity, due to its being genetically modified and therefore considered unfit for human consumption.

In addition, alcohol production can vastly improve animal feed. In making alcohol from grain, all of the starch is removed, but all of the protein and fat, some of the cellulose, and a wide array of vitamins and minerals remain, along with yeast from fermentation. This remaining substance is called **distiller's dried grains with solubles (DDGS)**, which occupies about one-third the volume of the original corn, since the starch has been removed.

DDGS has been used as superior cattle feed for more than 100 years. If you feed 33 pounds of DDGS to cattle instead of 100 pounds of corn, you get 14 to 17% more meat up to 30% faster, with a fraction of the veterinary costs.[19] (Since cattle evolved to eat woody brush, not grass and grain, up to 80% of the starch in grain goes through cattle undigested and unused, causing huge health problems in the process.)

So, fermenting the corn to alcohol results in more meat than if you fed the corn directly to the cattle.

Myth #5: Big Corporations Get All Those Ethanol Subsidies, and Taxpayers Get Nothing in Return!

Even TV shows like *The West Wing* have gotten into the act on this one![22] Everyone in the alcohol fuel industry would be completely in favor of a no-subsidy system, but only if there were no subsidies for any energy source. The subsidies for oil and other nonrenewable sources are massive and hidden in a myriad of ways in our income tax structure. What this does is make energy inexpensive for businesses

Fig. 2-9

PUBLIC TRANSIT WASTES MONEY. IT ALWAYS HAS TO BE **SUBSIDIZED!**

POLLUTION SUBSIDIZED BY HEALTH CARE PREMIUMS

PARKING SUBSIDIZED BY BUSINESS + TAXES

SUBSIDIZED POLICE, FIRE + PARAMEDICS FOR 4 MILLION ACCIDENTS PER YEAR

LOCAL STATE + FEDERAL TAXES SUBSIDIZE STREETS + HIGHWAYS.

COSTS OF CARS, GAS, MAINTENANCE + INSURANCE "SUBSIDIZED" BY CAR OWNERS.

SINGER

COPYRIGHT 1992-2002 ANDREW B. SINGER

POOR MEGAOILRON NEEDS HELP AGAINST TERRORISTS

Even though MegaOilron is making more money per year than at any time in history, it can't afford to protect its own facilities against "terrorists." In 2004, the U.S. Department of Homeland Security granted oil refineries $65 million for fences, cameras, and communications equipment. That's tax money the oil companies don't have to pay back.

"That makes no sense to me at all," said Bill Millar, President of the American Public Transportation Association. Maybe he thought that the corporations making huge windfall profits due to OPEC raising its prices ought to pay their own bills? That was certainly the viewpoint of the Project On Government Oversight. Its Executive Director, Danielle Brian, said, "They're taking absolute advantage of the situation, and the government is letting them get away with it," citing the Bush administration's close ties with the oil industry.[20]

In comparison, the entire U.S. Department of Energy budget for solar research was scarcely $80 million.[21]

(who don't pay income tax), since the cost they pay at the pump is not the fair cost of energy. Individuals pay for that in income taxes. Massive subsidies have kept gasoline in our lives for some time now.

So what is really the cost of a gallon of gas? The most detailed, inclusive, and documented study of this question has been done by the International Center for Technology Assessment in Washington, DC. The five primary areas of accounting were: 1) tax subsidization of the oil industry; 2) government program subsidies; 3) protection costs involved in oil shipment and motor vehicle services; 4) environmental, health, and social costs of gasoline usage; and 5) other important externalities of motor vehicle use. This totaled from $558.7 billion to $1.69 trillion per year. When added to the then-low price of gasoline, the result was a minimum of $5.60 per gallon, but a more complete accounting put the figure at $15.14 per gallon.[23] An updated report from 2005 raises the cost by 21 to 32 cents in 2003 dollars.[24] (Since 2003, you'd also probably be justified in adding all the costs of the Iraq occupation as oil-related.) By burying these costs in our taxes, the burden is shifted from corporate users of energy onto the shoulders of the citizenry.

A U.S. General Accounting Office study showed that, between 1968 and 2000, oil companies received subsidies (not counting military) of $149.6 billion, compared to ethanol's paltry $116.6 million.[25] Yet, oil companies have built no new refineries in the last 25 years. In fact, they have closed 50 refineries in the last 15 years.[26] After all, when you know there's not going to be much oil to refine, why build a refinery?

Today, even with very little government support, the cost of building much more energy-efficient and productive alcohol plants has dropped to roughly $1 to $1.15 per **annual gallon of production capacity**,[27] down from $2.50 per gallon in 1980. The alcohol is produced today for less than $1 per gallon, down from over $2 per gallon.

So it is clear that the subsidies alcohol did receive—small tax credits that various states provided and the 40 to 60 cents per gallon provided by the federal government starting in the early 1980s— have worked extremely well in bringing maturity to the industry. Farmer-owned cooperatives now produce the majority of alcohol fuel in the U.S. Those farmer-owners pay themselves premium prices for their corn and then pay themselves a dividend on the profit from the alcohol. And market share has been redistributed: Archer Daniels Midland, once the industry leader, is now only 28% of the ethanol market,[28] and alcohol fuel is only 5% of its billion-dollar-a-year business.

The increased economic activity derived from alcohol fuel production has turned out to be crucial to the survival of noncorporate farmers, and the amounts of money they spend in their communities on goods and services and taxes for schools have been much higher in areas with an ethanol plant than in areas without a plant. Moreover, since the capital generated by alcohol fuel sales and production recirculates within the U.S., $3 to $6 in tax receipts are generated for every dollar of ethanol subsidy. The rate of return can be much higher in rural communities, where re-spending within the community produces a multiplier factor of up to 22 times for each alcohol fuel subsidy dollar.

Let's look at a simple study done by LECG, a very conservative independent corporate analysis firm. Even though this study doesn't extend out the recirculation of capital as I describe above, it still shows the following figures for alcohol fuel plants' direct effects on taxes:[31]

Cost of tax incentives	−$1.8 billion
Farm program savings	+$3.2 billion
Increased tax revenue	+$1.3 billion
Net impact on treasury	+$2.7 billion

In other words, alcohol fuel subsidies actually increase government tax receipts, since fuel production is done by U.S. companies, not transnational corporations with massive tax breaks.

World Crude Oil Reserves, January 2006

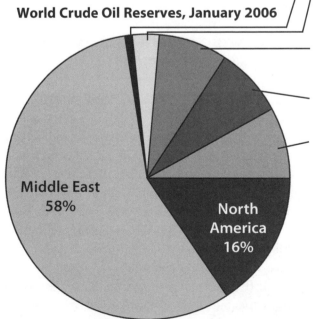

Western Europe 1%

Far East and Oceania 3%

Eastern Europe (former USSR) 8%

Africa 8%

Central and South America 8%

Middle East 58%

North America 16%

Fig. 2-10 World petroleum reserves.[29,30] *The Middle East clearly controls the majority of the planet's oil supply.*

In 1980, Employment Research Associates (ERA) did an analysis for the DOE to determine direct and indirect employment in the production of 12 billion gallons of alcohol per year.[32] The study considered a wide array of feedstocks in addition to corn, such as wheat, sorghum, citrus waste, and cheese **whey**. It concluded that the establishment of a six-billion-gallon-per-year ethanol industry would create 960,000 U.S. jobs.

The ERA study considered the 12 billion gallons to come from 48 fifty-million-gallon-per-year plants and 360 ten-million-gallon-per-year plants. It also assumed a $2 per annual gallon of production capacity for building costs.

We are now fast approaching the six billion gallon mark, 25 years after the study was done. The 48 big alcohol plants have certainly been built, but almost none of them with federal money. Over the years, the cost of building the plants has plummeted, and the dollar has depreciated to the point where the 1980 dollar is worth 2.4 times a 2005 dollar. So the cost of building alcohol plants has dropped by nearly five times since 1980.

Until recently, the economics for the smaller plants were not attractive, while the price of gasoline was artificially depressed by government policy and massive subsidies. But now, smaller plants making 500,000 gallons a year (or even making fewer than 100,000 gallons when integrated co-products are taken into account) look economically very rosy, while the costs that larger plants incur for transporting raw materials are now more significant.

Small-plant economics are especially attractive if the farmer can cut out the middlemen and sell the fuel directly to 500–1000 people locally in a community-supported energy (CSE) system.

The margin between production and selling prices is wide enough for a small plant to make a decent profit, especially when using some sort of low-cost niche feedstock, such as a waste product from another industry. Low feedstock costs offset the higher labor incurred in a small plant. Also, the volume of byproducts is not as massive as with a large plant, which makes the local marketing of them easier. Small-plant economics are especially attractive if the farmer can cut out the middlemen and sell the fuel directly to

500–1000 people locally in a **community-supported energy (CSE)** system (see Chapter 28).

So, based on the ERA study, replacing all of the United States' gasoline—160 billion gallons—with alcohol might generate as many as 26 million jobs. That would mean full employment for America. Full employment returns power to organized working people. The key to this employment bonanza would be the new proliferation of smaller plants with diverse co-products that can take advantage of market niches that large ones cannot. The multiplier effect of millions of employed people spending their money locally is the difference between the anxious state of the economy from the workers' point of view to one of robust energy.

In addition to subsidizing alcohol plants, the government should fund fueling stations that are independent of big oil. That might take a $20 billion revolving loan fund for stations, along with the current tax credits, to help stations get established.

Instead of allowing just the large-scale plants running primarily on corn to dominate, government funding or loan guarantees would permit all scales and many feedstocks to become part of the national energy system.

But even if the government doesn't get into the act, small-scale plants will start popping up all over the place in the next few years as the reality of **Peak Oil** will cause smart consumers to join community-supported energy businesses that will produce and distribute alcohol. Small-scale producers will find that alcohol will be as good a currency as whatever will be used instead of dollars in a world of diminishing oil supply. You'll be able to trade alcohol for nearly whatever service or locally produced goods you want. All it takes is for people to stop giving their dollars to oil companies and start giving them to local energy producers and distributors.

Myth #6: Ethanol Doesn't Substantially Improve Global Warming! In Fact, It Pollutes the Air!

Let's separate air pollution from the global warming caused by fossil fuel exhaust, and address them one at a time.

First, air pollution: Throughout the '80s and '90s, there was a lot of noise about ethanol blends having higher **evaporative emissions** than pure gasoline. (Evaporative emissions are what escape from under the fuel filling cap when some of the fuel becomes vapor.) This was often translated by reporters into, "Ethanol Dirtier Than Gasoline."

Alcohol fuel has been added to gasoline to reduce virtually every class of pollution. Adding as little as 5–10% alcohol can reduce **carbon monoxide (CO)** from gasoline exhaust dramatically. When using pure alcohol, the reductions in all three of the major pollutants—carbon monoxide, **nitrogen oxides (NOx)**, and hydrocarbons (HC)—are so great that, in many cases, the remaining emissions are unmeasurably small. Reductions of more than 90% over gasoline emissions in all categories have been routinely documented for straight alcohol fuel.

It is true that when certain chemicals are included in gasoline, addition of alcohol at 2–20% of the blend can cause a reaction that makes these chemicals more volatile and evaporative. But it's not the ethanol that's the problem; it's the gasoline. This is especially true because of the **natural gas condensates (NGC)** that now make up an increasing percentage of gasoline, although they once were considered too toxic and volatile. The increase of permeation emissions may be one gram per day under the right circumstances. But the reduction in the amount of the three major emissions of fuel during a daily commute is about 300 times the one-gram increase in permeation emissions. For alcohol mixtures above 20%, permeation emissions actually go down in comparison to straight gasoline.

Alcohol carries none of the **heavy metals** and **sulfuric acid** that gasoline and diesel exhausts do. And straight ethanol's evaporative emissions are dramatically lower than gasoline's, and are not any more toxic than what you'd find in the air of your local bar. You'll read more about all this in Chapter 14.

As for global warming, the two vehicle emissions that matter are carbon dioxide and water vapor. These gases increase global warming because they trap solar heat energy in the atmosphere that otherwise would escape into space.

Alcohol **combustion** and fermentation both emit carbon dioxide, and anti-alcohol propagandists are quick to allege that distillation and use of alcohol as a fuel aggravates the already high levels of CO_2 in the air. In fact, in simple terms, the production and use of alcohol neither reduces nor increases the atmosphere's CO_2. In a properly designed system, the amount of CO_2 and water emitted during fermentation and from exhaust is precisely the amount of both chemicals that the next year's crop of fuel plants needs to make the same amount of fuel once again.

Photosynthesis by plants takes carbon dioxide, water, and sunlight to make carbohydrates. Production and use of alcohol returns the carbon dioxide and water to the air, and produces work with the solar energy. It's essentially a closed cycle.

But the system, facilitated by alcohol fuel production, actually allows us to reduce carbon dioxide emissions.[35] How? The growing of plants actually ties up many times more carbon dioxide than is created in the production and use of the alcohol. The portion of the plant that becomes fuel is not all of the plant. All of the vegetative portions of a plant, from roots to stalk and leaves, are made of **sequestered** carbon dioxide and water, primarily the carbohydrate cellulose. Plants sequester up to ten times more carbon dioxide from the air, compared to just the balanced CO_2 recycling from the part of the crop (e.g., grain) that is used to make the alcohol (see Chapter 3). Plants also exude up to 80% of the carbon they **absorb** from the air through their roots, as sugar, to feed beneficial fungi and bacteria.

The implications are staggering. If a balance were struck (as it could be using alcohol as fuel), the oceans and plants would absorb our excess CO_2 over a period of 50 to 100 years. Increasing plant growth by converting from a hydrocarbon to a carbohydrate economy could reduce atmospheric carbon dioxide far more quickly. A crash program of

---

THE BIGGEST DANGER FROM GLOBAL WARMING

Contrary to popular belief, the most dangerous greenhouse gas is water vapor. It is 30 times more potent than carbon dioxide as a greenhouse gas, with a tremendous capacity to trap and store solar energy. The burning of fossil fuels creates huge quantities of new water vapor along with other emissions.

Carbon dioxide is a trigger that starts global warming, and once there is more heat, more water evaporates, causing a magnification of the original effect of the carbon dioxide. This feedback loop, among several, could eventually take on a massive momentum and continue the climate change effect far beyond what would be expected from carbon dioxide alone. Most scientists believe we haven't reached that tipping point just yet. But no one knows for sure just how close we are.

reforestation, energy crop planting, and ocean kelp farming could rapidly reverse global warming!

At present, it's not possible for the Earth's system to absorb all the CO_2 being emitted from fossil transportation fuel and power plant exhaust. This carbon dioxide is not recycled as in the alcohol system, because it is made of plants that were living millions of years ago and now have become fossil fuels. When they are burned and produce carbon dioxide and water, this adds a burden to our atmosphere. If we don't seek equilibrium, we can anticipate further massive changes in weather and a nearly irreversible heating trend in the environment, which could last centuries or even thousands of years.

Actual and Projected Motor Vehicle Petroleum Use

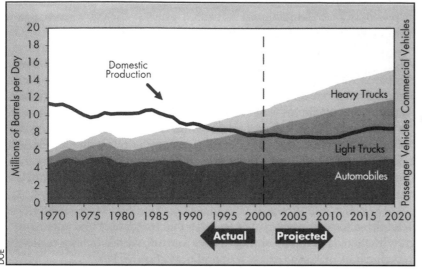

Fig. 2-11 Actual and projected motor vehicle petroleum use.[33,34] *This chart shows that vehicle use of petroleum will not rise significantly due to increased passenger vehicle adoption. The major use will be for transportation infrastructure, which will be led by trucks and buses in developing countries. Since the need for this infrastructure is driven by increases in population and income in product-producing countries, there is little hope of heading off the increasing demand for fuel. Much of this increased demand can be met by high-compression alcohol engines.*

World Vehicle Registrations

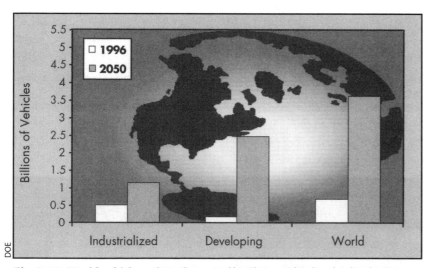

Fig. 2-12 World vehicle registrations. *Led by China and India, the developing world is experiencing explosive demand for motor fuels right when oil production is peaking. The effects on world economies and the climate depend on replacing fossil fuels with renewables at a much faster rate than simply making up for declining oil supply.*

"[Carbon] sequestration is difficult, but if we don't have sequestration then I see very little hope for the world. No one can be comfortable at the prospect of continuing to pump out the amounts of carbon dioxide that we are pumping out at present ... with consequences that we really can't predict but are probably not good ... [I'm] really very worried for the planet."

—LORD RON OXBURGH, SHELL OIL BOARD CHAIRMAN, JUNE 17, 2005

Myth #7: So If We Use Ethanol, We Can Still Keep Burning through Fuel Like the Irresponsible Consumers We Are?!

In the developed countries, there is a long, comfortable relationship with cheap energy, which has led to a culture of consumption and waste. The U.S. consumes the most energy in the world and is one of the greatest users per capita, just behind Japan. Even if we can produce all the alcohol we need, we still have to become more responsible with energy use.

Conservation and increasing efficiency are the most important actions we can take immediately to blunt the imminent effects of Peak Oil. But alcohol can play an important part in energy planning if it's taken seriously. For instance, a good start would be to take advantage of alcohol's high-octane qualities by switching to smaller, high-compression engines that squeeze much more energy out of each drop of alcohol.

The current energy use per capita in China and India is low, but as their standards of living continue to rise and as they emulate the West, there is the danger they will adopt profligate consumption, as well. Since their populations are large, they are destined to eventually be the largest energy users. The stakes are high, since massive increases in air pollution and greenhouse gases in the next 50 years from these two developing giants could be the final blow to our climate and to the quality of our air. Pollution wafting into the U.S. accounts for 30% of the nation's ozone, a major component of smog, according to the National Oceanic and Atmospheric Administration.[36] There is no way to dispose of pollution; each of us is downwind of each other.

As China and India continue to use more trucks and cars, they have the golden opportunity to avoid the failed path of petroleum—and produce their own high-compression alcohol engines, along with hybrid technology in large vehicles. The amount of energy saved by hybrid technology in cars is small, but quite large in buses and trucks. Since both countries have extensive rail systems, it would be easy to convert China's coal burners and India's diesel turbine locomotives to alcohol and/or biodiesel, keeping the cost of rail transit far below that of petroleum-fueled infrastructures. China is both building large alcohol plants and contracting for alcohol to be produced elsewhere for its use.

The renewable fuels movement is not about providing unlimited clean fuel for SUVs. What it's about is shaping policy now with our own individual and group actions to make sure the energy future we get is the one we want, not the one the Oilygarchy is planning for us.

Now that you have seen behind the curtain of propaganda against alcohol, you're better equipped to counter these arguments when they're unconsciously repeated by people in your own community. With the actual facts and figures to back you up, you can wage your own counter-propaganda campaign with letters to the editor and even in everyday conversations. Bit by bit, brick by brick, we can build public understanding of the solar path out of fossil fuel dependency.

To truly understand the implications of ethanol on our lives, we need to address issues that should be covered in our basic education but aren't. Agriculture, the science of feeding all of us, is woefully neglected in our basic education; surprisingly, the ecology of agriculture is hardly discussed even in our agricultural colleges. But to truly understand the revolution alcohol represents, it's important to understand the basics of agriculture. So let's delve into the life of plants and the soil and see how alcohol production can be used to chart a path away from the use of petrochemicals in the production of food and fuel.

Endnotes

1. "The Plow Boy Interview," *Mother Earth News*, March/April 1990 (www.motherearthnews.com/Nature_and_Environment/1990_March_April/The_Plowboy_Interview).

2. Isaias de Carvalho Macedo, "Greenhouse Gas Emissions and Energy Balances in Bio-Ethanol Production and Utilization in Brazil," *Biomass and Bioenergy* 14:1 (1998), 77–81.

3. Larry Rohter, "With Big Boost from Sugar Cane, Brazil Is Satisfying Its Fuel Needs," *New York Times*, April 10, 2006, Sec. A1.

4. *2002 Agricultural Census*, U.S. Department of Agriculture, www.nass.usda.gov/census/census2002/volume1/us/index1.htm (December 30, 2004).

5. National Agricultural Statistics Service, U.S. Department of Agriculture, http://usda.mannlib.cornell.edu/reports/nassr/field/pcp-bba/acrg0602.txt (June 28, 2002).

6. Author's calculation.

7. Barry Commoner, *The Politics of Energy* (New York: Alfred A. Knopf, 1979), 42–43.

8. *2002 Agricultural Census*.

9. Jeffrey S. Dukes, "Burning Buried Sunshine: Human Consumption of Ancient Solar Energy," *Climate Change* 61 (2003), 31–44.

10. Robert Shleser, *Ethanol Production in Hawaii: Processes, Feedstocks and Current Economic Feasibility of Fuel Grade Ethanol Production in Hawaii*, for the State of Hawaii Department of Business, Economic Development and Tourism (July 1994), 1–62.

11. U.S. Congress, Senate, Committee on Environment and Public Works, Subcommittee on Clean Air, Wetlands, Private Property, and Nuclear Safety, *Statement of Steven Gatto, Hearing 106-953*, 106th Cong., 2nd Sess., June 14, 2000.

12. Dukes.

13. Dukes.

14. Lawrence Wright, "Silent Sperm," *The New Yorker*, January 1996, 41–55.

15. *Infertility: An Overview* (Birmingham, AL: American Society of Reproductive Medicine, 2003).

16. D. Pimentel, et al., "Environmental, Energetic, and Economic Comparisons of Organic and Conventional Farming Systems," *BioScience* 55 (July 2005), 573–582.

17. John M. Urbanchuk, *Contribution of the Ethanol Industry to the Economy of the United States* (Renewable Fuels Association, January 2005), 2.

18. Author's calculation using *U.S. Department of Agriculture Feed Outlook*, January 2005.

19. *Distillers Feeds* (Cincinnati, OH: Distillers Feed Research Council).

20. "Taxpayer Dollars Diverted," *CBS Evening News*, May 11, 2005.

21. "Sci Tech: Hydro Fuel Faces Bumpy Road," *CBS News*, June 30, 2003.

22. David Morris, "West Wing's Ethanol Problem," *AlterNet*, February 2, 2005 (www.alternet.org/envirohealth/21147).

23. *The Real Price of Gasoline: An Analysis of the Hidden External Costs Consumers Pay to Fuel Their Automobiles*, Report No. 3 (Washington, DC: The International Center for Technology Assessment, November 1998), 1–43.

24. *Gasoline Cost Externalities: Security and Protection Services*, The International Center for Technology Assessment, www.icta.org/doc/RPG%20security%20update.pdf (January 2005), 1–7.

25. Jim Wells, *Tax Incentives for Petroleum and Ethanol Fuels*, GAO/RCED-00-301R (Washington, DC: U.S. General Accounting Office, September 2000), Table 1.2.

26. Mark Cooper, *Over a Barrel: Why Aren't Oil Companies Using Ethanol to Lower Gasoline Prices?* (Sioux Falls, SD: Consumer Federation of America, May 2005), 2.

27. Urbanchuk.

28. Informa Economics, Inc., *The Structure and Outlook for the U.S. Biofuels Industry*, for the Indiana State Dep. of Agriculture, October 2005, www.in.gov/isda/pubs/biofuelsstudy.pdf (March 1, 2006).

29. "Worldwide Look at Reserves and Production," *Oil & Gas Journal* 103:47, December 19, 2005 (www.eia.doe.gov/emeu/international/reserves.html). Oil includes crude oil and condensate. Oil reserve estimate for Canada includes 4.7 billion barrels of conventional crude oil and condensate reserves and 174.1 billion barrels of oil sands reserves.

30. Data for the United States are from *Advance Summary, U.S. Crude Oil, Natural Gas, and Natural Gas Liquids Reserves 2005 Annual Report*, DOE/EIA-0216 (Washington, DC: U.S. Energy Information Administration, September 2006).

31. Urbanchuk, 1–4.

32. Marion Anderson, *American Jobs from Alcohol Fuel* (Employment Research Associates, 1980), 1–13.

33. Stacy C. Davis, *Transportation Energy Data Book: Edition 21*, ORNL-6966 (Oak Ridge, TN: Oak Ridge National Laboratory Center for Transportation Analysis, October 2001).

34. Mary J. Hutzler, et al., *Annual Energy Outlook 2002*, DOE/EIA-0383, (Washington, DC: U.S. Energy Information Administration, December 2001).

35. *Biofuels for Transport* (International Energy Agency, May 11, 2004), 6.

36. Traci Watson, "Air Pollution from Other Countries Drifts into USA," *USA Today*, March 14, 2005, Sec. 1A.

COPYRIGHT 1992–2002 ANDREW B. SINGER

Fig. 2-13

CHAPTER 3

THE PERMACULTURE SOLUTION TO FOSSIL FUEL DEPENDENCY

As we showed in the last chapter, industrial agriculture and its components—oil-based fertilizers, pesticides, and herbicides—are harmful to the planet. A nationwide switch to organic farming is in order. But it can't work if we maintain a monoculture-based system, with its present emphasis on corn farming. Much of America's farmland is below 2% organic matter. At 2%, the soil biology collapses—and, with it, the fertility needed to grow crops. More and more chemical fertilizer is needed to prop up production on sterilized soil.

THE "GREEN REVOLUTION" AND PERMACULTURE

A lot of people think that the "Green Revolution"—marked by the advent of monocultures, pesticides, herbicides, and chemical fertilizers some 60 years ago—produces more food per acre than older methods of agriculture. It emphatically does not. In fact, a Mexican campesino using simple hand tools to grow a polyculture of corn, beans, and squash can produce, on a dry-weight basis, far more food per acre than the farmer of the most modern U.S. Midwestern cornfield—food worth far more money on a net basis.[1] Unlike subsidized Green Revolution farmers, the campesino would not survive if his farming were only 30% efficient in using its major energy input (sunlight) or if it were dependent on expensive consumable products (herbicides, pesticides, fertilizers) that damaged the operating equipment (soil).

More than a half-century of this so-called Green Revolution has created our current urgent need for permaculture ethanol systems to solve the many agricultural and energy woes left in its wake. My hope is that an appeal to farmers' bottom lines, as well as a national consensus for a cleaner environment, will motivate change toward permaculture.

As you will see later in this book, there are many crops that can produce much larger amounts of

TODAY'S CORN FARMERS WILL BEGIN TO ASK THEMSELVES, "WHY NOT GROW 800- TO 1000-GALLON-PER-ACRE CROPS AND STILL BE ABLE TO GROW A COVER CROP OVER THE WINTER, TO BE TURNED IN, PROVIDING FERTILIZER AND ORGANIC MATTER TO GROW THE NEXT CROP?" WHY NOT, WHEN IT NOT ONLY IS PRODUCTIVE AND PROFITABLE, BUT GIVES US A CLEANER ENVIRONMENT AND HEALTHY SOIL? WHEN FARMERS REALIZE THEIR CO-OP CAN GROSS $2300 PER ACRE JUST ON THE ALCOHOL, AND CAN ALSO DEMOLISH ALMOST ALL CHEMICAL INPUT COSTS—IT BECOMES DIFFICULT TO CONSIDER A REASON *NOT* TO DO IT.

Fig. 3-1

alcohol fuel than a monocultural crop such as corn. A key to the success of the permaculture system is crop rotation, where different nutrients are used each season and nothing becomes depleted. Right now, the only common rotation in the Midwest consists of corn and soybeans. If you had four to eight different energy crops rotating, the demands on the soil would be very much reduced and could even be complementary.

"Then why should the farmer pay an ever-increasing price for gasoline when he can produce all the alcohol he needs and a lot more on his own farm?"

—HENRY FORD, *DETROIT NEWS*, DECEMBER 13, 1916

As long as much of the organic matter from production is returned to the soil—as in permaculture—an agricultural system will increase in fertility each year. The present agricultural system destroys topsoil and, therefore, fertility. Once the level of organic matter in the soil reaches around 5%, organic farmers need only about five tons of compost per acre per year to maintain fertility. Spread evenly over an acre, this would appear as a light dusting. With more organic matter than that, farmers would build topsoil depth and soil biological activity.

For example, if you were to grow relatively shallow-rooted corn one year, the next year you might grow fodder beets that will go several feet deep, using their huge system of roots to bring potassium and phosphorus up near to the surface. When you harvest the massive, 15-pound beet, it is only the top of an inverted conical pyramid of roots that fan out to probably more than three feet in diameter at the soil surface, tapering to a point five feet down. The part we harvest is less than half the weight of this entire root system. Fungi and earthworms can feed on the many pounds of smaller roots left behind throughout the soil, freeing the phosphorus and potassium for the next crop.

With rotation of crops grown for energy, today's corn farmers will begin to ask themselves, "Why not grow 800- to 1000-gallon-per-acre crops like fodder beets, Jerusalem artichokes, or sweet sorghum—and still be able to grow a cover crop like fava beans over the winter, to be turned in, providing fertilizer and organic matter to grow the next crop?" Why not, when it not only is productive and profitable, but gives us a cleaner environment and healthy soil? When farmers realize their co-op can gross $2300 per acre just on the alcohol, and can also demolish almost all chemical input costs by returning (for example) the spent beet pulp, or the manure from animals eating the beet pulp, into the soil—it becomes difficult to consider a reason *not* to do it.

POLYCULTURE, PHOTOSYNTHESIS, AND PHOTOSATURATION

Polyculture is an advanced method of agriculture that obtains the multiple benefits of crop rotation by growing many crops simultaneously. Polyculture dramatically increases the productivity of photosynthesis by eliminating **photosaturation**, or solar saturation, the point at which a plant's photosynthetic machinery is shut down by excess sunlight.

Simply put, photosynthesis is carbon dioxide (CO_2) plus water (H_2O) plus sunlight (energy for the reaction), which creates six molecules of CH_2O **(glucose)** plus a molecule of oxygen (O_2). Glucose is the basis of starch and cellulose.

A wide cross-section of the plant world photosaturates at 30% of a day's total sunlight, at which point more sunlight will not increase photosynthesis. This means that most plants grown in full sunlight stop growing in the middle of the day due to photosaturation, and don't resume growing again until the afternoon. And even if they can take the solar stress, they "waste" two-thirds of the surplus sunlight falling on them. Why? Because if a plant species needs unobstructed full sunlight all day, and then some tree extends a branch and shades it, the species will go extinct.

But plants that have evolved to use a fraction of the available sunlight have also evolved to

RIGHT: Fig. 3-2 Photosaturation. Winter squash visibly demonstrate their saturation by excess sunlight. This well-irrigated plant is "wilting" at two p.m. in mid-September at a cool 60°F; in this way the plant deflects most of the direct sunlight. Note how much self-formed shade the leaf has created.

DAVID BLUME

cooperate in a natural polyculture of mixed plants. A good example is winter squash. On seed packets, gardeners are told to plant it "in full sun." Pumpkins, butternut, and Delicata squashes wilt in the middle of the day when exposed to full sun because their huge leaves have evolved to use the dim light in their original home, climbing through the trees in a tropical rainforest. They wilt in order to deflect the excess sunlight. Many people incorrectly assume that the plants are dehydrated and water them excessively. In the later afternoon, the leaves stiffen up again and look healthy. The gardeners pat themselves on the back for a job well done and then are incredulous the next day when the leaves wilt again.

TROPICAL BONANZA

Almost double the amount of sunlight falls per square foot on the tropical world as on the temperate U.S. Many equatorial countries could capitalize on their placement on the globe by producing surplus solar energy for export in the form of alcohol. By 2010, Brazil expects to export more than 9.4 billion liters of alcohol,[2] with none of this expected to go to the U.S. at present, because of our trade barriers.

Brazil is using only 8% of its land for cultivation.[3] Of this total, it uses only a small fraction, a little over five million **hectares (ha)** (12.35 million acres) for sugar and alcohol production (Brazil has not cut down rainforests to make alcohol, either![4]) A recent USDA report estimated that Brazil can open up an additional 420 million acres of cropland "without any additional deforestation in the Amazon basin."[5]

Instead of the small area of Middle East dictatorships controlling the world's energy, an alcohol surplus could be produced in a wide band of tropical countries. By selling to northern countries, they would democratize the production of fuel. The diffusion of capital from northern countries to purchase this value-added agricultural product, instead of low-cost commodities, would be a major economic boost for much of the developing world.

Photosynthesis and Light Intensity

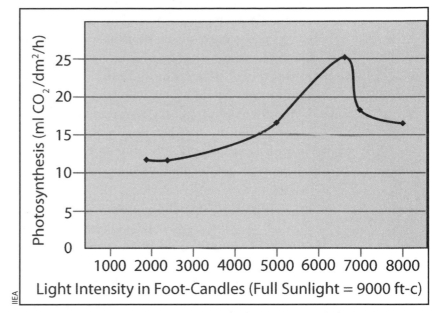

IIEA

Fig. 3-3 Solem saturation in sugarcane. *It's clear that as sunlight reaches its highest intensity in the middle of the day (9000 **foot-candles (ft-c)**), photosynthesis drops off dramatically and comes to a complete stop. Providing a light shade over the crop keeps carbohydrate production high throughout the day.*

Where Oil Originates and Who Consumes It

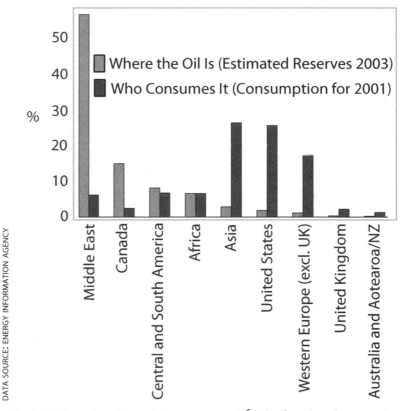

DATA SOURCE: ENERGY INFORMATION AGENCY

Fig. 3-4 Where the oil is and who consumes it.[6] *The fact that the countries that use the oil are far from its source is both energy-inefficient and the source of potential conflict over possession of resources. If all countries produced some or all of their energy from the sunlight falling on them, we'd have a lot fewer problems.*[7]

The first way to prevent photosaturation is to plant multiple canopies in a polyculture to cross-shade. For example, in a polyculture of corn, beans, and squash, the corn and beans shade each other, with the beans climbing up the corn, and both shade the squash growing underneath. In this way, each crop is harvesting about 30% of the sunlight, and little goes to waste. (Why doesn't corporate agribusiness do this? Because it's not a machine-friendly system and would result in higher dependence on labor, the increase in net profit notwithstanding.)

Another way to extend photosaturation limits (and also to initiate or extend photosynthesis in low-light conditions) is to add carbon dioxide, which alcohol fuel-makers can supply from their **fermentation tanks**. As you will see later, when yeast ferment carbohydrates to alcohol, they convert about half of what they eat into carbon dioxide, which they "exhale."

A small additional amount of carbon dioxide can mean plants will continue to photosynthesize down to 0.5% of average sunlight instead of being inactive below 5%.[9] Photosynthesis would therefore start earlier and end later in the day; or a denser overstory of crops and shade would be permissible.

POLYCULTURE A PROVEN PRODUCER

As we were going to press, a breaking study was released in *Science* Magazine.[8] In 1994, a degraded piece of farmland in Minnesota was laid out in 152 plots and planted with polycultures of prairie species consisting of mixed grasses, legumes, and herbs. The results after more than a decade consistently showed that those plots with the highest diversity of species were the most productive.

The conclusion was a mirror image of what we in permaculture have been routinely seeing with very basic polycultures of food crops. The study showed that the most diverse plantings had 2.38 times the potential biofuel output of a single-species planting.

What's more, the alcohol output should be half again as much per acre as corn starch alcohol production, without the use of any of the inputs used in corn agriculture. On an energy return basis, the prairie alcohol production was projected to be 17 times more energy-positive than corn alcohol.

This tripling of output over, say, a monoculture of switchgrass was achieved using only an herbaceous mixture of plantings, with no trees in the design, on degraded land, with no return of **stillage** or carbon dioxide. In other words, this was primarily a demonstration of overcoming photosaturation, along with soil biology being allowed to work without being poisoned by pesticides, chemical fertilizers, and herbicides. An even more diverse planting that included trees, on good farmland, and with all the surpluses of energy production being returned to the soil, would dramatically increase these already impressive yields.

This study also confirms my comments about plants sequestering far more carbon dioxide than ends up in the fuel. In this study, the prairie took 13.75 times as much carbon dioxide out of the air as was released in the production and use of the alcohol.

THE TASTY TEN PERCENT

While working on a project with the Nahuatl people of Mexico, growing oyster mushrooms on coffee pulp, I stopped to hang out with three campesinos for lunch. They welcomed my mycologist friend, Dr. Daniel Martínez-Carrera, and me to their lunch camp. They told us we could heat our beans, chiles, and tortillas using their *microonda* (microwave oven). I did a quick scan around the cornfield where we were sitting and realized they were having a little joke at our expense. The microwave was an old oil drum top sitting on three rocks with a fire under it.

While eating lunch, I asked if it got boring eating nothing but beans, squash, tortillas, and chiles all the time. (Academics had told me that was pretty much all that rural Mexicans ate.) The oldest of the three looked at me and asked me to follow him.

We went into the cornfield, and he started picking and naming plants, telling me what health benefits each had. To the untutored eye, all these plants would look like weeds. In reality, the fields had been weeded of all undesirable plants. The remaining plants were not weeds, but nutritious wild vegetable crops. We hauled a small armload back to the camp and wilted the veggies on the "microwave" and then stuffed our tortillas with different herbs. Each one tasted different and delicious. Although the corn, beans, and squash as primary crops used 90% of the sunlight, these wild greens used the last 10% of the sunlight falling on the field. That last 10% of the sunlight made for all the flavors and vitamins we could ever want.

With higher amounts of carbon dioxide, photo-saturation levels are increased, allowing unshaded plants to waste less sunlight in the middle of the day—resulting in up to triple the biomass produced during what normally would be a period of stress and no growth. Long-term study shows tripled yields of biomass and fruit in tree crops with carbon dioxide enrichment, even *outside* of greenhouses.[10]

A GREAT SYSTEM FOR FEEDING PLANTS

We were all taught in biology class that plants need various nutrients, such as nitrogen, phosphorus, and potassium. We were taught that plants absorb the water-soluble nutrients via the roots, and that all nutrients travel upward to make the building blocks of plant matter. The setup was portrayed as a one-way street, with the plants extracting what they need from the soil, and with each plant in competition with every other plant for scarce resources.

But, in reality, plants give back to the soil, through the roots, as much as or *more* than they receive. Furthermore, each plant is not a Lone Ranger, but a part of a larger community of below-ground organisms that provide sustenance to the plant and receive food in return.

Virtually every plant depends on special types of fungi to help it acquire food. These are called **mycorrhizae**; *myco* means fungus, and *rhiza* means root. The primary underground relationship is based on **mycorrhizal symbiosis**, a process that is key to understanding how plants are nourished.

At the heart of biological farming is organic matter, which is anything that rots. Leaves, twigs, manure, your jeans, and paper are all organic matter. Organic matter is consumed by both fungi and bacteria as the first step, at the bottom band of the soil's living biomass pyramid.

Most of us are familiar with the above-ground biomass pyramid. In rough terms, the base band of that pyramid is plants; the smaller band above that is plant-eating animals; and above that are carnivores, sometimes topped off by a top carnivore.

In the soil's living biomass pyramid, fungi and bacteria form the base; next up is a smaller biomass band consisting primarily of tiny microscopic mites that look vaguely like ticks and that eat the fungi and bacteria, devouring maybe ten times their weight. Go up the pyramid another step, and

you'll find a smaller population of slightly larger predatory mites, which eat ten times their weight in the smaller mites. Another level up, and you get into "large" but still microscopic creatures, such as springtails and nematodes, that generally eat both fungi and predatory mites. Finally, you get up to things you can see, such as earthworms and the like.

The bacteria and fungi, while not animals, are definitely not plants. Fungi and soil animals don't make their own food from sunlight or breathe carbon dioxide. They all eat or decompose something, they breathe oxygen, and they reproduce. And they do one other, very important, thing. They poop. Nine-tenths of all that eating ten times their own weight at each level ends up as poop. This microscopic creature poop is much of what plants need for food.

How it gets into the plants is another story. The fungi are after the same thing as any kid—mainly sugar. Unlike plants, fungi can't make sugar, so they need to take it from plants. Now, these fungi aren't generally parasites. They don't just take nutrients and kill the plant, which would be the road to evolutionary extinction. They engage in trade.

Fungi dissolve rocks, organic matter, microscopic creature poop, leaves, dead bacteria, and almost

---

MEAT-EATING TREES

Springtails are microscopic creatures that jump through the soil using a powerful tail; they are very common in the **rhizosphere** (root zone) of plants and trees. Scientists studying springtail feeding habits wanted to know how much of the mycorrhizal fungi springtails consumed. This experiment was analogous to seeing how much pizza a teenager could eat. But instead of the expected result, the pizza jumped off the table and ate the teenager!

The researchers radioactively tagged the fungi and were planning on capturing the springtails and analyzing how much radioactive fungus they had eaten. But instead, as a springtail would try to feed on the tip of the fungus, the fungus would squirt a chemical into the mouth of the springtail, paralyzing it. The fungus then grew into the mouth of the springtail and ate it from the inside out! Nutrients from the springtail were transported by the fibrous fungus and delivered into the tree roots in exchange for sugar. The researchers quickly adjusted their study, and they found that 25% of the nitrogen in the biomass of the study tree, symbiotically connected to the mycorrhizal fungus, came from springtails.[11] So, vegans, just remember all the little animals that your peach tree ate, with the help of a fungus, to bring you your fruit.

By the time the system has been operating for three years, there's enough water stored there to irrigate the trees lining the swale for 100 years, even if it never rains again…. One acre of integrated production like this would financially outperform 2000 acres of corn.

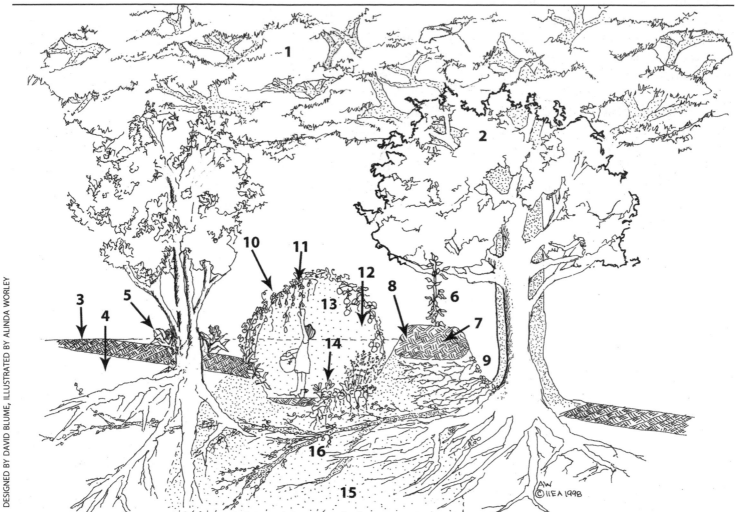

Fig. 3-5 Permaculture production system. *In order to make the most out of scarce rainfall, this permaculture design uses a* **swale** *(a dead-level ditch, the interior area between the grade behind and the berm in front) to collect water as its primary feature. Although this particular design is focused on food and flower production, it could easily be altered slightly to focus on energy crops. Bear in mind this is a fairly simple design; it could be much more dense, yielding even more crops and profit. Starting from the top is* **(1)** *a wide-canopied,* **nitrogen-fixing, Farmer's tree** *that provides light shade and bee forage. Understory trees* **(2)** *provide food or energy crops. Topsoil* **(3)** *is shown cross-hatched; subsoil* **(4)** *is shown white. Bulbs* **(5)** *are planted around bases of trees, generating a high-value flower crop while protecting trunk and roots from burrowing animals. Flowers or surface-harvested crops like nopales can be planted from the trunk out to the drip line. Drip line crops* **(6)** *such as trellised berries exploit the surplus moisture of this zone and are often planted on the well-drained part* **(7)** *of the swale berm* **(8)**. *Berm faces can be planted in deep-rooted permanent crops such as strawberries* **(9)**, *which root to six feet. Although shown here on only one face, such crops can be grown on all three faces of the swale (on the uphill side of the swale and the inside and outside faces of the berm). Over the swale, a trellis* **(10)** *increases cropping area. Beans or berries* **(11)** *or squash/cucumbers/melons/tomatoes* **(12)** *make good use of the trellis and provide shade* **(13)**. *The shaded bottom of the swale has rich soil for high-value vegetable crops or even energy crops such as fodder beets* **(14)**. *The swale bottom, with its air stabilized by the trellis crops, would also benefit from heavier-than-air carbon dioxide enrichment and irrigation with* **stillage**. *The shaded subsoil* **(15)** *is permanently hydrated by the water soaking in from rain. By the time the system has been operating for three years, there's enough water stored there to irrigate the trees lining the swale for 100 years, even if it never rains again. Nitrogen-fixing bacteria* **(16)** *on the roots of the Farmer's tree provide all the nitrate fertilizer needed by all the crops grown under its* **canopy**. *Mycorrhizal fungi in the top few feet of soil knit all the plants together, sharing the nutrients extracted from the soil and the carbohydrates provided by the upper canopies. One acre of integrated production like this would financially outperform 2000 acres of corn.*

anything they touch, by exuding powerful fluids such as acids or hydrogen peroxide. They then soak up the dissolved nutrients (nitrogen, phosphorus, and potassium) and pump them into the plant in return for the sugar. Both the plants and the fungi think they are getting a great deal. After all, plants make a huge surplus of sugar over what they need for growth, and fungi can dissolve and absorb much larger quantities of soluble minerals than they need to thrive. As with any permacultural system, surpluses need to be put to work—and so both the plants and fungi cooperatively trade their surpluses.

Plants pump 60 to 85% of the sugar produced in their leaves into their roots, and more than half of that goes to feeding mycorrhizae, or exuding the sugar solution out into the soil for bacteria and other microbes to eat.[12] For instance, strawberries can grow under a dense pine forest, where no direct sunlight penetrates. The mycorrhizae transfer sugar that originates with the pines to the strawberries in exchange for something the strawberry produces that the fungi want to eat. So, that sweet wild strawberry you pick is actually flavored by pine sugar.

So, unlike what we learned in school, each plant and tree is not a self-serving species. They are all a community connected via fungi, which collect and distribute the surplus resources cooperatively. By growing under the light shade, a crop not only avoids photosaturation and thus can photosynthesize all day, it can also receive surplus sugar from the top canopy via the fungi, if needed.

This whole cooperative system is based on a keystone: There must be sufficient organic matter in or on the soil. As I said, soil below 2% organic matter causes this system to collapse. Using soluble chemical fertilizers will kill most of the soil microlife. Insecticides and herbicides are toxic to almost all microlife, as well, although many bacteria and some fungi have been able to break them down. (Good thing, too, or life on Earth as we know it would have ended!) Which brings us back to the significance of organic polycultural farming.

DEALING WITH WEEDS AND PESTS

A well-designed agricultural system based on alcohol fuel-making can eliminate the need for Roundup herbicide, genetically modified Roundup Ready corn seed, chemical fertilizers, and even

Fig. 3-6

most insecticides. These are all losing propositions, anyway.

Insecticides rarely work very well. Biology responds to stimulus; biology "learns" and responds differently over time. Now, this doesn't apply only to individual organisms; it even more appropriately applies to populations. Let's say it's the late 1940s and you have a surplus of grasshoppers that are eating your wheat. The agricultural extension agent brings you this new nerve gas called DDT, and you spray it over your crop. Wow, 98% of your grasshoppers stop hopping. It's a miracle! You have more crop to harvest, and the stuff hardly costs anything. (Although you don't seem to hear any birds or toads anymore.)

"If we throw Mother Nature out the window, she comes back in the door with a pitchfork."

—MASANOBU FUKUOKA, AUTHOR OF *THE ONE-STRAW REVOLUTION: AN INTRODUCTION TO NATURAL FARMING*

ALCOHOL...THE SOLAR FUEL

RON HARPER

Fig. 3-7 Alcohol, the solar fuel.
Alcohol is really liquefied solar energy. Using sun, carbon dioxide, and water, plants make the carbohydrates that we turn into alcohol. When burned, the solar energy drives our engines, while returning the CO_2 and water back to be used by next year's crops.

Next year, the hundreds of eggs laid by the surviving insects the previous year hatch, not having been gobbled up by predators, who were killed along with the insects last year. So the grasshoppers come back even worse, since there are fewer predators to eat them. No matter, you just spray your DDT. But this year is not like last year. You need to use twice as much DDT, since maybe 20% (not 2%) survive your first spraying this year. Why? Because you killed all the hoppers that were easily killed by DDT last year, and only the ones with natural immunity have survived to reproduce. Without predators, which reproduce more slowly, you have selected the hardiest of the bunch to come back this second year. So it takes a lot more pesticide to kill these resistant bugs.

The third year comes around, and there aren't any birds, toads, fish, spiders, or snakes anymore—because the pesticide was in all the bugs they ate, and concentrated in their flesh, and killed them. So now you have to increase the use of DDT again, and this time you can kill only 20% of the hoppers. At this point, you generally lose the crop unless you drench it in DDT. By the fourth or fifth year, all the grasshoppers that have survived reproduce super-hoppers that not only are immune to DDT, but also excrete it through glands that produce the

"tobacco juice" that has been sprayed on nearly every kid who has captured a hopper. The hoppers learn to squirt the juice into the mouths of predator birds, killing them with the pesticide.

To continue using pesticides, you need to escalate every two or three years to ever more toxic ones. As mentioned in Chapter 2, some of the pesticides used today are 3000 times more toxic than the original DDT.

So, since Nature responds to stimulus, its response to the toxins will vary constantly. Nature will overcome any obstacle we put in her path due to the incredible diversity of genetics each species has.

In polyculture farming, a mosaic and rotation of primary crops both confuses pests and limits the amount of food they will find suitable. Monoculture is like an all-you-can-eat buffet for the pests that like that particular crop. So pests that like to eat corn and overwinter in the soil will be mighty disappointed when they awaken to find the field planted to sugar beets in the spring. Root crop pests that love beets will find little tasty fare in the tough fibrous roots of sorghum. You can grow strips of flowering plants, such as Jerusalem artichokes, to provide high-protein food to keep insect predators fed while they wait for more bugs to show up; it will also provide you with an energy crop at the end of the farming season.

Rather than being drained for corn, natural low swampy areas could sport energy crops, such as cattails, that make ideal habitat for birds or beneficial insects such as dragonflies and for countless toads—all of which would go out and decimate the pests in adjoining fields.

ORGANIC "WEED AND FEED"— PUTTING AN END TO MONSANTO IN AGRICULTURE

Monsanto is the industry leader in genetically modified seed and is the supplier of Roundup. It is in the vanguard of companies that wish to shackle farmers to patented herbicide-resistant seed—which allows farmers to use lots of Roundup herbicide to kill the weeds growing between the GMO crop. Monsanto prohibits farmers from saving the seed and forces them to buy new seed each year. I have a remedy for this strong-arm tactic.

In permaculture, we always think back to what happens in Nature for an explanation. So, let's observe corn in Nature to figure out how best to grow it without chemicals. When a cornstalk falls

"All the world is waiting for a substitute for gasoline. When that is gone, there will be no more gasoline, and long before that time, the price of gasoline will have risen to a point where it will be too expensive to burn as a motor fuel. The day is not far distant when, for every one of those barrels of gasoline, a barrel of alcohol must be substituted."

—HENRY FORD, *DETROIT NEWS*, DECEMBER 13, 1916

DAVID BLUME

DAVID BLUME

Fig. 3-8 Control group, plantings without DDGS. These rows are planted with the ten worst weeds found in cornfields. They are used for testing any new herbicide. In this flat, there was just plain fertilized soil, and in just a few days the weeds have come up rather nicely.

Fig. 3-9 Plantings with DDGS. This photo was taken on the same day as the control photo. You can see most of the weeds did not survive germination, and the rest are severely stunted. Within days, all the weeds were finished off. You can just see the remains of the bacterial gel that formed on the surface of the soil as a result of the DDGS addition.

over at the end of its life, the husk-wrapped starchy ears of corn plop onto the ground. Over the winter, a little decomposition happens to the husk. Birds might get to the kernels on the top half of the cob, pecking some open, scattering bits of corn, but leaving some of the grain on the cob, where rain washes it onto the ground. Come spring, three, four, maybe a dozen intact seeds sprout and come rocketing out of the ground. Very few or no weeds seem to be near the corn when it sprouts. So, in Nature, the corn seems to have an herbicidal effect that gives the clump of corn a big head start. To this day, indigenous people plant corn in clumps imitating Nature.

An experiment I performed shows how easy it is to take herbicides out of the picture. For a long time, organic farmers had known that **corn gluten meal (CGM)** was a very good pre-emergent herbicide. This means that it kills plants when they are just sprouting, as opposed to post-emergent herbicides, which kill plants beyond the seedling stage. (The high price of CGM has limited its use to organic gardeners; it is too expensive for most organic farming.) No one knew how CGM worked, however. USDA scientists said that the prevailing

theory was that weeds were nitrogen-poisoned by the high-protein gluten. It seemed a ridiculous theory, as most weeds I know suck up nitrogen better than most crops.

Based on my observation of corn in Nature, I conducted an experiment to see if the distiller's dried grains with solubles (DDGS), the byproduct of **dry-milling** corn, would have an herbicidal quality similar to CGM's. Remember, everything that came from the soil is in the DDGS. The only thing taken out of a crop to make alcohol is the solar energy. The plant carbohydrates contain only carbon dioxide, water, and sunlight. Nothing from the soil is used up in burning alcohol in cars. All the protein, fat, and soil minerals are still in the spent byproduct of the alcohol process.

I set up four flats with potting mix, and I put ten rows of weed seeds in each one. The seeds were for the ten worst weeds reported in cornfields. I reasoned that if anything would be resistant to the herbicidal effect of corn, it would be weeds growing in cornfields. I added nothing to the control flat, sprinkled whole organic corn meal (OCM) over the second flat, sprinkled CGM over the third, and sprinkled DDGS over the fourth. (I included

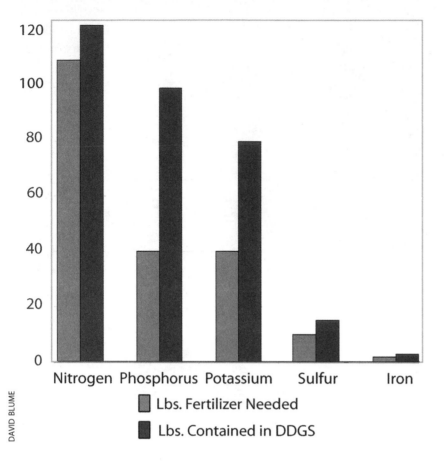

Comparison Between Corn DDGS Fertilizer Content and Recommended Fertilizer Levels for 160 Bushels/Acre Yield

- Lbs. Fertilizer Needed
- Lbs. Contained in DDGS

DAVID BLUME

Fig. 3-10 Corn DDGS fertilizer value. You'll note that the DDGS supplies considerably more nutrients than are needed to grow the next crop of corn, especially when it comes to phosphorus and potassium.

to soil particles in mats of webbing. This prevented soil erosion and seed washout by falling drops of water. It also created a breathable "seal" on the surface of the soil, limiting **evaporation**.

Next, there was an explosion of soil microlife eating the pioneering fungi, bacterial gels, and the DDGS. Within weeks, the mycelia, the bacterial gels, and the granules of DDGS disappeared. The fermented, ground seed in the form of DDGS had allowed things that like to eat germinating or broken seed coats to go through a powerful population explosion in the soil. These microbes really savaged the roots of sprouting weed seeds. Once this fungus and microbe explosion happens, it takes many weeks before any other seed trying to germinate in the bed has a chance to grow. Corn, however, was immune to the effect. The herbicidal control effect was biological, not chemical.

So how does this translate into a revolution in farming and an end to Monsanto's formula of selling patented seed and herbicides to match, and still dramatically cut production costs?

According to USDA records, the median corn yield in the very average state of Nebraska was

OCM in the experiment to make sure that any herbicidal effect was not from the residue of chemical herbicides; organic corn is never treated with chemicals. Later testing of the DDGS showed that it, too, contained zero residual **biocides**.)

So did DDGS act like an herbicide? Yes, it did. In fact, all three materials had significant herbicidal effects, but the DDGS results were the most pronounced. About half the weeds were killed, and the rest were stunted. Stunting was enough, though, since the corn grew right up and buried the puny weeds in darkness, where they withered.

I did another trial at the same time to see if herbicidal qualities would inhibit the germination of corn. I seeded four flats with corn instead of weeds, and then treated them with nothing, OCM, CGM, and DDGS. The various additives delayed germination one day.

The granules of OCM, CGM, or DDGS were attacked by bacterial and fibrous fungal **mycelia** right away, which grew into the grains from the soil. The bacteria frequently formed water-holding gels, which adhered to both the grains and the soil particles. The filamentous fungi knitted the grains

almost 160 bushels per acre.[13] I had the DDGS in my experiment analyzed for its composition of soil nutrients, and then calculated how much DDGS would come from 160 bushels, and what would be the soil nutrient value of that amount of DDGS. It was then a simple matter of comparing these figures to what the USDA says it takes to grow the 160 bushels of corn in that average field.

I found that there was enough fertilizer value in the DDGS "left over" from growing 160 bushels of corn to raise more than the next 160 bushels, since there was actually 10% more nitrogen and an even greater surplus of other critical nutrients, such as phosphorus, potassium, iron, and sulfur, than was required to grow the next corn crop. Everything that came out of the soil to make that corn was still there in the DDGS. So, from the soil's point of view, it was like returning last year's crop nutrients for use by this year's crop.

I found that this ratio of increased percentages of certain nutrients with repeated DDGS applications to corn held true even at yields of more than 200 bushels per acre. So, over a period of years, 160-bushel land would be producing 200 bushels. Plus, the soil organic matter would rise year after year after year. Combined with the soil-protecting effects of DDGS, the higher organic matter would mean more retained nutrients, increasing **humus** levels.

It would also mean relative drought-proofing of the crop, due to the mycelial sealing-in of moisture in the early phases of crop growth, and the sponge-like water-retention effect of organic matter.

What I had literally discovered in the experiment with DDGS was organic, drought-proofing, "weed and feed."

Now, how would this affect a farmer's bottom line and even free him from corporate dependence? Using very general numbers, you'll find that a corn farmer historically grosses something like $250 per acre on his crop. For that acre, he spends more than $50 on toxic Roundup herbicide and Roundup Ready genetically modified herbicide-resistant seed. He spends about $80–$100 in fertilizer per acre and a smaller amount on insecticides. With all his expenses totaled, a farmer will generally net, in a decent year, about $50 on an acre of corn.

But if the farmer produces alcohol instead of selling his grain into agribusiness—and takes the DDGS that results from his grain-growing, and applies it to his field during soil prep and

planting—his costs to produce that acre of corn will drop from $150 to about $50, tripling the net profit.

And remember, the farmer is now making alcohol from his grain, instead of selling it cheap for animal feed. If the farmer sells his alcohol to a community-supported energy alcohol distribution station, he can bring in, after deducting the cost of making the alcohol, $1.50 per gallon, or $588 per acre, instead of getting $250 selling corn for animal feed.

The net profit from his crop would be over $500 instead of $50. He wouldn't have to borrow money for fertilizer or GMO seed (since he could now save his own), and he would have, of course, no herbicide costs. This could be big news in Farm Country. Instead of corn being a soil-draining crop in rotation, it could be soil-building. Never again would a single year's crop failure bankrupt a farm.

Once a farmer starts to farm with DDGS, even if he only does it every other year or one in every three years while rotating through other energy crops, he is within an inch of being organic—which is better for the environment, his family's health, and his bottom line. The Oilygarchy—which sells more than one billion pounds of insecticides per year, as much as four billion pounds of herbicides per year, and an ungodly amount of highly energy-intensive nitrogen fertilizer, made from natural gas—would be deprived of a high-value market for its products. Monsanto would have no one to sell its proprietary GMO seed or chemicals to. Now, *this* would really be a farmers' revolution.

Fig. 3-11
Improved growth of corn using DDGS. *On the left is corn in fertilized potting soil, and on the right is corn in the same soil with DDGS added to the surface. Neither patch is nutrient-deficient, but clearly the addition of the DDGS has improved the growth of the crop on the right. In the original color photo, the difference is even more marked.*

So here is my contribution to the revolution. I was granted a patent in 2007 on the process to use DDGS as a combination fertilizer and herbicide. Now an agribusiness corporation cannot patent it and then prevent others from using it. I am going to handle use of my patent in a completely different manner than a large agribusiness corporation would. A program will be set up to let individual farmers and small collectives with small capacity license this patent for a nominal fee, perhaps only requiring registration (details are still being worked out as of the publication date of this book).

I ask the following three things of these small-volume users of my patent: 1) that they not use chemical herbicides and not use genetically modified seeds; 2) that they learn how to breed and save their seed from season to season; over time, you will end up with a variety tailored to your climate and soil (write me if you need help with this); and 3) in lieu of the patent royalty, donate what you think is fair to the International Institute for Ecological Agriculture so that I can keep doing this kind of work. Make sure that you take care of your family and workers first. I trust my fellow small farmers to do right by me if their profit increases. So take that, Monsanto!

Alcohol fuel production can result in more concentrated corporate farming of monocrops processed in giant plants, or it can save the family farm and provide farmers with markets for a wide variety of crops. The preferable route to a sustainable agriculture system looks to me like farmers cooperatively producing fuel and multiple co-products in small plants, using a wide range of feedstocks, from roots to grains to cellulose.

That sort of mosaic would go a long way toward eliminating the pest and soil problems that monoculture has created, as well as eliminating the use of toxic chemicals and fertilizers in an attempt to mitigate those problems. Such diversity would make for a more resilient farm economy with less chance of failure than the high-risk gambling involved in growing only corn and soybeans. Ideally, a sustainable agricultural system would be three-dimensional, harvesting three times as much sunlight as biomass. It would have mixed canopies of trees, and ground crops whose environmental needs closely match the local climate and **solar income**.

As you will see in Chapter 11, farms can become producers of many value-added human foods and products, rather than just the suppliers of low-cost raw materials to corporations. Alcohol fuel production sets us on the road to a permanently productive and fossil-fuel-free agriculture.

Endnotes

1. Stephen R. Gliessman, "Multiple Cropping Systems: A Basis for Developing an Alternative Agriculture," in Charles A. Francis, *Multiple Cropping Systems* (New York: Macmillan, 1986), 72–76.

2. Mark Avery, email communication, *Agência Estado* website, reprinted in *Grain Journal* and *Biofuels Journal*, www.grainnet.com (March 9, 2005).

3. Brazilian Department of Sugar and Ethanol, *Sugar and Ethanol in Brazil,* Ministry of Agriculture, Livestock and Food Supply Secretariat of Production and Commercialization, January 2005, 4.0.

4. Brazilian Department of Sugar and Ethanol, 8.

5. *Brazil: Future Agricultural Expansion Potential Underrated*, USDA Foreign Agricultural Service, January 21, 2003, www.fas.USDA.gov/pecad2/highlights/2003/01/Ag_expansion/ (September 21, 2003).

6. "Where the Oil Is," compiled from statistics in *Oil & Gas Journal.*

7. "Who Consumes It," compiled from U.S. Energy Information Administration tables on world petroleum consumption.

8. David Tilman, Jason Hill, and Clarence Lehman, "Carbon-Negative Biofuels from Low-Input High-Diversity Grassland Biomass," *Science* 314:5805 (December 8, 2006), 1598–1600.

9. S.B. Idso and B.A. Kimball, *Responses of Sour Orange Trees to Long-Term Atmospheric CO_2 Enrichment* (USDA ARS), 2.

10. Idso and Kimball.

11. John N. Klironomos and Miranda M. Hart, "Food-Web Dynamics: Animal Nitrogen Swap for Plant Carbon," *Nature* 410 (April 5, 2001), 651–652.

12. Elaine Ingham, Ph.D., communication with author, 2005.

13. Nebraska Agricultural Statistics Service, *Nebraska Agri-Facts* 17/2002, www.nass.USDA.gov/ne/agrifact/agf0218.txt (September 4, 2002).

CHAPTER 4

DARKER VISIONS OF OUR ENERGY FUTURE

The world produced 72,477,000 barrels per day of oil in 2004.[1] Half of that oil typically becomes engine fuel, which means we burned 555,536,205,000 gallons—over 500 billion gallons—in 2004 to fuel the world's internal combustion engines.

In 2005, the budget for the U.S. Department of Defense topped $500 billion. By mid-2006, the total expenditures for the occupation of Iraq also totaled over $500 billion.[2]

What is the connection between these figures? The cost of building an alcohol fuel plant conveniently comes to about $1 per gallon of annual production capacity; as more plants are built, the cost will decline. So, we can build enough alcohol plants to replace all the world's use of petroleum fuel with alcohol, essentially permanently—with less than one year's defense spending.

Building those plants around the globe would substantially reduce pollution, begin to reverse global warming, fully employ rural communities, guarantee an end to world hunger, and spread wealth based on solar energy collected and soil properly conserved. Would we need a defense budget if the world were fed, employed, and permanently supplied with locally grown energy and food? What would be left to fight over?

You'd think all of us would want all of this. But we live in a world of skewed priorities. Our system runs on privatizing profits for the few while socializing the costs for the rest of us.

As the state of the planet makes clear, we can no longer afford the fantasy that there is a river, an airspace, or a landfill in which to dump inconvenient filth; we can no longer deny that proper disposal is a cost of production. We are all downwind of each other; there is no anonymous fence over which we can pitch our wastes without them coming back to haunt us.

Corporations can no longer be allowed to have the rights of persons without the responsibilities. We need to look squarely at big ideas that expand

MEGAOILRON'S VISION OF NONRENEWABLE RESOURCES

FOR THOSE WHO CAN PAY HAS BEEN CATASTROPHIC FOR

ALL OF US TO DATE AND WILL GET FAR WORSE.

IF WE WANT A FUTURE POWERED BY RENEWABLE FUELS

AND DEMOCRATICALLY GOVERNED PEOPLE,

WE HAVE TO WORK TO MAKE IT SO.

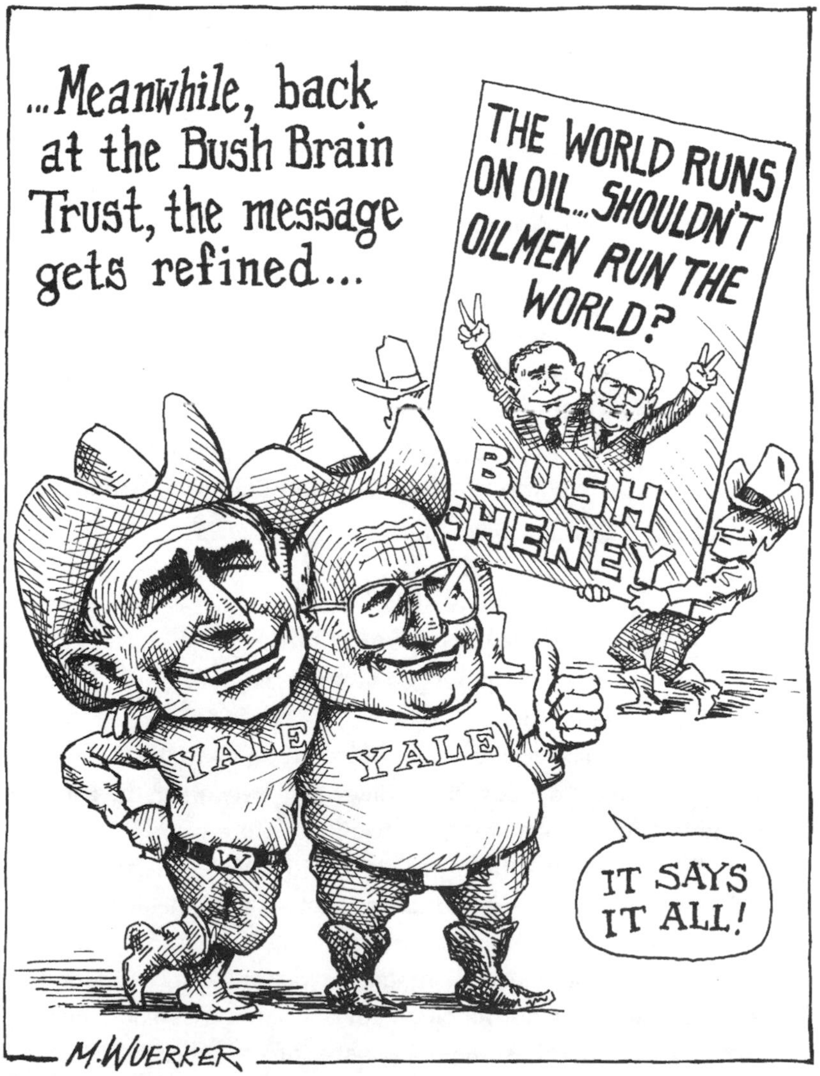

MATT WUERKER

Fig. 4-1

Fig. 4-2

SEE IF IT RIDES ANY SMOOTHER, NOW!

EMISSION STANDARDS

US AUTO INDUSTRY

MICHAEL KEEFE

the boundaries of consideration in solving the world's ecological and economic crises.

There are those who actually look forward to oil production peaking and then declining. They believe the end of oil will automatically trigger a transition into a future of renewable energy, a return to localization and community, and *poof!* corporate power will vanish from our lives. Would it were only so!

Using greed, blind faith, speculation, and generation of fear, a conglomerate of energy, high-finance, and government forces I refer to from time to time as "MegaOilron" is already moving toward a far different vision of the "post-petroleum" future. This vision of nonrenewable resources for those who can pay has been catastrophic for all of us to date and will get far worse. It's not an option to simply do nothing and wait for a social utopia. If we want a future powered by renewable fuels and democratically governed people, we have to work to make it so.

TAR SANDS

We are scraping the bottom of the tar barrel now that all the good oil has been used. Today's newly minted "conventional oil" is extracted primarily from Canadian **tar sands**. Deposits of tar sands also exist in central Asia and in Venezuela (which also has a lot of **oil shale** deposits). Although most Americans don't know it, imports of "oil" from tar sands exceed what we import from Saudi Arabia, OPEC's largest producer of crude oil.

But tar sands are not convenient black gold that is pumped out of the ground. They're a tarry goo mixed with sand, currently surface-mined in Alberta, Canada, and transported to processing centers.

It takes five barrels' worth of natural gas energy to provide the steam to process and harvest four barrels

---

TANKERS ARE MAJOR POLLUTERS

Oil and liquefied natural gas tankers use bunker oil, which although related to diesel, is barely usable as fuel. Tankers travel about 11,000 miles round-trip bringing crude oil to the U.S. Russell Long, project director of the Bluewater Network at the Earth Island Institute in San Francisco, testified before the U.S. Senate that, "The oil shipped from these areas [the Middle East] to the U.S. travels in an aging world tanker fleet—a fleet that is responsible for 14% of all global nitrogen emissions, and 16% of all sulfur emissions from petroleum. Carnegie Mellon University concluded in 1999 that the world shipping fleet is actually changing global climate as a result of its staggering emissions over oceans."[3]

of oil-like **kerogen** (which yields petroleum products after further distillation) from mined sands. Each barrel of oil from sand requires two to five barrels of water and carves up four tons of earth.[4] Two tons of sand yield only one barrel of the tarry fluid, but the rest of the sand and oily residues are flushed with water into huge holding ponds that will be highly toxic for thousands of years.

At the Alberta sites, the holding ponds are located right in the flight path of one of the biggest waterfowl migration routes. Untold millions of birds are killed in the oily ponds when they try to land to rest. The oil companies had to put up electrified scarecrows and detonate cannons to try to keep the birds from landing. Did it work? Well … they tried.

Over the next few years, many oil companies will find that the mineable sands on the surface will be gone. They will then have to extract the kerogen from deposits deeper in the earth, using steam to melt the underground tar, and sucking up the barely liquefied goo with a deeper well. That process is expected to take four or five times the natural gas energy that the current process takes to get one barrel of oil—20 barrel equivalents.[5]

This 20:1 ratio of energy input to energy output is staggeringly bad. The sheer volume of carbon dioxide produced from the natural gas will likely be the single highest source of greenhouse gas in the

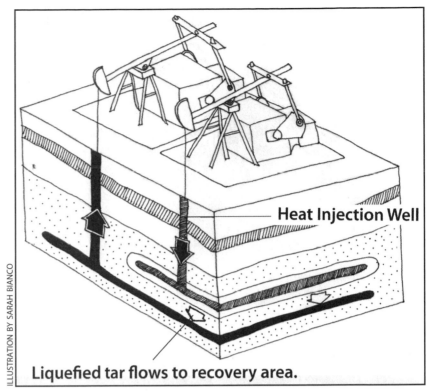

ILLUSTRATION BY SARAH BIANCO

Heat Injection Well

Liquefied tar flows to recovery area.

Fig. 4-3 Deep tar sands extraction. *Tar sands in an upper layer are heated using natural-gas-derived steam until the tar melts and flows down to a porous layer, from which it is pumped. This is a massively energy-wasteful process, resulting in enormous production of greenhouse gases.*

Fig. 4-4 Years of supply of fossil fuels. *This General Motors projection done in 1978 clearly shows that industry and government were aware that Peak Oil was likely to occur shortly after 2000. No significant governmental steps were taken to reduce dependence on oil for energy.*

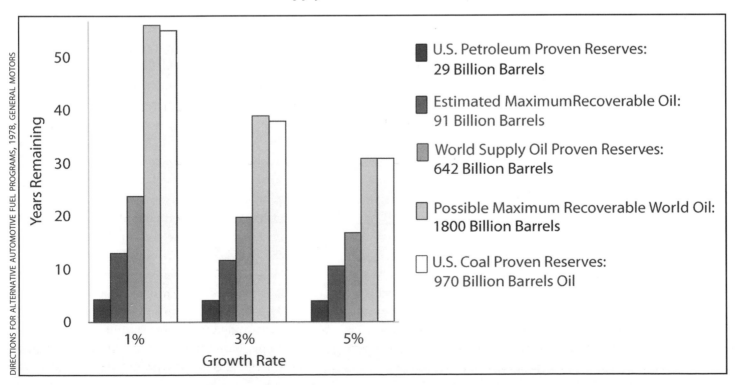

Years of Supply of Fossil Fuels (Base Year 1978)

DIRECTIONS FOR ALTERNATIVE AUTOMOTIVE FUEL PROGRAMS, 1978, GENERAL MOTORS

Years Remaining / Growth Rate

- U.S. Petroleum Proven Reserves: 29 Billion Barrels
- Estimated MaximumRecoverable Oil: 91 Billion Barrels
- World Supply Oil Proven Reserves: 642 Billion Barrels
- Possible Maximum Recoverable World Oil: 1800 Billion Barrels
- U.S. Coal Proven Reserves: 970 Billion Barrels Oil

world—that is, until they start in on oil shale (see next section). Proposals are already on the table to build a nuclear power plant next to the tar sands facility in Alberta to produce steam for heat instead of using natural gas.[6] The justification for this is to reduce the CO_2 emissions from natural gas!

This truly borders on madness. Replacing natural gas waste (carbon dioxide) with nuclear waste is certainly a case of out of the frying pan and into the fire.

The foul stuff, kerogen, is so loaded with sulfur that some attempt has to be made to get some of the sulfur out of the oil so it doesn't become sulfuric acid rain when burned in an engine. Near Alberta, huge sulfur pyramids comprise five million tons of the yellow acidic sludge and grow larger each day. Surplus sulfur from synthetic petroleum production is becoming a worldwide problem, with Chevron/Texaco and Exxon/Mobil sulfur mountains in Kazakhstan topping 7.8 million tons. These piles are supposedly so reflective of sunlight that shuttle astronauts have seen them from space. Fines of $72 million were levied against oil companies for their pollution of the Kazakhstan desert, later whittled down to $7 million.[8]

In Canada, where Exxon/Mobil and Shell operate the majority of the plants, 1700 tons a day of sulfur are added to the Alberta pile, which leaches rivulets of battery acid into the Earth and waterways. These huge semi-solid mountains flake and powder with the changes of temperature, releasing glittering clouds of potent sulfur particles to blow in the wind until they encounter moisture and become liquid acid. Canada, the United States' largest supplier of oil, "can't afford to" clean up its mess; Syncrude Canada Ltd. whines that it can't afford to move the sulfur the 930 miles to where it might be sold.[9]

What's more, China has already acquired large stakes in the pipelines that will be moving the tarry crude to Canadian Pacific Coast refineries. These pipelines are being built over the vociferous objections of indigenous people whose land must be crossed. Everyone expects China to outbid the U.S. for the tar sands operations of the future. "We are looking for profitable projects, which could include everything from minority stakes to full ownership of oil sands companies," states Hou Hongvin, Vice President of one of China's largest oil and chemical companies, Sinopec.[10]

To sum it up, tar sands will eventually create a 2500% negative energy returned on energy invested (EROEI), while putting out enormous greenhouse gases. And the U.S. will be fully dependent on them—if we let it happen.

OIL SHALE

Oil shale production takes the madness of prolonging petroleum use even further. In the 1980s, Harvard and other universities estimated that producing petroleum from oil shale would only be cost-effective once oil reached $100 per barrel. In 2005, oil company executives and stock analysts were warning that people should expect prices of $105 per barrel in the near future.[11] So oil shale was to be called "conventional oil," as well.

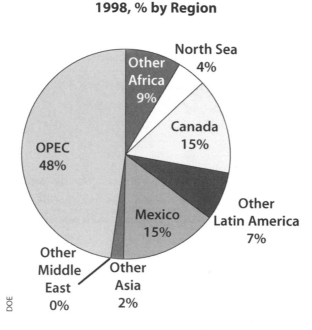

Fig. 4-5 U.S. crude oil imports by country. Note the large contributions by both Mexico and Canada. Due to the commercialization of tar sands, Canada's percentage has continued to grow since 1978, and Canada is now our largest foreign source of so-called "conventional oil."

U.S. Crude Oil Imports 1998, % by Region

- Other Africa 9%
- North Sea 4%
- Canada 15%
- OPEC 48%
- Other Latin America 7%
- Mexico 15%
- Other Asia 2%
- Other Middle East 0%

DOE

THE UNFROZEN TUNDRA

Canada has always assumed that the Arctic plant matter, the tundra, would absorb enough carbon dioxide to allow it to meet its Kyoto carbon dioxide emissions reduction target—a full 40% of the world's Arctic vegetation is in the Canadian tundra. Canada assumed that the carbon dioxide extracted from the atmosphere by respiration of tundra plants would cancel the CO_2 output from the production of tar sands.

However, a recent study shows that an ominous feedback loop is occurring. As the climate warms, the deep rich organic matter and peat of the tundra soil is decomposing at an unnaturally fast rate, liberating far more carbon dioxide than is being absorbed. It is calculated that when the top meter of Canadian tundra organic matter decomposes, atmospheric carbon dioxide will increase a full 25% worldwide, dramatically accelerating global warming.[7]

Oil shale is a hard, porous rock that also contains a tarry substance similar to tar sands. But unlike tar sands, simple steaming won't extract the oil. Over time, hundreds of small operators tried their hand at making oil from shale, but all went bust when West Texas oil was discovered in 1925.

The U.S. Bureau of Mines operated a small research facility for shale from 1925 to 1929. During World War II, Congress passed the Synthetic Liquid Fuels Act, authorizing the building of demonstration plants for $87 million, subsidizing research for the oil companies once again. The government privatized the project in 1972, when it leased the Anvil Points plant, a government-built demonstration plant in Rifle, Colorado, to a group of 17 private companies. Eventually, all

---

"I'd put my money on solar energy … I hope we don't have to wait till oil and coal run out before we tackle that."

—THOMAS EDISON, IN CONVERSATION WITH HENRY FORD AND HARVEY FIRESTONE, MARCH 1931[12]

---

interest petered out without a single commercial plant being built.[13]

Then came the energy crisis of the 1970s, and the oil companies once again bellied up to the bar. They convinced the Carter administration to put up $14 billion in 1980 for the Synthetic Fuels Corporation, to be used to support private synthetic fuels projects from oil shale and other nonrenewable sources (e.g., coal to liquid fuel). Once again, no commercial plant came out of that pork barrel.

If oil shale and tar sands are developed, hundreds of millions of tons of waste contaminated with acridine (a coal tar derivative) will have to be dumped in U.S. Western states. *Every barrel (42 gallons) of oil produced will generate a ton of toxic waste.* To reach the minimal economically effective pilot plant level of 500,000 barrels a day, the experimental plant would have to dispose of *200 million tons* of waste every year. Such waste contains not only acridine, but a whole spectrum of toxic hydrocarbons, voluminous heavy metals, and salts. And the production of 500,000 barrels a day is less than oil refineries lose in steam leaks—a mere drop in the bucket as far as supply is concerned.

Fig. 4-6

MICHAEL KEEFE

Science Magazine reports that acridine causes curious defects in insects. Minute amounts of acridine were added to the sand in which cricket eggs were laid. Twelve days later, when the eggs hatched, the crickets had a variety of mutations, including two or more heads, extra eyes, and extra or branched antennae. The amount of acridine that caused the mutations was miniscule.[14] No specific tests have been performed on higher animals, but such a powerful **mutagen** should be suspected of having negative effects on all manner of animal life.

A ton of oil shale produces only 25 gallons of oil-like goo. To capture the oily vapors for refining, large amounts of water have to be sprayed over the burning shale. In the U.S., massive pipelines would have to bring water from Canada to supply this inferno. To produce 50,000 barrels per day of this sticky precursor to petroleum—one-tenth of the amount slated for the pilot plant described above—would take enormous water resources and emit mind-boggling air pollution. Just making the 50,000 barrels a day would consume 4000 acre-feet (1.3 billion gallons) of water per year, and produce atmospheric pollution of three tons of sulfur oxides, up to 20 tons of nitrogen oxides, two tons of carbon monoxide, two tons of largely carcinogenic hydrocarbons, 20,000 tons of carbon dioxide,[17] and 110,000 tons of solid waste, *every single day.*[18]

In the early 1980s, it was estimated that if oil shale replaced our imported oil (at the time about half our current imports), it would produce more than 1000 pounds of toxic waste per person per year, doubling the amount of toxic waste of all other industries combined at that time. The waste would be so voluminous that it would require a dumping ground the size of every canyon in Wyoming, Utah, and Colorado. That was in 1980. Nothing, I repeat nothing, would be likely to grow on these wastes for thousands of years, while they leach carcinogens, mutagens, and heavy metals into our water supplies.

It gets worse. One oil company recently announced that it was successful in lowering "heaters" into wells drilled into the shale, thus getting the rock hot enough so that the goo melted and drained down to an old **aquifer** where it could be pumped out. Something about this didn't seem right to me, so I asked a physicist friend to figure out what would be the energy balance of using natural gas to provide the heat for this. He got back to me and said, "The good news is that they can't be using natural gas as the heat source. There'd be no way to get that much heat down there with gas. The bad news is that they are almost certainly using large plugs of hot nuclear waste to heat the rock. Nothing else comes close to making this scenario fly." Halliburton is purported to have designed an identical system for NASA, using nuclear waste plugs to melt the Martian ice cap to provide water for a Mars base.

LIQUEFIED NATURAL GAS

Liquefied natural gas (LNG) is a supercooled, compressed liquid that starts off as natural gas. The energy it takes to compress the gas is a large percentage of the amount of energy you recover from the gas, making it a very inefficient method of transportation. Natural-gas-powered **compressors** do the work—spewing carbon dioxide, making this energy source a huge contributor to the **greenhouse effect**. But unless someone figures out how to run natural gas pipelines across the ocean, LNG is the only way that natural gas can be moved from where there is a surplus (Siberia and the Middle East) to North America.

LNG is very tricky stuff and causes metals to become brittle and susceptible to failure in ways

---

CHINA—BICYCLES OR TWO-STROKE MOTORCYCLES? WHY WE SHOULD CARE

In the mid-1990s, I read an article about the huge effects that any major decision in China has on the entire world. The story was an interview with a Sinologist from the University of California, Berkeley, who was initially talking about the largest mass migration of people in history coming from the rural areas of China to the cities. He then digressed and talked about bicycles.

He said that if every family (not every person) who lived along the coast of China were to replace one of its bicycles with a two-stroke motorcycle, the consequences would be significant. When asked what those would be, he said, "Well, to start with, the air in the five westernmost states of the U.S. would become unfit to breathe."

In 2004, it was estimated that 10% of the air pollution in California comes from China. By 2010, scientists project that a third of the smog-forming ozone in California air will originate in the booming economies of Asia.[15] Many families in the relatively prosperous coastal cities of China aspire to have a motorbike.[16] Alcohol-powered two-stroke bikes using biodiesel for a lubricant would prevent the environmental disaster.

that wouldn't normally happen. Someone firing a large-caliber bullet at the super-pressurized tank could cause the tank to shatter instead of just be perforated—or the massive stresses on an LNG tanker during a violent storm could cause failures wherever the tank is connected to the hull.

An LNG loading terminal, or a supertanker carrying LNG, contains as much explosive power as (or more than) a good-sized atom bomb. Some estimates predict a blast radius of ten miles from an exploding LNG tanker or land terminal. And it's bound to happen sometime as an industrial accident. In 1944, in Cleveland, a small LNG tank ruptured and exploded, killing 130 people and injuring 200. This was a tiny tank compared to those planned at terminals today. The Marine barracks that were destroyed in Lebanon in 1983 were devastated by the detonation of a truck carrying compressed natural gas bottles, which are far less energy-dense than LNG.

In the 1970s, the U.S. Federal Power Commission projected that an LNG explosion at the New York City (Staten Island) LNG terminal would kill up to 100,000 people.[19] Since then, planned LNG terminals and tankers have ballooned in size due to the increased dependence on natural gas by utility companies. Due to the public sensitivity of living next to a very attractive terrorist target, current LNG terminals are being planned at the borders in Mexico and Canada, where American protesters can't complain. But with domestic natural gas production about to reach its peak, LNG terminals are expected to proliferate all around the world.

As I write, there is legislation pending to take the authority for siting LNG terminals away from counties and states and preempt it with federal authority. That means that citizens would no longer be able to challenge corporate plans to site LNG terminals near where they live; an LNG terminal could be sited without regard for local or state planning, zoning, environmental, or any other regulations.

COAL

Coal is considered the United States' ace in the hole for energy. The U.S. has very large reserves and uses a phenomenal amount of coal to produce electricity.[20] Like oil, coal is a fossil fuel, so its releases of carbon dioxide are continuously added to the Earth's atmosphere.

CHARLES WELCH, <WWW.SOLCOMHOUSE.COM>

Fig. 4-7 Act of a terrorist in the U.S.? *Hardly. A disgruntled American shot a hole in the Alyeska pipline in Alaska. Hundreds of thousands of gallons of oil spewed out until pressure reduced enough to plug the leak. More than 100,000 miles of pipeline around the world are at risk of sabotage. A shot like this, using a 75-cent bullet, could take several days to repair, while a sabotaged pumping station might take weeks.*

U.S. DEPARTMENT OF THE INTERIOR

Fig. 4-8 Four Corners coal strip mine land. *This coal strip mine, which fuels the power plants at Four Corners, is one of many in the Southwest. It may be centuries before anything will grow on these spoil piles, which are all the while leaching toxins into the water table.*

Coal emissions are sulfurous and therefore acidic. They cause acid rain, which has largely sterilized thousands of lakes in the Eastern United States. In America's manufacturing heyday, when the coal-fired steel mills in the Rust Belt were running, the air was so acidic that you couldn't leave your clothes to dry on an outdoor line because the acid would eat holes in cotton. Rainwater acidified by sulfur emissions would also leach toxic aluminum, cadmium, copper, and lead from the soil that lined the reservoirs where the water was stored. The water was so acidic that it leached lead from the solder used to join copper pipes in homes, and the longer the water was in contact with the solder, the more lead was released into the water. Due to the prevalence of this phenomenon, solder from plumbing has been reformulated to eliminate almost all of the lead.

For years, health officials in the Eastern U.S. cautioned people to run their taps for several minutes to dispose of this water contaminated by their own plumbing. Cities such as Boston must dump millions of pounds of **alkaline** lime into reservoirs to neutralize the acid. This is another clear example of corporations privatizing the profits while socializing the costs resulting from their choice of fuels.

Coal also contains mercury, one of the most potent neurotoxins known. The saying "mad as a hatter" came from 18th-century hat manufacturers using mercury in the processing of felt material; over time, hatters poisoned their entire neurological systems.

Coal companies can now emit more mercury than ever, since the Bush administration weakened rules, allowing them to trade mercury credits with utilities that don't use coal. This bizarre system allows plants that don't emit mercury, for instance hydroelectric or windmill-based electrical plants, to garner credits that are saleable to coal-burning plants so they can meet their mercury reduction goals—on paper. After all, it's much cheaper to buy credits than it is to actually scrub mercury from coal emissions.

The small amount of mercury in each pound of coal takes on enormous significance when billions of tons of coal are burned and the mercury from the plant's flues falls or is washed into the ocean. There it is absorbed by plankton, and makes its way up the food chain—until the level of mercury contained in your can of tuna concentrates in breast milk and damages the nervous systems of children. We are all downwind in some way.

In a related way, there are small amounts of radioactive particles in coal, too. These radioactive seeds of cancer rain down silently all around us and in our food chains, since most radioactive particles are water soluble. Far more radioactive particles are released into the air by coal burning than leak from nuclear power plants, since such massive amounts of coal are burned.

The particulates that coal burning generates are small, which not only makes them more harmful to natural systems—such as your lungs—but also provides abnormally tiny condensation points for water vapor to become droplets on. In the dry air of the U.S. Southwest, droplets formed on the coal emissions are too small to form up into large enough drops to rain. So, downwind of the Four Corners coal-burning plants (where Arizona, New Mexico, Utah, and Colorado meet), a near-permanent drought is in effect for hundreds of miles.

These particulates are carried all over the world by the wind. Polar wind currents deposit untold tonnage of these tiny black specks onto our icecaps. So not only do we have ice melting because of warming from the greenhouse effect, we also have ice melting because of tiny particles of soot. As the particles absorb sunlight, they melt down into the ice, conducting the heat. Although it's not visible to the naked eye, the icecaps are no longer mirror white, but a shade of grey.

NUCLEAR POWER

Amazingly enough, nuclear power plants have come back up for discussion. Due to solid public opposition, none have been built since the 1970s. When they were built, their cost overruns were

---

A PERMACULTURE SOLUTION TO SULFUR RELEASE

It is possible to stem the issuance of sulfur into the atmosphere from coal plants. From a permaculture perspective, that sulfur represents a surplus that is not designed into a system to use it. By spraying alkaline lime into the sulfur-filled exhaust, the sulfuric acid is converted into gypsum, a very useful **pH**-neutral (not acid or alkali) soil amendment that is also a primary ingredient in sheetrock. This system, used at the Archer Daniels Midland plant in Decatur, Illinois, is still more costly to operate than simply venting sulfur into the atmosphere, but it largely solves the problem. So there is an answer, but it would just cost corporations a bit more money to implement.

so enormous that the capital used to build them drained much of the total investment capital out of our entire economy at the time.

Since the covert reason nuclear power plants were built was to provide the military with the enriched uranium and plutonium our weapons programs needed, the nuclear industry received subsidies that dwarfed even those of the oil companies. Since the plants were part of "national security," there was never any questioning of their unprofitability and totally tax-subsidized nature. By 1980, nuclear power plants had already received over $50 billion, which today would be worth more than $100 billion.

In a report made public in 1980, the Energy Information Administration (EIA) of the DOE determined that the $37 billion (1979 constant) in subsidies given to the nuclear industry had seriously distorted the true cost of nuclear energy. Without our tax dollars to support the industry, the EIA figured our nuclear utility bills would have been conservatively 66 to 100% higher than they are now. The subsidy was the equivalent of a $28 to $42 per barrel subsidy on oil. This study did not even take into account the nuclear industry's favorable tax arrangement, nor the limitation on its liability in the event of an accident (provided by the 1957 Price-Anderson Act). Since the study was done, the subsidy level has increased dramatically, while government monitoring of radiation surrounding nuclear plants ended under Reagan and Bush, so as to frustrate any attempt to sue reactor owners for radiation-related cancers. Ask some of the Pennsylvanians around Three Mile Island how successful they've been at seeking redress.

Faulty reactors (or reactors on faults) cost several billion dollars apiece to build. Uranium—a limited, nonrenewable source of fuel, is following the same exponential cost growth curve that oil is traveling now, but at a much faster rate because of its extreme scarcity. The mining of uranium presents environmental risks too numerous to detail.

"Coal and uranium mines can deduct up to 50% of their taxable income. In both instances, total deductions can frequently exceed the original investment costs of buying and preparing the land for resource extraction."[21]

—GREEN SCISSORS, WASHINGTON ENERGY MONITORING ORGANIZATION, 2002

But forget the technological problems of reactor design, limited supply, and nonrenewability. Forget the cost. Forget that nuclear power presently has no near-term impact on our transportation system (we'll talk more about that later in

Fig. 4-9 Nuclear transportation accident. This overturned truck sent drums full of nuclear waste through the roof and across a major highway in Kansas.

Fig. 4-10

HE'S GROWN A FOOT SINCE I SAW HIM LAST....

this chapter). And forget—if you can—the threat of nuclear terrorism (it takes under ten pounds of plutonium to construct a bomb of more force than the one that fell on Hiroshima), and the accident potential. The bottom line is that after nuclear fuel has expended its economic life, you've got to get rid of the waste. It will be extremely toxic for a quarter-million years, and still very dangerous for up to four billion years. Now, this is a *real* problem. There are over a million tons of depleted uranium from nuclear reactors. It is mind-boggling how much toxic waste this is. So far, the only use for this toxic waste has been **hardened munitions**.

One could argue that nuclear power has already caused enough misery in the name of Peak Oil. The 3000 tons of depleted uranium weapons the U.S. used against Iraq to acquire its oil equal the radiation release of over 250,000 Hiroshima-sized atom bombs, but concentrated largely within one country, rather than blown all over the Earth. The massive incidence of radiation-caused birth defects in Iraq is already a horror beyond words.

Many are making the argument that CO_2 reductions will occur if we replace our coal-fired plants with nuclear ones. This argument has no merit. Mark Diesendorf, a senior lecturer at the Institute of Environmental Studies at the University of New South Wales in Australia, points out that a 1000-megawatt nuclear plant would cost at least $3 billion to build—two times as much as a coal-fired power plant—and would cost much more to operate than fossil fuel plants. Building many nuclear plants over, say, 20 years would create

Fig. 4-11

SO WHAT'S WRONG WITH RELEASING HARMLESS AMOUNTS OF KRYPTON GAS INTO THE ATMOSPHERE?..

so much greenhouse gas to produce them that it would take 40 years to "break even" in terms of CO_2 "savings."[22] And 40 years is longer than the typical design life of the plant.

If we instead spent that same $3 billion on 3000 integrated, one-million-gallon-per-year alcohol fuel plants (roughly one per U.S. county), I calculate that we'd eliminate more than 300 million tons of carbon dioxide normally produced from petroleum fuel during those same 20 years. We'd also sequester several times that amount as increased soil organic matter.

EXCURSIONS INTO MADNESS: THE FLAT EARTH GROUP

New books are making Pollyanna pronouncements about our supposedly inexhaustible oil supply; their writers are members of what I call The Flat Earth Group. In this section, we look at some of their proposed alternatives for powering our society, which are so inherently frightening that they don't fit in the same league with fossil-fuel-based energy sources.

Abiotic Oil

These charter members of The Flat Earth Group say that we will never run out of oil because oil is not the product of old solar-generated algae, but is produced abiotically, without a formerly biological source, at very great depths of more than 15,000 feet. The very great depth is supposed to be why we haven't found all this oil yet.

The studies that posit the existence of abiotic oil came from the Soviet Union, where publishing studies that were politically unpopular was decidedly bad for your health. Petrogeologists, in order to keep their heads firmly attached to their shoulders, wrote papers for Stalin theorizing that oil could be produced at great depths from nonbiological deposits of carbon. These theories have no basis in fact.

The ultimate physical argument for why these claims are impossible is that organic matter buried below 15,000 feet is under such pressure that it cannot survive as a liquid and becomes natural gas.[23] So the claim that if we only dig deeper, we will find all the oil we need is sheer fantasy. Oil companies have been routinely drilling deeper than 15,000 feet since as early as 1938,[24] which has led to our understanding about the depth at which natural gas begins and oil ends.

Methane Hydrates

The Flat Earth Group promotes **methane hydrates**, also known as methane clathrates, as an energy "solution." Methane is natural gas. When natural gas is under tremendous pressure, at low temperatures, and in the presence of water, a weak association of gas and water forms into a slushy ice-like substance: methane hydrates. These are found under the ocean floor in very deep parts of the ocean, and also in massive quantities below the permafrost of the Arctic tundra. Sci-fi author John Barnes wrote a thriller called *The Mother of All Storms*, in which a deep-ocean-floor deposit

Fig. 4-12

of methane hydrates is suddenly released to the atmosphere, where it sheds its icy nature and becomes methane gas. In the thriller, the effect of the release of this ancient methane was to accelerate global warming 500 years within days.

"This much is certain: No initiative put in place starting today can have a substantial effect on the peak production year. No Caspian Sea exploration, no drilling in the South China Sea, no SUV replacements, no renewable energy projects can be brought on at a sufficient rate to avoid a bidding war for the remaining oil."

—KENNETH S. DEFFEYES, AUTHOR, *HUBBERT'S PEAK*

There are many thousands of cubic miles of this slush, so the Flat Earthers point to it, saying, "See, there's no energy shortage." They say, "It's simple to recover the stuff—all you have to do is drill a hole into it and warm it up a little while bringing it to the surface. The methane gases right out of the slush, and you've got nice clean natural gas. What could be simpler?"

First of all, it's important to remember that methane is 23 times as potent as carbon dioxide as a greenhouse gas over 100 years—but 62 times as potent in the first 20 years it reaches the atmosphere.[25] Second, whereas oil shale has its kerogen trapped in tiny pores in hard rock, methane hydrate is just the opposite—it has no more physical integrity than a blended margarita. Like the ice slush in a margarita, it will float if it isn't held down by something. In the case of the ocean floor, that's about 800 feet of mud on average.

But unlike a margarita, the methane slush does **Fig. 4-13**   not melt; it sublimates. This means it goes from

solid to gas in one step. Things get more complicated, too, since ocean-floor thermal sources can gasify some of the methane, which then lies trapped as a giant gas bubble under the slush. Unstable doesn't begin to describe the situation. So, puncturing the ice with a drill could result in an explosive release of untold millions or even billions of tons of methane.

The Permian-Triassic extinction event (which eliminated all but fungal life from land for *a million years* during an extreme global heating period) and also the Paleocene-Eocene Thermal Maximum (when heating was so great that no ice was to be found on Earth) are both thought to be the result of massive methane hydrate releases.[26,27]

Back during the Paleocene or Permian periods when these trillion-ton bursts out of the ocean occurred, the primary effect was to dramatically increase global warming. But if we were to have such a release today, once the gas reached the first water heater pilot light along the coast … well, the tsunami rolling out from such a blast would rival a large meteor strike. A 1000-foot-tall wave is not outside of the calculations.

"For what's the use of a house, if you haven't got a tolerable planet to put it on?"

—HENRY DAVID THOREAU

Global warming by itself could be the cause of massive methane explosions, with no need for us to muck around with drills.[28] What has kept the methane slush at the bottom of the oceans, so far, is that it is capped by a few hundred feet of deep-ocean-bottom mud, on average. As long as the temperature stays dead cold and the mud cap holding the gas down doesn't change, this massively explosive greenhouse gas is somewhat safely entombed.

But not forever. How warm is too warm when it comes to methane release? Already, water in the deep oceans along continental shelves has warmed three-tenths of a degree **Celsius (°C)** in the last hundred years. A temperature rise of two degrees would mean an additional *800 feet* of mud would be needed to provide enough pressure to keep the hydrates from blowing off the bottom of the ocean.

Methane hydrates are also trapped under hundreds of feet of permafrost in Canada and the former

"Exploitation of Spaceship Earth's inventory of integral atoms either by fission or fusion will probably violate the integrity of the complex cosmic design for successful maintenance of human life aboard Earth."

—R. BUCKMINSTER FULLER, IN THE 1983 FOREWORD TO THIS BOOK

Soviet Union. As permafrost warms, as it is doing today, more and more methane hydrates have the potential to seep or explosively release. One drilling team has already drilled specifically and "successfully" for permafrost methane hydrates, ignoring all the risks. Warming of the tundra is already causing millions of tons of methane to seep into the air right now.

According to climate change scientist Jeremy Leggett, "Think[ing] of burning methane hydrates is like opening a Pandora's box, knowing a murderous and quite genocidal genie lurks within it."[30]

Hydrogen Power and Nuclear Energy

Nuclear power is also a prominent feature in the so-called "hydrogen economy," where it is also strange bedfellows with the coal industry. Here's how it works:

The coal industry, a big owner of hydrocarbons, after failing for years now to come up with a safe and economical process to convert its rocks to synthetic gasoline, thinks it has finally hit the jackpot with hydrogen. Quite a number of processes have been developed over the years to extract hydrogen or convert the coal to vehicle fuel. These processes use various configurations of high temperature and pressure, and lots of water, and generate huge quantities of toxic slag. If the coal industry is able to get the right subsidies, and if carbon dioxide continues to be unregulated, it would be a viable business plan to simply generate more electricity, using coal, to electrolyze water (split it into hydrogen and water)—if you ignore the fact that taxpayers need to subsidize it.

So, how could coal or nuclear contribute to a viable hydrogen infrastructure? Peak electricity production happens during the workday, which is when industry is in full production. "**Peakers**" (generators) kick on at the height of air-conditioning hours, to prevent blackouts. But at night, when there are no users, the opposite is true. Then the producers of expensive electricity must throttle back their contributions to the grid, since the grid cannot accept more electricity than it distributes. Electricity producers must idle their plants, waiting for those electricity-using workers to get back up and start up the world each day.

It's a complex dance for computers and utility controllers to match use with demand. But the bottom line is, if there's no place to sell the power, then the grid doesn't want it, and you can't put it into the grid.

A shift to hydrogen power would give a market to all the thousands of coal plants and scores of nuclear plants across the country that stand mostly idle every night. Instead of costing utility companies money nightly, these already existing plants could produce hydrogen all night. This reserve capacity means that virtually no capital is needed

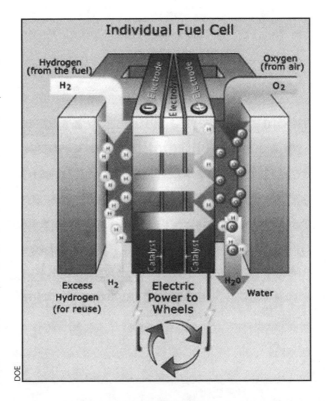

Fig. 4-14 *Individual fuel cell.* **Fuel cells** *produce electricity, which then must be used by an electric motor—not an internal combustion engine.*

to increase the electrical output of the nation to begin hydrogen production.

Making a small but steady profit at night also means a massive increase in pollution and radio-activity in the environment, from both coal and nuclear plants.

The Solar-Hydrogen-Fuel-Cell-Engine

I'm only going to briefly debunk the **solar-hydrogen-fuel-cell-engine** here. You've probably gathered that what I'm calling "solar-hydrogen-fuel-cell-engine" is the combination of several different technologies in one mouthful.

Fuel cells are very old technology, even predating the internal combustion engine (ICE). Developed in the 1800s, fuel cells strip electrons from hydrogen and produce *electricity*. So they are fundamentally different than the ICE, which takes chemicals, explodes them, and produces *mechanical energy* directly—turns a driveshaft, for example.

Since it's mechanical energy that propels our cars down the road, we need a different technology to transform the electricity of fuel cells into a spinning driveshaft. That mysterious device is none other than our familiar electric motor. The electric motor turns electricity into a spinning driveshaft, which takes the car down the road. This two-part, more complex, system replaces the ICE—which, remember, directly converts fuel into a turning driveshaft. So a fuel-cell-engine combination is

not a way of *producing* energy, but it is instead an alternative way of *using* energy.

Although little-discussed in the press, hydrogen gas can propel an ICE just fine without the involvement of fuel cells, using technology similar to what's used to run a car on propane or natural gas. And, like a fuel cell car, an ICE using hydrogen gas doesn't produce any carbon dioxide in the exhaust. Of course, an ICE made of inexpensive cast iron costs a fraction of the sticker price and **embodied energy** cost of the fuel-cell-engine, which uses two expensive systems (platinum-laced fuel cell and copper electric motor).

So why use a fuel-cell-engine combo instead of our well-understood and mass-produced ICE? Because, in theory, converting hydrogen to electricity and powering an electric motor is more efficient than simply burning hydrogen and turning an engine.

But that calculation leaves out what it takes to make the hydrogen. Although there are a number of ways to do it, two of the most common ones are to use electricity to break water apart into hydrogen and oxygen, or to use very high-pressure steam to tear the hydrogen loose from either natural gas or water. Either way, you need to overcome vast obstacles in transporting, compressing, and storing hydrogen gas.

Another method of producing hydrogen, known as **severe reforming**, converts natural gas, CH_4 (methane), into carbon dioxide and H_2 (hydrogen)

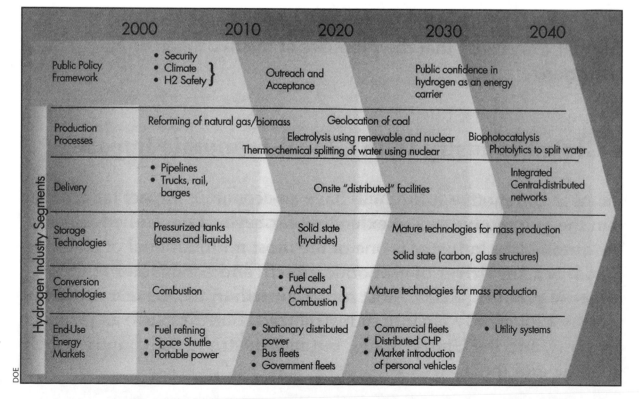

Fig. 4-15 Overview of the transition to the hydrogen economy.[31,32]

In this optimistic scenario, hydrogen does not become a major fuel until long after oil production goes into a prolonged nosedive.

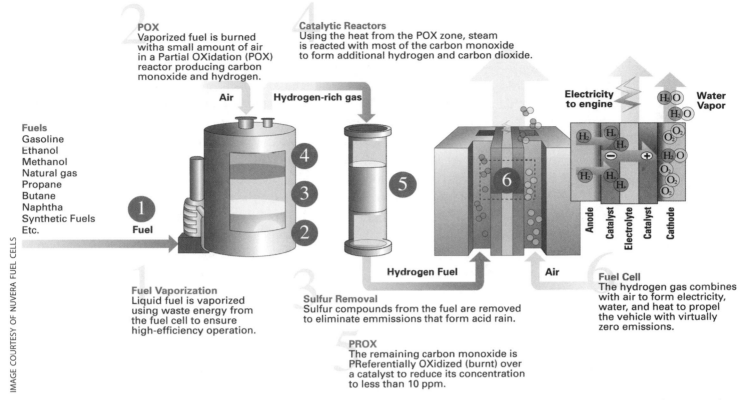

POX
Vaporized fuel is burned witha small amount of air in a Partial OXidation (POX) reactor producing carbon monoxide and hydrogen.

Catalytic Reactors
Using the heat from the POX zone, steam is reacted with most of the carbon monoxide to form additional hydrogen and carbon dioxide.

Air

Hydrogen-rich gas

Electricity to engine

H_2O Water Vapor

Fuels
Gasoline
Ethanol
Methanol
Natural gas
Propane
Butane
Naphtha
Synthetic Fuels
Etc.

Fuel

H_2 H_2 H_2 H_2

O_2 O_2 O_2 O_2

Anode Catalyst Electrolyte Catalyst Cathode

Hydrogen Fuel

Air

Fuel Vaporization
Liquid fuel is vaporized using waste energy from the fuel cell to ensure high-efficiency operation.

Sulfur Removal
Sulfur compounds from the fuel are removed to eliminate emmissions that form acid rain.

Fuel Cell
The hydrogen gas combines with air to form electricity, water, and heat to propel the vehicle with virtually zero emissions.

PROX
The remaining carbon monoxide is PReferentially OXidized (burnt) over a catalyst to reduce its concentration to less than 10 ppm.

Fig. 4-16 Hydrogen fuel cell. Hydrogen fuel for transportation must be converted to electricity to run an electric motor. It does not reduce carbon dioxide emissions if the fuel being converted to hydrogen is a fossil fuel. At each step of the conversion process, there are huge energy losses.

molecules at temperatures that require **supercritical steam**—which is, of course, produced with more fossil fuels or nuclear power. This is not a solution for the long run, with natural gas, like oil, about to enter a decline. All that's accomplished by using natural gas to make hydrogen is to move the emissions from the driver's tailpipe to the flues of the hydrogen plants: no greenhouse gas reduction here.

Another way to make the hydrogen needed to power a fuel-cell-engine is via on-board conversion of a liquid fuel to hydrogen, as you need it. If you convert a liquid fuel to electricity using an on-board **reformer**-fuel cell combination to run the car, you can approach 45% efficiency, compared to the gasoline engine's 20%.

It turns out there is one way to make hydrogen out of sunlight without increasing greenhouse gases. That is to use ethanol as a liquid fuel for on-board reforming to hydrogen. The huge advantages that ethanol has over gasoline for reforming are that the fuel is renewable and the catalyst does not contain platinum. Ethanol will reform at a tepid 550°F over a nickel catalyst, or more efficiently auto-reform over a rhodium/cerium catalyst at 1200°F.

Now we finally come to a renewable light at the end of the tunnel. The tailpipe emissions from a reformed *ethanol* fuel cell vehicle are only 13% of the emissions of a gasoline vehicle.[33] So we are faced with the unexpected prospect that the solar-hydrogen-fuel-cell-engine is actually a practical concept, if it runs on ethanol as its solar hydrogen source. This is clearly University of Minnesota chemical engineering professor Lanny Schmidt's answer, and it exists right now. This energy solution can be implemented immediately, with a small investment in development costs— ethanol-sourced hydrogen can run the internal-combustion-powered cars of today, as well as the fuel cell cars of the future, if they are ever made.

So am I advocating going to fuel cells for cars? No. Because even if on-board reformers are about 45% efficient, **vaporized** alcohol in an ICE has already been proven practical at 43% efficiency (see Chapter 13), so the added complexity and expense of the solar-hydrogen-fuel-cell-engine solutions are not worth it.

But fuel cells for other uses could make sense. By the time you read this, companies should be marketing small fuel cells that use **methanol** and ethanol to run laptop computers and cell phones.

Alcohol contains far more energy per cubic centimeter than any known battery. Current estimates by makers of small fuel cells calculate that you can fit 18 times the alcohol energy in the same space as a battery would occupy, and that cell phones so equipped would have to be refueled only once per month![34] And, of course, you wouldn't need much time or an outlet to recharge; you'd just refill the cell with more high-power alcohol from a flask in your briefcase. (I like that. Business meetings would be a lot less tense if everyone had a flask in their briefcase.) New biological fuel cells promise even higher efficiency for small devices.

A FRAGILE PLANET

How many humans will survive the certain massive ecological changes coming is a question no one can answer. Bucky Fuller once described to me just how much of a miracle it is that we are here at all. We live on a very thin crust of inhabitable land on a planet surrounded by a very thin veil of gases, which makes life here possible. If you were to take a 12-inch stainless steel sphere, polish it to a high shine, and then breathe on it, you'd have an accurate representation of how fragile is the Earth we live on. The condensed steam of your breath on the ball is the thickness of the Earth's crust. Under that slip of steam, the Earth is liquid molten rock. The veil of atmosphere that envelops us is less than half as thick as the Earth's crust. The highest points on Earth are only some tens of thousands of feet from the bleak vacuum of space. Outer space is not "out there" as much as it is just "over there."

Given the miracle of our existence on this wisp-like, balloon-shaped spaceship of land and atmosphere marking the boundary between the minus 200°F vacuum of space and a 2000°F molten mass of fluid metal at our planet's core—given the delicacy of that miracle, I think it's time we realized that we can't keep pouring hundreds of millions of tons of garbage into the beautiful film in which we live and expect that we will survive. That wisp of breath, which might contain all the life there is in the Universe (since we have yet to hear of any other), is worth saving. It's worth fighting for.

So let's get started.

Fig. 4-17

"Don't ask yourself what the world needs—ask yourself what makes you come alive, and then go do it. Because what the world needs is people who have come alive."

—HAROLD THURMAN WHITMAN

Endnotes

1. Michael Dabrowa, "The Energy Equation: China's Thirst for Oil Alters World's Dynamics," *Atlanta Journal Constitution*, May 29, 2005, Sec. C.

2. Richard Wolf, "Military May Ask $127B for Wars," *USA Today*, November 16, 2006 (www.usatoday.com/news/washington/2006-11-16-iraq-costs_x.htm).

3. U.S. Congress, Senate, Committee on Environment and Public Works, Subcommittee on Clean Air, Wetlands, Private Property, and Nuclear Safety, *Statement of Dr. Russell Long*, in S. Hrg. 106–953, 106th Cong., 2nd Sess.

4. Doug Struck, "Canada Pays for U.S. Oil Thirst," *Washington Post*, May 31, 2006, Sec. A.

5. Robert Collier, "Fueling America: Oil's Dirty Future," *San Francisco Chronicle*, May 22, 2005, Sec. A.

6. Ian Wilson, "Brand Nuke Idea," *Calgary Sun*, April 13, 2006.

7. Peter Calamai, "Tundra Test Stuns Scientists— Carbon Dioxide Could Be Dumped Into Atmosphere, Raises Spectre of Accelerated Global Warming," *Toronto Star*, September 23, 2004.

8. "OGJ Newsletter: Quick Takes," *Oil & Gas Journal* (April 21, 2003), 8.

9. Alexei Barrionuevo, "A Chip Off the Block Is Going to Smell Like Rotten Eggs—Sulfur Is Piling Up in Alberta, Millions of Tons Nobody Needs or Can Get Rid Of," *The Wall Street Journal*, November 4, 2003, Sec. A.

10. Robert Collier, "China Moves Fast to Claim Oil Sands," *San Francisco Chronicle*, May 22, 2005, Sec. A.

11. *World Oil Prices Surge after Study Tips US$100 a Barrel*, Channel News Asia, April 1, 2005, www. channelnewsasia.com/stories/afp_world_business/ print/140299/1/ (December 15, 2005).

12. James B. Newton, *Uncommon Friends: Life with Thomas Edison, Henry Ford, Harvey Firestone, Alexis Carrel, and Charles Lindbergh* (San Diego: Harcourt, 1987), 31.

13. Jonathan Lash and Laura Boynton King, *Synfuels Manual: A Guide for Concerned Citizens* (New York: Natural Resources Defense Council, October 1983), 16.

14. B. T. Walton, "Chemical Impurity Produces Extra Compound Eyes and Heads in Crickets," *Science* Magazine 212:4490 (April 3, 1981), 51–53.

15. Traci Watson, "Air Pollution from Other Countries Drifts into USA," *USA Today*, March 14, 2005, Sec. 1A.

16. Andrew C. Revkin, "A Far-Reaching Fire Makes a Point About Pollution," *The New York Times*, July 27, 2004, Sec. F1.

17. Lash and Boynton King, Fig. 11, 289.

18. Lash and Boynton King, 113.

19. Peter Van der Linde, *Time Bomb: LNG, the Truth about Our Newest and Most Dangerous Energy Source* (Garden City, NY: Doubleday, 1978).

20. Electric Power Annual, U.S. Energy Information Administration, www.EIA.doe.gov/cneaf/electricity/ epa/epa_sum.html (November 9, 2006).

21. *Running on Empty: How Environmentally Harmful Energy Subsidies Siphon Billions from Taxpayers*, a Green Scissors Report by Friends of the Earth, Taxpayers for Common Sense, and U.S. Public Interest Research Group Education Fund (January 31, 2002), 6.

22. Wendy Frew, "Nuclear No Cure for Climate Change, Scientists Warn," *Sydney Morning Herald*, May 2, 2006.

23. Kenneth S. Deffeyes, *Hubbert's Peak: The Impending World Oil Shortage* (Princeton, NJ: Princeton University Press, 2001), 22.

24. Deffeyes, 8.

25. Climate Change 2001: Working Group 1: The Scientific Basis, Intergovernmental Panel on Climate Change, www.grida.no/climate/ipcc_tar/wg1/248.htm (December 13, 2006).

26. Paleocene-Eocene Thermal Maximum, Answers.com, www.answers.com/topic/paleocene- eocene-thermal-maximum (December 13, 2006).

27. Permian-Triassic Extinction Event, Wikipedia, http://en.wikipedia.org/wiki/Permian-Triassic_ extinction_event (December 13, 2006).

28. Article in Reuters India, http://in.today.reuters. com/news/newsArticle.aspx?type=worldNews&story ID=2006-07-21T011013Z_01_NOOTR_RTRJONC_0_ India-260413-1.xml (2005).

29. R. Buckminster Fuller, communication with author, 1983.

30. Sonia Shah, "The New Oil," *Salon*, March 16, 2004 (http://dir.salon.com/story/tech/feature/2004/03/16/ methane_hydrates/index.html).

31. *Fuel Cell Vehicles: Race to a New Automotive Future* (Washington, DC: U.S. Department of Commerce Office of Technology Policy, January 2003), 11.

32. *A National Vision of America's Transition to a Hydrogen Economy—To 2030 and Beyond*, U.S. Department of Energy, www1.eere.energy.gov/ hydrogenandfuelcells/pdfs/vision_doc.pdf (December 13, 2006).

33. Jeffrey Bentley and Robert Derby, *Ethanol & Fuel Cells: Converging Paths of Opportunity*, for the Renewable Fuels Association (2002).

34. *Fuel Cell Vehicles.*

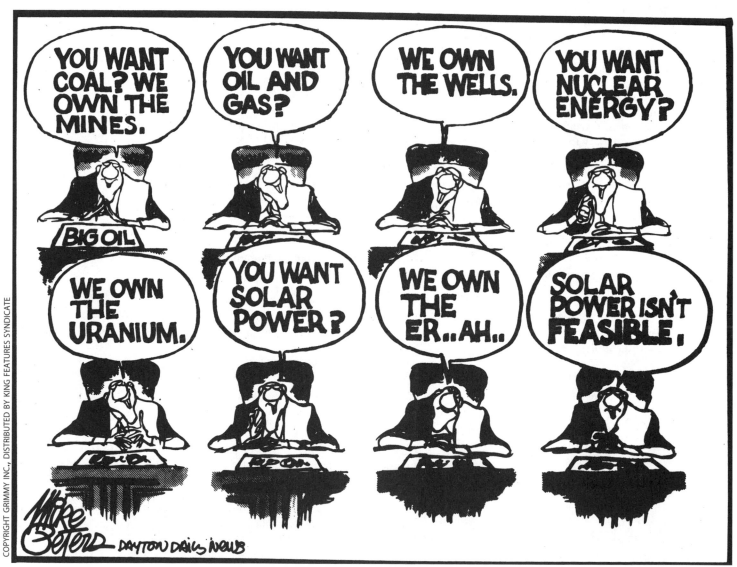

Fig. 4-18

In January of 2004, I made a trip to Brazil along with filmmaker Tom Valens, photographer Bob Fitch, and translator Alessandra Ramos Caiado. Our goal was to penetrate the wall of silence concerning alcohol fuel use in Brazil. In the U.S., it was almost impossible to find anyone truly knowledgeable about the Brazilian program, so we decided to get the story firsthand.

Brazil is around the same size as the United States, and there are central areas where alcohol is produced, somewhat like the Midwestern U.S. One can scarcely avoid the conclusion that if a program is successful on a national basis here, one of the largest countries in the world, it will be transferable almost anywhere.

In Brazil, we interviewed Secretaries of Agriculture and State, owners of large-scale sugar/alcohol plants, people waiting in line for fuel, aircraft manufacturers, university professors who had devoted their lives to renewable energy studies, cooperative agronomists, representatives of new-technology companies for automobiles, veterans of the battle to keep alcohol on the table as a national priority, and manufacturers of the largest alcohol fuel plants in the country.

We found surprises—and learned lessons that could be duplicated around the world. The attention to technology there was much greater than in the U.S., and was very practical; the attitude of the alcohol fuel industry in Brazil is so much more open and collaborative than the competitive intellectual-property mode of the U.S. The Brazilian people were proud of what their country had done about fuel, and it really showed.

A BRIEF HISTORY OF BRAZIL'S ALCOHOL FUEL PROGRAM

When Brazil's alcohol fuel program started in the 1970s, it was in response to the same oil shocks that rocked America. Embargoes of oil and sky-high prices motivated the rulers of Brazil to begin

IN 1980, LESS THAN SIX MONTHS AFTER CAR COMPANIES TOLD THE GOVERNMENT IT WAS IMPOSSIBLE TO MAKE A CAR RUN ON ALCOHOL, 60% OF ALL CARS COMING OFF BRAZILIAN ASSEMBLY LINES WERE SPECIALIZED ALCOHOL-ONLY VEHICLES. THEY HAD HIGH-COMPRESSION ENGINES (WITH 12.5 TO 1 COMPRESSION RATIOS), AND THEY RAN FINE AFTER AUTOMAKERS WORKED OUT THE KINKS. VERY QUICKLY, THE PRODUCTION OF BOTH ALCOHOL AND THE CARS USING IT TOOK OFF, REACHING A PEAK IN 1984, WHEN 94.4% OF ALL CARS PRODUCED RAN ON ETHANOL.

BOB FITCH PHOTO, <WWW.BOBFITCHPHOTO.COM>

*Fig. 5-1 Gas pump hydrometer. The author is looking at the **hydrometer** that is part of every alcohol pump in Brazil. The driver can verify at a glance that the alcohol has not been watered down.*

to wean the country of dependence on oil. The military dictatorship of the era ordered the major manufacturers of cars to produce dedicated alcohol-fueled vehicles, and ordered sugar producers to shift away from table sugar to fuel production.

In a democracy, it would have taken years to get action on this front, but when the dictators said, "Read my lips—we want alcohol-fueled cars now," the corporations were forced to act. In 1980, less than six months after car companies told the government it was impossible to make a car run on alcohol, 60% of all cars coming off Brazilian assembly lines were specialized alcohol-only vehicles. They had high-compression engines (with 12.5 to 1 **compression ratios**), and they ran fine after automakers worked out the kinks. Very quickly, the production of both alcohol and the cars using it took off, reaching a peak in 1984, when 94.4% of all cars produced ran on ethanol.

Production was managed by the government to stay in close connection with demand. The national oil company, Petrobras, which had full control of the distribution of fuel, complained loudly of surplus gasoline that it had to get rid of on the open market internationally. The group of elites that controlled the national oil company were a different political faction than the military government.

Then, in the late 1980s, military rule came to an end in Brazil. In the first elections, heavily influenced by Washington and big money, a rightist, "free-market fundamentalist" government was elected. All subsidies and departments involved with alcohol fuel were essentially dismantled. Also, the sugar growers were released from having to supply the nation with alcohol, and they immediately began producing sugar for export, cutting the amount of alcohol roughly in half. Suddenly there was a huge shortage of fuel, which the producers exploited by raising prices. The public was the victim of these changes.

It's important to note that many in the United States allege that alcohol fuel in Brazil "failed" due to removal of government subsidies. But the reality is that there was a sea change in the way elites ran Brazil. The new bosses included the large sugarcane plantation owners, who wanted to make more profit by selling table sugar.

In 1989, following this debacle, sales of alcohol vehicles dropped to 10% of the new auto market there. During the 1990s (a global era of artificially cheap oil), and until oil prices rose at the end of 2001, alcohol-fueled vehicle manufacture stayed at about 1% of the total number of Brazilian vehicles produced per year. Everyone thought the fuel alcohol movement would die.

But it didn't die. It changed and expanded. Brazil's auto fleet continued to dramatically increase, and gas-powered vehicles needed alcohol for the 25% blending with gasoline. Also, many people kept their old alcohol cars going, and they made up a stable market for the wet alcohol (96% alcohol/4% water) sold at more than 31,000 locations all over the country.

FARMING EFFICIENCY IN BRAZIL

When I went to Brazil to investigate alcohol fuel production, I expected to see lots and lots of sugarcane. After all, it is the largest alcohol crop in the world. I was also prepared for environmental destruction, soil erosion, a mega-monoculture doomed to collapse, and lots of water and air pollution—what the opponents of alcohol have been

---

FUEL ALCOHOL HAS BEEN USED AROUND THE WORLD

The world is no stranger to fuel alcohol. China used alcohol as a farm fuel after its revolution, with rice and marine algae as some of its feedstocks. When other countries, for political reasons, refused to sell diesel fuel to China, the Chinese used soybean oil in their tractors.

Before there was any widely developed oil tanker system in operation, petroleum was hardly available in the Philippines or New Zealand, and those countries were almost entirely alcohol-powered early in the 20th century. The Philippines made alcohol from nipa palms and sugarcane; New Zealanders used beets and other feedstocks. Neither country had an automobile industry, but their alcohol fuel was used to run imports from Studebaker and Chrysler—whose automobiles were specifically designed for alcohol fuel.

In the early 1900s, France had a well-developed alcohol fuel system and did everything possible to forestall oil dependence and trade deficits. French fuels were a blend of anything that could be produced within French borders. Benzol/alcohol, benzol/ether, benzol/ether/alcohol, and benzol/alcohol/gasoline were common.

Today the European Union is moving to renewables far more quickly than the United States. Sweden already has a national network of **E-85** (85% ethanol, 15% gasoline) pumps at service stations.

moralizing about for 30 years. I have seen all that I can stand of cornfields and grape vineyards laid out with the rows running up and down hills, swiftly eroding the soil with each rainfall. So I braced myself for what might be a gruesome sight.

I really enjoyed being wrong here. What especially impressed me when I first saw cane country in Brazil was the obvious influence of permaculturists. The fields were all tilled to **keyline** (close to contour) with periodic swales (on-contour, dead-level ditches that hold water, letting it gradually sink in) made by bulldozed berm walls of about three feet. This was done on hills that were sloped about 15 degrees! We watched the results of a downpour just pool up behind the swales and soak in. No runnels took soil downhill with each rain. Each sugarcane row was also a contour berm that soaked the water in evenly. This water collection technique is very important later in the season when the canes form, since they shoot up during the dry season and need to draw upon water stored underground.

This keyline layout went on for scores of miles, on a massive industrial scale. I've never seen modern swales done like this in the U.S. The last time was in the 1940s (before my time) by the Civilian Conservation Corps, to stop soil erosion in the Southwest and the South.

Not only does the new Brazilian layout design significantly charge the aquifers with surplus rain, it also conserves the soil. The swales also prevent nutrient and chemical runoff into surrounding streams.

I spoke with Brazilian agronomists who told me that this system of field layout started in the early 1990s and has spread throughout the country. They explained also that the shading of the soil by the closed canopy of the sugarcane went a long way toward preventing water evaporation from the soil. The Brazilians could teach a few things to our Midwestern agribusinesses, whose farming methods have resulted in the Ogallala Aquifer dropping many feet per year and our topsoil going down the Mississippi.

Some plantations in Brazil are using beneficial insects to control cane pests. Octavio Lage, a large grower I visited, told us his costs were maybe 30% higher with organic practices, but the higher price brought by organic sugar more than made up for the increased costs.[1] Organic acreage there is increasing rapidly, but is still considered to be quite small. Lage told me that he had "only" 4600 hectares

dedicated to organic production. That's as much land as the largest organic farm in California.

COGENERATION

With the continuing solid demand for alcohol in Brazil, mid-sized distilleries consolidated to form bigger companies, and the planting of cane, with its dual markets of sugar and alcohol, continued to grow.

Due to slim profit margins, it was uneconomical for smaller distilleries to compete with the bigger operations, so many of them converted to beverage production. This was sad news for our team, since we had hoped to visit them.

The deciding issue was process energy. The bigger plants could afford to invest in cogeneration equipment—which turns the bagasse (the fibrous byproduct left over after crushing cane to extract the juice used in making sugar or alcohol) into steam and electricity to run the plant.

The larger plants' ability to invest in cogeneration gave them an edge over the smaller operations that were common at the start of the alcohol movement, which had to buy natural gas and electricity to run

Fig. 5-2

their plants. The larger plants were producing alcohol at about 40 cents a gallon, plus they had the profits from sugar, and now also from selling electricity to the grid.

The average Brazilian alcohol plant uses about 23% of its self-produced electricity in its own sugar and alcohol operations, and sells 77% back to the grid. Having a positive energy balance has never been a problem for Brazilian alcohol producers.

When Brazilian energy corporations pulled an Enron-like scam in 2001, complete with rolling blackouts to frighten people into paying more for power—and pressuring the government to give huge incentives to energy interests to build new natural gas electrical plants—it was clear this was total chicanery. Brazil has a huge system of dams and generators, and more than 90% of its electricity comes from hydropower. Petrobras wanted the government to sponsor a natural gas pipeline from Bolivia; the government instead started making loans to all the big sugar/alcohol plants so they could build much bigger and more efficient **cogenerators** that would export electricity to the grid from the biomass-fired boilers.

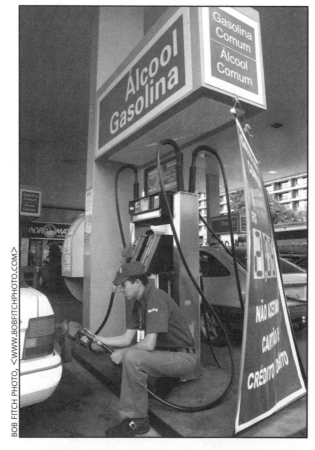

Fig. 5-3 Alcohol pumps in Brazil. *Brazilian alcohol pumps are appropriately color-coded green versus gasoline's red. Alcohol is offered at over 30,000 stations across the country. Brazil wisely pumps 96% alcohol/4% water, which costs much less energy to produce than 100% alcohol and is compatible with all automotive materials.*

BOB FITCH PHOTO, <www.BOBFITCHPHOTO.COM>

These boilers start up on farmed eucalyptus wood to get a nice hot core going; then the moist bagasse is fed in. The flues are fitted with spray recovery, so that smoke and particulates are trapped and not released to the air. The "cake" from the recovery system is comprised of the mineral content of the crop and is returned to the soil as fertilizer.

Since the electricity shortages experienced by Brazil earlier this decade, there has been a lot of investment in cogeneration. The average Brazilian alcohol plant uses about 23% of its self-produced electricity in its own sugar and alcohol operations, and sells 77% back to the grid. Having a positive energy balance has never been a problem for Brazilian alcohol producers.

Sugar/alcohol plants are now being looked at as primary utility providers during the driest season when hydropower is at its lowest level. Someone with a good ecological sense took a look from a systems point of view and figured out that the wet season hydropower could be significantly increased if it didn't have to fill huge reservoirs that were later emptied for dry season power. With biomass fuels providing power during the dry season, there would be less need to siphon off water for storage when hydro was roaring; thus, even more power could be produced during the wet season, making virtually all other sources of electricity irrelevant for several months each year. This was the argument that finally convinced the government to back the cogeneration model. The loans from the government are for ten years, but the sales of electricity from the bagasse allow the plants to pay off the loans in only three years if they choose.

In some ways, the move to burning bagasse for electricity is too bad. There are so many wonderful things that can be made from bagasse (see Chapter 8) that would make quite a profit and have ecological benefits, but the simplicity of burning the biomass to make electricity suits the owners of the giant plants. Perhaps since there is such a massive, and always increasing, volume of bagasse, some of it can be siphoned off for these other uses.

The efficiency of bagasse-to-electricity conversion could be much higher. Before cogeneration, getting rid of the mega-tonnage of bagasse was a serious problem, so boilers were designed to operate more like incinerators than efficient steam-boilers-to-power generators. These older plants usually produce low-pressure steam. Newer plants are using boilers designed to increase efficiency;

ETHANOL'S IMPACT ON THE RURAL ECONOMY

Ethanol's rise has had far-reaching effects on the economy. Not only does Brazil no longer have to import oil, but an estimated $69 billion that would have gone to the Middle East or elsewhere has stayed in the country and is revitalizing once depressed rural areas. More than 250 ethanol mills have sprouted in southeastern Brazil, and another 50 are under construction, at a cost of about $100 million each. Driving to lunch at his local churrasco barbecue spot in Sertãozinho, the head of the local sugarcane growers' association points to one new business after another, from farm-equipment sellers to builders of boilers and other gear for the nearby mills. "My family has been in this business for 30 years, and this is the best it's been," says Manoel Carlos Ortolan. "There's even nouveaux riches."[2]

Fig. 5-4 First dedicated alcohol aircraft. Our team had the good fortune to be on hand for the assembly of the first four aircraft manufactured by Embraer to run solely on alcohol. These cropdusters are sold to farmers who emphatically have demanded aircraft that operate on the fuel they produce themselves. As a result, Embraer no longer makes a gasoline version. (See Chapter 13.)

they produce high-pressure steam, especially useful in the evaporators used in sugar-making, and in electrical cogeneration. Researchers at the University of São Paulo in Piracicaba, Brazil, are designing more energy-efficient fluidized bed boilers (where ground bagasse is combusted with a powerful airflow up through the burning fuel).

Brazil's exclusive focus on huge plants is changing. Advocates at universities and research institutes, who have long been ignored about the need for widespread small-scale production, are perking up again with revised studies showing the viability of smaller, integrated alcohol plants that also make other high-value products (such as fish, poultry, and cattle) and conduct efficient smaller-scale cogeneration and/or produce methane. Entrepreneurs are starting to ask pertinent questions about scale and production. New models of an approximately 200-acre operation are being analyzed and found to be very profitable, as long as production is diversified, using all parts of the cane.

Brazil's switch to making alcohol instead of table sugar will put a lot more alcohol on the market—for export, and for domestic consumption, where the additional alcohol can be a political problem. Brazilian alcohol producers already make 40% of the nation's nondiesel transportation fuel. The

Brazilian state oil company, Petrobras, is in charge of marketing the nation's alcohol as well as its oil, and it has traditionally done everything in its power to limit alcohol's domestic market share. As far as Petrobras is concerned, alcohol has already made gasoline (the toxic waste from refineries) a surplus commodity, which it has to dump on the open market. I would expect Petrobras to redouble its efforts to export alcohol to countries like China and Thailand.

DIESEL AND BIODIESEL

More than half of the vehicles in Brazil (this includes trucks) are run on diesel, so the government is becoming very interested in growing palm oil for biodiesel. Biodiesel made from palm oil is a bit of an issue ecologically, since it involves a barren monoculture under a dense stand of nut palms. This is a hot political football in Brazil, as I found during my visit. Although Brazil has not needed to import any oil for gasoline, until recently it had to import oil for about 15% of its diesel fuel. Petrobras favors using oil for diesel, but has apparently given the nod to biodiesel in order to make up the 15% imports, since the national policy is to reduce dependence on foreign oil.

Brazil does not have a lot of natural habitat in which oil nut palms can grow well. There are a few coastal areas, but the main area being looked at for biodiesel production is, you guessed it, the Amazon. It would, of course, be a public relations disaster for biodiesel if Brazil starts ripping out rainforest to put in nut palms. So, in Brazil, we are potentially faced with a difficult battle in which alcohol must vie with its natural ally, biodiesel, to prevent destruction of the jungle.

Fig. 5-5 Brazilian Flexpower.

Taxis like this one led the initial purchases of flexible-fuel vehicles in Brazil. Now, due to public demand, four out of five new cars in Brazil are flex-fuel, and some companies have phased out gasoline-only vehicles completely.

Even though Brazil pioneered many alcohol-based diesel alternatives in the 1980s, one of our interviewees put it this way: "If the rest of the world is going to biodiesel, we feel like we should go that way, too. It's not that alcohol won't work; it's just that we don't want to be alone again, like we have been with alternatives to gasoline." Also, Petrobras keeps trying to prevent alcohol from taking any more market share. Letting alcohol into the diesel market is particularly discouraged. As you will see in Chapter 25, biodiesel can be made in small quantities to blend alcohol with normal diesel, which could eliminate the shortfall of diesel in Brazil overnight and make Brazilians fully self-sufficient.

BRAZIL'S FLEXIBLE-FUEL CARS

In 2005, the sales of **flexible-fuel vehicles (FFVs)** in Brazil raced ahead of gasoline-only models. More than 70% of all vehicles sold were able to run interchangeably on either alcohol or gasoline. "Demand has been unbelievable," said Barry Engle, President of Ford Brasil. "I am hard pressed to think of any other technology that has been such a success so quickly."

This change is happening overnight, with Peak Oil creeping up. The new higher prices of oil have woken up the Brazilians, who are rushing to buy FFVs, convert their cars to dual fuel, or simply mix alcohol and gasoline at the pump. Gas now costs about 45% more than alcohol at the pump. The cost to produce alcohol in Brazil from sugarcane is about 40 cents per gallon,[3] a little less than half the cost of alcohol produced from corn.

The surprising jump in FFV sales has said it all for the future of alcohol in Brazil. There is no going back now. The public demand for flex-fuel vehicles is several years ahead of predictions. In fact, several manufacturers, such as Fiat and Volkswagen, have stopped making gasoline engines and only offer the flex-fuel option. "There is something curious that we are just starting to see," says Alfred Szwarc, an ethanol consultant with São Paulo's sugarcane association. "Gasoline-powered cars lose more of their resale value than flex cars. People know that oil is finite and that it is going to get more and more expensive. They think that a gasoline-powered car is going to be more difficult to sell. They see flex cars as the car of the future."[4]

MegaOilron has always known that if alcohol becomes widely available at the pump, its ability to control the auto fuel market will evaporate overnight, and Brazil has resoundingly confirmed its worst fears. Our job is to imitate Brazil's success: We need to end the control of distribution of fuel by the Oilygarchy and to ratchet up production of alcohol. Brazil's experience has demonstrated that the public can quickly influence automakers to abandon gasoline-only vehicles. MegaOilron has every reason to fear what will happen when the public finally understands the reality of alcohol fuel.

Endnotes

1. Octávio Lage, President, Jalles Machado S/A Brasil, communication with author, January 2005.

2. Adam Lashinsky and Nelson D. Schwartz, "How to Beat the High Cost of Gasoline—Forever!," *Fortune*, 53:2, February 6, 2006, 74–87.

3. José Luiz Olivério and Antonio Geraldo Proença Hilst, "DHR–DEDINI Rapid Hydrolysis: Revolutionary Process for Producing Alcohol from Sugarcane Bagasse," *International Sugar Journal* 106:1263 (2004), 168–172, Fig. 4.

4. Andrew Downie, "Brazil Fights Oil Prices with Alcohol," *Christian Science Monitor*, October 7, 2005, Sec. World.

BOOK 2

MAKING ALCOHOL: HOW TO DO IT

Although making alcohol has been termed humankind's second-oldest profession, that doesn't mean everyone knows how to do it. While it's relatively simple to make alcohol, the devil is in the details. I've tried to walk a line between making sure you understand everything that's important for producing alcohol on the small scale, and being too obsessive with trivia and minutiae.

I've also resisted the temptation to outline only one way of doing things. Instead, I've tried to give you the principles and tools to design a plant that fits your site and goals. I think this is a far better approach than to give you every nut and bolt of a single specific plant design. It permits you to bring creativity and ingenuity into the process so that the plant you build will truly be your own.

In this section, you will learn about many different feedstocks and waste products that can be made into alcohol. My goal is to give you an overview of not only traditional crops but, more importantly, unusual crops that are often considered to be weeds.

Much land that isn't considered agriculturally valuable can be used for growing. Usually, humans have damaged or swamped those soils or turned them into deserts.

Some of the crops I discuss could be used as part of an ecological restoration of the soil, as we routinely do in the field of permaculture. Alcohol fuel production can be a vehicle to repair land, while simultaneously providing relief to our overstressed, yet still productive soils.

You should come away from the discussion of energy crops with hope and the certainty that we can design our way to an organic agriculture that provides both food and energy while improving the soil every year.

People often ask me, "What is the best thing to make alcohol from?" But there is no such thing as an ideal feedstock for fuel. The choice is always specific to the site and operator, whether you are a farmer or someone exploiting a surplus of something that's going to waste.

FEEDSTOCKS FOR FARMERS

Feedstock selection can make or break the small-scale alcohol business just as quickly as an obstinate bureaucrat can. There's a lot more to consider than what crop produces the most alcohol per acre. The farmer needs to consider the entire process: from soil preparation to cultivation to harvest, the loading of crops onto the truck, storing and preprocessing them, moving them from the shredding/grinding area into the still, separating them out (if possible) for animal feed, moving and storing the byproducts after drying, etc.

A farmer who grows an energy crop first narrows down what does well in his particular locale. He looks at all the crop's possible uses in his operation and in his farming community. He asks himself, *How many byproducts can be made from the original feedstock?* Then he checks the market for alcohol, and for all potential co-products both within his own operation and within his general farming community. Then it is time to get out paper and pencil to calculate which of the various crops under consideration seem to have the best cost-benefit ratios. And, of course, the farmer should pick crops that he really likes growing.

Weather and geography are the two biggest considerations. A Minnesota farmer isn't going to try to grow sugarcane; the Arizona farmer won't find corn to be his best crop.

A key question might be: Can the crop be dried or stored to be processed over the whole year? If a farmer must process all his feedstock in three months (as with sorghum), he'll have to build a bigger plant than he would if he had a year-round

THERE IS NO SUCH THING AS AN IDEAL FEEDSTOCK FOR FUEL. THE CHOICE IS ALWAYS SPECIFIC TO THE SITE AND OPERATOR, WHETHER YOU ARE A FARMER OR SOMEONE EXPLOITING A SURPLUS OF SOMETHING THAT'S GOING TO WASTE. NO MATTER WHO YOU ARE, IT PAYS TO DO PLENTY OF RESEARCH AND TALK TO A LOT OF PEOPLE BEFORE HAULING OUT THE WELDING EQUIPMENT AND STARTING TO BUILD YOUR DISTILLERY. FIGURING OUT WHO YOUR NATURAL ALLIES ARE AND WHAT THE MARKETS ARE FOR ALL YOUR CO-PRODUCTS IS SOMETHING TO DO IN ADVANCE.

Fig. 6-1

Fig. 6-2 Alcohol yield per ton of feedstock. *Notes:* [A]*Based on several sources, probable yield of 199+-proof fuel calculated from the percentage of* **fermentables** *per 2000 lbs. of feedstock.* [B]*In this case "dry" means* **all** *water removed rather than the average level of* **dehydration** *used in commercial feedstock processing.* [C]*Roots and foliage combined, sugar only, no cellulose.*

Alcohol Yield per Ton[A]

Feedstock	Yield (gal.)
Wheat	85.0
Corn	84.0
Buckwheat	83.4
Raisins	81.4
Grain sorghum	79.5
Rice (rough)	79.5
Barley	79.2
Dates (dry)	79.0
Rye	78.8
Mesquite	76.0
Sago palms (fresh)	75.5
Prunes (dry)	72.0
Molasses (blackstrap)	70.4
Sorghum cane	70.4
Oats	63.6
Lichens (reindeer moss)	60.0
Figs (dry)	59.0
Marine algae (dry)	55.0
Cassava (U.S.)	48.0
Manure (dairy cattle)	40.0
Cassava (Brazil)	39.0
Sweet potatoes	34.2
Buffalo gourd	32.0
Plantains (Costa Rica)	29.6
Bananas	28.4
Yams	27.3
Chili peppers	27.2
Papayas	27.2
Jerusalem artichokes	27.0
Fodder beets	27.0
Mangos	27.0
Onions	24.2
Prickly pear	24.0
Garlic	23.1
Cattails (starch only)	23.0
Potatoes	22.9
Sugar beets	22.1
Forage crops (dry weight)[B]	21.1
Nipa palms	21.1
Figs (fresh)	21.0
Oranges (whole)	21.0
Pineapples	15.6
Sugarcane	15.2
Grapes	15.1
Apples	14.4
Apricots	13.6
Pears	11.5
Peaches	11.5
Plums (nonprune)	10.9
Carrots	9.8
Comfrey (whole plant)[C]	9.0
Whey (per 225 gallons)	6.7
Marine algae (wet)	6.0

feedstock source to produce the same amount of alcohol. He might find it helpful to grow more than one kind of crop. On the other hand, the farmer might choose to run his plant only three months out of the year and lease it during the other nine months to farmers with other feedstocks that are able to be run during his downtime.

Local insect or disease problems can make growing a crop like beets impractical without long-term rotation and/or expensive chemicals. The ground may be too wet for too long for tuber crops like the Jerusalem artichoke, encouraging various kinds of rot. Farmers in arid areas may wish to experiment with some of the crops designed for dry climates, although some of these require specialized equipment.

Naturally, a feedstock crop's most desirable characteristic is a high alcohol yield per acre, balanced against the cost of growing. Look for the lowest capital input per produced gallon of alcohol. Also important is the crop's resistance to disease or catastrophe such as drought or flood. For most farmers, the crop should be available over a long period of time, or storable for use year-round.

Farming for energy is a little more complex than farming for food. Those farmers who combine both—harvesting food/fuel crops—have many considerations, including water, cultivation, soil preparation, costs for fussy crops like beets, fertilizer costs, harvest costs, and more. You need to ask yourself if you can reduce costs by growing organically—since half of all pesticides are used for cosmetic reasons, and cosmetics aren't an issue in fuel crops. I address cutting chemical input costs in several scenarios later in the book.

Switching from a corn crop to a crop with a higher alcohol yield, such as fodder beets, may require careful planning, since different equipment and growing techniques are required. Spreading out the equipment expense among a few growers may be a good solution. On the other hand, you might have an easier time convincing a nervous banker that sorghum—which can use corn equipment with little modification—is not risky. If the growing season is long enough, it can yield two crops (sometimes three) per year, surpassing the yield from corn.

If you're a farmer in arid or marginal grazing lands, you might do well to plant some energy-producing experimental crop, like pimelon or buffalo gourd. These crops produce fuel and food, and add some organic matter back to the soil by returning their spent mash to the Earth. Alternatively, the mash can be processed into biodiesel and seed meal for feed.

I knew a farmer in California who grew watermelon for seed. Until he started making alcohol, he hadn't used the melon pulp for anything. It used to be pumped into a big stinky pond during harvest season. His watermelon crops have much more value since he began using what was thought of as the "byproduct."

Ask yourself if there is a market for an essential oil that can be distilled from the crop (see Chapter 8). Or what about the carbon dioxide market in your area? Is there industrial demand for the gas? Would it be better used to encourage your crops' growth? Is there a business that could be set up next door that could use the excess hot water from your plant? Is there a business next door, such as a packing house with big chillers, that has excess heat energy you can use?

Water is another consideration in choosing feedstocks. An energy crop may need to use extra water to give a good alcohol yield. In that case, you can cut your irrigation costs by using alcohol as a pumping fuel. Or perhaps water is becoming too expensive in your area, and switching to an arid-area crop would save production costs?

If you raise livestock, there are additional considerations. For dairy operations, and to a lesser degree, cattle operations, alcohol may be considered a beneficial byproduct of its own "waste" product—distiller's dried grains with solubles (DDGS)—since upgrading feed grains to a higher value using DDGS produces more milk or meat from the livestock (see Chapter 11). Operations that grow feed for their animals may well decide to go with a crop like Jerusalem artichokes or beets, since their total feed and fuel yield would be greater than grains and might require lower capital input. Moldy or spoiled grain that shouldn't be fed to livestock can be turned into alcohol and the grain byproduct used to raise earthworms.

Another consideration is that some crops need specific fertilizers. Petroleum/natural gas fertilizers are expensive and going up in price, but you can substitute your mash byproduct for fertilizer and avoid those costs. But, if you do that, you lose potential animal feed byproduct revenues. As you can see, it's quite a juggling act to balance all the factors in feedstock selection.

Alcohol Yield per Acre[A]

Feedstock	Yield (gal.)
Cattails (single crop, managed, starch only)[B]	2500
(in sewage, including cellulose)	10,000+
Sorghum (including cellulose)	3500
Nipa palms (Phillipines, managed)[E]	2140
Cassava (U.S.)[C]	1662–2045
Cassava (Brazil)	585–1440
Cattails (wild, approximate)[D]	1075
Fodder beets (Monrosa)	940
Sugarcane (22-month crop)	900
Buffalo gourd	900
Nipa palms (wild)	650
Sago palms (wild, New Guinea)	650
Jerusalem artichokes	550–750
Sugarcane (U.S.)	555
Prickly pear (cultivated)	500–900
Sorghum cane	500–1000
Comfrey[F]	500
Pimelons (managed)[E]	450
Sugar beets	400–770
Mesquite (managed)[E]	341
Potatoes (starch only)	299–447
Corn	214–392
Prickly pear (managed wild)[E]	200–500
Sweet potatoes	190–255
Rice (rough)	175–230
Forage crops (Lucerne)	145
Apples	140
Dates (dry)	126
Grain sorghum	125–256
Carrots	121
Raisins	102
Yams	94
Grapes	91
Peaches	84
Barley	83–133
Prunes (dry)	83
Wheat	79
Pineapples	78
Oats	57
Rye	54
Pears	49
Apricots	41
Buckwheat	34
Figs (fresh)	32
Figs (dry)	30

Fig. 6-3 Alcohol yield per acre of feedstock. *Notes:* [A]*Average yield (annual unless otherwise noted) of 199+-proof fuel based on several sources. When range of yields is known to vary, high and low yields are indicated.* [B]*Based on cattails grown in a wastewater-fed, one acre test plot, in Northern California.* [C]*Calculations based on University of Arizona test plots.* [D]*Calculations made based on papers by Boyd and Jenkins, single crop, starch only, no cellulose.* [E]*Indicates results of plants usually regarded as non-agricultural, or weeds managed as an agricultural crop.* [F]*Roots and foliage combined, sugar only, no cellulose.*

"There are a lot more stills in this country than gasoline stations."

—HENRY FORD, WHEN ASKED WHY HE PERSISTED IN MAKING CARS THAT WOULD RUN ON ALCOHOL

FEEDSTOCKS FOR URBAN/ SUBURBAN DISTILLERS

If you don't grow your own feedstock, you will need to buy it or find a free source for it. On a farm, the mash byproduct has a variety of beneficial uses, but in an urban plant it can be a nuisance to dispose of. A farmer spends little, if anything, on transportation to bring his crop to his distilling area, but urban distillers have to arrange to collect their raw materials and have them moved to their plant. Such expenditures are critical in determining final fuel costs.

GILROY GARLIC

After hosting me on his talk show, the amiable country-western DJ at KFAT radio in California's Salinas Valley suggested we mosey on down to the local watering hole. Since he was buying, I went. We put a quarter in the jukebox and got to talking about Hank Williams. A burly farmer on my left punched me on the shoulder and said, "I knew Hank. He lived with me and the wife up near Sacramento." He patted his wife's hand. "She gave ol' Hank a pretty bad time—*Your Cheatin' Heart* was written for her."

Well, one thing led to another, and we discovered that this fellow grows the best garlic in Gilroy—garlic so good he doesn't even bother competing in the annual Garlic Festival. His garlic bulbs are exported to Japan, where they make garlic pills out of them. He said he dumped four to five tons a day of substandard garlic during harvest.

By this time, I was pretty well-lubricated, so I asked him, since he knew so much about garlic, did he know how much sugar it contains? Without hesitation, he told me that it's usually around 16%, but that his garlic contained over 18% sugar. My eyes bugged out, and I told him that he should be fermenting his waste, and making alcohol to run his farm machinery.

The farmer laughed. "Run my machinery on it? Hell, it's too good for that. I make my own 190-proof—and, like my garlic, it's the best in its way, too. You want to try a taste? I've got some out in the truck."

His wife was getting nervous. People in the bar began to stare at us. The disc jockey was amazed. We do a show about alcohol, and here's a man who's been moonshining longer than we've been alive—with garlic.

That same night at the bar, I met the owner of the local dump where the packing houses take all their throwaways. He introduced me to a member of the city council who wanted to know if they could run their city vehicles on alcohol.

So, if you live in a rural area and are looking for feedstock, go to the bar where the farmers go and put a quarter in the jukebox.

Ideally, you would want to be able to charge a **tipping fee** to take the material off the producer's hands. Transportation fuel is so expensive nowadays that, if you use a pickup truck, it's rarely worthwhile to go as far as 75 miles to collect your feedstocks. It's almost out of the question at 100 miles. Clearly, feedstocks closer to home have a big advantage over those an hour away.

A major factor in choosing a good feedstock is how many gallons of fuel per ton you'll be able to make. This is quite a different calculation than gallons per acre. For instance, since fruit is mostly water, fruit waste (and most other sugar crops) yields only around 20 gallons of alcohol per ton. You might not want to drive 75 miles for fruit waste when a one-ton pickup load uses as much as six or seven gallons of fuel, leaving you only a 11- to 12-gallon net yield for your effort.

On the other hand, most dry, starchy crops yield 85+ gallons of fuel per ton, and some food processing waste, such as noodles or candy, top out at over 115 gallons per ton. Driving 75 miles for a ton of starchy material would net you about 77 gallons of fuel. If you're thinking in terms of a semi-large cooperative, you may find it advantageous to look for a site in an industrial area where food is processed. The cooperative I joined in the 1980s in the San Francisco Bay Area was located within a 30-mile radius of over 80 major food processing plants and canneries.

There's a lot more waste available than you'd ever imagine, but if you do *buy* a feedstock, be careful of its cost. Ideally, you want your final price per gallon (calculated to include your byproducts) to cost you less than today's price of fuel.

"Say Max, I found this station that's open 24 hours and has all the gas you want...."

Fig. 6-4

Here are some further considerations for urban distillers: If the waste you want is being hauled away under contract by a disposal service, can you arrange for that company to deliver it to your land rather than to a commercial dump site? If the waste hauler agrees to deliver it, will they pay you to dump it there (woo-hoo!), will you take it off their hands for free (high five!), or will you have to pay a little to reimburse them for coming out of their way?

If you're hauling the waste material yourself, do you need to supply containers for it, such as 55-gallon drums or a tank for liquid waste? How are you going to lift or pump the stuff into your truck? How often do you have to come and take it away? If you are not responsible about reliably collecting the feedstock, you will lose that source. Is it a regular, dependable source? Is your need for fuel well matched to the quantity of feedstock that you need to pick up? If your feedstock is not available regularly and you get it all at once, do you have proper storage space at your site? If you do, you'll be able to handle a greater variety of feedstocks.

"As a country, we need to look inward for the answers to the energy of the future. We need to bring down our demands for oil, rebuild some bridges and highways, and allow the farmers to grow something that replenishes the soil. Those who don't know what that is, should do some research. The problem is not in Iraq, and the answers are not in Iran."

—MERLE HAGGARD, JUNE 2003

Will your feedstock yield saleable byproducts? If not, will you be able to handle your own waste disposal?

How much alcohol do you need in a month, and how many trips do you have to make to produce it? Is convenience more important to you than low price? If so, you may want to purchase an easily handled feedstock, molasses for instance, and have it delivered to you.

Finally, what are your zoning laws? Who are your neighbors? If you're experimenting with some smelly material as a feedstock, you may not be the most popular kid on the block. The list of

Fig. 6-5

MIKE PETERS

feedstocks in the next chapter includes several possibilities that might be less likely to offend your neighbors.

No matter who you are, it pays to do plenty of research and talk to a lot of people before hauling out the welding equipment and starting to build your distillery. Figuring out who your natural allies are and what the markets are for all your co-products is something to do in advance. When I was a farmer, I had a policy of never putting a seed into the ground until I had the crop sold. Use the same care in choosing your feedstock.

Fig. 6-6

CHAPTER 7

FEEDSTOCK PREPARATION AND FERMENTATION

Let's start our look at feedstocks with a quick primer on plant biology. In the process of photosynthesis, plants use solar energy to make sugar (glucose) from the carbon dioxide that they breathe in from the air and water. They make a dizzying array of substances from glucose, and these substances form the plants' structure, as well as store solar energy for later use in growth or reproduction. Glucose is the basic sugar building block, and all the more complex things made from it are known as carbohydrates.

Cellulose, or basic plant fiber, is made of tens of thousands of glucose molecules strung together in chains that intertwine and can even take a crystalline shape. The glucose found in cellulose is called beta glucose and is subtly different from the alpha glucose that makes up starch. Starch (grains are a typical example) is made of dozens of sugar molecules. Table sugar—**sucrose**—is made of two glucoses joined in a particular way.

Sugar is Nature's form of immediately usable energy for growing, metabolizing, and doing all the other things plants do. Of course, bacteria and animals love sugar, too. Sometimes a plant will create a surplus of sugar in its fruit, surrounding its hardened seeds (e.g., cherries) to attract animals to eat it. The animals then deposit the seeds far and wide in their poop. Other plants may pursue a different strategy for propagation, for instance having the seeds go through winter in the ground and then sprout in the spring.

In many cases, the sugar molecules are joined together in short chains of starch, a form more resistant to attack from bacteria and grazing animals. Starch is a more stable form of energy. The seeds initially store their carbohydrates as starch and then change them into sugar in the sprouting process to power the seedling until it can start making its own sugar from sunlight.

Sometimes, instead of providing energy for sprouting seed, starch is a method of temporarily

THE ENZYMES WE'LL BE PRIMARILY USING—ALPHA AMYLASE AND GLUCOAMYLASE—WORK AS CATALYSTS, CAUSING A REACTION WITHOUT GETTING INVOLVED. TECHNICALLY, GLUCOAMYLASE CAN ACTUALLY BREAK STARCH DOWN TO SUGAR WITHOUT THE HELP OF ALPHA AMYLASE, BUT IT TAKES TOO LONG. USE ALPHA AMYLASE FIRST TO SPEED THINGS UP FOR A PRACTICAL PRODUCTION RATE.

RON HARPER

Fig. 7-1 Enzymes at work. *These* **enzymes** *chop your long-chain carbohydrates into smaller* **dextrins** *and then sugars. Other enzymes can be used for dissolving protein, fats, and pectin in specific feedstocks.*

safeguarding energy in the fleshy part of a plant until it's time to convert to sugar. Many root or root-like crops—for instance, cassava, cattails, or potatoes—will store their energy as starch until conditions are right for sprouting and growing.

Fermentation is the process of converting carbohydrates to alcohol, usually using yeast. Although there are hundreds of types of yeast, the one that we use in alcohol production is usually **Saccharomyces cerevisiae**. Yeast literally eat the sugar, excrete alcohol, and "exhale" carbon dioxide. Yeast generally can eat only **simple sugars**, so more complex carbohydrates must be broken down for the yeast to be able to eat them.

Before we begin making our mash, let's look at terminology—you need to know what pH is before you can adjust it. You need to understand what yeast and enzymes do, and how temperature, sugar content, and quantities of ingredients affect your **wort** (the pulpy mush you bring to the fermentation party) and your mash (what the wort becomes after yeast is added and fermentation occurs).

"Yeast does not live by sugar alone!"

—DAVID BLUME, AUTHOR AND PERMACULTURIST

TERMS TO KNOW FOR FERMENTATION

Yeast: The key player in the process of fermentation, yeast changes sugar into alcohol and carbon dioxide. Yeast acts like both a plant and an animal. In the capacity of fuel-maker, it is loyal and hardworking. The easier you make it for yeast to get at the sugar inherent in whichever feedstock you've chosen, the faster fermentation will occur. Since yeast don't have hands or mouths, it's necessary to deliver the sugar to them mixed with water so they can absorb sugar from solution.

Enzymes: Enzymes are found in Nature in every plant and animal. They not only break down a substance into other substances, but can also build **compounds** by combining separate molecules. Each enzyme creates the conditions in which one specific reaction can take place. All enzymes are proteins and operate best at specific temperatures and pH levels.

pH: The pH is a measurement, expressed in a range of 1–14, of how acidic or basic (alkaline) a material is. A range of 1–7 indicates acid; 7–14

indicates alkaline; pH 7 exactly is neutral, neither acid nor base. Pure distilled water has a pH of 7. The farther away in either direction from 7, the stronger or more intense a material is in its **acidity** or alkalinity. And each successive pH unit of measure is ten times stronger than the preceding unit—not unlike the Richter scale. We'll talk about ways to measure pH later on. Levels have to be checked at many different steps in whatever process you employ. Knowing your pH is a key to getting enzymes to work properly, keeping your yeast producing in top form, detecting contamination, and, in some cases, breaking down a feedstock using just heat.

Temperature: We'll be doing a lot of cooking and holding the wort or mash at particular temperatures in the fermentation process. For effective preparation, maximized enzyme action, and strong fermentation, temperature control is key. It's important in determining how viscous a batch is, which affects what you have to do to mix or agitate it.

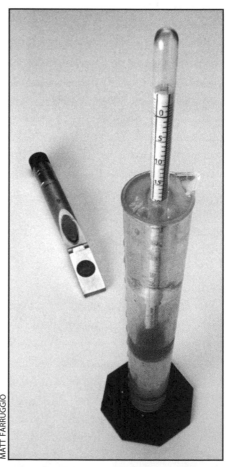

MATT FARRUGGIO

Fig. 7-2 Saccharometer and refractometer. *The saccharometer at right is reading 19% sugar. The refractometer on the left measures sugar when light is shone though a sample of the liquid.*

Dosage: You'll need to know how many enzymes to use. How much yeast? How much acid or alkali? What levels of nutrients need to be added? What happens when dosages are off? How do you measure?

Sugar content: The key to knowing how much alcohol fuel you're going to get in your batch is its sugar content. We'll look at tools, the **refractometer** and **saccharometer**, that help you measure, and the importance of precise adjustments.

Nutrition: No, we're not trying to stuff leafy green vegetables into our wort/mash. But we'll look at nutrient shortages in some feedstocks and what yeast need to thrive in your wort.

"Nature is trying very hard to make us succeed, but nature does not depend on us. We are not the only experiment."

—R. BUCKMINSTER FULLER

THE SUGAR METHOD

Because sugar is immediately absorbed and digested by yeast, some people consider making alcohol from sugary crops an easier way to make fuel. The method for fermenting sugary materials has fewer steps than fermenting a starchy material, since starch has to be broken down into sugar before fermentation.

This recipe for sugary feedstocks is a general one; specific processing for specific plants is discussed further in the next chapter. We'll also look at a variety of techniques for getting sugar from a feedstock into a solution where the yeast can feed on it.

In deciding which type of feedstock to use, consider that sugary feedstocks are seasonal; few plants store their energy as sugar for long periods, and sugar sources are most abundant in temperate climates during summer and fall months. Sources include waste or culled fruit, peas, sugar beets, fodder beets, molasses, candy waste, syrup from sorghum stalks, and sugarcane.

Preparation of sugary feedstocks for fermentation is relatively easy, involving five steps: 1) juicing, or shredding and crushing; 2) cooking; 3) cooling; 4) making nutritional adjustments; 5) checking sugar **concentration** and adjusting if necessary; and 6) correcting the pH balance.

RON HARPER

Step 1: Juicing, or Shredding and Crushing

Immediately, a conundrum is put before us. Should we juice pulpy feedstocks or ferment them with their pulp?

To Juice ...?

Advocates of juicing (and **diffusion**—see next section) point out that solids limit yeast movement and interfere with yeast's ability to get at the fermentables, since much of your sugar will be locked inside shreds of pulp, rather than floating free in solution.

The sugar concentration is often quite a bit higher in mash made of juice rather than whole fruit, because water has to be added to thick pulp in order to be able to work with it. So, in theory,

Fig. 7-3 Small juice press. This small juice press can juice up to a ton per hour of fruit at high efficiency.

PHOTO COURTESY OF GOODNATURE PRODUCTS, INC.

juice yields you more alcohol per batch, which saves energy in distillation.

Sometimes the convenience of fermenting a pure liquid is worth losing the 20 to 30% of the sugar that stays trapped in the fruit pulp—for instance, when waste material is only available for a short season. If you have an essentially unlimited amount of waste fruit available seasonally, it may make sense to juice it in order to get maximum yield faster from your fermentation tanks, and distill around the clock. In this case, your limitations are time and season, so sacrificing some sugar may make business sense.

Juicing is a good choice if you have livestock that will eat fruit pulp—in which case potential sugar loss becomes your animals' gain. The yeast at the bottom of our **fermenters** is a very high-grade protein feed. Leftover mash from juice fermentation can be ideal feed for **single-cell protein**, too, for all you bacteria rustlers out there.

Residual sugar in the pulp can be recovered efficiently as methane in a digester. The bacteria that produce the gas just love the easily digested carbohydrate, and the rate of production is pretty spectacular when it's sugar rather than cellulose being converted.

If you decide to get into juicing, you'll need a juice press. For maximum juice and better sugar extraction, I recommend a hydraulic pump and **piston** rather than the old-fashioned cast-iron screw that you see in illustrations of traditional wine presses. If your operation is a small one, you may want to consider a 30-ton hydraulic truck jack (hand-operated variety) instead of a fancy hydraulic system.

Small-scale commercial juice presses have come down in price in the past 20 years, and a little diligence could turn up a used one. In a larger operation, an auger press often makes the best sense; these can often be found used.

Many homesteading books and self-sufficiency magazines advertise plans for building a cider press using jacks. If you're building your own, make it more heavy duty than the plans indicate, since you'll be putting it to harder use than the designers probably had in mind.

Fig. 7-5

Diffusion

There is another route that allows nearly complete extraction of the sugar, fast fermentation, and an all-liquid mash. This method does away with the need for a press. It's called diffusion.

Early in the 20th century, major alcohol producers used diffusion to get sugar from thick feedstocks into solution so as to have an all-liquid mash. It's also the way that sugar beets are processed to make table sugar today. It's a very good alternative to juicing, since there's very little loss of sugar. Diffusion can be worth doing if you are working with

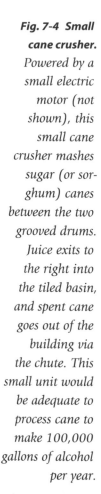

Fig. 7-4 Small cane crusher. *Powered by a small electric motor (not shown), this small cane crusher mashes sugar (or sorghum) canes between the two grooved drums. Juice exits to the right into the tiled basin, and spent cane goes out of the building via the chute. This small unit would be adequate to process cane to make 100,000 gallons of alcohol per year.*

BOB FITCH PHOTO, <www.bobfitchphoto.com>

a dilute feedstock like coffee pulp or pulpy fruit waste, since it can end up modestly concentrating sugar above the level in the fruit.

In 2002, when I visited Jalapa, Mexico, I was told tales of woe about how the bottom had dropped out of the pineapple market and the crop was going to waste. Due to the fibrous nature of pineapples, diffusion would have been a good technique to convert the wasted crop to piña colada fuel. So if you are planning on using diverse feedstocks in your plant, even if it is a **batch plant**, diffusion might be in the cards.

In a solution, diffusion is the movement of concentrated dissolved solids into areas of lower solids concentration. For instance, when a solution with no salts is suddenly exposed through a permeable **membrane** to a substance or solution high in salts (or dissolved solids), the initial rate of infusion into the saltless solution is quite rapid. This is called having a high **osmotic pressure**. As the solution begins to take on salt, the rate of diffusion from the high-salt substance slows, and when the two solutions approach equal amounts of salt, diffusion comes almost to a stop.

This is also true of sugars. But since there are more solids in beets than just the sugar, it is possible to efficiently transfer all the sugar to the solution, and even end up with a higher percentage of sugar in the solution than started off in the beets. If the solids were only sugar, you'd reach an equilibrium where there would be the same percentage of sugar in the solution as there would be in the **substrate**. But since a substantial portion of the beet solids are not sugar, the sugar continues to diffuse into solution.

You need special equipment for diffusion—four tall, cylindrical tanks with large bottom **hatches** for emptying beet pulp (for example) when the process is over. Fill all tanks with coarsely sliced beets; chunks can be a bit larger for diffusion than the grated material described in the whole fermentation method below.

Add 185+°F spent mash liquid or water to the first tank and steep for 45 minutes. Approximately 35 to 45% of the beets' sugar will dissolve into solution. Add heat to bring this solution back up to 185°F, and pump it into the second vat full of beets. The solution should sit in tank #2 for 45 minutes, while it picks up even more sugar. Not quite as much sugar dissolves out of the beets in tank #2, though, because of the sugar already in the solution.

Pump the contents of tank #2 to tank #3, after heating to 185°F. You've got a strong sugar solution now, which picks up even less sugar in tank #3, and which has reached or approached the sugar concentration of fresh beets. By the time you move the solution to the fourth tank of fresh beets, it's very high in sugar and only absorbs the littlest bit more to bring it to or even above the percentage of fresh beets—18 to 24% or so. This sugar-enriched juice can go directly to the fermenter. This ends the solution's trip through diffusion's starting cycle.

MICHAEL KEEFE

Fig. 7-6

Once tank #1 has had four charges of hot water, the beets are sugar-free. They are then emptied and the tank refilled with fresh beets. The #2 tank is now the "first tank," and once it's had four charges, it gets emptied. Another way to visualize this is that the solution is advancing in one direction and the spent beets are "advancing" in the other direction, counterflow to the solution. In fact, some sugar plants use one very long pipe, instead of four tanks, pumping beet shreds in at one end and hot water in at the other end. Pulp leaves the pipe just after the water inlet, and the solution leaves just before the introduction of the fresh beets.

... or Not to Juice?
(Whole Fermentation)

If you are going to use a continuous distillery, whole-pulp fermentation would be impractical. But the small-scale producer who uses a batch still—which doesn't require the separation of unfermentable pulp before distillation, as is common in commercial continuous distilleries—may come out ahead. Whole fermentation will yield a higher quantity of alcohol per ton of feedstock in comparison to juicing, with no penalty in time, as long as a greater amount of yeast is used at the beginning.

For smaller-scale fuel production, pulping or shredding feedstock to prepare it for "whole" fermentation is the preferred method of fermenting sugary materials—where all of the feedstock, not just the juice, is going to be fed to the yeast. A ton of soft fruit should yield 200+ gallons of pulp and juice. Even crops that are normally crushed or juiced, such as sugarcane, can be shredded and fermented whole to convert almost all the sugar to alcohol.

A common 5- to 8-horsepower-capacity garden shredder (alcohol-fueled, of course) is all you need. Stan, one of my students from Fresno, demonstrated to me that he could shred six to seven tons of peaches an hour, pits and all. If you're using fruit with pits and intend to use the mash for feed, be sure to grind the fruit fine enough so that pieces of pit can't hurt your animals. Apparently, cattle can swallow pieces up to ¾ inch, but other animals should have pits ground to less than ³/₁₆ inch.

"We need to change the culture, that is absolutely right. We have had a culture which says, 'The more energy you use, the more successful you are.' We need to have a culture that says, 'The less energy you can use to be comfortable, the better off you are and the better you should feel about yourself.'"

—REPRESENTATIVE ROSCOE BARTLETT (R-MARYLAND)

If you are using a feedstock that is limited or non-seasonal, whole fermentation might be a good choice. Australian researchers K.D. Kirby and C.J. Mardon point out how yeast benefits from whole fermentation, as opposed to juicing.[1] Their study focuses on beets, but can apply to many other feedstocks.

Say you've diffused or juiced a batch of beet shreds to yield a 20+% sugary juice. As we will see later in this chapter, a high sugar percentage can initiate the **Crabtree effect**, slowing the initial reproduction rate of yeast. Yeast has to really stretch to accommodate higher sugar concentrations and will actually alter its membrane structure to "toughen" itself, and that toughening reduces the rate of sugar uptake.

In whole fermentation, all the beet shreds are mixed with the least amount of water to immerse them. The dissolved sugar concentration starts off fairly low. Initially, only a small portion of all the fermentables dissolves into solution; most of the sugar is still in the feedstock shreds. So yeast can metabolize the available sugar in solution without the osmotic pressure problems it faces in pure juice. With whole fermentation, osmotic pressure in the beets is sufficiently high to allow a steady diffusion of sugar from the pulp into solution, and yeast in turn consumes it at a steady rate.

Kirby and Mardon indicate that there is no point at which the diffusion of sugar into solution is too slow to keep up with yeast's ability to absorb it.[2] In fact, yeast tend to stick to the feedstock, absorbing the sugar as soon as it's released, and excess sugar floats off into solution around them. Periodic agitation allows suspended yeast to absorb the sugar in solution and introduces it to new stock not yet exhausted of sugar. Using relatively high inoculation rates (one part yeast to 100 parts feedstock by weight), Kirby and Mardon were able to ferment a batch of whole beet pulp in 10–16 hours with exceptionally high efficiency.[3]

Their research agrees with my experience. I've seen references in older texts demonstrating good results putting thinly sliced beets in burlap bags and fermenting them in successive tanks until all the sugar was exhausted. Whole fermentation of any feedstock I've tried depends on one thing only—particle size. Kirby and Mardon's study finds that particles of ¹/₈ to ³/₁₆ inches are fine, and cause no impractical slowdown in sugar diffusion.[4] The problem with really big chunks is that sugar has a hard time getting through all those layers of feedstock cells. It does eventually diffuse completely, but not before the yeast and I have both run out of patience. You may find, depending on the feedstock, that a fine grind and heating to over 200°F for at least ten minutes, to burst cell walls and pasteurize the batch, will do the job quite well.

Step 2: Cooking

The wort should cook at between 200°F and a boil, for up to an hour, until the lumps have partially dissolved into a coarse sauce. Cooking does the double duty of breaking down the pulp's cell walls more fully and sterilizing the wort to eliminate bacteria that could interfere with your fermentation.

The sauce can be fairly thick (a little thinner than applesauce). You don't want to have to add

any more than a minimal amount of water in the cooking process, to avoid diluting the sugar content. Most fruit naturally contains 12 to 18% sugar, which is already slightly lower than ideal for our purposes, so definitely avoid diluting it much further. Beets, tubers, and stalky feedstocks like sorghum or sugarcane should be shredded as fine as possible and cooked with just enough water or **backslop** (the liquid left over after distillation) to get the sugar into solution.

Instead of cooking the wort, some fuel-makers simply bring their shredded feedstock up to 100°F and add extra-large dosages of yeast. They count on the extra yeast outstripping the bacteria in a race for food. It's a trade-off: Higher and faster yields are gained by cooking for an hour, at the expense of energy. I generally feel the risk of contamination (see later in this chapter) is too great to recommend skipping the cooking step.

Step 3: Cooling

Once you have a thick sauce of sugar and pulp, cool the mixture to 90–100°F as quickly as possible, using a **cooling coil** or **cooling jacket**. Rapid cooling is essential in preventing ubiquitous airborne bacteria from getting a head start eating your sugar before you add your yeast. You generally don't add water for cooling, since it would dilute the wort.

Step 4: Making Nutritional Adjustments

To do its job, yeast requires nitrogen, phosphorus, calcium, and trace minerals. Most crops have most of the needed nutrients already present, unless you juice them. Fruit is typically a little low in nitrogen, so it can be supplemented by adding **barley malt**, or diammonium phosphate (a common fertilizer). These materials take care of both nitrogen and phosphorus needs.

To make sure that all needed micronutrients are available, I recommend using either liquid or powdered kelp. A few ounces of kelp solution are all you need per 1000 gallons of wort. There are also good commercial yeast foods made specifically for this purpose, although they're a little more expensive.

Barley malt (see more below) is normally associated with starch **conversion**, as it contributes a lot toward breaking larger molecular weight sugars down into simple **fermentable sugars**. Barley malt is generally used in fruit-to-alcohol conversion for its nutrients. But there may be enough starch present in some sugary crops (e.g., unripe fruit, sugar beets, fodder beets, legumes, and squashes) to actually get some benefit from the malt enzymes.

Check the yeast manufacturer's recommendations for nutritional adjustments. If you're using malt, the ratio is generally 10–15 pounds of malt per 500 gallons of fruit mash. You would generally use 2–4 pounds of diammonium phosphate per 500 gallons[5] to provide sufficient nitrogen in fruit worts. If your feedstock is low in calcium, adding a few pounds of common agricultural gypsum (calcium sulfate) will ensure sufficient calcium (see the section on the starch method later in this chapter).

In addition to the mineral nutrients, it is important that you add some form of yeast extract to provide other critical micronutrients. Adding five to ten pounds of dried (dead) **brewer's yeast**, the same stuff you can buy at health food stores, will do the job. (After all, what could be better for feeding yeast than dead yeast?)

For Jerusalem artichoke tubers or other drier fibrous sugary materials, some liquid must be added. This will also be true for fruits with a lot of **pectin**, which thickens the wort.

Backslop can be used in the cooking process instead of water; it provides some of the necessary nutrients, requiring little or no pH adjustment after cooling. If you're using backslop instead of water, it's critical to use it immediately after distillation to prevent contamination. Never use backslop that has cooled for any length of time or sat overnight. You shouldn't have to add any extra brewer's yeast as a nutrient if you are using backslop, since it contains the dead yeast from the last batch.

Pectin is a gelatin-like substance that is found in a lot of fruit, where it "glues" together a matrix of cellulose and **hemicellulose** fibers, holding the fruit together. Its gel-like nature traps the sugars and makes it difficult for the yeast to get at them. Cooking helps break them down, but in some cases where a fruit has a lot of pectin, you may want to employ a pectin-dissolving enzyme, **pectinase**, to thin an otherwise thick wort.

The pectinase would be added as the temperature is rising; you might hold the temperature at the optimum for the enzyme (around 130° F) for 10–30 minutes before resuming heating to cook the pulp. The pulp should thin out quite a bit after the addition of the pectinase. This will shorten the cooking time, reduce the amount of additional water needed, and speed the fermentation rate by helping sugar dissolve out of the pulp more easily, which could justify the extra expense. For instance, apple juice makers sometimes use pectinase to soften the shredded apples before pressing, to extract a higher yield of juice. You'll learn more about enzymes and how to use them in the starch recipe.

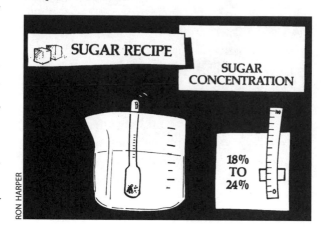

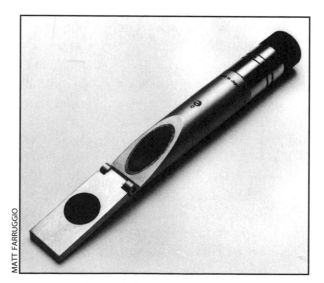

Fig. 7-7 Refractometer.
A drop of filtered wort is placed on the glass disc at bottom left. The main body of the unit is closed on the drop. Pointing the instrument into a light, look into it as if it were a telescope. The light refracting through the solution reads as a shaded area on a scale etched on the viewing field to indicate sugar percentage. Refractometers are available with automatic temperature correction and are accurate to less than 1%.

Step 5: Testing Sugar Concentration

The sugar concentration of your mash is obviously one of the most important measurements you'll take in the course of fuel production. An accurate measurement is the only way to predict alcohol yield (see later in this chapter) and the only way to make sure no sugar is wasted. It can be done either of two ways.

A saccharometer (sometimes called a **Brix** or **balling hydrometer**) is cheap and reasonably accurate. As all hydrometers do, it measures the density of a liquid—for our purposes, the density of liquids containing sugar, up to about 35% usually. The level at which the saccharometer floats tells you how much sugar is in your liquid—sort of. Saccharometers are usually calibrated for a 60°F solution. Higher temperatures cause the saccharometer to read as if there's less sugar than there really is, while colder temperatures give you a false higher reading.

Any nonfermentable solids floating around in the liquid will cause a false high reading, as well. Saccharometers actually read the total amount of dissolved solids in a liquid, rather than the sugar alone. Before measuring sugar content, I always filter the wort through two coffee filters at least once to remove the crud. Even then, your reading can still be inaccurate because of salts, unfermentable sugars, and excess minerals in the liquid. Molasses used in your recipe is famous for giving wrong readings, since molasses contains a lot of galactose, other unfermentable sugars, and alkali salts. A saccharometer invariably reads a molasses mix as if there's a lot more sugar in solution than there really is—sometimes by as much as 6%. (You can, of course, correct for false readings by comparing a before and after reading on a test batch that you know has been totally fermented.)

The other tool that measures sugar content is called a refractometer. Although saccharometers are a good cheap way to start experimenting, once you start producing steadily, a refractometer will pay for itself many times over in greater yields and less waste of valuable sugar. The hand-held variety costs around $200. In many areas, it can be difficult to buy refractometers from September to November because of the grape harvest.[6]

A refractometer measures the bend of light through a droplet of filtered mash, and gives a reading of the sugar content accurate within 0.1–1%. Wineries use refractometers to get an indication as to when their fruit is becoming ripe, squeezing a drop of juice onto the tool's ground glass from a grape right off the vine. Holding the meter up to the light gives you the sugar reading, with the shaded area indicating the sugar level.

Temperature and salts have an effect on a refractometer, but less so than on a saccharometer. Slightly more expensive refractometers can be purchased that have temperature correction built in so they will be accurate over a reasonably wide range of temperatures.

If you are a grower, refractometers are very useful for testing the health of growing plants.[7] You squeeze a wad of plant matter with a pair of modified vise-grips—or normal vise-grips in a pinch, so to speak—to extract a drop or two of liquid and monitor its sugar content. Low relative sugar levels are a reliable indicator of plant stress or over-fertilization with nitrogen.

The goal in making your wort is ideally to end up with a solution that is roughly 22–24% sugar. Most sugar-containing materials will have less than this amount when mixed with a small amount of water. But using molasses or waste candy can result in higher concentrations.

Usually, you will check your sugar concentration at the end of the process after adding your nutrients and correcting the pH. If you know that the sugar concentration is likely to be higher than 22–24%, then you will check the sugar concentration twice, once right after cooking after you've added the yeast food, and once before cooling.

Refractometers are more accurate than hydrometers in thicker solutions.[8] If, after cooking, by some wonderful chance your sugar level is above recommended levels (24% usually), add water before you start cooling. But be careful not to add too much; if you need to dilute your batch with water, this can also cool your wort.

Step 6: Correcting pH

Once you've provided a balanced diet and proper temperature for yeast, check the pH balance. If it's between 4.0 and 4.5, as it is naturally in many fruits, you don't have to do a thing. If it's outside this range, you must adjust it. The pH can be adjusted by adding either acid or base (alkali) to a material, depending on the direction you want the pH to go. Good old basic lime is useful for raising pH level, while any number of acids can be used to lower it.

Fig. 7-8 Sugar testing. Before testing the sugar content, filter the wort with a single or double coffee filter to remove as many suspended solids as possible to increase the accuracy of your reading.

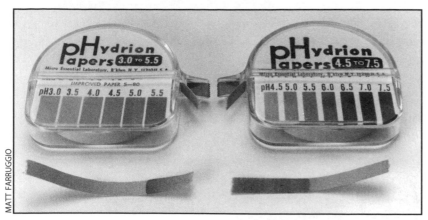

Fig. 7-9 Litmus paper. *Although this black and white photo doesn't do it justice, the color distinctions are pretty accurate. Rather than using one wide-range pH paper, you will find your readings more precise when you use two narrow-range papers.*

Fig. 7-10 pH meter. These come in a variety of accuracies and sizes. Some are even pocket-sized. You generally get what you pay for, so a cheap one is not going to be as accurate or last as long as a more expensive one.

Enzymes and yeast used in making alcohol are very sensitive to pH levels in wort or mash. The worts we use while producing alcohol are slightly acidic, in the range of pH 4.0–7.0, depending on the step. The wort mixture's pH must also be held to a fairly narrow range for each enzyme step in starch processing and for yeast to do their work in fermentation. Although letting the pH go a little outside the specified range won't knock the enzymes out altogether, their efficiency will drop dramatically. The actual range varies some from manufacturer to manufacturer, but **alpha amylase** is effective at pH 5.5–7.0 (ideally 6.5), and **glucoamylase** operates well at pH 4.0–4.5. **Cellulase** and other enzymes have their own pH ranges to pay attention to, if you are going to experiment with them.

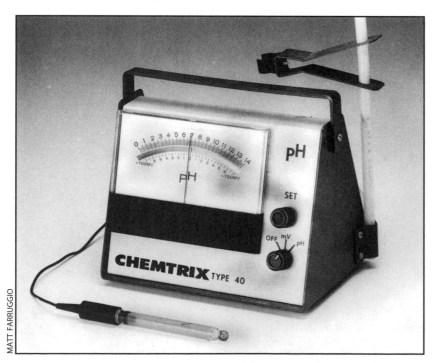

To a lesser degree, calcium levels affect many enzymes. If you're working with grain, it's likely you'll be using some lime (calcium hydroxide), an alkali, to adjust the pH at the start of the process, which tends to correct any calcium deficiencies. When using other materials that have a calcium deficiency, such as cellulose, it may be necessary to also add lime to bring the pH up for enzyme use, and then later correct the pH back down with a little acid before fermentation. Alternatively, you can use gypsum (calcium sulfate) if you need to supplement with calcium but don't want to raise the pH.

Of course, before you adjust the pH, you have to measure it, and this is generally done in one of two ways. The most inexpensive way is with litmus paper, a strip of paper impregnated with a special dye, which you may remember from fifth grade science or chemistry sets. Dipped into your mash, the paper turns a distinct color. Each color corresponds to a particular pH number. Litmus paper comes in different ranges—the narrower the range you work in, the more accurate the measurement will be. For making fuel, use two paper ranges that focus on pH ranges of 4.0–4.5 and 5.0–7.0, which correspond to the requirements of enzymes and yeast. Litmus paper is available at swimming pool supply stores, nurseries, and chemical supply houses.

The second, more expensive but more accurate, method is a pH meter, which only needs to be dipped into your mixture for a few seconds and it obligingly reads out the appropriate pH number. A minor inconvenience inherent is that it must be calibrated every so often. The easiest way to do this is to wash the probe and then immerse it in distilled water; your reading should be 7.0. If it isn't, calibrate until it is. Some meters calibrate at pH levels other than 7.0, in which case you'll immerse the probe in a prepared standardized solution available wherever you purchase the meter. For best accuracy, you should use calibration solutions that are in the ranges that you commonly encounter in use. (Soil pH meters work differently and are not accurate for use in fuel production.)

Lime can be added to bring the pH back up to where it should be in a too-acidic wort. If the pH is too alkaline, lower it with sulfuric, **phosphoric**, or **lactic acid**.

To be accurate in adjusting the pH for a large quantity of wort, say 500 gallons, remove five gallons (1%) from the batch. Measure how many grams or ounces of lime or milliliters or liquid ounces of acid you need to adjust the pH to the right level. Multiply this amount by the amount of mash left in the tank (495 gallons), and divide by five. This gives you the right figure for adjusting the entire batch.

Here's an example, using a 500-gallon batch of alkaline mash:

1. Remove five gallons.

2. Determine pH: If it tests low, start adding lime, stirring, measuring carefully. Let's say that three ounces of lime will correct five gallons of mash to the proper pH in this case.

3. Multiply the three ounces by the remaining 495 gallons, then divide by five:

$$3 \times 495 = 1485, \div 5 = 297 \text{ ounces}$$

You need 297 ounces of lime to adjust the rest of your wort. Mix well and double-check the corrected pH.

The same steps are taken when adding acid if a wort's pH is too high, but with acid, measured in milliliters or liquid ounces instead of grams.

The sample adjustment of five gallons is to prevent overcorrecting a wort. If you had added too much lime and then had to add a lot of acid, you'd initiate a reaction that would create inorganic salts, which would contaminate your batch and might inhibit yeast in its fermentation process. Rest assured, however, that out of your first few batches, you'll destroy at least one in this manner. It's much less painful to learn this lesson with five gallons than 500. You will be surprised at how little lime or acid is needed to make a big change in pH.

One last thing: When correcting the temperature at different stages of your recipe, do so with a cooling coil rather than by adding water. Adding water in the wort changes the pH in most cases, generally increasing the amount of lime or acid needed to do the job and making it necessary to repeat the adjustment.

ENZYMES

Before we move on to the starch method, let's talk about enzymes. The better an understanding of their functions and limitations you have, the more fuel you'll get out of your more complex carbohydrate feedstocks. Fermentation expert Easter Trabulse says of enzymes, "They are like a marriage counselor; they bring two pieces together

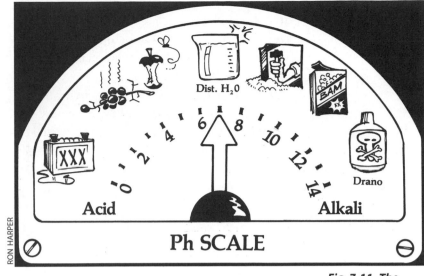

Fig. 7-11 The pH of common materials. *This figure shows the range of common materials from acid to alkali.*

or break them apart, but they are not part of the relationship."[9]

The enzymes we'll be primarily using—alpha amylase and glucoamylase—work as catalysts, causing a reaction without getting involved. They won't combine with starches or dextrins, but they will speed up the processes of starch-to-sugar conversion. Technically, glucoamylase can actually break starch down to sugar without the help of alpha amylase, but it takes too long. Use alpha amylase first to speed things up for a practical production rate.

Liquid commercial enzymes have a limited shelf life of around six months at room temperature (about a year and a half at 34°F). They don't go bad at the end of their stated shelf life, but their potency begins to drop, so you need to use more. Enzymes derived from barley (dry malt) will lose

COOKING IT RIGHT

Nature uses starch for long-term energy storage because it has a very stable chemical structure. It's much more resistant to decomposition by bacteria than sugar. In the case of corn, while the ear is on the stalk, its kernels are chock-full of sugar and contain little if any starch. The moment an ear is separated from the stalk, however, its sugar begins converting to starch, an enzymatic process that links all the sugars together, which occurs within 90 seconds or so.

Cooking stops the conversion of sugar to starch by deactivating the assembly enzymes. If you've ever boiled water near a cornfield, and picked, husked, and dropped an ear into a boiling pot inside 30 seconds, you know what a difference enzyme conversion makes. If you haven't been lucky enough to eat corn that fresh, take it from me—it tastes as sweet as candy.

potency quickly over a period of about eight months to a year. If you're using malt, I recommend sprouting and drying a new batch every two to three months.

There are enzymes for use with such carbohydrates as pectin and **inulin**, and there are now new enzymes on the market suitable for converting cellulose and hemicellulose. Some large companies are already starting to experiment with cellulase when processing corn to try to increase alcohol yield by converting the cellulose that's in the corn kernel to sugars along with the starch.

Theoretically, if you could somehow get the enzymes back out of your pot of mash, you'd be able to use them over and over again.

With high-protein mashes like those from grain, you can eliminate most or all of the nitrogen supplementation by using a **protease** to break down the protein, releasing nitrogen and other nutrients. As demand for these less common enzymes grows, we can expect the prices to drop.

Theoretically, if you could somehow get the enzymes back out of your pot of mash, you'd be able to use them over and over again. From time to time, there have been experiments to attach the enzymes in some way to paddles so that during agitation the enzymes contact all the feedstock but can be re-used. No really practical way to do this has been accomplished.

In less than favorable operating conditions, enzymes will act inefficiently or become completely inactive. The factors that need to be critically controlled are temperature, pH, agitation, and dosage relative to the starches or dextrins to be broken down.

Temperature

Enzymes use heat energy in the wort as part of the process of breaking the bonds between molecules. In general, the more heat available, the better, until you reach upper heat tolerance. While too high a temperature causes the protein structure of the enzyme to fall apart, too low a temperature will bring any reaction to a near halt, since there isn't enough energy for the enzymes to do their job. In general, the higher the feedstock-to-water ratio (the thicker the mixture), the higher the temperature tolerance of the enzyme.

With alpha amylase, the acceptable range is usually 170–190°F (in some cases 200+°F). Both **beta amylase** or glucoamylase work best between 120–140°F. Cellulase enzymes and pectinase enzymes operate at these temperatures, as well. The general temperature ranges vary in their specifics from manufacturer to manufacturer, so always refer to the company's spec sheet.

The above temperatures are accurate for enzymes extracted from bacteria. It's important when you buy alpha amylase to make sure it's bacterial in origin. Alpha amylase extracted from fungal sources or malt has much lower temperature tolerance. Ideally, your alpha amylase enzymes should come from bacteria of the *Bacillus* genus. Some of the best species are *B. subtilis*, *B. amyloliquefaciens*, or *B. licheniformis*. Each manufacturer uses its own variety of one of these species.

Glucoamylase enzymes are always bacterial in origin and come from *Aspergillus niger*, *A. awamori*, and certain strains of *Rhizopus* (all common bread molds). Before glucoamylase was available, moonshiners used beta amylase (now less common, from barley malt or sweet potatoes) to do the conversion/**saccharification** step.

Dosage

Only a very small amount of enzymes is needed to convert a huge amount of starch or dextrins. As long as conditions are right and your enzymes are continuously being introduced to fresh substrate (the material the enzyme is acting upon), they will continue to break the substrate down. The variables to be controlled are pH, temperature, and agitation.

Agitation speeds breakdown, causing more and more substrate to make contact with the enzymes. For speeding up a reaction, it's actually more beneficial to agitate than to add more of the expensive enzymes. If your tank is fitted with a good **agitator** or pump to recirculate the wort, I recommend heat exchange coils or a jacket for cooling, rather than adding water. The exception is if your mixture is very thick with solids not affected by the enzyme. In that case, some dilution may actually yield a better breakdown.

Manufacturers of enzymes produce them in their own carrier medium, and therefore each has its own particular recipe. Dosages are specified for the dry-weight basis of the material in question. Refer to nutritional handbooks for the dry starch content of your material to calculate proper enzyme dosages. If you are going to be using a particular feedstock for some time, it's worthwhile to have it chemically analyzed; it can save you a lot of money on enzymes in the long run.

MATT FARRUGGIO

As an alternative, you can zero in on the correct enzyme dose in the following way: Test containers can be one-quart mason jars. Begin by making a three-gallon basic batch that's two-thirds water, one-third dry feedstock. If your original feedstock comes to you wet, dry a known amount of it in an oven so you know how much water to account for. Cook the batch by slowly raising the temperature to a boil without causing **gelatinization** (see starch recipe in next section). I recommend one and a half hours of simmering—clearly an excessive cooking time, but it eliminates the possibility of under-**hydrolyzing** the starch. Adjust the pH of the three-gallon batch and cool it to the temperature range of the alpha amylase you are using.

Pour it into ten one-quart mason jars. Use any and all information you have at hand to make as good a guess as possible as to suspected starch content. Then refer to your enzyme manufacturer's recommended dosage. In one jar put in the amount of enzyme that should be right based on your guess of starch content. This is your test-case jar. In four jars, use progressively less enzyme than is called for. You could, for instance, cut the amount by 20% for each jar. In the five remaining jars, increase the amount in each jar by 20% compared to the test jar.

Keep all the jars in a water bath at the optimum temperature for the enzymes. Stir each one of them for 30 seconds, five times in an hour. At the end of

one hour, run cool water into the bath to stop the enzyme action.

Add five drops of iodine to each of the samples. Iodine will turn starch purple. If your guess at the starch level was good, all samples from five through 10 should show no remaining starch, so no purple. In the four jars with less enzyme than the test jar, there should be progressively more and more purple. Filter all the samples that do not turn purple through a coffee filter and check the sugar level in each one with a refractometer or saccharometer. The one with the highest Brix reading indicates the best dosage.

This would be a perfect world if you always guessed perfectly. You may find that the first jar that doesn't purple might be jar #9, not #5. If that's the case, do the test over with a new identically made batch, using the level found in #9 as the lowest dosage, and increase the amount in the remaining nine jars by 10% each until you zero in on the best dosage.

Obviously, having a chemical analysis done is much easier, but the experience you gain from performing this experimental testing yourself is quite valuable. A self-built testing "lab" also allows you to check effects on potential feedstocks of cooking time, agitation, particle size, and other variables.

THE STARCH METHOD

Starch takes on several forms. Simple starch is 35 glucose molecules linked together in a chain. Corn and a few other crops have what's known as **flinty** starch, where the chains are bunched together in semi-crystalline masses of 1000 or more glucose molecules.

Yeast need to eat bite-sized bits of sugar, not the longer, more complex forms. Our simple six-step starch method boils down to chemistry that breaks up those 35-link chains, or their more complex crystals, into smaller chains of three to seven glucose molecules called dextrins, and then breaks the dextrins down even smaller into one- or two-molecule sugars like glucose and **maltose**.

To achieve a simple sugar from corn or any other flinty starch, there's one additional step, before breaking apart the chains, called **premalting**, which is required to untangle and loosen the knotted crystals. Premalting is for flinty starches only.

Popular starchy crops include grains, potatoes, peas, manioc (also known as cassava or tapioca root), and some tropical fruits such as plantains,

LEFT: Fig. 7-12 *Iodine test.* *Two drops of iodine turned the starch in the top of this beaker quite purple. The reaction continued until the entire contents of the beaker became purple.*

breadfruit, and unripe bananas. But in the United States, more abundant than any of these is corn—so even though that particular feedstock is not among the most desirable of energy crops (in part because it requires lots of water, fertilizer, time, and sophisticated equipment), it's often favored for making fuel due to its abundance, ease of storage, and high concentration of starch. So we will use it as a model. Besides, if you can master the corn recipe, any other starch product will seem easy.

A bushel of corn weighs 56 pounds. Of that, 60 to 75% or more is starch. The **batch method** described below allows you to convert starch to sugar in six easy steps: l) milling; 2) **slurrying**; 3) premalting; 4) **hydrolysis**; 5) **liquefaction**; and 6) conversion. The batch method isn't used in large plants, but for our scale, batch makes the most sense.

Fig. 7-13 Starch recipe model. *Starch starts off as a granule of tangled chains. Cooking and premalting separate them further into single 35-link chains of individual starch molecules. In liquefaction, starch gets broken down into dextrins (medium-length chains). In conversion, the dextrins are broken down into one- and two-link simple sugar molecules.*

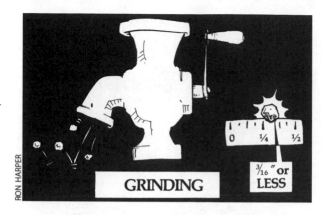

GRINDING   ³/₁₆" or LESS

Step 1: Milling

To facilitate the breaking down of starch molecules, grind corn into pieces small enough to be accessible to the chemicals and enzymes that you will be using. For making alcohol on a small scale, use a coarse grind, #20 to #60 screen (1/8 to 3/16 inch).

Intuitively, you'd think that the smaller the particle, the better the breakdown, so you might be wondering why I don't suggest a finer grind, right? After all, some large-scale alcohol producers use a #100 screen. But for small-scale use, screens that are between #20 and #60 are preferred because a coarse grind is easy to separate from the liquid after distillation.

The recovered grain can be sold as animal feed or used in other ways. If you mill the grain finer, you'll need a **centrifuge** or a sophisticated filter system to separate the **spent grain** from liquid in the end.

Milling is done by **gristmill** or **hammermill** (see Chapter 10). Small hammermills are not expensive and can often be found used, or you can have your grain ground for a nominal fee. You can even buy grain already ground at most feed stores; most coarse animal feed is around #60 screen.

In very large operations, the grain is cleaned with strong detergent after milling, rinsed, decontaminated with strong disinfectant, then rinsed again with cold water to flush out the disinfectant. These steps, designed to prevent contamination, are not generally done on the small scale.

Step 2: Slurrying

At everyday temperatures, starch granules are relatively impervious to water or enzymes. It's part of what makes starch such a good energy storage medium. So we are going to have to raise the temperature of the water to get it inside the starch granules. This process is called slurrying.

SLURRYING

RON HARPER

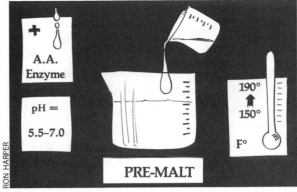

PRE-MALT

RON HARPER

So you've got a bushel of #60 screen grain. Next, you tie on an apron and whip up a 28- to 30-gallon batch of cornmeal mush. Any cornmeal mush chef worth his salt (no, don't add salt) is aware of two critical hazards: lumping and burning.

The worst is lumping, which will occur if the mash gets hotter than 155°F, the temperature at which corn starch gelatinizes and becomes a gooey mess. Gelatinization is the process in which the outer layers of the granules of corn start to absorb water and lose their crystalline structure.

Always add your cornmeal *before* the water reaches the gelatinization temperature, which varies with the grain. You can test your corn on the stovetop to determine its gelatinization temperature. Corn ranges from 155 to 191°F, depending on how much amylose (a type of starch) it contains. Gelatinization points are 126–139°F for barley; rye 135–158°F; rice and sorghum (milo) 154–171°F; and wheat 136–191°F. You'll find that if you can master corn, the other grains will be much easier.

Start your slurry with water well below 150°F. You could use fresh cold water or warm waste water from your distillery **condensers** or even hot liquid spent mash mixed with cold water. Spent mash is synonymous with backslop, the solution left over after distillation. It contains a mixture of unfermentable materials and dead yeast.

How much water should you use for one bushel (56 pounds) of corn? If you are using standard yeast, start with ten to 15 gallons per bushel. By the end of the fermentation process, you will use 28 to 30 gallons, which will yield an 18 to 20% sugar solution.

It's important not to burn the wort, which is easy to do. Constant agitation will prevent burning and sticking, and is also necessary for the infusion and circulation of chemicals and enzymes you'll add later.

Step 3: Premalting

Remember, this step is only for flinty starches. At this point, your nice smooth corn slurry is approaching 150°F (danger! danger!), and its starch molecules are still tied up in hard little clusters. It's time

MATT FARRUGGIO

BELOW LEFT: Fig. 7-14 Gelatinized corn. *This corn has been cooking at about 165°F for ten minutes without the addition of alpha amylase. It has become so thick that the spoon is standing straight up in it. If this happened in your **cooker**, your agitator would probably stop dead.*

BELOW: Fig. 7-15 Starch recipe materials. *Alpha amylase, gluco-amylase, sulfuric acid, yeast, malt (if no liquid enzymes are used), iodine, lime, and a feedstock (ground corn in this case).*

MATT FARRUGGIO

for premalting, which does two things: It prevents sticking and lumping above 150°F, and begins breaking starch down into dextrins. Premalting means adding a small amount of the enzyme alpha amylase to the wort. Since bacterial alpha amylase is only efficient at pH levels between 5.5–7.0, test the slurry and adjust the pH appropriately.

When the slurry is at the correct pH, add alpha amylase equal to up to one-quarter of one percent of the dry weight of the starch in the grain (0.25% **w/w**), and mix well. Within five to ten minutes after the initial thickening, the mixture should thin and begin to give off a pungent odor. As the temperature continues to rise, the mush will continue to thin, until it has reached a temperature of 195+° F. At that point, common alpha amylase enzyme is generally deactivated (cooked) by the high temperatures, after bringing the slurry through the premalting stage.

Step 4: Hydrolysis (Cooking)

Only a small portion of your corn's starch has so far been cut up into dextrins. In the next step, we are going to use heat and the small amount of acid in the water to loosen the corn starch chains from one another.

By the time the wort is boiling, the first hit of alpha amylase has "gone up in smoke," and your cornmeal is popping and bubbling. The moisture in each granule of corn is boiling and bursting as well, and is thereby breaking down the bundles of chains (which can number as high as 10,000 glucose molecules) into smaller bundles and individual chains. As cooking progresses, the wort will become very cloudy with released starch.

Hydrolysis takes 45 minutes to an hour at a boil. Don't try to shorten the cooking process—complete hydrolysis ensures a higher yield of alcohol in the end, because the more broken-up the starch,

the more effective the succeeding steps will be. Cooking times longer than an hour won't hurt the mixture, but will use more energy.

Once the wort has finished hydrolysis, you won't need any more heat. Depending on your plant size and layout, you will do the next step in the same cooker, or you may choose to transfer your wort from cooker to fermentation tank (see Chapter 10).

There are other ways of hydrolyzing starch into individual chains. Industrially, corn is sometimes cooked under pressure. Optimum alcohol production is obtained at about 150 **pounds per square inch (psi)**, 350°F, with steam. Under these conditions, grain can be almost completely hydrolyzed in three minutes. Another industrial method is continuous cooking under pressure, in which a small amount of material is subjected to a very-high-pressure jet of much hotter steam, instantly cooking and blowing the material into a fermenting tank or plug flow cooker. Of the pressure methods, continuous cooking uses the least amount of energy. Unfortunately, the sophistication and expense of the equipment needed make neither of these methods practical for the small-scale producer.

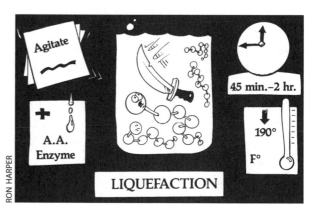

Step 5: Liquefaction

Liquefaction (turning the thick mush into a liquid) completes the process of breaking down tangled starch chains into **complex sugars**, and employs our old friend alpha amylase, at the same temperature and pH you used before. But now the temperature is on its way down.

Time is of the essence during liquefaction because conditions are ripe for bacterial invasion and infection. Three factors determine how fast liquefaction can take place: temperature, pH, and agitation.

To facilitate cooling at this stage, add cold water to the wort, or use a cooling coil, to bring the temperature down to 190°F. A cooling coil is easy to make out of a copper or stainless steel tubing spiral through which you run cold water—see Figure 7-20. Alternatively, you can use an external **heat exchanger** to speed the cooling process.

Keeping the temperature at close to 190°F encourages a furious-fast reaction. A temperature of 170°F is still acceptable, but lower than that will slow liquefaction to nearly nothing. In wintertime, it's a good idea to insulate your tank: 1000 gallons of hot corn mush is going to resist rapid cooling anyway, but insulating is easier and cheaper than adding heat to keep the tank temperature up in cold weather. If you do have to add heat, use your "cooling" coils and hot water, or low-pressure steam, since copper tubing can't take much more than 10 to 15 pounds of steam pressure.

Once the wort is at 190°F, check the pH to make sure it's still between 5.0–7.0. The pH level can vary during the liquefaction, and it's better to make an adjustment or two during the process than to let it wander from the optimum range. To fine-tune for the fastest possible reaction, keep the pH at 6.5; actual optimum point varies slightly with different manufacturers. Different companies' enzymes will be optimum at different temperatures, usually between pH 6.0–6.5; in some cases, pH is optimum between 5.5–5.7. The variation is mostly due to whether the enzyme comes from a plant, fungus, or bacteria. Being precise and following the manufacturer's specifications will give you the best results.

Typically add enzymes equal to about ½ of one percent of the weight of the starch in the grain (0.5% w/w). Mix thoroughly. You'll see a distinct change over the next 30 minutes to two hours. The wort becomes very pungent and very fluid, and the grain color darkens. Your starchy corn will have become a dextrin-rich soup.

Agitation has the largest influence on reaction time. When enzymes constantly contact new starch during agitation, the reaction is speeded up by at least four times over an immobile liquefaction. Be careful, though, not to mix a lot of air into the wort during agitation, since air contains bacteria, which could reduce alcohol yield.

If everything has gone according to schedule, your batch should be liquefied in as little as 30 to 45 minutes. If you don't agitate, the process will take closer to two hours. Your batch is done when the grain is fully brown, the odor of the enzyme has all but disappeared, and there's no stickiness

Fig. 7-16 Liquefied corn. *Two minutes after the addition of alpha amylase, with some vigorous stirring, the corn has thinned out to become quite fluid.*

MATT FARRUGGIO

at all in the mash. You can try to check the starch content of the wort by taking a cup of it and adding a couple of drops of iodine (see Enzymes Dosage section earlier in this chapter).

Don't dawdle once the mash has thinned and the iodine test looks good. If you leave the batch

SOFT STARCH

Soft starches are found in cassava, wheat, barley, potatoes, and buffalo gourd. These feedstocks can save time and energy over the starch method outlined for corn; it's not necessary to boil a soft starch mash.

Basically, you'll want to combine premalting, cooking, and liquefaction in one step by adding all the required alpha amylase immediately after your initial feedstock is combined with cool water; or by adding fungal amylase first and then bacterial alpha amylase as temperature rises above 160°F. In either case, because it can tolerate high temperatures, bacterial alpha amylase is the active liquefaction enzyme. This enzyme can actually stand temperatures just above 200°F for short periods without getting fried, even though their highest level of activity is at 190°F.

It's possible for liquefaction and cooking to occur simultaneously at 200–205°F for 60 to 80 minutes with a full alpha amylase dose. Alpha amylase will survive these temperatures for one to one and a half hours if your pH is perfect, the temperature is controlled to below 205°F, and the mash is vigorously agitated. You can save an hour or more of heat-up/cooking time with this technique in batch production, due to water's high **latent heat of vaporization**. The mixture will heat quickly to 200+°F, but takes a long time to begin boiling. Soft starches, especially wheat, will liquefy well at temperatures below 200°F.

But if your pH, calcium, and feedstock-to-water ratio are not optimal, you won't get an efficient starch-to-dextrin breakdown. So I recommend using and mastering the basic starch recipe to learn good process control before trying this energy-saving method.

Batch Cooking of Dried Sago

Mash Concentrate	20% Dissolved Solids				
Enzyme Dosage	0.5 g/kg Starch				
Maximum Cooking Temperature (°C)	65	75	85	95	139
Relative Viscosity of Cooked Mash, Centipoise	5000	1500	1300	1200	1350
Alcohol Yield, Liters per Metric Ton Starch (3-day, 30°C Fermentation)	450	610	640	630	640
Dosage of (Pressed, Moist) Yeast g/kg Mash	2.0				

at liquefaction conditions too long, a carbohydrate called maltulose can start to form, and it will not break down in the next step and will not ferment.[10]

To my knowledge, there is no simple definitive chemical test to distinguish between large sugars and starch. Practiced moonshiners do a taste test until the batch has reached its maximum sweetness—not your most scientific method, but surprisingly accurate, with enough practice. In any case, have patience. Don't go to the next step in less than 30–45 minutes with agitation, or less than two hours without. Experiment with reducing times once you've got some experience.

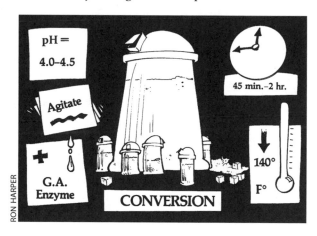

Step 6: Conversion (Saccharification)

Home stretch. The starch has been hydrolyzed, then liquefied by way of a protein biocatalyst. You started with a starchy cornmeal mush, and you now have a strange sugar molecule soup. You can almost hear the yeast call, "Is it soup yet?" Well, almost, but not quite.

Yeast, to whom we want to feed this mess, only eat small sugar molecules such as glucose, maltose, sucrose, and **fructose**. To convert the larger sugars of your liquefied wort into simple ones, you'll do the process of conversion, also called saccharification, using glucoamylase or beta amylase.

Glucoamylase can break down some of the more unusual types of dextrins that beta will not. For glucoamylase, the pH requirement is 4.0 to 4.5, with an upper temperature limit of 140 to 150°F.

Use cooling coils to reduce the temperature. Or, in the case of a concentrated feedstock such as corn, you could add water to get the temperature down to 140°F. Remember, you might use up to 30 gallons per bushel before the process is completed, although we started with only 15 gallons.

In conversion, always adjust the temperature *before* adjusting pH. This is not because temperature and pH are directly related, but because some cooling procedures can affect pH. For instance, you may have cooled your wort by adding water. Water is around pH 7.0. If you adjusted the pH first and then added water, the added water would raise the overall pH, and you'd just have to adjust it again.

The other reason to reduce the temperature first involves safety. Adding acid to a 190°F wort (to lower the pH) produces a much more violent boiling reaction than adding it to a 140°F wort. Reduced water temperature helps avoid being injured by spattering acid (see Acids and Bases section later in this chapter).

Now that we have the right temperature and pH, we add four ounces of glucoamylase enzyme per bushel (0.6% by weight (w/w)), depending on the manufacturer. Over the next half-hour to two-hour period, keep the wort close to 140° F, and definitely not below 120°F, to ensure rapid action by the enzyme. It's too bad that glucoamylase can't tolerate 170°F, since that is the temperature necessary to continue to kill *Lactobacillus*, our most dangerous potential contaminant at this stage.

Agitation is as important now as it was in liquefaction.

Conversion requires sufficient calcium. In general, most grains have plenty, but if conversion seems sluggish, your feedstock may not be supplying enough on its own. Check the manufacturer's specifications for how much calcium your enzyme needs. It can't hurt to throw some calcium in if you suspect a deficiency. If you used lime to bring the pH up in the liquefaction step, you can be pretty sure that you'll have enough calcium in the wort. If you need more calcium, adding gypsum (calcium sulfate) at this stage instead of lime will cause a much smaller pH change, which saves you having to add more acid.

Now try the iodine test again. This time your sample should show no color at all, or at most turn a very faint pink. I always wait another 20 minutes after the last trace of red has disappeared before going on—a shortcut here can slow fermentation. If your sample has a tint of blue or purple, conversion is not yet complete. Give it more time. If the purple persists, you didn't do an efficient job back in liquefaction. Your residual starch is unfermentable, and therefore wasted.

As soon as conversion is complete, add enough cold water to yield 18–24% sugar—the remainder of the 30 gallons in the starch recipe should do it. Some newer yeasts can tolerate up to 40% sugar and produce over 20% alcohol. If you are using one of these yeasts, you will need to add less water. Generally, though, on the small scale, it is tough

to maintain the conditions needed to attain the higher alcohol content. Use a saccharometer or refractometer to measure.

Water not only dilutes your sugar content, it lowers the temperature of your brew. When using a standard yeast, you want it down to 90–95° F before going on to fermentation. If water doesn't lower the temperature sufficiently, use a cooling coil. If you do use water, check the pH again, and if it's too high, bring it back down to pH 4.0–4.5.

In the last 20 years, there has been a move to eliminate conversion as a separate and distinct step. Large alcohol plants move directly from liquefaction to fermentation while simultaneously saccharifying the mash. What has facilitated **simultaneous saccharification and fermentation (SSF)** has been the breeding of yeasts that can tolerate temperatures of over 100°F, and glucoamylases and complexes of enzymes that remain active at lower temperatures than those of the past.

The advantage with SSF is that you are not creating a huge tank of bacteria food in advance of adding your yeast. Once you add the yeast in fermentation, it takes several hours for it to multiply fast enough to saturate the mixture, which gives the faster-reproducing bacteria a long head start on the smorgasbord of sugars you've created, perhaps contaminating the batch. In SSF, the glucoamylase produces sugar at a rate the yeast can easily consume. There is a lot less free simple sugar for bacteria to eat. You save a step and help eliminate contamination problems. It has been found that the enzymes can always keep up with the demands

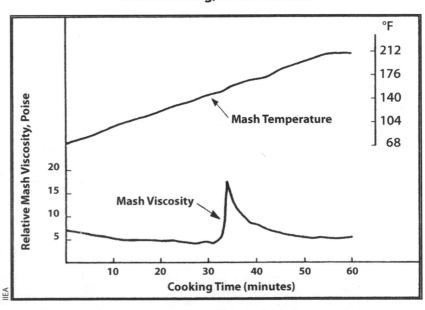

Fig. 7-18 Batch cooking of fresh cassava. This chart shows the sharp thickening effects of gelatinization at temperatures below optimum action of bacterial alpha amylase. Poise is a measurement of viscosity. At just a few degrees hotter, the enzyme radically thins the mixture. Semi-continuous cooking and/or use of fungal alpha amylase, which is active at lower temperatures, would largely eliminate the spike in viscosity.

Batch Cooking, Fresh Cassava

RON HARPER

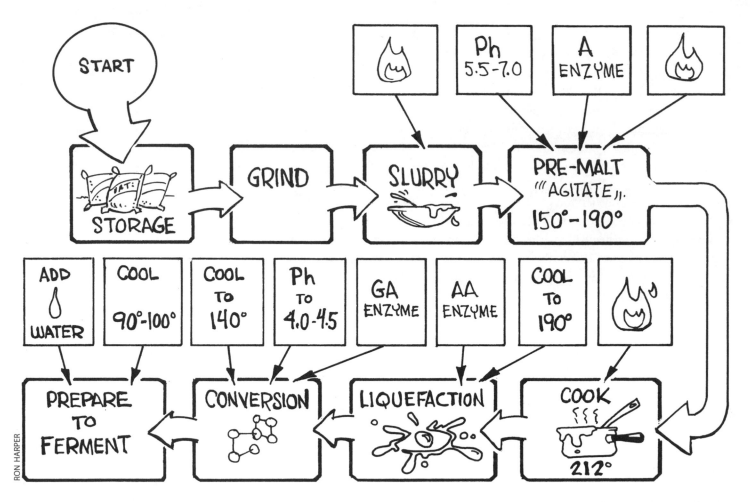

Fig. 7-19 Starch cooking flow diagram.

of the yeast and do not become a limiting factor. This new development is a boon to small-scale producers, as well as the big boys.

THE KEY TO FERMENTATION: YEAST

While you may feel justifiably proud of having mastered two enzyme steps in converting starch to sugar, yeast is going to sail through 11 enzyme and co-enzyme steps in converting sugar to alcohol, without even giving it a thought. And the little creatures are going to love every minute of it. From the moment you introduce them to your wort until a moment one to three days later when they die, drunk and exhausted—all they do is eat and reproduce.

During their short lives, yeast go through two distinct phases. The first is called the aerobic phase, during which yeast breathe in oxygen, eat sugar, reproduce like crazy, breathe out carbon dioxide—and don't really produce any alcohol to speak of. Once they've used up all their oxygen, they shift to the **anaerobic** phase, in which they breathe in carbon dioxide, eat sugar, excrete alcohol, and breathe out carbon dioxide—without reproducing. (You might say we run our cars on yeast manure.)

It's important to know just how much alcohol your yeast can stand, in order to figure out how much of a sugar concentrate to feed them. Yeast will use a little over half the sugar to produce alcohol, and almost half to produce carbon dioxide. Different strains of yeast will tolerate different strengths of alcohol before they die. Baker's or brewer's yeast, for instance, can stand a mash of only about 8% alcohol. Distiller's yeast holds its liquor better, tolerating up to 12% alcohol, and some special breeds can handle up to 15%. The top of the line, as you might expect, are the champagne yeasts, which tolerate (under perfect conditions)

17% alcohol. A few of the newer yeasts will go up to 23% alcohol under special conditions, but those conditions are hard to provide in a small plant. As a practical rule, you can expect most good distiller's yeasts to handle about 10–12% alcohol.

The critical factors to control in order to end up with a 10–12% alcohol mash are (does this sound familiar?) temperature, pH, nutrition, agitation, initial yeast dosages, and contamination.

Temperature

Optimum operating temperature depends on the strains produced by the yeast company. Most yeast prefer 80 to 90°F. Some good newer yeasts can tolerate 100–105°F. Below 75°F, yeast acts so slowly that it seems as if nothing's happening at all. Bringing the temperature back up often revives yeast activity. Temperatures above their optimum for a long period of time will kill yeast. Clearly, some method of temperature control is in order.

The problem of keeping yeast warm is not nearly as critical as keeping it cool, since its frantic orgy of eating and reproducing generates quite a bit of heat all its own. In fact, it's common for vigorously fermenting batches to get so hot that much of the yeast dies halfway through the process.

Cooling the mixture can be accomplished with cooling coils or cooling jackets. Some low-tech producers have gotten away with spraying the outside of the tank so that evaporating water cools it. If you're going to try external spraying, it's a good idea to wrap the tank with burlap first to enhance the evaporative effect.

Even better is to pump the mash through an external heat exchanger. With an external heat exchanger, you can inject pH-correcting chemicals into the mash as it passes though the heat exchanger, or bubble in oxygen for the first half hour of fermentation to further stimulate yeast growth.

The main disadvantage of internal cooling coils is that they are harder to clean and decontaminate between batches. The external heat exchanger is disinfected by pumping from a tank full of disinfectant back to the same tank, which is a form of **cleaning-in-place (CIP)** system. Caustic soda is the cleaner commonly used.

Temperature changes in fermentation are generally quite slow and really only need to be checked once or twice a day. The highest rate of heat production will happen between ten and 30 hours. This is when you want to be sure your agitation and heat exchange are at their optimum. If the fermenting mash is agitated, temperature throughout the mash can be kept more or less even throughout the tank, and temperature changes made less often.

If you want to make cooling or heating automatic, it's easy to use a temperature-controlled twin-inlet **solenoid valve** to regulate tank temperature. Plumb hot and cold water to each of the solenoid valve's supply inlets. Set the upper limit

Fig. 7-20 Fermentation cooling coil. This Brazilian alcohol producer used a one-inch-diameter stainless steel coil to absorb heat from the hotter center of this non-agitated fermentation tank.

of your thermostat for a couple of degrees below the yeast manufacturer's maximum recommended temperature, and set the lower end a couple of degrees below the optimum temperature. The hot or cold water admitted to the cooling coils controls the temperature nicely.

In practice, you will rarely use hot water, and seasonally you may wish to use insulation if too rapid a cool-down becomes a problem. A thermostat will also work to automatically start and stop the pump and **valves** of an external heat exchanger.

The new higher-temperature-tolerant yeast can make quite a difference in plant operation. Depending on air temperature, fermentation tanks may naturally cool themselves enough just through radiation of heat to stay within the temperature tolerance of 100+°F yeast.

pH

Yeast used in making alcohol are very sensitive to pH levels in mash. To operate at top capacity, yeast prefer a pH of 4.0–4.5. This low pH inhibits the growth of most butyric and lactic acid bacteria. But too-low pH will inhibit yeast and reduce your yield. Yeast can tolerate pH between 3.0 and 4.0 for a few hours, and pH 2.0 for less than two hours. Tolerance of low pH drops as the alcohol level in the batch begins to rise. A pH level above 4.5 will cause your yeast to make acids and glycerol out of some of your sugar, rather than alcohol, to bring the pH down to its optimum level. It's always cheaper to adjust the pH yourself than to let the yeast do it.

The pH should always be the last thing you check before you add yeast. You should check it during fermentation, as well, to see if corrections are necessary.

Nutrition

Yeast need a balanced diet of proteins and minerals. Grain mashes almost always provide this balance. Perhaps the best yeast nutrient source is backslop—what you've got left after distillation, the spent mash full of cooked dead yeast—which can provide most of the nutrients that the new living yeast need. Using backslop instead of water for about a third to half of your liquid when preparing your wort takes care of most all of yeast's nutritional requirements.

Backslop is often not a sufficient source of nitrogen, however. For instance, 25% backslop in your mash will add around 15 milligrams of nitrogen per liter, with the corn itself adding about 71 milligrams per liter. The goal is to reach approximately 100 milligrams of nitrogen per liter, so a small amount of fertilizer or malt would need to be added to get optimum results.[11] The batch will still ferment with most grains and backslop without nitrogen supplementation, but it will slow down earlier than if it has enough.

Do not use urea as a nitrogen source, since yeast will make a dangerous byproduct, ethyl carbamate (**urethane**), from it.

Instead of adding fertilizer, you can extract more usable nitrogen from **amino acids** in the mash by using protease enzymes to break protein down into amino acids, which the yeast can use as food. Some companies combine proteases with their glucoamylase (the enzyme used for saccharification of starch just before fermentation), e.g., Rhizozyme.

You can get into problems with not having enough protein or minerals for the yeast when you use straight starch, or waste fast foods, which are relatively devoid of anything other than fat and starch. (Just think—if it doesn't have enough nutrition to support yeast, what are we doing eating it ourselves?)

The type of water you use in fermentation can sometimes affect the mash. Water containing

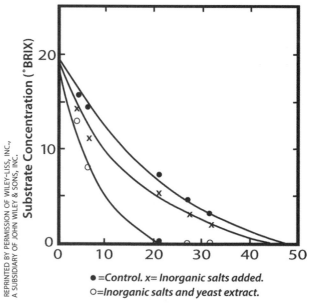

Effect of Nutrients on Fermentation Rate

●=*Control.* x= *Inorganic salts added.*
○=*Inorganic salts and yeast extract.*

Fig. 7-21 Effect of nutrients on fermentation rate.[12]

As you can see here, the combined benefits of fertilizer (inorganic salts) and yeast extract (which is contained in backslop) result in the least number of hours for fermentation.[13]

sodium, selenium, cobalt, or lots of chlorine kills yeast. Plutonium, strontium 90, cesium 137, and cadmium are also toxic to yeast—but if this is what's contaminating your fermentation water, you have bigger problems (living too close to a nuclear reactor) than getting your yeast to reproduce!

In a few areas, water that has a very high ratio of calcium to magnesium may need extra magnesium. When using waste junk food, make sure you have at least 0.3 parts per million (ppm) of zinc, which is often missing from highly processed foods. However, more than 1.0 ppm of zinc will inhibit yeast. In supplementing junk foods, a handful of calcium-magnesium-copper-zinc pills ground up will handle these nutrient needs. (They are good for you, too, if you eat too much junk food.) I often find it helpful to neutralize chlorinated water with hydrogen peroxide before using it.

If you use malt instead of backslop, ten to 15 pounds per 500 gallons of mash is plenty. With diammonium phosphate, five to eight pounds per 500 gallons of mash is sufficient. If you don't use backslop, add a few pounds of dried brewer's yeast from the health food or feed store to provide the trace substances that your yeast use.

Agitation

Understandably, once yeast actually starts producing alcohol as a digestive byproduct, it experiences some difficulty in recognizing its food source. Basically, the yeast is staggering around in a cloud of alcohol and can't make contact with its food. To some degree, the carbon dioxide bubbles that the yeast make will help them travel upward, until they lose the bubble and drift down again. But for fastest fermentation, doing some agitation yourself is helpful. This way, the yeast is constantly faced with fresh sugar to absorb. With some of the thicker mashes, where yeast mobility is limited, agitation can cut fermentation time in half.

Mashes of Jerusalem artichoke tuber pulp or sugar beet pulp require slow, constant agitation to avoid the additional problem of trapped carbon dioxide. Carbon dioxide gets trapped under a heavy cap of solids in thick mashes; its release resembles the boiling and plopping of a giant bowl of oatmeal. There's the potential for plugging your **fermentation lock**. Another danger in thick mashes is stranding your yeast above the liquid by a high-floating cap of solids, fatal to the majority of the little critters.

Agitation during fermentation doesn't take a lot of energy; it can be done with a low-power gear-reduced motor (six rpm is fine). Your agitator paddles should rise up from above the cap and slap down to continually break up floating solids. For 1000-gallon tanks and smaller, if you don't want to build an agitator right away, the periodic circulation of the mash by pump is adequate. If you use a pump, beware of contamination. Lines and pump will have to be flushed and disinfected between batches to avoid introducing lots of bacteria to the tank.

As an extra added attraction, continuous agitating allows you to fill your tank just a little higher with mash. Usually, you'd fill a tank to 80% capacity to allow space for the cap and foaming. If your agitator works well, it may be safe with many feedstocks to fill the tank to 90%.

Dosage

The amount of sugar to feed your yeast depends primarily on how much alcohol it can tolerate. For instance, if you have a 10%-alcohol-tolerant yeast, you'd expect to feed it a 20% sugar solution—

Fig. 7-22 First stages of fermentation. Added dry, yeast go right to the bottom of the tank and sit for a few minutes, eating sugar and beginning reproduction. Notice the trapped carbon dioxide in the yeast mass and the miniature explosions of escaping carbon dioxide and yeast. Initially, yeast may actually pull a little air in from the outside (as seen in the fermentation lock). A few more minutes, and almost all the yeast will be floating on top of the mash and pushing out remaining air with their carbon dioxide exhalations.

MATT FARRUGGIO

RON HARPER

cerevisiae (the scientific name for distiller's yeast) can't regulate their internal osmotic pressure. A high concentration of sugar outside the yeast tries to force the yeast to burst, since it creates high osmotic pressure inside them. In effect, the more dilute fluid inside the yeast wants to mix with the concentrated sugar outside. Under these conditions, fermentation can get off to a sluggish start. Consequently, high sugar content gives an advantage to many osmotically self-regulating bacteria. Since they multiply much faster than yeast does, even low levels of initial bacterial infection can greatly impair your alcohol yield.

Furthermore, if you start with a very high sugar concentration, you can trigger what's known as the Crabtree effect. In this case, the yeast switch over from reproductive phase to alcohol production phase very quickly. This, too, would reduce the quantity of yeast, and slow fermentation, while giving bacteria an advantage.

One way to avoid the problems related to high initial sugar concentration is to use simultaneous saccharification and fermentation. This allows the

remember, yeast only turns half your sugar into alcohol, so 20% sugar should yield 10% alcohol.

Providing an over-rich diet of sugar (e.g., 24% sugar for 10%-alcohol-tolerant yeast) will kill the yeast before it uses all the sugar. If your sugar level is lower than ideal, your yeast will actually survive fermentation, but you'll have to use a little bit more energy in the distillation step to extract alcohol from the weaker mixture in your batch.

Since fermentation often slows quite a bit as the yeast get close to their alcohol limit, I prefer to use a yeast with a slightly higher tolerance than what I'm aiming for, so as to have strong fermentation right to the end. You can also consider live yeast recycling with liquid feedstocks, if the fermentation stops before reaching levels toxic to the yeast.

High initial sugar concentrations inhibit alcohol production for another reason. *Saccharomyces*

BELOW: Fig. 7-23 Structure of yeast.

This yeast is reproducing by budding a new cell. Yeast store their energy as glycogen—the way animals store energy—instead of as glucose like plants do. Yeast have some qualities from both kingdoms.

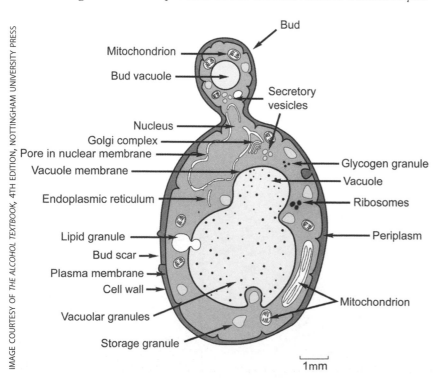

- Bud
- Mitochondrion
- Bud vacuole
- Secretory vesicles
- Nucleus
- Golgi complex
- Pore in nuclear membrane
- Vacuole membrane
- Endoplasmic reticulum
- Glycogen granule
- Vacuole
- Ribosomes
- Lipid granule
- Periplasm
- Bud scar
- Plasma membrane
- Cell wall
- Vacuolar granules
- Mitochondrion
- Storage granule

1mm

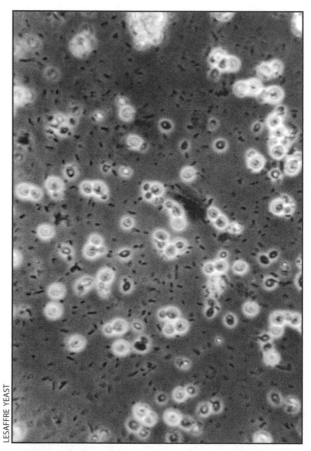

LESAFFRE YEAST

ABOVE: Fig. 7-24 Bacteria-infested mash. *The small dark specks are bacteria. Note the difference in size between the yeast and bacteria. Bacteria can double every ten minutes, compared to yeast taking an hour.*

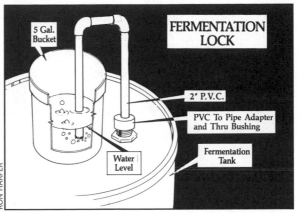

RON HARPER

RON HARPER

Fig. 7-25 **Fermentation lock.** *This simple device used on large tanks replaces the fancy little fermentation lock you saw on the five-gallon bottle.*

yeast to get off to a great start with lower sugar, but with a continuous release of sugar over time.

Now that we are seeing the first cellulose-to-alcohol processes coming on line, high sugar content may become more of an issue. Unless you are using a GMO yeast to consume the half of the sugar which is not normally fermentable, a normal yeast will have to contend with higher osmotic pressure initially. The likely solution will be to make a larger volume of **pitching solution** to prevent sluggish fermentation (see next section). In the case of cellulose fermentation without GMO yeast, there will be plenty of excess sugar that is not fermentable. But it doesn't have to go to waste, as you'll see later.

With sugar concentration at the optimal level for your particular yeast, you're ready to inoculate your batch with the little fellows. For a 500-gallon batch, add two to five pounds of dry yeast. Adding more usually results in a drop in lag time to less than six hours before fermentation becomes visibly vigorous. The thicker the mash, the closer to five pounds you should add. This higher inoculation rate in thick batches is needed because the yeast requires extra time to reproduce and spread throughout a thick batch.

To add yeast directly to the wort, first dissolve it in a few gallons of 104°F nonchlorinated water[14] with a few ounces of sugar or malt for about 15 minutes. This conditions the yeast, gives it some practice using this particular wort, and helps it spread throughout the wort in the initial stages. Once you've added yeast, the wort is called mash.

A more advanced and effective method of inoculating your batch is by preparing a pitching solution—making a solution of wort and yeast a few hours before the mash is ready to start fermenting. Doing this is also an effective way to avoid the problems of high sugar concentration.

Fermentation takes from one to three days. You'll know the process is complete when you sample some filtered mash with a saccharometer or refractometer and get a reading of less than 2% sugar. Dissolved solids and other substances in the mash give you the impression that there's still a small amount of unfermented sugar left. Some mashes, which are relatively free of dissolved solids, will actually register below zero when complete because alcohol has lowered the liquid's density. As you become familiar with your feedstock, you'll be able to determine your own correction factor.

Your first clue that fermentation is over comes from your fermentation lock. This is a water trap that releases the yeast's exhaled carbon dioxide but prevents oxygen and bacteria from entering the tank. The lock also captures alcohol vapors. For this last reason, water in the lock should be added to the mash just before distilling, since that water will contain some alcohol. If your site dictates that you must control odor, you may want to add

Fig. 7-26 **Magnified yeast.** *Here's how distiller's yeast looks under a powerful microscope. Note the new yeast bud and bud scars.*

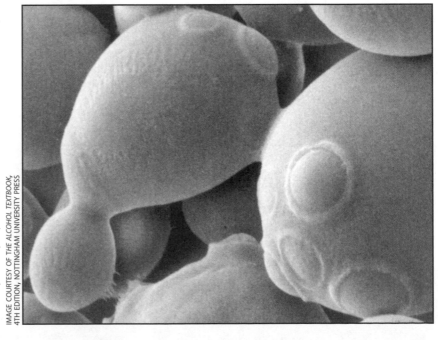

IMAGE COURTESY OF *THE ALCOHOL TEXTBOOK*, 4TH EDITION, NOTTINGHAM UNIVERSITY PRESS

an activated charcoal filter following the lock to deodorize the CO_2. This won't be necessary if you are collecting, cleaning, and compressing the gas.

Activity in the tank and lock goes through definite cycles. In the beginning, while yeast is still inhaling oxygen, there is very little CO_2 bubbling around. It takes a few hours for some slow bubbling to evolve. But as fermentation continues, quite a large volume of gas is expelled from the lock. For every pound of ethanol produced, the yield is just under a pound of carbon dioxide. At the peak of fermentation, a 500- to 1000-gallon tank can produce over two cubic feet of carbon dioxide *per second*. At the end of fermentation, the bubbles will have slowed to a virtual stop.

Another visual indication that you're done fermenting is when the cap of solids that has been suspended on top of the mash falls to the bottom of the tank, signifying that the yeast is dead or in a drunken stupor. At this point, the mash is as ready to be distilled as it will get.

Pitching the Yeast

One way to make sure fermentation gets off to a running start is to make sure that your yeast are primed for the "party" that will ensue. If dry yeast is added directly to the mash tank, even if it's dissolved in water just before addition, there's often a lag of six to 12 hours before the fermentation really gets going (see Figure 7-27). During this period, bacteria can do a great deal of damage. Eliminate this lag time by making a large starter batch of yeast called the pitching solution.

I like to start my pitching solution hours before I even begin to make the wort. Using a sterilized portion of a previous batch (backslop) as your starting solution and breeding yeast for several hours is quite a bit cheaper than adding lots of fresh dry yeast, and accomplishes the same thing. It also acclimates your yeast to the kind of mash you are using.

Pitching will multiply two pounds of dry yeast to the equivalent of eight to 25 pounds of dry yeast. A massive inoculation of yeast from the pitching solution reduces lag time to an hour or two, eliminates problems of high sugar concentrations, and cuts losses to bacteria considerably. What you are doing in the process of pitching is accomplishing most of the aerobic reproductive phase of the yeast's party first, so it doesn't have to happen in your fermentation tank.

Pitch can be spent, sterilized mash left over from a previous batch and mixed half and half with water. Avoid using backslop from a problematic batch for making a pitching solution. Add enough molasses or sugar to bring the mixture up to 18 Brix (sugar percentage). The volume of your pitching solution should be in the range of 7 to 10% of the volume of the fermentation tank.[15]

If you don't have any leftover mash, add two to three pounds of malt and a pound of dried powdered nutritional yeast to 50 gallons of dechlorinated water, to provide nutrients.

Check the pH (4.0–4.5), and bring this solution up to 104°F. Add the amount of dry yeast that the manufacturer specifies for the whole batch. Be careful to avoid clumping. Make sure you have enough headspace in the container to allow for foaming. You can refine the process of pitching by adding the molasses or sugar slowly, rather than all at once, to avoid the Crabtree effect. Keeping the sugar solution at about 1% or so during pitching means frequent additions of sterile sugar solution or molasses.

Hold the temperature at 90°F over the next six hours while you prepare your batch for fermentation. During those hours, yeast will multiply up to ten times in biomass during three to five generations of budding, providing you with a highly concentrated active yeast culture.

Fermentation Rate at Two Temperatures

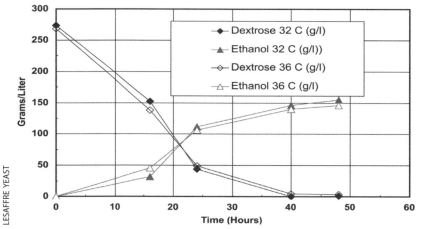

Fig. 7-27 Fermentation rate at two temperatures. *Rounding up, the two temperatures are 89° and 99°F. You can see the early reproductive stage as sugar concentration goes steadily down while alcohol concentration goes steadily up. Then the second phase kicks in, and the rate of fermentation speeds up. It then slows again as the fermentation finishes. With this yeast, you can see the advantage in yield by keeping the temperature at the lower optimum level. At the lower temperature, more alcohol is produced, and you can see that the higher temperature inhibits the yeast using the* **dextrose.**

Supply your yeast with lots of oxygen to encourage reproduction instead of alcohol production. The best way to supply yeast with oxygen is to slowly bubble (**sparge**) it in through a tube at the bottom of the tank. The smaller the pinholes in the sparging tube, the smaller the bubbles, the better. Oxygen sparging further prevents the Crabtree effect and stimulates the yeast to reproduce rapidly. Oxygen is available through welding supply houses. Older chemistry books can teach you how you can "make" oxygen by purifying it from the air if you are the tinkering sort. By the way, smoking around the oxygen-rich **yeast breeder** could result in a flash flame, which could make a real mess of your beard, eyelashes, or hairdo.

Don't shortcut this step, since if you do a good job propagating a vigorous batch of yeast, it can make a big difference in the yield of alcohol and resistance to contamination. The shorter fermentation can have logistical advantages, too, in getting batches ready for distillation earlier.

The more exotic yeasts *have* to be bred. For instance, *Kluyveromyces* yeast, used with Jerusalem artichokes or whey, is not commonly produced commercially, due to lack of demand. You'll have to breed these yeasts from a slide, to test tubes, to quart-sized flasks, and finally to your pitching solution. Your local agricultural university should be able to get you slide samples of unusual strains if you wish to experiment. Alternatively, you could join the **American Type Culture Collection** and order starter yeasts.

To make the yeast propagation less work, you can make twice as much pitching solution as you need and store half of it to start your next pitching solution. Simply refrigerate the excess yeast-rich pitch at close to 34°F between uses. Next time you need a starter batch, add an equal amount of sterile mash to the pitch, and slowly bring the temperature up to 100°F. Within two to five hours, you'll have a new double batch of yeast pitch. You can again store half of it for the next preparation.

To avoid bacterial contamination, though, you should make your pitch from scratch every fourth or fifth batch. It's difficult to keep all the bacteria out, and after a few batches the bacteria level can become too risky. Keeping several flasks of the original propagation from the slide in a refrigerator, to start from each time, will keep your strain fairly pure. But even these should be re-bred fresh once a month if you're using an unusual yeast. Standard yeasts should be pitched fresh each time.

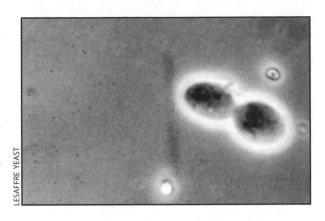

Fig. 7-28 Yeast reproduction. An almost complete budding of a yeast cell. Each yeast cell will reproduce itself each hour, until the oxygen in the mixture begins to be depleted.

A microscope is useful to determine if your yeast starters are contaminated (see Figure 7-24). If you want to save a contaminated yeast flask because of difficulty in replacing the strain, drop the pH to 2.0 with phosphoric acid for one to two hours, then bring the pH back up. This should kill almost all the bacteria and not weaken your yeast too much. As I mentioned, the pH should never be lower than 2.0 and never for more than two hours.

Contamination

Following completion of fermentation, don't let your mash sit for more than a day at the most. In its finished form, mash is an invitation to dangerous bacteria, which can turn your alcohol, and remaining proteins and sugars, if any, into lactic acid, **butyric acid**, or, more commonly, **acetic acid** (vinegar).

During fermentation, yeast protects your mash naturally. Its carbon dioxide exhalations are heavier than air, and the gas creates a thick CO_2 blanket over the mash, effectively sealing the tank against butyric or acetic acid bacteria.

Butyric acid bacteria (identified by long, ropey strands in the mash) love sugar but hate cleanliness—an ounce of prevention is the appropriate measure of defense.

In the course of fermentation, if your mash's pH suddenly drops to 2.0, you'll know that lactic

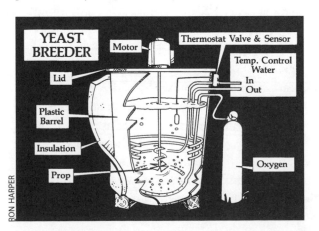

Fig. 7-29 Yeast breeder. A breeding chamber like this one provides everything yeast need for reproduction: good food, lots of fresh oxygen, gentle circulation, and tight temperature control. Not pictured is a fermentation lock.

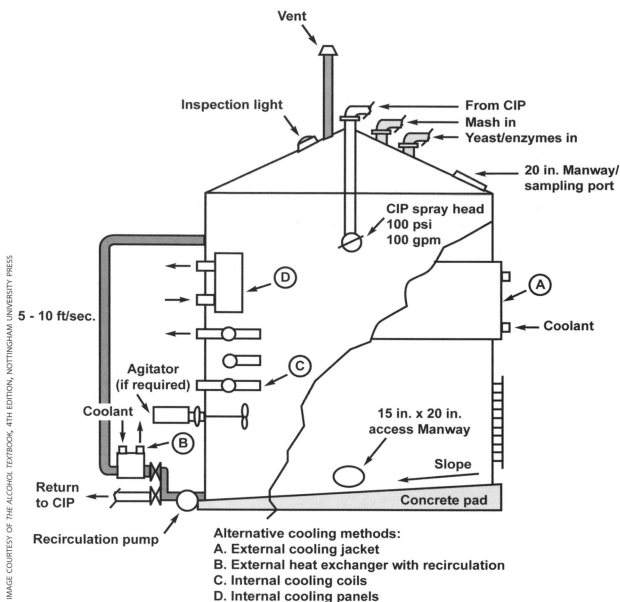

Fig. 7-30 Fermentation tank. This illustration shows all the different methods of temperature control and agitation. Heat exchangers internally are both coils or waffle panels. External are jackets, and separate heat exchanger (not shown but indicated). Both mechanical agitation and pumped recirculation are shown. The vent shown would lead to a fermentation lock. Note the cleaning-in-place "sprinkler head" for disinfection. In a real tank you would be unlikely to use all these features, just the ones that fit your operation.

Vent

Inspection light

From CIP
Mash in
Yeast/enzymes in

20 in. Manway/ sampling port

CIP spray head 100 psi 100 gpm

(D)

(A)

5 - 10 ft/sec.

Coolant

Agitator (if required)

(C)

15 in. x 20 in. access Manway

Coolant

(B)

Return to CIP

Recirculation pump

Slope

Concrete pad

Alternative cooling methods:
A. External cooling jacket
B. External heat exchanger with recirculation
C. Internal cooling coils
D. Internal cooling panels

acid bacteria have somehow crashed the yeast's party. If this happens, after cleaning thoroughly, in your next batch, use lactic acid instead of sulfuric acid to balance the pH. A little lactic acid bacteria contamination *late* in the fermentation actually lowers the pH a little more, which helps protect against other bacteria—but early contamination uses up valuable sugar.

Acetic acid bacteria will turn your mash to vinegar. As of this writing, I don't know any way to make your car run on vinegar.

The best way to avoid contamination is to keep all tanks and equipment bacteria-free, but this is impossible. The best you can hope for is to keep bacteria at a minimum. Clinical sterilization isn't necessary, but cleanliness is. After each **run**, all solids should be washed from the fermentation tank, especially from any nooks and crannies, with a

pressure washer. The inside of the fermentation lock and areas around temperature probes are important—but often overlooked—areas to clean. If you've used cooling coils, or cooling panels, they should be washed, too. After every second or third run, wash your tank down with a disinfectant—chlorine solutions are relatively safe to handle and effective. Remember to rinse well after disinfecting, or neutralize any remaining chlorinated water with hydrogen peroxide, since chlorine is toxic to yeast as well as bacteria. Of course, ethanol itself is a top-notch disinfectant!

As mentioned above, CIP using caustic soda is pretty foolproof. Larger plants can use a CIP system. Basically, these cleaning-in-place systems consist of a separate tank with a strong solution of sodium hydroxide (lye), which is sprayed or circulated through the tank to be cleaned, to kill everything. The solution recirculates back into its storage tank for later use, through a filter to capture particles. The pH of the CIP solution is checked from time to time to see if more sodium hydroxide is needed. CIP also works to clean out external heat exchangers.

If you use backslop instead of water in making your wort, use it immediately after distillation, when the mash is virtually sterile. Don't let backslop sit overnight to cool, because it will be seriously infested with bacteria by morning.

There are certain **thermophilic** (resistant to high temperatures) bacteria which can survive the distillation process, but unless they've got a big head start, they'll be killed off by a healthy yeast inoculation.

If all this seems like a lot of hassle, it really sounds worse than it is. The minor inconvenience of preventing bacterial infestation of your batch doesn't compare at all to how it feels to dispose of 500 gallons of smelly, soured mash.

CALCULATING ALCOHOL YIELD

Knowing your potential alcohol yield tells you if a feedstock is worth transporting to your plant, helps you narrow down problems in fermentation when actual yields don't measure up to projected ones, and helps determine enzyme dosages for starchy materials. Once you know the carbohydrate content and composition, you'll be able to calculate your potential yield.

There's a shortcut for making your calculations when perfect accuracy isn't important. For approximate yields of gallons of alcohol per ton, for either starch or sugar, multiply the percentage of fermentables by 151.5. That figure is usually close enough.

But let's say your creative instincts provide you with a feedstock that you figure no one else is using. How do you know it'll work? Well, you try it, of course. But don't make 500 gallons; make a three- to five-gallon test batch. And before you even attempt the test batch, there are steps you can take to figure out if it's really worth trying.

Is your carbohydrate sugar, starch, or both? The simple way to really know is to send it off to be analyzed. Laboratories with experience will give you an analysis of the carbohydrates in your feedstock; not every lab is able to do this.

There are less formal ways to get an approximate fix on carbs. You should be able to determine levels by referring to a booklet assembled by the USDA, *The Nutritive Value of Foods*,[17] or

Laurel's Kitchen,[18] which give a breakdown of many materials you'll run into. You can also look up the standard nutrition label for the food product, which should give you a rough idea of the amount of carbohydrates in the material but not what they are. Sometimes the label will tell you what percentage of the carbs is sugar.

If your particular feedstock isn't listed, and you suspect that there is some starch, start off testing with iodine. Iodine will dye most starch compounds purple and has no effect on sugar. Try a drop or two in a few different locations on your feedstock. With bananas, for instance, a drop on the surface of the fruit will yield little reaction, but the cross-section indicates some starch in the fruit's core. Unripe fruit will sometimes store carbs as starch until it ripens and converts them to sugars (e.g., apples).

Sugars

If your material is known to be primarily composed of simple sugar with no appreciable starch, initial testing is easy. If it's liquid or very juicy, test some filtered juice with a saccharometer for a reading of dissolved solids—most of these will be sugar.

To determine how much of the dissolved solids is sugar, ferment your test batch completely, distill off the alcohol, and add back to the batch an amount of water equal to the amount that left with the alcohol during distillation. Take another

---

ETHANOL, THE BYPRODUCT OF SCRAPING THE BOTTOM OF THE OIL BARREL

Oklahoma's first ethanol plant is making alcohol as a byproduct of getting oil out of the ground. Unlike in Saudi Arabia, you can't cheaply pump saltwater into the ground to displace the oil in Oklahoma. You can't pump air into the well to pressurize it, either, since mixing oxygen with ever-present natural gas is a good way to cause an explosion and make an oil well into a big, big crater.

So Chaparral Energy, an oil and gas company, co-ventured with the Oklahoma Farmer's Union to build an ethanol plant in Enid. The eight million cubic feet of carbon dioxide produced per day by the plant will be pumped into old wells to pressurize them safely with the nonflammable gas. The pressurized oil-bearing strata then help drive oil to the surface through the oil wells.[16] This type of oil recovery is normally conducted by burning natural gas to make carbon dioxide to pump back into the well. That not only contributes to global warming, it's a waste of natural gas.

sugar reading. If your fermentation is complete, the remaining solids are unfermentable sugars, salts, and other dissolved materials. The difference between the initial reading and the final reading gives you a reasonably accurate correction figure for testing the sugar percentage.

A fairly dry feedstock has to be mixed with water, and possibly cooked, for an analyzable solution. Add water in measured proportion to the volume of dry feedstock to correctly be able to account for the additional liquid. For instance, if you add two parts water to one part feedstock, the sugar concentration in solution will only be one-third of the actual sugar concentration in the feedstock.

"This hour in history needs a dedicated circle of transformed nonconformists. Our planet teeters on the brink of atomic annihilation; dangerous passions of pride, hatred, and selfishness are enthroned in our lives; and men do reverence before false gods of nationalism and materialism. The saving of our world from pending doom will come, not through the complacent adjustment of the conforming majority, but through the creative maladjustment of a nonconforming minority."

—DR. MARTIN LUTHER KING, JR.

Once you know the sugar percentage, you can calculate your yield. Let's take a feedstock like plums or hot chili peppers, with an 18% sugar reading. The plum is probably about 75% water. To find alcohol yield per ton, multiply 2000 pounds (one ton) by 0.18. You get 360 pounds of sugar. Yeast uses a touch over 50% of that for alcohol conversion, so multiply 360 by 0.50 to determine the *pounds of alcohol* that yeast will produce. That's 180 pounds. Since most of us are used to thinking in terms of gallons rather than pounds of fuel, we'll convert pounds per ton of feedstock to gallons per ton. Alcohol weighs 6.6 pounds per gallon. So, 180 pounds of alcohol divided by 6.6 is about 27.3 gallons per ton.

Starches

Calculating yield works a little differently with starchy materials. Starch is stored in plants chemically in a much drier form than sugar, usually resulting in a higher percentage of carbohydrates. Since starchy plants are often prized for their food value, their composition is usually recorded in books like *The Nutritive Value of Foods*.

Corn, depending on the variety, contains around 60–75% starch. A pound of starch produces more

than a pound of sugar. When starch is hydrolyzed, it chemically picks up molecular weight from the water you're using and ends up producing more sugar. The final alcohol concentration after fermenting a pound of starch is theoretically 0.568 pounds of alcohol.

To figure the amount of corn starch in a ton of grain, multiply the starch percentage—let's say 70%—times 2000 pounds, in this case equaling 1400 pounds of starch. Multiply the weight of the starch by 0.568 to get the pounds of alcohol. That comes out to 795.2 pounds of alcohol, which divided by 6.6 pounds per gallon gives us a yield of 120 gallons per ton. In practical terms, you should figure that you'll get 90% of the theoretical 120 gallons, so you'll probably end up with 108 gallons.

ADVANCED COOKING TECHNIQUES

Instead of simply adding all your feedstock and then cranking up the heat, there are some refinements you can employ to make your starch cooking process proceed more smoothly.

Semi-Continuous Cooking

The usual approach for corn or starch recipes is to add all the feedstock to your starting water, and begin raising the temperature while agitating and adding your enzyme. At about 150°F, corn starts to gelatinize, and the agitator is pushed to its limits to keep the thickening mash from becoming unmanageable, until the temperature gets a bit higher and the bacterial alpha amylase begins to cause the mash to thin. The extra energy your agitator will need to perform this tough job cuts into the energy you save by using less water in cooking.

While the mash is thickening, if alpha amylase isn't well distributed to speed the process of thinning the mash, then you're likely to burn the corn. There's a fairly sophisticated way to avoid burning and lumping in a thick mash—a sort of semi-continuous cooking.

Use as little starting water or backslop as possible—ten gallons per bushel of corn or even a little less should do—and a quarter or less of the corn ration you're processing. Bring the temperature up to about 165–175°F (you'll already have added your premalting enzyme with some preliminary pH and calcium adjustment). At that temperature, such a small amount of corn thins pretty well in a few minutes. Then start adding the rest of the feedstock, pausing a minute or two in between additions so that each batch has a chance to thin. Adding feedstock can even be automated with an auger controlled to turn on and off intermittently. Temperature rise is slower this way, but you'll be able to add more and more feedstock as the temperature increases and the enzymes become more active.

I don't recommend this process for beginners. It takes time to learn at what rate to add whatever pH-adjusting chemicals you're using to the growing increments of feedstock. Once you get the recipe worked out, repeating it should be relatively easy—you'll have determined a ratio of ounces of lime or acid to so many pounds of feedstock.

This technique requires a few more minutes before you can start cooking, but it should speed up subsequent steps since you'll need less water and less process energy; your final alcohol yield will be higher, too. It's a worthwhile technique once you feel confident you've mastered the basic process.

Combining Fungal and Bacterial Enzymes

An easier alternate technique to prevent gelatinization, or one to use in combination with semi-continuous cooking, is combining the use of fungal and bacterial alpha amylase. Depending on your feedstock, you can reduce costs, speed up the beginning of fermentation, and do less pH adjusting in your typical or semi-continuous starch recipe by using a combination of fungal and bacterial alpha amylase.

Fungal enzymes operate at a lower pH and at a much lower temperature than bacterial alpha amylase. While lower temperature tolerance makes fungals inferior enzymes for liquefaction after cooking, they work better in premalting, which occurs initially at lower temperatures. Fungal alpha amylase starts being effective at 125°F, quite a bit lower than bacterial alpha amylase.

To make your mash palatable for *bacterial* enzymes, you brought the pH up to around 6.0. *Fungal* alpha amylase will operate well at pH 5.1.

Being substantially cheaper than the bacterial enzymes, fungal enzymes allow you the choice of using more of them in premalting than usual for the same final alcohol production cost; and they cut down premalting time, as well. Bacterial enzymes are still used for liquefaction, and pH will require less adjustment after a strong premalting step.

When fungal alpha amylase starts working on starch, it breaks some of it down to maltose rather than all to dextrins, and yeast metabolizes maltose directly. Providing more simple sugar right at the beginning of fermentation allows yeast to begin reproducing much faster than they normally would. Maltose-fed yeast compete better against invading bacteria by reproducing faster and smothering them. If there is no contamination, fungal/bacterial or all-bacterial enzyme batches finish at the same time with the same yield, since the fermentation curves of the two enzymes catch up with each other within a few hours.

Barley Malt as a Source of Enzymes

There are advantages and disadvantages in using malt instead of bacterial enzymes. Barley's biggest advantage is that you can grow it, or at least sprout it, yourself in practically unlimited quantities. For the self-sufficiency-minded farmer or homesteader, barley is the only way to self-produce necessary enzymes for making alcohol. Of course, barley contributes to the alcohol yield of your starch process. Since barley, like most grains, is largely starch, the alcohol produced from it will somewhat offset the higher apparent cost of barley over bacterial enzymes.

The disadvantages are that it takes time and effort to produce barley malt, it can be quite expensive if you don't grow it yourself, and it doesn't do quite as good a job as glucoamylase in the conversion step. It also varies in quality. If you aren't getting a good

*Fig. 7-31 Semi-continuous starch cooking. You can clearly see that **viscosity** rises each time more feedstock is added at temperatures below the enzymes' operating temperature. Spreading out the peak of viscosity makes the work of the mixer/agitator easier and uses less horsepower.*

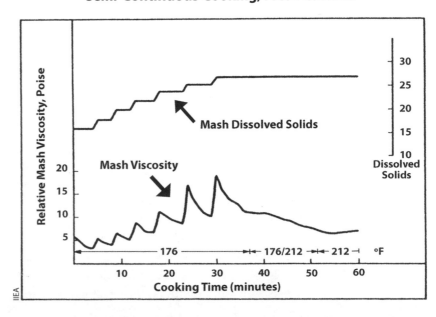

Semi-Continuous Cooking, Fresh Cassava

Net Cooking Energy Consumed

PROCESS/RAW MATERIAL	MASH DS %	COOKER FEED TEMPERATURE, °F (RAW MATERIAL/ WATER)	MAXIMUM COOKING TEMPERATURE, °F	NET ENERGY CONSUMPTION MJ/LTR ALCOHOL
Batch/Dry Sago	20	86/86	266	3.8
	20	86/86	185	2.2
"Semi-Continuous"/Fresh Cassava	25	86/176	203	1.1
Continuous/Cassava Flour	23	86/194	203	0.52
Continuous/Ground Corn	30/35	86/194	221	0.77/0.71

Fig. 7-32 Net cooking energy consumed (in megajoules per liter). This shows how important preheating the feedstock/water mix is in cooking. Semi-continuous cooking, a batch method, results in a higher amount of dissolved solids with a much-reduced, but slightly higher than continuous cooking, energy input. Batch energy level can be reduced to nearly continuous levels by good energy recovery and storage. Although best energy efficiency is achieved using low-pressure steam over 212°F, small plants will generally not save enough money to warrant using a boiler.

breakdown of starch to sugar, you may find it necessary to increase the malt ratio from 10% to 15%.

In Nature, most seeds that lie through winter before germination store the solar energy they will need for sprouting in the form of starch. To recover that energy, Nature uses enzymes just like we do, to break down starch into usable sugar for the seed to begin its life.

Imagine you're a wild barley grain in the soil (no, not a wild oat you've sown), spending the winter underground, trying to stay comfortable. Suddenly, spring is sprung. The sun comes out and starts warming the soil, which sets off some of your chemical triggers. Then it starts to rain. After you've absorbed enough water and soaked up enough energy in the form of solar heat, you start releasing alpha and beta amylase, and the conversion of starch to sugar begins—which allows you to start growing. (This explains why sprouts are so sweet, when the grain itself is bland.) Since you've been lazing around all winter, you've got a lot of catching up to do. You release ten times as many enzymes as you need to get yourself growing. Our process employs those excess barley enzymes to break down starch to sugar in mash, the same breakdown as in the seed for growth.

Making Barley Malt

There are two ways to sprout barley for malt, depending on how much you need. If you only need 70–80 pounds, use a 50-gallon plastic garbage can as a sprouter, with holes punched in the bottom and the lower third of the sides. Line the can with a garbage bag and fill it a quarter full of barley. Then fill the can with water and let it soak for several hours. Barley will absorb as much as four times its own weight in water. Pull up on the bag, ripping it, and

let the water drain completely out through the holes in the can. Rinse the sprouts three times a day.

If you intend to make a larger quantity of malt (for instance, to dry it and store it for use for some weeks), pile your barley on a concrete slab that can drain. Allow the barley to stand in the shade at room temperature (65–85°F). Continue to water it at least twice a day, and once or twice a day turn the grain over to keep it aerated to keep anaerobic bacteria from getting a foothold and rotting the barley. Always handle the grain carefully—use a pitchfork, not a shovel. After three to five days, your barley will have produced a sprout up to 1/2 inch long. You can grind it to use now, or dry it first and then grind and store it. Some of my former students recommend a small garbage disposal for grinding the sprouts.

Barley malt contains both alpha and beta amylase enzymes. To store barley malt, you have to dry it at a temperature of up to 120°F—but no higher than that, or you'll lose some of malt's beta amylase. Circulating warm air (like the waste heat recovered from your flue) is effective for drying. The waste hot water that we've been collecting can also be effectively used for drying the malt. Using a large version of a food dryer is a simple way to get the job done. To make sure that your hot air is just below 120°F, pull the air through a vehicle radiator supplied with the hot water flow that is controlled by a thermostatic valve. Just put the sensor end of the valve in the airflow entering the dryer, and it will keep the radiator just the right temperature for drying by opening and closing the hot water valve as needed.

Effective drying is really more a function of good steady airflow rather than heat, so using a high-volume fan will make drying fast and efficient.

Using Your Barley Malt

Ten percent of the weight of a given batch of grain is as much barley malt as you'll need: for 1000 pounds of corn, use 100 pounds of malt. The weight percentages listed in the following procedure are the same for dried malt or fresh malt! Fresh or green barley malt is more potent, with more enzymes in it than dry barley, but there are more enzymes in dry barley per given weight.

Fig. 7-33 Sprouting barley. *These sprouts are ready to be dried and ground for malt.*

Premalt and liquefaction steps in fermentation are slightly different if you use malt instead of bacterial enzymes. For a bacterial enzyme, the pH is 5.5–7.0 for both steps. When you use barley malt, the pH should be 5.6–5.8. Malt's alpha amylase is less temperature-stable than bacterial enzymes.

In the premalting stage, add about 10% of the malt called for to your slurry. The premalting stage should include a period during which the batch is held at 160–165°F for 15 minutes, which thins the mash, and then brought to a boil. When the wort has finished cooking and cooled to 165°F (instead of the normal 190°F for bacterial alpha amylase), and you're ready to go to liquefaction, substitute 40% of the total barley malt. We've now used half of the malt. The remaining 50% is used in the conversion step, replacing glucoamylase with barley malt's beta amylase. The pH and temperature in conversion are the same as they were for bacterial glucoamylase.

It's important to note that although malt always contains both alpha and beta amylase enzymes, only one or the other is actually operating at any one time. In the first two steps, the temperature is far too high for beta to survive, and in the last step the pH is too low for alpha to survive. Alpha amylase only works on starch, and beta amylase only works on dextrins.

ACIDS AND BASES—HANDLE WITH CARE

Probably the most dangerous part of making alcohol fuel is handling acids and bases. We use both when correcting pH for enzyme and cooking purposes, as well as when neutralizing mash at the end of fermentation, before distillation. Although not as instantaneous as burning your hand on a hot part of the still, injuries from acid or alkali can be more serious—permanent and even deadly.

You need to pay close attention to what you are doing with these substances. There are distinct procedures for this, which we discuss further below. Properly handled, you will never have a problem using them.

The three acids used for adjusting pH are sulfuric, phosphoric, and lactic acids. Sulfuric acid is the one most commonly used; its chief advantage is that it's the cheapest of the three, and it's available everywhere. Unfortunately, it reacts the most violently of all when added to hot water. Phosphoric acid is more expensive but less violent in its reaction with water, and it even gives the yeast a little nutrition and later makes a fertilizer if mash is composted. Lactic acid is rather expensive and is used only as a preventative, if lactic acid bacteria destroyed your last batch. Avoid nitric or hydrochloric (muriatic) acids, both of which can either inhibit or kill yeast.

Fig. 7-34 Small farm-scale dehydrator. *A dehydrator of this size is adequate for micro- and small-scale plants in making barley malt or even drying mash for feed. Trays with whatever is being dried would be inserted in the slots and blown dry by the fan on the right.*[19] *The building plans are public domain.*

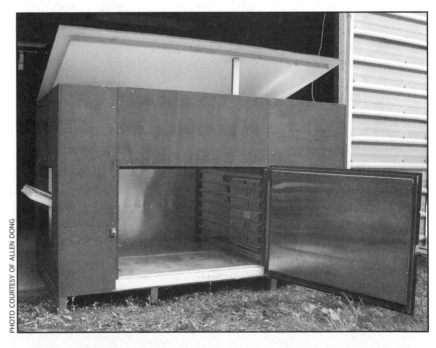

PHOTO COURTESY OF ALLEN DONG

EMERGENCIES WITH ACID

If you should spill some concentrated acid on your skin, flood the area with large quantities of water (not baking soda) for at least 15 minutes.

Wherever acid and bases are present, an emergency first-aid eyewash should be readily available. Eyewashes contain a **buffered** neutralizing solution that acts more quickly than plain water. If eye injury occurs, flood the eye with eyewash or 15 minutes of plain water, then immediately get medical attention. When flooding an eye with water or wash, always flood from the bridge of the nose to the outside of the face. If you flood the other way, you may spread acid into the other eye.

Baking soda is used to neutralize spills, not to treat acid splashed on you.

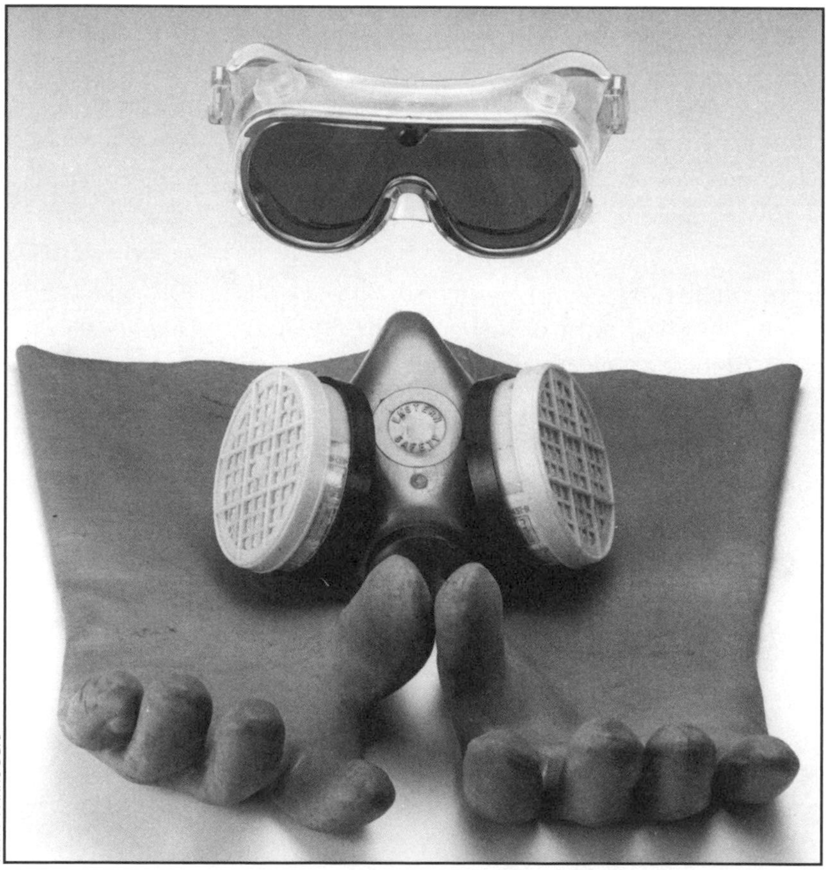

Acid can be purchased in a wide variety of grades and prices, but larger amounts need to be bought from chemical supply stores. When buying acid, you are generally looking for "Technical Grade." You should check the impurities analysis to see if there are significant quantities of heavy metals (over 0.1%), especially in sulfuric acid. Some of these acids are collected as byproducts from industrial processes and repackaged for you to buy. But heavy metals aren't good for cattle, your soil, or yeast, even in the minute quantities we're adding to 500 gallons of mash. So if any feed or food production use is to be made of the byproducts, then you should use the more expensive "Food Grade." "Reagent Grade" is expensive and unnecessary.

Acids should be respected. Handled carefully, they're safe enough, but handled sloppily, they're extremely hazardous. Make sure that you have multiple boxes of baking soda (not baking powder) on hand to neutralize any spills on the floor, counters, or equipment.

For your own personal protection, when transferring acid to a storage tank or adding it manually to your batch, always wear elbow-length rubber gloves, a long-sleeved shirt, long pants, boots,

Fig. 7-35 Minimum safety equipment. *Acid-proof gloves, indirectly vented goggles (not numerous small ventilation holes passing directly through the plastic), and a good filter mask are all necessary to the smallest of fuel plants. Long pants, with boots and long-sleeved shirts, should be worn at all times in an alcohol plant.*

MATT FARRUGGIO

Fig. 7-36

RON HARPER

tight-fitting chemical splash goggles with indirect ventilation, and a rubber apron. Some people like to use a full-face shield in addition to goggles. You can't be too careful. Do not take shortcuts with safety gear—your continued eyesight and naturally charming mug are especially at risk.

Only transfer acid by using a special acid pump or gravity flow. Never pressurize an acid container with compressed gas or air. Always add acid to water; *never add water to acid.* Either way, mixing will cause some bubbling and splattering—but when adding water to acid, the reaction is potentially explosive. Avoid adding concentrated acid to liquids above 150°F; the reaction is greatly accelerated by heat. Always add acid to water slowly, with a slow, steady agitation. Adding acid quickly builds up a lot of heat at the entry point, which aggravates the reaction.

Lime, the chemical base (alkali) we use to raise pH, is much safer to handle than acid, but a few precautions should be noted: Lime is available as hot lime (calcium oxide) or as slaked (hydrated) lime, sometimes called agricultural lime (calcium hydroxide). Hot lime in its crystalline form is an irritant to eyes and mucous membranes and will cause skin rash and burns if left in contact with the skin for very long. These problems are far less of an issue with slaked lime. When handling lime, you should use protective goggles as described above, rubber gloves, a chemical mask, and a long-sleeved shirt to protect your arms.

Before you add lime to your main batch, it should be wetted in a pail and dissolved. A rubber pail (the kind used in horse barns) lasts a lot longer than a metal one for this purpose. You can add lime without wetting if you add it slowly, while agitating to prevent lumping.

Now that you've learned the basic processes for fermenting the classes of material commonly used to make ethanol, you are better equipped to evaluate which feedstocks you will want to use in your own plant. You now know that fermentation is not rocket science—it's ecology. Just as with most agriculture, it is our job to provide the right conditions to encourage Nature to do all the hard work.

As a farmer, I became acutely aware that my job was "simply" to take care of the soil so that everything my seeds or transplants needed was available. The plants really grew themselves without much direct help from me. The same is true

Fig. 7-37

for fermentation—we provide the right conditions for our yeast, and they do all the heavy lifting in making sugar into alcohol.

Now, when it comes to distillation, that's where our intelligence as humans comes into play. But, before we go there, it's time to consider the abstract concepts of fermentation in the light of the crops and materials we will actually be using to make fuel. As you will see, feedstock choices are far more complex and interesting than the corn and cellulose the media promote.

Fig. 7-38

Endnotes

1. K.D. Kirby and C.J. Mardon, "Solid Phase Fermentation for Intermediate Scale Ethanol Production," *Proceedings of the IV International Symposium on Alcohol Fuels Technology, Guarujá, Brazil,* 1 (October 1980), 13–19.

2. Kirby and Mardon.

3. Kirby and Mardon.

4. Kirby and Mardon.

5. Easter Trabulse, Enologist, communication with author, June 7, 2005.

6. Trabulse.

7. Arden B. Anderson and Arden A. Anderson, *Science in Agriculture* (Acres USA, 1992).

8. Trabulse.

9. Trabulse.

10. Ronan F. Power, "Enzymatic Conversion of Starch to Fermentable Sugars," in K.A. Jacques, et al., eds. *The Alcohol Textbook: A Reference for the Beverage, Fuel and Industrial Alcohol Industries,* 4th ed. (Nottingham: Nottingham University Press, 2003), 29.

11. Inge Russell, "Understanding Yeast Fundamentals," in K.A. Jacques, et al., eds. *The Alcohol Textbook: A Reference for the Beverage, Fuel and Industrial Alcohol Industries,* 4th ed. (Nottingham: Nottingham University Press, 2003), 114.

12. S.L. Chen, "Optimization of Batch Alcoholic Fermentation of Glucose Syrup Substrate," *Biotechnology and Bioengineering* XXIII (1981), 1827–1836.

13. Chen.

14. Trabulse.

15. Russell.

16. *The News On 6,* KOTV, Tulsa, Oklahoma, August 29, 2005.

17. Susan E. Gebhardt and Robin G. Thomas, *Nutritive Value of Foods,* Home and Garden Bulletin no. 72 (Beltsville, MD: USDA Agricultural Research Service, Nutrient Data Laboratory, 2002).

18. Laurel Robertson, Carol Flinders, and Brian Ruppenthal, *The New Laurel's Kitchen: A Handbook for Vegetarian Cookery and Nutrition* (Berkeley, CA: Ten Speed Press, 1986).

19. Allen Dong, Farm-Scale Food Dehydrator, http://agronomy.ucdavis.edu/LTRAS/itech/dehydrator.htm (November 10, 2006).

CHAPTER 8

INFORMATION ON VARIOUS FEEDSTOCKS

The debate about alcohol sources is generally narrowed to a choice between sugarcane in the tropics and corn in temperate areas. Both crops require good-quality agricultural soils, which are a small fraction of the dry land, or for that matter agricultural land, on our planet.

Many of the potential energy crops in this chapter can thrive in arid or swampy lands, artificial marshes used for sewage treatment, or brackish water that interfaces with land; there are even crops that require no soil at all. In many countries, these areas cut a much wider swath than those used for corn or sugarcane and are not usually considered cropland or farmland.

Other crops described in this chapter are already being grown in smaller acreages than corn, but due to their much higher yields from alcohol and alcohol co-products, they could find themselves in rotation with corn and soybeans rather quickly—with modest backing from government in the form of financing for crop-specific equipment and expertise.

Some crops considered weeds or useless already cover vast areas. They represent a current resource that, by my calculations, can be managed sustainably to produce billions of gallons of ethanol and co-products per year right now.

Some would say that it takes too much effort to get farmers to change what they grow. To that I would respond that 60 years ago, once soybeans became backed by the USDA, they replaced half the acreage of corn within a few years. On a more ominous note, over 65% of corn today is genetically modified, a massive shift in a very few years. So if the need and the profit are there, farmers (and their bankers) can and do shift their crop choices.

As for cellulose—now being hyped as ethanol's most promising feedstock—most of the crops mentioned in this chapter are not specifically sources, although some of them also contain it, and I mention potential yield where possible. A

MANY OF THE POTENTIAL ENERGY CROPS IN THIS CHAPTER CAN THRIVE IN ARID OR SWAMPY LANDS, ARTIFICIAL MARSHES USED FOR SEWAGE TREATMENT, OR BRACKISH WATER THAT INTERFACES WITH LAND; THERE ARE EVEN CROPS THAT REQUIRE NO SOIL AT ALL. SOME CROPS CONSIDERED WEEDS OR USELESS ALREADY COVER VAST AREAS. THEY REPRESENT A CURRENT RESOURCE THAT, BY MY CALCULATIONS, CAN BE MANAGED SUSTAINABLY TO PRODUCE BILLIONS OF GALLONS OF ETHANOL AND CO-PRODUCTS PER YEAR RIGHT NOW.

PHOTO COURTESY OF ERIK VAN BOCKSTAELE

Fig. 8-1 Turbo Klein variety fodder beets. Modern varieties of fodder beets like these may become a cornerstone of temperate-climate alcohol fuel production. They rival the production from sugarcane but do so in a fraction of the time, yielding valuable byproducts.

treatment of cellulose strategy would be a small book in itself.

Just so you know, the crops I profile here are a good cross-section of potential energy crops, but this list is by no means exhaustive, especially with respect to cellulose sources.

This chapter also touches upon some kinds of waste materials that could form the basis of viable small-scale businesses. Although I describe only a few, my hope is that I stimulate your thinking of waste as simply a surplus material for which a use has not yet been designed.

Fig. 8-2 Young buffalo gourd plant. Planted from seed, the buffalo gourd rapidly sends a deep root into the soil to drought-proof it against infrequent rains.

BUFFALO GOURD

Buffalo gourd, *Cucurbita foetidissima* (also called Missouri gourd, chili coyote, and the fetid gourd, for its vine's pungent odor), is a fairly new fuel feedstock, although it has been around for over 9000 years. The plant migrated with the movements of indigenous peoples in North America. Buffalo gourd has a history of use as a medicine and a detergent. Americans in the Southwest still use the mature fruit pulp and root for cleaning and scouring. Except for the seeds, the plant is quite bitter and must be processed to be a food product.

Test plots of buffalo gourd operated by the University of Arizona projected yields of 12,024 pounds per acre of starch. That would put the alcohol yield in the vicinity of 900 gallons per acre.

The plant's taproot can achieve a record 200 pounds after four years of growth. More commonly, the roots are about 100 pounds; they are 18 to 20% starch on a wet-weight basis, but up to 52% on a dry-weight basis. Due to the root's great mass, buffalo gourd can survive winter air temperatures to 25°F below zero, especially if the ground is covered by an insulative layer of snow. The pungent vines are killed off as soon as the temperature goes below freezing.

Buffalo gourd reproduces by seed, but more commonly by sprouting **adventitious** roots from its vines, like the roots that sprout from strawberry vines, which attach themselves every few feet to the ground. Typically, one plant will set nine new rootings in a season, so 500 plants per acre become 5000, or roughly one per square meter. Once established, the vine between the original plant and the new rooting can be cut without harm to either.

The vines are prolific, laying a thick carpet of vegetation that helps conserve moisture in arid lands. A single root may send out from six to 20 vines. Each vine grows to an impressive 20 feet under good conditions. In five months, a single plant can put out over 700 feet of vine. For gathering solar energy, a single plant may have 15,000 sandpapery leaves, about five inches wide and almost a foot long.

In addition to a huge starchy root and massive foliage, each plant produces around 200 three- to four-inch-diameter gourds in a single season. Each gourd contains 200 to 300 seeds, a nutritional wonder for a **xerophytic** (low water use) plant.[1] The seeds yield an oil comparable to safflower oil in flavor, and comparable to corn oil or soybean oil in fatty acids. The seeds' edible oil content is 30 to 40%, and the protein level is a whopping 30 to 35%. As an animal feed, buffalo gourd's seed meal contains three times the protein of cottonseed, and five times as much as sunflower seeds. The seed meal is very close in nutritional value to standard **soybean meal (SBM)**, which is astounding since soybeans are legumes.

Seeds account for up to 712 pounds of oil potential per acre—enough energy to provide for 149% of the process energy at 40,000 Btu per gallon of alcohol produced. Alternatively, the oil could be processed into more than 110 gallons of biodiesel, far more than soybeans at 48 gallons per acre, while growing in lands that would never support a thirsty crop.

Incorporating some of the buffalo gourd seeds in the mash can benefit the alcohol fermentation process, since they are very high in fatty acids, especially **linoleic acid**. High amounts of fatty acids enable yeast to tolerate higher percentages of alcohol, and speed alcohol production.

The potential for increasing the yield of both alcohol and vegetable oil from buffalo gourd is excellent with selective breeding. Great genetic range has been observed in the field in oil content, number of fruits, root size, and other key characteristics, which generally indicates a great potential for crop improvement.

The buffalo gourd is not thought of as a cultivated plant, although it grows in the arid American Southwest and has been spotted as far north as Illinois and South Dakota, and as far south as Central Mexico. Its western limits are the coastal mountain ranges of California. Curiously, the plant seems to have migrated around the country following railroad lines, picked up, perhaps, as a souvenir because of its unusual beauty.

When cultivating buffalo gourd, consider the crop's perennial growth cycle, its asexual reproduction, and its three distinct products (gourds, roots, and foliage). The crop is direct-seeded in hills, spaced eight to 16 inches apart in rows about six feet apart. The first year, gourds are harvested to recover the seed, roots are left in the ground, and vines stay in the field; dead and dried vines can be returned to the soil or be grazed by sheep or cattle. The harvesting of the dried vines as "hay" or their removal by grazing is beneficial to perennial growth, stimulating early growth of all the new rooted sections as "new" individual plants the following season. As winter approaches with freezing or near-freezing weather, vines will begin to die off. As long as temperatures stay below 36°F, the roots remain dormant. The gourds can be harvested well after the vines are dead.

In the second year of cropping, the stand will be up to ten times as dense as the first planting because of all the additional rootings. Foliage and gourds are again harvested, and now root harvesting begins. During the dormant season, alternating 40-inch-wide strips have their roots harvested. The digging depth of the root harvester should be about 16 inches. The starchy roots are used for alcohol production. The viny stalks of the remaining plants are trained across the harvested patch to send down new roots.

The next year, each alternating 40-inch strip that wasn't brought in previously is harvested, and the vines that replenished the previous year's strip are trailed back across the newly harvested area for re-rooting. As long as you're getting good yields, keep alternating without planting new seedlings.

Although you will get some pollination from common honeybees, you won't get maximum yield of seed if that's the only bee in town. Normal honeybees find other plants more attractive.

Buffalo gourd has evolved alongside specialized pollinators. Only the solitary squash bee *Peponapis* and the gourd bee *Xenoglossa* prefer buffalo gourd to posies and petunias and so forth. Another caution: Your cropland must be well drained. Soil saturated for more than 24 to 36 hours can cause a physiological collapse in buffalo gourd.

Your cropland must be well drained. Soil saturated for more than 24 to 36 hours can cause a physiological collapse in buffalo gourd.

For alcohol, buffalo gourd root is processed like a cassava root (see next section), but buffalo's starch granules are much smaller than either cassava's or corn's. Its starch gelatinizes sooner and at a slightly lower temperature, and that can be a sticky problem when bringing the batch up to temperature. The semi-continuous premalting used with cassava or corn is also necessary in preparing buffalo gourd for fermentation (see Chapter 7).

The feed value of buffalo's seeds and foliage is excellent for **ruminant** animals, but the protein found in the seed coat is bound up within a **lignocellulose** structure, and is not digestible by humans and other nonruminants. Luckily, most of the seed's protein is concentrated in its embryo, which can be digested by nonruminants.

Fig. 8-3 Buffalo gourd mature fruits. *Buffalo gourds are prolific in arid climates.*

Fig. 8-4 Buffalo gourd canopy. *A dense foliage canopy, typical of a managed field, is an efficient collector of solar energy. It also dramatically cools soil temperatures and conserves water.*

UNIVERSITY OF ARIZONA DEPARTMENT OF PLANT SCIENCES

The leaves of the buffalo gourd plant, with a dry-weight protein content of about 13%, could provide a major protein source for animals. Harvest the foliage prior to frost. The leaves are almost 60% digestible, but have a bitter flavor from the chemical cucurbitacin—they're repulsive to humans, but cattle don't seem to mind them after they're dried (probably because drying oxidizes the bitter chemical).

Seed protein and fats, mixed with root fermentation byproducts and yeast's own proteins, help balance the seeds' amino acids, making an excellent feed blend.

CASSAVA

Fig. 8-5 Mature cassava root.

This mature root is 18 months old. The tall plant in the background is Calotropis, *also under study as a hydrocarbon source.*

You probably know cassava root as tapioca; it's also called manioc. Brazil grows many millions of tons annually, the largest harvest in the world. Cultivation has been dramatically increased in both Asia (Indonesia) and Africa (Ghana). In addition to its use as a starch, cassava has been employed in the textile industry, in the explosives industry, for dyes, linoleum manufacture, and as a dextrose glue. During the early 1950s, cassava accounted for up to 86% of the United States' starch imports.

Many developing countries are now being encouraged to grow cassava for export to industrialized countries for use as cattle feed. It is shipped as dry chips, and is considered a low-quality feed due to its low protein content. It's an extremely hardy plant even in bad soil (e.g., high acidity), and it requires low amounts of mineral fertilizers and, better still, low capital investment.

Cassava test yields at the University of Arizona offer encouraging results for alcohol production:

Roots at 32% starch yielded from 1637 to 2015 gallons of alcohol per acre, based on test plots over an 18-month growing season.[2] Optimal water requirements have not been fully analyzed, but these yields were realized with irrigations of three inches of water every two weeks, for a total of eight to ten times a year. It's likely that less water will do the job in more humid, lower-altitude areas. Drip irrigation would further reduce water usage and would be very important to avoid soil salt buildup by evaporation of irrigation water from flood or sprinkler irrigation.

The plant has a pretty, leafy top growth three to six feet high at maturity. The roots often are a foot to 16 inches long and around six inches in diameter, resembling a huge, elongated turnip. Root meat consists of little else than starch and some fiber. Be aware that some varieties of cassava have a percentage of retrograde starch that is impervious to cooking and enzymes, and will produce a lower yield than other varieties.[3] If you have a positive iodine test after you've finished liquefaction (see Chapter 7), you may want to consider varieties with softer starch. The retrograde starch makes a good feedstock for methane production.

The cassava tops, which resemble a small tree, are used as animal feed, and have a higher content of protein than hay. In poor soils, it may be best to return the tops to the ground, or to return the manure from the animal fed the tops to the field, since half of cassava's nutrients and a disproportionate amount of minerals reside in the tops. The roots—mashed, peeled, and granulated (ground)—are used to make alcohol.

Cassava is grown throughout the tropics as a fuel crop. Studies in developing countries show that the plant can have an important socio-economic role where living conditions are poor, providing jobs and a source of revenue. A colleague from Indonesia described to me a proposed small cassava alcohol processing plant, which would produce five million gallons of alcohol per year. He said the plant would provide employment, above subsistence level, for 4000 farmers and related personnel. In this case, the plant was being designed to be labor-intensive instead of capital-intensive, due to the high unemployment in the region.

Although the cassava is primarily a tropical plant, it's grown commercially in Florida and has definite potential as a food/fuel crop in the Southwestern U.S., along the Gulf, and possibly into Texas. The plant generally needs 18 months to reach maturity. New varieties that mature in short time spans of three to six months have a lower starch content, but might be a good rotation crop with other fuel crops in areas that have relatively frost-free growing periods during this shorter growing season. Since these varieties have higher cellulose content, they may yet become very desirable to grow in the U.S.

Although Brazilian cassava growers claim little need for fertilizer, their yields have been less than half what they might be with good cultivation—a dressing of the soil with composted stillage or other organic fertilizer. Standing water and soils with very slow drainage may cause the roots to rot, so be sure your field has good drainage.

Since cassava is essentially an underground root, its soil should be more or less free of rocks and sticks. A slightly modified fodder beet harvester works well for harvesting cassava, as does a potato harvester, or better yet, a modified carrot digger. Cassava roots are considerably larger than fodder beets, and grow deeper. Harvesters must reach a 24-inch depth for the largest roots. In developing countries, they are frequently harvested by hand.

Given its properties and growth requirements, cassava might be a good crop to make use of beet sand, even to replace sugar beets, in warmer southern areas, since many such fields presently go to pasture in the absence of a strong beet sugar industry. Cassava could even replace soybeans and cotton. The cassava uses a lot less water than beets, although optimal water levels for the U.S. have not been established. In Australia, where yields would be higher than in the U.S. South, yields of over 50 tons per acre are attained commercially. Many yields in South America are 25 to 30 tons per acre, using no fertilizer and less soil preparation than in Australia. Returning stillage to the field should increase yields in such fields to rival those in developed countries using chemical fertilizer.

Cassava is planted not from seed, but by insertion of a short piece of basal stalk. Stalks often are passed through a spray of fungicide, nutrients, and sometimes root growth hormones as they are planted (with a cane planter) in nitrogen-enriched hills. Each stalk segment rapidly sprouts roots and becomes a new plant.

CASSAVA: AN INDIGENOUS LEGEND

The beautiful daughter of the indigenous chief gave birth to a child she named Mani. According to tribal custom, the chief killed his daughter, since she was unmarried, but kept his grandchild, who died a few years later of no apparent disease.

Some weeks later, a strange plant grew from the ground in which the mother had been buried. The plant had large green leaves, and its huge roots were white inside. After eating its fruits, birds became dizzy and drunk. The Indians kept the plant and propagated it, and experienced no more hunger. In honor of Mani, the plant was named Manioc.

UNIVERSITY OF ARIZONA PLANT SCIENCES DEPARTMENT

Fig. 8-6 Cassava. This tall, leafy plant was photographed in a test plot of the University of Arizona. Its vigorous growth, moderate water requirements, and high yields make it a potentially valuable crop for the arid states.

Sugarcane billet harvesters can be used to cut stalks before root harvest, separating the leaves and upper stalks for feed, while cutting and saving the lower stalks for replanting. Presently though, it's still more practical to use good old-fashioned people and machetes to cut three- to six-foot stalks by hand and feed them into a cutter/planter.

The 30- to 40-inch water requirement for cassava is not considered high for a commercial crop, but a great deal of care must be taken when irrigating in arid lands. Unless there is a good insulating organic matter layer, the evaporation of mineral-laden irrigation water could leave behind substantial salt, eventually making the land unusable. This has certainly been the case with cotton in many parts of the world.

Presently, cassava should be planted three feet apart in three-foot rows to yield 4800 plants per acre. Tests are needed to determine if closer spacing can reduce water requirements and increase starch yield. We do know that too much watering reduces root yield and increases foliage growth.

Arid climates should help reduce the incidence of cassava mosaic virus, which plagues Floridian cassava growers. Northern cultivation is limited by cassava's sensitivity to cold, although several cold-resistant strains grow in the U.S. Researchers are trying to create new cold-tolerant hybrids that combine the high starch yield of common cassava with the tolerance for cold of high-altitude Southwestern varieties.

Cassava foliage dies off in winter and re-sprouts the following spring. It's smart to harvest the tops for animal feed just as the first cold weather hits. Dry-weight protein content of the foliage can be as high as 35%, almost all of it located in the leaf. Protein content of fresh foliage is about 5%, for a protein yield of over 5000 pounds per acre. This is far more protein than can be produced by an acre of corn.

In a Chinese study, the DDGS-like byproduct of cassava alcohol production had a 27% protein level instead of the low level normally found in whole cassava. This has major implications for export price and shipping costs, if all the feed byproduct isn't used internally in the country producing it.[4]

Analysis of Fresh Cassava Roots

Root Dimensions (Length/Maximum Diameter, in cm)	20–30/6–9
Root Weight, in grams	500–1000
Moisture Content	Approx. 56%, by Weight
Starch	35%
Fiber	4.5%
Calcium	0.1%

Fig. 8-8 Analysis of fresh cassava roots.

Prussic acid (in essence, cyanide) is natural in cassava roots and leaves—the material should not be fed fresh. Many plants, including many pasture grasses like sudan grass, produce this chemical to protect young growth from being eaten. **Ensiling** the foliage (anaerobically fermenting it in lined pits or silos) breaks down cyanide. Even sun-drying destroys the cyanide.[5] Toxic components in the roots are eliminated in the cooking process for alcohol production.

The foliage is quite high in the protein **lysine**, but very low in **methionine** and weak in tryptophan. Blending the ensiled foliage with grain DDGS or spent oyster-mushroom-growing substrate makes a well-balanced feed.

Peeling of the roots is critically important when making alcohol. A substance known as linamarin is in the peel and will deactivate or inhibit your enzymes.[6] Most of the protein is in the peel, and it is useful as cattle feed.

Granulating the roots allows optimum enzyme action (see Figure 8-7); as with corn, you'll use a #20- to #60-mesh screen size. A 2mm size has been found useful in small-scale plants that have a centrifuge to separate out the solids at the end of the process.[7]

Fig. 8-7 Effect of particle size on efficiency of cassava fermentation. Cassava shows 90% or better yields with particle size of #20 mesh or greater (finer). Coarser grinds leave much of the starch untouched by enzymatic action.

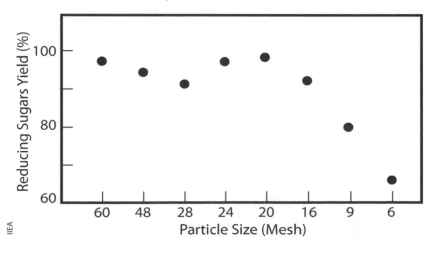

Effect of Particle Size on Efficiency of Cassava Fermentation

IIEA

Run fresh cassava root through your shredder twice—once coarse and once on a $1/8$-inch screen. If you're using dried cassava chips, a hammermill will do an even better job of grinding the feedstock into small particles.

Cassava root is handled like corn, which we talk about in detail later in the chapter. Premalting is essential to keep the stuff from gelling into a thick, unmixable goo. In studies with about 18% dissolved solids in cassava mash, its starch gelatinizes at 149 to 158°F; the mash gets very thick for five to ten minutes at this point, and then begins to thin to its original viscosity as alpha amylase takes effect.

In practice, your mash should contain about 25% dissolved solids to give you a 10 or 11% alcohol yield at the end of the process. Such a mixture will be much thicker than the 18% mash.

You might want to start with an estimated 14% dissolved solids wort, bring the temperature above the gelatinization temperature, and slowly add shredded root until you reach the desired concentration. After each addition of shredded cassava, the mixture will thicken, and then thin out. Although this technique takes extra time to get the wort up to a full 25% filtered wort saccharometer reading (75 minutes instead of 60 minutes), it will result in lower agitator energy usage and higher alcohol yield.

Hydrolysis is shorter for cassava than for corn. Bring it to a boil for ten minutes, instead of 45 minutes to an hour in the case of corn. Continue to liquefaction and conversion just as if it were corn starch, calculating enzyme dosages based on dry starch weight of the cassava.

CASTOR BEANS

The castor oil plant, otherwise known as *Jatropha curcas L.*, has spread throughout the tropical and semi-tropical world. Originally from Eastern Africa, the *Jatropha* grows in arid, inhospitable places and stores most of its solar energy as oil instead of carbohydrates. It has over 700 industrial uses and an established market, even without being considered for fuel.

You wouldn't normally expect to find it in a list of alcohol fuel crops, due to the perception that its only value is in producing oil, but it is very versatile. For instance, castor bean oil is the only unprocessed vegetable oil I know of that will blend wet alcohol with normal diesel fuel, which is known

PHOTO COURTESY OF G. DAIDA

Fig. 8-9 Castor beans. *Castor beans are formed in a spiny case.*

for its strong incompatibility with water. Mixtures of castor oil and alcohol have been used, although not extensively, for 20 years in Brazil in diesel engines. But as you will see, in addition to being used as an alcohol blending agent, this plant can be used to make alcohol.

The castor bean plant can live in dry, unirrigated lands that most plants wouldn't consider; it would do well in the arid U.S. Southwest. It has a deep taproot and a thick fibrous surface root system that allows it to thrive where other plants wilt.

Although the castor bean plant is in the grass family, the "tree" varieties will grow to ten or 15 feet high in a single season and are often perennial.[8] Dwarf varieties are annual in Nature and produce smaller seeds. In more tropical areas, it's possible to get two crops per year of short varieties. In temperate climates, the plant will grow only as an annual.

Castor beans probably should not follow cotton as a rotation, since they share a few of the same bacterial root diseases. The plant does do well following corn, however. Most varieties need a minimum 140-day growing season.

Recent selective breeding programs have increased the oil content of the seed from 24% to 40–60%, while favoring lower-height varieties more amenable to mechanical harvest.[9] In Brazil, yields of 120 gallons of oil per acre per crop cycle are common, but yields of twice this are possible in lands with greater rainfall or irrigation.

When heated for an hour and put through a **screw press**, five gallons of pure castor oil is obtained from only 100 pounds of seed.[10] In developing countries, the castor bean seed is first roughly crushed by a roller press and then squeezed between hemp mats with a screw press, much like olive oil. But if the seeds are sprouted, the oil quickly turns into sugars and other fermentable carbohydrates at a ratio of 2.7 to 1;[11] thus, the alcohol yield from the sprouted castor beans should be 324 gallons per acre.

The spiny hulls of the castor bean are said to be comparable to barnyard manure and can be spread as fertilizer.

CATTAILS

Cattails *(Typha sp.)* live in marshy places where the water is rich in nutrients. They can tolerate a relatively high amount of salt, and they grow very fast and thick, outstripping bullrushes *(Scirpus sp.)* and most other marsh plants. Its **rhizome** (root-like structure) accounts for over 65% of a cattail's wet weight,[12] and it's the fast-growing rhizome that makes the plant such a pest in the waterways and ditches of the world. Rhizomes don't root very deeply, tend to grow laterally, and send up new foliage as the rooting network develops.

ABOVE: Fig. 8-11 Cattail experiment. This section of a cattail marsh has been picked clean of rhizomes. The experiment will determine how quickly rhizomes will invade the area from the edges. The surface is covered with duckweed, a nitrogen-fixing floating plant. Researchers find that a couple of ducks are able to keep the various plots clear, which is a good practice in treating sewage since the point of the treatment is to oxidize the nitrogen. But in an energy marsh, co-culturing with duckweed would capture free nitrogen from the air to fertilize the cattails. On the other hand, duck eggs sell for double what you can get for organic chicken eggs. Decisions, decisions.

Fig. 8-10 Comparison of wild and wastewater-grown cattails. The wastewater-grown plant on the right is at least three times heavier than the one on the left. Note the much larger starch-filled base of the plant.

American Indians found the plant's rhizome a good source for a crude, starchy flour from which they made bread, and cattails are once again being recognized as a useful crop. Alcohol is being made from cattail roots in the former Soviet Union. Cattail foliage makes a good cellulose pulp source for the paper industry and a pretty good animal feed.

A crop of cattails is an excellent candidate for natural marshy areas in which few other energy crops thrive. And they can provide a profitable way to clean up rivers, streams, and oceans. It has been conservatively calculated that 35 acres of cattail marsh could treat five million gallons of secondary sewage a day.[13]

While cleaning up sewage better than any known crop, cattails can yield several high-quality byproducts from the process, one of them being alcohol. Cattail productivity in sewage liquid is incredible. Its nutrient uptake and biomass production is several times higher than corn's. A 1982 study produced results that were as high as 130,000 to 150,000 pounds of biomass *dry weight* per acre. Almost 100,000 pounds of that was rhizome mass. Cattails grown in the wild only produce between three to 30 tons per acre dry total weight.[14] Rhizome yield per acre of wild cattails would be about 15 tons average.[15]

Wild cattail protein content is regularly 6.9% dry weight. It's thought, although not verified, that protein content of wastewater-grown plants is higher.

Assuming the 6.9% figure to be average, protein production per acre would be over two tons. Cattail protein stays fixed in its solids through fermentation, and crude protein readings from fermented wild cattails in dry stillage show 19.1%.

Cattail rhizomes are credited with having a higher starch content than potatoes. One 1975 study[16] and one in 1981 recorded 40 to 60% starch content (dry weight, rhizomes only). Such yields suggest a conservative per-acre alcohol estimate of 2500 gallons from cattail rhizomes fed by wastewater. This calculation is based on a mean yield of only 34.8 tons dry weight of rhizomes at 45% starch in a single crop cycle.[17] These results were achieved in relatively cold northern areas—in warmer climates, the researchers feel that a second crop or even a third crop could be harvested, depending on the length of the frost-free growing season.

A group of my alcohol fuel students, led by Dave Hull and Steve Wilbur, designed and built one of the first cattail marsh secondary sewage treatment facilities in the world, for the city of Arcata, California. Primary sewage treatment settles out or digests solids, but leaves a high level of nutrients in solution. Most rural plants stop processing at that point and release the nutrient-saturated sewage into a waterway or land disposal. Small plants typically cannot afford secondary or tertiary treatment, which typically use sunlight and microorganisms in pools to lower the **biological oxygen demand (BOD)**, as well as nitrate and ammonia levels.

Growing crops in nutrient-rich, secondarily treated sewage to absorb dissolved nutrients and chemicals works as an alternative to part of the secondary and all of the sophisticated tertiary treatment. Cattails excel at this. Not only do they remove solids, but cattails are powerful detoxifiers of chemicals dissolved in water; mercury, for instance, is taken up by the plant and evaporated out of the leaves. They do more than remove chemicals and nutrients, too. Pores on the lower portions of the plants actually capture bacteria and "eat" them.

Grown in wastewater, the plants may be said to be living almost hydroponically. In fact, they don't root any more deeply in wastewater pond soil than is necessary to keep them from falling over. This characteristic makes it possible to grow cattails anywhere trench ponds can be built to simulate aquatic conditions. Agricultural land need not be used at all. Plants grown this way in wastewater reach 12 feet in height, compared to their usual five

LEFT: Fig. 8-12 Cattails, mature foliage. *The seed heads pictured are what give rise to the common name of the plant. Seeds are attached to a light fluff that allows them to be carried for miles on the wind until deposited on some new wet place.*

Fig. 8-13 Cattail rhizomes. *Note how the main plant on the left spreads by adventitious sprouting of new foliage a foot away.*

to seven feet. Using liquid stillage to fertilize cattails, instead of sewage, produces similar yields.

There is potential for closed loop fuel production, where the stillage left over from distillation goes back into the cattail marsh, enabling the growing plants to use those nutrients to trap more solar energy in the form of starch in a relatively small area. Matching up a cattail system located alongside an animal feeding operation or dairy, with alcohol production—where both the stillage and manure are cycled through the marsh—might make for a very dynamic, productive system.

The ten feet or so of cattails' top growth (leaves) should also be considered in energy planning.

LET THE ROADSIDES PROVIDE THEIR OWN FUEL?

In another implementation of "the problem is the solution," how about using the roads to provide the fuel for the cars that use them?

Roadsides have to be mowed or sprayed by counties at considerable expense. Since roadsides are where water gathers, their runoff spreads herbicides and automobile excreta (oil, antifreeze, tire dust, etc.) for miles downstream.

If each county were to convert only 1000 miles of county-maintained roadsides so that a five-foot-wide strip of cattails was cultivated on each side of the road, boom mowers could shred and harvest up to three crops a year of starch and cellulose, as if it were **silage**. At the production level of wild cattails or a polyculture of other high-output cellulose crops, we'd be looking at four pounds of starch and cellulosic plant fiber per square foot from an average of two harvests. This could in theory produce 61 billion gallons of fuel (40% of U.S. gasoline use)—without using a single acre of farmland and while thoroughly detoxifying road runoff water. Planting energy crops in the nation's unused median strips in divided highways would generate additional billions of gallons.

Not everyone wins under this scenario. Monsanto and Dow would be deprived of a lucrative market for their herbicides.

Fig. 8-14 **Experimental solar vacuum still.** *Used by Wetlands Engineering and Technology for test batches. Heat energy comes from the panel in the rear.*

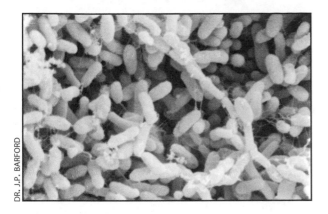

DR. J.P. BARFORD

Depending on the time of year, the sugar content in the leaves averages 10 to 15%. You could use the diffusion process to extract the sugar for fermentation or ferment it whole. But the alcohol yield would be pretty low for the mass of material handled. Combining the sugary foliage with cheese whey to increase the overall sugar content may be an attractive way to produce alcohol from the tops in several cuttings a year. Toward the end of the growing season, you would cease cutting the tops, to allow the sugar to sink down into the rhizomes for winter energy storage so the plants can sprout early in the spring.

On the other hand, now that cellulose technology is becoming economically feasible (see next section), the 25 dry tons of largely cellulosic leaves and the nearly equal amount of cellulosic byproduct left over after rhizome starch fermentation lead me to some provocative speculation. Using the state-of-the-art rate of 180 gallons of alcohol per ton for cellulose conversion, it is possible to project more than 10,000 gallons of alcohol per acre, based on very real cattail biomass production figures, when combining cellulose and starch. Bear in mind that these biomass yield figures do not take into account the potential of stimulation with fermentation carbon dioxide.[18]

At 10,000 gallons per acre, we'd need only about 6367 acres per U.S. county[19] to treat all of our sewage and to replace our entire 200-million-gallon fuel demands. That amounts to only 1.46% of our agricultural land.[20] The energy to run the plant would come from the **lignin** recovered from the cellulosic part of the crop.

Until cellulosic processes become practical on the scale of small municipalities or counties that process sewage, the cattail leaf matter could be dried using solar or waste heat, after diffusion of its sugars. Then it could be used like sugarcane bagasse to cogenerate all the process heat and electricity for the alcohol plant, plus sell surplus electricity back into the grid (see Chapter 11). All ash from the biomass boiler would go back into the marsh as fertilizer.

Remember, too, that if this tertiary treatment plant is, almost by definition, part of a sewage treatment plant, then methane from the primary sewage processing could also be available to provide all the process energy for the alcohol plant.

CELLULOSE BIOMASS TECHNOLOGY

After hearing about the potential from cattails in the previous section, you are really primed to hear about cellulose technology. Until recently, the various methods of producing alcohol from cellulose were only possible with large capital investments for heavy pieces of equipment. But intensive research has been under way over the last 20 years to find an economical method to produce ethanol from our planet's most voluminous carbohydrate. The varied approaches to the questions posed in pursuit of cellulosic alcohol have become a blizzard of research projects spanning more than a dozen universities and many government agencies.

Cellulose is a distant relative of starch and inulin. Like starch, it's essentially a series of linked sugar molecules; unlike starch, which is typically only 35 glucose units, the chains of cellulose are often more than 10,000 glucose molecules. Cellulose is made up of beta-glucose instead of the common alpha glucose that makes up starch. Plus, the chains of cellulose link up side-to-side, forming fibrous crystal-like units with their hydrogen atoms bonding to each other. These cellulose structures are very

LEFT: Fig. 8-15 Scanning electron micrograph of Zymomonas mobilis. *This bacteria has been used extensively by genetic engineers working on the direct consumption of hemicellulose- and cellulose-producing alcohol without an intermediate sugar-producing step.*

"The extreme events to which climate change appears to have contributed reflect an average rise in global temperatures of 0.6°C over the past century. The consensus among climatologists is that temperatures will rise in the 21st century by between 1.4° and 5.8°C—by up to ten times, in other words, the increase we have suffered so far. Some climate scientists, recognizing that global warming has been retarded by industrial soot, whose levels are now declining, suggest that the maximum should instead be placed between 7° and 10°C. We are not contemplating the end of holidays in Seville. We are contemplating the end of the circumstances which permit most human beings to remain on Earth."

—GEORGE MONBIOT, *THE GUARDIAN*, AUGUST 12, 2003

stable and not easily hydrolyzed (broken down by water or acid).

The penetration and separation of cellulose's sugar molecules is further complicated because cellulose is often all bound up in lignin—the virtually indestructible "glue" that provides strength to the fiber of growing plants. Lignin around cellulose is like the hardened resin binding fiberglass cloth. Cellulose has great strength in regard to stretching or tearing, but trees made of just cellulose would be pretty floppy. Lignin, like fiberglass resin, is very hard and strong, but brittle. The combination of cellulose and lignin is incredibly strong, flexible, and decay-resistant.

Waste paper, sawdust, municipal green waste, and biomasses such as leafy materials and grass clippings are all potential sources of cellulose suitable for producing alcohol. Grass clippings alone are a staggering source.

Sources of Cellulose

Switchgrass is bandied about as the best cellulose crop around, and it has become synonymous with cellulosic alcohol. This is what happens when the press picks up a story and it becomes part of the culture. Switchgrass was originally merely the recommendation of one crop that could be grown in certain climates, such as semi-arid plains. It has been treated as a magic bullet by policy makers, venture capitalists, and farm organizations—as a new monoculture crop to rival corn.[21] In trials, it has demonstrated that it might produce a bit more fuel per acre than corn with reduced inputs—less fertilizer and agricultural chemicals, fewer passes though the field with equipment for cultivation weeding, and so forth.

Switchgrass is a native plant of the Canadian Plains, which could yield about 4000 liters per hectare, or 427 gallons per acre.[22] But switchgrass does not exist as a monoculture in Nature, and once planted in this way will require substantial inputs over time—just like any monoculture. Its problems will then outweigh its bright promises. Many of the other crops described in this book for arid regions will produce a higher yield of alcohol, without the added complication of cellulose processing.

Yes, in some cases, switchgrass might be just the right thing to incorporate into a regional energy agricultural system *in rotation with other crops*.

Monocultures of such subtropical and temperate crops as sweet sorghum, eucalyptus, and *Leucaena* could yield 4000 to 5000 gallons per acre, with sugar and cellulose taken into account.[23]

Better still would be a multiple-canopy strategy with shorter trees, shrubs, bushes, and grasses. Polycultures of multiple crops maximize the absorption of sunlight and produce more carbohydrates and other valuable products per square foot. Soil structures designed to capture water runoff in large tracts of the American West, where up to 87% of rainfall runs off, have been known to triple biomass, and therefore cellulose output. This sort of water collection would skew yields in favor of polycultures over switchgrass even further. Many of the crops (e.g., eucalyptus) produce valuable byproducts such as essential oils, too.

Some coppiced trees (trees that are radically pruned each year to stimulate new growth) can produce pretty astounding amounts of biomass. One study showed that clear-cutting a dense five-year growth of river red gum (*Eucalyptus camaldulensis*) would yield more than 40.3 tons dry weight per acre.[24] But coppicing would yield about 12 tons per acre dry weight *each year* from cutting a fourth of the canopy annually—producing over 2000 gallons of alcohol per acre (at 180 gallons per ton).[25]

The study that produced those figures showed that adding nitrogen chemical fertilizer to the woodlot had no measurable effect on yield. How could this be, since the soluble fertilizer was certainly taken up by the trees? Also, the nitrogen fertilizer certainly would have severely damaged the mycorrhizae. Decimation of mycorrhizal fungi by chemical fertilizers, resulting in lowered yields, has been amply demonstrated in other crops.[26] Since return of stillage or stillage compost to the forest would substantially stimulate the population of fungi and other critters in the root zone community, it is entirely possible that the potential for tree-based yields could be dramatically higher than what was observed in this study, using other fast-growing trees.

A less studied tree, the temperate, nitrogen-fixing *Tipuana tipu*, is known to have annual ring production of more than three inches per year. A six-year-old *Tipuana* can have a 100-foot-diameter canopy and reach 90 feet in height. Coppiced *Tipuana* might generate three to five times the alcohol possible from river red gum.

Throughout the tropics, there are extremely fast-growing leguminous and nonleguminous trees. In researching a fast-growing "hardwood" tree to use as feedstock for **shiitake** mushroom production, Dr. Daniel Martínez-Carrera introduced me to *Bursera simaruba*. It is used as a living fence in Mexico—a cut branch shoved into moist earth will sprout and become a massive tree in a few short years, with rings almost as thick as *Tipuana*. I've seen two-year-old *Bursera* attain a height of 30 feet and a diameter of eight inches. Coppicing seems to just make it grow faster. Florida residents know the *Bursera* as the gumbo limbo tree or the tourist tree (due to its flaky red "skin").

Investigation of tropical trees suitable for cellulose production would be a formidable task. The diversity in tropical trees is almost beyond imagination. While working in Costa Rica in 1982, I discovered that this small country had more varieties of trees than all of North America and that *new ones* were being discovered and named on a weekly basis. So the field of tree-based cellulose energy plantations is in its infancy.

Waste paper, sawdust, municipal green waste, and biomasses such as leafy materials and grass clippings are all potential sources of cellulose suitable for producing alcohol. Grass clippings alone are a staggering source. Each acre of lawn produces on average 4206 pounds dry weight, which is almost all cellulose.[27] Alcohol yield ought to be about 350 gallons per acre, about the same as corn![28]

Not only is this a pretty good yield, but grass clippings are reputed to be the United States' largest irrigated crop by dry weight.[29] In 1978, lawns covered 16 million acres and used about 20% of the total fertilizer of the nation (at that time equal to the amount of fertilizer used by the food agriculture of India).[30] During this period, lawns consumed 44% of California's domestic water use.[31] In California, an arid state that is not a major producer of grass clippings per acre, clippings can account for up to 10% of urban solid waste production.[32] Georgia puts out some serious tonnage of clippings, and Florida is off the charts. If California grass clippings alone were converted to alcohol, they would produce 1.5 billion gallons of alcohol, or roughly 15% of the state's auto fuel needs.[33]

A calculation by NASA in 2005 shows that lawns have swelled to 29.7 million acres.[34] This ought to yield over 11.2 billion gallons of alcohol, or more than twice as much as the alcohol produced from corn in 2005.

Of course, there is a lot more green waste than just grass clippings, but this gives you some idea of cellulose that could be processed that's already available within an existing collection infrastructure.

Breaking Down Cellulose

Biochemical engineers look to Nature for clues to frame their research. Obviously something digests cellulose and lignin, or we'd be up to our armpits in slowly decaying plant materials. Anyone looking out a window at a lush pastureland where livestock graze is seeing mobile cellulosic digesters at work. After all, cattle consume cellulose-loaded feedstock all day long and convert it to sugars for their energy.

We will talk about the issue of freeing the cellulose from the lignin a little later (although there are many cellulosic materials that don't have much

---

HERBICIDES SELL INSECTICIDES

On my farm, the most effective nighttime bug control was performed by a brigade of toads. Each little toad eats about 167 bugs per night.

In a recent test that any organic farmer would immediately verify, Roundup, Monsanto's popular herbicide, proved to be deadly to amphibians. One tablespoon of Roundup was added to 250 gallons of water in which tadpoles were living. Within three weeks, 98% of the tadpoles had died. "It's much deadlier than we thought," said ecologist Rick Relyea.[35] In a terrestrial experiment, Roundup killed 79% of the toads and frogs in just one day.

To kill weeds, genetically modified corn and soybean crops are sprayed with huge quantities of Roundup (to which they've been given immunity). A following rainstorm can wash up to 99% of the herbicide immediately into low spots and wetlands, where the toads would congregate. (Toads live on land but breed in water, whereas frogs stay in water most of the time.) Much of the Midwest farmland is a checkerboard of wetlands with slightly higher areas that are drained to make farming possible in what would otherwise be a marsh.

Monsanto's response to the study: "Roundup isn't meant to be used near water, and its directions clearly say so."[36] Remember that next time your country road maintenance crews spray your roadside ditches, which then drain into your creeks and kill all the amphibians for miles downstream. And, of course, without the bug protection provided by amphibians, the use of insecticides would without a doubt have to rise dramatically.

RIGHT: Fig. 8-16 USDA researcher Nancy Ho. *Spending over 20 years on its development, Nancy Ho has created a genetically modified distiller's yeast that can use many formerly unfermentable cellulose-derived sugars.*

lignin). Let's look now at the two basic methods currently recognized for breaking down cellulose once it's free of lignin. The most well-known method is **acid hydrolysis** in water, and a more recent technique is **enzymatic hydrolysis**.

Acid Hydrolysis

In acid hydrolysis, biomass is put into a reactor. High temperature and high pressure are then used, along with exposure to weak sulfuric acid; or the biomass is put under little to no pressure and exposed to a stronger acid. The heat, acid, and pressure break the cellulose into separate sugar molecules. These methods have been effective at making alcohol from wood or other cellulose for a century. They have not been in common use for some time, since it has been more expensive to make alcohol this way than from starch or sugar. But there is nothing mysterious about these processes. They work just fine.

Weak acid hydrolysis uses little acid, around 1%, but high temperature and pressures (400°F and around 400 psi). **Strong acid hydrolysis** uses a very high percentage of acid, but does the reaction using no pressure and temperatures near the **boiling point** of water. The trade-off is that weak acid treats the acid as a consumable input, which is neutralized after making the sugar. In strong acid, an acid recovery step is required that is technologically a bit complex but avoids having to use pressure vessels.

Fig. 8-17 Zymomonas mobilis. *This bacterium actually digests cellulose and produces alcohol in one step. Since bacteria have low tolerance for alcohol in comparison to yeast, bacterial fermentation has never been commercialized, as the distillation energy would be significantly higher. But gene splicers continue to use it as a source of genetic code to try to make alcohol-tolerant yeast with cellulose-eating capabilities.*

PURDUE NEWS GROUP

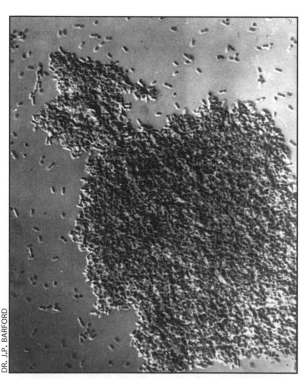

DR. J.P. BARFORD

The reactor for weak acid hydrolysis is sophisticated and expensive. Paper manufacturers use this very high-pressure weak acid technology in their pulping process, which turns trees to newsprint. In this case, the conversion to sugar is not desirable, and once the cellulose is sufficiently pulped, the reaction is stopped earlier than it would be if alcohol production were the goal. The paper mill's process waste, sulfite liquor, contains about 5% sugar and can be directly fermented to alcohol—requiring only a pH adjustment (with lime) and some nutrients (ammonium salts or dead yeast). In plants designed specifically for handling sulfite liquor, sulfur dioxide is steam-distilled out of the liquor, which makes neutralizing the remaining acid pretty easy.

If there's a paper plant in your area, you may want to examine its byproduct, but it's highly toxic and acidic. In many parts of the world, it's actually still legal for paper companies to dump this stuff into rivers, where it sterilizes the water for miles. Processing waste liquor could solve a hideous pollution problem.

During World War II, the U.S. had a sizable number of plants producing alcohol from cellulose (wood fiber). These plants used the strong acid hydrolysis method described above. Following the war, the plants were not cost-effective compared to alcohol made from grain, so they were either converted to paper production or dismantled. The reason there is little emphasis on the strong acid method today is that it destroys all the hemicellulose (primarily **five-carbon (C5) sugars**) and, if it's not carefully controlled, it produces toxins such as **furfural**, which inhibit fermentation yeast, resulting in low yields of alcohol. There is currently a lot of interest in converting the hemicellulose to alcohol, as you'll see below, so strong acid is not generally being considered these days.

Enzymatic Hydrolysis

Nature doesn't use severe pressure or concentrated acid to decompose cellulose, so there must be a biochemical way to do it. The primary alternative approach to acid hydrolysis is enzymatic hydrolysis, generally using the enzyme cellulase.

Cellulase was at one time available only from bacteria that live inside one-celled animals (flagellates) that live in the digestive track of termites (an animal inside an animal inside an animal). Smashing up a lot of termites, however, was not found to be a practical source for cellulase.

Until recently, the U.S. Army had focused its research on a top-notch cellulase producer, a mutant strain of a wind-disseminated soil bacterium, *Trichoderma viridae*, also called jungle rot. This bacterium was discovered eating World War II surplus ammunition belts that were made of cotton (cellulose fiber) on GIs in Vietnam. Unfortunately, *Trichoderma* turned out not to be the miracle bacteria the Army hoped it would be. They bred and mutated it, and still could not get it to produce cellulase economically.

Later, other researchers began to mutate strains of bacteria such as *Zymomonas* and *Clostridium* or gene-modifying *Klebsiella*, to produce ethanol directly from cellulose, bypassing sugar fermentation.[37] Although this research met with modest success, with hemicellulose often being digested— none of the bacteria could do the job without at least some help from some sort of acid hydrolysis process. This was also a potentially catastrophic use of genetic engineering, especially in the case of altered *Klebsiella*, since it is an aerobic organism that dwells in the soil and would not die if released, unlike the anaerobic bacteria discussed above.

In the 1980s and 1990s, we kept hearing that economically attractive cellulosic alcohol via enzyme hydrolysis was five years away. It almost became a standing joke at alcohol fuel conferences. Lo and behold, that day has finally arrived, due to the price of gasoline rising to a level that makes the advances in cellulose technology competitive.

One barrier had been the cost of cellulase enzymes needed to take fibrous plant material through several steps to provide fermentables. In the last few years, the cost of cellulase has finally dropped to a tenth of what it was, due to financial incentives offered by the DOE to get private industry to focus on reducing the cost. As a result, cellulase enzymes from two companies now cost about 30 cents per gallon of alcohol produced, and the price is expected to drop even further. These enzymes cut up the cellulose into both five-carbon and **six-carbon (C6) sugars**. Common distiller's yeast eats most of the C6 sugars, but C5 sugars need less-common yeasts, or GMOs, to produce alcohol.

This wasn't exactly the outcome alcohol fuel supporters had been hoping for. Much of the result of enzymatic hydrolysis is the production of a larger quantity of five-carbon sugars (mostly **xylose**). We had all dreamt of a cellulose enzyme process that would yield all C6 fermentable sugars that normal yeast could ferment. Although there are some natural organisms that ferment C5 sugars to ethanol, the organisms that researchers focus on today are proprietary, patented, gene-altered bacteria like *Klebsiella*[38] or modified *E. coli*.[39] Although these organisms can convert hemicellulose-derived sugars to alcohol, they don't have the alcohol tolerance developed over eons by yeast. These GMO bacteria

---

HEMP'S POTENTIAL

There is much work to be done as far as cultivating plants with high cellulose content to be used for fuel. The much-maligned hemp plant (a fibrous industrial version of marijuana) has been known to provide cellulosic stands 14 feet tall when irrigated in good soils. With cellulose yields of five tons or more per acre, hemp could be a new contender in the energy field, possibly yielding 900 to 1000 gallons per acre in six months. Some species of annual *Crotalaria*—which goes by the name Sunn Hemp but is not actually related to true hemp, and is currently used for summer cover crops—can reach similar heights in half the time, less than 90 days, all the while enriching the soil with nitrogen.

typically would die at 3% alcohol concentrations. This makes distillation far more costly and energy-intensive than the 12 to 20% alcohol produced by yeast from C6 sugars.

Enzymes need intimate contact with the cellulose to cause it to break down into sugars. Somehow the cellulose fibers need to be physically broken down so that the enzymes can do their job. Various pretreatments are needed to physically break apart the long chains of cellulose. Most of the pretreatments still require pressure and/or acid to expose the cellulose. Newer pretreatment alternatives for enzyme use might permit economical building of smaller reaction chambers that fit the farm cooperative scale.

Some of the promising research techniques for pretreatment are steam explosion or ammonia explosion of the cellulose fiber. The feedstock absorbs the liquid ammonia, swelling the cellulose, decrystallizing and releasing it. The ammonia flashes to a gas, exploding the cellulose fibers apart,

Fig. 8-18
Removing solvent.
*The black liquid is the dissolved cellulose. The author is removing the solvent by **vacuum distillation** in order to end up with a solution of C5 and C6 sugars.*

making them very well pretreated for hydrolysis.[40] In a similar process, water under high pressure and temperature is forced into the cellulosic material; when the pressure is released, it becomes steam, tearing the fibers apart.

So, since we are still required to do acid hydrolysis or other radical processes to expose the cellulose to the enzymes, why use enzymes? Some would say that there is the promise of enzymatic hydrolysis giving a higher yield of sugars while producing fewer toxins. There may be a modest savings in energy over weak acid hydrolysis, depending on the pretreatment strategy.

But isn't our goal to get sugars from cellulose to feed our yeast to make alcohol, not to obsessively focus on creating conditions for relatively expensive enzymes? The radical pretreatments described above are about as difficult as the tried and true weak acid hydrolysis that has worked for 100 years without the expense of enzymes. The focus on enzymes smacks of the tail wagging the dog. After millions of dollars expended on building the first pilot plant, operated by Canadian biotechnology firm Iogen Corporation, using radical pretreatments, GMO enzymes, and GMO yeast—the effort to go forward seems stalled on the need for hundreds of millions of dollars to build the first commercial plant. Although oil companies have spent tens of millions of dollars to purchase much of Iogen's stock, they seem unwilling to provide the capital to prove the process works in the commercial world

On the other hand, Brazilian commercial interests, with a relatively small amount of capital, have improved on weak acid hydrolysis and actually beat Iogen, being first to operate a successful pilot plant converting sugarcane cellulose to alcohol for less than 50 cents per gallon—just fermenting the C6 sugars coming out of the process. No GMO enzymes or yeast are required in their rather straightforward plant. Their original pilot plant was a 100,000-gallon-per-year unit. They are now operating a scaled-up, 4.6-million-gallon-per-year pilot plant.

This second so-called pilot plant scale is not only working well, but I feel is already sufficiently proven as reliable enough to be a practical farm-scale commercial proposition. This scale would produce more than four million gallons per year, which would efficiently use the cellulose that could easily be produced on five to ten average-sized Midwest farms, depending on the selected mix of crops. It would produce all its own process energy via methane from the unfermentable sugar yield. The original 100,000-gallon-per-year pilot plant would serve a farmer who has just 200–300 acres.

So what can we do with the C5 sugars that normal yeast cannot ferment? To some degree, the fantasies of fuel producers have been answered by a couple of standard yeasts that have been genetically modified to use C5 sugars derived from woody matter. A USDA team has come up with a *Saccharomyces* (common distiller's yeast) with three altered genes that allow it to consume both normal C6 sugar and many of the C5 sugars that are produced in enzymatic or weak acid hydrolysis. For 20 years, Dr. Nancy Ho of Purdue University's Laboratory of **Renewable Resources** Engineering led the team that finally produced a stable yeast which efficiently eats C5 sugars such as xylose, arabinose, galactose, and cellulose-derived beta glucose. The new yeast will operate at normal temperatures of about 92°F and at normal pH and requires no special conditions.[41]

The new enzymes, which should be on the market for public sale by the time you read this, break down cellulose to make a banquet for Nancy Ho's new yeast. This combination of enzymes and

RAIN DANCES

All us have seen it in old Western movies: indigenous Americans doing a dance to encourage the rain gods to bless their land with life-giving rain. "How quaint," we think. "These superstitious people believe they can change the weather, when with a little education they would understand that Nature isn't under the control of rain gods." After looking down our noses at these permaculture land experts, we find that rain dances were usually held on distant mountaintops upwind from their land. The participants brought healthy boughs and herbs that were loaded with *Pseudomonas syringii,* and dancing on these plants raised a bacteria-filled cloud of dust to fan out on the winds, seeding the clouds. And yes, it rained, sometimes hundreds of miles downwind.

Is it a little dry where you live? Now you know what to do.

GMOS AND THE END OF RAIN

I often tell a story about GMOs and how close they have come to ending the world. It turns out that strawberry growers typically lose the last few weeks of their harvest to light frosts. In studying what makes frosts occur, it was discovered that the strawberry plants were covered with bacteria that like the sugar on the leaves and fruits. Each of these bacteria, *Pseudomonas* or *P. syringii,* has a protruding sharp needle sticking out of it (i.e., a syringe). This needle is so sharp that it provides an excellent place for water to condense out of the air into a droplet; or, if it's cold enough, the water condenses and freezes into an ice crystal, i.e., frost. This bacteria is thus able to get water out of the air when the climate is otherwise dry. The droplets or melting frost wash over the bacteria and onto the leaf, where sugar dissolves and the bacteria can absorb it.

Biotech firm Calgene was contracted to find a solution. They studied the bacteria and found that a few of them had faulty genes (a recessive trait) for making the needle and instead made little stubs. These defective bacteria typically die out in nature since they can't condense water as well. So the gene-splicers identified the stubby gene and inserted it in normal healthy bacteria, making this a *dominant* genetic trait. Their plan was to spray this on the strawberries, and it would overwhelm the normal bacteria, and would stop the frost from forming, and boom! the strawberry growers could get two or three more weeks of crops.

But *P. syringii* don't grow only on strawberries. They also love growing on pigweed (*Amaranthus sp.*). Pigweed has a taproot that goes three-quarters of the way to hell and puts out trillions of really tiny, very durable seeds. As far as farmers are concerned, the only good thing about pigweed is that frost kills it. So if this GMO bacteria were sprayed on strawberries, it would then be permanently in the environment, reproducing and blowing in the wind. Pigweed, with the protection of the GMO bacteria, would survive winters and become a serious problem for other farmers. So, thanks to a lawsuit by the Center for Food Safety, this GMO bacteria was banned due to its potential to make pigweed a super-weed.

But what never came up was the truly big threat. *P. syringii* is most commonly found in trees. The bacteria feed on tree sugars that dissolve when its sharp point condenses water at night and out of fogs onto trees. When there is a storm and the warm air of the forest rises up into clouds, it carries the *P. syringii* with it. In coastal regions, most raindrops have condensed on tiny salty mineral crystals at sea. But when you get away from the coast, up to 80% of the condensation of raindrops or formation of snowflakes is on the very sharp point of a *P. syringii.*[42] Had the Center for Food Safety not won the battle on behalf of the farmers who would be affected by the pigweed, rain inland from coasts would have stopped falling, due to these dull-pointed GMO bacteria. It could have meant the end of life as we know it on planet Earth.

tailored yeast is the basis for Iogen's plant. Iogen's first feedstock is wheat straw. But this combination of enzymes and yeast should handle any sort of woody waste, including municipal green waste after weak acid hydrolysis. I have had conversations with the developers of both the enzymes and the yeast, who say that neither product will be exclusively licensed; both products are expected to be made commercially available to those wishing to make alcohol. This may be good news for aspiring small producers.

Another new GMO yeast has just been announced in Europe, where they took a different approach. Nature's most powerful decomposers are fungi. In developing this GMO yeast, very powerful genes were spliced in from fungi instead of from bacteria.[43] Fungi are experts at breaking down cellulose; examine any fallen tree.

Use of C5-sugar-digesting yeast is controversial, although not nearly as controversial as using GMO bacteria. Both cellulose and hemicellulose are basic building materials of plants. GMO yeast only eat sugars that result from a pretreatment that breaks plant fiber down into various sugars.

But the proposed use of GMO bacteria to directly digest cellulose/hemicellulose and produce alcohol in one step is a dire hazard. The risks of these organisms getting out in the world is unknown, so there are reasons for concern. For instance, almost every piece of fruit has yeast naturally occurring on its skin. As soon as the skin is broken, the yeast can start fermenting the sugars in the fruit, which can result in birds flying intoxicated after dining on them. But if GMO bacteria can partially eat hemicellulose, will it just eat through the skin of plums, peaches, and so forth? Unlike yeast, even GMO yeast, it may now have *Fig. 8-19* the capacity to do so.

Some of the bacteria (*Kelbsiella*) used in experiments have been proven, once in the soil, to continue to thrive. Once there, they've been shown to kill virtually all seedlings that attempt to grow in the contaminated soil, by a combination of ethanol toxicity and probable cellulose digestion of root tips. We don't know for certain what the effects of escaped cellulose- or hemicellulose-eating GMOs will be. But one thing we do know for certain is that GMO bacteria, once made commercially available, will escape into the wild. There is a risk such an occurrence would be an extinction-level event.

Perhaps the best use of the C5 sugars is not conversion to ethanol but as an input to a methane digester. The methane, instead of ethanol, produced from these sugars should easily provide all the process heat and surplus electricity to sell back into the grid from the alcohol plant. This solution avoids all the risks associated with GMO organisms. The energy to run the plant has to come from somewhere; why not from the cellulose turned into methane/electricity?

Xylose, the major C5 cellulose sugar, can be converted to xylitol. Xylitol is what is known as a **sugar alcohol**, based on its chemical structure. It can substitute for table sugar in almost any product requiring sugar's sweetness. The advantage, however, is that humans don't digest **xylitol**, and so diabetics can use it with abandon. The negative effects of high-sugar diets can be reduced the more we replace table sugar with this safe alternative.

Costs of Ethanol from Cellulose

The cost of making ethanol from cellulose is a fuzzy projection at this point. Novozymes, the world's largest biotech company specializing in enzymes, figured that the projected price in 2004 would be about $2.30 per gallon.[44] This estimate

is probably on the very high side because of the uncertainty as to what kind of pretreatment will be used on the cellulose before using the enzymes.

The pretreatment process could easily be the major expense in building an enzymatic cellulosic alcohol plant. Since this major expense in the building of the facility has not been codified by experience, as it has for starch-based plants, prices are still speculative. On the other hand, Brazilian weak acid hydrolysis, as mentioned above, is currently producing alcohol from C6 sugars alone for about 50 cents a gallon from sugarcane bagasse. No attempt has been made to use the C5 sugars that the process generates, as well. The Brazilian process was operated successfully for two years at a 250-gallon-per-day output, which means it's applicable to the farm scale.

With corn, feedstock is often 75% of the final cost of the alcohol. With cellulosic ethanol, feedstock costs can actually be minus 15 to 20 cents per gallon if you run the landfill that is charging people to dump yard waste; five cents if you buy bulk used newsprint; or far less than $1 per gallon for a wide variety of other materials such as corn **stover** or bagasse.[45]

As recently as 20 years ago, the price of building a dry-mill corn-to-ethanol plant was over $2.50 per annual gallon of production capacity. Now, a far more sophisticated and energy-efficient plant is $1.50 per gallon lower. An even faster decline in the price of building cellulosic ethanol plants could be expected.

Currently, it is possible to economically use waste paper with the new cellulase enzymes, which work at normal air pressure, pH 5.0, and 132°F. Since paper has already been pretreated in its manufacture, if you simply pulp it with water, adjust its pH with acid, and add a little bit of yeast nutrients, the enzymes should yield a good fermentable mixture. Yields might be well over 100 gallons per ton.

Solvents for Cellulose

Curiously enough, almost everyone looking to produce cellulosic alcohol has assumed it should be done in water. A few thinkers have gone outside of the box and challenged this assumption. More than 20 years ago, Canadian scientist and professor Laszlo Paszner demonstrated on the pilot scale that when the organic solvent **acetone** was used instead of water in weak acid hydrolysis, lignin quickly dissolved (and could later be recovered), and cellulose

was readily exposed and easily broken down to mostly C6 and a lesser quantity of C5 sugars. Many easily extracted useful byproducts, such as pure xylose, were worth far more than the alcohol.

Brazilian researchers have experimented with more than a dozen solvents and developed a similar process using ethanol (cheaper and less toxic than acetone) as the solvent. Instead of a big reactor, the process is carried out in a tall column (pipe). Into one end of the pipe, they pump a mixture of bagasse and ethanol. In the other end of the pipe, they pump a weak acid/ethanol solution. This reaction column is operated at high temperature. The two streams mingle in the center of the pipe. In about ten minutes, the lignin is fully dissolved and the cellulose converted into a fermentable liquid. When the column is operated in a continuous fashion, the fermentable C6 sugars leave the column reactor through a pipe at one point, and the C5 sugars largely leave through another pipe at a different point along the column.

"Ethanol is not hurting our water, it's the MTBE. In fact, even though I am not a drinker, I know that ethanol is little different than corn whiskey, so if ethanol gets in the water, the worst that can happen is that you might want to add ice, tonic, and soda."

—SENATOR CHARLES GRASSLEY (R–IOWA)

This technique preserves much of the hemicellulose-derived sugars that otherwise would be destroyed in a lengthy acid process, making process control less touchy. It also permits the separate processing of C5 sugars by either GMO or natural hemicellulose-fermenting organisms, to alcohol, or to methane in a separate digester.

I like the simplicity of this process. It can be adapted to the smaller scale, and uses no proprietary GMO organisms or expensive enzymes (although GMO yeast could be used to ferment the C5 sugars). There is a potential downside to using solvent-based technology, since the solvents must be boiled off before the sugars can be fermented. Of course, the solvents can be reused over and over again, but the energy cost is high.

It would be critical for the plant to produce its process energy renewably, and to aggressively conserve and reuse process energy to begin to justify the higher energy demand over other processes. The Brazilian pilot plant, which makes 5000 liters per day, already produces alcohol at a price of less than 50 cents per gallon, just from the C6 sugars. Its energy source, however, is biomass-fired boilers, and energy cost is not an issue. The price should drop further when the process is implemented on a much larger scale and the C5 sugars are either fermented or extracted for other use.

After decades of saying that cellulosic alcohol is five years away, it is no longer true. The future has technologically and economically caught up with the present due to the higher price of oil. All that remains to make cellulosic alcohol a current reality are courageous investors to begin to build plants of a wide range of capacities, using the various technologies we have right now. Short-term tax incentives to lure private money into this new field would certainly speed its growth.

The hard shells from one acre's worth of hazelnuts would weigh over a ton and burn like hardwood (7000 Btu per pound), to produce approximately 14 million Btu, enough to distill 466 gallons of alcohol in a small-scale distillery.

CHESTNUTS AND HAZELNUTS

Permaculture designs are often heavily based on trees. If you have sloped land that would not be suitable for row crop production, you might think about using starchy or even cellulosic products from certain trees to provide your fuel.

Chestnuts are not actually nuts but starchy fruits, and they have a very short shelf life and season. Imported chestnuts take a week to get to the U.S. by ship and a week to get through customs and into stores, leaving only a week to eat them before they start to become moldy. So domestic producers have a distinct advantage in marketing chestnuts locally. The cultures that value chestnuts most are Asian and Italian. But if you live far from the population centers of these ethnic groups where you could sell organic chestnuts for $10 a pound, you might consider using them for fuel.

Chestnuts are a good overstory crop, and new Chinese/American hybrids can produce up to 4000 pounds per acre, or about 2000 pounds dry weight.[46] Subtracting the shells (which are largely cellulose and therefore usable in a cellulose process, or as a boiler fuel), you end up with about 1600 pounds of kernels per acre.[47] These contain between 80 to 85% carbohydrates, which

are mostly sucrose and starch.[48] The sucrose can be as high as 25%![49] You would end up with 133 gallons of alcohol per acre from the kernels alone,[50] as well as high-protein, low-oil mash for feed. Sprouting the nuts could convert most of the starch to sugar and, as a bonus, convert the small amount of oil (about 1 to 4%) into sugars as well, which would simplify making alcohol. Both the chestnut prunings (made to keep an open canopy over a hazelnut understory) and shells would provide plenty of fuel for the process heat used to make the alcohol. To extend the processing season, it's pretty easy to dry chestnuts; the process doesn't result in an edible product but would be fine for making alcohol.

Hazelnuts are another possibility for a perennial crop that can yield some fuel. When intensively grown as an undercanopy shrub instead of pruned into tree shape, hazelnuts can yield up to 3.95 metric tons per hectare, or 1.76 tons per acre, in the shell.[51] Newer varieties have been driving up the yields. When you include the nuts that aren't large enough to be considered marketable, the yield can be as much as 25% more. In general, you can calculate nutmeat at 40% of the nut;[52] based on the yield above, you are looking at 1408 pounds of nutmeat per acre. Since, unlike chestnuts, hazelnuts are true nuts, they store most of their energy as fats (triglycerides), so you can expect approximately 16% carbohydrates and 68% oils.[53] If you were to cook this up as mash, you would get a tiny alcohol yield of perhaps 19 gallons per acre. However, you would get 150+ gallons of oil to use for boiler fuel or biodiesel, triple what you'd get from soybeans.[54]

But plants don't use triglycerides to grow. When the seed sprouts, each gram of triglyceride is converted into 2.7 grams of usable carbohydrate as sucrose.[55] Rapidly growing seedlings use the sugar for energy. So, if the hazelnuts were sprouted before grinding and fermentation, you might theoretically get 335 total gallons of alcohol per acre.[56] At their peak, the sprouts will actually be 40% sugar![57] If the best-case scenario of 335 gallons could be reached, it would be better than corn yield per acre (in most places), but without all the annual soil work, need for water and chemicals, and soil erosion. I have seen one paper that suggested that some of the carbohydrate will not be fermentable, depending on the stage of sprouting. The older the sprout gets, the more quickly it converts triglycerides to cellulose.

Cellulase may then be needed to convert the cellulose in both the sprout and the unfermentable part of the nutmeat.

The hard shells from one acre's worth of hazelnuts would weigh over a ton and burn like hardwood (7000 Btu per pound), to produce approximately 14 million Btu, enough to distill 466 gallons of alcohol in a small-scale distillery.[58] Pruning the hazelnuts to keep production high might yield another ten tons of dense, oil-enriched firewood, the equivalent of five cords per acre, especially if used in a **downdraft "rocket stove"** design. This would provide a large surplus of energy for process heat. So no fossil fuel would be needed for heat in this alcohol plant.

Hazelnuts should be considered as an understory crop, under another crop—pruned chestnuts, or a nitrogen-fixing leguminous tree like mesquite in the South, or perhaps black locust in the Northern U.S.

CITRUS FRUITS

The sugar concentration of oranges (including the peel) hovers around 14%. The peel itself is only about 5% fermentable, so including peel drops the overall sugar level of your wort. Pure orange juice from small, sweet oranges may contain sugar concentrations as high as 18%.

There are some difficulties processing oranges. The main one is dealing with the colored layer of peel, which contains tiny sacs of volatile oil. D-limonene is the oil's major component, and it's quite toxic to yeast. Although D-limonene has a high boiling point (in the vicinity of water's), I've successfully fermented small batches of whole orange pulp by first cooking the batch at a boil for an hour, causing the volatile oil to be evaporated out of the mash.

Citrus peel oil is valuable and could be worth selling. When I looked at this possibility for a cooperative of citrus growers in the 1980s, we determined that, for a given quantity of lemons, the small quantity of lemon oil was as valuable as the alcohol. Condensing the vapor arising from the cooking process would contain the oil. Running the condensate through your **fractional still** would result in much more concentrated oil. Higher concentrations of oil in the condensate are not water-soluble, so they float and are easily decanted. Crude lemon oils are usually worth several times the value of orange oils.

COFFEE

Although coffee seeds are often referred to as beans, they have no relationship with any true bean or other legume. Coffee fruit is roughly like a cherry, and the pit is the savored "bean" that makes over 430 million cups of coffee a day in the U.S.[59]

In processing the coffee "cherry," the pulp is removed and most often flushed into rivers. Coffee pulp contains 14.4% fermentable sugar and is available in huge quantities seasonally.[60]

The pulp also contains about 6.5% pectin.[61] Pectin is a carbohydrate similar to starch, but instead of being made up of lots of glucose, it's mostly made up of a similar substance called galacturonic acid. The difference is that when it is joined together in a chain like starch, one of the carbons has a carboxyl group attached (COOH) instead of CH_2-OH.[62] Pectinase is an enzyme that can break the bond between the #1 and #4 carbons that hold one pectin molecule to the next, as well as breaking the pectin itself down into simpler sugars. As with the breakdown of starch, you actually get more sugar by weight than you get from the original pectin that you started with. This is due to the enzyme grabbing water, splitting it apart, and attaching the hydrogen and oxygen to the resulting simple sugars. Home jelly makers use this water-binding quality to thicken their jams.

The catch-all term *pectinase* actually refers to three enzymes: polygalacturonase, pectin lyase, and pectin methyl esterase.[63] The three enzymes break the pectin molecule in different places. Commercial pectinase comes from common bread mold, *Aspergillus niger*, and operates best

Fig. 8-20

Fig. 8-21

at a temperature of about 130°F and a pH of 4.8 to 5.0.[64]

Some new GMO bacteria can ferment pectin to alcohol, but can only tolerate about 5% alcohol before keeling over, and so are not really useful. Some yeasts appear to have a limited ability to do direct pectin conversion, also.

Caffeine is made as a pesticide by many plants. It works by jamming nerves into the "on" position, causing convulsions and death to insects that bite the coffee fruit.

Once the coffee fruit (pulp) has been removed from the pit, there remains mucilage that covers the seed. The coffee beans are fermented to break down the adhering mucilage and to expose the thin, cellulosic, paper-like layer (silverskin) covering the pit (bean). The mucilage surrounding the seed has about 4.1% sugars, about 9% protein, and 1% pectin, the rest being water.[65] The fermentation of the mucilage is done in an uncontrolled manner with airborne wild yeast and bacteria. Any potential alcohol that might be incidentally produced quickly turns to lactic acid or acetic acid. There might be an economic opportunity in carefully controlling this fermentation and recovering the alcohol via vacuum distillation to avoid heat damage to the pit.

The coffee seed (bean) is 68.2% nonfiber carbohydrate.[66] This carbohydrate turns out to be mostly fermentable sucrose, fructose, and glucose![67] This composition has major implications for coffee as a sustainable, ecologically sound fuel source, when the crop is grown under organic conditions.

In some countries, alcohol is produced from the pulp and then fermented further into vinegar.[68] In an integrated alcohol multiple product processing plant, the profit from the co-products along with the alcohol make processing the pulp worthwhile, even if gasoline is under a dollar per gallon.

What would alcohol yields be like if we looked at the whole fruit and seed? Using USDA figures of 1500 pounds of coffee seed and 5000 pounds of pulp, we might be looking at 150 gallons of alcohol per acre as a rough figure from normally fermentable sugars.[69] In countries where gasoline is over $4 a gallon, this value of the fuel would make the alcohol more profitable than selling the coffee beans to processors.

But more can be done with coffee pulp than simply making alcohol. The environmental benefits and profitability of building an integrated facility to produce alcohol and co-products are substantial. For instance, percolating and extracting the caffeine from the pulp would give farmers an organic, safe pesticide to use on a wide variety of crops.

You didn't think coffee made caffeine to get you high, did you? Caffeine is made as a pesticide by many plants. It works by jamming nerves into the "on" position, causing convulsions and death to insects that bite the coffee fruit. Since we are considerably larger than these bugs, we only get a mild buzz. In 2005, the India division of Coca-Cola was faced with a public relations disaster when farmers there began using caffeine-loaded Coca-Cola as an insecticide.[70]

Using caffeine as an insecticide would be a step up from trying to use coffee pulp as animal feed, since the caffeine makes the cattle nervous and lose weight. Kind of defeats the purpose of feeding it to them, wouldn't you say?

Other co-products could be earthworms or mushrooms. In Mexico, they raise worms on nothing but waste coffee pulp. It seems that caffeine has an aphrodisiac effect on worms. The castings are then used as potting mix for higher-value vegetable and fruit crops, such as melons.

There has been excellent work done on cultivation of mushrooms grown on coffee pulp alone or in combination with straw. In the case of oyster mushrooms, the flavor of the mushrooms grown on pulp was found to be superior to mushrooms grown on pure straw.[71] Mushroom cultivation on the mash byproduct of whole-fruit alcohol fermentation should be even better. Mushrooms efficiently use the amino acids in the coffee seeds to make a very balanced protein for human use (see Chapter 11). Yields of mushrooms are typically one pound of fresh mushrooms per pound of dry substrate.

Caffeine and other toxins are broken down into minerals by the mushroom mycelium, leaving the byproduct quite usable for animal feed or fertilizer.[72] Of course, the byproduct would be an excellent material to work into intensive vegetable garden soils. Coffee growers find it an excellent organic fertilizer to put back under their plants as **mulch**, since it contains everything needed to grow the next crop and helps stabilize soils on slopes.

Nicaraguan researchers have also found that spent mushroom substrate makes an excellent food source for worm production. Since there is no market for earthworms in Nicaragua, the researchers fed the high-protein worms to cattle and chickens, thus converting the red wigglers into valuable animal, and eventually human, food. It does take some getting used to watching cattle chew on a mouthful of worms.

Presently, most production of coffee results in environmental degradation. Each pound of coffee pollutes about 30,000 gallons of water, since the waste water from manufacturing plants is usually dumped into rivers.[73] During the three-month processing season, the rivers run with pulp and pectin, making the water a flow of ropy, gelatinous goo. The pulp in Mexico alone is conservatively estimated by the National Water Commission of Mexico to equal 80,000 tons of biological oxygen demand per year.[74] High BOD uses up all oxygen in the water in the process of decomposing, and all aquatic life is then suffocated. Over that three-month season, those 80,000 tons would be the equivalent of the sewage of nearly two million

people a day.[75] Since coffee is the third largest commodity shipped, after oil and water, this scenario is repeated all around the globe.

By making it a condition of their loans, the World Bank and other lenders have made sure that coffee is nearly always produced in surplus quantities. It wouldn't do to have a shortage of the world's most commonly consumed addictive drug and the second-leading drink, after water. (It's one thing to have an oil cartel, but quite another to have a coffee cartel.) But between the pulp's sugar, the sugar in the mucilage, and the fermentables in the seed, it could make sense to convert this very sustainable crop into fuel in surplus years. Alcohol production from coffee would give coffee producers an alternate market and therefore a floor price that the World Bank-designed surpluses would be unable to undercut. Having a price floor under coffee, created by having an alternative market for fuel,

---

PULP MORE VALUABLE THAN THE COFFEE?

In 2002, near Jalapa, Mexico, I was visiting a *finca* (ranch) which was de facto organic. Its main crops were coffee as an understory crop, with macadamia grown as the overstory. Under shade trees, its owner had also developed large foot-deep bins, using broken concrete for the borders, and filled them with coffee pulp. His red worms processed the pulp into fine castings, which he then harvested. Some of the harvest went into making fine potting soil for melon growers who were begging for him to produce more. But many of the castings were soaked in water and then filtered to make a potent **compost tea**, very high in phosphorus and humic acids, which the farmer had no trouble selling to various flower and vegetable growers in the area. The pulp operation made much more money than the entire seed part of the coffee crop.

CARTOON COURTESY OF PHIL FRANK

Fig. 8-22

would stabilize agricultural employment and prevent exploitive corporations from victimizing the growers.

The hidden value here is in employment. In years where coffee is in oversupply, which by now is a permanent affair, Mexican growers do not consider it cost-effective to pay even ten cents per pound to coffee pickers. Coffee harvesting employs an enormous percentage of rural people for several months and is a major source of revenue for all the small businesses in the coffee-producing regions.

In a permaculturally designed tropical or subtropical orchard, coffee as a multiple-use crop is a sustainable understory plant, which can have a valuable overstory of trees, such as macadamia, and nitrogen-fixing legumes.

Comfrey has extremely low alcohol yields per ton compared to fruit and grain sources. However, its total alcohol yield per acre is nearly 500 gallons, due to its enormous and efficient growth. The small-scale producer will find this an ungainly amount of material for solely making alcohol, but it might seem more worthwhile if alcohol were looked upon as a byproduct of making large quantities of liquid fertilizer and mulch/compost for other high-value crops.

COMFREY

Comfrey, a prolific leafy weed with enormous medicinal properties, grows wild throughout the Western United States. A common ingredient in many shampoos, cosmetics, and herbal preparations, it is also known as boneset. It has powerful effects when used externally to treat fractures, strains, thrombophlebitis, and bruises, and has also been used internally for stomach and respiratory ailments.

Comfrey produces about 80 tons of leaf and five tons of roots, wet weight, per acre.[76] Organic farmers often build their soil by turning under comfrey's leaf, which is an excellent source of carbon and minerals. It is a perennial crop that propagates from pieces of root, since it has nearly all sterile seeds. Gophers take root pieces through their network of tunnels, and it sprouts wherever they leave it. Once you've planted comfrey, you have it forever. You can literally dump a six-inch-deep layer of shredded yard waste over it, and it will send new shoots right up through it.

What's more, this plant is strongly mycorrhizal, both feeding and receiving huge quantities

of nutrients via symbiosis with soil fungi. Its rhizosphere (root zone) is a very dense community loaded with fungi and cooperative bacteria, so any rough organic matter used to mulch the crop is rapidly digested and incorporated into the soil life and eventually the comfrey roots. This means that comfrey requires no finished compost or special fertilizers when grown in most soils. A comfrey field could efficiently be used to process huge quantities of liquid or solid byproducts of distillation, if no other use takes priority.

"What is a weed? A weed is a plant whose virtues have not yet been discovered."

—RALPH WALDO EMERSON, 1878

Comfrey has very deep roots, which store huge amounts of energy for growth. These powerful, **hardpan**-puncturing roots dredge up nutrients from deep in the soil, usually below the level that most crops can reach, routinely penetrating up to ten feet.[77] The nutrients are then pumped into the leaves, which yield two to five cuttings per year. When the leaves are used as a mulch, they decompose on the surface, and these deep-mined nutrients become available to shallow-rooted plants. This makes the leaf matter of comfrey an excellent combination of weed-suppressing mulch and wide-spectrum fertilizer for other crops, when it is harvested and used for vegetable or row crop production.

Since the leaves have little lignin and a fair amount of nitrogen, they decompose quickly to a blackish goo and readily release their nutrients. Total nitrogen from comfrey leaf is 64.4 pounds per ton of dry matter. This would mean, in general terms, that one-quarter to one-third of an acre of comfrey leaf would provide enough nitrogen to fertilize an acre of other crops. Organic farmers have been successful in using shredded leaf naturally fermented in water for a few days as a foliar (leaf-related) fertilizer. On my farm, I have combined this with a general compost tea, with spectacular results.

Comfrey as an energy source doesn't initially appear all that attractive. It is a poor source for sugar or starch; its leaves contain a little less than 1% starch, and only 3.19% sugar. Its roots are even poorer in fermentable material, with 1.80% starch content and 1.06% sugar. Total gallons of alcohol per ton are 6.18 for the leaf and 2.82 for the roots. These are extremely low alcohol yields per ton compared to fruit and grain sources. However, comfrey's total alcohol yield per acre is nearly 500 gallons, due to its enormous and efficient growth.

The small-scale producer will find this an ungainly amount of material for solely making alcohol, but it might seem more worthwhile if alcohol were looked upon as a byproduct of making large quantities of liquid fertilizer and mulch/compost for other high-value crops. Large farm systems or industrial plants can run the crop through a protein extraction process to recover its valuable leaf protein, and then ferment the de-proteinized juice for easy alcohol production.

Comfrey will become more attractive as cellulose-to-ethanol conversion becomes cost-effective. Its low lignin content and highly available cellulose make it easy to process compared to more woody materials like sawdust and dried crop residue. The cellulose content of comfrey is about one-third of its dry matter;[78] that comes out to 3.5 tons per acre. This ought to yield about 630 gallons of alcohol per acre.[79]

But there is a tantalizing potential to double comfrey's annual biomass production and its various product yields. Since comfrey is a perennial crop with a closed, dense canopy and is harvested by mowing, it may be one of a handful of crops that can ideally use a drip irrigation/carbon dioxide system, which could be installed semi-permanently. Its dense upright growth pattern would work perfectly to trap carbon dioxide for increased plant growth. Combine this aspect with comfrey's ability to process thickly applied rough organic matter, and it may be a major contender in many regions.

It also does well under a canopy of trees, so it could be a companion crop under hazelnut, chestnut, or (if there is sufficient water) mesquite.

CORN

Corn is native to the Americas, spreading from its original home in Mexico over an enormous range, wherever people have gone over thousands of years. In some Latin American cultures, the ability to grow corn even extends to mountainsides so steep that farmers have been known to fall to their deaths by stumbling in their fields. Corn has been carried all over the world and planted nearly everywhere at least once. These adaptations have given the crop an enormous genetic range, from high-altitude arid plains to nearly swampy lowland farms.

Corn was once noted as being a terrible crop for soil erosion, but in recent years, returning the stalks to the soil instead of burning has helped to retain more organic matter. Precise use of fertilizer has also helped, and, if corn is used as one of a number of rotation crops, it will continue to have an important place in agriculture.

As a crop, corn has a poor ratio of top growth to seed. For every pound of corn produced, there are two and a half pounds of stalk, compared to the one-to-one ratio of wheat.

Feeding cattle corn wastes the starch, makes them sick (which costs plenty in veterinary bills), and shuts down their absorption of other nutrients.

The yield of corn varies between 200 to 392 gallons of alcohol *per acre*, on average. This is in the medium range for most of our energy crops. Corn's big advantage is that its yield *per ton* is 80 to 90 gallons; each kernel of corn can contain more than 70% starch. This yield will go up in the future as we convert the cellulose present in corn kernels.

So, corn is a top crop choice for handling the least material to get the most alcohol. Some might see that as a disadvantage, though, because it means less valuable byproduct tonnage to market.

High alcohol yield may not be the primary goal in future processing of corn. A pound of corn DDGS replaces one kilogram of corn and 0.823 kilograms of protein-rich soybean meal in feeding animals.[80] As human population increases, it is conceivable that potential high prices for animal feed could cause farmers to choose low-starch/high-protein varieties to emphasize DDGS volume.

Corn needs a lot of fertilizer and (as with any crop grown in massive quantities) needs a lot of pesticides and other chemicals. So why does the U.S. base its alcohol fuel industry on corn? Well, because right now we have lots and lots of it. The alcohol fuel industry is an outgrowth of trying to find a market for all that corn. This is a bit like the tail wagging the dog. But facts are facts. The U.S. plants, on average, over 70 million acres of corn per year (recently up to 78.7 million acres), which is typically alternated with soybeans. This yields 10.1 billion bushels, or 2.85 billion tons, of corn a year.[81]

In the U.S., corn was originally a crop with multiple uses: for human food, for whiskey as a value-added cash crop, and as animal feed. Over time, corn has lost out to wheat as a human food crop. Much of it is used for high-fructose corn sweetener (HFCS), which isn't exactly food. In fact, almost all U.S. corn is used for something other than human food, primarily for animal feed. Eighty-seven percent of corn goes for animal feed here and abroad, leaving only 13% for all other uses. Of that 13%, HFCS makes up 5%. All other uses—including cereals, corn chips, commercial starch, industrial products, liquor, and seed—make up the last 8%. Eleven percent of corn is used to make ethanol; however, all of this corn is processed for both alcohol and DDGS fed to livestock, so it's part of the 87%. This means the alcohol industry is a long way from having to worry about affecting the human food uses of corn in their conversions of available corn to better-quality animal feed.

What is not widely acknowledged is that cattle have not evolved to eat corn. They do not have to graze, as sheep do, nor do they follow an omnivorous diet like pigs. Instead, cattle prefer browsing, eating woody shrubs. Unlike prairie-evolved grazers, which have divided cloven hooves so they can get traction in tough sod, cattle have wide, flat feet, which are great for soft forest soils. Cattle's browser teeth are optimized for pruning brush, not eating grass.

Feeding cattle corn wastes the starch, makes them sick (which costs plenty in veterinary bills), and shuts down their absorption of other nutrients. Up to 80% of the starch in corn can go largely undigested through a cow's four stomach compartments, which contain 45 gallons of mixed organisms. Perhaps the most serious thing that can go wrong with corn-fed cows is feedlot bloat. The **rumen** (first stomach) is always producing copious amounts of gas, which is normally expelled by belching during rumination. But when the diet contains too much starch and too little roughage, rumination all but stops. A layer of foamy slime that can trap gas forms in the rumen, and it inflates like a balloon, pressing against the animal's lungs. Unless action is promptly taken to relieve the pressure (usually by forcing a hose down the esophagus), the cow suffocates.

A corn diet can also give a cow acidosis. Unlike our own highly acidic stomachs, the normal pH of a rumen is neutral. Corn makes it unnaturally

PHOTO COURTESY OF ERIK VAN BOCKSTAELE

acidic, however, causing a kind of bovine heartburn, which in some cases can kill the animal, but usually just makes it sick. Acidotic animals go off their feed, pant and salivate excessively, paw at their bellies, and eat dirt. The condition can lead to diarrhea, ulcers, bloat, liver disease, and a general weakening of the immune system that leaves the animal vulnerable to everything from pneumonia to feedlot polio.[82]

FODDER BEETS

Fodder beets, or mangel-wurzels, are not a well-known crop to American livestock growers, but they are often used in European, Scandinavian, Australian, and New Zealand agriculture. The first fodder beets were specialized varieties of mangel-wurzels and some sugar beets. Today they are a hybrid, a combination of both. They were developed in the 1920s to replace imported grain, and, self-sufficiency being the mother of invention, they are becoming one of fuel producers' most important crops.

Fodder beets are huge in size, dwarfing the sugar beet, each weighing as much as 15 pounds. Although they often have a lower percentage of sucrose (table sugar) than sugar beets, they're much heavier producers per acre. Some varieties of fodder beets have up to 20% more total sugar than sugar beets, as well as the higher per-acre tonnage. Fodder beets do not contain as much crystallizable sucrose as sugar beets, and they contain a small percentage of starch (up to 8%), which interferes with crystallization of table sugar, so they are unsuitable for the sugar industry. Their other sugars are perfect for feed and fuel in the U.S. agricultural economy.

Root crops generally produce higher yields of usable carbohydrates than grains do, and beets are no exception. Unlike grains, beets store their energy primarily as sugar in the roots, not in the seed or foliage. Older varieties of fodder beets have produced 850 to 950 gallons of alcohol per acre,[83] compared to sugar beets at 400 to 770 gallons per acre. Some of the newer European varieties of fodder beets are reporting well over 1000 gallons per acre.

Fodder beets store longer with less sugar loss than sugar beets and are less prone to being stricken with curly top virus, a devastating disease of sugar beets. In the Northern United States and Canada, fodder beets would be a good rotation crop with Jerusalem artichokes and sweet sorghum. You could also rotate soybeans, but their yield is reduced in extreme northern areas.

Fodder beets are to cold temperate regions what cassava and sugarcane are to the tropics. They like an alkali soil with pH 6.5 to 8.0, so lime is needed. Fodder beets don't grow on marginal lands. They need deep, well-drained soil and lots of water, especially during the driest months of June, July, and August. Usually, no irrigation is done after the last day in August, so as to allow the sugar to set and concentrate out of the remaining tops and into the roots.

Beets of this size will send roots down five feet, helping to break up hardpan in many cases. Hardpan is caused by tractors or animals compacting the soil several inches below the surface. This compaction can make the soil impervious to water and impenetrable for many crops' roots. In fact, I often recommend using fodder beets as a cover crop after ripping (the deep loosening of the soil to break up hardpan). The beets exploit all the new cracks in the soil and prevent them from recompacting, and leave behind important organic matter later relished by worms. Turning the whole beet into the soil dramatically raises organic matter content.

I often recommend using fodder beets as a cover crop after ripping (the deep loosening of the soil to break up hardpan). The beets exploit all the new cracks in the soil and prevent them from recompacting, and leave behind important organic matter later relished by worms.

Planting a fodder beet crop is facilitated with a precision-type seed drill, and requires three to six pounds of seed per acre. If you start with about 40,000 seeds per acre, ideally 28,000 plants will survive. Rows should be spaced 20 to 22 inches apart. The seeds are planted at a shallow depth of $^3/_4$ to one inch, and benefit by having the soil tamped or compacted immediately following planting. By the time the second set of leaves is present, you should thin plants to seven to eight inches apart; any closer, and the sugar yield will drop.

High levels of fertilizer salts inhibit sprouting and starting growth, so it's best to fertilize in the fall prior to spring planting. This sinks the salts to a depth that's valuable to the plant as it grows taller and reaches deeper. Of course, it's best to avoid these issues with fertilizer salts and use organic methods. Begin the cropping with about half your nitrogen already in the ground and add the rest later.

Fig. 8-24

Fodder beets are resistant to fall frosts and can be harvested in late fall.

Don't store the tops with the beets. Fodder beets are stored in clamps (heaps), either outside or in partially underground buildings. Beet tops create a lot of heat in a clamp pile or pit due to decomposition, which encourages rotting. A possible strategy would be to have cattle come in and defoliate the tops in a rotational grazing pattern prior to harvest.

Clamped beets preserve their sugar better if they're clamped dry. The crop stores best over the winter at about 34 to 36°F. When clamping, wait for a couple of days of freezing temperatures and then insulate the beets. Your pile should be five feet high or so to generate enough metabolic heat (the beets are still alive) to stay a few degrees above freezing. The simplest clamping procedure is to cover the pile first with a clear plastic sheet, then with two to three feet of straw, and finally with a black plastic sheet held in place with old tires. Less insulation is needed in less frigid climates.

For alcohol production, beet tops are usually chemically defoliated in the weeks before harvesting to raise the sugar yield 10% and the overall beet weight about 15%. The use of defoliants has numerous agricultural and ecological drawbacks. If you're running a livestock operation, and alcohol is simply a valuable byproduct of fermentation-improved animal feed, the animals can eat the tops instead. The root sugar lost by not chemically defoliating is not as important as the weight gain from the cattle eating the high-nitrogen tops. Since beets need soft, friable (easily crumbled) soil with good **field capacity** (water retention) and reasonable drainage, it's a good idea, in my opinion, to suffer a slightly lower alcohol yield and return much of the foliage to the soil, via the process of passing it though a cow, to keep the carbon and humus level high.

Unfortunately, it's not practical to grow beets in the same ground more than once every four years. After several years in a row, the soil becomes saturated with nematodes, and it's difficult, if not impossible, to grow a crop for several years after infestation has established itself. Rotating fodder beets with other diverse food and energy crops should prevent a nematode problem and maintain soil health.

A few words on fodder beet byproducts: Their value varies since they're priced in comparison to the animal feed crop they're replacing. In the U.S.

West and Northwest, dry unfermented beet pulp (from sugar extraction industries) is considered the ideal replacement for barley. The prices for this beet pulp as a byproduct of the sugar industry are lower than for pulp from fermentation. Fermented fodder beet pulp is used by the dairy industry, since it produces a higher milk yield in the first third of a cow's lactation cycle. The beef industry generally prefers to use fermented beet pulp instead of DDGS from corn. After distillation, an acre of sugar beets yields approximately a ton of cattle feed. The same acre of fodder beets gives you close to two tons, with higher protein. East Coast producer cooperatives may find their best market is Europe, where fermented beet pulp is a high-priced, undersupplied commodity.

Fig. 8-25

FORAGE PLANTS

Forage plants are crops typically grown in pastures. They often are a collection of many types of green leafy plants mixed with grasses. Many very young green leafy plants often start off with 65% sugar and/or starch (dry weight) and don't need a strong support structure, so they haven't started to produce lignin. In these early growth stages, there's often a surplus of free sugar. A few days later, much of the carbohydrate production shifts toward cellulose production, which is not as easy to process on the small scale yet, compared to sugar or starch.

Many of these forage crops can be "green chopped" to grow right back again, allowing several cuttings in a year. If these plants are juiced, the juice contains most of the plant's sugar and protein. The protein can be extracted, and the sugary juice fermented. Extracted protein can be dried and sold as animal feed, with its price historically double that of soybean meal.

The experience of the New Zealand Ruakura Agricultural Research Station offers a guideline for the amount of protein possible in leaf protein extraction. The station processed a pasture mix of ryegrass (*Lolium perenne*) and clover (*Trifolium repens*). These forage plants contain only 20% total dry material; the rest is water. The dry weight of the forage was as high as 7.2 tons per acre per year. The percentage of protein per dry weight of forage is about 25% to 26%.[84]

The study focused on the carbohydrates produced in the process of making **leaf protein concentrates (LPC)**. When the forage crops were pressed for juice, the soluble part of the carbohydrates comprised only 2% DM (dry material) of the liquor, but about a quarter of this was readily fermentable fructose, glucose, and sucrose. The rest of the carbohydrates were insoluble. When the nonsoluble carbohydrates in the juice were treated with sulfuric acid and 220°F heat, the hemicellulose broke down into the sugars arabinose, galactose, and xylose, along with some glucose.[85]

Thus, based on 7.2 tons of dry material per acre of forage crops, alcohol yield from juiced forage (not counting the cellulose) would be about 100+ gallons per acre. But the fermentation would have to be done by the new GMO yeasts that use C5 sugars.

Even though the amount of alcohol is low, it would be worthwhile recovering as a byproduct of

LPC production. This is because there are several tons of the biomass, which is mostly cellulose. The amount of potential alcohol from this easy-to-pretreat cellulose would be several hundred gallons per acre. So adding the de-proteinized sugary juice back into a processing chain for cellulose would be a bonus.

JERUSALEM ARTICHOKES

The Jerusalem artichoke (*Helianthus tuberosa*), a truly American plant, is a promising new crop for alcohol production. Pilgrims traded with the Indians for this nutritional tuber; like buffalo gourd, it migrated with the natives from camp to camp. It has nothing to do with the Middle East, and it's a sunflower, not a thistle like the true artichoke.

Agriculturally, the plant is considered a vigorous pest in some areas, a delicacy in others, and, more and more lately, a good cash crop. The variety used

Fig. 8-26 Freshly dug Jerusalem artichoke tubers. *This mass of not quite mature tubers came from two small pieces of tuber planted a few months prior.*

COURTESY OF LYNN ALAN WILLIAMS, UC DAVIS DEPARTMENT OF VITICULTURE AND ENOLOGY

Fig. 8-27 Flowering Jerusalem artichokes. *This field is nearly ready for harvest.*

Fig. 8-28 Processing Jerusalem artichokes. *The fermentables in the tuber on the left produce the amount of alcohol in the long-stemmed beaker on the right. The remaining pulp is dried into animal feed in the small tray on the lower right. Note the viscosity of the tuber mash.*

in North America for fuel is called the French white mammoth, which actually originates from Swedish breeding programs of the 1950s, in which the native North American Jerusalem artichoke was crossed with an improved sunflower. The Swedes hoped to produce sugar from the plant's stalk in a manner similar to sorghum and sugarcane, and to achieve yields as good as sugar beets. The plan was to cut the stalks, and leave the tubers in the ground. The result was a hybrid that grew more vigorously than either of its parents, but the stalks never did act like proper sugarcane.

Fig. 8-29 Jerusalem artichokes, mature tubers.

Jerusalem artichokes are an excellent food for diabetics. They contain inulin (not insulin), a sweet but nondigestible carbohydrate that can't upset a diabetic's metabolic balance. About 75 to 80% of the carbohydrates in Jerusalem artichokes are inulin.[86]

The tall, stalky Jerusalem artichoke plant has broad dark green leaves and grows dozens of small yellow flowers, unlike sunflowers grown for seed, which bear one big sunny bloom. In the wild, even under poor conditions, Jerusalem artichokes can grow six to eight feet tall. Fertilized and watered, they can double that height. Since the dynamics of the choke's soil nutrient use have not been thoroughly studied, no real formulas for optimal fertilization have been tested. Good results have been achieved using well-composted manure, pH-corrected with lime before addition, with and without stillage. The plant also responds well to heavy mulch applied in the fall after planting, where the tubers are protected from freezing until they sprout in the spring.

Chokes grow in a wide variety of climates and soil types. Avoid mucky, swampy, or poorly drained soils, which can expose the tubers to infection and cause them to eventually rot. The chokes thrive in sand as long as they get some water on a regular basis; in fact, sandy or light soils high in organic matter make harvesting easier. Tubers will be smaller and more numerous in clay soil, larger in loose soils. Given present harvesting equipment, the larger the tuber, the better.

Jerusalem artichokes tolerate a wide range of pH factors, growing vigorously in ranges of 4.4 to 8.6. Use a pH of 6.0 to 8.0 for maximum sugar yield. If you have to raise your soil's pH, you can use pH-corrected liquid stillage, which contains a whole complement of minerals. Raising the pH of your soil with lime will work, but best yields are

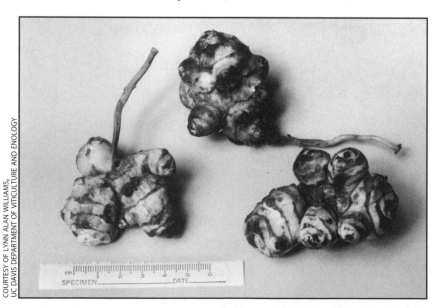

realized with balanced levels of calcium, potassium, sodium, and magnesium, along with a healthy dose of trace minerals.

Although the Jerusalem artichoke has yielded profitable crops with as few as 20 inches of rain, it does best with at least 30 to 40 inches, or an equivalent amount of moisture in irrigation. Amounts over 50 inches seem to reduce yields in heavier soils. The plants will tolerate huge amounts of rain (and do, in the Pacific Northwest) as long as the soil is well drained.

Good results have been obtained with overhead sprinklers for irrigation, and by normal hilling and irrigation. Depending on your location and soil capacity, the irrigation interval is from seven to ten days. In drip irrigation, much less water is needed. Adequate water is especially important while tubers are in the midst of filling (sugars descending from the stalks down into the tubers), just before and during the plants' flowering. Once flowering is complete, water can be reduced to a minimum or withheld to help the tuber drain all its sugar from the stalks.

Planting is done with a potato planter or by hand when cheap labor is available. For the densest foliage canopy, make rows 24 or 30 inches apart, depending on your equipment. Wider plantings, up to 34 inches, have also yielded good returns. Place pieces of tuber about four inches deep every 12 inches. If you intend to cut your foliage early for feed, some literature recommends planting every six inches. You'll be using about 600 to 800 pounds of tubers to the acre.

The seeds produced by the numerous flowers are, for the most part, sterile. The plant primarily reproduces from tubers transported by animals— who either eat them and poop them out, or, like gophers, transport pieces underground.

The larger your seed pieces, the better your top growth. Pieces as small as half an ounce will produce reasonably healthy plants, but your biggest sizes should weigh about one and a half ounces. Although even larger pieces work fine, they don't produce a significantly higher amount of tubers, and do increase your initial seed cost.

Planting is best done in the fall. It can also be done in the spring or as soon as the ground has thawed and can be worked. Be careful, though, not to try a fall planting in soils which stay soaked or which flood in winter. Snow cover is fine.

Fig. 8-30
Sunflower beetle damage. *Occasionally, such damage has been found in isolated stalks of Jerusalem artichokes, but no major infestations of this pest have been reported.*

The dense foliage is an efficient collector of solar energy. Foliage will form a closed canopy 30 to 40 days after sprouting. Such vigorous growth chokes out weeds and is reported to overrun and kill the normally invincible Johnson grass. You won't need repeated mechanical cultivation to control weeds with Jerusalem artichokes, as you do with other fuel crops like beets.

A modified potato or onion harvester is preferred for harvesting tubers (once the tops have been cut and removed). Equipment modifications vary from machine to machine; the primary changes are to increase the agitation of the soil as the tubers are dug, and to add rods to the elevator/screen so tubers are retained (most of them are smaller than potatoes or onions).

Rotate nitrogen-fixing crops in marginal soils, or in warmer areas plant a nitrogen-fixing cover crop over the winter after the tubers are planted. Using Jerusalem artichokes as a rotation crop to smother weeds before replanting with other food/fuel crops will both reduce weed pressure for the next crop and break any pest cycles from the prior crop.

Jerusalem artichokes have no serious insect pests. Researchers have reported isolated cases of sunflower beetle, and it's possible that intense cropping may necessitate pest control in the future.[87] Occasional corn borers find themselves an artichoke crop, but they don't appear to affect yields.

No major plant diseases have been found to affect the Jerusalem artichoke. Sclerotinia wilt will occasionally infect and kill isolated plants or patches of crop, but it has never been epidemic. Harvesting the tops immediately after tuber filling reduces the incidence of sclerotinia, since it usually strikes after the plant has flowered.

Top growth should be cut once or even twice early in the plant's growth. Cuttings done five to six weeks after the plant develops its second set of leaves have yielded 15 tons of forage at 10 to 14% protein. This makes great animal feed, but will reduce tuber yield by 25+%. Cut shoots will recover and rebuild their canopy within a month, but the plant must operate with much less photosynthetic efficiency per acre until closure occurs. Tops cut after the tubers have filled have less protein than young shoots, but the tubers are the destination of the missing protein, and feed value can be recovered in the byproduct mash. Fermented tuber pulp yields a whopping 26+% protein feed high in lysine, an amino acid usually lacking in grain or green forage.

The whole idea of cutting tops may have to be revisited as cellulose processing gets more common. It may turn out that three cuttings of tops, at more than 40 tons per acre, will yield more alcohol than harvesting the tubers. If so, then this crop would become a perennial semi-permanent crop, avoiding all the costs of harvesting and then replanting tubers.

Realistic yields for tubers grown as annuals are 15 to 25 tons per acre; exceptional yields are 30 to 40 tons per acre; and a few experimental yields report 60 tons per acre. Based on the reported range of 15 to 27.7% fermentable carbohydrates,[88] it's reasonable to expect yields between 550 and 750 gallons per acre.[89]

Tubers can be stored in several ways. You can slice and dry them into chips of 10 to 12% moisture similar to sweet potato or cassava chips. Investigators in the 1930s reported that loss of carbohydrates is negligible and that the chips last indefinitely. Circulation of dry warm air at 100°F has proved adequate for drying, or you can go as high as 200°F. Chips should be stored in a cool, dry place.

Tubers can be stored underground in the field during winter, weather permitting, while you harvest them as needed. Storing the tubers at 36°F in an environment filled with carbon dioxide or nitrogen will preserve them with minimal losses.

Be sure to wash your tubers in a 0.025% sodium hypochlorite solution to sterilize their surface; then air dry and cool rapidly.

Comparison of Three Inulin-Fermenting Yeast

PARAMETER	K. MARXIANUS UCD (FST) 55-88	K. MARXIANUS SU (B) 79-267-5	S. ROSEI UWO (PS) 80-38
Maximum specific growth rate, μ_{max} (h^{-1})	0.41	0.19	0.1
Growth-Associated Constant	12.5	9.4	6.8
Ethanol Yield, $Y_{p/s}$ (g EtOH/g Sugars) (% Theoretical)	0.45 (88%)	0.43 (85%)	0.44 (88%)
Biomass Yield, $Y_{x/s}$ (g Dry Wt/g Sugars)	0.04	0.05	0.05
Initial Sugars, S_0 (g/L)	100.5	100.2	100.3
% of Initial Sugars Utilized	92.0	90.0	91.0
Maximum Ethanol Concentration, P_{max} (g EtOH/L)	44	42.8	42.5
Maximum Biomass Concentration, X_{max} (g Dry Wt/L)	3.9	4.7	4.6

Fig. 8-31 Comparison of three inulin-fermenting yeast. All do a very good job converting inulin to ethanol.

IIEA

Processing for Alcohol

The tuber's carbohydrates, called inulin, have been assumed to be a single compound composed of fructose and a few glucose molecules, with a total of 35 sugar links in all. It now appears that in artichoke tubers the inulin family may include several shorter chains, as well. Although inulin can be as large as a starch molecule, it's much easier to hydrolyze. Once that's accomplished, normal yeasts or slightly different varieties ferment inulin's fructose and glucose. An alternate process makes use of some of the more obscure varieties of yeast that directly ferment inulin to alcohol.

The old-fashioned way to hydrolyze the inulin to convert it to fermentable sugar uses heat and acid, but there is also the new commercial enzyme, inulinase. Regardless of the method used, steps must be taken to prepare tubers for hydrolysis. It's a good idea to wash the tubers to remove mud, pebbles, and bacteria. Tubers float pretty well, so agitating them will shake and wash loose most of the crud.

Using either method, tubers are first shredded and pumped to your cooker. The pulp is viscous, so you'll need a high-quality **positive displacement pump**, or better yet, a **progressive cavity pump**. Cook with as little water as possible, just enough to prevent burning. The pulp thins out a bit as it cooks. Using sulfuric acid, the pH should be dropped to 2.0 for hydrolysis. Then cook at a boil for an hour to break down all or almost all of the inulin to simpler sugars that a variety of yeasts can handle.

Enzymatic hydrolysis wasn't available 20 years ago, but it is now. Enzyme magnate Novozymes, for instance, makes a product called Fructozyme which hydrolyzes inulin. The total amount of enzyme needed is based on the total inulin in your feedstock on a dry-weight basis. You'll need 1.48 pounds of Fructozyme for every ton of dry inulin in your tubers, which will be somewhere between 15 and 28% inulin, based on their wet weight.

The pH needs to be between 4.0 and 5.0. As you begin heating the mixture, you can use a small quantity of enzyme to help liquefy it, and bring it up to a boil for maybe ten minutes. Use the rest of the enzyme after you have lowered the temperature to about 160°F, and hold the mixture between 140–160°F for ten to 12 hours.

At 12 hours, you'll get about 90% conversion of the inulin.[90] You can then ferment the wort with standard yeasts. This method does increase the

possibility of bacterial contamination, since you are holding the wort for so many hours before adding yeast. Constant slow agitation can slash this time during enzyme hydrolysis and reduce the contamination risk.

If you plan to use the alternate "specialized" yeast method of processing, you'll add less acid, only bringing the wort down to pH 4.0 instead of 2.0. You'll still cook your pulp for an hour at a boil. Since much of the inulin will remain unconverted, you'll want to use an inulase-producing yeast such as *Candida pseudotropicalis*, *Candida kefyr*, *Torulopsis colliculosa*, *Kluyveromyces fragilis*, or *Kluyveromyces marxianus*. Canadian trials have rated *K. marxianus* the fastest acting yeast with a high enough alcohol tolerance to metabolize the carbohydrates. Most of these yeasts are selective, metabolizing small sugars first, and then the more complex carbohydrates like inulin. There is evidence that most inulase production occurs early in fermentation.

So why would you go to the extra trouble instead of using a lower pH and normal yeast? The specialized yeast method is a useful technique if you have access to cheese whey (discussed later in this chapter), which can replace the water in the process. Since special yeasts can also eat the **lactose** in cheese whey, you end up with a higher concentration of alcohol. The quality of the protein recovered in the byproduct mash is also much higher than if you use a lower pH.

Minimum inoculation with *Kluyveromyces* should be 50% above the level listed in Chapter 7 on fermentation. And because inulase-producing yeast is not made industrially, you'll have to breed

Fig. 8-32

your own from sample slides obtained through your local agricultural college's culture collection or the American Type Culture Collection. This may be changing now that more cheese factories are using this yeast to turn their whey into alcohol.

You'll get better results inoculating your batch with more than the usual amount of yeast. Breeding extra yeast requires a little extra time and money, but decreases fermentation time, helps avoid contamination, and increases alcohol yields. But even if you use the enzymes, it is probably still worth putting in some *Kluyveromyces* to pick up any non-converted inulin.

"Thousands of sites around Australia are still contaminated with serious concentrations of DDT. Over 1600 of these highly polluted sites now have hope of being cleaned up due to seaweed. A three-university team discovered that with enough seaweed worked into a contaminated site, 80% of the DDT vanishes in six weeks. The sodium in the seaweed opens up the clay soil, and the rich nutrients feed soil microbes (probably fungi) that breakdown the persistent nerve poison."

—RICHARD MACEY, IN *THE AGE* (AUSTRALIAN NEWSPAPER), MAY 10, 2004

Fermenting a Jerusalem artichoke's thick pulp tends to trap lots of carbon dioxide gas, resulting in a lot of pulp and yeast rising out of the batch, possibly plugging your fermentation lock and creating a potential for bursting your tank. This will be less likely with enzymatically processed mash. Fill your tanks less than you otherwise would, to account for the head of solids. Floating mash also immobilizes a lot of yeast above the solution, killing it, and lengthens your fermentation time. Thorough, slow-speed mechanical agitation prevents all these problems.

Whole-pulp fermentation requires that you add some fluid to make the pulp workable. Of course, diluting the 18% fermentable content of your tubers will reduce alcohol yield to around 6 to 7% in the final mash. A partial solution to the low final alcohol level is to dilute the tuber pulp with whey instead of water. Whey's 4 to 6% lactose reduces the dilution effect of the liquid, and whey protein increases the value of your mash byproduct as animal feed. Ideal yeasts for inulin are generally good for lactose, as well. A combination of *Kluyveromyces marxianus* and *Candida pseudotropicalis*, in particular, are worth trying on whey and tuber pulp mixes, since the pulp byproducts make

an excellent dairy feed. The same tank truck that picks up fresh wet pulp for feed can deliver cheese whey.

We've been discussing what to do with fresh wet tubers. If you have a waste heat source available, you can slice and dry the tubers in a method similar to cassava chips. Dry chips can be hammermilled to a fine-grained meal, and a lot of the fermentation and hydrolysis issues become more manageable.

Studies done in Germany in 1959 comparing feed values for hogs demonstrated 274.5 kilograms of Jerusalem artichoke tubers yielding the same 100-kilogram weight gain as 286.4 kilograms of potatoes, or 315.5 kilograms of grain.[91] Compared with corn or barley feed, tubers are 1.15 times as valuable. In France, the feed has been a dairy cattle staple for years. Jerusalem artichoke feed is palatable and an excellent boost for milk production. The fermented pulp should outperform the above figures for raw tubers.

Once planted, Jerusalem artichokes can take over a field. If you decide to replant with another crop, dig up the majority of tubers once or even twice, and sun and wind will do most of them in. Then allow some hungry hogs to occupy the field for a month. The pigs, with their excellent sense of smell, will efficiently root out just about every tuber.

LICHENS

Massive bogs of lichen grow in far northern lands such as Iceland, Sweden, Siberia, and Alaska. Iceland moss and reindeer moss were tried early in the 1900s as a source of alcohol. Their highest yields were over 60 gallons per ton of moss.

Lichens contain a complex carbohydrate, difficult to break down, called lichenin, that is related to common starch or sugar. There are no commercially available enzymes for reducing lichenin; the best results are achieved by cooking the entire lichen with steam and acid under modest pressure, about 50 pounds per square inch, adding either 2.5% hydrochloric acid or 6% sulfuric acid. Most small-scale users will need to avoid processes that involve expensive pressure vessels.

Twenty years ago, I theorized that if global warming continued to widen the growing season for tundra plants in the Arctic, lichen might become an almost sustainable feedstock. But then I concluded that if the weather changed enough to make lichen a more practical feedstock, we would be in deep trouble. Now, a recent study has confirmed my nightmare, and that accelerated growth is already a fact in the tundra.[92]

Tundra and bog ecosystems are quite delicate, and regrowth of lichens is slower than would be practical for a sustained industry. This feedstock should not be harvested massively, since it could seriously damage populations of migrating deer, reindeer, and a whole variety of other life forms directly or indirectly dependent on a tundra ecosystem. It may become necessary to reforest the tundra to immobilize and prevent rapid melting of winter snow, in order to prevent the decomposition of its organic matter to greenhouse gases. If so, the process of berming and tree planting may temporarily make available huge quantities of lichen.

MARINE ALGAE

A few species of algae are commercially grown for human consumption; these are also excellent sources for energy production.

Laminaria digitata was used in the early part of the 20th century for alcohol production. Currently, Chinese **mariculture** uses *Laminaria japonica* (kelp). Laminaria and several other genuses contain **laminarin**, a complex carbohydrate comparable to starch in land plants. The plants use this laminarin to store solar energy, then hydrolyze it to glucose (probably with an enzyme). Further plant metabolism converts it to the so-called "sugar alcohol" **mannitol**, which is a common short-term energy storage chemical in marine plants, just as sugar is in a land plant. Although technically an alcohol, by chemical definition mannitol takes the form of a white powder. It has a number of uses, including as the active ingredient in some baby laxatives, and has long been a favorite of cocaine dealers, who use it to dilute the concentration of the drug.

Laminarin can comprise 33 to more than 35% of algae's dry matter,[93] which may be as low as 10% by weight. At those rates, your yield is about 55 gallons of alcohol per ton of dried seaweed. By 1962, Chinese production reached 4.8 dry tons per acre. Reports from later in the '60s were already showing 70% increases in productivity.[94] Chinese breeding programs are ongoing, and varieties that will live in warmer water and produce more biomass continue to be developed.

Other marine algae are appropriate in warmer water, while kelp prefers cooler water. Intensive production of *Kappaphycus alvarezii* in Vietnam has been quite remarkable, since central Vietnam has an extensive system of shallow lagoons and estuaries where shrimp and algae can be grown. Yields as high as 4.5 tons per acre dry weight have been achieved in six months (October–March), with the balance of the year used for shrimp production.[95] Both crops have higher yields with this rotation. This species of marine algae has been cultivated in ponds in Israel, as well.

Other algae with potentially high amounts of recoverable carbohydrates are *Chondris, Rhodymenia pertusa, Joculator maximus, Corallina officinalis, Porphyra* (nori), *Fuchus, Palmaria,* and especially *Rotomeria palmaria* (the last two are both dulse).[96] Some of these genuses contain laminarin, and some contain what is called Floridian starch, a chemically close relative of the starch produced by land plants. Many of these algae are currently cultivated along various U.S. coasts. Many of the carbs in these species, for instance in *Dilsea edulis,* are almost identical to corn starch.[97] One of the handling differences between these carbs and corn starch is that corn starch gelatinizes at a higher temperature and after longer cooking (see Chapter 7).

There has not been enough of a market thus far to support commercializing the production of enzymes to easily convert the laminarin to fermentable sugars. Good success at the laboratory scale has been obtained by using enzymes extracted from Helix snails.[98] (Mashing an infinite number of snails does not at first glance appear to be a constructive solution.) Although some success has been reported with enzymes from wheat, barley, hyacinth bulbs, and marine bacteria, the breakdown isn't all fermentable. So although some

Fig. 8-33
Cultivated kelp.
In Nature, kelp anchors itself to the bottom in water less than 100 feet deep. Kelp cultivation on nets can cover hundreds of acres of ocean without regard for depth and makes for easy harvest, fertilization, and carbon dioxide enrichment.

PHOTO COURTESY OF <WWW.ALGAEBASE.ORG>

Fig. 8-34 Modern net cultivation of kelp. This method is more efficient than single rope cultivation.

of these enzymes, such as barley malt, may do part of the breakdown, to really get all the sugar out we are stuck with acid hydrolysis for now.[99] But given the mild conditions required, it is a viable method.

Hydrolyzing laminarin to fermentable sugars can be as simple a matter as heating the shredded seaweed to a boil in a solution, using sulfuric acid to lower the pH to about 2.0. Early researchers found that cooking the seaweed under pressure (15 to 20 psi), which raises the boiling point of the

ALGAE AND BIODIESEL

There has been some discussion about producing biodiesel using algae in constructed ponds. Algae can convert sunlight into fats more efficiently than trees under the right conditions. But the capital and operating costs are much higher than they are for land-based crops.

As I detail here, marine algae is a rich resource for alcohol, much more productive and cost-efficient than algae produced for biodiesel. In the long run, mixtures of alcohol that contain 1% biodiesel and **cetane**-improving chemicals made from biomass will very likely be the diesel fuel of choice. This means the market for biodiesel will be limited to its use as a lubricant in these fuel mixtures.

Under these conditions, algae-produced biodiesel may be at a distinct financial disadvantage compared to biodiesel derived from nuts or castor beans; we already produce enough vegetable oil in our currently inefficient ways to make up the 1% lubrication additive we need.

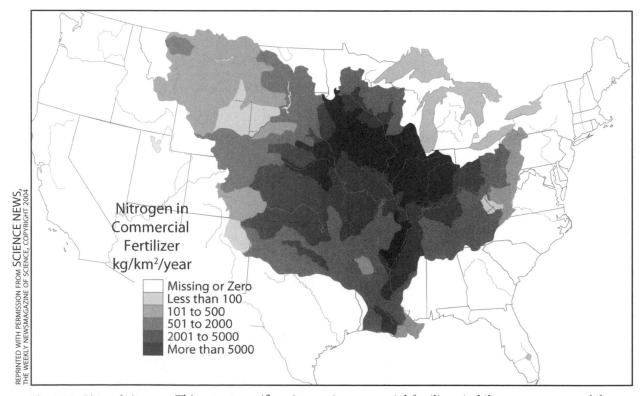

Nitrogen in Commercial Fertilizer kg/km²/year

Missing or Zero
Less than 100
101 to 500
501 to 2000
2001 to 5000
More than 5000

Fig. 8-35 River of nitrogen. This map quantifies nitrogen in commercial fertilizer, in kilograms per square kilometer per year, applied by farmers to their fields throughout the Mississippi River watershed. Typically, 15% of the total drains into rivers that feed the Gulf of Mexico.

solution to over 250°F, yields the highest final fuel output. Cooking with 250+°F live steam at **atmospheric pressure** would have a similar effect.

If you don't use high temperatures or 15-psi live steam, and simply boil the mixture, you can still get close to 90% conversion. Cooking the acidic mixture at a simmer without pressure for five hours will do the job nicely. Pressure-cooking reduces the time to minutes, however.

Resulting sugars are glucose and fructose, which ferment easily with standard yeast. Before shredding or cooking, be sure to rinse away residual sea salt, since salt inhibits or kills yeast.

In North America, the best harvest time for algae is fall. The algae must be used quickly after harvesting, or actively dried, for instance using waste heat from your alcohol plant. Laminarin is not a stable compound like starch, and air-drying laminaria results in large losses of carbohydrates.

Laminarin takes two forms: one quite soluble in cold water, the other only soluble in hot. The hot-soluble ("insoluble") form will precipitate out in the presence of ethanol, without much or any heat. *Laminaria digitata* contains mostly the insoluble form, and *Laminaria hyperborean* has the more easily soluble form.[100] *Laminaria cloustonii* has been reported as the best source of the insoluble form.[101]

Using Marine Algae for Alcohol

Although there are several species of marine plants with high enough concentrations of carbohydrates to be used for alcohol, to my knowledge there are no comprehensive studies as to which algae have the highest amounts of usable fermentables. In the early 1900s, a few small commercial enterprises produced alcohol from marine algae. In the 1940s, some algae-to-alcohol operations appeared in Norway and Sweden.

It's not that it's difficult to make alcohol from algae; the problem is that there are so many more valuable products to produce from it, such as carrageenan, agar, and dozens of valuable compounds. Alcohol is a low-priced commodity in comparison.

Economically, laminaria is not currently practical for production in North America because it requires high human labor input. This is primarily because insufficient industrial uses for marine algae are present in the U.S., with petroleum prices being so low. In developing countries with high unemployment, a labor-intensive industry such as kelp cultivation provides jobs, food, alcohol, fertilizer, high-value industrial substances, and methane.

China supplies the world's demand for algae. China's cultivated marine algae production is now hundreds of times the size of the United States' wild algae harvesting industry. Carbohydrate-rich kelp is commercially grown in China and Japan as a food source for bread and other valuable products.

Currently, laminaria is largely limited to use as an ethnic food and in a small industry producing agar and other industrial chemicals. With the ever-spiraling price of oil, cultivation of seaweed will not only become economical, but will become necessary as the effects of Peak Oil make availability of inexpensive chemical feedstocks a thing of the past.

As energy prices continue to rise, or when oil-based industrial material that could be replaced with algae-sourced material becomes sufficiently expensive, a domestic algae mariculture industry will boom. Engineering and ecological study should be put into this field. This research could increase the density of yield, lowering the cost of machinery to harvest and tend the algae. It's very possible that the energy return on this crop could be more that 15 to 1, much higher than even sugarcane—with virtually no fossil fuel used in the

Fig. 8 36 Suffocating stretch. This map depicts 20,700 square kilometers of the dead zone in the 2001 Gulf of Mexico. The zone probably extends farther west, but researchers ran out of money before they could finish charting that area. This dead zone is by no means the planet's largest.

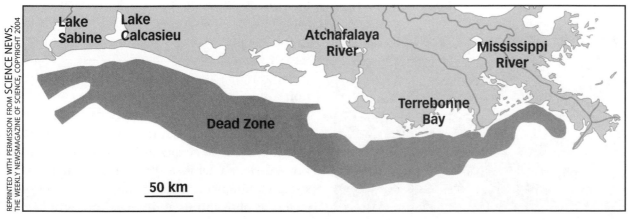

Lake Sabine

Lake Calcasieu

Atchafalaya River

Mississippi River

Terrebonne Bay

Dead Zone

50 km

process, since methane (natural gas) production from kelp is a proven process.

The Untapped Potential

The potential impact of a crop such as algae can't be ignored. The attraction of marine algae over land plants is that algae don't have to fight gravity. As a result, the cellulose-lignin structures that give land plants their structural integrity and ability to stand up aren't needed. Thus, much more of the plants' energy goes into growth and carbohydrate production.

Kelp can grow inches or even more than a foot per day! Another big advantage is that in many locations no fertilizer is necessary to produce it. Kelp lives in a "hydroponic" solution, also known as the ocean. Nutrient-limiting factors on growth evaporate where kelp is cultivated near river outflows containing sewage entering the ocean.[102]

Kelp's foot-a-day growth is primarily limited by the level of dissolved carbon dioxide in the water. If coastal kelp-to-alcohol plants were built, kelp farms could return the fermentation carbon dioxide to themselves, bubbling it through the kelp, increasing growth. This would generate more oxygen and cool the water further.

So let's design an energy system to work as a crash program of kelp farming for energy. There are major ecological reasons to do so. The United Nations has concluded that there are 150 intermittent or permanent dead zones in the world today.[103] These are areas of the ocean where the elevated nitrogen causes a population boom and then decomposition of microscopic algae. In the process of decomposition, all the oxygen in the water is consumed, killing off sea life. Some of the dead zones, like the one in the Baltic, are over 62,000 square miles in size! Although no one has

fully measured the extent of the Mississippi/Gulf dead zone, it is at least 20,700 square kilometers (7992 square miles). All along the East Coast and at river mouths on the West Coast, there are dead zones or areas with very elevated nitrogen levels.

It may already be *necessary* to start these kelp farms for their water-cooling function—in order to save the Pacific fisheries. Due to warming of the water along the California, Oregon, and Washington coasts, **krill** have disappeared. Although they are called zooplankton, krill get to be one to two inches long. Animals from birds to whales depend on them for food.

"It's the krill that drive the food web dynamics off this coast," said Ellie Cohen, the Executive Director of the Point Reyes Bird Observatory in California. "Their absence has tremendous implications for everything out there, right up to the humpback and blue whales. We don't know if this is a result of global warming ... but without the krill, you could be looking at a food web collapse."[104]

Water temperatures along the Gulf of Alaska are the highest they've been in 50 years.[105] This effect does not match the usual patterns of El Niño and seems to be the result of global warming. If the water doesn't cool, then phytoplankton and krill that eat it cannot survive. So massive kelp farming might *have* to be implemented to locally absorb solar energy and cool the ocean surface so that plankton can survive and feed the food chain.

In the process of growth, kelp produces oxygen while absorbing carbon dioxide dissolved in the water. So, kelp farms would be oxygen-rich oases for sea life in the dead zones. Putting massive seaweed farms in the Gulf, for instance, would dramatically cool the surface water, since the solar energy would be turning into kelp carbohydrates instead of heated water. This would serve as a buffer against hurricanes, causing them to cool and stumble down a couple of categories before hitting land. We could convert the oil platforms to plants that process seaweed for alcohol, and pipe it to shore.

The liquid stillage remaining after distillation would resemble the kelp solution currently used by organic and other farmers as a natural wide-spectrum fertilizer. In Norway and China, kelp is dried in large quantities for kelp meal or kelp solutions as fertilizer. If we adopted a national strategy to implement kelp farms, the amount of chemical phosphorus and potassium fertilizers

Fig. 8-37 Marine algae pond. *Using inexpensive liners and low-power paddlewheel circulators, marine algae has been grown intensively in places like Israel and Vietnam.*

PETE BRITZ, RHODES UNIVERSITY

used by farmers would dwindle to zero, since the forms found in kelp solutions are superior. This plan would also go a long way toward eliminating the toxic, and in some cases radioactive, chemicals released into the environment as byproducts of current production of commercial fertilizers. So alcohol production could be seen as a byproduct of producing nontoxic, petroleum-free fertilizers for the nation's agriculture.

Do you think I am proposing an outlandish scheme? In looking at kelp for methane production, the American Gas Association, hardly a wild-eyed utopian group of tree huggers, estimated somewhere near 23 quads (23 quadrillion Btu) a year of methane from kelp just from the California Coast.[106] If the kelp was first fermented to make alcohol and the remaining mash was then fermented a second time for methane, to be used primarily for alcohol plant energy, about a third of that energy would be recovered as alcohol. This might be almost 90 billion gallons of fuel from the California Coast alone.

The remaining two-thirds of the energy as methane would provide all the alcohol plant process energy plus a huge surplus of gas/electricity. That's roughly half of the transportation fuel the U.S. currently uses per year. Add to this the potential production from the Oregon and Washington Coasts, the nutrient-saturated dead zone of the Gulf of Mexico, and possibly the outflow from Chesapeake Bay. Looks like we've replaced all the transportation fuel for the U.S. just from marine algae, as well as the lion's share of natural gas and electricity, as well. All without using a square foot of farmland.

So then all we'd have to do would be to nationalize the now-useless oil pipelines to send some of the alcohol and all of the digested liquid kelp to fertilize our nation's agricultural heartland. Of course, building such kelp farms would be a massive undertaking, but if building 41,000 miles of highways to carry our vehicles or mounting a $500 billion war for oil in Iraq doesn't intimidate our Congress, then neither should a project like this—which neatly solves many problems in one stroke.

MESQUITE

Given its potential, mesquite may become as permanent a crop in the more arid parts of the world as almond or walnut trees.[107] Mesquite is currently found on over 70 million acres of the American Southwest;[108] mesquite seed meal was a major part of the diet of the area's indigenous people. It can be grown in a much greater area, including Hawaii, for instance.

Mesquite is well known for growing on arid mountainous land and in places where nothing else grows well—just another good-for-nothing weed, or so we've been told. The best stands of mesquite in arid country are on valley floors, where

UNIVERSITY OF ARIZONA

Fig. 8-38 Wild mesquite tree. A nitrogen-fixing tree able to live in both dry and wet regions, mesquite produces huge quantities of seed pods chock full of fermentable carbohydrates. Cattle ranchers call it a weed tree and clear-cut it from their land to raise cattle. It would yield far more profit than the cattle that replace it!

Fig. 8-39
Mesquite flowers.
These flowers will become the pods that contain the plant's fermentable sugars.

its astounding roots have reached more than 100 feet to find water. Mesquite is a good source for alcohol and for limited wood burning because it grows under arid conditions where other large trees won't, and it's very dense and long-burning. As many of us know, mesquite is used to make commercial charcoal.

Importantly, mesquite (*Prosopis*) is a nitrogen-fixing genus, related to beans and peas, and provides critical shade. This means that neither the tree nor what grows under it is likely to need any nitrogen fertilizer, since bacteria on its roots convert nitrogen from the air into nitrate fertilizer. In an arid area with some ability to irrigate, the potential for growing other crops underneath the canopy of mesquite is quite substantial. Such arid energy/food crops as buffalo gourd, pimelons, or nopales, forage crops for animals, or even a range of human food crops thrive under the protected canopy, in comparison to the open conditions beyond the trees. So there is a tantalizing opportunity to plant a complementary energy crop or two in conjunction with this self-fertilizing tree.

Mesquite pods are up to 50% sugar, of which 35% is sucrose, and they might be bred to even higher concentrations. Up to four and a half tons of pods per acre can be harvested every year, yielding as much as 341 gallons per acre and 76 gallons per ton.[109] The **hydrophilic** soft fiber, which makes up much of the rest of the bean, could be processed for a cellulosic alcohol yield as well, bringing the annual production to a much higher figure.

Even without the cellulosic alcohol potential, mesquite equals corn's starch-to-alcohol yields per acre without expensive inputs or annual tillage. In fact, if we harvested only the pods from the current 70 million acres, we could be looking at 23.9 billion gallons of alcohol, a substantial proportion of our national usage, without even dealing with conversion of the pod's cellulose.

Just a good-for-nothing weed. But what if we started to plant this weed? Not counting the cellulosic alcohol yield or understory crop, yields of 5.5 billion gallons of ethanol could be produced in the Texas Panhandle alone.[110]

THE TEXAS PANHANDLE

I frequently use the size of the Texas Panhandle to illustrate points about yield of alcohol or land needed for a particular example. The Texas Panhandle, a small but easily visualized patch on the map of the U.S., is about 26,000 square miles. This works out to be an area equal to just a touch over 1% of the agriculturally rated land in the United States.

Mesquite can survive quite nicely down to 0°F, so the potential of planting scores of millions of acres of this tree on simply constructed, water-collecting swales across the entire Southern U.S. is staggering. Since many of these arid lands were former prairies that have been desertified by overgrazing, we practically owe it to the land to restore it using trees like mesquite.

Residual mash from fermented mesquite pods is 13 to 16% protein, a high enough quality to be considered excellent human food, which, of course, mesquite has been for thousands of years. Unlike most plants, mesquite is high in the amino acid lysine. Animal feed has to be supplemented with lysine, which is expensive, to make the digestion of protein more efficient.

In recent tests, mesquite meal was found to be one of the top ten foods in controlling diabetes. Its glycemic index is only 25, and that's with the whole bean.[111] A product made from mesquite mash should serve as an excellent high-protein source of food for diabetics.

MOLASSES

A byproduct of the sugar industry, molasses is one of the easiest of all materials from which fuel can be made. Molasses is what's left when sucrose is crystallized out of sugarcane juice to make white sugar. It's a thick syrup of uncrystallized sucrose and a whole range of fermentable sugars. Molasses is often a very cheap feedstock in tropical sugar-growing regions. It is also a byproduct of sugar production from beets and sorghum. When you read about Brazil or India making alcohol from sugarcane, remember that, unless sugar prices are extremely low, half of the sugarcane crop is turned into much more valuable table sugar and only the molasses is turned into alcohol fuel.

In its favor, molasses has a very compostable, highly nutritious "mud" left over after fermentation and a high sugar content (generally 35 to 55%), it's easy to store, and it's available year-round. The chief disadvantage is economic. There are no direct saleable end products, and, depending on the cost of sugar, the price of molasses fluctuates widely and can be high. When I was teaching alcohol fuel seminars in the 1980s, the price of molasses varied from $80 to $110 a ton. At this writing, it's down to $54 per ton. At this lower price, alcohol made from molasses will cost you less than a dollar per gallon. One ton of molasses yields 65 to 80 gallons of fuel, depending on sugar content.

Molasses is a heavy substance: 170 gallons weigh over a ton. You'll almost certainly have to use a positive displacement pump to transfer the syrup into your storage tank. And keep in mind that pumping is easier on a hot day. On cold days, molasses thickens to the degree that you can open the valve at the bottom of the tank and slice off slabs as it crawls out, slower than molasses in January! For regular production, it's wise to install a hot water heating coil just inside the tank and around the outlet. When you circulate very hot water through the coil, the molasses near the outlet will thin

enough for pumping. Seventy-five feet of ³/₄-inch copper tubing formed into a one-foot-diameter coil, with about an inch between each coil, does the job nicely.

Molasses processing is accomplished by dilution, acidification, and fermentation. Dilute to 14–18% fermentable sugar (very hot, spent mash liquid is preferred for dilution and dissolving the molasses), then add cool water, adjust the pH with sulfuric or lactic acid to 4.0–4.5, and inoculate. When taking your sugar reading, be aware of the high levels of alkaline salts, unfermentable sugars, and minerals in molasses, which can give you a 4 to 5% false high reading.

You can determine the true sugar content by making a five-gallon test batch, starting with a batch reading of 15% on your hydrometer. Let it ferment all the way through for up to three days, or until the batch is completely finished. Glucose test strips will not give you an accurate reading with molasses since they don't read sucrose, so watching the fermentation lock is the best bet. Test your filtered final mash with a hydrometer or refractometer for apparent sugar content. You won't have any fermentable sugar left in the mixture, so whatever this final reading is constitutes the correction factor. Your reading should indicate 2 to 6% remaining sugar. In determining the starting sugar concentration of your batch, add this correction to the actual sugar percentage you desire.

It's not a good idea to push standard yeast to its theoretical alcohol tolerance limit when using molasses. Unfermentable sugars, salts, and other minerals inhibit yeast in molasses, causing them to start dying off before they reach their advertised full alcohol tolerance. Shoot for a couple of percent less than the manufacturer-specified alcohol yield, for the most time- and energy-efficient processing. **Fig. 8-40**

It helps to breed your yeast on a molasses diet, to prepare them for fermentation. Bred on molasses, yeast can "learn" to tolerate the feedstock's salts and minerals to a better degree. Some yeast manufacturers provide special yeast strains for molasses fermentation that permit a higher final alcohol percentage.

Inoculate with a pitching solution of about 5% of the main mash. Fermentation should be rapid and complete in 50 hours or less, sometimes as little as 24 hours. If fermentation is slow and the temperature and pH are all right, then you may need to add some ammonium sulfate, ammonium phosphate, or commercial yeast food to supple-

ment nutritional deficiencies. This is often standard procedure with molasses. Using backslop for dilution instead of water reduces the need for additional nutrients.

Spent molasses mash is particularly good for single-cell protein production (see Chapter 11), which could easily make molasses a cost-effective feedstock if you have animals to feed, even when sugar prices drive the cost of molasses up.

PALMS

Although sugarcane is the crop most identified with tropical alcohol production, several species of palms may come to rival cane alcohol in the years to come. Like cane, palms are a perennial crop; once established, most species can produce indefinitely. Very little study has been directed to palm alcohol, since the environment in which palms thrive bears little resemblance to that of monoculture crops like grains or cane. Most of the species we'll discuss in this section are coastal, and there are many thousands of square miles of wild stands already in existence that can be pruned back into managed stands.

Co-production of shrimp, crabs, and fish is highly complementary in the shade provided by the overstory "trees." Living inside or along the edges of marshes, these palms have a nearly hydroponic relationship with nutrients and can efficiently use solar energy to produce abundant carbohydrates. In some cases, yields can be more than double the per-acre yield of sugarcane—even without recycling of liquid nutrients in spent mash, or use of fermentation carbon dioxide. Best of all, several of these species are so hardy they are considered weeds, my favorite kind of crop.

Nipa Palms

Nipa palms, *Nipa fructicans*, grow naturally on tropical and semi-tropical Pacific islands and on the Pacific Coast of the U.S. But they have spread like weeds around the world's tropical coastal wetlands. They've been used as a major feedstock in the past, especially in the Philippines, where their alcohol fueled up to half of the island's vehicles early in the last century. In the early 1900s, commercial operations there produced from the palms, in an uncultivated state, an annual alcohol yield of 217 gallons per acre.

The nipa palm grows especially well in estuaries and sometimes is considered a weed tree when it

Fig. 8-41 Nipa palm fruit.

PHOTO COURTESY OF MARY ANDREWS, COPYRIGHT 2003 MONTGOMERY BOTANICAL CENTER

PALM OIL AND BIODIESEL

There are about a dozen trees, primarily nut trees, that can make around 100 to 250 gallons of biodiesel per acre. African nut palms are at the top of the plant world in producing oil; their nuts and fruit both have a lot of oil to extract, and 500 or more gallons per acre is not unusual from a mature stand.[112] With the food market for palm oil now all but gone in the developed world, the price recently dropped to somewhere near 20 cents per pound,[113] less than wholesale crude oil. So palm oil now has come full circle, back to being the lower-priced diesel fuel it was always intended to be.

---

infests mangrove swamps. It is a tough survivor and is credited with being able to survive and mitigate damage from tsunamis. It has a large underground "trunk" which sends up a new "tree" as enough food resources accumulate to make it possible. This feature allows the plant to quickly shoot back up after severe storms break off the aboveground trunk.[114] The seeds of nipa float and even germinate in water, establishing themselves in any waterside location, especially when washed inland by storms.

The tree has not been harvested like other biomasses; sap has been the only part of the plant used for making alcohol, collected in a manner similar to maple sugar collection in the United States. The sap's sugar content from wild trees is 12 to 14%; trees must be at least four years old before tapping. The nipa generally produces two or three basketball-sized fruits per tree. It's been observed that removing all but one fruit increases sap flow, and the fruit itself may be tapped three times a month to further increase stalk sap yield.

Modern techniques tap the plant's peduncle (the stalk supporting the fruit) for sap collecting, rather than the trunk as in the early days. Sap yield varies from roughly *0.5 to 2 liters a day!* Harvesting of wild trees' sap occurs between March and May, and again in September through November.

A natural nipa stand may have over 19,750 trees per acre. Such overcrowded conditions reduce sap yield. In 1913, the Philippine output used a fraction of available forest to produce over two million gallons from 75 farm-sized distilleries. The

same year, North Borneo's nipa forests could have produced 67 million gallons of alcohol. For such quantities, a mechanical collection/plumbing system, similar to large maple sugar operations, would have to be developed to produce alcohol cheaply. On the other hand, if world fuel prices are high enough, it would probably be worth sacrificing efficiency for increased rural employment with daily sap collection.

Nipa palm resources in New Guinea are estimated at over 70 million gallons a year. Recent studies there show possible yields, in partially managed forests, over three times those recorded in the early 1900s. But manual labor costs make the extraction of over 650 gallons per acre slightly prohibitive if fuel prices are in the vicinity of $1 per gallon. At higher prices, the additional yield and labor costs work out.

In the Philippines, however, where nipa grows much better, a conservative yield of highly managed nipa forests should yield an awesome amount of 190-proof alcohol: 2140 gallons per acre per year!

In their 1980 study, the National Institutes of Biotechnology and Applied Microbiology at the University of the Philippines Los Baños, in cooperation with the Philippine National Alcohol Commission, based these yields on a planting density of about 4940 trees per acre and continuous tapping for eight months. They calculated that the fruit-bearing trees would yield a very conservative average of one liter of sap per day per fruit at 14% sugar. These calculations were based on an 80% efficient fermentation and distillation. Biologists

PHOTO COURTESY OF MARY ANDREWS,
COPYRIGHT 2003 MONTGOMERY BOTANICAL CENTER

Fig. 8-42 Nipa palm trees growing right along brackish shores. *Since nipa spreads laterally, the entire clump of shoots you see here may all be part of one "tree" joined underground.*

Fig. 8-43 Nipa palm roofing. *This woman is making "shingles" of woven waterproof nipa palm leaves. They are durable, cheap, and can be made all over the tropical world.*

PHOTO COURTESY OF JIM RICHTER, <WWW.GOTOURING.COM/RAZZLEDAZZLE>

evaluating this project said that it might be possible to optimize nipa fuel production to boost annual yield another 25%.

In the case of damaged trees or periodic thinning, the interior **pith** of the trunk may be used. It is an amazing mixture of soft punky wood and thick flakes and veins of starch. Once the trunk interior is coarsely ground and added to water, the starch sinks and the wood floats, while free sugars remain in solution. The starch can be strained out, dried, and stored indefinitely for processing in the off-season of sap production. From this point, a normal starch recipe can be executed.

The waxy fibrous nipa palm fronds make excellent rainproof roof thatch.[115] They are more woody than the rest of the tree, and their generous production would provide all the biomass needed to fire distilling operations. Regular pruning is already done for the nipa roofing "industry."

From a permaculture point of view, in order to mitigate permanent loss of farmland to saltwater, the integration of salt-tolerant sago palm production with shrimp production might make a very good diversified design, providing useful shade as part of a dike system to compartmentalize the shrimp marsh. The high level of nitrogen in the water due to shrimp poop would dramatically stimulate palm growth.

Sago Palms

The sago palm, *Metroxylon sagus*, is a pithy tree rich in soluble starch. It grows in swampy environments, and in coves and valleys opening on the ocean. It reaches the less than majestic height of a few feet. Starch content is highest just before flowering and fruiting, which occur only once

during its life. Managed as a crop, sago palms should yield 650+ gallons per acre.

Harvest is not a one-time affair, since once the entire above-ground portion is harvested, new sprouts rise from the trunk, and pinching off all except two or three will continue the production.

Plans for using sago for alcohol go back many decades in New Guinea; its potential yield there is about 60 million gallons of fuel yearly. But sago stands are spread out, and only farm-sized plants could succeed in light of transportation difficulties. Still, there are huge areas in the Pacific tropics where this tree could be cultivated profitably.

Over time, low-lying sugar plantations have been (regrettably) converted to shrimp production worldwide. From a permaculture point of view, in order to mitigate permanent loss of farmland to saltwater, the integration of salt-tolerant sago palm production with shrimp production might make a very good diversified design, providing useful shade as part of a dike system to compartmentalize the shrimp marsh. The high level of nitrogen in the water due to shrimp poop would dramatically stimulate palm growth.

Fresh sago contains approximately 50% starch and dried sago about 75% starch. That's as high as corn. In **batch hydrolysis**, you use a saccharometer after the conversion step, when the starch has been converted to sugars, to determine when enough water has been added to a sago mash to read 25% sugar for a 9% alcohol yield. Semi-continuous cooking (see Chapter 7) may be necessary to attain alcohol yields higher than 9%. Ideal cooking temperature seems to be 185°F. During cooking, to prevent gelatinization, a maximum temperature rise of 5°F per minute is allowed.

Not enough work has been done to quantify yield per acre, but this is such a promising feedstock that trials should be conducted to work out the optimum growing conditions.

Other Palms

The sugar palm, *Arenga pinnata*, is like a cross between sago and nipa. In Southeast Asia and India, this palm is actually cultivated for semi-commercial table sugar production. It is tapped at its male and female flower stalks for sugar. Larger than the others, this palm has a huge trunk packed with starch for periodic or thinning harvests. Optimum planting density and alcohol production levels have not been well studied as of yet, but given the large amount of starchy biomass generated per year in this large palm, it may turn out to be a significant species.

In West Africa, various species of *Raphia* palms can be tapped like nipa, or the trunks can be processed for starch like sago. Some species have the additional benefit of providing a useful oil, which is edible or can be used for biodiesel production.

PIMELON (WILD WATERMELON)

Throughout the southern arid portions of America lies land considered "useless" for agriculture. Pimelon is a weed generally found in those arid lands, all wound up on the cotton harvester, a real pest. In 1981, it was reported that the melon's pulp yields about 450 gallons of alcohol per acre,[116] with valuable byproducts.

The melons are extremely seedy; one acre of pimelons produced over 6000 pounds of seed, which yielded 1300 pounds of safflower-like oil and 1100 pounds of high-quality protein for livestock and possibly for people. Like the oil from buffalo gourds, the oil yield from pimelons could be used to provide at least one and a half times the process heat needed for the alcohol plant. If directed to biodiesel production instead, the oil would produce over 200 gallons—quite a bit more oil per acre than canola and far more than the paltry 48 gallons per acre from soybeans.[117]

If we grew pimelon on an area the size of the Texas Panhandle, we'd be able to produce about 17 billion gallons of alcohol, well over 10% of annual U.S. motor fuel usage.[118] And we'd have about two and a quarter million tons of high-quality animal feed to boot. In the Texas Panhandle, where it takes 100+ acres to range-feed one beef steer, feeding pimelon mash after processing makes a lot more sense. Of course, we have literally millions of acres of arid lands where pimelon would thrive.

Fig. 8-11 Pimelons. *Yet another pesky weed that can turn arid land into a carpet of solar collectors.*

As an understory companion crop under mesquite, pimelon's yield should go quite a bit higher—due to the mesquite's nitrogen-fixing capabilities fertilizing the soil, and its shade preventing solar saturation of the "weed" melon crop. My own experiments on my farm, as well as the experience of subsistence farmers around the world, indicate that growing squash and melon species under the shade of corn increases yield. The melon yield underneath mesquite would roughly double the per-acre yield of mesquite alone. Once again, the potential is in the billions of gallons.

PRICKLY PEAR

Prickly pear (*Opuntia polycanta*, *Opuntia ficus-indica*, and others) is an extremely hardy cactus found all over the world. It is also called beaver tail cactus and nopales.

Fig. 8-45 Prickly pear.

LET ME KNOW IF YOU SEE A PRICKLY PAIR OUT HERE....

RON HARPER

Fig. 8-46

Although it's tempting to think of all cacti as denizens of blazing hot deserts of the South, the prickly pear has a much wider range. Various closely related species of *Opuntia* grow as far north as Alberta and Saskatchewan, and all the way down to Texas.[119] Cactus is grown commercially in California and Arizona for jams and dried fruit, and is a common cultivated food crop in Mexico.

In their wet state, the prickly pear's fleshy, segmented leaves may contain 16% fermentables or more. In the arid North African countries where prickly pear grows over millions of acres without cultivation, sustained harvests of over 500 gallons per acre are possible. In Brazil, the plant covers vast areas and grows densely enough, as a weed, to yield over 900 gallons per acre annually.

Prickly pear should be considered a permanent crop, and only new leaves and a few whole branches should be harvested every year, so the main plant can continue producing. Yields in such a nonirrigated, managed *wild* crop might be 200 to 500 gallons per acre. Irrigated managed crops can be triple this figure.

Cultivation of prickly pear is limited, and so is information on optimum growing. What is known is that the key to a large sustainable yield is surface mulch and water. Returning stillage to the land, along with any leftover water from cooling, should support high yields.

Cacti like prickly pear do produce flowers and seed fruit. But most of the time the plant counts on asexual reproduction. Any largish piece of a pad (leaf) which falls to the ground can sprout roots and become a new plant. One reason why cacti have thorns is for reproduction! A passing animal encounters a thorny pad, which becomes attached to the animal and breaks off from the plant. The animal often carries it for days before the thorns work themselves free and the pad falls to the ground, to sprout into a new plant. New commercial production is created by planting a pad halfway in the ground. A single leaf I planted in California at my farm had 100 pads in three years, without irrigation.

A co-planting of mesquite and prickly pear makes a good energy crop system for arid lands. Mesquite fixes nitrogen in the soil, and helps to maintain a moister microclimate, since its canopy provides helpful light shade for the prickly pear. Prickly pear is highly efficient in its photosynthesis and makes full use of whatever shade is available.

Mesquite prunings supply wood heat for cooking and for distilling the alcohol made from prickly pear's leaves and mesquite's pods. This combination of valuable plants is a natural occurrence, and the mycorrhizal connection between the mesquite and nopales greatly increases the yield of the cactus, since it receives substantial surplus carbohydrates from the mesquite.

In Mexico, prickly pear has been used as a rotation crop for a couple of years after land had been exhausted from corn cultivation. The water-managing and fungal-feeding aspects of the cactus/fungus symbiosis dramatically regenerate the soil by feeding beneficial soil microorganisms. This recharges the soil with nitrogen and carbon. After as little as two years of cactus production, with cattle allowed to graze on about half the growth and with all the grass symbiotically growing around the cactus, the soil would be ready for several more years of corn cultivation.[120]

If you are planting a new field of *Opuntia*, I recommend one of the Burbank spineless varieties. These are common in gardens and have escaped to grow prolifically in many places. Legend has it that Luther Burbank developed this **cultivar** by going to his cactus and telling them that they were loved and didn't need their spines anymore. They eventually agreed with him!

This changed their flavor somewhat, at least to cattle, which like it less than wild varieties. Cattle have eaten nopales, thorns and all, ever since they came to the New World.[121] When mixed with mesquite pulp, nopales become an even more nutritious animal feed.

In remote locations without a feed market, the mash should be treated as fertilizer to be returned to the desert crop-growing area. New markets in *Opuntia* extract for treatment of diabetes could be a valuable co-product.

SUGAR BEETS

A favorite fuel crop in temperate regions is sugar beets. Yields of 770 gallons of alcohol per acre of beets have been achieved. During the late 1990s and early part of this decade, many beet sugar refineries closed in California and other states. There is a lot of current discussion about reopening them as alcohol production facilities.

Beets are 16 to 18% sugar. They can be shredded and then pressed to remove their juice. A garden compost shredder does the job. Any apple press will provide the juice, and many presses come with a built-in shredder.

Beet juice is acidified to a pH of 4.0; 1 and ³/₄ ounces of yeast is added per 22 gallons of juice, at 82°F. Fermentation takes no longer than 24 hours! Juicing leaves behind a great deal of unfermented sugar, and many small producers prefer alternate methods of fermentation that yield more alcohol per ton of beets (see Chapter 7).

Beets can be fermented whole if they're shredded or hammermilled to the consistency of coarsely grated carrots. Add enough water to get the beets

Fig. 8-48 Sugar beet harvester. *Harvesters easily lift the beets out of the ground and load them in trailers. Note that the beets have been topped prior to mechanical harvesting.*

into a thick slurry; cook for up to an hour. Depending on the kind of beets you're using, there may be enough starch to warrant a liquefaction and conversion step while the wort is cooling.

Diffusion has been a popular method for getting beet sugar into solution (see Chapter 7), but the small producer will probably want to go the route of whole fermentation.

Sugar beets are normally considered a temperate crop, but a new variety that will grow with irrigation in dry tropical areas is being used in India and a few other countries. Its yields are still in the vicinity of 700 gallons per acre (France gets 714 gallons); its advantage would appear to be that it matures much faster than sugarcane. The beets are ready in six months, while cane takes more than a year. Tropical sugar beets use less water by half than sugarcane, so beets might be a better choice in areas marginal for cane.

Sugar beets can be used as a rotation crop with cassava and sugarcane, allowing formerly sugarcane-only alcohol plants to spread out the harvest and processing season, leaving the plants idle for less time. In a temperate twist, a project is being planned in southern California where beets will be grown in the winter half of the year, and a fast-growing, short-season sugarcane will be grown in the summer.

For information on the culture and byproduct value of sugar beets, see the section on fodder beets.

Fig. 8-47 Tropical sugar beet. *This selectively bred sugar beet grows in much warmer climates than its northern sisters.*

*Fig. 8-49
Sugarcane.
It's a grass!*

*Fig. 8-50 Sugar-
cane seedlings.*

*Seedlings of
many variet-
ies are grown to
be transplanted
into the field
in an attempt
to prevent the
catastrophic crop
failure that would
happen if all the
plants were of a
single variety and
encountered a
disease or pest.*

SUGARCANE

Sugarcane is a perennial grass, much to most people's surprise. Perennials are crops that you plant once and harvest for several years. Blackberries are a perennial vine, for instance.

Perennial grasses tend to store much more than half their biomass in their roots, unlike trees and other plants that have about half their biomass above ground and the other half below. But as the plant ages, the root-to-shoot ratio dramatically changes, reaching as high as 22%, meaning most of the biomass is in the leaves and stalk.[122]

Sugarcane can have some roots go as deep as 18 feet, but 42% of the root mass is between two to eight inches deep, and another 23% is between eight to 16 inches. Different varieties will stray from these percentages, especially depending on soil type, but this is a good rough set of figures.[123] Sugarcane roots really knit the topsoil together in a tight weave along with their attached mycorrhizal fungi, which can virtually eliminate soil erosion in a well-designed field.

Cane is definitely a subtropical crop. It starts to sustain damage at 46°F, and at 37°F the whole plant can die. But for all its apparent tropical credentials, it needs a cool winter. Both **chill days** in winter and **degree days** in summer are used to calculate the potential yield of sugar content very accurately. A degree day for sugarcane measures distance above a minimum of 70°F,[124] so a day that reaches 71°F would be one degree day. A 100°F day would equal 30 degree days. Chill days are similar in that they describe the number of degrees below a fixed temperature in the cool season. Constant breeding of sugarcane is slowly producing more cold-tolerant cultivars. The cooler-climate varieties mean sugarcane can be grown in some parts of the Southern U.S. and frost-free coastal areas.

When perennial grasses go to flower and set seed, they draw on that underground storage of carbohydrates in the roots. They put all their current photosynthetic products into sending up a seed stalk (known as the "cane" or **ratoon**), rather than pumping sugars into the roots in the ground. The roots no longer exude sugar to feed the bacterial population, which decomposes, providing a boost of nutrients to further the reproductive effort.

Planting

With a perennial like sugarcane, a thriving rhizosphere is maintained, and yields get bigger each year, even if they start off more slowly. Tilling up the soil for annual plants is often destructive to fertility levels and oxidizes a lot of valuable humus, which is key to holding nutrients in the soil. In addition, freshly tilled soil in a rainy season is easily eroded and washed downhill. So the more time we can keep the soil covered with a fibrous rooted crop of some sort, the happier the soil tends to be.

A well-managed sugarcane field can live for ten years without replanting, but renewal is often done in five years. Modern cane farmers plant a rotation crop, usually a food crop, for the six months between removal and the time for planting new cane. Cane plants are big, and the rows are often about five feet apart. Normally, the first cane is cut after the plant has been growing for at least 12 to 14 months.

But cane can be grown as an annual (a plant which must be planted every year) instead of as a perennial, depending on variety. This makes it possible to use cane in areas that are not tropical

but still have a long growing season between early and late frost dates.

Part of an energy-focused permaculture design for a sugarcane farm would be to plant many cultivars of cane that mature at different times. This would allow the CO_2 from processing the earliest cultivars to be used to enrich the atmosphere in the canopy for later-maturing cultivars. The planting of diverse varieties of cane is already a standard practice in Brazil, so implementing this change there would simply involve emphasizing varied maturation dates as a primary criterion.

To create optimum conditions for photosynthesis, broad-canopied leguminous trees could be planted on the swale berms to provide light shade and to fix nitrogen in the soil. Using a combination of shade and CO_2 enrichment strategies would augment yields of both sugar and biomass.

Growing

Sugarcane is among the most efficient photosynthesizers of all land plants. In theory, plants can photosynthesize about 11% of the total sunlight falling on them. The highest levels are realized by algae. Trees can sometimes reach half of the theoretical efficiency. But most land plants average 1 to 2% conversion of sunlight to biomass. Sugarcane is always over 2.6% and sometimes closer to 3%. This maximum photosynthetic rate on the conversion of sunlight to chemical energy is most significantly limited by the amount of carbon dioxide found in the atmosphere.

There are two basic categories for how plants photosynthesize, referred to as C3 and C4 types. Grasses and other C4-type plants, like sugarcane, are more efficient in their use of CO_2 and sunlight than the simpler, C3-type.[125]

But there comes a point where even sugarcane becomes saturated by light and can no longer make more carbohydrates. Once a leaf is saturated, any further sunlight goes to waste, and the additional stress associated with overexposure to sunlight can actually reduce the amount of carbohydrate produced. Some high-producing varieties of sugarcane saturate at about 4000 foot-candles (ft-c), which is quite a bit less than half of the 9000 ft-c of peak sunlight that falls on the cane in the middle of the day.[126]

There are two ways in which a grower can get around the photosaturation limit. The first is by

PHOTO COURTESY OF UNICA BRAZIL

Fig. 8-51 Manual sugarcane harvesting. *Harvesting cane in an unburned field increases rural employment and permits the harvest of leaves, which are one-third of the crop's biomass. Manual harvesting also prevents the seasonal air pollution caused by the usual practice of burning fields in order to reduce labor costs, so that laborers need only cut down the cane.*

IMAGE COURTESY OF CANEHARVESTERS

Fig. 8-52 Mechanical harvesting of sugarcane. *Capital-intensive, mechanical harvesting of sugarcane permits recovery of the leafy biomass.*

In walking around farms in Brazil, I was constantly amazed at how much taller the cane was every time I found it under a tree.

increasing the carbon dioxide levels in the field, and the other is by reducing the amount of sunlight via shade to below the saturation point (see Figure 8-53).

Increased CO_2 levels in the field raise the saturation point and allow more sunlight to be converted into sugar, especially in midday light intensities. In fact, the growth rate can be doubled if the rate of carbon dioxide goes from the background rate of around 360 ppm up to 900 ppm.[127] In a well-designed integrated alcohol farm, fermentation CO_2 would be returned to the field during the day to be absorbed by the next crop.

Without additional CO_2, high rates of photosynthesis can be maintained throughout the midday if the cane is grown under a light shade (30 to 50% shade), as shown in Figure 8-53. At 9000 ft-c—direct full sun—photosynthesis would drop and then stop. It would resume as the weaker afternoon sun came back into the optimum photosynthetic range, which would not fully saturate the leaf. As the afternoon wore on, the photosynthetic rate would increase again, eventually reaching a peak, before dropping off due to lowering temperatures and the sun setting.

The light shade from a diffused canopy, however, would maintain peak photosynthesis throughout the midday. In walking around farms in Brazil, I was constantly amazed at how much taller the cane was every time I found it under a tree.

Solar Saturation in Sugarcane

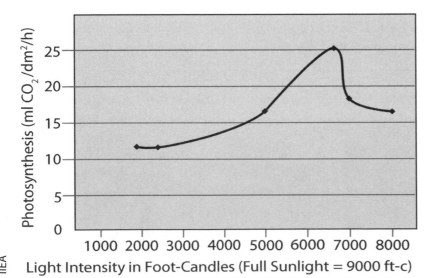

Fig. 8-53 Solar saturation in sugarcane. *It's clear that as sunlight reaches its highest intensity in the middle of the day (9000 foot-candles), photosynthesis drops off dramatically and comes to a complete stop. Providing a light shade over the crop keeps carbohydrate production high throughout the day.*

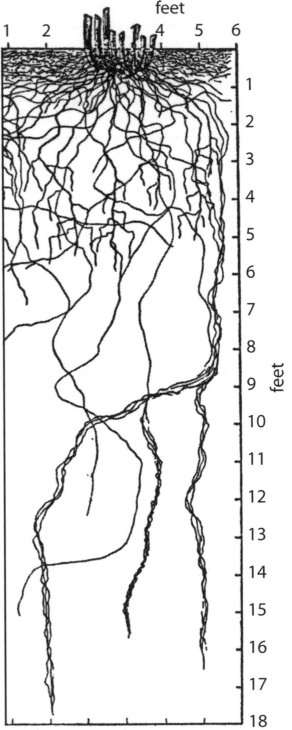

Fig. 8-54 Sugarcane root system. *Cane has a dense mat of roots just below the surface but can reach down to 18 feet, bringing up water and mineral nutrients to feed the whole plant.[128]*

Harvest and Crop Management

Sugarcane harvesting is done in one of two ways, either manually or mechanically. The traditional manual way requires the burning of the field to remove the leaves. Once the leaves are removed, the average cane harvest worker can cut and stack ten tons of cane per day. If the leaves have to be removed by hand, the rate of harvest drops to two tons of cane per day.[129]

Mechanical harvesting has started to be used on the larger plantations; it avoids the environmental issue of burning and enables the harvesting of the leaves as another biomass product. The leaves are becoming an important issue, since they are about a third of the biomass and have been found to burn well in cogenerators. But mechanical harvesting deprives a large part of the rural population of the best-paying work in their region, hand harvesting. This is a really tough situation for Brazil to wrestle with, but it is only a real issue on the biggest farms, since a mechanical harvester costs over half a million dollars. A large portion of the sugar in Brazil is grown on small farms, each with about 20 hectares in cane, which they sell either to a mill or through a cooperative.

Whenever you cut back a perennial plant, as when the cane is harvested, a portion of the extensive root system dies back, too, without the leaves to feed it. The decomposing roots become a nutrient source for the remaining plant. It's been calculated that for every 100 tons of cane harvested, decomposing roots comprise at least five tons DM (dry material).[130] This amount of root matter is almost exactly the amount needed to maintain fertility in an organic field. Of course, leaf trash decomposes also, adding additional organic matter to the field.

Twenty-five years ago, the **vinasse** (spent mash liquid) left over after distillation was disposed of in the closest stream. But this practice has changed due both to enforcement of environmental laws and to recognition of the stillage as an excellent source of all of the micronutrients (most importantly, potassium) needed for cane production. It turns out that using vinasse as fertilizer results in superior yields—which is not surprising, since the vinasse contains all the nutrients from the soil that produced the cane.

The new practice is to either flood-irrigate the vinasse back into nearby fields or use it as sprinkler **fertigation**, sending it by pipeline to more distant fields. When the vinasse and the ash from burning bagasse are returned to the soil, the nearly closed cycling of nutrients is completed, with only the solar portion (carbohydrates) being removed, as alcohol.

Once again, good permaculture thinking has turned a waste into a resource. As a result, little or no potassium needs to be added to cane fields today. With this nearly closed loop nutrient system, fields fertilized with vinasse are far more easily certified organic. In fact, Brazil leads the world in biomass-fertilized fields, with over three million hectares.[131] These fields not only yield more cane, but the vinasse strongly controls nematodes, soil worms that can sap the strength of the plant.

BOB FITCH PHOTO, <WWW.BOBFITCHPHOTO.COM>

Fig. 8-55 Vinasse fertilization. *These pumps send the vinasse through irrigation lines (stacked behind them for storage) to sprinklers, which distribute the nutrient-rich liquid back to the fields.*

PHOTO COURTESY OF GERALD HOLMES, NORTH CAROLINA STATE UNIVERSITY, RALEIGH

Fig. 8-56 Sweet potatoes. *These were thinned out because they don't meet a certain size or shape standard to go to market. They would likely be used for animal feed, canning, or baby food.*

At least one distillery in Brazil, Jalles Machado S/A, is planning to first run the vinasse through a methane digester to produce methane from the unfermentable sugars, and then spread the digested vinasse on the fields.

Yields

Sugarcane yields have been increasing in Brazil since the 1980s, when they were about 40 metric tons per hectare of cane from primarily one variety. These days, a large field may have up to 40 varieties, and Brazilian law dictates that no variety can take up more than 20% of a field. This diversification of cultivars was done to partially mitigate the problems associated with the monoculture of cane. So, if a disease comes through, it will be less likely to affect more than a couple of the varieties, leaving most of the crop intact. This single-species diversification is not a great solution to the issues with monocultures, but it is a step in the right direction.

Brazilian sugarcane yields have risen to 80 metric tons per hectare on average, with 100 tons becoming more and more common. Interestingly, the higher yields are often on the smaller farms. A family on 20 hectares of cane can take more care than can be taken on a farm of 20,000 hectares. There's an old saying in China, "The best fertilizer is the farmer's shadow." The more time you spend in the field, the more likely you are to spot problems with pests, soil deficiencies, or disease early and take action.

In Brazil, alcohol yields from sugarcane used to be down around 3500 liters of alcohol per hectare. Now, with the diversity of new cultivars and better processing technology, the national average comes close to 7000 liters per hectare (748 gallons/acre).[132] These alcohol yields are just from the cane juice. There are new details in the section about cellulose, above, on converting the bagasse and leaves—which combined make up two-thirds of the crop by weight—to alcohol. So the potential increase in yield is just now beginning to be dramatically realized.

A new study claims, too, that a profusion of smaller processing plants would reduce transportation costs of the cane, while requiring additional labor, which is a positive impact on the rural economy.[133]

Although the Brazilian alcohol industry may sound large, sugar/alcohol production takes up less than 8% of the *cultivated* land in Brazil. For comparison, Brazilian soybean cultivation takes up more than double the acreage of sugarcane.[134]

Brazilian sugarcane growing practices take advantage of the ways that sugarcane benefits from mycorrhizal fungi. Before being planted, new sugarcane plants are inoculated with these fungi, and the fungal mycelia quickly colonize the whole field. The fungi dissolve and transport nonsoluble phosphorus to the roots of the cane plants. This plant-fungi symbiosis eliminates the need for phosphorus fertilizer. This biological approach to fertility is a big advance in Brazilian agriculture, which is just now being fully appreciated. The advantages of this symbiosis are being observed by Brazilian farmers and have led to a reduction in the use of herbicides and pesticides, since these toxic chemicals are the nemeses of the beneficial fungi.

Fig. 8-57 Tilby Separator system. *The Tilby process derives several useful products from sorghum or sugarcane.*

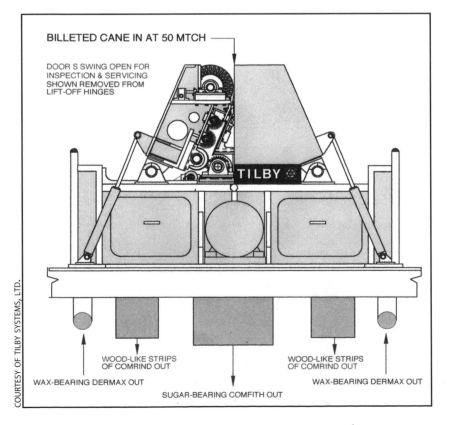

BILLETED CANE IN AT 50 MTCH

DOOR S SWING OPEN FOR INSPECTION & SERVICING SHOWN REMOVED FROM LIFT-OFF HINGES

TILBY

COURTESY OF TILBY SYSTEMS, LTD.

WOOD-LIKE STRIPS OF COMRIND OUT

WOOD-LIKE STRIPS OF COMRIND OUT

WAX-BEARING DERMAX OUT

WAX-BEARING DERMAX OUT

SUGAR-BEARING COMFITH OUT

Although producers add about 50 kilos of nitrogen per hectare, studies show the plants contain more than 100 kilos of nitrogen per hectare.[135] Normally, you would look to a nitrogen-fixing bacteria to be the source of such a boon. But sugarcane has no such bacterial associates on its roots; it does not have any nitrogen-fixing bacteria of the sort found on beans or peas, for instance, consisting of round colonies attached to the roots, which turn atmospheric nitrogen into nitrate fertilizer.

Here's what is actually going on with the extra nitrogen: A third of the sugar made in the sugarcane plant is pumped through the roots into the ground, where it feeds the mycorrhizae and general bacteria, whose populations explode in response to this rich food source being introduced into the rhizosphere. The bacteria decompose and provide nutrition for the plant but do not fix nitrogen from the air. Some mycorrhizal fungi actually capture soil microorganisms and eat them, transporting the nitrogen to the plants. The sugar pumped into the soil represents a huge amount of carbon dioxide being taken out of the air. When the micro-life excrete or die, nitrogen and other nutrients are released into the soil, where plants can take them up.

Making Alcohol

Until recently, growers found it more profitable to produce sugar using the crystallizable portion of the cane juice. But with the recent rise in petroleum prices and the anticipated removal of sugar crop subsidies, making alcohol from the whole-cane juice is now becoming the rule rather than the exception. When sugar prices are high and petroleum prices are low, you'll find most of the industrial plants make alcohol solely with the molasses left over after crystallizing the sugar.

There are two ways to make alcohol from the cane, if that is your primary goal. The first is with the use of crushers, which come in many sizes all the way down to personal production scale. The theory of the crushers is that the juice is forced through the fibers of the cane to squirt out the ends, but, in reality, the cane juice does not flow very well through the fiber, so it ends up rupturing the strands, and the juice is released. Done this way, 97% of the juice is recovered.[136]

But for the small plant that doesn't have the capital for a big crusher, diffusion turns out to be the best method (see Chapter 7). Using a slicer or shredder to make three-millimeter-wide coins of cane, diffusing the sugars into solution, and then fermenting it recovers more than 98% of the juice.[137] The resulting chips of bagasse are yeast-impregnated and make a usable animal feed that is easily handled by the small producer.

Fermentation of cane juice can take anywhere from four to 24 hours, depending on the amount of yeast used to start and the quality of agitation and temperature control.

SWEET POTATOES

In the early part of this century, sweet potatoes were often dried into chips after harvesting and stored until they could be used for conversion to fuel. This spread out the processing season far beyond the harvest.

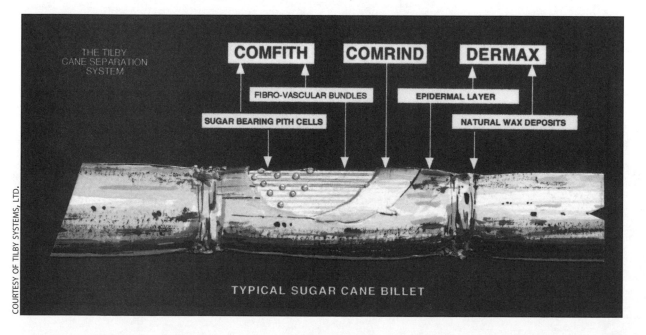

Fig. 8-58
Typical sugarcane billet. *Although this illustration is of sugarcane, sorghum cane is very similar.*

Fig. 8-59 Tilby strandboard. *The tough fibers recovered from the canes of sugar or sorghum can replace many traditional uses of wood, helping reduce deforestation.*

Fig. 8-60 Modular building panels. *These building panels, derived from cane fiber, combine insulation with structure for low-cost building techniques, minimizing needs for timber.*

Fig. 8-61 Sweet sorghum spacing. *Close row spacing, with about 18-inch centers, gathers in much more solar energy than conventional wide spacing.*

The following recipe comes from an old alcohol fuel book, with malt references converted to bacterial enzymes. It's a variation on the starch recipe from Chapter 7.

1. Heat 35 gallons of water to 120°F.
2. Prior to cooking, add ⅔ ounce alpha amylase per bushel.
3. Add 56 pounds (one bushel) dehydrated sweet potatoes, ground or in small pieces. Correct to pH 6.0–6.5.
4. Bring temperature up to 190–200°F and hold it there for half an hour at pH 6.5.
5. Adjust pH down to 5.3 and cook mash at a boil for one hour.
6. Cool mash to 195°F and correct pH back to 6.0–6.5.
7. Then add 1-½ ounces alpha amylase per bushel and agitate for 45 minutes.

The conversion step with glucoamylase is the same as for corn, but the temperature should not be allowed to drop below 135°F, and conversion time should be extended to an hour with good agitation. Cooling should be done with a cooling coil.

Your yield is about 6.5 gallons of fuel per bushel of dehydrated sweet potatoes, or 20 gallons per ton with fresh potatoes. Fresh potatoes require only five to ten gallons of water per bushel, allowing you to eliminate pH reduction during liquefaction. If you're using dry potatoes and have a good agitator, you can start with less than the 35 gallons of water.

An Alabama sweet potato farmer tells me that when a crop is thinned, and again when it's harvested, finger-sized and larger undeveloped (ugly) sweet potatoes attached to the plant stalk are considered waste, or allowed to be eaten by grazing livestock. I'm told they are also used for deer bait, and, if they aren't too bad, they get used for baby food. But the point is that there is a lot of waste. Pound for pound, they should be almost as good as whole sweet potatoes.

SWEET SORGHUM

Although only a few thousand acres of sweet sorghum are grown in the U.S.,[138] it is an excellent feedstock. Sweet sorghum can grow anywhere that corn, soybeans, or sugarcane grow, and can rotate successfully with any of these crops. Sorghum is sort of the temperate world's version of sugarcane. In fact, I used to grow a sweet sorghum variety on my farm called Apache sugarcane, which could be eaten, well, like sugarcane.

Sorghum matures faster than corn and will tolerate more stressful conditions. In some Southern areas in the U.S., it is possible to get two or sometimes three crops per year. Since sorghum has a very high sugar content (approximately 40%) and is a prolific plant, the per-acre yield is anywhere from 500 to1000 gallons of alcohol per acre in one crop cycle.[139]

There are often seasonal or market-based surpluses in tropical areas. Mangoes, for instance, glut the market each year, leaving huge quantities unpicked.

In the U.S. and other temperate areas, sorghum makes a natural replacement for tropically grown sugarcane in its culture and harvest. In the extreme Southern U.S., sugarcane's yield is about 10% higher than sorghum's, but cultivation costs are considerably higher. Unlike sugarcane, sorghum has been grown economically in the North-Central Plains of the U.S., and in places like Ohio. Several factors affect yields: planting date, variety, date of ratooning, height at cutting, and fertilizer applications.

There are two basic types of sorghum grown for energy. The first has a lot of sugar in its stalk; the second stores its energy as starch in the plant's grain head. The second kind produces small round seeds on a comb, like corn.

Crossbreeding has produced sorghum plants that have both a grain crop and a syrup yield. To extend sorghum's alcohol production season, you could use the stalk varieties first for sugar-to-alcohol conversion and save the grain crops for starch processing later. You'd get a lot of burnable biomass from the cane fiber and leaves that could be used to fuel most of the operation, in a manner similar to sugarcane.

As an alternative to simply burning the non-sugar biomass, there are several valuable byproducts worth considering. Sorghum cane consists of a central pith, which contains the plant's fermentable sugars, and a perimeter rind, which is composed of long, sturdy cellulose fibers perfect for "plywood" or composition board manufacture. The fibers of the sorghum plant are the traditional

I COULD TELL YOU, BUT I'D HAVE TO KILL YOU

In 2003, I attended a national fuel ethanol workshop with over 1000 other people. One of the problems with these sorts of conferences is that in the current cultural hysteria of everyone trying to patent intellectual property (IP), you find a lot of speakers talking about stuff without really telling you anything specific, since they are all tied in knots by their patent attorneys. I don't know about you, but this sort of thing really ticks me off.

In a plenary session with all of us there, we were treated to a presentation from a sorghum growers association. We were hearing all about advances in varieties, harvesting, and uses of the byproduct. The speaker said, "And we are even finding valuable uses for human food products with major health benefits."

So when the question and answer session started, I was handed the microphone and posed the question, "Well, since you've removed all the starch from the grain, you have a pretty nearly carbohydrate-free material to make into breads, etc. As a diabetic, I am pretty alert to the need for foods that we can eat. Since there might be 30 to 60 million people with diabetes, have you thought about just how large this market might be for your sorghum DDGS?" It got so quiet you could hear a pin drop. Being caught out with his IP exposed in front of a thousand people, he said, "I could tell you, but I'd have to kill you."

BANANAS (WHILE THEY LAST)

Scientists are warning that bananas could die out in the next decade because of rampant fungal diseases. But meanwhile they make a good waste crop source for fuel. Although bananas are a little starchy when ripe and very starchy unripe, fermentation tests I performed in Costa Rica went without a hitch. In a 1982 visit to Costa Rica to study alcohol fuel feasibility, I learned that one out of every three bananas was culled due to not meeting market aesthetics. These were flushed into rivers, which flowed into the ocean.

Fig. 8-62

THAT'S ODD...THEY'RE COOKED ALREADY...

broom fibers used for sweeping. There is even an industrial wax that can be extracted from the outer rind of the plant.

Syrup sorghum varieties are selected for their high content of crystallizable sucrose or for their content of high-sucrose syrup (food-grade molasses). Presently, syrup processors prefer varieties with low fiber content, which corresponds to small-diameter canes. For alcohol production, higher yields could be obtained by using thicker-stalked cultivars. Such stalks, when encased within heavier rinds (with a higher volume of sugar-rich pith) allow for more efficient processing of multiple products. Slicing the cane into "coins," as Brazil has done,[140] or shredding it into coarse pieces and then using either diffusion or whole fermentation methods (see Chapter 7), will convert 98% of the sugar from the stalks.[141] Although this is only 1 or 2% greater sugar conversion efficiency than crushing, it is logistically much easier for small producers.

In medium- to large-sized operations, the Tilby Separator seems to be a practical solution

for separating pith, wax, and rind economically, without traditional crushing of the entire plant. Before the Tilby system can be used efficiently, modifications will have to be made to cane and forage harvesters in use on the sorghum-producing acreage in the U.S. The modified machinery will have to handle 18-inch row spacings, rather than the more common 40-inch row spacings. This type of harvesting will maximize high-value byproduct recovery.

Although it's not ideal, a modified silage harvester, with several blades removed and blade speed reduced, could be used to gather sorghum stalks. You might want to make some adjustment in the harvester's gearing to speed the transport of stalks through such a machine. Rather than harvest the whole plant, a modified sugarcane harvester could strip the seed heads, leaves, and other trash, and then bale it for boiler fuel or cellulosic alcohol conversion. When harvesting the whole plant and converting the cellulose to alcohol, it has been projected that yields will be 3500 gallons per acre.[142]

Planting in 18-inch row spacing is very important. Dense planting causes a canopy to form rapidly and early in the season, choking weeds and making use of the maximum amount of solar energy. Dense planting does not decrease the sugar content of each stalk—in many cases, it slightly increases it. This net increase is due to the mycorrhizal fungi knitting all the plants together underground and providing storage of the sugar, which is then transported back to the roots during formation of the cane. The early closure of the canopy also reduces water needed for the crop, since the soil is not exposed to the evaporative effects of sun and wind.

Some sorghum varieties are better than others for ratooning, and more selective breeding could improve this characteristic. The primary benefit of ratooning is the extension of the harvest season. Second and third cuttings yield less than the first, so in warm areas, you might choose to plant a second crop rather than cut a third ratoon. Storing the carbohydrates as concentrated juice (e.g., molasses), or separating and drying the pith, are both possibilities for lengthening the processing season.

Sweet sorghum can be a primary or a rotation crop. The Battelle Institute suggests that sorghum be rotated with sugar beets and cane during fallow seasons to increase alcohol yields per acre. In fact, in Brazil, sweet sorghum is now being planted as a rotation crop every five to ten years when sugarcane is pulled out. Brazilians use the new varieties of sorghum that have both seed head and syrup stalk.[143] This rotation reduces pests in sugarcane and, more importantly, extends the processing season from eight months to 11–12 months.[144]

In temperate areas, sweet sorghum may be rotated with winter wheat to help prevent soil erosion between sorghum crops. Sorghum pulp and leaves can be plowed back into the ground for stabilizing easily wind-eroded wheat soils. Of course, the vinasse (fermentation liquid after distillation) can be returned to the fields for fertilization.

TROPICAL FRUITS

The equatorial plant community receives the most solar energy and produces papayas, mangos, bananas, plantains, chayotes, pineapples, and a variety of other fruits prized in northern temperate countries. The tropics also produce a variety of plants with byproducts that are great for alcohol, but that aren't being used.

Citrus crops grown closer to the equator have higher sugar content than those grown in temperate climates. Soft fruits such as papaya and mango have sugar percentages in the 14 to 20% range and ferment readily. There are some notable exceptions: Plantains are not sugary at all; they contain 20% or so starch instead. The chayote, a semi-perennial, prolific, squash-like crop well adapted to growing in the deep shade under trees, is primarily starch, and its periodically harvested roots are also a valuable fermentable (although the roots taste too good to waste by making them into alcohol).

As in the U.S., a large percentage of the citrus crop worldwide doesn't meet the cosmetic standards required for the export market. Also, there are often seasonal or market-based surpluses in tropical areas. Mangoes, for instance, glut the market each year, leaving huge quantities unpicked. While consulting for macadamia growers in Mexico, I was approached by pineapple growers who had seen the bottom drop out of their market due to globalization, leaving enormous quantities of the fruit with nowhere to go.

Cashews are an interesting case, since the "nut" is formed on the outside of the fruit. The fruit,

Fig. 8-63

which weighs many times the cashew seed, is disposed of in huge quantities during the nut processing and would be a great feedstock for alcohol production. Regional, multi-feedstock-capable distilleries would make a lot of sense in these areas.

WASTE SOURCES

As a permaculturist, I find I almost never use the word "waste." Whatever surplus material you are considering is a resource, not a problem. Something is a waste product when it's a surplus that doesn't have another use it can be put to; there are few things in this category. (Okay, spent nuclear fuel rods are serious waste, I'll grant you that.)

Use your imagination. A small produce store will dump tons of fruit and starch produce each week. As much as 25% of produce goes to waste before consumers buy it. Some of my students have found that waste soda pop syrup from local bottling plants compares favorably with molasses in yield; another student processed leftover bar drinks he collected in a special drum. Bread too old for day-old stores is bulky, but pound for pound is nearly as good as pastry droppings. Tortillas, broken pasta, lima beans, peas, frozen waffles, starch for gluing boxes, almond hulls, **lees** from wineries, and rejected seeds and bulbs from flower seed growers—these are all feedstocks that my students have used. Look for seasonal products. The squash family is a good source in winter; you can make a lot of alcohol from unwanted pumpkins after October 31 when the price plummets. In some cities, thousands of tons of starchy chestnuts fall to the ground in just a few weeks each year. Just remember, it's not waste; it's a resource needing to be put to work.

You may find surplus fruit the easiest source for alcohol. It's been conservatively estimated that culled fruit in California could yield 10+ million gallons of alcohol.

Fruit Waste

You may find surplus fruit the easiest source for alcohol. It's been conservatively estimated that culled fruit in California could yield 10+ million gallons of alcohol. Visit the areas in your location that handle food—industry, packing houses, and canneries. If it isn't the right color, or shape, or size,

or ripeness, canneries pay to have it hauled away. Farmers can ask their local canning association for a list of canneries in their vicinity. Haulers' needs for fruit waste may be too small or infrequent to be worth the bookkeeping for a cannery to pay them, but almost without exception the cannery will give it away free. The bigger the truck you've got, the happier they are to see you.

If you're having stock delivered, you'll have to have a storage facility. Since a ton of peaches yields a little over 200 gallons of pulp and juice when it's shredded for the fermenter, you can anticipate needing ten to 30 tons for a few hundred gallons of alcohol. A concrete holding pen will allow the trucker to dump his load quickly; once they start dumping, they most likely won't be able to stop, so your container size should conform to the size of the load you're receiving. You won't have to deal with long-term storage as long as you schedule your runs around delivery time.

Although fruit waste is bulky, it's cheap and plentiful in many places. Your water requirements for processing are less with fruit, since it's got plenty of its own juices. You don't need expensive enzymes to ferment fruit, but the energy input required for cooking is sometimes higher than with other mashes. Fruit ferments quickly with fewer complications than other stocks, but leaves you with quite a bit of pulp to dispose of. The pulp is usually a combination of cellulose, hemicellulose, and pectin. Depending on the fruit, you may be able to apply an enzymatic cellulose hydrolysis or pectin-dissolving enzyme step to increase your yield and reduce your pulp. However, pulp makes a good animal feed component for methane digestion or compost. Livestock operations in your area should be delighted to take the stuff off your hands if you can **dewater** it with a centrifuge or screw press; you may even make a little money selling it.

Fermenting fruit breaks down many of its residual pesticides and adds yeast to the mash, which raises the nutritional content considerably. Fruit mash is an excellent fertilizer, especially if it's returned to an orchard. There's a limit to how much mash you can put in any one spot, though. You don't want to put any more pulp and/or liquid under a tree than would naturally collect there as fallen fruit. Gardens and orchards appreciate your pulp for composting with some kind of clippings, chipped branch prunings, etc.

Pastry Droppings

Bakeries provide another good source of waste for the urban distiller. Pastry droppings are processed like wheat, but their yield is much higher than raw wheat's because of the incredible amount of white sugar in bread. White sugar may not be healthy for people, but it's terrific for yeast. Yields can be well over 100 gallons of alcohol per ton for pastry droppings. Some farmers are willing to pay $50 or more per ton of bakery droppings to use as animal feed. A bakery near the co-op plant we had in a San Francisco suburb in the 1980s produced pastry droppings totaling over 25 tons of dough a week.

Although a certain amount of oil helps keep foaming down, your dough shouldn't be excessively oily. You may want to use a commercial antifoam agent for really foamy batches—ones with high amounts of wheat. I used to use doughnuts as a feedstock; when we cooked them down, the fat they were fried in would float to the top. We used that thick layer of fat as a resource. I calculated there was a lot of energy in the fat, and we used it as heating fuel for the still—providing 80% of the distilling energy we needed, cutting down on wood use. (Of course, we could have made it into biodiesel.)

While consulting in Costa Rica, I was taken to a small factory making ice cream cones. The cones were pressed and cooked into shape with waffle-iron-like machines. All the excess was trimmed off and would have made a perfect feedstock.

Alternate feedstocks should be considered if you have to pay more than $30 to $40 per ton for dough products. Have a potato chip factory nearby? There's a lot of starchy waste from operations like that. But if you're set on dough, you can try selling your liquid byproduct to dairies or hog farms to help balance your costs. The liquid byproduct also makes a good base for production of single-cell protein production or methane.

Candy Waste

If it's available, candy waste (broken or dented pieces of candy) is an excellent source for alcohol. Because of its nearly pure sugar content, yields can be as high as 125+ gallons per ton. Storing candy is easy, even for long periods, because its sugar content is so high that even bacteria can't live in it. In fact, nothing can live in it. I sure have used a lot of it; once in a while we'd find a mouse nest in the

Storing candy is easy, even for long periods, because its sugar content is so high that even bacteria can't live in it. In fact, nothing can live in it.

candy, with the baby mice looking really sick from a constant diet of the stuff.

Also, making candy often results in burned or caramelized sugar, which is scraped off the bottom of the cooker and thrown out for you to put to use.

WHEAT

Historically, wheat has often been considered a core food. Old-fashioned varieties of wheat tended to have high protein and gluten content. In the 1800s, a baker could go to jail for making bread from wheat that contained less than 12% protein!

Our increase in yield of wheat tonnage per acre over time has really been only in the starchy part of wheat, with most modern varieties having a fraction of the protein of the older wheats. So the amount of protein produced per acre hasn't actually gone up very much. These soft wheats are used to make junk food, crackers, pita bread, French bread, pasta, and flatbreads, for starters. Hard wheats, which more closely resemble the old-fashioned grain, are used to make yeast-raised loaves.

You'd save money and hassle by using cheaper, softer wheat, which has less protein, but you'd end up with a less valuable byproduct. So it does matter if you are going to use the DDGS as animal feed.

Wheat's yield in northern climes can average 40 to 45 bushels per acre. That's just a little over a ton per acre.

Like corn, wheat is a grass; it generally requires the addition of about 60 pounds of nitrogen per acre, significantly less than corn. Wheat has a little bit less starch, slightly depressing its yield of alcohol per ton. Wheat's production of high-quality DDGS is higher than corn's, 38% of the original grain instead of corn's 32%, with about 38% protein content, compared to corn's approximately 30% protein. One pound of wheat DDGS replaces a kilogram of corn and 1.04 kilograms of soybean meal in feed rations.[145]

Wheat is processed like corn, with a few exceptions. The same recipe also works with softer (non-flinty) grains like barley, millet, milo, and sorghum grain. The grind should be a bit coarser than corn; a #20 screen is about perfect. Softer grains tend to gelatinize a bit earlier than the flinty ones, so it's important initially to mix your grain with cold water (below 100°F) to prevent lumping. Less water is used with wheat than with corn; the final dilution of water to grain should be about 22 gallons to one bushel.

In cooking, add all your alpha amylase right at the start. Bring the temperature up to about 190°F with vigorous agitation, and hold it there for an hour or so. (If you go over 190°F—even for only a few minutes—the enzyme becomes inactive, unless your enzyme source is bacterial.) Optimum pH really makes a difference to your enzymes when maintaining this high a temperature for an extended time. What you will have done is combined premalting, cooking, and liquefaction steps into one. The conversion (saccharification) step is identical to corn's.

Wheat's high protein content makes it foam a lot during cooking. You can add a bit of vegetable oil to the mash to overcome excessive foam; a commercial anti-foam agent works much better. Sometimes wheat is cooked in a mix with corn, using the corn recipe. When the amount of corn is over half the mixture, foaming is minimal.

WHEY

When milk is made into cheese, most of its fat and protein end up in the cheese, and the milk sugar (lactose) is left in a watery byproduct called whey. There isn't a lot of sugar in whey (only about 6%), and roughly 2% protein. Whey solids are considered a fairly nutritious human and animal food. The cheeses that produce the highest lactose levels in the whey are cheddar and Swiss, as well as some Italian cheeses.[146]

Fig. 8-64

A tank truck is really the only way to move the large quantities of whey necessary for alcohol production. Practically, you'd be better off making the alcohol adjacent to the source rather than moving it anywhere else. One thousand gallons of whey will only yield 30 gallons of fuel, so you'd need a tank truck for transporting and large tanks for processing. If you have a surplus of biomass energy, you can evaporate some of the water out of the whey, producing a lactose concentrate, which would be easier to handle. Modern alcohol plants concentrate the whey on-site to produce a solution of over 24% lactose, using reverse osmosis.

Cheese whey is a huge, expensive disposal problem for the cheese industry. If you're willing to take the stuff off their hands, they'll go out of their way to accommodate you. One cheese factory near the San Francisco Bay area produces 50,000 gallons of cheese whey a day and can't get rid of it fast enough. In the tradition of permaculture, some dairies have stopped thinking of whey as waste, and instead see it as a resource to be used. Golden Cheese Company of California produces over 2.7 million gallons of alcohol per year from its whey.[147]

Lactose is not metabolized by *Saccharomyces cerevisiae*, so a specialized yeast must be used for fermentation. One yeast that works is *S. fragilis*, but it is not the most efficient at converting the sugar. The best yeasts available today for this purpose are *Kluyveromyces fragilis, K. lactis,* and, to a lesser degree, *Candida pseudotropicalis*. *Kluyveromyces* is selective in the sugars it absorbs, going for the simple sugars first, then lactose and inulin.[148] It's not as alcohol-tolerant as *Saccharomyces* yeast, which means alcohol levels higher than 10% will be quite a strain. Since *Kluyveromyces* is not commonly cultured commercially, you'll

MUFFET AND THE SPIDER ARGUED LIKE THIS FOR HOURS

probably have to propagate it yourself. Starter colonies should be available through your local agricultural college at little or no charge, along with recommendations as to optimum pH and temperature to ferment whey.

Another useful process uses *Torula cremoris* yeast, which is very efficient. Heat the whey up to the boiling point and adjust pH to 4.7–5.0. Sulfuric acid is best at curdling the protein, which is then filtered out. Cool the mash to 93°F. Add yeast at the rate of one pound of yeast per 120 gallons of whey. Fermentation times average about 20 hours.[149] Either centrifuge the yeast to recover it for animal feed first, or go straight to distillation.[150]

Since whey uses lactic-acid-based fermentation, such as you might find in yogurt or kefir, many normally troublesome bacterial contaminants are inhibited. This helps whey fermentations achieve an 88% out of a 92% theoretical rate of conversion of sugar to alcohol.[151]

As I mentioned in the *Torula* recipe above, whey is acidified first to bring down its pH. This curdles the protein, which precipitates out of the mash and can be easily removed. Curdled protein, though, is not as digestible and therefore not as valuable as uncurdled protein. Modern plants use ultrafiltration to remove the protein altogether before acidification, leaving the producer with a higher-quality protein solid, especially when mixed with the yeast separated at the end of the process.

Protein filtration is followed by reverse osmosis through a membrane to reject some of whey's water and to concentrate its sugar up to 15–23%. This economizes on space and lowers distillation energy requirements, since the resulting mash has a higher alcohol content. Although a reverse osmosis concentrator is expensive to buy new, you may be able to pick one up at a plant auction and have it fitted with the right membrane for whey.

A potentially valuable byproduct of the whey process is a lactase called beta-D-galactosidase. This enzyme is made by a lactose-eating yeast, which breaks lactose down into the simpler sugars glucose and galactose. This enzyme can be commercially valuable in processing dairy products to make them edible for people with lactose intolerance. Extracting the enzyme is something a larger plant can consider before distillation.

Whey can be used in place of water when processing other feedstocks. Low-sugar feedstocks that need dilution to make them workable benefit from mixing with a liquid that already has some sugar in it. You can only use whey to dilute feedstocks that have a low salt/mineral content, since whey is high in these very things. Too-high salt/mineral content will make your yeasts unhappy.[152] You'll have to experiment to determine the right amount of whey versus water.

To simplify the use of whey for dilution, you can convert whey's lactose into simpler sugars using the enzyme lactase. This way, normal distiller's yeast *Saccharomyces cerevisiae* can be used in fermentation instead of the lactose-eating yeast discussed above, so you won't need to use two kinds of yeast to do your fermentation. The simpler sugar solution can be used as a liquid for diluting fruit pulp, corn, beets, grain, or other normal feedstocks. Alternatively, you can use two yeasts to handle the different sugars, if it's too expensive to use lactase.

Or you can use whey's lactose just as it is, with Jerusalem artichokes or other inulin feedstocks (such as *Asphodel*) that use *Kluyveromyces* for direct fermentation. It's not uncommon for Jerusalem artichokes to start with only 14% fermentables, after minimal dilution with water. The milk sugar in whey can bring the sugar concentration up to 20%, for a respectable 10% alcohol yield.

The liquid byproduct from whey fermentation is very high in phosphorus, and this should either be used in compost operations, or diluted and added in irrigation of fields. Although it's tempting to think that the fermentation process has dramatically lowered the biological oxygen demand, the presence of so much phosphorus precludes discharging the liquid as "treated" waste. Besides, why give away all that phosphorus?

As you can see, I depart from the popular wisdom that cellulose is the only path to follow when addressing our transportation energy needs. I believe that the choice is more varied and multi-layered than that. For every type of climate and soil condition, from arid lands to marshes, and even in the ocean, there's something we can use to make alcohol. The ability to produce fuel is not limited to using one of the world's big four crops—rice, wheat, corn, potatoes—or even to sugarcane.

Although I have been thinking through energy crops for 30 years, I've had people force me to reevaluate preconceived notions. While I was editing this chapter, a young man called me and said, "What about turnips and rutabagas?" My first reaction was that although fresh turnips have as much sugar in them as apples, they couldn't amount to much per acre. But then I realized that the majority of the turnip is cellulose. A few minutes of quick figuring, and I realized that big turnips, with their multiple crops per year and huge growth, are serious contenders down the road when their cellulose is included.

So now that you know that we can grow plenty of feedstocks under almost any conditions, it's time to distill them to moonshine.

Endnotes

1. W.P. Bemis, et al., "The Buffalo Gourd: A New Potential Horticultural Crop," *HortScience* 13:3 (June 1978), 235–240.

2. M. Porto and V. Marcarian, "Studies on Cassava (*Manihot esculenta*) as a Potential Crop in Arizona," *Forage and Grain: A College of Agriculture Report* P:57 (1982), 73–74, Table 1.

3. Olaoluwa T. Bamikole, Fermentation Engineer, communication with author, October 2004.

4. Cheng Zhang, et al., "Life Cycle Economic Analysis of Fuel Ethanol Derived from Cassava in Southwest China," *Renewable and Sustainable Energy Review* 7:4 (2003), 353–366.

5. Bamikole.

6. Bamikole.

7. Bamikole.

8. Yuli Tri Suwarni, "Jatropha Oil: A Promising, Clean Alternative Energy," *The Jakarta Post*, July 4, 2005.

9. Dovebiotech Ltd., *Castor Bean (Ricinus communis): An International Botanical Answer to Biodiesel Production & Renewable Energy*, 6, www.dovebiotech.com/technical_papers.htm (July 28, 2005).

10. Dovebiotech, 14.

11. H.B. Pierce, Dorothy E. Sheldon, and John R. Murlin, "The Conversion of Fat to Carbohydrate in the Germinating Castor Bean," *The Journal of General Physiology* 17 (August 1933), 311–325.

12. S.J. McNaughton, "Ecotype Function in the *Typha* Community-Type," *Ecological Monographs* 36:4 (autumn 1966), 311.

13. David Hull, Wetland Engineering and Technology, communication with author, 1982.

14. J.F. Morton, "Cattails (*Typha sp.*): Weed Problem or Potential Crop?" *Economic Botany* 29:1 (1975), 7–29.

15. Robert A. Jervis, "Primary Production in Freshwater Marsh Ecosystem of Troy Meadows, New Jersey," *Bulletin of the Torrey Botanical Club* 96:2 (March–April 1969), 209–231.

16. Morton.

17. David Hull, Karl Klingenspor, and Steven Wilbur, of Wetland Engineering and Technology, communication with author, 1982.

18. Amory B. Lovins, et al., *Winning the Oil Endgame: Innovation for Profits, Jobs, and Security* (Snowmass, CO: Rocky Mountain Institute, 2004), 104, www.oilendgame.com/ReadTheBook.html (November 10, 2006).

19. 3141 counties in the U.S.

20. Combined USDA-defined cropland and farmland.

21. Oak Ridge National Laboratory, *Biofuels from Switchgrass: Greener Energy Pastures*, for U.S. Department of Energy Bioenergy Feedstock Development Program, http://bioenergy.ornl.gov/papers/misc/switgrs.html (July 16, 2005).

22. Susanne J. Brown, *Biomass Crops Seen as an Opportunity for Future Energy Markets*, McGill University Ecological Agriculture Projects, http://eap.mcgill.ca/magrack/sf/FAll%2094%20D.htm (June 22, 2005).

23. Robert Shleser, *Ethanol Production in Hawaii: Processes, Feedstocks, and Current Economic Feasibility of Fuel Grade Ethanol Production in Hawaii*, for the State of Hawaii, Department of Business, Economic Development and Tourism (July 1994), 20, Fig. III-8.

24. Stephen T. Cockerham, "Irrigation and Planting Density Affect River Red Gum Growth," *California Agriculture* 58:1 (January–March 2004), 43.

25. Author's calculation.

26. R.L. Miller and L.E. Jackson, "Survey of Vesicular-Arbuscular Mycorrhizae in Lettuce Production in Relation to Management and Soil Factors," *Journal of Agricultural Science* 130:2 (1998), 173–182.

27. M. Ali Harivandi, Victor A. Gibeault, and Trevor O'Shaughnessy, "Grasscycling in California," *California Turfgrass Culture* 46:1,2 (1996), 5.

28. Lovins.

29. Bill Mollison, *Permaculture: A Practical Guide for a Sustainable Future* (Washington, DC: Island Press, 1990), 434.

30. Mollison.

31. Mollison.

32. Harivandi, Gibeault, and O'Shaughnessy, 4.

33. Author's calculation based on Lovins.

34. Earth Observatory, *Ecological Impact of Lawns*, http://earthobservatory.nasa.gov/Study/Lawn/lawn3.html (November 10, 2006).

35. Eric Hand, "Roundup Is Killing Off Amphibians, Ecologist Says," *St. Louis Post-Dispatch*, August 10, 2005, Sec. Sunday, P1.

36. Hand.

37. T. Han, et al., "Engineering a Homo-Ethanol Pathway in *Escherichia coli*: Increased Glycolytic Flux and Levels of Expression of Glycolytic Genes during Xylose Fermentation," *Journal of Bacteriology* 183:10 (May 2001), 2979–2988.

38. Helen Golias, et al., "Evaluation of a Recombinant *Klebsiella oxytoca* Strain for Ethanol Production from Cellulose by Simultaneous Saccharification and Fermentation: Comparison with Native Cellulose-Utilising Yeast Strains and Performance in Co-Culture with Thermotolerant Yeast and *Zymomonas mobilis*," *Journal of Biotechnology* 96:2 (June 26, 2002), 155–68.

39. Han, et al.

40. Shleser, 33.

41. Nancy Ho, Biologist, of USDA, communication with author, October 2004.

42. Bill Mollison, Permaculture Design Course lecture, Cross Timbers, April 1996.

43. Kevin Wenger, of Novozymes, and Nancy Ho, Biologist, of USDA, communication with author, October 2004.

44. Kevin Wenger, of Novozymes, communication with author, October 2004.

45. Shleser, 24, Table III-7.

46. Greg Miller, of Empire Chestnuts Company, communication with author, June 14, 2005.

47. Miller.

48. Miller.

49. Miller.

50. Author's calculation.

51. Shawn Mehlenbacher, "The Hazelnut Situation in Oregon," for the Hazelnut Marketing Board, *VI International Congress on Hazelnuts* (March 2005), Table 1.

52. Mehlenbacher.

53. M.A. McCarthy and R.H. Matthews, *Agriculture Handbook No. 8-12: Composition of Foods—Nut and Seed Products*, (Washington, DC: USDA Human Nutrition Information Service, 1984).

54. Author's calculation.

55. Cyberlipid Center, *Fats and Oils*, www.cyberlipid.org (November 10, 2006).

56. Author's calculation.

57. Pierce, Sheldon, and Murlin, 1.

58. Author's calculation.

59. Alice Waruguru Kamau, *Coffee 101*, Batian Peak, http://batianpeakcoffee.com/coffee_101.html (January 28, 2005).

60. Jan C. von Enden and Ken C. Calvert, "Limit Environmental Damage by Basic Knowledge of Coffee Waste Waters," *Post Harvest Processing: Arabica Coffee* (May 2002).

61. von Enden and Calvert.

62. Enzyme Services & Consultancy, *Questions About Enzymes*, www.enzymes.co.uk/answer24_pectinase.htm (June 20, 2005).

63. Enzyme Services & Consultancy.

64. Enzyme Services & Consultancy.

65. von Enden and Calvert.

66. James A. Duke, *Handbook of Energy Crops* (1983), 3.

67. H.G. Walker, Jr., and R.M. Saunders, "Chromatography of Carbohydrates on Cation-Exchange Resins," *Cereal Science Today* 15:5 (May 1970), 142.

68. Kamau, 2.

69. Duke, 4.

70. John Vidal, "Things Grow Better with Coke," *The Guardian* [UK], November 9, 2004.

71. Daniel Martínez-Carrera, communication with author, July 2005.

72. Daniel Martínez-Carrera, et al., "Commercial Production and Marketing of Edible Mushrooms Cultivated on Coffee Pulp in Mexico," in T. Sera, et al., eds., *Coffee Biotechnology and Quality* (Dordrecht, Netherlands: Kluwer Academic Publishers, 2000), 471–488.

73. Martínez-Carrera, communication with author, March 1990.

74. Eugenia Olguín, Gloria Sánchez, and Gabriel Mercado, "Cleaner Production and Environmentally Sound Biotechnology for the Prevention of Upstream Nutrient Pollution in the Mexican Coast of the Gulf of México," *Ocean & Coastal Management* 47:11–12 (2004), 645.

75. Author's calculation.

76. T.M. Teynor, et al., "Comfrey," in *Alternative Field Crops Manual*, University of Minnesota Center for Alternative Plant and Animal Products and the Minnesota Extension Service, www.hort.purdue.edu/newcrop/afcm/comfrey.html (November 10, 2006). The dry weight yield of comfrey in California is 10.7 tons per acre.

77. Lesley Bremness, *The Complete Book of Herbs* (London: Dorling Kindersley, 1988).

78. J.M. Wilkinson, "A Laboratory Evaluation of Comfrey (*Symphytum officinale L.*) as a Forage Crop for Ensilage," *Animal Feed Science and Technology* 104 (2003), 227–233.

79. Lovins, et al.

80. (S&T)² Consultants, *The Addition of Ethanol from Wheat to GHGenius* (Ottawa: Natural Resources Canada, Office of Energy Efficiency, 2003), 9.

81. Darrel Good, "Corn: A Record Large Crop," *Grain Price Outlook* 7 (October 2003).

82. Michael Pollan, as quoted in John Robbins, *What about Grass-Fed Beef?*, The Food Revolution, www.foodrevolution.org/grassfedbeef.htm (July 29, 2005).

83. F.J. Hills, et al., "Comparison of Four Crops for Alcohol Yield," *California Agriculture* 37:3/4 (1983), 17–19.

84. S.R. Vaughan, et al., "The Co-Production of Fuel Alcohol and Leaf Protein Concentrate—Towards Total Biomass Refining of Green Herbage," *Abstracts of the Fifth International Alcohol Fuel Technology Symposium, Auckland, New Zealand* (1982), 54.

85. Gerald N. Festenstein, *Leaf Protein Concentrates* (Avi Publishing Company, 1983), 215.

86. D.R. Cosgrove, et al., "Jerusalem Artichoke," in *Alternative Field Crops Manual*, University of Minnesota Center for Alternative Plant and Animal Products and the Minnesota Extension Service, www.hort.purdue.edu/newcrop/afcm/jerusart.html (October 12, 2004).

87. Roy M. Sachs, et al., "Fuel Alcohol from Jerusalem Artichoke," *California Agriculture* (September–October 1981), 4.

88. Z. Duvnjak, et al., "Production of Alcohol from Jerusalem Artichokes by Yeasts," *Biotechnology and Bioengineering* 24:11 (February 18, 2004), 2297–2308.

89. W. Pilnik and G.J. Vervelde, "Jerusalem Artichokes as a Source of Fructose, a Natural Alternative Sweetener," *Journal of Agronomy and Crop Science* 142 (1976), 159.

90. Novozymes, *Uso de Fructozyme en la Producción de Fructosa a partir de Inulina*, Technical Paper 2002-19779-01.pdf, 1–5.

91. Duvnjak, et al.

92. Peter Calamai, "Tundra Test Stuns Scientists," *The Toronto Star*, September 23, 2004.

93. Johan A. Hellebust and J.S. Craigie, eds., *Handbook of Phycological Methods* (Cambridge: Cambridge University Press, 1978), 135.

94. Tien-Hsi Cheng, "Production of Kelp—A Major Aspect of China's Exploitation of the Sea," *Economic Botany* 23 (1969), 215–236.

95. Alan T. Critchley, et al., eds., *Seaweed Resources of the World* (Japanese International Cooperation Agency, 1998), 68.

96. W.D.P. Stewart, ed., *Algal Physiology and Biochemistry (Botanical Monographs)* (Berkeley: University of California Press, 1974), 206–214.

97. Ralph A. Lewin, ed., *Physiology and Biochemistry of Algae* (New York: Academic Press, 1962), 293.

98. M. Quillet, "Sur la Nature Chimique de la Leucosine, Polysaccharide de Réserve Caractéristique des Chrysophycées, Extraite d'*Hydrurus foetidus*," *Compte rendu hebdomadaire des séances de l'Académie des Sciences, Paris* 240 (1955), 1001–1003.

99. Lewin, 294.

100. Hellebust and Craigie, 139.

101. Lewin, 293.

102. Cheng.

103. United Nations Environment Programme, *GEO: Global Environment Outlook Year Book 2004/5*, www.unep.org/GEO/pdfs/GEO%20YEARBOOK%202004%20(ENG).pdf, as referenced in Janet Ralof, "Dead Waters," *Science News Online* 165:23, June 5, 2004 (November 10, 2006).

104. Glen Martin, "Sea Life in Peril—Plankton Vanishing; Usual Seasonal Influx of Cold Water Isn't Happening," *San Francisco Chronicle*, July 12, 2005, Sec. A1.

105. Martin.

106. Robert Hodam, *Energy Farming* (California Energy Commission, 1978).

107. Liz DeCleene, *Mesquite, the Rediscovered Food Phenomenon,* http://chetday.com/mesquiteflour.htm (November 10, 2006).

108. Jane E. Brody, "To Preserve Their Health and Heritage, Arizona Indians Reclaim Ancient Foods," *The New York Times,* May 21, 1991 (http://query.nytimes.com/gst/fullpage.html?sec=health&res=9D0CEFD71F31F932A15756C0A967958260).

109. W.P. Bemis, *Potential of Buffalo Gourd and Other Arid Land Plants as Food Resources* 17 (Tucson: University of Arizona Dept. of Plant Sciences).

110. Author's calculation.

111. DeCleene.

112. Joshua Tickell, *From the Fryer to the Fuel Tank: The Complete Guide to Using Vegetable Oil as an Alternative Fuel,* 3rd ed., (New Orleans: Joshua Tickell Media Productions, 2003).

113. Ahmad Faizal Abd Hajat, "Minor Impact on M'sia's CPO Prices Despite De-Pegging of Ringgit," *Malaysian National News Agency,* July 28, 2005 (www.bernama.com/bernama/v3/news_business.php?id=146666).

114. Franklin W. Martin, *Multipurpose Palms You Can Grow: The World's Best,* Agroforestry Net, http://agroforestry.net/pubs/multipalm.html, October 15, 2004).

115. Jeffrey Dukes, "Burning Buried Sunshine: Human Consumption of Ancient Solar Energy," *Climate Change* 61 (2003), 31–44.

116. Karl Kessler, "Bad Plants Go Straight," *The Furrow,* North Plains Edition (March 1981), 15.

117. Kessler.

118. Kessler.

119. *"Distribution and Occurrence of the Species Opuntia polyacantha,"* USDA Fire Effects Information System, www.fs.fed.us/database/feis/plants/cactus/opupol/distribution_and_occurrence.html#GENERAL%20DISTRIBUTION (October 12, 2004).

120. Marco Antonio Anaya-Pérez, "History of the Use of Opuntia as Forage in Mexico," in Candelario Mondragón-Jacobo and Salvador Pérez-González, eds., *Cactus (Opuntia spp.) as Forage* (Rome: Food and Agriculture Organization of the United Nations, 2001).

121. Liliana Rodriguez, "Opuntia," in Gena Fleming, ed., *Plant by Plant: Your Gateway to Traditional Food and Healing Plants,* www.plantbyplant.com/pages/opuntia77.htm (November 2, 2004).

122. C. van Dillewijn, *Botany of Sugarcane* (Waltham, MA: The Chronica Botanica Company, 1952), 144.

123. van Dillewijn, 138.

124. van Dillewijn, 177.

125. Harry F. Clements, *Sugarcane Crop Logging and Crop Control Principles and Practices* (Honolulu: University of Hawaii Press, 1995), 198.

126. Clements.

127. Constance E. Hartt and George O. Burr, *Factors Affecting Photosynthesis in Sugarcane* (Honolulu: Hawaiian Sugar Planters' Association, Experiment Station), 597.

128. van Dillewijn, 177.

129. Angelo Bressan, Director, Department of Sugar and Alcohol, Ministry of Agriculture, Brazil, communication with author, January 12, 2005.

130. Edgar Beauclair, of Dept. of Vegetable Production, Universidade de São Paulo, Escola Superior de Agricultura "Luiz de Queiroz," Piracicaba, Brazil, communication with author, January 2005.

131. Octávio Lage, President, Jalles Machado S/A Brasil, communication with author, January 2005.

132. *Sugarcane: Technological Advance,* São Paulo Sugarcane Agroindustry Union (UNICA), www.unica.com.br/i_pages/cana_tecnologia.asp (July 29, 2005).

133. Geraldo Lombardi, Pedro Antonio Rodrigues Ramos, and Romeu Corsini, *Potencial Econômico, Social e Ambiental da Produção Integrada de Álcool, Electricidade e Alimentos* (São Paulo, Brazil: Universidade de Sao Paulo, 2003).

134. Bressan.

135. Beauclair.

136. Lombardi, Ramos, and Corsini.

137. Lombardi, Ramos, and Corsini.

138. Morris Bitzer, of National Sweet Sorghum Producers and Processors Association, communication with author, July 2005.

139. Shleser.

140. Lombardi, Ramos, and Corsini.

141. Lombardi, Ramos, and Corsini.

142. Shleser, 20.

143. Lombardi, Ramos, and Corsini.

144. Lombardi, Ramos, and Corsini.

145. (S&T)^2 Consultants.

146. Elmer H. Marth, "Fermentation Products From Whey," in Byron Horton Webb, ed., *Byproducts from Milk,* 2nd ed. (Westport, CT: Avi Publishing, 1970), 43–82.

147. Golden Cheese Company of California, *Alcohol,* http://ourworld.compuserve.com/homepages/gccc/alcoholp.htm (October 4, 2004).

148. *An Interesting Article on Kluyveromyces and Whey Fermentation,* www.pbf.hr/cabeq/Grba.pdf (September 11, 2004).

149. D.M.L. Sandbach, "Production of Potable Grade Alcohol from Whey," *Cultured Dairy Products Journal* 16:4 (1981), 17–19, 22.

150. M. Rogosa, H.H. Browne, and E.O. Whittier, "Ethyl Alcohol from Whey," *Journal of Dairy Science* 30:4 (1947), 263–269.

151. Marth.

152. Sandbach.

"Given how many of us had parents or grandparents who came here … to the United States—looking for nothing else but the opportunity to work and have decent lives, I think we also grew up believing this country really does have a unique role to play in this world—and that America really does stand for something better, something honorable. That's why I think it causes so many of us so much pain to know that people in this world who once respected us, people who once loved us, now fear and even despise us.

"The values we share are fundamental to who we are as a people—it's what defines us as Americans.

"Now, what does all that have to do with energy policy? Everything…. Let me share just a few statistics with you.

"The United States today has slightly less than 3% of the world's oil reserves, but we use 25% of the world's oil production.

"We import 51% of our oil. Two and a half million barrels per day come from the Persian Gulf. Of this, over 1.7 million barrels comes from Saudi Arabia alone. Sixty-five percent of the world's (known) oil reserves are in the Persian Gulf.

"And where will oil come from in the future? Eighty-five percent of the increase in oil production between 2010 and 2020 is going to come from the Middle East.

"That's why … the leverage of Middle East and Persian Gulf regimes on the United States not only isn't contracting—it's growing.

"Our message to the powers that rule there has essentially been: We don't care who gets rich, we don't care about corruption, we don't care about poverty, we don't care how you treat women, we don't care that there's no democracy, we don't care that there are no free trade unions, we don't care if dissidents are imprisoned and tortured and killed. Just make sure we get our gas. Now, that's not the kind of foreign policy that's consistent with the values I talked about a moment ago."

—JOHN PODESTA, PRESIDENT, THE CENTER FOR AMERICAN PROGRESS

Congratulations—you've already accomplished the hardest part of the process of fuel-making: brewing the alcohol. The next step, distillation, is actually very easy; it's harder to explain than to actually do. I'm going to give you an operational understanding of distillation rather than an engineering explanation, since that's what will be most useful for our purposes.

Distillation comes down to a few simple principles, which anyone can understand with the right vocabulary. Simply put, distillation is the separation of two or more materials—in this case, two liquids: alcohol and water. The way we separate the two is by taking advantage of the differences between alcohol and water.

One big difference is that alcohol boils at about 173°F, and water boils at 212°F. So alcohol will boil first. Alcohol also has a higher **vapor pressure** than water, so it takes less energy to turn it from liquid into gas. It takes three times the energy to boil water compared to alcohol.

However, just boiling off alcohol does not constitute a complete distillation. A complete distillation not only boils the alcohol/water vapor, but also cools it back to a liquid. You can then measure the strength of the liquid—the percentage of alcohol—with a "proof" hydrometer. Proof is a unit of alcohol measurement; one proof equals one half of a percent. Pure, 100% alcohol measures 200 proof; half alcohol, half water equals 100 proof.

The run is the length of time it takes to complete one distillation, not the distillation process itself. Sometimes you'll hear "run" used to mean both things within a single distillation.

A PRIMER ON ENERGY AND HEAT TRANSFER

It's useful to outline some of the core concepts around how energy works before talking about specific technology. Since we need to have an understanding of these basic concepts for all aspects of

SIMPLY PUT, DISTILLATION IS THE SEPARATION OF

TWO OR MORE MATERIALS—IN THIS CASE, TWO LIQUIDS:

ALCOHOL AND WATER. THE WAY WE SEPARATE THE TWO

IS BY TAKING ADVANTAGE OF THE DIFFERENCES

BETWEEN ALCOHOL AND WATER.

IMAGE COURTESY OF *MOTHER EARTH NEWS,* <WWW.MOTHEREARTHNEWS.COM>

Fig. 9-1 Mother Earth News' *hot-oil furnace/distillery.* *This is a woodstove surrounded by an oil-filled chamber to provide thermal transfer. The heating unit is teamed with a vacuum still on the left.*

alcohol production and engine conversion, I'd like to spend a little time making sure we are all on the same page.

Heat and Temperature

Heat and temperature are not synonymous. Heat is a description of the energy contained in a substance. Imagine yourself looking under a powerful microscope at a glass of water until you can see individual water molecules. These molecules are not just sitting still. They jiggle around quite a bit and bump into each other. If they have little energy, they bump into each other infrequently and don't vibrate very far on each jiggle. If energy is added to the water, they move more forcefully in all directions, bashing into each other with greater impact and more often.

One of the little-discussed effects of global warming is the expansion of the oceans. Even if the icecaps didn't melt, sea levels would rise quite a few feet just because slightly warmer oceans would jiggle more and take up more room than colder oceans.

In fact, this is precisely how a microwave oven heats your food. The microwave radiation causes the food to vibrate at about 2.5 billion times per second. This causes lots of friction, which heats the food and also tears apart cells with the motion. Mobile phones use virtually identical frequencies of microwave energy, due to the frequencies' penetrating power, so they also do this sort of vibration to human cells. (Hence, they are called cell phones?)

The more violently molecules slam-dance, the more intense the energy level. The more intense the energy level, the greater the natural tendency for the molecules to spread further away from each other. When things spread apart, they become less dense. So warm air or water takes up more room (expands) as it takes on energy. One of the little-discussed effects of global warming is the expansion of the oceans. Even if the icecaps didn't melt, sea levels would rise quite a few feet just because slightly warmer oceans would jiggle more and take up more room than colder oceans.

When things heat up and become less dense, it is commonly, and incorrectly, said that hot air rises. Although hot air does end up on top, it's because the colder, and therefore denser (heavier), air sinks and displaces the hot air. So it is gravity moving

denser air below the fluffier warm air that causes the warm air to "rise." This is why a ceiling fan can't push lightweight hot air down into the denser and colder air near the floor.

It's also important to note when talking about energy that there is no "cold" energy. When things seem cold, it's because they have lost their energy. Cold is not a substance or force. So you can't let the "cold" out of a refrigerator when you open the door. You are letting warm air in because the cooler air is denser and falls by gravity out of the fridge. Heat always moves to a colder area.

Now, temperature is a measure of how hard the molecules are slam-dancing. It is indirectly related to heat, but it really depends on how tightly packed the molecules are when you add the energy pressure. In the English/American system, the standard developed long ago to measure heat is called the British thermal unit (Btu). It represents the amount of energy it takes to make one pound of water (about a pint) raise its temperature one degree Fahrenheit. The amount of energy you need to make a pound of water go from 60°F to 212°F is 152 Btu.

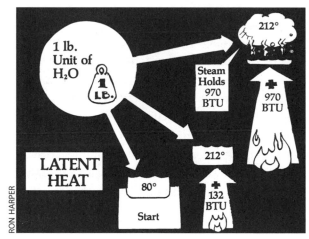

Fig. 9-2 Latent heat of vaporization. *The amount of extra heat that is required to make liquid water into steam.*

Phase-Change Energy

Energy and temperature stay pretty well linked as long as the water remains liquid. But distillation requires us to change a liquid into a gas. To change a pound of water that is 212°F into 212°F steam requires over 970 additional Btu. This huge extra input of energy is called the latent heat of vaporization, or the **phase-change energy**, to describe the change from a liquid phase to a gaseous phase (or from solid to liquid).

The reason "a watched pot never boils" is because it takes a lot of time to pack in those extra 970 Btu. Once those units have been absorbed by your water, though, every additional Btu has a huge impact on the amount of gas produced.

It's easy to see why steam is a more efficient heating source than hot water. When a **pound of steam** condenses in a room radiator (a heat exchanger), it gives up at least 970 Btu to become a liquid again. This immense amount of heat is absorbed by the cooler metal and then radiated into the still cooler room. Hot water—the same temperature as steam at 212°F—has less than 100 Btu of easily exchanged heat to be radiated into your cold room.

Pressure

When water molecules absorb enough heat to boil, they become very agitated and need a lot of room. At atmospheric pressure, steam needs 1700 times more room than its original liquid water. This is the basis of steam engines. The idea is that you put in enough energy, and the water expands mightily, driving pistons or a turbine. When it gives up its energy by condensing back to a liquid, it contracts 1700 times its volume.

This brings the next important fact about heating and gases into focus. It's all relative to pressure. Our first examples were all at atmospheric pressure in open containers. Things get interesting when you boil things and they try to expand in a closed container. Since there is nowhere to expand to, the pressure in that container rises instead. In an open container, at sea level, both sides of the container are at about 14.7 pounds per square inch (psi), due to the miles of air pressing down on everything. In a closed container, the outside pressure stays at 14.7 psi, but the inside pressure climbs when heat is added. If the pressure inside the container drops—by, say, vapors condensing back to liquid or the atmosphere being removed from the inside by a pump—the outer air pressure pushes in and can collapse the container.

You might remember a science trick in high school where water was boiled to steam in a metal can with a screw lid. If you seal the can and let it cool, the condensing steam will contract to a liquid, and the can will be crushed by the outside air. It's not the contracting steam that sucks the can in, although it may seem that way!

In making alcohol, we may wish to make use of pressure or vacuum. It's important to use a proper,

stronger container for either pressure or vacuum. Why would we need to have pressurized or evacuated tanks? At different pressures, the boiling point of a liquid will change. We can take advantage of that in our various designs. The higher the air pressure, the higher the boiling point. In the case of water, putting the tank under pressure to 15 psi raises the boiling point of water to about 250°F. In practice, some of the water turning to steam in the closed container will supply the pressure. Dropping the pressure and creating a vacuum will lower the boiling point. We can make use of the relationship between pressure and boiling point in different aspects of distillation, cooking, and later in car conversion.

Rate of Heat Transfer

Another important principle in heat transfer is that the bigger the difference between the temperature of a substance you're heating and the heat source, the better. Nature abhors a vacuum and always tries to equalize differences. The greater the difference, the faster the initial rate of equalization. Conversely, as the temperature of heat source and substance begin to draw closer, the transfer of heat slows down.

Say you're using 180°F solar-heated water circulating in coils that are immersed in your mash to preheat it. A 60°F mash will go to 80°F very

---

SOLAR ENERGY: A POOR HEAT SOURCE FOR DISTILLATION

Obviously, it's important to match quality of heat (its temperature and quantity) to its use. Consider solar energy as a heat source: Solar collecting panels will generally heat water—quite a bit of water—up to 180°F. Each panel may absorb 40,000 Btu per day, more than enough energy to distill a gallon of alcohol. With the same amount of surface area, and therefore the same amount of potential heat energy, you can use a parabolic reflector to focus all the heat at one point. This can generate temperatures as high as 2400°F at the focal point—which could be used to generate a small quantity of 350°F high-pressure steam. But there's no more heat in the concentrating collector steam than in the solar panel water. There are roughly the same number of Btu in each, even though the temperatures are different.

It's as inappropriate to use 350°F steam for domestic hot water as it is to use 180°F hot water to try to boil mash. Temperatures produced by solar panels are not high enough for distillation (except in the case of vacuum distillation, discussed later in this chapter).

quickly, but going the rest of the way takes progressively more and more time. So, in general, you want to have a wide difference between the temperatures of your heat source and whatever is being heated, for the fastest heat exchange.

As boiling water expands and turns to steam in a closed container, the pressure rises. As the pressure rises, so does the temperature at which the water will boil. At 15 psi, the heat exchange rate into the food is much higher at 250°F than at 212°F. Conversely, when we want to cook something more slowly, at a lower temperature, we use a double boiler, which doesn't get any hotter than 212°F, in comparison to the 1200°F temperature of a gas flame.

Raising the temperature of the heat source is one way to increase the difference between the heater and the heatee. Another way to go about this is to reduce the pressure on the heatee, thereby lowering the boiling point. So a lower-temperature heat source that wouldn't have been useful if the heatee had a higher boiling point is now perfect. The ability to use 180°F water to heat and boil liquid in a tank that's under vacuum is a good example, which we'll look at more when we discuss vacuum distillation.

Fig. 9-3

Storing Surplus Heat

In designing your alcohol plant, it's important that you design uses for any surpluses the system generates. That means matching the surplus resource to a need elsewhere in the system. When working with heat, it's best that you use the surplus in real time, as you produce it. This is practical in a system that runs 24/7, but small-scale producers are more likely to do things in batches. This means that we have to do the intermediate step of storing surpluses until they can be used.

When it comes to heat storage, we have a few options. In essence, what we want is to store heat in some sort of "battery," the way we store electricity. Our heat battery will hold the heat in water or oil stored in a tank for later use, either directly or as a carrier of heat to exchange into something that needs the energy.

As we mentioned above, the high latent heat of vaporization represents just how much extra heat we can store in water at its boiling point without actually turning it into steam. As long as the surplus heat is in this middle range of useful temperatures (130–212°F), water makes a good

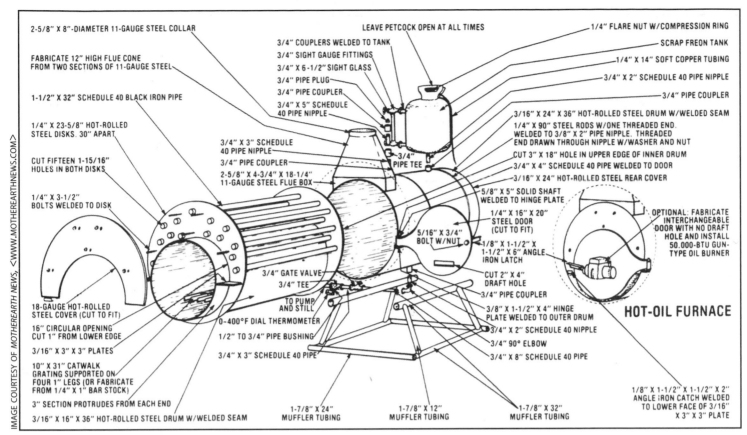

2-5/8" X 8"-DIAMETER 11-GAUGE STEEL COLLAR

FABRICATE 12" HIGH FLUE CONE FROM TWO SECTIONS OF 11-GAUGE STEEL

1-1/2" X 32" SCHEDULE 40 BLACK IRON PIPE

1/4" X 23-5/8" HOT-ROLLED STEEL DISKS. 30" APART

CUT FIFTEEN 1-15/16" HOLES IN BOTH DISKS

1/4" X 3-1/2" BOLTS WELDED TO DISK

18-GAUGE HOT-ROLLED STEEL COVER (CUT TO FIT)

16" CIRCULAR OPENING CUT 1" FROM LOWER EDGE

3/16" X 3" X 3" PLATES

10" X 31" CATWALK GRATING SUPPORTED ON FOUR 1" LEGS (OR FABRICATE FROM 1/4" X 1" BAR STOCK)

3" SECTION PROTRUDES FROM EACH END

3/16" X 16" X 36" HOT-ROLLED STEEL DRUM W/WELDED SEAM

LEAVE PETCOCK OPEN AT ALL TIMES

3/4" COUPLERS WELDED TO TANK
3/4" SIGHT GAUGE FITTINGS
3/4" X 6-1/2" SIGHT GLASS
3/4" PIPE PLUG
3/4" PIPE COUPLER
3/4" X 5" SCHEDULE 40 PIPE NIPPLE

3/4" X 3" SCHEDULE 40 PIPE NIPPLE
3/4" PIPE COUPLER
2-5/8" X 4-3/4" X 18-1/4" 11-GAUGE STEEL FLUE BOX

3/4" PIPE TEE

5/16" X 3/4" BOLT W/NUT

3/4" GATE VALVE
3/4" TEE
TO PUMP AND STILL
0-400°F DIAL THERMOMETER
1/2" TO 3/4" PIPE BUSHING
3/4" X 3" SCHEDULE 40 PIPE

1-7/8" X 24" MUFFLER TUBING

1-7/8" X 12" MUFFLER TUBING

1-7/8" X 32" MUFFLER TUBING

1/4" FLARE NUT W/COMPRESSION RING
SCRAP FREON TANK
1/4" X 14" SOFT COPPER TUBING
3/4" X 2" SCHEDULE 40 PIPE NIPPLE
3/4" PIPE COUPLER

3/16" X 24" X 36" HOT-ROLLED STEEL DRUM W/WELDED SEAM
1/4" X 90" STEEL RODS W/ONE THREADED END. WELDED TO 3/8" X 2" PIPE NIPPLE. THREADED END DRAWN THROUGH NIPPLE W/WASHER AND NUT
CUT 3" X 18" HOLE IN UPPER EDGE OF INNER DRUM
3/4" X 4" SCHEDULE 40 PIPE WELDED TO DOOR
3/16" X 24" HOT-ROLLED STEEL REAR COVER

5/8" X 5" SOLID SHAFT WELDED TO HINGE PLATE
1/4" X 16" X 20" STEEL DOOR (CUT TO FIT)
1/8" X 1-1/2" X 1-1/2" X 6" ANGLE IRON LATCH
CUT 2" X 4" DRAFT HOLE
3/4" PIPE COUPLER
3/8" X 1-1/2" X 4" HINGE PLATE WELDED TO OUTER DRUM
3/4" X 2" SCHEDULE 40 NIPPLE
3/4" 90° ELBOW
3/4" X 8" SCHEDULE 40 PIPE

OPTIONAL: FABRICATE INTERCHANGEABLE DOOR WITH NO DRAFT HOLE AND INSTALL 50,000-BTU GUN-TYPE OIL BURNER

HOT-OIL FURNACE

1/8" X 1-1/2" X 1-1/2" X 2" ANGLE IRON CATCH WELDED TO LOWER FACE OF 3/16" X 3" X 3" PLATE

battery, and we can store this heat until we need it later.

As we will see a little later, lower-temperature, water-based storage is an option for use with **vacuum distilleries**. But in **atmospheric distilleries**, which operate at normal air pressure, high temperatures are needed to boil and distill the mash, roughly 212°F.

If we want to store or capture energy at a higher temperature, then water alone won't serve our purposes. Flue gases from a still are well over 1000°F, and condenser temperatures from heat pumps are over 300°F. Capturing this premium high-temperature energy requires something with a much higher boiling point than water. There are any number of vegetable, petroleum, or synthetic oils that can serve this purpose.

Although these oils will allow you to store energy at higher temperatures, they don't have water's ability to store a lot of heat per pound. This means that to store the same number of Btu at these higher temperatures, you will need a bigger tank.

It's important to realize that using a hot-oil-type product is more efficient than steam, electric, or direct-fire heating methods[1] in heat transfer, especially when you want to boil water. The boiling point of average everyday vegetable oil, motor oil, or similar lubricating oils is in the vicinity of 350–400°F. Although oil doesn't have the high latent heat of vaporization of water, its ability to resist boiling makes it efficient at storing heat temporarily.

Specially designed petroleum or synthetic oils can boil at much higher temperatures; boiling points over 750°F are possible. The wide difference between the temperature of the oil and whatever is being heated makes heat transfer very fast and effective. For our purposes, it's not really worth the extra expense to go higher than 600°F. Typically, 600°F oils can still be pumped at below-zero temperatures, whereas higher-temperature oils can be too thick to pump on those cold winter mornings. Even 350°F vegetable oil is very effective at boiling water, is inexpensive, and can use the less expensive pumps.

Oil-based heating has a lot of advantages. For one thing, the oil heating system operates at atmospheric pressure. Steam heating operates at a variety of higher pressures, depending on how high a temperature you want to use to get fast heat exchange. Pressurized steam systems have to be built of heavier materials and require several levels of safety equipment.

Hot-oil heating can be done by circulating it through coils in the tank or in a jacket around the tank. The jacket can be made of channel steel bent

Fig. 9-4 Cutaway of hot-oil furnace. *This hot-oil heater was developed for use primarily with vacuum distilleries in the 1980s. It is adequate for production of up to ten gallons per hour of alcohol fuel.*

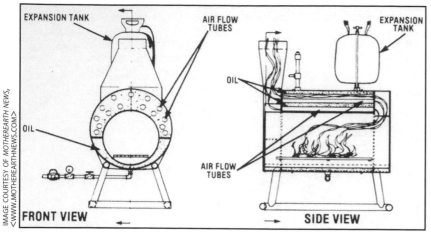

Fig. 9-5 Heating with hot oil. A wood-burning furnace like this is useful to generate hot oil if you don't have a waste heat source to capture with the oil.

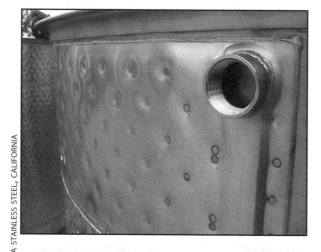

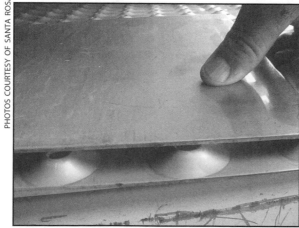

Fig. 9-6 Jacketed tank. Hot fluid is pumped into the waffle-pattern jacket through the two-inch fitting shown here. The fluid circulates through the jacket and exits through a similar fitting at the other end of the jacket.

Fig. 9-7 Cutaway view of jacketed tank.

to the shape of the tank and welded on to flow the heat transfer fluid through the channels. A common design in the food processing business is the waffle-pattern heat exchange jacket, with numerous dimples in the jacket to force the fluid to flow around the dimples.

In storing either water or oil heat in tanks, insulation is extremely important because the heat of the stored liquid will do its very best to heat the air surrounding the tank. My favorite way of insulating these high-temperature tanks is with straw bales. A three-string bale is two feet wide and rated at about R-50. With this much insulation, it's possible to store hot water heat for up to a week with only moderate losses. Doubling the straw-bale wall thickness can extend storage for a few more days.

For hot oil storage over 450°F, wrap the tank with mineral wool first to prevent scorching of the straw at very high temperatures, and then use one to two bales of insulation thickness to reduce the amount of heat lost from the higher-temperature oil. In any case, the corners need to be stuffed with loose straw to make up the difference between round tanks and square bales.

Energy Loss in Conversion

One last concept on energy and heat transfer: Whenever you change from one form of energy to another, you're guaranteed to lose something in the process. When natural gas is converted into electric energy, there's a huge energy loss. It takes about three Btu of gas energy to run the generator to make one Btu of electricity.

Obviously, it's crazy to use electricity or hydrogen for heating water. Electric heating starts with the stored energy of gas, converts it to heat, to steam, then to mechanical energy (spinning the turbine), then to electrical energy (in a generator), then transports it through transmission lines—losing energy the whole way—and then back to heat, which transfers to a pot on the stove, and finally affects the water in the pot (and the air around it). The only thing more insane is to turn the electricity into hydrogen so that you can run a fuel cell car (turning the hydrogen back into electricity and heat).

Burning gas to heat water directly, instead of making electricity, cuts out an enormous amount of waste. Match your energy use to the energy source least processed to do the job. And, more importantly, use all system surpluses first, before generating any new energy to do a job.

Boiling and Proof

Since alcohol boils at 173°F, you might assume that the alcohol should begin boiling off when the mash reaches 174°F. In actuality, the more alcohol in the mixture, the lower the boiling point. The more water in the mixture, the higher the boiling point.

So the boiling point of a 10% mash might start off at about 199°F. In the first third of the run (the **foreshot**), the vapors rising from the mash are very high proof, 180 or more. As this first portion of alcohol boils off and is cooled by the condenser into a liquid, the boiling point in the **mash pot** rises. Why? Since it's mostly alcohol that has vaporized in the foreshot, the mixture now has more water in proportion to alcohol than when you started.

Remember, the more water, the higher the boiling point. Now that the boiling point is higher, it's possible for more water to reach the vapor phase.

During the second third of the run, the alcohol produced is not so strong, averaging 80 to 100 proof. The boiling point has continued to rise, and the last third of the run might average only 10 to 20 proof until all the alcohol has been vaporized. In one simple distillation, you can expect to get an end product of about 100–110 proof.

DISTILLERIES

Distilling uses the differences between water and alcohol, in terms of their latent heats of vaporization and boiling points. A distillery may serve only the function of separating the alcohol from the mash, or it may incorporate multiple functions. In some cases on the small scale, it's possible to use the same piece of equipment to operate as a cooker and fermenter before operating it as a distillery. Larger operations will have separate pieces of equipment to do each of these functions.

The Moonshine Still

In the old days, we'd begin boiling our fermented mash in a primitive moonshine still. As the mash began to boil, vapors would begin to form and rise.

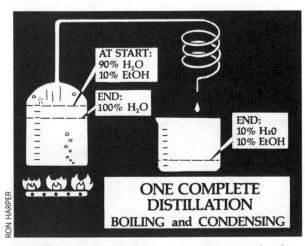

Fig. 9-8 One complete distillation. When completed, the alcohol (EtOH) will be removed from the mash. The condensed liquid will be half alcohol and half water (100 proof).

The simple still existed for thousands of years in different forms. Pre-Christian Arabs and later Greeks are thought to have used stills to make rose water and other concentrated herb concoctions. This simple still has a major drawback, though—on the first distillation, it can only produce 100-proof alcohol. For auto fuel, you need at least 170 proof and preferably 185 to 190 proof.

A simple still requires a series of distillations to get the proof up. A second distillation would yield 140 proof, and the third a little over 160 proof, depending on how carefully the still is controlled. In effect, the increase in purity is less and less with each distillation. Up to ten simple distillations are required to reach 190+ proof.

I don't advise buying or building a simple still. Old-style moonshine stills have an alarming tendency to "explode." Contrary to popular belief, the usual reason they explode has nothing to do with the alcohol at all, but is exclusively a design failure. Imagine all that corn porridge bubbling and popping in the mash tank. While you're

Practical Boiling Points

PROPORTION OF ALCOHOL IN THE BOILING LIQUID IN 100 VOLUMES	TEMPERATURE OF THE BOILING LIQUID	PROPORTION OF ALCOHOL IN THE CONDENSED VAPOR IN 100 VOLUMES
92	171.0	93
90	171.5	92
85	172.0	91.5
80	172.7	90.5
75	173.6	90
70	175.0	89
65	176.0	87
50	178.1	85
40	180.5	82
35	182.6	80
30	185.0	78
25	187.1	76
20	189.5	71
18	191.6	68
15	194.0	66
12	196.1	61
10	198.5	55
7	200.6	50
5	203.0	42
3	205.1	36
2	207.5	28
1	209.9	13
0	212.0	0

Fig. 9-9 Practical boiling points. The greater the proportion of alcohol, the lower the boiling point of the mixture. The boiling points shown here are practical boiling points rather than theoretical, since the impurities in fermented mash are different than in a pure water/ alcohol mixture. (Throughout this section, however, we refer to the temperatures that would be found in distillation of a straight water/ alcohol mixture.)

watching the tank, there's no problem. But as soon as you turn your back, a piece of corn jumps up and clogs the condenser coil. With no place to go, steam pressure builds up, and the still bursts. (**Slag boxes** have helped to solve this problem, but not 100%.) In many cases, the only indication you have of an impending explosion is when the steam pressure begins rocking the still back and forth. So, whenever the still starts to walk, you'd better start to run!

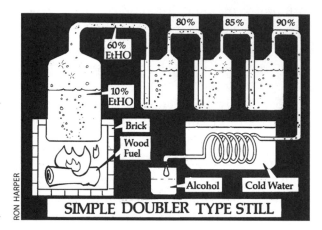

Fig. 9-10 Simple doubler still. By using three "doublers," alcohol vapor is condensed in the liquid and then re-evaporated three times to automatically produce high-proof alcohol in one run.

The Doubler Still

In the 1800s, a fellow named Adams was credited with making a real advance in distillation, the **doubler still**—a way of raising the proof automatically without having to stop the still, empty it, and refill it with alcohol for every distillation. To understand his development, think of distillation as the *leaving behind of water*. Remember, in each additional distillation on a simple still, some of the water is left behind. Less water, therefore, travels with alcohol vapor through the condenser.

The mash pot of Adams's still is the same as in a simple moonshine still, but between the main tank and the condenser is an intermediate tank which runs vapor under the liquid (see Figure 9-10). This tank, known as the doubler, works best when it starts off one-half to two-thirds full of water.

As vapor rises from the mash pot, it begins bubbling through the doubler, but before it can rise through the liquid, it condenses and mixes with the rest of the water. As the process goes on and the alcohol is vaporizing and condensing, the mash pot's boiling point continues to rise toward 212°F. The boiling point in the doubler, on the other hand, is going down as the alcohol enters and condenses in it. Eventually, hot steam from the mash pot reboils the alcohol-rich liquid in the

RIGHT: Fig. 9-11 Enrichment. Each time the alcohol is condensed and revaporized, the vapor contains more alcohol.

doubler, and the reboiled vapor moves on to the condenser, where it's cooled to a liquid of about 140 proof.

The alcohol has been completely distilled twice: one complete boiling, a condensation, another boiling from the doubler, and then another condensation. In both steps, water is left behind, and a purer product is vaporized each time. Carried to its logical conclusion, if there were ten of these doublers in a row, each would be successively higher in proof and lower in boiling point.

Note that the difference in boiling points narrows with each successive doubler. On a good set-up during the first run, the mash pot reaches a temperature of 212°F, and the last doubler is just above the boiling point of pure alcohol (173°F). Accordingly, the proof of the mash tank is close to zero, while the proof leaving the last doubler should be over 190.

On the first run, an appreciable amount of the alcohol remains in the doublers. On subsequent runs, the doublers will all be pre-charged to the correct intermediate proofs (see Figure 9-10). Almost all of the alcohol in your second batch is distilled as high proof, leaving the doublers with approximately the same amount of alcohol in them as when they started.

When you let alcohol enter a liquid, thus lowering the liquid's boiling point, and then allow the mixture to reboil, you are practicing **enrichment**, one of the two principles of distillation. (**Countercurrent stream** is the other one; see next section.)

In the old days, since it was illegal to moonshine, the longer you ran a still, the more likely you were to get caught by the government agents. By using doublers and the enrichment principle, moonshiners were able to cut down the length of the run to end up with high-proof liquor. But doublers created another problem—they took up a lot of

space, making it much more difficult to hide the still.

There were a couple of other problems. After each reboiling, you'd have all that leftover water, which has a tendency to continue to build up in the doubler. A couple of full doublers recreate the same pressure buildup and bursting potential that corn pieces do in a simple still, plugging up your condenser. Overflow drains were thus added to some doublers to collect the spillage for re-distillation with a next batch.

Another negative aspect of the doubler still is that it uses an awful lot of energy. You only apply heat to the mash pot, but each tank and line is constantly radiating heat from the process. On a very cold day, it might be impossible to put enough heat into the mash tank to keep the process going.

The Bubble Cap Column Still

To save energy (chopping wood) and space, moonshiners came up with the first primitive **bubble cap column still**, consisting of a tube or a **column** filled with a series of doughnut-shaped plates stacked one on top of another, spaced a few inches apart. A small stub of pipe, called a **riser**, extends out an inch or two from each plate's center hole, retaining a small reservoir of liquid on each plate. A cap with a wider diameter than the riser sits over the stub of pipe, dips below the liquid, and directs vapor that is going up the riser back down. As vapor bubbles back under the liquid, it condenses into alcohol, and the boiling point of the liquid on the plate declines as the boiling point in the tank increases.

Responding to the higher temperatures, steam, with its high water content, now boils through the liquid. The alcohol and a little water are reboiled up to the next plate, where the solution is again condensed, enriching the liquid on that plate, revaporizing, and leaving water behind. Each successive plate holds a higher proof, with a corresponding lower boiling point.

Although this new still solved a lot of problems in terms of space and energy, it left one last problem to solve—what to do with the leftover water. As the alcohol rose from plate to plate, becoming purer and purer, something had to happen to the water. Left to its own devices, that water would overflow the plate and begin flowing back down

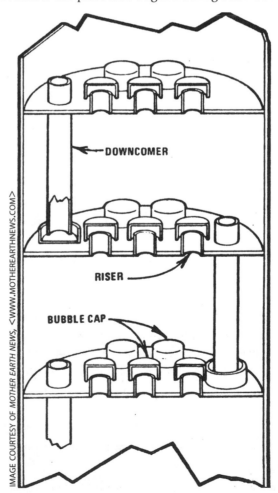

MATT FARRUGGIO

IMAGE COURTESY OF MOTHER EARTH NEWS, <WWW.MOTHEREARTHNEWS.COM>

DOWNCOMER

RISER

BUBBLE CAP

FAR LEFT: Fig. 9-12 Bubble cap distillery model. *These distilleries often use many small caps on the plate. Alcoholic liquid fills each plate to the height of the* **downcomer**.

Fig. 9-13 Bubble cap plate design. *This design was a logical extension of the concept of doublers but was more compact, used less energy, and automatically controlled liquid levels on each plate.*

the riser, defeating all our efforts to maintain separate levels of proof and boiling point.

The solution was an overflow drain at one side of each plate (each drain being situated on the opposite side of the plate above or below), so excess water and a little alcohol can follow the drain down to the next lower plate. Each subjacent plate is hotter because it's a lower proof, which allows part of the alcohol to revaporize out of its water until that plate overflows and the mixture drips down to the next plate in line.

To ensure maximum stripping of excess alcohol, some advanced bubble cap designs used channels on their plates so that down-coming liquid was directed in a long serpentine route, encountering dozens of small bubble caps (instead of a single large cap) on its way to the next lower level.

The bubble cap column illustrates the second principle of distillation—countercurrent stream—based on the down-flowing action of water versus the up-flow of alcohol vapor.

The Perforated Plate Column Still

The next type of plate still came on the scene around the same time steam technology was gaining in sophistication. The use of steam instead of direct firing as a heat source for distillation changed the design of the column still. Now, with a steam boiler, a steady, reliable pressure could be maintained in the column.

DISTILLATION PRINCIPLES

The concepts of enrichment and countercurrent stream govern all modern distilleries. Basically, these principles say that alcohol rises as a vapor and water drops as a liquid; and that since alcohol has the lower boiling point, the top of the column is always cooler than the bottom. In fact, the top plate of a properly working column with relatively pure alcohol will be just under173°F, and the bottom will be around 212°F.

Although alcohol at the top of the column may boil at 171°F (due to lower-boiling-point impurities in the initial mix) during the start of the run, it will boil at somewhat higher temperatures than 173°F during the last third of the run (because of **fusel oils** and other higher-boiling-point compounds). Water is also subject to a fluctuating boiling point. Salts and other dissolved solids affect it, as does elevation. Proper top and bottom plate temperatures are site-specific and more or less feedstock-specific, within a degree or so.

The new plate was nothing more than a sheet of metal with hundreds of tiny holes drilled through it in a particular pattern. The holes covered a specific amount of the surface area of the plate (often around 8%). There was still a downcomer on each plate, and liquid on each plate, as in the bubble cap still. But with the advent of steam, risers and caps could be eliminated as liquid retainers. Now the liquid stayed on the plates by means of the pressure of alcohol/water vapor shooting up through the holes.

Picture keeping a Ping-Pong ball afloat in the air above your face by blowing on it: The pressure of your breath holds up the ball. Like a Ping-Pong ball, the **perforated plate column still** is very sensitive to changes in pressure. If you stop blowing on the ball, it falls back in your face; if the pressure drops in a perforated plate column, the liquid falls back through the holes. The falling back is called a **dump** in distillation terminology.

When all those plates with all their separate proofs and boiling points dump back into the tank (or into the **reboiler** in a continuous still), you have to start all over again. Obviously, very careful attention must be placed on maintaining pressure in these stills. Directly fired coal or wood heat don't work well with perforated plate stills, since the intensity of a fire changes continuously as it burns—so steam was the answer.

Perforated plates further streamlined the development of industrial-level **continuous distillation**, where all of the mash, rather than just its vapor, is flowed through the still. Because you are not processing one batch at a time, but instead steadily pumping mash into the still from fermentation tanks, the process is continuous, instead of using sequential batches.

Although bubble cap columns can be and are used for continuous distillation, thick mashes

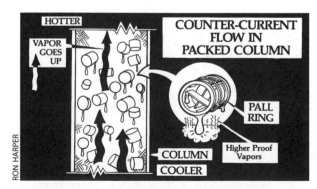

Fig. 9-14 Countercurrent flow. In a **packed column**, in a process similar to a plate distillery, high-proof vapor rises and water is left behind to flow to the bottom.

generally require perforated plate design. In bubble caps, thick mashes can gather around the caps and partially plug the flow. Perforated plate distilleries—in which the vapor jets up through the plate through small holes, while the mash solids and surplus liquid go over the **weir** dam or down the downcomer—work better with thick mashes in continuous distillation.

There are several advantages in using a perforated plate still even for **batch distillation**: Less energy is used when steam is the source of heat, and a smaller-diameter distilling column containing the perforated plates will still produce more alcohol than a larger-diameter bubble cap continuous distillery.

But perforated plate stills are expensive, and somewhat difficult to design and build. The best compromise for the batch producer is the packed column type, since it saves space and energy, and is the least expensive and easiest to construct.

The Packed Column Still

Packed column stills are an offshoot of oil industry technology for distilling or cracking oil into its component parts for use in different industries and materials. While the oil industry distills several chemicals at once, we're only concerned with alcohol and water.

The packed column still uses the same enrichment and countercurrent flow principles that other stills do. But instead of having separate plates with distinct and separate proofs and boiling points, the packed column allows a gradual, continuous rise in proof and lowering of boiling point, as you go higher.

Plates are replaced inside the column with a packing material to give the alcohol/water vapor sites on which to condense

Fig. 9-15 Bronze wool packing. *Far superior to marbles, bronze wool is okay to use in three-inch-diameter columns. In six-inch columns, the wool uses a lot more energy than pall rings to overcome greater back pressure, and the lower pads tend to load up with particles of mash in a pretty short time.*

Fig. 9-16 Pall ring packing. *Pall rings come in a variety of materials, but the plastic ones are the least expensive and work well in alcohol/water distillation.*

Fig. 9-17 Lathe chip packing. *Stainless steel lathe chips are an economical packing if pall rings are too expensive initially. As far as energy economy goes, they rank between pall rings and bronze wool.*

and revaporize. Remember, each time alcohol and water boil, the alcohol rises, leaving a little more water behind. Now, instead of that overflow running down a tube to a hotter plate to have its alcohol stripped, a droplet of condensed vapor falls a short distance through the porous packing in the column until it reaches a sufficiently hot zone for the alcohol to revaporize. It's a dynamic process—the vapor is condensed and revaporized literally thousands of times on the way to the top of the column. Each revaporization leaves a small water deposit, which ultimately makes its way to the bottom of the column and back into the mash pot.

Alcohol/water vapor requires a lot of space to rise through, but the vapor needs to encounter a lot of surfaces on which to condense on the way up. So your packing material should provide plenty of surface area and still leave a great deal of free gas space. While rocks, marbles, or pieces of glass packed into the column may provide a lot of surface area, they don't allow enough free gas space to be a good packing material. Too little free gas space causes a loading-up effect in the column, in which there's so much condensed wash in the passages that the vapor has a hard time getting through. To compensate, you'd need higher pressure, a taller column, and more energy to run the still.

I can recommend several excellent materials for packing. Bronze or stainless steel **scrubbing pads** or **lathe chips** made of stainless steel or any other nonferrous metal fit the still requirements. Even better are **pall rings**, which were computer-designed for distillation columns. They provide an enormous amount of surface area and occupy only about 5% of the volume of the still column. Since pall rings vaguely resemble shotgun wads, some people have tried (not too successfully) to use such wads for packing.

Even though pall rings are more expensive than some materials, the 20+% energy saved justifies

their use, so invest now to save later. Enough 5/8-inch diameter pall rings to fit a six-inch diameter column still (approximately 2-3/4 cubic feet loosely packed) costs about $130. Lathe chips cost about $10 to $15. Scrubbing pads are about $100 purchased in bulk. Many of my students started off with one of the cheaper packings, then switched to pall rings once they'd sold some fuel and made some extra money.

Another commercial packing material to consider is **ceramic saddles**. Although they are cheaper than pall rings and almost as efficient, they seem to break down over time, whereas pall rings are nearly indestructible. While newer saddles are being made from plastic, the cost still isn't sufficiently cheaper than pall rings.

Lathe chips are a byproduct of the metal machining industry. When a cylinder of metal (or some plastics) is turned rapidly on a lathe for cutting or reshaping, the material removed comes off in spiral chips. These chips (the coarser the better) are relatively efficient packing material in stills eight inches in diameter or less. Of course, your chips should be of a material that won't rust. Don't worry about any cutting oil that may remain on lathe chips—it'll drop to the bottom during the distillation process the first couple of times you use the still.

Since alcohol can reach 190+ proof in a packed column, after condensing and revaporizing many times on the way to the top, it follows that there is a definite relationship between the height of the column and the proof of the alcohol. The farther it has to travel up the column, the more times it condenses and revaporizes.

There is, of course, a point past which your alcohol can't get any purer, no matter how many more revaporizations occur. This point is approximately 192 proof. The distance your alcohol vapor must rise to get to this point is expressed as a ratio of 24 units of height to one unit of diameter. Any higher wastes energy. Any lower and you get a lower-proof yield, because your vapors don't get to condense and revaporize enough times to reach the 190+ proof level—they'd still be carrying too much water. A height-to-diameter ratio of 16 to 1 yields 160 proof at best.

Your choice of height to diameter should be determined by how you plan to use the alcohol. A fuel oil burner converted to alcohol requires a minimum of 150 proof. If you're going to generate

Fig. 9-18 Height-to-diameter ratio for packed columns. These ratios are based on pall ring packing and 192-proof distillation. Building columns taller will not raise the proof of the alcohol and will use more energy.

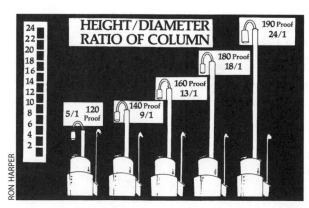

HEIGHT/DIAMETER RATIO OF COLUMN

24
22
20
18
16
14
12
10
8
6
4
2

5/1 120 Proof

140 Proof 9/1

160 Proof 13/1

180 Proof 18/1

190 Proof 24/1

RON HARPER

electricity with alcohol, you may find 170 proof more than adequate to run your generator. Most people, though, are planning to use this alcohol to fuel their cars, or at least want the option to do so. Although you can run a car on as low proof as 170, it's preferable to use a higher proof (see Chapter 14). A column capable of producing 190 proof can be "detuned" easily to produce a lower proof.

In a batch still, the diameter of the column determines the output in gallons per hour, but the relationship between diameter and output isn't as straightforward as the height-to-diameter ratio. A column still six inches in diameter yields about eight gallons per hour of 192-proof alcohol, an eight-inch still yields 16, and a 12-inch gives you about 35.

Matching Tank Size to Column Size

COLUMN SIZE	OUTPUT (GAL/HOUR)	RECOMMENDED TANK SIZE (GAL)
2"	¼ to ¾	10–50
3"	½ to 1-½	30–150
4"	1–3	50–250
6"	5–10	200–1000
8"	8–20	500–1500

IIEA

Packed column stills are clearly the most cost-effective up to 12-inch diameters. Once you get above 35 gallons per hour, it begins to make sense to consider going to a continuous distillation design for ease of automation and energy savings. Going much smaller than a six-inch still is hardly worth it: A three-inch still, for example, produces only about a gallon per hour.

Packed column stills require a lightweight support plate of expanded stainless steel into which the plastic pall ring packing is poured. Packed column stills need minimal maintenance. After a year of heavy use, the packing will accumulate a coating of proteins and other guck. You can clean the packing by removing it from the column and washing it, or by pouring a biological solvent down through the top of the column, to dissolve the protein off the packing. If you use the solvent method, do it every four months or so, and still fully disassemble the column and wash the packing every couple of years.

Recirculate the cleaning solvent several times by pumping it through the column from top to bottom and back through a filter.

When you build your still, it's important to scale the tank to your column correctly. Putting a huge column on a tiny tank is obviously not practical, nor is a tiny column on a large tank. What is practical is distilling alcohol at a rate that allows you to complete a run in a normal working day or shift, somewhere between five to ten hours. It will take about six and a half hours to distill alcohol from the contents of a 500-gallon tank through a six-inch column; 500 gallons of mash yield 50 gallons of alcohol.

Column Control Factors in the Packed Column Still

Two factors are under your control in a packed column: temperature gradation and the pressure of the vapor entering. And both these phenomena are related. The more pressure there is, the more vapor enters the column, and the more temperature control is required. If your pressure is too high, you'll continually be fighting to keep the column temperature in line.

First, let's look at temperature. To ensure that the top of the packing contains high-proof alcohol on its way to the condenser, the temperature should be close to 173°F. If we keep this point at a controlled 173°F, only relatively pure alcohol can pass it. At the start of the run, the column won't need controlling; the vapors will be high enough in alcohol content to distill automatically at very high proof.

But as alcohol is depleted from the main batch, the temperature of those vapors rises (due to the higher water content) and begins to heat up the column. When this happens, you'll want to begin condensing vapor above the packing to 173+°F.

On the small scale, the usual method for cooling is by flowing water through cooling coils placed

LEFT: Fig. 9-19 Matching tank size to column size. Too small and the run time is short and hard to control. Too large and you run into logistical issues, planning on supervising the still for too many hours out of a day.

SHOULD WE GET USED TO GLOBAL WARMING?

Reporting on the White House admission that global warming was due to mankind—by extension MegaOilron—Terry Moran of ABC said, "Today's report paints the starkest pictures yet when scientists say what could be the impacts in the U.S. of global warming, from the diminishing snowpacks in Western mountain ranges, which could cause real water problems in California and elsewhere, to the disappearance of barrier islands on the Atlantic Seaboard." When asked by Peter Jennings what the Bush administration thinks mankind should do, Moran answered, "Get used to it."[2]

Fig. 9-20 Liberator 925 six-inch distillery. *This is the distillery my employees and I designed and operated in the 1980s. The thin tube to the left is the condenser, the center tube is the distilling column, and the right-hand tube is the flue coming from the* **firebox**. *Not visible is the cone bottom, which provided lots of surface area for the wood fire to transfer heat to the mash. The dials are connected to remote thermometers in the column and tank.*

MATT FARRUGGIO

above the packing. Using a heat pump (air conditioner) to chill the top of the column is a good idea if you have limited water.

If a temperature of 173+°F is maintained at the top of the column above these coils, then the bottom of the column will always be about 212°F (as hot distilled water leaves the bottom of the packing), and the center will be about 200°F.

Your cooling coils won't really affect the whole column's temperature. Alcohol condensing off the coils and flowing partway back down the packing is what cools the overheated column, by absorbing heat in order to revaporize—like giving the column an alcohol rub. Alcohol has a high latent heat of vaporization, and using that property allows us to get a good even temperature spread from top to bottom.

Direct refluxing is a superior, but more expensive, method of controlling temperature. Instead of cooling some of the alcohol at the top of the column and waiting for it to flow back through the packing, you can actually pump and spray alcohol back into the column.

Of course, with refluxing, you'll have to double or triple your condenser capacity, since you won't be using a **control coil** to absorb heat. A special alcohol pump returns cool alcohol back to the column a foot or two below the top of the packing and through a perforated ring of tubing which delivers a spray of alcohol over the packing below it. A column responds to temperature change much faster this way than by condensing alcohol on the coils and then letting it dribble back down into the packing. The **reflux ratio** (the amount of alcohol returned, compared to what is being produced net) varies between 2.0:1 in top-notch stills, to as much as 8:1 in very inefficient stills that operate at too high a vapor pressure.

How do you determine the amount of control at any given time? When using either cooling coils or direct refluxing, watch the vapor temperature, reading it at the top of the column, preferably in the narrow tube that the vapor whistles through to the condensers. As the temperature tries to go up, slowly allow more pumped cooling water through the coils, or heat-pump refrigerant, or more alcohol reflux. As the temperature decreases, slowly reduce the delivery rate by closing the valve on your cooling water or reflux alcohol.

With either method, a temperature-sensitive **flow-control valve** is a valuable tool for automating

this control factor. Temperature controls for heat pumps (refrigeration) are easily available, and that fact alone might make it worthwhile to use a heat pump for your control coil. If you are handy with electronics, you could experiment with a pump to send condensed alcohol back to the top of the column for reflux control.

The flow-control valve has a liquid- or gas-filled bulb inserted into the vapor flow right at the point where the vapor leaves the column. The high turbulence and speed of the vapor at that point give you a faster heat exchange, and a better reading with the bulb. The bulb's fluid or gas contracts or expands to drive a valve which opens and closes at an infinitely variable rate, regulating the amount of cooling water through the coils, or alcohol pumped back into the column. A flow-control valve is much more sensitive and accurate than an on/off solenoid valve.

Your flow-control valve should have a sensitivity of 1°F *at the most.* Compare response rates of valves you're shopping for and get the fastest one you can afford. The inlet and outlet ports of the valve should be no larger than a half inch if you are using water, or even the slightest motion of the valve's gate could allow tremendous changes in flow and you'd lose the sensitivity you're paying for. It is important, too, to buy a valve with a narrow temperature range in which 175°F is the approximate middle. A 160°F to 195°F valve, for instance, would be great, but it's not a standard size. Many large valve companies will be happy to give you a custom range at a properly exorbitant price.

Fig. 9-21 Coil rolling tool. This is a good tool for coiling copper tubing in a single plane.

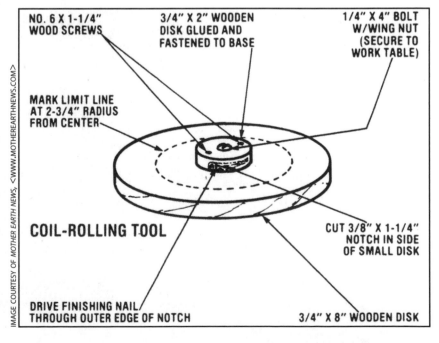

NO. 6 X 1-1/4" WOOD SCREWS

3/4" X 2" WOODEN DISK GLUED AND FASTENED TO BASE

1/4" X 4" BOLT W/WING NUT (SECURE TO WORK TABLE)

MARK LIMIT LINE AT 2-3/4" RADIUS FROM CENTER

COIL-ROLLING TOOL

CUT 3/8" X 1-1/4" NOTCH IN SIDE OF SMALL DISK

DRIVE FINISHING NAIL THROUGH OUTER EDGE OF NOTCH

3/4" X 8" WOODEN DISK

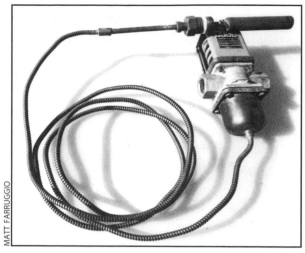

Fig. 9-22 Temperature-sensitive flow-control valve. This inexpensive valve (about $300) is suitable for columns from three to six inches in diameter, producing 180- to 192-proof fuel. It uses cold water to chill the coil in the top of the distilling column. It is all mechanical, requiring no electricity nor computer. Larger units are available from boiler parts suppliers.

When you put in a flow-control valve, you need to plumb a manual override system. At a minimum, you can get away with a T in front of and a T behind the automatic valve in the water line. An alternate line should be plumbed between the two Ts, with a quarter-turn **ball valve** installed in-line. Place a **gate valve** up-line between the T and the automatic valve inlet.

If the valve malfunctions, turn off the gate valve in front of the inlet. The water will have to flow through the turn in the up-line T and through the quarter-turn valve, which you can open so water bypasses the automatic valve. You can use the quarter-turn valve to adjust your cooling coil or reflux spray until the end of the run. Then, once everything has cooled down, you can figure out what went wrong with the automatic valve. An additional gate valve following the automatic valve outlet can be installed to give you the ability to isolate a defective leaking valve.

If you don't have a safety bypass, a couple of things can happen. If the valve sticks closed, you get no cooling in the column, and lots of hot alcohol vapor ends up billowing out of your overworked condenser. If the valve sticks wide open, you may not get any alcohol in your collection tank; it'll just keep flowing back down the column and into the tank until you realize what's going on. It's rare that valves malfunction, but they sometimes do, and you should be able to override the system manually. Having a manual override allows you to perform

TEMPERATURE
SENSOR

HOT
OUT

CONDENSER

PACKING
TO
BOTTOM
OF COIL

ISOLATION
VALVE

THERMOMETER

FLOW CONTROL
VALVE

ISOLATION
VALVE

MANUAL
OVER RIDE
VALVE
(NORMALLY
CLOSED)

STILL
COLUMN

COLD IN

Fig. 9-23 Installation of coolant flow-control valve. Isolation valves are closed and the manual override valve is opened in the event of valve failure.

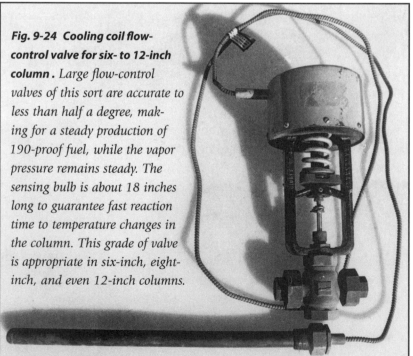

Fig. 9-24 Cooling coil flow-control valve for six- to 12-inch column. Large flow-control valves of this sort are accurate to less than half a degree, making for a steady production of 190-proof fuel, while the vapor pressure remains steady. The sensing bulb is about 18 inches long to guarantee fast reaction time to temperature changes in the column. This grade of valve is appropriate in six-inch, eight-inch, and even 12-inch columns.

diagnostic tests on your column, and it's a convenient way to shut down your controls completely.

In a vacuum still, the flow-control valve is a pressure-sensing, not a temperature-sensing, device. If your vacuum still uses hot water, hot oil, or steam heat exchangers as its energy source, use a flow-control valve similar to the one in the column's cooling coil control.

Remember, the other factor you can control in a packed column, besides temperature, is the pressure of the vapor entering the column. Pressure control depends on your heat source. An *uninsulated* still experiences surges in pressure, due to condensed alcohol on the tank's ceiling periodically dropping back into the tank and flashing into vapor.

The pressure in an *insulated* still is more or less directly related to the amount of heat put into the tank. Natural gas or fuel oil burners (fossil fuel heaters can use restaurant fryer oil here) can be set to deliver an even amount of heat continuously. If your tank is well insulated, you'll check it no more than every hour or so, and adjustments will be slight.

You can fully automate your still with off-the-shelf automatic pressure-reading controls to vary the amount of heat delivered in relation to the pressure. For six-inch stills, burners and controls capable of putting out a maximum of 500,000 Btu per hour are appropriate. An eight- or ten-inch column calls for a rating of 1,000,000 Btu per hour (most pool heaters need this much). It's unlikely you will require anywhere near this much energy, except for the initial heating of your mash to a boil. At that point, activate your automatic valve (or release your manual override), and the flame will throttle back down to an appropriate setting.

Alternatively, you can use the steam boilers used by small commercial laundries. These small boilers provide an easily controlled heat source and would fit our needs nicely. They can be picked up for a song at auctions of cleaners' equipment. They generally run on natural gas, but also can run on propane. If you are producing methane—which is essentially natural gas and works with natural gas equipment.

Small biomass boilers are now available, and can run on wood or sawdust. I've also seen ads for small boilers that run on fuel oil, which means they should also run well on used, heated vegetable oil.

Another option is to make use of low-pressure waste steam, if you're lucky enough to have such a source available. Pressure-sensitive, automated steam flow-control valves are common at industrial heating or boiler supply houses.

Chances are, though, that if you're an average home distiller or small business anywhere except

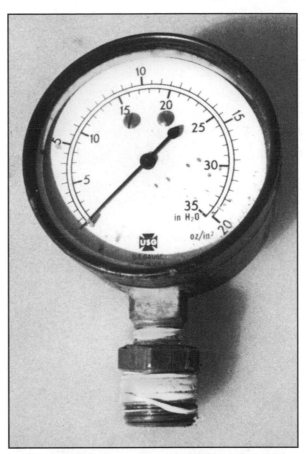

Fig. 9-25 Pressure gauge. *Note that the gauge registers a maximum of 20 ounces psi. It should be installed wherever vapor enters the column.*

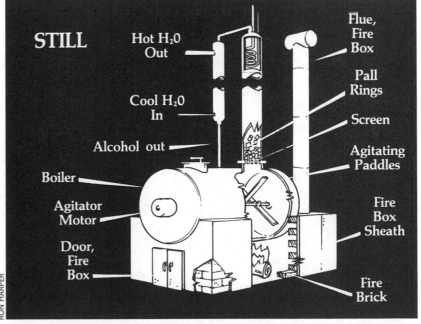

Fig. 9-26 Generic wood-fired distillery. *Shows major components of a direct-fired distillery.*

in the heart of the Midwestern plains or the desert, you'll be using direct-fired wood, or waste-oil heat as your process energy source. Unfortunately, compared to any of the "neat" sources of energy mentioned previously, a direct wood-fired still is more difficult to control, since fires are erratic in their burning intensity.

The biggest factor in controlling a wood fire is the amount of oxygen available for combustion. In a common household, wood stove oxygen is indirectly controlled by a flue damper, which

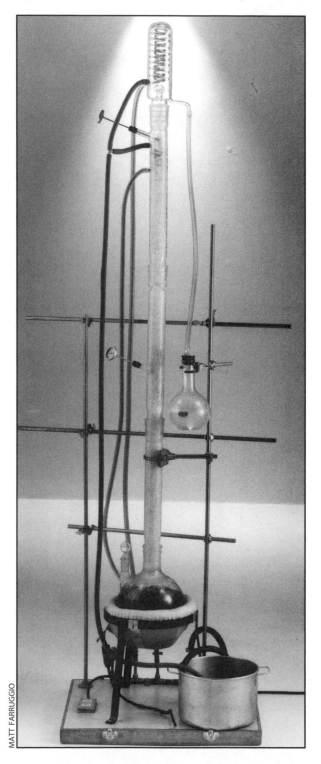

Fig. 9-27 Packed column laboratory distillery. *The packing is short lengths of glass tubing (5mm). Vapor travels up the column length, and the temperature is monitored at the top and middle. Just below the top thermometer, you can see a control cooling coil regulated by a* **needle valve** *behind the still. Cold water is recirculated from the pot (kept cold with ice). The condenser at the top is an immersion-type, coil within a coil, design. Condensing alcohol collects in the condenser's sump and then drains down into the flask.*

MATT FARRUGGIO

determines the amount of air drawn in through the air inlets. Adjusting the inlet dampers sets the upper limit for airflow. There's always a slight lag between adjustment and response when a flue damper is moved. Some wood stoves are able to use a temperature-sensitive automatic damper to stabilize the temperature of the stove. A similar device can work on a still.

Controlling the inlet air is more effective than controlling the flue damper. Most well-designed stills have a fan that forces air into the firebox to ensure a really hot fire up to boiling. After the mash starts to boil, the fan should be adjusted to a lower air input. It can be powered by a motor controller to slow down or speed up air input, as necessary. Don't attempt to adjust the speed of a normal fan motor by using a dimmer switch, as it will burn the motor out after a few months. One inexpensive solution is to provide an air bypass vent between fan and firebox to bleed off fan-forced air when less is needed. Never restrict air from entering the fan—restricted air will damage it.

Probably the easiest way to use wood or biomass fuel would be to use a large rocket stove to heat oil to be circulated through a jacket, or even to direct-fire the still. I recommend using a rocket stove or firebox to heat hot oil and then using that oil to heat the still tank. This gives you the ability to use a pressure-regulated flow-control valve, and that could be the easiest way of all. It's certainly going to be the way I build a still for myself.

There's a good test to determine maximum operating pressure and alcohol output for your still while it's under control. Start with a 1.5–2% alcohol mixture in the still (make this with pre-measured water and alcohol before you begin). Manually open the cooling coil valve all the way and then move it back a hair. Start increasing the temperature by whatever heating method you've chosen.

As the pressure begins to rise in the column, keep an eye on the temperature at the top of the column. You'll reach a point at which, if the pressure is great enough to overcome the cooling control coil, the thermometer will alert you to the unacceptably high temperature of vapor leaving the still, because the temperature will begin to rise. Calibrate your heat input to the pressure range just below the point where pressure is about to overcome the control coils.

Toward the end of the distillation process (when there's 1 to 2% alcohol left in the mash), you will generally turn off the controls to save energy and time by collecting your last alcohol as **low wine** (see next section). Since you are doing this test with only 1–2% of alcohol in your tank, you will be simulating the point in the run where the column controls get turned off. At this point, the vapor pressure will be at its highest, and you still want to control the proof of the alcohol.

So, set your pressure controller at a maximum of an ounce or two below the breakpoint to control heat input such that the system is never quite "maxed out." This will give you the highest output and good sensitivity for controlling your proof.

A new technology to explore would be pressure sensors and controlling software on a PC. The array of equipment to make this sort of thing possible continues to multiply and should get some attention by the geeks among us who are good at this stuff.

As with temperature control, you should always have the ability to manually override whatever system you've devised. A couple of things can

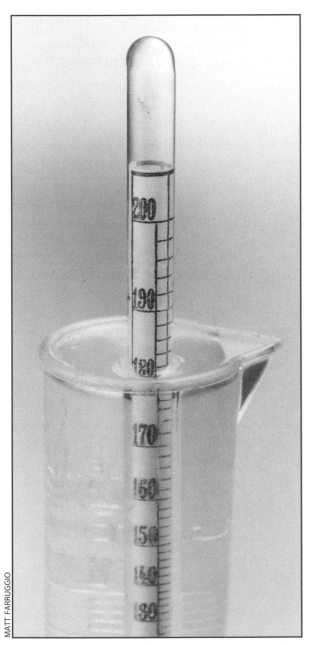

Fig. 9-28 Proof and Traille hydrometer. *This alcohol is 180 proof. A proof hydrometer must be designed for the proof range you expect to be working in. As you can see, the accuracy of this hydrometer would be considerably reduced for lower proofs.*

Fig. 9-29 Liberator 925 control panel. *A hydrometer sump is shown on the left. The three thermometers are remote types with capillary lines leading to sensing bulbs at the top of the column, at the initial vapor entry point, and in the tank itself. Handles at the right are for manual override of cooling controls and condensers, and for filling the reboiler with water.*

You should always have the ability to manually override whatever system you've devised.

happen if your pressure-sensing/controlling system malfunctions. The worst is that the tank builds up enough pressure to set off the relief valve on the tank, wasting your valuable alcohol. If the air inlet **blower** turns off or if your air vent bleed-off door sticks open, your fire, starved for air, slowly dies down, and your still stops producing alcohol. Partial sticking causes irregular distillation pressure changes, which can exceed your temperature valves' response rate and result in a low-proof product.

Low Wines

As you get toward the end of your run, it becomes harder and harder to control the column, since there's so much less alcohol and so much more water in the vapor. When you get to the point at which it takes a great deal of control energy to get just a little alcohol at high proof (1 to 2% alcohol in your mash in a good batch still), it's best to simply run off the remaining low-proof alcohol—called low wine—without reflux controls. This lets your vapor go rapidly through the still column and condense (basically just what the moonshiners did with their simple stills)—allowing the whole column to become approximately one temperature and one proof.

Low wines run off at 30 to 60 proof. Condense these off into a drum separate from the rest of your alcohol. Then when you start your next run, add those low wines to the batch to begin distillation.

It works like this: The first run starts off as a 10% alcohol mash; you distill 8% of the alcohol as **high wine** (fuel) and the remaining 2% is low wine. On the second batch, you again start with 10% alcohol mash, but now you add your low wines, bringing the alcohol level to roughly 12%. At the end of the second batch distillation, you'll get 10% high wine and 2% low wine, which you'll add to your third batch before distillation, and so on. Low wine is like the deposit on a soda pop bottle.

Although automation makes it possible to carry a run all the way to the end without switching over to low wine, it takes as long to get that last 2% as it took for the entire first part of the run. Besides, you get the added bonus of a somewhat faster average output when you start with a 12% batch rather than 10%. So in the long run (so to speak), your average gallons-per-hour is higher when you recycle the low wine.

Chemicals with high boiling points (acetic acid, some **aldehydes**) enter your fuel at the end of a run; so do **esters**, which have a tendency to deposit a little gum or carbon on valves. The acids will enhance corrosive electrolysis of metals in an engine's fuel system when using less than 190-proof fuel.

It's wise to save the last part of your run, the low wines, in a separate tank until you have enough to run a full batch of them off. Add plenty of lime to them before distillation to neutralize acids and bring the pH to 8.0. Only extract alcohol for fuel during the first 60 to 70% of the run; use the surplus twice-concentrated low wines as heating or boiler fuel. This is an excessively conservative approach, but is a useful technique for beginning distillers whose first few fermentations will be less than perfect, with more noxious byproducts than later, better-fermented batches.

Insulating Columns

Your column (and tank) should, of course, be well insulated in order to conserve energy—heat is continually radiated from exposed metal, especially on windy days. But there's a more important reason: Heat radiating from the column cools it and causes vapors to condense on the column walls. A cooled alcohol/water solution will run down the inside walls of your column without properly revaporizing the alcohol. In fact, depending on how cold the outside temperature is, alcohol will run all the way back to the mash tank and will have to be reboiled all over again—a major defeat for the principle of countercurrent stream. If the outside temperature is cold enough, your alcohol may never reach the top of the column. The condensing effect will be much more severe at the bottom of the column than at the top because of the difference between vapor temperature and outside temperature, causing strange "bumps" in the distillation curve, which affect the quality of your output.

To insulate your column, wrap it with fiberglass batting, and then with either tarpaper (ugly but cheap) or 30-gauge aluminum sheets (nicer looking, but more expensive). There are several excellent but very expensive spray-on urethane-foam-type insulations that work in the right temperature range. Your best bet might be a snap-on, aluminum-backed, molded insulation designed for outdoor steam lines, the same diameter as your column. Seal the seams with silicon caulk, even if the insulation is rated for outdoor use. Good sources for this insulation are overstocked nuclear power plant construction suppliers or boiler supply houses. Insulation choices increase if you house your distillery indoors.

Control Coil Construction for Six-Inch Distillery

I found that the following control coil design worked exceptionally well for regulating alcohol proof leaving the column. If you aren't using a countercurrent condenser, you may use this control coil configuration to construct an immersion-type condenser. It should be at least equal in size to the control coil. Here's what you'll need.

1. Condenser housing
2. Vapor tube
3. Three $1/2$" Ts
4. Cold water inlet
5. $5/16$" lead lines from coils
6. Eight $3/8$" × $5/16$" compression through fittings
7. Eight $3/8$" half pipe couplers
8. Condenser flange
9. Welds on bottom of couplers
10. Small top coil
11. Leads from lower coils
12. Large lower coil
13. Small lower coil
14. Large lower coil
15. Flange bolt holes
16. Cold water through fittings (only half couplers shown)
17. Vapor tube hole
18. Hot water through fittings (only half couplers shown)

Step 1: Drill or cut with a cutting torch eight $7/8$-inch holes (#16 and #18) in the top flange (#8 which connects to the condensers). The holes should be $2^5/8$ inches from the perimeter and spaced equally as shown. Cut four $3/8$-inch pipe couplers in half. Slip them (#8) into the $7/8$-inch holes to about half their depth, with the cut side facing down (towards the inside of the column). Weld them in place on the bottom side of the flange (#9). Welds must be pit-free, since they have

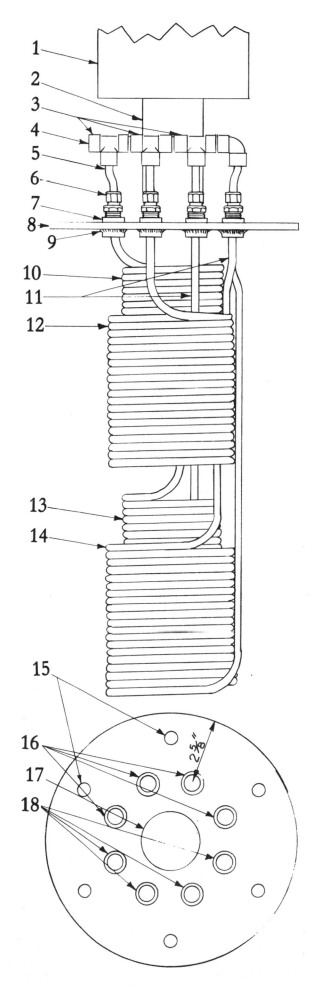

Fig. 9-30 Control coil construction.

to prevent alcohol vapor leakage. If in doubt, silver solder over the welds.

Step 2: You need to construct four separate coils. Two of the coils will be formed by wrapping

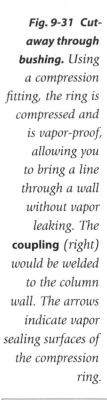

Fig. 9-31 *Cutaway through* **bushing.** *Using a compression fitting, the ring is compressed and is vapor-proof, allowing you to bring a line through a wall without vapor leaking. The* **coupling** *(right) would be welded to the column wall. The arrows indicate vapor sealing surfaces of the compression ring.*

MATT FARRUGGIO

$5/16$-inch copper tubing around a two-foot length of four-inch pipe (which is actually $4\frac{1}{2}$ inches outside diameter). Each coil should be 18 wraps. Leave some excess tubing on each coil to extend through the top of the flange. On one coil, the two lead lines should be 24 inches (bottom lead) and 18 inches (top lead). The other coil should have leads of 14 and eight inches.

In addition, you'll need to form two smaller-diameter coils. Bend 22 wraps of $5/16$-inch copper around a two-foot piece of three-inch (outside diameter) tube for each coil. Leave lead lines on these two coils of the same lengths as the larger coils. For each pair of coils (one large and one small), you'll need approximately 50 feet of soft copper tubing.

Step 3: Assemble the coils as shown, running all cold water inlets (bottom leads) through one group of couplers (#16) and all hot water outlets (top of coil leads) through the other group (#18). The leads will not only run through the coupler, but also through the $3/8$-inch × $5/16$-inch compression through fittings (#6), which seal the tubes against vapor leaking around them.

Plumb and manifold the two groups of lead lines as illustrated into the three $1/2$-inch Ts and elbow (#3). Each lead line is connected to the branch of the T by $1/2$-inch × $5/16$-inch sweat (solder) **bushings**.

CONTINUOUS DISTILLATION

Continuous distillation is used on the industrial level for production of alcohol beverages and fuel. It is clearly cost-effective to use continuous distillation in plants that produce 50+ gallons per hour. Small-scale producers—who use a mix of feedstocks, run on different schedules depending on the season, generally don't run around the clock, and don't need to extract trace organic chemicals from their fuel—won't find any advantage to this technology.

The principles that govern continuous distillation are the same as those for batch distillation, namely enrichment and countercurrent stream. The continuous distillation system was developed as a more labor-efficient process and to facilitate the removal of valuable trace chemicals. It is primarily controlled by computer or some other fairly advanced system, and makes more efficient use of heat exchangers to constantly recycle heat energy.

In a batch still, alcohol is slowly but surely extracted from the main batch until there's

---

ALCOHOL FOR INDUSTRIAL USES

Low wines come into play if you're going to sell your alcohol to industrial sources rather than use it as fuel. In this case, your process for selling the least contaminated alcohol will be very similar to the old moonshiners' methods.

Industrial-grade alcohol from your still comes from the **middle cut** of the distillation. There are higher levels of aldehydes, **ketones**, and other volatile trace components in the foreshot. The end of your run contains the fusel oils (higher alcohols, etc.), esters, acetic acid, and various lower-boiling-point contaminants. These materials have little effect on a car engine, but may be considered unacceptable by the industrial market.

What you do is collect the first 15% of your alcohol run and throw it in with the low wine collected at the end of the run. Once you've got a sufficiently large quantity of low wine, distill it again to further concentrate the impurities, and use the alcohol middle cut for auto fuel. The middle cut of redistilled low wine is somewhat higher in impurities than normal fuel but still quite acceptable. Eventually, after the very concentrated loss of several low wine distillations, you'll have a product too concentrated with contaminants for fuel or industrial sale, but one that you can use as fuel in an oil burner or even to burn under your still. In the future, you may be able to take a year's supply of concentrated low wines to a larger alcohol plant to separate out valuable fusel oils, etc., for sale.

---

practically nothing left. At the end of the process, you have a tank full of boiling nutrient-rich water and solids; you empty the tank, refill it, and start the distillation process over again. The new batch has to be heated, which, depending on how large the tank is, can take hours. Energy lost during mash boiling is not as easily recovered as in the continuous system, and heat storage of some sort must be used to hold heat (as hot water) until it's needed.

In a continuous still, mash is introduced into the center of the column. And when I say mash, I mean liquid, gloppy, corny mash—right into the middle of the column. One hundred percent of your mash can be handled by a continuous column, while only 20% runs through a batch column, as vapor.

To handle the excess water and obtain maximum proof, your continuous column must be much taller than a packed column batch-type still. A perforated plate continuous still often uses a 42 to 1 height-to-diameter ratio, compared to a batch packed column still's 24 to 1 ratio.

Although alcohol goes up and water goes down, the continuous process takes a lot more control and monitoring than batch processing. Incoming mash must be preheated close to its particular boiling point, and since alcohol content varies from batch to batch, the boiling point differs as well. Alcohol content of your mash must be monitored, along with rate of flow of the in-feed. And watch the column temperature very closely, throughout the *entire* column.

A batch still's heat source is at the bottom of the tank, where the liquid is heated either by direct firing, steam, or heat exchangers. In a continuous still, the mash pot is actually replaced by a short extension of the column, where the reboiler water is boiled by steam-heated coils or injected live steam. In a continuous still, the heat comes from two places: Preheated mash enters the column at its mid-point, and the reboiler that sits at the bottom of the column provides primary heat for the lower half of the still, as well as additional process steam to vaporize incoming mash.

Column pressure must be constantly and accurately maintained in a perforated plate continuous still, or water and solids flowing down will suddenly enter the reboiler and overflow, wasting alcohol.

If everything is working right, each individual plate is held continually at a constant boiling point and proof. Getting there takes about an hour and a

MATT FARRUGGIO

Fig. 9-32 Perforated plate. *This weir-dam-type plate was made from scratch by Floyd Butterfield (see Chapter 27). Up to 45 of them are needed in a continuous distillery.*

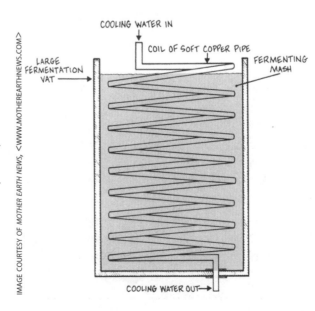

IMAGE COURTESY OF *MOTHER EARTH NEWS*, <WWW.MOTHEREARTHNEWS.COM>

COOLING WATER IN

COIL OF SOFT COPPER PIPE

LARGE FERMENTATION VAT

FERMENTING MASH

COOLING WATER OUT

Fig. 9-33 Immersion coil heat exchanger. *This can be used as a beer/stillage heat exchanger, cooler for cooked mash, fermenter temperature controller, or preheater for distillation. Usually, the coil can return to the top of the container instead of going through the bottom.*

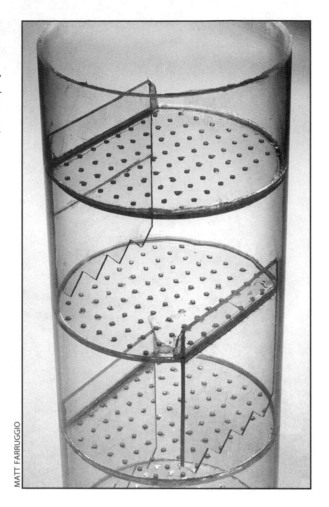

Fig. 9-34 Weir dam perforated plate distillery. *Used in continuous distilleries only.*

MATT FARRUGGIO

half and produces irregular proof until it achieves its continuous state.

Once the still has reached equilibrium, you can pull some neat tricks in the further purification of alcohol for making beverages or perfume. There are certain valuable trace contaminants/chemicals in your alcohol, such as fusel oils, and their removal and recovery are desirable. Since these chemicals also have specific boiling points, they tend to accumulate on specific plates in the continuous still.

For example, a chemical like fusel oil in the mash mixture boils at a temperature between alcohol and water's boiling points. As you go up the column, there's a point at which the plate above is cool enough to halt the fusel oil vapor's upward flow, and the plate below is hot enough to revaporize it back up. That point, on that particular plate, is where most of your fusel oils will congregate, and every few hours you can siphon that plate off through the column wall. On other plates, there are other chemicals that can be removed. Obviously, you can get very pure alcohol from a continuous still.

Fortunately, trace chemicals—which are terrible to drink—have little to do with the use of alcohol as a fuel, so your operation need not be as complex as this. There is the tantalizing potential for fusel oils to be used as **cetane improvers** for diesel engines running on alcohol; if more research in this area proves this to be practical, it may become important to recover these chemicals.

The big energy bonus of a continuous still is its ability to recycle heat while it's running and to preheat incoming mash before it enters the column. You can use this mash as your alcohol vapor-condensing medium. But more commonly the voluminous hot water leaving via the reboiler is used to heat the cold incoming mash, which enters the tube of your heat exchanger, which is absorbing heat from the shell containing the hot reboiler water. The mash absorbs the heat, recovering it from the reboiler water, and leaves the tube of the heat exchanger as hot mash ready to enter the column. This saves a considerable amount of water and heat, since separate cooling is reduced, and separate heating of the mash (to bring the temperature up) is reduced, as well.

Alternatively, the hot reboiler water can be used for the water needed for the next batch of grain to be cooked. Since reboiler liquid contains most of the heat that the condenser does not, little additional heat is needed to complete the cooking of a next batch. When operated continuously, the system is fairly economical and more energy-efficient than a batch still. Run sporadically, operating costs can double. In a (noncontinuous) batch system, some of these energy savings can be achieved, but most often you'd have to store much of the heat you had accumulated in an insulated water tank, part as high-quality high-temperature heat and part as low-quality warm water.

"Saudi Arabia's interest may appear to be served by lower production rates and higher prices, irrespective of the outcome. Let me remind you ... whenever prices go down, consumption goes up and vice versa. Whenever oil prices increase, large amounts of capital are invested in search of alternative sources of energy and in a search for oil in different areas. If we force Western countries to invest heavily in finding alternative sources of energy, they will. This would take no more than seven to ten years and would result in reducing dependence on oil as a source of energy to a point which will jeopardize Saudi Arabia's interests. Saudi Arabia will then be unable to find markets to sell enough oil to meet its financial requirements. This picture should be understood."

—SHEIK ZAKI YAMANI, FORMER SAUDI OIL MINISTER, JANUARY 31, 1981

Continuous stills cost a great deal more to build than batch stills; you need high technology to run them. But if your operation warrants one, be aware that this generally calls for a perforated plate column. Bubble caps plug up in a continuous still, running on grain, but are still used with solids-free mashes (e.g., sugarcane). When continuous distillation first became a workable technology, all kinds of presumably nonobstructive caps were invented—and discarded. Packed columns are rarely used for continuous distillation, since they'd gum up in a minute if the solids weren't completely filtered out before introduction to the column. Perforated plates rarely plug up, since solids drain with the excess liquid by way of a downcomer or spill over the edge of the weir (see Figures 9-32 and 9-34).

In some systems, though, the majority of solids are filtered out before beer is injected into the column. Sometimes a perforated plate can be used in the lower half of the still (the **stripper**), while a packed column system is used in the upper half (the **rectifier**), where only vapors enter after reboiling. In general, though, continuous stills use perforated plates throughout.

It's interesting to note that small-scale continuous stills were manufactured briefly in the 1980s. Some were even built by enterprising farmers. At rates below 35 gallons per hour, though, batch stills tend to be more practical. On the other hand, if your alcohol production plans call for 24-hour-per-day running for months at a time, you might consider a continuous system controlled by computer.

VACUUM DISTILLATION

There's been a lot of promotion of vacuum distillation for small producers. Vacuum stills offer conditions and opportunities that will definitely benefit certain producers on a site-specific basis, but they are not a panacea.

Have you ever tried to boil beans at 8000 feet? It takes all day, because there's less air pressure at higher elevations. At sea level, there's about seven miles of air pressing down on everything, and that seven miles weighs about 14.7 pounds per square inch (psi). At higher elevations, less air presses down.

The boiling point of a liquid is directly related to air pressure: the less air pressure, the lower the boiling point. People in high elevations use pressure cookers to get the pressure, and therefore the temperature, up high enough to cook their food. Beans won't cook unless you get the temperature well over 200°F so that the protein will cook.

Boiling water has very little to do with the 212°F (100°C) temperature we learned in school, except at sea level. A boiling point is the temperature at which water changes from a liquid to a vapor. Changing the pressure on the water also changes the boiling point.

Conversely, we can lower the pressure in a still to perform distillation at lower temperatures than 173–212°F.

Although it is possible to simulate the moon's zero air pressure in the lab, it's virtually impossible to remove all of the air in a still. A perfect vacuum is 29.929 **inches of mercury**. The practical limit for an industrial application like ours is 25 to 26 inches of mercury, in which environment alcohol boils at about 104°F and water boils at about 120°F.

So, if you start with an 80°F mash, you have to go up 40 Btu to get to a new lower boiling point. You'll have saved 90+ Btu between 120°F and 212°F—a savings of about 10%. There are more savings—in reduced heat loss due to radiation, etc., but in a well-insulated plant, it isn't huge.

Fig. 9-35

THE GREENHOUSE EFFECT

It's been estimated that if all the waste process heat that was practical to recover with present technology was used in industry, we'd be able to decrease energy costs by 40%.

Lowering the boiling point obviously reduces the amount of energy needed to vaporize a liquid, but the drop is not as dramatic as it would first appear. From the boiling point, it still takes over 970 additional Btu to vaporize a pound of water (that's 416 **kilojoules** per kilogram of water) (see Figure 9-36). The reduction in boiling point only reduces the amount of energy it takes to begin vaporization.

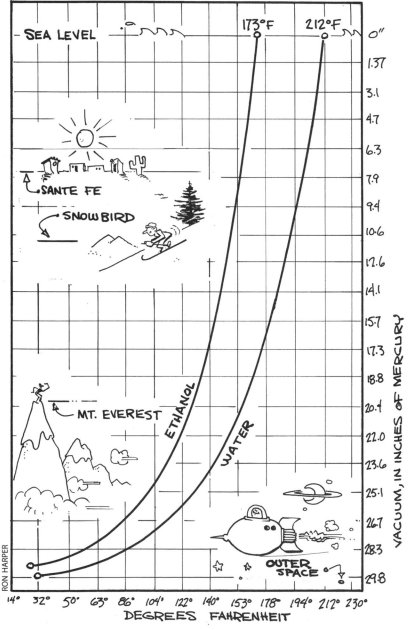

Fig. 9-36 Vacuum and boiling points. *Effect of vacuum on boiling points of alcohol and water.*

Useful Opportunities from a Lower Boiling Point

Lowering the boiling point of mash does create some useful opportunities. At a lower boiling point, you can transfer heat from mediums that were not high enough in temperature to use with atmospheric distilleries. Lowering the boiling point makes four important energy sources possible for distillation that are impractical at higher boiling points. These are direct-solar-heated water, the alcohol itself, low-grade waste heat from other industrial processes (cogeneration), and heat pumps.

Direct-Solar-Heated Water

The most seductive of alternate heat sources is solar energy. In many parts of the country, active plate collection systems can accumulate a reasonable quantity of water up to 180–190°F. This temperature is just barely hot enough to effectively transfer heat to the mash. Heat is transferred through heat exchange water coils in the still tank. With a 60–70°F difference between the mash and hot water, your coils have to be quite long to exchange the majority of the heat (450 feet of 3/4-inch tubing for a 1000-gallon tank, for example).

You need a lot of panels or a lot of surface area for a concentrating collector to gather enough Btu to do the job. Let's say it takes 20,000 Btu to distill a gallon of alcohol, and the average good-quality panel only gathers 40,000 Btu per day. A run of 100 gallons of alcohol a day will take at least 50 average solar panels. You could run your still every third day and store the energy from about 20 panels over the three days in a well-insulated hot water tank. If you're only going to make a run once a week, ten panels may work, with some hot water left over for your home. But the panels, if purchased retail instead of being homemade, will be pretty expensive and require a long payback period.

Liquid Alcohol for Heat

Alcohol itself can be used for heat as a liquid fuel. You can burn the fuel in a boiler to heat your water to steam, which then exchanges its heat into the mash through heat exchange tubing. It's important in a vacuum still to use a pump to agitate during distillation to make the heat exchangers more efficient.

This exchange will work in an atmospheric still, as well, but the energy return is very poor. In a good plant you can get, optimistically, from 5:1 up to 10:1 energy return with alcohol-based heating of your still. Since vacuum stills use less energy to accomplish the distillation, it is just barely acceptable to use alcohol to provide the heat of distillation. In a normal atmospheric distillery, it is almost always better to use some lower-quality form of heat to do distillation. Liquid fuel is much more valuable than most fuels that can be used to provide heat, so using alcohol as boiler fuel is pretty much a last resort.

Low-Grade Waste Heat: Cogeneration

It's been estimated that if all the waste process heat that was practical to recover with present technology was used in industry, we'd be able to decrease energy costs by 40%. This level of

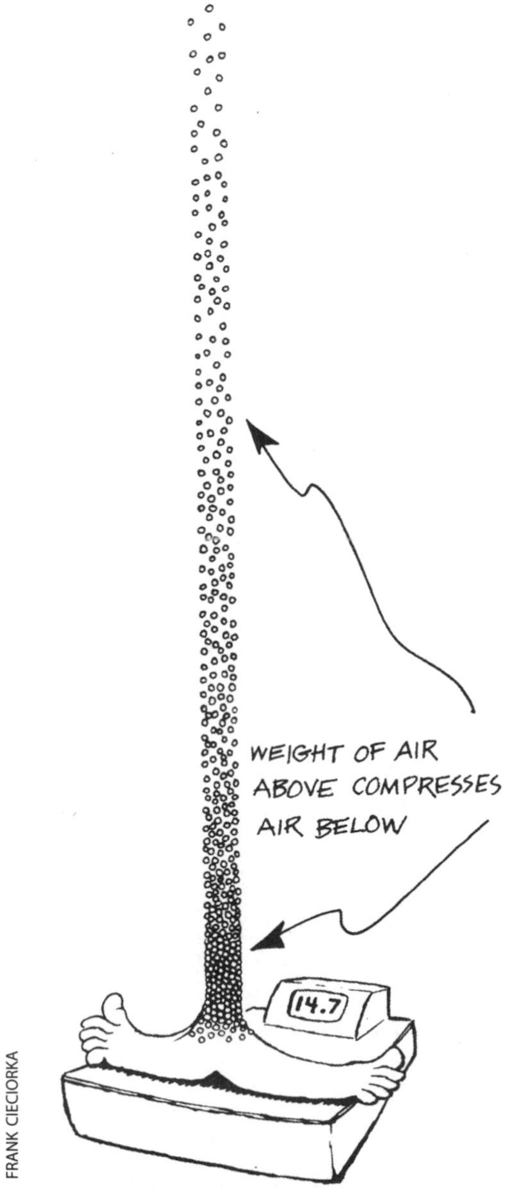

WEIGHT OF AIR ABOVE COMPRESSES AIR BELOW

14.7

FRANK CIECIORKA

Fig. 9-37

savings has been approached in some European countries.

Thousands of businesses worldwide have **low-grade steam** available for process heat. Low-grade steam may be the condensed byproduct of a high-pressure steam application for any number of industrial operations. The temperature of this kind of steam can be 250–300°F, although some inefficient operations can leave even higher temperatures. If you know someone who owns a business that has waste steam, you may wish to design your plant around it. If the steam is not a steady source, it might be easiest from a control point of view to store it by heating oil, which can then be used at your leisure.

Such steam is good for cooking and distillation, but may still be 15 to 70 psi, which requires heavy-duty heat exchangers—normal copper water pipe shouldn't be used with more than 10-psi steam pressure. If you are tempted to do your cooking with alcohol-produced steam, I'd caution that a less premium fuel, such as wood, is more appropriate.

Handling steam is not kid's stuff. It requires proper engineering and is subject to regulation by building and safety regulations in your area.

Heat Pumps

A somewhat more involved method of energizing heat exchange oil is by using an open-compressor-

---

DELICIOUS VACUUM STILL MASH

One of my first students worked for Gallo Wines in California. If equipment was needed to do a special job and there was no machine built to do it, Bob created it. Two weeks after he'd taken my class, I went to see a bench-scale vacuum still he'd constructed. It was a one-inch column hooked up to an air conditioner.

More interesting than his still was his chosen feedstock—he was using sherry from the winery. Since he knew exactly how much alcohol he had in his "mash," it was easy to measure alcohol recovery and know how efficient the still was. When he had finished his distillation, he gave me a taste of the sherry, without alcohol. It was delicious! He never allowed the liquid to go above 122°F, so heat couldn't hurt the flavor. Normal distillation would have left it horrible to drink.

He concluded rightly that distilling fruit and sugarcane mashes by using a vacuum unit gently heated by an air conditioner/heat pump would allow a lot of temperature-sensitive nutrients to be retained in fruit and sugarcane mashes.

ABOVE: Fig. 9-38 Earl Webb. In 1982, Earl was a full-time California school-teacher and a fuel producer.

type heat pump to heat mash and condense alcohol. A regular heat pump has an electric motor, which runs a compressor from inside a sealed container in the system. **Open compressors** run on an external motor, which can be an alcohol utility engine or an electrical motor. The external motor drives the compressor by a pulley and belt (like the kind for air conditioning in automobiles).

Larger heat pumps can transport or "pump" almost ten times as much energy as they use to operate, although smaller ones still pump a respectable three times the energy they use. An open compressor heat pump takes a low-temperature heat source, makes it into high-temperature heat, and then moves it to where it's needed. The pump can absorb the heat from such sources of energy as room-temperature air, alcohol in the condenser, warm or hot water/mash, or the exhaust of the engine powering the heat pump. It can even absorb the heat from the soil.

This concept may seem confusing if you've never dealt with heat pumps—it took me three tries to figure it out (see Heat Pumps, further below). The gist of it is that you can use a very small amount of alcohol, or self-produced methane, to power an external motor-type heat pump to move a lot of free or nearly free external heat to power your still. At that point, you've got a truly self-sufficient system.

A lot of heat pump waste heat is available from any business that has major refrigeration needs. After all, a refrigeration heat pump is moving heat from inside the building to the outside heat exchanger. If you recover that waste heat, which is normally blown into the air, it makes an excellent heat source for vacuum stills.

There is an easily scavenged, high-quality vacuum pump that anyone can afford: the refrigerant compressor of any good-sized refrigerator (or automobile air conditioner). Such a compressor used as a vacuum pump can create up to 29+ inches of mercury (vacuum). It doesn't do it very fast, but the extra few minutes it takes to create the vacuum is not a serious problem. You only have to do it

Fig. 9-39 Earl Webb's mobile vacuum distillery. Earl wanted to take this trailer-mounted unit to a nearby orchard during fruit season and make alcohol from the culls. On the road, the distillery is powered by a propane hot water heater. At home, Earl uses solar-heated hot water for power.

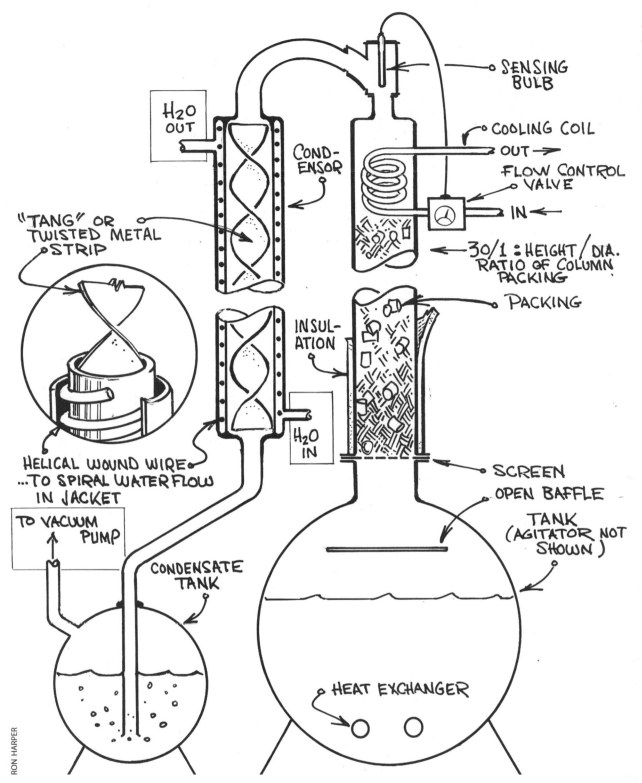

Fig. 9-40 Basic vacuum distillery.

SENSING BULB

COOLING COIL

OUT →

FLOW CONTROL VALVE

IN ←

30/1 = HEIGHT/DIA. RATIO OF COLUMN PACKING

PACKING

H₂O OUT

COND-ENSOR

"TANG" OR TWISTED METAL STRIP

INSUL-ATION

H₂O IN

SCREEN

OPEN BAFFLE

TANK (AGITATOR NOT SHOWN)

HELICAL WOUND WIRE ...TO SPIRAL WATER FLOW IN JACKET

TO VACUUM PUMP

CONDENSATE TANK

HEAT EXCHANGER

RON HARPER

once, and then shut off the pump as you begin heating the mash for distillation. Start it up again when you begin distilling, pulling the vacuum from the receiving tank.

Of course, several of these inexpensive little pumps could be hooked together to speed evacuation of the tank. At 28 inches of mercury (vacuum), the boiling point of water is close to 90°F, and alcohol will boil at about 60°F. Most people distill at 25–26 inches of mercury. Boiling points at these settings should be approximately 120°F and 104°F, respectively.

In any of these heat pump schemes, you will probably find it easier to engineer if you transfer the refrigerant heat to a tank of oil and then use the 300°F oil as your heating media. For a very easy-to-control system, use a French fry fryer fat pump or fuel oil burner pump to circulate the hot oil through heat exchange coils in your vacuum still. Many oil pumps can tolerate up to 300°F.

Advantages and Disadvantages of Vacuum Stills

Vacuum stills eliminate the greatest danger of the atmospheric-type stills: Put your hand on a regular exposed tank at 200°+F, and you'll have yourself a good burn. A vacuum still tank never gets over 140°F, about the temperature of hot household water, and is therefore a bit safer. A vacuum tank can't leak alcohol vapor, either, because a leak would be sucked inwards.

There are a few other advantages. Vacuum stills can reach a slightly higher proof than atmospheric stills—about 194 proof compared to 192 proof. And a four-inch-diameter vacuum still will put out about the same amount of alcohol per hour as a six-inch-diameter atmospheric still.

But there are a few definite disadvantages to vacuum stills, one being the cost. Also, the force the vacuum creates is powerful enough to collapse a tank suitable for a normal still. A vacuum still tank must be spherical or cylindrical, with domed ends to resist collapse. A vacuum tank must be very thick. I recommend ³/₈-inch thick tanks in the 500–1500 gallon range. A new vacuum tank can cost more than an entire atmospheric still costs for materials.

All the welds should be done by an ASME-approved welder, because a vacuum still is considered a pressure vessel. If you do your own welding, be very careful or you'll have tiny invisible leaks, which can make the still unable to hold a vacuum until you've found them. One way to avoid leaks is to silver solder over all your welds to seal most, if not all, pinholes.

You'll need a good vacuum pump or aspirator to create your vacuum. A vacuum pump rated at ½ to one cubic foot per minute and at 27–28 inches of mercury costs a couple hundred dollars. Piston pumps are more efficient than **vane**-type pumps for this job. Used belt-driven vacuum pumps on older Caterpillar diesel engines can be a good bargain.

The heat source for a vacuum still has to be external. No firebox-under-the-tank designs are allowed, because vessels under vacuum or pressure tend to act pretty weird when they're heated or stressed unevenly at high temperatures. One of my students saw a vacuum still collapse simply because someone dropped a hammer on it from a few feet up. You'll need a small boiler, heat pump, hot oil, waste steam, or solar–powered system to heat it with internal heat exchange tubing.

Although a vacuum still has low heat requirements, you'll have to build a separate cooker for your feedstocks, since you need high temperatures for cooking. The alcohol also has to be received in a vacuum-tight tank (about 15% of the size of the mash tank), since it obviously won't flow out of the still as long as you've got a vacuum sucking back in.

Another disadvantage to the vacuum still is that there's less of a spread between the temperatures of your alcohol vapor and cold water. The closer the temperature of the vapor to be condensed and the condenser liquid, the less effective your condenser will be. Most groundwater is 50°F, and the alcohol vapor is just over 100°F (rather than 173°F, as is usual in atmospheric stills).

So, a great deal more water must be used to cool and condense the alcohol to a liquid. (If you're using icy cold stream water, you may not need an excessive amount.) An average output

*Fig. 9-41 Vacuum pump and alcohol receiver. Vacuum stills deliver their **distillate** into an evacuated tank. A small vacuum pump is all that's necessary to evacuate a six- to eight-inch distillery.*

MATT FARRUGGIO

of ten gallons per hour in an atmospheric still can require as much as 20 gallons of water per minute at peak production. A ten-gallon-per-hour average output vacuum still, using 50°F groundwater, can use as much as 200 gallons a minute. It's practically impossible to use a **cooling tower** to get sufficiently cold water for reuse in a vacuum system.

Using a heat pump refrigerator condenser is going to be the way to go. With a heat pump, you won't need any water at all, since the refrigerant does the condensing for you. This is certainly the best way, especially if you recover the condenser heat with the refrigerant and deliver it into oil stored in a tank, which can then heat the still, or boost the temperature of your hot water storage tank. It is possible to heat the oil to over 300°F with the hot side of the heat pump.

Another problem with vacuum stills is cleaning them. In the simplest designs, you're not likely to be able to get into the tank for this job, so you have to separate the solids out before pumping lumpy mash into the distillery tank. Since you lose some alcohol in the separated solids, there is at least a 10% drop in alcohol yield. Even after separating the solids, you'll have to periodically boil caustic soda through your still to clean it. Welding an

expensive pre-fabricated vacuum-capable hatch would allow you to clean the tank more effectively and get a higher yield.

With all these disadvantages, if you have the extra capital up front, is it cost-effective to use a vacuum still? It all depends on energy costs. If you were using waste wood in a direct-fired still, it would take a long time to pay off the extra expense. But if you were buying wood at $250 per cord , you would be spending 30 cents per gallon on energy. That would be $300 annually on a 1000-gallon production, and you only save a part

Fig. 9-42 Filling port for small vacuum still. Earl Webb welded a six-inch coupler to his distillery tank and uses it to pump in his fermented mash. He used a six-inch pipe plug to close it up and has fabricated this wrench to make use of the large hexagonal area, rather than the small square plug end, for easy opening and closing.

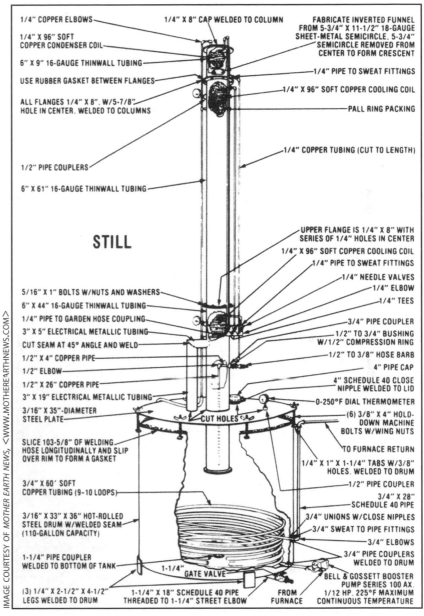

Fig. 9-43 Cutaway view of vacuum distillery. This diagram shows the basic components of a small vacuum distillery. It was developed to operate on hot oil or could even operate on very hot water. This design can be simplified by simply mounting the column directly on the tank and doing away with the reboiler and vapor input. Vapor would then simply enter by the bottom of the column.

of that by doing vacuum distillation. If that's all you're saving in energy costs, it would take a long time to pay off the separate cooker you'd need to do vacuum distillation (a vacuum still normally does not permit you to combine cooking and distilling).

A vacuum still does become more worthwhile if you are operating daily, as in a **small plant** versus a **micro-plant**. If all you are doing is making 1000 gallons per year, it probably is not worth the extra expense and complication to go vacuum, but if you are making 60,000 gallons per year, the savings start to add up, and it is worth the extra expense to save on energy costs.

Some plans recommend heating liquid in the reboiler by immersing it in boiling mash. This doesn't really work.

OUR TECHNOLOGICAL UTOPIA

SINGER

"If a society opts for high energy consumption, its social relations must be dictated by technocracy and will be equally degrading whether labeled capitalist or socialist."

— Ivan Illich, "Energy and Equity," 1973

Fig. 9-44

Cutting Costs

It turns out there *are* ways to cut capital costs in building a vacuum still. One of my students beat the high tank cost by coating a weather balloon with several layers of fiberglass to make a super-strong spherical tank. Another used a gigantic, spherical, steel World War II submarine net mooring buoy as his tank.

Another good source for vacuum tanks, especially the 500-1000 gallon size, is a retired butane tank. Butane causes tiny stress cracks in metal over a period of years, and the tanks are retired when they're considered unsafe for the very high pressure used in butane storage. At the relatively low pressures we need, they work just fine. Check with your local liquid petroleum gas (LPG) distributor or utility company for sources for these tanks.

Some of my former students have experimented successfully with a swimming pool chlorinator/mixer as an aspirator. You can create the required vacuum by running an unlimited water source, from an irrigation ditch, say, through the aspirator to pull the vacuum.

A few other distillers have solved their cooling problems by creating a gigantic ice mountain in the winter (usually in an inexpensive "dough-boy" above-ground pool—the kind that are about 20 feet across and use a round sheet-metal side assembly with a liner inside). The distillers spray water on inactive, empty heat exchange coils and let sub-freezing weather make ice out of the spray. When summer comes around, they just run their water through the 32°F coils embedded in the ice mountain, and then on to the condenser. If you're willing to do some scrounging, some hustling, and some tinkering, you may be able to save $300 a year in energy costs per 1000 gallons of product.

Occasionally, there are secondhand propane tanks available at more reasonable prices than new ones. But unless every bit of propane has been removed from the tank, when you get down to welding on it, you can get yourself blown up. Filling the tank with carbon dioxide to prevent explosion *may not be enough*—the tank's metal pores may still hold trapped propane. Filling the tank most of the way with water and topping it off with a head of carbon dioxide where you're working may be the only solution to an explosive problem.

REBOILERS

Reboilers allow you to introduce alcohol vapor at the midpoint of your column rather than at the bottom. A reboiler system lets you extract a greater amount of the total alcohol in a batch as high proof. The vapor proof, which is quite high at the start of distillation, won't need to travel through as much packing as it otherwise would when entering from the bottom of the column.

The biggest advantage in using a reboiler is that it allows you to preheat your column before the mash begins to boil. This forces all the cold air out of the column. When the alcohol does begin to distill, it takes much less time for the column to settle down, reducing low-proof foreshots and irregular running.

The disadvantages are higher construction costs and more monitoring.

Using a reboiler will not mean you can run a batch still continuously. Keep in mind that a batch still doesn't put out alcohol at a steady rate. The first hour might yield as much as three or four times the output toward the low-wine end of the run.

Reboilers are generally used on continuous stills, but they can be fitted to batch stills, as well. Reboilers on batch stills are worth considering with mashes that are 9 to 12% alcohol. If you routinely run 8% or less alcohol, a simple column may actually be superior.

In a batch distillery, reboilers allow you to introduce alcohol vapor at the midpoint of your column rather than at the bottom. Notice that *vapor* is introduced in the center of the column — remember that a continuous still processes 100% of its mash as a *liquid*, while a batch still only processes 20% of the mash as vapor.

Introducing vapors in the center of your column leaves the bottom half essentially unheated, containing only condensed liquid from the packing above the center point. But the reboiler provides steam and heat for the bottom of the column, in order to create the temperature gradation necessary through the whole length of the column.

In large continuous systems, steam is either injected into the reboiler or circulated through coils to boil that portion of the liquid on the bottom. But let's forget about steam; we have other good alternatives.

Some plans recommend heating liquid in the reboiler by immersing it in boiling mash. This doesn't really work. Reboiler water can only get as

The biggest advantage in using a reboiler is that it allows you to preheat your column before the mash begins to boil. This forces all the cold air out of the column. When the alcohol does begin to distill, it takes much less time for the column to settle down, reducing low-proof foreshots and irregular running. The disadvantages are higher construction costs and more monitoring.

hot as the boiling mash, and since the boiling mash has alcohol in it, it can't reach 212°F—the absolute minimum temperature necessary to properly evaporate or boil the reboiler's nearly pure water. Consequently, the reboiler will always have some alcohol in it, which is wasted when the reboiler overflows. Although the waste is small, an underheated reboiler takes a very long time to produce very little steam. It's inefficient during start-up and useless for preheating the column.

A preferred method is to introduce a separate heat supply to the reboiler: hot gas from the firebox surrounding it, or a separate burner if the still is powered by natural gas or propane. The outlet gases of a rocket stove are usually between 200 to 300° and might work well for a reboiler, too.

Better yet would be heating the reboiler with heat exchangers filled with hot oil or heat pump refrigerant. This guarantees the reboiling of alcohol and a sufficient amount of steam to carry on distillation efficiently. If you use hot gas from a firebox, the amount of heat can be controlled by a

Fig. 9-45 Fuel oil tank distillery. The ubiquitous, inexpensive 275-gallon fuel oil tank is a favorite of home builders. Its large surface area makes for fast efficient heating. But be sure to make it easily replaceable, since it doesn't last for too many years in its unconventional role.

TRICKLE DOWN ECONOMICS

BIG OIL™ GOUGES YOU AT THE PUMP...

...BIG OIL'S HUGE PROFITS THEN ALLOW THEM TO MAKE EVEN BIGGER POLITICAL CONTRIBUTIONS

...ENSURING THE RE-ELECTION OF THEIR CANDIDATES WHO THEN FURTHER "GREASE THE SKIDS" FOR BIG OIL™

VOTE BIG AL

MATT WUERKER

M. WUERKER

Fig. 9-46

separate damper that controls the amount of flue gases diverted to surround the reboiler. That's what I used on my old Liberator 925 distillery.

Nowadays, I would favor hot oil or **freon** coils; they're easier to automate. Hot oil is heated by flue exhaust gas to a temperature of 300 to 700°F (depending on the boiling point of the oil) and stored in a highly insulated tank. Reboiler heating coils using freon (approximately 300°F) can be really useful, if the heat source for the freon is the distillery's condenser. You would be recycling heat removed from the alcohol vapor by the condenser into the reboiler, before finally scavenging the rest of the heat for your hot water storage. Both oil and freon coils are amenable to being controlled automatically by temperature or pressure sensors in the column, with solenoid valves to open and close the flow of heating fluid through the coils.

Reboilers for wood-fired stills should be heavy stainless steel to avoid burning out if exposed to flue gases directly. The center vapor feed tube that extends from mash pot to column should be about

half the diameter of the column. A vacuum created from the condensers will pull the vapor in through the smaller tube quite easily, and a smaller tube means less metal to have to keep hot. The inlet tube should be just as well insulated as the column. If you're using a pressure-sensitive controller to operate a firebox air bleed, the controller should have its pressure sensor reading at the vapor entry tube.

HEAT PUMPS

We've discussed these to some degree in the vacuum distillation section above. Heat pumps aren't for every plant, but in the right situation they're nearly miraculous in terms of what they do with so little energy.

As a source of heat, a heat pump is safer than most. There's no open flame, there are no regulations to comply with, and no smoke to give you away to neighbors. You can use the refrigerating side of your heat pump to condense alcohol at a very low temperature. Alcohol below 50°F cannot possibly explode or even evaporate significantly.

As usual, though, there are a few drawbacks. Unless you can salvage a compressor and parts, a heat pump system is expensive. But salvaging is not difficult. I salvaged a good-sized $100 heat pump from the refrigeration unit for a walk-in refrigerator at a restaurant auction. The cost of heat exchangers is minor compared to the cost for the heat pump itself.

A heat pump is basically an air conditioner in reverse. An air conditioner cools a room by absorbing the room's heat and depositing it outside of the area being cooled. There are few basic components to an air conditioner: low-pressure coils for absorbing heat, a compressor to squeeze heated gas into a liquid, and heat disposal coils with a fan to remove heat from the liquefied gas.

An air conditioner also has an **expansion valve** to allow liquefied freon to flow into the heat absorption coils, where it expands into a gas. The freon's original heat thus expands over a much larger area, and so the gas becomes "cold" (cold being the intense ability to absorb heat, or the absence of heat). The coils become very cold (–10°F), and your room's air gets blown over them. The cold coils absorb heat, which is transferred to the freon.

Another example of expanding gas becoming very cold is when liquefied propane becomes gaseous propane. Its original heat expands over a much larger area, and the gas becomes "cold" as its heat is dissipated.

The absorption of heat, which "boils" the propane to a gas, causes the metal valve to become so cold that air moisture condenses and becomes ice as the water vapor gives up its heat to the expanding, evaporating propane. In an air conditioner, compressed cool freon flows into coils through an expansion valve and boils; the coils become very cold (–10°F), and your room's air gets blown over them. The cold coils absorb heat, which is transferred to the freon.

The freon is now perhaps 30°F. It is then compressed, which sends the temperature up to 300+°F. The compressed fluid freon next goes through another radiator/heat exchanger that has outside air blowing over it. The outside air absorbs the heat, reducing the temperature of the liquefied freon to perhaps 100°F. The process then starts over with the liquefied freon expanding to absorb heat from the indoor air.

An air conditioner throws away the heat collected from the room. But in alcohol production, you can use hot compressed liquefied freon in the coils in your mash tank or jacket to help boil the mash or reboiler liquid. In a vacuum distillery, as heat is exchanged into the mash, the liquefied freon cools from 300°F to 120°F while bringing the mash up to a boil. As in an air conditioner, cooled freon travels back to an expansion valve, expanding into a gas to become quite cold so it can absorb heat again.

One way you can use this cycle in distillation is that, after freon's heat has been taken up by your mash, the cooled freon is then allowed to expand and therefore chill your distillery's condenser coils or jacket. The freon will absorb heat from the hot alcohol vapor, chilling it to cold liquid.

If you're with me so far, we can see other uses of this cycle, especially for heating. Cool freon gas can be used to absorb heat—from hot flue exhaust, spent mash, overly hot air at the top of greenhouses, or even hot air in the upper reaches of your

Fig. 9-47

A HOME MADE HEAT PUMP

HEAT SOURCE Ⓐ (POLITICAL RHETORIC) WARMS AIR AROUND COILS Ⓑ WHICH RAISES TEMP. OF GAS IN COILS. SENSITIVE MAN FROM MARS Ⓒ FLEES FROM PAINFUL NOISE TOWARD SPACE SHIP, TURNING TREADMILL WHICH POWERS COMPRESSOR Ⓓ. COMPRESSED GAS CONCENTRATES HEAT IN ITS DENSER, LIQUID STATE. HOT LIQUID IS USED TO HEAT MASH Ⓔ. LIQUID NOW GOES TO EXPANSION VALVE Ⓕ WHICH ALLOWS IT TO RETURN TO GAS STATE, ENABLING IT TO ABSORB HEAT AGAIN FROM Ⓐ, A SEEMINGLY INEXHAUSTIBLE SOURCE OF HOT AIR!

"The real problem is the gasoline market. We do not produce enough gas. The refineries are running at very high rates—and if one goes out, we have a shortage and prices spike. This puts the large sellers in a position to artificially create price spikes by withholding supplies. Are they doing it? It's very hard to know. But clearly, they can make more money by selling less gasoline."

—SEVERIN BORENSTEIN, DIRECTOR OF THE
UNIVERSITY OF CALIFORNIA ENERGY INSTITUTE, 2003

distillery building. Lower-quality heat from any of these sources can be increased in temperature to be used wherever heat is needed in the process.

Beautiful system, isn't it? Furthermore, the heat you initially pump into your still comes from the air. The sun heated the air; you merely absorbed some of it, compressed it to increase its temperature, and then pumped it into your still. Since the warmth in the air is solar in origin, several U.S. states consider these to be active solar heating systems.

The main drawback to the freon system is that it doesn't easily store heat. Unlike water or oil, the refrigerant and the pressure vessel that would be required to contain it are too pricey to make sense as heat storage. If your freon system is producing

Fig. 9-48 Firebox construction. *Firebrick layer removed to expose Kaolwool insulating backing (see arrow) in the cone-bottomed Liberator distillery.*

Fig. 9-49 Firebrick-lined door. *Note the openings in the brick for entry of forced air. Brick is supported and retained by an angle-iron frame. This keeps the door well protected against burnout or warping.*

surplus heat, then it might be best to use the freon to heat a tank of hot oil to 300+°F. You would then use the oil instead of the freon to do your heating chores.

Freon as used back in the 1980s was nontoxic to the mechanic or user but dangerous to the ozone layer. New types of refrigerants, including the variants of freon, are now safer for the environment but toxic if inhaled. Get professional help installing your freon system if you don't have the skills to safely handle this material.

Before we leave heat pumps, which you're probably anxious to do, I'd like to point out a couple of subtleties in the system. Once your mash is boiling, and the column is heated, and distillation actually begins, an equilibrium is reached—the amount of heat removed from the alcohol in the distillery condenser is pretty close to exactly the amount of heat it took to distill the alcohol. So, if you return the condenser heat back to your mash tank, you need little additional heat to keep the distillation going.

A heat pump using an alcohol-powered utility engine lets you move from three to ten Btu of heat energy for every Btu of electricity expended in running your heat pump compressor. This same ratio will hold with alcohol or methane used as the compressor power source instead of an electrical motor, if you absorb the waste heat from the engine using the heat pump!

I recommend buying a few hours of a refrigeration engineer's time to calculate the exact specifications for your needs. If you're fortunate enough to set up a plant adjacent to a food processing plant or cannery that uses massive refrigeration equipment as part of its operation, a refrigeration engineer can devise a safe way for you to tap their super-hot freon (or CO_2 refrigerant) and run your plant.

Using a heat pump is the most practical way to run a batch-type vacuum still. Although freon is hot enough to boil mash at atmospheric pressure, it is much more time-efficient and cost-effective in the long run to use a vacuum still rather than an atmospheric still. With the lower boiling point under vacuum, the heat transfer rate from 300°F freon coils is very rapid. Since freon or carbon dioxide boil at such low temperatures, far below 0°F, you can actually absorb heat from 32°F snow to power a still. There have been a few multi-million-dollar stills designed with these principles in mind for areas as far north as North Dakota and

Canada, which were expected to run almost completely on heat pumps using solar energy.

WOOD AS A HEAT SOURCE

In rural areas around the San Francisco Bay Area, oak firewood could recently be had for about $300 per cord. (For you city slickers, a cord is a stack of wood four feet wide, four feet tall and eight feet long—128 cubic feet.) In Los Angeles, about 400 miles south, the same cord goes for $350. Four hundred miles north, near the Oregon border, a cord costs $225. Another 500 miles north, and the price drops to $200. In the Midwest, which is far from any large sources of wood, firewood is high-priced. But the Northwest, West, Southeast, East, Northeast, and some North-Central areas have plenty of wood at prices that are attractive to the distiller.

More wood is used to heat homes than nuclear power, even now. While "too cheap to meter" nuclear power keeps getting more expensive, the renewability of wood has kept its price relatively low.

Build-it-yourself wood stove books have many good designs for fireboxes, but there are a few wrinkles to burning wood to power a still. In a simple firebox, most heat goes right up the flue—very little makes it into the mash.

Controlling the flame, which controls the pressure of vapor in your still's column, is a little more difficult with wood than with other fuels. The problem is that a wood fire is always changing in intensity. You've got to control the available air that's being used for combustion.

When designing a firebox, consider first the mash pot—the more of that receptacle's surface area that gets exposed to flame, the more heat it can absorb. If you have a cylindrical still tank, set it up horizontally, not vertically. This allows you to build a long firebox with a lot of tank surface area exposed to the flame. A cooker or still set up the wrong way (with its small end down) won't distill alcohol from a full tank no matter how much wood you burn—there will almost always be too much liquid in the tank if enough surface area isn't exposed to the flame. Of course, there is a limit to how much heat you can transfer through a square foot of metal.

Another thing to consider is that a great deal of heat is wasted through the walls of the firebox, unless it's insulated with firebrick or **cob**. Lining

a steel firebox with firebrick can save as much as 50% of wood's energy. Backing the brick with Kaol-wool or other types of high-temperature insulating fiber increases savings.

Fig. 9-50 Small Brazilian wood-fired distillery. *The firebox around the mash tank is concrete. It just goes to show you can dispense with all the fancy firebrick and even firebox doors if you are creative. The tank on the stand behind the owner is the condenser. An immersion coil is inside the tank, which uses the tank's water to condense the alcohol. The column is short, since this is used to make a Brazilian form of rum at about 100 proof. Note the sophisticated firebox air control damper.*

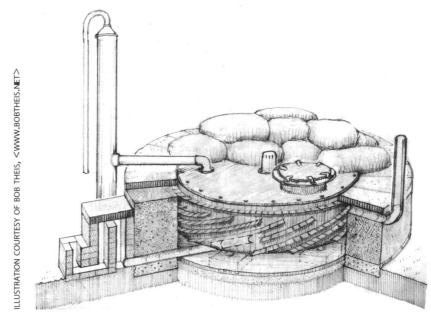

Fig. 9-51 Rocket still. *In this concept drawing, the cob rocket stove is on the left, and the hot wood gases are directed under the center of the still. Thick cob walls surround the still, and spiral tabs are tacked onto the tank to make the flue gases rise along the tank walls before going out the flue to the right. The "pillows" on top are bags filled with rice hulls or chopped straw for insulation.*

DAVE BROWN

Fig. 9-52
Liberator 925
firebox. *Designed*
with oversized
doors for easy
loading and
cleaning out of
ash. The interior
walls are molded
refractory
material. Heat
fins are visible
under the conical
tank.

MATT FARRUGGIO

Fig. 9-53 Squirrel cage fan. *A common squirrel cage fan like this one is used to pump air into a distillery or cooker firebox.*

When I was briefly in the distillery manufacturing business, I used a material known as **castable refractory**. You could mix this material like mortar or cement, and pour it into forms inside the sheet-metal walls of the firebox, and it would harden into a monolithic ceramic slab with even better characteristics than firebrick. It's surprisingly affordable.

Temperatures as high as 2500°F can be generated in a firebox insulated in this way, while uninsulated fireboxes generally go up to only the 1600–1800°F range. This lower temperature results in incomplete combustion and thus more smoke. Also, a lot (30+%) of the wood's energy is not utilized, since particles of smoke, which don't actually burn, do contain a lot of burnable hydrocarbons.

The higher the temperature, the more completely the wood is combusted, and the more heat transfers to the mash. Super-hot fireboxes also burn a good quantity of what would become creosote and smoke, giving you almost invisible flue gases (which is important if you have nosy neighbors), far less pollution, and more usable energy.

If your firebox doors are metal (as many popular plans recommend), they'll warp and then be eaten away by corrosive hot wood gases from 2500°F fires. Firebox doors should be designed like kiln doors, with firebrick set in a frame. Or build a door similar to an incinerator's—brick in a frame door that's raised by cables and pulleys. Special grooved firebricks are made just for this kind of construction (see Figure 9). Their grooved bottoms and sides sit in the angle-iron frame of the door. If you have trouble finding such firebricks, you can groove your own with an abrasive cutting disc on a table or radial arm saw. Alternatively, you can cast refractory material in place on the door.

If you insulate your firebox with brick and high-temperature mineral wool, the metal on the outside of the firebox should stay below 300°F. If so, the outside of the firebox and the tank above it can be insulated with any of a number of commercially available urethane spray foams. Such insulation can have a protective waterproof coating sprayed over it, which eliminates any need to paint your still. An inch thickness of hard foam should be plenty. Foam insulation comes in a variety of

Fig. 9-54 David Ange's distillery. *This personal-sized still tank is a used cyclone funnel (upside down) from a dust collection system.*

hardnesses; the harder, more durable consistencies insulate a bit less effectively than softer, easily dented foams.

Applying foam requires specialized equipment and expertise. Foam companies will send out a truck and charge you by the square foot of foam applied, plus a set-up fee. If you can't find a spray-on foam company in your area, or if you don't have the funds for a commercial insulation, you should be able to find standard mineral wool to wrap around the firebox and tank. Cover the wool with 30-gauge aluminum sheet metal. This should reduce skin temperature below 200°F.

If you have small children (or clumsy adults) around, you'll need to protect them from the hot still. Since the tank and firebox are not safe to touch, you can attach an expanded metal screen about an inch from the hot surfaces, with metal tabs welded to the tank/firebox and bolted to the screen.

Just when I thought I had seen everything, a new potential way to heat with wood or biomass came to my attention. Cob is a material made of sand, clay, and straw, which can be used to build houses. Ianto Evans, master cob builder, has used cob to make large versions of the tin-can rocket stoves that efficiently burn wood for cooking. The key part of the design is that the flue is inside the firebox, which ensures high heat to fully combust the biomass.

Fig.9-55 Firebox door. To retain brick in a firebox door, use an abrasive cutting wheel to groove the bricks. One edge of a half-inch angle iron can then be used to hold the brick in place. Note the half brick at the top of the door. The angle iron will also run around the bottom of the door. I filled my door channel with Kaolwool rope to seal it against escaping smoke.

Fig. 9-56 Walter Wendt's personal still. Walter lives near a dairy and plans to use whey and perhaps Jerusalem artichokes in addition to food processing waste. He made good use of normal brick to build a firebox. His property is in a suburban area.

We can learn a few things about gas-to-liquid heat transfer from looking at the type of old wood-burning kitchen stove that had a built-in water heater, which worked in a remarkably simple but efficient manner. Waste heat fled directly up the flue, but if you wanted to boil water, you'd close a special damper and detour hot gases through passages all around the water tank before they slipped up the chimney.

To work right, the flue had to have a good strong draw. And since pulling gases the extra distance around the water tank cooled them considerably as heat was exchanged into the water, the flue had to be taller and larger in diameter in such a stove (cooler gases draw less effectively than very hot ones). Still, the process was more energy-efficient than heating water on top of the stove, since flue gases are waste heat anyway. Saving energy isn't always measured in cents per **kilowatt-hour (kWh)**; sometimes it's measured in how many cords of wood you have to split.

Gas-to-liquid heat transfer is not super-efficient unless the gas has enough time to exchange its heat. Hot gas directed around a tank allows it more time in which to exchange its heat and more surface area through which to accomplish the exchange, and results in less waste. Baffling the firebox directs the heat even more efficiently against the bottom of the tank, instead of it going up the flue. Welding tabs on the bottom of the still tank widens the heat absorption surface area and significantly increases the turbulence of the combustion gases crucial to increasing heat transfer.

If you want to squeeze every bit of energy out of your wood, you can get a bit more sophisticated. In the 1980s, Corning came up with a **catalytic combustor**, a wonderful device for wood-burning stoves. This is a ceramic honeycomb with catalytic metals inserted in the passage from the firebox to the area to be heated. The catalysts in the ceramic matrix force smoke particles to burn at a much lower temperature than they otherwise would. Smoke's combustion temperature is usually about 1400+°F, but smoke passing through a catalytic honeycomb will spontaneously burn at temperatures as low as 400+°F.

The extra heat you get from the secondary combustion of smoke helps economize on your burnable biomass. As much as 30% more energy can be extracted from wood, with a very significant drop in **polycyclic organic materials (POMs)** and other toxic pollutants. No matter how organic it is, wood smoke is definitely a pollutant.

A grate lets you burn smaller pieces of waste wood, rather than cordwood, by improving air circulation and making ash clearing easier, but building a grate for your still can be tricky. Usually, you can't use **mild steel** because that metal succumbs quickly to intensely hot coals and oxidation.

You can build a good grate which will last (although not as long as cast iron) by using ¼-inch mild steel **flatstock**, about two inches wide and as long as your firebox. Cut enough flatstock so that when you set them on edge they're roughly three inches apart. Weld heavy ³/₄-inch **barstock** or **roundstock** spacers between them. To prevent the top edge of the flatstock from being eaten up by the coals, place inverted ⁵/₁₆-inch **channel stock** over them. The channel stock should be made of

MATT FARRUGGIO

type 316 stainless steel. Stainless top channels will protect your grate pretty well and serve you for at least a hundred batches.

Remember to put short, heavy legs on your grate so you can clear the ash from underneath. Also, account for the space the grate will take up when you design your firebox.

AZEOTROPIC DISTILLATION

In normal distillation, there's a point at which you can't get the proof any higher, no matter how many times you condense and revaporize the alcohol. Under atmospheric conditions, that point is about 192 proof; under vacuum conditions, 194 proof. An alcohol mixture distilled to its highest point of proof is referred to as an **azeotropic** mixture—one in which alcohol and water are attracted to each other too much to be separated by simple heating. To remove the last bit of water requires special techniques which collectively can be referred to as **azeotropic distillation**.

Most of us who produce alcohol prefer to use our alcohol straight, as a "neat" fuel. It'll normally range from 190–192 proof. However, you may want to sell alcohol to someone who requires the alcohol to be water-free (dry or **anhydrous alcohol**). Or perhaps you live in a very cold climate, and are using alcohol along with gasoline to run an unmodified gasoline car—in this unusual situation, dry alcohol can help you avoid phase separation problems.

Phase separation is where the alcohol and water separate from the gasoline in distinct layers in your fuel tank; low-lying alcohol/water would be pumped into the system almost pure. Your car would sputter, cough, and die, since the fuel system in an unmodified car is set too lean for pure alcohol.

On a warm day, gasoline will tolerate quite a bit more water than on a cold day. Alcohol at 192 proof will mix with gas without separating in all but the bitterest of winter temperatures. If you expect temperatures below –22°F, then you will want to produce anhydrous alcohol if you anticipate switching back and forth with gasoline.

Of course, if you have a flexible-fuel car, there is no need whatsoever to use anhydrous alcohol. In

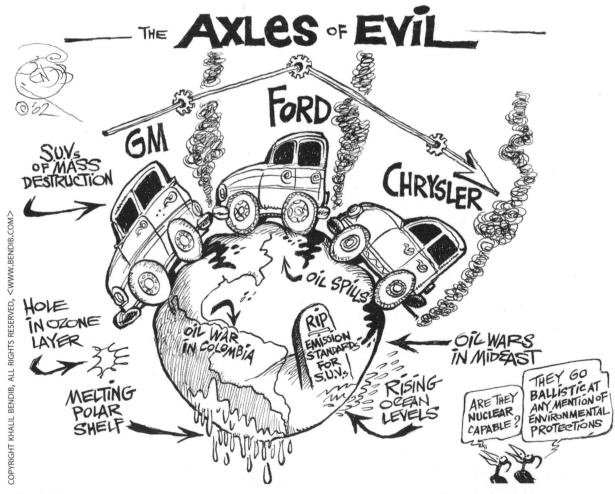

Fig. 9-57

Fig. 9-58 Energy for distillation. You can see that the amount of energy needed to distill alcohol to higher proofs skyrockets after 80% (160 proof). It's far more energy-efficient to switch to adsorbent methods above 160 proof.

fact, all Brazilian flex-fuel vehicles run on 192-proof alcohol containing 4% water. For FFVs, if phase separation at extreme temperatures occurs, it is not a problem, since the vehicle is already designed to run on the neat alcohol at the bottom of the tank, as well as the gasoline in the upper phase.

But neither 192 nor 194 proof is high enough to reliably stay mixed with gasoline if the air temperature is far below –22°F.

Nowadays, it's common to use **molecular sieves** or cellulose **adsorption** when removing the last few percentages of water from alcohol (see next section). The older method commonly used was **benzene** tertiary distillation. Benzene, a powerful carcinogen, is no longer used to dry out alcohol but is still a major component in modern gasoline.

This very high-proof anhydrous alcohol must be stored and handled with extreme care. When 200-proof alcohol is exposed to the air, it will not be 200-proof for long. With a few hours of open exposure in humid air, enough moisture is absorbed to bring the alcohol back down to 194 proof. Once alcohol reaches that proof, absorption slows dramatically. Dry fuel should be stored in sealed tanks. Any air that may enter your tank (due to expansion and contraction of the tank's

atmosphere) should be dried first with a **desiccant** (silicate beads, etc.) to preserve the alcohol's anhydrous nature. In actual practice, a **conservation vent** limits the exposure of the alcohol to outside air and effectively prevents dilution.

There is a simple way to make dry alcohol on the small scale, based on the fact that certain metallic minerals like calcium (Ca) and hot lime (CaO) adsorb water, but not alcohol. The safest, most energy-efficient way to accomplish such a solid/liquid distillation is with a vacuum tank and heat pump. A discarded 50- to 100-gallon gas water heater makes an ideal tank for this enterprise. You'll have to cut away some of the top for loading the lime—I recommend an airtight hatch that you can close between uses.

Fill the tank two-thirds full with coarse (unslaked) hot lime, and then up to the top with 190+-proof alcohol. Let it sit overnight, protected from cold nighttime temperatures. In 24 hours, the lime will have selectively adsorbed the water from your alcohol. But there'll be small specks of lime, so drain the alcohol into a separate tank, boil it, and then condense it. This is a simple one-time vaporization and condensation. All the calcium dust or lime flakes will collect in the bottom of your evaporator, and you should be left with 199+-proof alcohol.

Alternatively, it is possible to take the vapor directly from the column and run it through additional columns of lime pellets to pull out the water, leaving perfectly dry alcohol to go to the condenser. This avoids the energy-wasting evaporation step.

How do you know when the lime has adsorbed all the water? You can purchase lime that has been treated with cobalt chloride, a chemical that turns blue when the lime has adsorbed its fill. If you put an inch or two of this special lime in the top of your column and weld a port with a window in it so you can look inside, you can see when it turns blue. Drierite is a common brand of treated lime.

When making dry alcohol this way, you'll need at least two tanks (columns) of lime, so that you can keep adsorbing in one tank while the other is being regenerated. The lime in your drying tank must be dried before it can be used again. Once dry, wet lime (calcium hydroxide) has regenerated to hot lime (calcium oxide), so the tank should be sealed immediately, since the lime will absorb water from the air.

You can make good use of your still's waste flue heat by directing the hot flue gases through self-installed **fire-tubes** running through your lime

Energy for Distillation

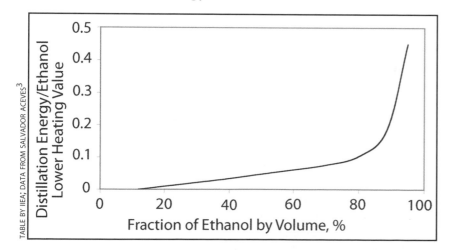

TABLE BY IIEA; DATA FROM SALVADOR ACEVES[3]

tank. The moisture can be vented through the filling hatch. Or, better, use 350+°F compressed CO_2 blown through the tank to carry away all the moisture. If you are using oil as a surplus heat storage system, you can use the oil to preheat air or CO_2 to over 350°F, a good way to really dry out the lime.

ADVANCED FORMS OF AZEOTROPIC DISTILLATION

One advanced way to separate alcohol and water using little energy is with molecular sieves. These are cylinders filled with a synthetic **zeolite** material that allows alcohol to flow through but has tiny pores that trap water. Unfortunately, the material can't hold much water, and it's very expensive. Molecular sieves are still appropriate for removing the last bits of water to dry alcohol, without having to go through a benzene distillation.

Back when they were first developed, molecular sieves were used in tandem. When one sieve became saturated with water, the flow of wet alcohol switched to the second sieve. The sieves were flushed with 450°F CO_2 to drive all the water off, and then cooled by CO_2 that had been allowed to expand from liquid just before entering the sieves. If you are regenerating your sieve material on the small scale, you can keep it simple by cooling with incoming air that has been drawn through silica desiccant, ensuring that you'll keep your regenerated sieve material dry.

Nowadays, you can advance, even on the small scale, to the clever technology of **pressure swing vacuum regeneration**. When one sieve column is saturated, a smart way to evaporate the water is to immediately pull a hard vacuum on the sieve column, and, voilà! it becomes a de facto vacuum still. The energy needed to get the water from its very hot, barely condensed state in the sieve back into vapor state, is small. When a vacuum is pulled in the sieve, the energy requirements for vaporization drop, and the residual heat is more than enough to get the water to evaporate back out of the sieve.

Molecular sieve material, when purchased retail from a professional supplier, can cost plenty, but there is one sieve material that is mass-produced and cheap. It turns out that several brands of kitty litter are actually molecular sieve material, and, as you can imagine, do a great job of adsorbing urine in a cat box. (They do a great job adsorbing the water from alcohol, as well.)

The most exciting form of azeotropic distillation makes use of cellulose, instead of lime or molecular sieve material, as a water adsorbent. It's a great example of the permaculture principle of "the problem is the solution." Like many great discoveries, this one came by accident. A researcher at Purdue University spilled a few gallons of 160-proof alcohol over some oven-dried ground corn.[4] After he had sopped up the spill, he re-measured the proof of the alcohol he had drained from the

LABORATORY OF RENEWABLE RESOURCES ENGINEERING (LORRE), PURDUE UNIVERSITY, WEST LAFAYETTE, INDIANA. RESEARCH AND ENGINEERING CARRIED OUT IN LORRE AND DEPARTMENT OF AGRICULTURAL ENGINEERING; VISUAL DESIGN (ESTHETICS) IN UNIT SHOWN BY BRUCE PAPIER. PHOTOGRAPH BY DAVE UMBERGER.

Fig. 9-59 Corn grit alcohol dehydration unit. *Developed at the Laboratory of Renewable Resources Engineering, this prototype* **alcohol dryer** *is capable of yielding 10,000 gallons of 199+-proof alcohol, starting with 190-proof, if it's run continuously. The electronic controls are visually keyed to guarantee nearly foolproof operation.*

corn, and found that it had, amazingly, gone to 198 proof. Apparently, corn, or more precisely, the cellulose in corn, adsorbs water and lets alcohol pass through. It does seem strange that moonshiners making alcohol from corn for thousands of years have never reported such an obviously simple, low-energy way to raise proof.

In the 1980s, this research was considered a novelty, but now Archer Daniels Midland uses ground corn to dry its alcohol. To use corn, or cellulose, as a selective water adsorbent, fill your column with cracked corn, or even cellulose fiber, and allow 180+-proof alcohol to vaporize up through and condense at 99+% purity.

With standard distillation, getting alcohol up to 160–180 proof is a low-energy input process, but distilling from 180–192 proof requires a proportionately greater amount of energy: Energy input needed to go to 180 proof is only $1/11$ of the energy needed to go to 192 proof. As you push toward 192 proof, energy consumption soars, and you end up with a very poor energy balance.

But energy sources are much more diverse at the lower temperatures used for regenerating cellulose, versus lime or zeolite. Many sources that are considered waste heat are available at these lower temperatures.

Using adsorption for the vapor after distilling to 180 proof avoids the energy-intensive distillation of 180–192 proof. A solid adsorbent can give you a ten to one energy return, including the energy for regeneration. The process is still energy- and cost-efficient even when using nonwaste energy to regenerate the dehydrant. It can be even less energy-intensive than molecular sieves.

In the original lab-scale trials, Purdue University reported that drying alcohol vapor, as it exits the distillery at 190–199 proof, uses only 1500 Btu of energy per gallon. Starting with liquid 190 proof, energy requirements are only 5000 Btu per gallon. Modern distilleries using these methods along with pressure swing regeneration, like ADM's plants, are getting energy use for dehydrating alcohol down to 500 Btu per gallon. This compares favorably to the 15,000–25,000 Btu for old-fashioned, dangerous benzene extraction.

Corn and cellulose can be dried at temperatures between 140 and 220°F by running dry, heated nitrogen or carbon dioxide through the adsorbent at atmospheric pressure. The hot carbon dioxide or nitrogen should be pumped in at the top of the column and exit through the bottom. Obviously, your column should be heavily insulated for the most effective adsorption or regeneration.

Using corn, you can expect water to be adsorbed at a ratio of 2% of the weight of the adsorbent. Because of this low retention ratio (compared to molecular sieve material), the researchers found that four corn-filled columns made a workable system. Three of the columns are regenerating, while one column dehydrates alcohol vapor.

For going from 190 to 199 proof in a corn grit dehydration column, an approximate height-to-diameter ratio of about 12 to 1 seems to work adequately. The **pressure drop** in such a column is sufficiently low, considering that the column is full of #20 corn grits, to be connected directly to the distillery (accepting high-proof alcohol vapor directly, rather than allowing it to be condensed and then revaporized separately before it's sent through the corn). The Purdue pilot plant used four columns, each four inches in diameter, capable of drying at least 10,000 gallons of 190-proof alcohol per year. Contrary to expectations, corn does not degenerate rapidly and seems to have a working life of several months of daily use.

Would-be distillers have proposed simply distilling alcohol to 100 proof without a packed reflux column, and then using the adsorbent technique to bring the proof up to whatever level is desired. This approach is inefficient with time, energy, and capital. On the small scale, it makes the most sense to shoot for between 160 to 180 proof with relatively imprecise reflux control, and then use larger-volume corn or cellulose adsorption for however dry you want to make the alcohol. Remember that it does not have to be completely dry to mix fully with gasoline, if that's your worry. There's lot of opportunity to experiment with this approach.

Again, it is not necessary to dry alcohol all the way out to 199+ proof unless you are mixing it with gasoline in an unmodified car at bitter temperatures. The process adds unnecessary complexity, needs extra energy, and complicates storage of the finished alcohol.

REDUCING THE HEIGHT OF YOUR COLUMN

In order to get the kind of proof you need, your still's packed column has to conform to the 24 to 1 height-to-diameter ratio shown in Figure 9-18. However, for cosmetic purposes, or to deal with the obnoxious neighbor forever calling officials to complain about whatever thing, there is a way to reduce the overall column height without losing high-proof output.

Very simply, you can cut your column into two or three sections, place the sections next to one another, then connect them so they act like one continuous-length column. To keep the divided column working as one, you need to provide a way for the alcohol vapor to travel from one column to the next, and you have to make sure the liquid runs back down from the top of each column through all their lengths. The following description of a divided column is based on dividing a column in half, but the same rules apply with more than two sections.

The section of column taking vapor directly from the still tank is referred to as the lower section; the second section, mounted next to the still, is the upper column. They don't have to be equal in length—in fact, the upper section is usually longer than the lower, since the lower column is often mounted on top of the still tank. A vapor transfer line runs from the *top* of the lower section to the *bottom* of the upper section, which allows vapor to rise through the lower section, travel through the vapor tube, and then through the upper section, where it reaches the condenser. The transfer tube should be one-third to one-fourth the diameter of the column. It should be well insulated to avoid any condensation on the way to the bottom of the upper column.

Once the vapor has condensed in the upper column, part of it starts its liquid descent. A divided column uses a small **sump**, sort of like a reboiler, at the bottom of the upper column to collect liquid as it descends, and a fuel pump to transfer the fluid up to the top of the lower column. The sump should have a **float switch** to engage the pump as soon as the liquid level has risen in the upper column. The pump's outflow should be restricted and adjustable, allowing the pump to operate more or less constantly, rather than requiring you to pump a large amount of fluid, empty the sump, turn everything off until the sump refills, and start again, which would result in a pretty irregular distillation.

Your liquid should enter the lower column through automotive **fuel injectors** that reduce the alcohol to a fine mist, mounted a foot below the top of the packing. I like the simple injectors from old Bosch CIS fuel injectors, since they are designed for continuous operation. Use a screen above the **foggers** to support your packing, creating a space in the column so the foggers have a couple of inches of free space. Overall height of the separated column should be taller than 24 to 1. The packing above the sprayers should be about 2–2.5 to 1, and I recommend the extra height of the upper column be an additional 2.5–3 to 1.

If you've got some extra money, you can use a temperature-operated flow-control valve to control the amount of liquid reaching the injectors. The temperature-sensing bulb should be installed at the top of the lower column. In this case, your sump should be larger, perhaps a foot deep or

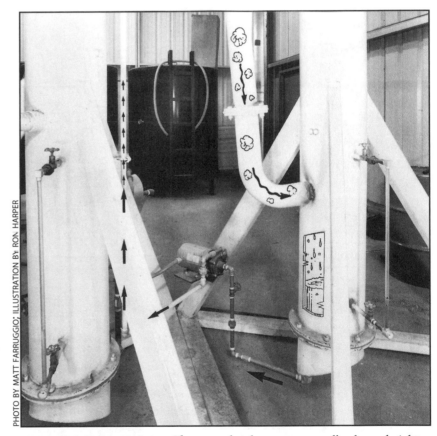

PHOTO BY MATT FARRUGGIO; ILLUSTRATION BY RON HARPER

Fig. 9-60 Splitting a column. *If you need to keep your overall column height low, use Jim Hall's distillery as an example for splitting the column. The lower section is on the left of the photo. Vapor travels from the top of the lower section to the bottom of the upper column (right), through the vapor tube (center). The pump takes countercurrent liquid descending through the upper column and delivers it to the top of the left-hand column. The bottom of each **sight tube** (difficult to see here) is a fitting containing a thermometer available through hydraulic parts catalogs to help you to monitor the temperature of your sump.*

more, with a float switch to turn the pump off in case the sump somehow empties itself.

Distillation has always had an aura of mystery surrounding it. More than once in the history of alcohol fuel, lack of good information on distillation stymied farmers and others from taking charge of the transportation fuel system. As with most things, once the veil of mystery is pierced, the actual details are quite understandable and practical. Rather than simply give you the detailed design of one column, I hope I have succeeded in giving you the tools to design the distillation system that fits your needs and site.

But what are the needs of your project/site? It's easy to focus on the amount of alcohol you may want to produce in a year, month, or hour. But once you get a sense of how much you'd like to make, it's time to get real about the practical aspects of

Fig. 9-61

operating your plant. There's a lot more to a plant than just the still, and making good choices about all the details can make a world of difference in how efficient your operation will be. So, next we'll be looking at what it takes to put together efficient micro- and small plants.

Endnotes

1. Michael R. Damiani, *Selecting a High Temperature Heat Transfer Fluid Synthetic or Hot Oil?*, Radco Industries, www.radcoind.com/TechTips2.html (June 16, 2005). [Original source: *Process Heating* (October 1998).]

2. *ABC News with Peter Jennings*, Story 3, June 3, 2002.

3. Joel Martinez-Frias, Salvador M. Aceves, and Daniel L. Flowers, "Improving Ethanol Life Cycle Energy Efficiency by Direct Combustion of Wet Ethanol in HCCI Engines," *Proceedings of the ASME International Mechanical Engineering Congress and Exhibition* (2005).

4. Michael R. Ladisch, et al., "Cornmeal Adsorber for Dehydrating Ethanol Vapors," *Industrial and Engineering Chemistry Process Design & Development* 23:3 (1984).

CHAPTER 10

DESIGNING YOUR FUEL/FEED PLANT

Before I launch into laying out your **fuel/feed plant**, I want to make a pitch for thinking about the operation as a diverse entity producing multiple yields, not just alcohol and a couple of byproducts. We'll be talking in much greater detail about this subject in the next two chapters.s It is frequently more profitable to examine each component of a system as having multiple uses, and to be obsessive about squeezing every possible purpose out of any energy we use. By doing things this way, we can make an enterprise that is not just sustainable, but highly profitable.

It's important to get used to the idea that alcohol is just one product created from biorefining plant matter into multiple products. So are you going to use the byproducts to operate an off-season greenhouse; to raise earthworms, fish, shrimp, mushrooms, or high-value vegetables; to set up an organic **community-supported agriculture (CSA)** project; to produce and market organic combination fertilizer/herbicide; or simply to make animal feed?

The suggestions in this chapter for plant layout apply to producers making 1000 to 15,000 gallons of alcohol per year, which I refer to as *micro-plants* (see Chapter 12 for a model for such a plant); or those turning out 15,000–250,000 gallons per year, which I call *small plants*. More sophisticated plants with a greater capacity than 250,000 gallons are somewhat beyond this book's scope, although many of the integrated design concepts will still apply (up to 5,000,000 gallons).

THE MICRO-PLANT

Most beginners start with a system that uses one tank for cooking, fermentation, and distillation. Such an operation distills alcohol every third or fourth day, since the unit is tied up as a fermentation vessel for several days. This kind of system is quite easy to run, simple in design, and can accommodate several vehicles' fuel requirements.

IT IS FREQUENTLY MORE PROFITABLE TO EXAMINE EACH COMPONENT OF A SYSTEM AS HAVING MULTIPLE USES, AND TO BE OBSESSIVE ABOUT SQUEEZING EVERY POSSIBLE PURPOSE OUT OF ANY ENERGY WE USE. BY DOING THINGS THIS WAY, WE CAN MAKE AN ENTERPRISE THAT IS NOT JUST SUSTAINABLE, BUT HIGHLY PROFITABLE.

Fig. 10-1 **Three Meter Island cooling tower.** *This cooling tower works basically like a swamp cooler. At the top of the unit, a fan blows air down into the structure (made of 2-inch × 2-inch frame) and through the burlap stapled onto it. The small pump takes hot water to be cooled and pumps it through many greenhouse-type foggers inside the frame. The evaporation of water from the burlap cools the rest of the water, which collects in the pool at the bottom.*

MATT FARRUGGIO

Since the single-tank plant is often used by a single owner or several neighbors on a time-share basis, the still should produce enough alcohol in one run to supply one or two people with a full month's supply. Most people with one car use about 50–100 gallons of fuel a month. To produce this amount in one batch, your distillery tank should hold 750–1500 gallons—remember, you'll need some head space to allow for foaming, etc., while you're using the distillery for cooking and fermenting. You should need to run it only once per month to produce the fuel you need for one or two people.

The alcohol percentage in your mash may be less than 10%, depending on your feedstock, so it's best to use a slightly larger tank than would actually allow you enough mash for a month's supply. If you are working with a seasonal feedstock, you could make your year's supply in about a month, running your still every few days. The alcohol does not go bad in storage. Alternatively, if you have a steady source of feedstock, 10 to 20 people could share the same distillery to get their 50 to 100 gallons of fuel per month.

Use either a six- or eight-inch column to efficiently distill all the tank's alcohol in one run lasting a few hours. Most micro-producers opt for the six-inch column to keep height and building costs down.

You'll also need a good agitator system to keep the mash from burning during cooking and distillation, and if you're using chain drives on the agitator, it helps to have a second set of pulleys or sprockets so you can agitate faster during cooking, and slower during fermentation and distillation. When using an electric motor, a **motor-speed controller**, which looks like a big dimmer light switch, lets you safely vary the speed. In a pump powered by air from a compressor, varying the speed is accomplished with a gate valve to regulate the **cubic feet per minute (cfm)** of air reaching the pump. Similar regulation can be made with hydraulically driven pumps by controlling the amount of hydraulic fluid that goes to the motor.

You should install a two-inch gate valve with a PVC fermentation lock above it (see Figure 10-2)

Fig. 10-2 Fermentation lock valve. This two-inch gate valve is opened when the distillery is used as a fermenter, and closed when you're distilling alcohol. The fermentation lock for this still (not shown) is constructed from PVC pipe, as illustrated in Chapter 7.

MATT FARRUGGIO

MATT FARRUGGIO

ABOVE: Fig. 10-3 Hatch closure. *Les Shook's hatch closure hardware is simple and effective. He has welded a nut on the perimeter of the hatch frame. The bolt holds down a piece of ³/₈-inch flatstock, which has a one-inch diameter section of barstock welded on the end. The barstock bears against the top of the hatch. It takes Les about 30–40 seconds to unscrew all the bolts holding down his ¹/₄-inch-thick hatch. The hatch frame is ¹/₄-inch thick angle iron with the "leg" pointing in and up. On top of the leg, a piece of ¹/₂-inch square barstock is welded all the way around the frame to stiffen it. The hatch is cut to fit inside the perimeter formed by the ¹/₂-inch stiffener.*

LEFT: Fig. 10-4
Top view of
hatch. Note the
½-inch stiffener
surrounding the
hatch plate, and
the position of the
hold-down bolts.

Most builders of micro-plants are tempted to choose not to spend $600–$700 on the parts for an external **counterflow heat exchanger**. Instead, they consider mounting cooling coils in the tank itself. But it is penny-wise and pound-foolish to not build the external heat exchanger. A heat exchanger can deliver "free" hot water to your house water system, after extracting heat from your mash—using an insulated hot water tank for hot water storage. A well-insulated tank holding several hundred gallons of hot water loses only a few degrees a day. This hot water is a valuable resource that should be put to use, as we will see.

You may find a shredder or hammermill useful for opening up feedstock possibilities that require grinding or pulping prior to cooking. (When a hammermill is used for grain, it's usually called a hammermill grinder; when used for garden feedstocks, it's called a hammermill shredder.) But, for some people, the most attractive feedstocks for a micro-plant are those that require no grinding or preprocessing before cooking: waste bread, doughnuts, ground grain, molasses, soda pop syrup, candy waste, mill screenings, air raid shelter crackers, etc.

In addition to your still and testing equipment (refractometers, thermometers, pH paper or meters, etc.), you'll need a pump to move liquids— for instance, to transfer mash from your tank to screens for drying animal feed byproducts, or to a compost pile area. If you're selling wet mash for feed, the pump will transfer your byproduct to a farmer's tank truck. You can use the pump to transfer water from your hot water storage tank to your still or to pump mash through the heat exchanger.

A micro-plant can require a lot of water, as much as 20 gallons a minute during peak distillation for a water-cooled condenser alone. You'll also need a small amount of electricity—for lights at night, to run a fan for your firebox, for an agitator motor, and for a water pump if you don't have municipal water pressure. (If you live where water is scarce, see the suggestions on cutting your water use in the next section, The Small Plant.)

If wood is your heat source, storage space for a few cords is another consideration. One to two cords should be enough for 1000 gallons over a year's time. For greater outputs, to get the best price, buy enough wood in the spring to last the entire year.

on top of a single-tank system. The valve is closed during distillation and left open for fermentation to allow the release or collection of carbon dioxide.

When the **micro-still** serves multiple uses, i.e., as a cooker and fermenter, you need to be able to get inside it, so it requires a sturdy leakproof hatch. Buy a commercial **tank head** (hatch) or build one like Les Shook's (see Figures 10-3 and 10-4).

You'll also need a **pressure relief valve** set at four psi or a little less, as well as a **vacuum breaker**. Your distillery ought to be operating at about one psi normally. After distillation or cooking, if the hatch remains closed and the alcohol outlet tube is still immersed in fuel, condensing and therefore contracting vapor in the tank, a vacuum will be created, and the tank will collapse. I learned this the hard way. It requires an attention to detail not normally found in human beings to not allow the alcohol outlet line to get immersed in liquid, shut off by a valve, or plugged up in any other way.

To automatically protect against unintended vacuum, a vacuum breaker should be installed on top of the tank. This allows air into the system in the event a vacuum is inadvertently created, yet does not permit any vapor or carbon dioxide to escape through it.

The single-tank system's major separate accessory is its heat exchanger to cool mash after cooking, and perhaps after distillation. Rapid cooling is essential for prevention of bacterial infection.

THE SMALL PLANT

Small plants use more equipment than micro-plants, since cooking, fermentation, and distillation are performed in separate vessels. This permits you to run the plant every day, since your distillery is not tied up for days doubling as a fermenter, as it is in a micro-plant. These larger plants run at least one distillation every day.

Although the process of making a lot of alcohol is somewhat more involved than in a micro-plant, when you calculate the increased number of vehicles that can be served by a larger system, the cost per gallon of alcohol per dollar invested is quite a bit lower. This is especially true if extra revenue benefits are realized through the collection and sale of a steady production of various byproducts or the operation of related enterprises.

Larger plants usually grow out of smaller ones, adding one component at a time as capital is available and a working knowledge of the rest of the plant is acquired. Trying to incorporate all the systems described herein from the start would be a hectic, confusing affair for the first-time alcohol producer. But knowing what it takes to expand will allow you to design your initial plant without duplicating costs or causing a bottleneck in production.

A shopping list for a small plant might include a **mash cooker**, three fermentation tanks (each with the same volume per batch as the still), a semi-automated distillery, a hot water storage tank, a warm water tank, one or two counterflow mash heat exchangers, a hammermill, one or two **mash pumps**, animal feed separation screens, and a feedstock **lifting auger** and gristmill for grain-based plants. You may also want or need a cold water tank, water circulation pump, cooling tower, heat pump, double-sized fermentation tanks for two runs a day from the three or four tanks, a yeast breeder, **auger dewatering press** and/or **drum dryer** for grain or thick-type feeds, a cogenerator, **carbon dioxide compressor**, and perhaps an alcohol dryer. If you are including animals in your design, you may want to consider a methane plant to use both manure and liquid stillage to produce gas for system heat and electricity.

Your independent fermenters should be equipped with agitators to help speed the conversion of sugar to alcohol if you're using thick feedstocks. Or you might choose to build a **separator** to remove solids before fermentation if you want to perform an essentially all-liquid fermentation. Some of the side enterprises will require either

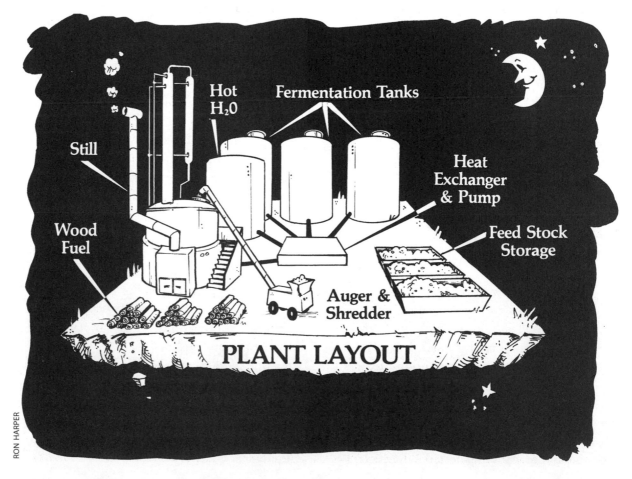

Fig. 10-5 Plant layout. This shows a basic layout that minimizes difficulties in filling and draining tanks.

RON HARPER

greenhouses or small buildings for growing high-value plant crops, fish, earthworms, or mushrooms, depending on climatic conditions.

Your cooker and still should be of similar design, and each must have a good agitator. In a small plant, time is the scarcest resource. It's worth the energy expended by your agitators to ferment a thick mash a day or two faster than you would without them.

In an entirely grain-based plant, your cooker can be about half the size of your fermentation tanks or distillery, since you can postpone adding the last half of the necessary water till after the wort has been pumped from cooker to fermenter. This transfer can occur either between cooking and liquefaction, or more commonly between liquefaction and conversion, with the wort going through a heat exchanger to cool it for simultaneous saccharification and fermentation (SSF) in your fermenting tanks.

Figure 10-5 shows a basic layout that minimizes difficulties in filling and draining tanks. For hot water storage, you'll need two tanks: one for very hot water, and another for warm water, both well insulated to retain heat.

If you run more than one batch a day, a heat exchanger is essential to collect hot water from spent mash or cooked wort. Position the exchanger so it can serve several functions: Mash from your cooker should be able to go through a counterflow heat exchanger for cooling before fermentation, and the water heated by the mash will be moved to your hot water tank. After fermentation, preheat the mash for distillation by running it through the heat exchanger, with hot water circulated through the jacket and exiting to a warm water tank. Preheated mash goes to the still if it's your first run of the day, or to an insulated holding tank adjacent to the still if it's a second or third batch. When doing two batches in a day, following spent mash removal through a heat exchanger, you would immediately pump preheated mash from your holding tank into the still.

Two heat exchangers allow you to run hot water recovered from either cooked wort or spent mash instantly into a second heat exchanger for heating a batch before distillation. It would seem that transferring heat from spent mash to incoming mash directly makes the most sense, but mash-to-mash counterflow heat exchangers are difficult to build, and commercial ones are often too expensive to warrant their use in plants under 250,000 gallons a year.

If you live where water is expensive or scarce, and you are using water for condensing the alcohol vapor, you'll probably need a cooling tower to reduce water temperature to around 60–70°F;

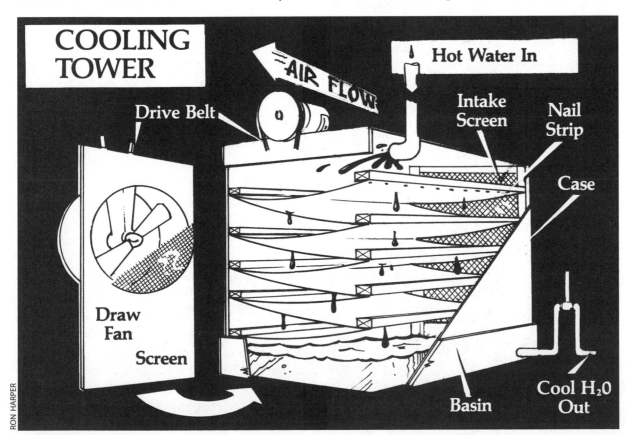

RON HARPER

Fig. 10-6 Cooling tower. Using evaporation of some of the water to cool the rest, a cooling tower can reduce water use for condensing the alcohol.

the tower uses water's high latent heat of vaporization to draw heat out and transfer it to the air by evaporation. A tower allows you to recycle coolant and save tons of water. But that extra pump and fan cost energy—both need electricity to run. Plus a cooling tower throws away the surplus heat by transfer to the outside air.

You could choose instead to use a heat pump (air conditioner) to operate your condenser, chilling the alcohol vapor to liquid and transferring the heat to your hot water tank. Using a heat pump can upgrade the heat recovered from your condenser from low-grade warm to high-quality hot water.

For general mash transfer, both micro-plants and small plants use a mobile pump. In a 50,000-gallon-per-year plant, it may be necessary to use a separate positive displacement mash pump just for the heat exchanger. An additional pump is required if you use two heat exchangers.

Fig. 10-7 Feedstock elevator.
This efficient setup takes feedstock (apples) from the washer up into a shredder.

MATT FARRUGGIO

At each fermenter and at the still, you'll want a water line plumbing just behind the drain valve to back-flush your drains if they plug up, and to rinse the hose when you finish draining a tank. If you'd rather not have a **hard plumbing** setup as your back-flush system, a hose connection and gate valve allow back-flushing—with the slight inconvenience of having to screw on the hose every time you want to use it.

You'll find it cost-effective to use a hammermill or shredder for preprocessing a wider variety of raw materials. If your feedstock is a dry material like grain, you may want to transport it to your cooker's hatch by way of a lifting auger. This consists of a spiral flight turning within a tube. The turning of the auger lifts the feedstock up until it flows out of the top. Shredded soft fruits that are watery won't go up an auger very well; you'll need a positive displacement **sump pump** under your shredder to pump the stuff into your tank.

One good alternative method, if you have sloped land, is to arrange the distillery below the feedstock storage area, using gravity to do all the hard work. It can even be worth building a truck dock and ramp next to the distilling area to bring a truckload of material in at the loading height of the still hatch or shredding area.

Drying malt or animal feed byproducts may be necessary if there isn't sufficient local demand for wet feed. Drying allows you to sell to more places, including granaries where feed is blended. A dryer requires a bit of space and quite a bit of energy, because of water's very high latent heat. Use solar or waste heat for drying wherever possible. If you're processing tons and tons of feed, it's probably most effective to use a screw press to dewater your mash to 50–60% water. This extends storage life to several weeks in cool weather as well, and makes further drying with hot air much easier.

Small quantities can be dewatered by putting wet mash in fine mesh bags and tossing them in a washing machine on spin cycle. This setup can actually get the moisture content down to around 65–70%. If you live in a dry climate, you can then spread the mash out on black-painted cement, compacted clay, or an asphalt surface to dry very quickly. Roughly one square foot of surface area is needed to dry one pound of feed per day on average. Covering the area with a hoop house and blowing air through it will maximize the benefit of solar drying.

If you intend to use a screen for draining off your main liquid, and an auger press to squeeze the feed down to 65% moisture, there are some common-sense design factors to incorporate into your plans. Since wet feed can be pumped, it's a good idea to position your **shaker table** or screens and auger press above the storage bin and pump wet feed up to it. The drained liquid from screens will flow into a storage tank by gravity, as will liquid squeezed from the feed by the press. Dewatered feed then leaves the press and falls into the storage area.

Mash without enough solids to separate may be dewatered by using it as a growth medium for single-cell bacteria, which is quite valuable as 50%-protein animal feed. Growing bacteria, however, is a whole other ball of wax. You'll need to build an essentially airtight greenhouse-like structure to house shallow ponds or trays of liquid mash and bacteria. Filtered air has to be pumped in to prevent unwanted bacteria from slipping through and contaminating the culture (see Chapter 11).

Drying is not necessary after dewatering if you can match your production to feeding livestock, whether it's cattle, chickens, earthworms, or fish. The same would be true of using it as a soil amendment or substrate for mushroom growing. Avoiding the drying step by using the byproducts immediately should be seriously considered in a plant design that is intended to be essentially full-time. It saves energy and extra handling.

Another byproduct is liquefied or compressed carbon dioxide. For every pound of alcohol you produce, you also produce nearly a pound of carbon dioxide. To harness it as a byproduct requires plumbing the fermentation locks to a water column **scrubber**, compressor, heat exchanger, dryer, and finally a propane-type tank for storage. This way, you can store carbon dioxide produced at night for use in greenhouses or algae ponds during daylight hours (see Chapter 11).

If you are using the alcohol yourself in a co-op, you won't generally need to dry the last 4% of water out of it. Alcohol-drying equipment is an additional expense, but for ease of marketing, it's often necessary. Drying equipment is usually underused, so if you can collaborate with other alcohol plants in your area to dry their alcohol, the payback is much quicker. Individual plants may find corn dehydration columns quite cost-effective, if used to bring the alcohol from 160–180 to 190–200

proof, avoiding the high energy cost of distilling in this range (see Chapter 9).

LAYOUT

There are a few common-sense designs that seem to have been repeated by my students in laying out their plants. The cooker and distillery can be placed at the ends of an aisle with fermentation tanks on either side of the aisle, or all on one side, with the feedstock preparation area, water storage, mash holding tank, and cleaning-in-place tanks on the other. If your tanks are on both sides of the aisle, then the feedstock preparation area should be at one end, along with space for byproduct processing and storage. There should be room between the cooking/distilling area and the fermenters for your heat exchanger(s), but the hot water tank can be almost anywhere near the distillery or cooker.

Positioning of tanks and distilleries makes a big difference in good layout and ease of operation. The fermentation tanks, cooker, and distillery should be placed so that each one's drain is oriented to a central point between the various tanks. At this central area will be the mash pump(s) and heat exchanger(s). This way, a minimum amount of hose-bending is required to pump material from the cooker to the fermenters, and from fermenters to either the distillery or predistillation holding tank, with each transfer likely to go through the heat exchanger to either heat or cool the mash. It's often a good design feature to place your cleaning-in-place tank near the central pump area so that you can circulate your disinfectant through lines and tanks as necessary after they have been used, to prevent bacterial contamination.

Although moving carbon dioxide long distances to greenhouses, or gas back from methane digesters, is easy, don't make the common mistake of placing these components of your design far from the distilling plant. It's difficult to move hot water to where you need it over long distances. Even with good insulation, every foot of pipe causes some loss of heat. So place the various co-product facilities close together and put the high-value vegetable growing just beyond these facilities, then outdoor crops, fuel crops, and biomass forests beyond that.

A good basic rule from permaculture design is to cluster the facilities that make products for each other close together, and put those enterprises that require the most supervision or most visits per day close to the alcohol plant.

Site Preparation

Some forethought about site preparation can save lots of headaches later. In many cases, proper site preparation is the key to economizing on labor costs. This is not such an important consideration for micro-plants, so much of this advice is directed to the small-scale plant as run by a co-op, farm, or small business.

To avoid problems, before you place any equipment on an outdoor site, you should level the site. The entire area should have a minimum of four inches of **road base** (rock or recycled concrete that compacts a surface for you to place your equipment on) spread over most of the site so water will drain quickly without pooling. You can compact

the road base with a rented tamper/packer that vibrates and pounds the gravel in places where big compacting equipment won't fit around your equipment. Compacted road base also helps discourage weeds from sprouting up around your cooker and still. It's quite cheap and can be delivered to your site via dump truck by a grading or landscaping contractor.

While you're moving earth around, it helps to build a simple earthen ramp from clay four or five feet high for backing up a truck above your feedstock processing area. The top of the ramp should have a level space as long as your truck's wheelbase, with the ramp tapering down from there. The end and sides should have a concrete or heavy wood retaining wall rising a few inches above the level platform, so you can back up to it without fear of rolling off the back. This makes loading a shredder a breeze, since all you have to do is open a gate at the end of the raised, inclined concrete holding pen. This allows the feedstock to flow by gravity into the **hopper**. A larger holding pen and taller ramp make it possible for a co-op-sized plant to receive 8–25 tons of feedstock from a debris box or dump truck.

A concrete holding pen should be covered to keep out rodents. Venting carbon dioxide into the holding area will kill any insects or rodents, as well as keep the bacteria population depressed. Building in a slight grade toward one end of the

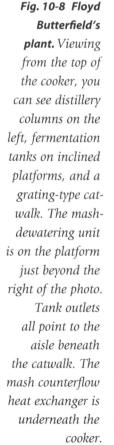

Fig. 10-8 Floyd Butterfield's plant. *Viewing from the top of the cooker, you can see distillery columns on the left, fermentation tanks on inclined platforms, and a grating-type catwalk. The mash-dewatering unit is on the platform just beyond the right of the photo. Tank outlets all point to the aisle beneath the catwalk. The mash counterflow heat exchanger is underneath the cooker.*

MATT FARRUGGIO

concrete pen—say, an inch or more drop per foot of length— helps drain juicy feedstocks. A screened three-inch drain set in the end of the pen allows you to collect the last juices and later to drain off rinse-down water.

Consult any how-to book on concrete work for help in building forms for a holding pen and retaining walls for an earthen ramp/platform.

In the tropics or in other hot climates, site preparation is particularly critical. High air temperatures tend to make the cooling of fermentation tanks a continual problem. Building your fermentation area partially or totally underground keeps temperatures around the tanks cool. If the building is partially underground, the above-ground portion can be bermed with earth to insulate it.

Be sure to check the water table and flood potential of your site before attempting to build underground. It's possible to build below the water table, but special techniques must be employed to make floors and walls watertight. Consult texts on underground construction for details.

Directly under the still, there should be an inch or two of clean white sand. Sand tolerates heat better than the rocks in road base, which can explode or crack if they absorb a significant amount of water. If you need to lift and move your still, it's much easier to slide forklift blades underneath through the sandy bottom. Some people prefer a four-inch slab of reinforced concrete under the still with anchor bolts to keep the still from vibrating out of place during agitation.

If you have to run a water line underground to your site, it's a lot smarter to do so before you start moving in any equipment. In areas where it routinely freezes in winter, underground plumbing below the frost line between warmer facilities is the safest way to get water to different parts of your site.

For facilitating cleaning in a flat-bottomed tank, such as a plastic fermenter, provide a slight grade under it (half an inch per foot minimum) to allow it to drain to an outlet on the low side.

If you have a concrete area for mash preparation, you'll need a drain so wash-down water can be carried away. Although codes in your area may call for drainage lines to drop only a quarter-inch per foot, I recommend *at least* a half-inch per foot, since you'll have much more particulate matter in your wash-down than standard wastewater does.

At the cooker, fermentation tank, or distillery tank outlet, there will often be a small spill of mash each time you disconnect the pump hose. It helps to dig a pit two feet in diameter and two feet deep in front of the outlet valve. Pour several inches of lime into the bottom of the pit and fill the rest with coarse wood chips mixed with lime. In a spill, just hose down the area, and any particles of mash will drain through the chips. The lime in the base of the pit controls odor, and the organic matter will break down in the chips. Once every few months, depending on how messy you've been, renew the chips and toss the old ones in your compost pile.

Grinders and Shredders

Unless your feedstock is whey, molasses, or some other all-liquid material, you'll probably need some preprocessing equipment. Grains require milling or grinding to reduce them to a coarse meal. Most fruits require shredding to crush and chop the feedstock into pulp and small pieces for cooking. Pulp can be pumped into a still or cooker, rather than be carried to the top hatch of a tank and dumped. You may want to experiment with juicing some of your material after shredding.

Fig. 10-9 Hammermill shredder. MacKissic, Inc. manufactures the 12-PT Series hammermill shredder. The hopper toward the back where material is being loaded is the place to load feedstocks with this ten-horsepower shredder.

PHOTO COURTESY OF MACKISSIC, INC.

On the back side of the process, you can use your shredder to prepare dry biomass for mixing with stillage to make compost, a valuable and marketable byproduct.

The most commonly used grain grinders are the hammermill variety. Hammermills work by crushing grain between a metal plate and rapidly spinning hammers. In grain grinders, the metal plate is perforated—a heavy screen of sorts; the size of the perforations determines how fine a grind you get. Such equipment is available at farm equipment dealerships, or by mail order in many publications for small farms and ranches.

Gristmills do an even better job than hammermills, which break grain into random-sized pieces, from powder to chunk; gristmills produce a consistent grind of more or less equal-sized particles. DDGS byproducts are much easier to recover if the grain is ground by gristmill, and yield more feed, since less is lost as flour. Small grain mills are manufactured for people who want to grind their own grain for animal feed. A grinder that can do 10–15 bushels of grain an hour is practical—and costs about $750. Gristmills can grind *only* grain, however. Hammermills can serve a dual purpose, as you'll see.

Shredders are used for chopping fruit, stalks, and other material into little pieces before cooking or pressing. Shredders operate on a system of spinning blades, or on the hammermill principle.

Blade-type shredders (similar to lawn mowers) are okay for stalks and dry materials, but gum up pretty quickly on soft stuff. The hammermill type also gets gummy after shredding one-third to one-half of a cubic yard of biomass with 40–60% moisture content. To clear a hammermill shredder of semi-wet fibrous pulp, grind in a dry material like straw, or stalks from Jerusalem artichokes, sorghum, or corn. Fruit or biomass with moisture content over 70% or less than 30% will not gum up a hammermill shredder. Soft fruits will wash themselves through your mill without additional clearing.

Hammermill shredders cost more than blade shredders, but they're worth it. An eight-horsepower (alcohol-fueled) hammermill can shred around seven tons of soft fruit an hour, and a ton of soft fruit yields about 200 gallons of pulp and juice. If your hammermill uses a screen down to about 1/8 inch, you can also use it as a quasi-grain mill.

If you're not using grain at all, and won't be shredding any great quantity of woody orchard prunings, etc., for compost, you can easily get away with a three- to five-horsepower shredder, although it is slower going.

On a well-built shredder, if you aren't going to be shredding things like big tree branches, you may be able to use a larger motor than the factory has specified, thereby increasing its ability to handle more tons per hour.

Fig. 10-10 Hammermill conveyor and delivery chute. *Les Shook was able to purchase this very labor-efficient used 25-horsepower hammermill at a very low price. It self-feeds with the conveyor and blows milled material into his still.*

MATT FARRUGGIO

Running grain through a shredder twice grinds it at a rate of about 20+ bushels an hour. When shopping for a shredder, look for heavy cast-iron castings (as opposed to aluminum castings), heavy **bearings** and shaft, a large, well-built hopper (you may want to make it bigger), and good clearance under the hammermill for placing a sump to catch feedstock to be pumped out or augered up into your cooker. Some manufacturers do a better job than others in arranging the screen assemblies for easy removal so you can switch feedstocks without a lot of hassle.

Shredded pulp may be suitable for juicing. Pulp is usually fermented whole, but there are some considerations in further processing by juicing. Juicing makes pulp much easier to handle, but causes some sugar loss—about 30% of the total sugar. On the other hand, juice can be fermented in as little as 24–36 hours. Whether this is really faster than fermenting whole shredded feedstock, is subject to debate (see Chapter 7).

Selecting Tanks

The tank requirement in a micro-plant system may be a single receptacle for cooking and distillation, and possibly an additional tank for hot or cold water storage. By contrast, systems for a small plant may need tanks for fermentation, hot, warm, and cold water storage, a cooling tower reservoir, wet spent mash storage, cooking, distilling, and perhaps a preheating tank or holding tank.

Some considerations in making the appropriate tank selection are the type of material your tanks should be made of, what shapes they should be, what size, what kind of access you'll have to the inside of the tank, what plumbing fittings you'll need, and, of course, the price.

CHINA'S VORACIOUS APPETITE

In 2004 and 2005, China's demand for steel of all kinds was felt worldwide. Not only did China buy, dismantle, and ship home idled steel mills from around the world, but it vacuumed the globe of scrap steel. U.S. East Coast scrap yards that might have had several acres of used stainless steel tanks were bought in their entirety and closed, and the tanks exported to China.

Stainless steel tanks will last three lifetimes. It's good they last that long, since it'll take you about that long to pay them off.

Tanks are usually mild steel (common, plain steel that rusts), stainless steel, or plastic. New mild steel tanks are the most common, and cost about $1–2 per gallon capacity. Depending on their thickness, mild steel tanks have a limited life of 3–10 years. They can last longer as fermenters if the interior is painted with winery paint or some nontoxic (to animals and yeast) epoxy paints. Used in the cooking or distilling processes, mild steel has the additional advantage of better heat transfer than stainless steel. In cooking or distillation, the tanks should be a minimum 12 gauge; as fermentation tanks, the minimum gauge is 14. Used mild steel tanks can be found from 25–90 cents per gallon capacity. Keep your eyes open as you drive around—once you start looking for used tanks, you won't believe how many there are.

Stainless steel tanks will last three lifetimes. It's good they last that long, since it'll take you about that long to pay them off. Look for used tanks at dairies: Growing dairies often replace their milk coolers with larger models. Milk coolers make ideal fermenters. They often have agitators and heat exchanger jackets built in (the agitators usually have to be modified for the thicker materials you'll be using, and the heat exchanger jackets are usually high-pressure refrigerant types). Many used coolers develop leaks in their cooling jackets,

Fig. 10-11 500-gallon plastic tank. Note square pads on the top and bottom designed for the installation of an agitator and/or through-wall fittings.

DAVID BLUME

which is often why they're being sold. Those of my students who've bought milk coolers have found the leaks are so tiny that neither hot water nor low-pressure steam seep through. They can quite successfully use hot or cold water rather than refrigerant for heat exchange.

Within a few years, I anticipate distiller's feeds will become a prime ingredient in a new generation of fast foods, breakfast cereals, and Third-World exports. That being the case, food-grade plastic tanks will become even more valuable.

Stainless steel is easy to clean, which keeps bacterial invasion to a minimum. There's also something aesthetically pleasing about stainless steel. Type 304 is durable enough without getting into exotic alloys. Some fabricators are of the opinion that type 316 stainless is better for cookers or distilleries. Stainless is stronger than mild steel, so you can get away with lighter gauges: 14 or 16 gauge is adequate for fermenters, with 14 or 12 gauge for cookers and distilleries. But never let a stainless cooker be heated without liquid in it. Its expansion and contraction are quite different from mild steel's, and the stainless steel will buckle.

Like stainless steel, plastic tanks have a long life (because they won't rust), and like mild steel they're reasonably priced. New plastic tanks run anywhere from 60 to 80 cents per gallon capacity. They're easy to clean and are available in many shapes and proportions. Plastic's main disadvantage is its intolerance to fire. Plastic tanks make very good fermenters or water tanks, but aren't used for distillation or cooking.

A plastic tank's strength is often rated in terms of how dense a liquid it can hold. Tanks for holding water should be rated for 12.5 pounds per gallon. If you're going to install agitators in them, the tanks should be rated for 16 pounds per gallon or more. The heavier tank rating also applies when water is above 160°F. Most plastic tanks are not rated for continuous use above 140°F. Hot water reduces their life to maybe 20 years, as opposed to 40 years. **Cross-linked polyethylene** seems to be the most durable tank plastic for use with mash. These are often referred to as **medium-density polyethylene** tanks.

Many otherwise high-quality tanks have deficient hatch designs. Before you buy a tank, request a guarantee that the tank can sustain at least two pounds of pressure for an hour. I had no problem securing this type of guarantee, in writing, from better tank manufacturers. If your tank won't hold

Fig. 10-12 1000-gallon tank. *A thousand gallons may sound like a lot, but a tank like this doesn't take up a lot of ground space.*

DAVID BLUME

DAVID BLUME

ABOVE: Fig. 10-13 Spherical tank. *Easy to pump, agitate, and ferment in; and the ideal shape for vacuum stills. Spherical tanks are more expensive because of their additional support structure.*

MATT FARRUGGIO

light pressure, it's because the hatch won't seal. Leaking hatches allow bacterial invasion and prevent the fermentation lock from operating. Carbon dioxide would rather leave by way of a leak than bubble through water in the lock, making it difficult to monitor fermentation. If you're collecting and compressing your CO_2, you'll be losing some of that valuable byproduct.

Plastic tanks are safe for animal feed byproducts. For a slightly higher cost, special **polyolefin plastics** (a form of cross-linked **polyethylene**) are FDA-approved for human food quality. Within a few years, I anticipate distiller's feeds will become a prime ingredient in a new generation of fast foods, breakfast cereals, and Third-World exports. That being the case, food-grade plastic tanks will become even more valuable.

There are a few other tanks worth mentioning. Copper tanks, which will last practically forever, are priced right through the roof. Their only real use is in the production of beverage alcohol. Copper requires frequent cleaning and polishing to prevent oxide poisoning, which forms naturally between uses.

LEFT: Fig. 10-14 Cone bottom fermentation tank.

GALLONAGE OF COMMON-SIZED CYLINDRICAL TANKS

	2'	3'	4'	5'	6'	7'	8'	9'	10'
2'	47	106	188	294	423	576	752	952	1175
3'	70	159	282	441	634	864	1128	1428	1762
4'	94	211	376	587	846	1151	1504	1903	2350
5'	117	264	470	734	1057	1439	1880	2379	2937
6'	141	317	564	881	1269	1727	2256	2855	3525
7'	164	370	658	1028	1480	2015	2632	3331	4112
8'	188	423	752	1175	1692	2303	3008	3807	4700
9'	211	476	846	1322	1903	2591	3384	4283	5287
10'	235	529	940	1469	2115	2879	3760	4759	5875
11'	258	582	1034	1616	2326	3167	4136	5234	6462
12'	282	634	1128	1762	2538	3454	4512	5710	7050
13'	305	687	1222	1909	2749	3742	4888	6186	7637
14'	329	740	1316	2056	2961	4030	5264	6662	8225
15'	352	793	1410	2203	3172	4318	5640	7138	8812

Read across for diameters and vertically for heights. Rounded to the nearest gallon.

Fig. 10-15 Gallons in a tank. For more accurate calculation, use the following formula: Multiply the square of the tank's radius by 3.14159 (pi) and then by the tank's height. That gives you the number of cubic feet in the tank. Multiply that figure by 7.48 for the number of gallons in the tank. Example: The tank in question is 5 feet in diameter, which means the radius of the tank is 2.5 feet. The radius of 2.5 squared (times itself) is 6.25. Multiply that times pi and you get 19.6349, times the tank's height (6 feet) = the cubic foot volume of the cylinder: 117.8084. To figure gallonage, multiply the volume times 7.48 = 881.207 gallons. To calculate the capacity of rectangular tanks, simply multiply length times width times height. Then multiply volume times 7.48, as above, to get the number of gallons in the tank. To calculate the volume of cones, use the same formula for determining the volume of cylinders, radius squared times pi times height. Divide your answer by 3, then multiply by 7.48 for your gallonage. For larger tanks, see Appendix C.

**Fig. 10-16
Stainless steel
tank head.** *This
large (20-inch)
hatch was
purchased and
welded onto my
distillery tank.
It provides easy
access for clean-
ing and loading.*

heat, and because they encourage yeast circulation. Yeast naturally float and sink hundreds of times during the three-day fermentation process. The longer and farther they travel up and down, the better. Many experimenters found no need for auxiliary cooling systems with tanks of this shape. And they take up less ground space.

Conical bottoms are easiest to clean. The bottom naturally empties itself and actually creates its own **vortex** that carries almost all the solids out easily. A bottom slope also helps circulate yeast and speeds up the entire fermentation process.

If you're stuck with short, fat tanks with flat bottoms, build a five- to ten-degree slanted platform underneath them, and install permanent sprayer fittings at the tanks' upper ends to help rinse out sludge after fermentation or cooking. They can be part of a cleaning-in-place system for sterilizing between uses.

If you're using your still as a cooker *and* fermenter, use long narrow cylindrical tanks, but lay them horizontally. This gives you more surface area to expose to your flame during cooking and distillation. Many homebuilt stills suffer from not enough surface area to absorb the heat necessary for a consistent production of vapor. Agitating horizontal tanks is done with a slow-speed (20- to 200-rpm), paddle-type agitator. The height of your access hatch is manageably low with a horizontal tank. Access can be very convenient if the tank is heated with a jacket, eliminating the firebox under the tank.

It's quite helpful to have a big enough hatch to actually go inside your tank (our first production still doubled as a hot tub). If you do manage to burn some stuff on the bottom—and rest assured you will—you can actually go inside and scrape it out. The minimum size for easy access is 18 inches; you need enough space to drop a ladder down and crawl in after it. If there's no way to enter your tank, you'll have to periodically boil caustic soda through it for cleaning.

Without a manhole or man-sized hatch, you'll have to mix your feedstock with sufficient water in a vessel outside the still to pump it inside. Diluting, of course, reduces your fermentable sugar content.

If you take this advice and build a manhole into your tank, make sure it seals tight: Alcohol vapor leaks are hazardous if the tank is used as a **still pot**, and if the tank is used as a fermenter, a leaky seal will prevent the fermentation lock from operating

Wooden fermentation tanks are the traditional vessels for fermenting wine and other beverages, but they're not practical for regular fuel production. The main problem is keeping the tanks reasonably free of bacteria. Wood's high porosity traps tiny particles of mash and bacteria that are difficult to clean without powerful disinfectants and plenty of elbow grease. If wood is to remain leak-free, it must be maintained with oil or pitch. In situations where you're using the tank only a few weeks a year, as in an annual grape crop fermentation, you can get away with wooden tanks. Wood tanks are often as cheap as and more durable than steel tanks for storing water in coastal areas where salt-laden air can corrode steel.

Concrete is similarly bacteria-prone as a tank material. And *never* use aluminum or galvanized tanks. Aluminum, in particular, leaches metal into mash when it's heated in an acid condition, and the results are toxic to most animals (including humans). To a lesser extent, zinc from galvanized tanks also causes toxicity problems. Welding on a galvanized tank can expose you to toxic gases.

Tank shape is important, too. In a perfect world, fermentation tanks should be tall, narrow cylinders with rounded or conical bottoms. Tall, narrow tanks are preferred, both because you get a better volume-to-surface ratio for radiating excess

properly. You may decide to buy a commercially produced tank hatch (tank head). I was glad I did when we built our stills. Some people have gotten away with welding half a 55-gallon drum with a removable top onto the top of the tank. Lock-top drums should withstand the normal operating pressure of up to two psi. These are hard to get to seal reliably, though, since it's easy to dent either the lip or the cover.

Often-overlooked design items on tanks are the fittings to access the tank for instruments, and for filling and emptying. The smallest pipe fitting that you should consider for filling and emptying a 500- to 2000-gallon tank is two inches, and life is a lot easier if the fittings are three inches **fip (female internal pipe thread)**. The best way to clear a clogged outlet is to back-flush with the water or fluid that was just pumped out. It helps to have quick-releases on your hoses, too (see further below). In plastic tanks, large fittings can be bolted to the side of the tank; couplings or **nipples** are usually welded to metal tanks. Through-wall (also known as through-hull) fittings work on both.

Fig. 10-18 Three-inch tank drain and valve. *A 3-inch street ell is welded to the bottom of this metal fermentation tank as a drain. A quick-release* **camlock fitting** *is screwed into the valve.*

They resemble a pipe nipple that extends through the wall; each side has a **gasket** and a threaded ring, which compresses the gaskets when tightened.

Couplings that are ½-inch or ¾-inch fip can be welded onto the tank where you want to screw in a thermometer, **thermocouple probes** for valves, or a garden faucet for sampling mash in various tests. You will also weld ¾-inch or 1-inch fittings, through which you'll put the inlet and outlet of your cooling control coils.

Another nice touch is a sight tube so you can determine the level of your mash. A sight tube fitting should be ¾-inch **mip (male internal pipe thread)**, with elbows to which ¾-inch barb hose can be attached.

You need to consider how you plan to cool the tank if it starts overheating. One way is by

Fig. 10-17 Two-inch gate valve. *Valves are attached on plastic tanks to through-hull fittings. Since these fittings are installed at least an inch from the bottom of the tank, a greater slope is called for in tank supports.*

Fig. 10-19 Common thermometer. *The threads on the bottom go into a common ½-inch pipe coupler that can be welded to your tank or column.*

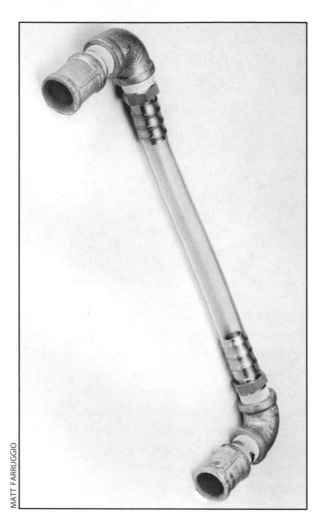

**Fig. 10-20
Sight tube
construction.**
*¾-inch couplers
are welded to
your tank. The
elbows are
connected to
the couplers by
a short nipple.
Brass ¾-inch
mip × ¾-inch
hose barb fittings,
threaded into
the elbows, allow
you to attach the
¾-inch Tygon
tubing between
the top and
bottom fittings.
Fluid level will
be clearly visible
in the tube. Such
sight tubes are
far less likely to
be broken than
standard glass
sight tubes and
are easier for the
inexperienced
person to install.*

³/₅-inch (or preferably 1-inch) internal cooling coils attached in such a way as to be reasonably easy to disinfect. If the tank is going to be fitted with an agitator, the coils should be set up well out of the way so they won't be damaged by agitator action.

Another way to cool a tank is from the outside with some form of cooling jacket. This can be as easy as wrapping the tank with burlap and soaking it slowly with cold water. The evaporating water takes away heat. Commercial tank-building companies can add a dimpled metal jacket to the outside of the tank that cooling fluid can be pumped through. External cooling works best with tall, cylindrical metal tanks. With either method, agitation while cooling will increase its effectiveness.

Agitators

Even distilleries that use an all-liquid feedstock, such as whey or molasses, benefit from slow-speed agitation in fermentation, and mixing in cooking. Agitation is useful in all three steps of the process—in cooking, in fermentation, and even in distillation.

Agitation is the moving around of material in your tank by means of such techniques as high- or low-speed mechanical motion, bubbling gas, tank shape design, or recirculating pumping. Depending on which alcohol-making process you're engaged in, one or more of these techniques is appropriate.

Agitation in Cooking

Cooking often demands the most vigorous agitation. If you're using fruit or tuber pulp, it often starts out as a thick and fibrous or chunky, applesauce-like mixture. Agitation is necessary here to keep the mash from burning on the bottom of the cooker, and also to help transfer heat by continually bringing new material in contact with the heating surface—turbulence in the mixture accelerates the heating rate considerably.

Agitation also affects the chemicals and enzymes you add as part of the cooking process. With corn, or other similar stocks, you use enzymes to break starch down into sugars. Vigorous agitation speeds starch breakdown by up to ten times as much as an unagitated batch. And when you add a pH-adjusting chemical to the mash, rapid homogenization makes subsequent pH readings more representative. The more effective your agitator, the less water you'll have to use to dilute the mash. The less water that has to be boiled with your feedstock, the less energy you need for all the steps.

Higher costs for a powerful agitator motor may pay off in time savings, although it will require a bit more energy to run. Certainly, this is an extremely expensive component in a small plant, but you can build your own agitator inexpensively with about the same skill level needed to build a distillery.

High-speed agitators (500+ rpm) are most often used in tall, narrow tanks. The blade can be a three-bladed medium-pitch boat propeller placed along the axis of the tank a foot or so from the bottom. High-speed prop-type agitators work okay in vertical tank cookers, but not so well in horizontal tanks, because you don't get a vortex. The natural curved shape of the tank's bottom makes circulation easy enough that you don't need a higher-speed motor for the mash.

The shaft should have a bearing at each end. On one end, a through-bearing with a **packing gland** allows the shaft to go through the tank wall without leaking; this is the end that is attached to, and driven by, the motor. The other end of the shaft can either go into a similar bearing with the end of the shaft extending outside the tank, or it can go into

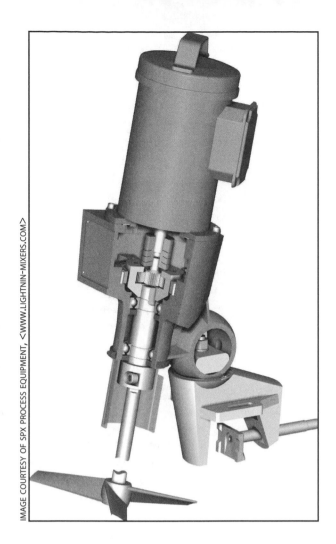

IMAGE COURTESY OF SPX PROCESS EQUIPMENT, <WWW.LIGHTNIN-MIXERS.COM>

expensive and work in high-speed applications. Chain-drive or high-torque belts are mandatory for lower speeds. The commercial version of a high-speed agitator has a **transmission** with automatic clutch instead of pulleys and belts. Such a transmission is generally much more expensive than all other components of the agitator.

Experiment to find the appropriate rpm setting for your tank—the setting will vary with the size prop, diameter, and depth. A six- or eight-inch prop works just fine in a 1500-gallon or smaller tank. The rpm range should be anywhere from 820 to 3200, depending on the tank's height-to-diameter ratio, the viscosity of the mash, and the diameter of the prop. You'll know your rpms are quite a bit too high if you start getting a wave action around the perimeter of the tank with plain water. You can vary the speed by changing pulleys at the end of the shaft and/or the end of the motor.

You will use an eight- to ten-horsepower (alcohol-converted) utility vertical shaft engine for cookers of 500 to 1500 gallons, or a three-horsepower electric motor for all but the thickest mashes. To engage and disengage your agitator from the motor, you can use a device to add tension to the belts (see Figure 10-23). Mini-bike clutches are great for slowly

Fig. 10-21
Clamp-on tank mixer. *Clamp-on mixers like this one work well for keeping feedstock moving while cooking. The mixer can then be removed, leaving no in-tank cleaning chores.*

a blind-ended bearing inside the tank supporting the end of the shaft. Many tank manufacturers sell agitator shafts, bearings, and through-tank fittings as a kit for horizontal tanks.

Since your shaft may have to take some shock loads or operate under high-**torque** conditions with thick mash, it should have some support. It should be housed in a heavy tube or pipe, with bearings in the tube supporting the shaft every two feet or so. If the tank is tall, the shaft should have a bearing supporting it at the bottom of the tank. The top of the shaft should exit the tank in a packing gland, or reasonably tightly through a bushing. Bushings may leak some vapor during distillation, so adequate ventilation is imperative. In cookers of 2500 gallons or less, the shaft should be one inch in diameter (one inch is a common size for boat props). The tank should be thickened where the shaft and packing gland come through the top, with a 1/4-inch steel plate extending at least a foot in every direction.

Attach one or more pulleys on the end of your shaft, coupled to your motor by fan belts. Chain drives are an option, but pulleys and belts are less

TUBES IN NUCLEAR REACTORS

Regardless of its heat source, any boiler uses many small tubes to absorb heat and boil water to steam. Retubing a normal boiler fired by coal, natural gas, or wood is a straightforward job that needs to be done every few years. If a tube leaks, it's no big deal, and there's no heavy downtime.

In a nuclear reactor, it's a bit more complex. There are roughly 17,000 tall, thin tubes in this kind of boiler, carrying 600°F pressurized radioactive water at 2000 pounds per square inch, making steam for a plant's turbines. The combination of radiation and water chemistry literally eats at the reactor's slender tubes until they crack or spring leaks.

Repairing worn tubes on a reactor can cost hundreds of millions of dollars. Needless to say, refitting tubes is put off as long as possible, and the tubes are simply plugged in the meantime.

The American Physical Society reported that failure of as few as six tubes in a reactor will prevent emergency cooling of the core, and virtually guarantee a meltdown. Six tubes is only 1/28 of 1% of the number of tubes in a reactor. In 1983, the Michigan Palisades plant was operating with *22% of its tubes plugged. Most plants now have a large number of tubes plugged as they reach the end of their useful life.*

CHAPTER 10 Designing Your Fuel/Feed Plant **247**

engaging pulley or belt drives. A motor controller varies the speed of an electric motor.

A horizontal-paddled shaft running down the axis of a vertical tank makes an efficient and low-powered agitator. The paddles can have a T on the end to which a thick piece of **neoprene** cut into fingers is attached for wiping the walls of the tank. This refinement is not really needed, but it helps prevent burning the mash. If the paddles turn at approximately 20+ rpm, your mash should be well agitated.

Fig. 10-22 Standard milk cooler agitator. *This type of blade configuration is okay unless you're using a thick mash.*

Powering such an agitator is best done with a one- to three-horsepower gear-reduced electric motor or a **hydraulic motor**. For turning at 100+ rpm, a two-horsepower motor is the minimum recommended size. For cooking grain-type feedstocks, the optimum rpm rate is 200, which shortens gelatinization time considerably. (Higher speeds don't improve that time significantly.) A three-horsepower electric motor (three-phase) is the preferred minimum size for cooking grains; two-horsepower motors may have to be shut off in the middle of gelatinization for 10–20 minutes, or until the thickening peaks and alpha amylase can thin the mash a bit.

At speeds between 4–200 rpm, belts and pulleys are ineffective, not having enough friction to turn the agitator shaft. But chain drives work well here, and are reasonably priced. A little more expensive are high-torque belts, which use less energy than chain drives and are easier to change for different speeds. Easier than changing belts is to use a motor-speed controller appropriate to your motor's size.

The **sheaves** for high-torque belts, although they look like gears, can be mounted in several pairs on the agitator and motor shaft. If the motor mount is left adjustable in a manner similar to an automobile's generator or alternator, the speed of the agitator can be altered by reducing tension on the belt and moving it to the next set of sheaves.

The main problem with slow-speed agitators is that a thick mash can end up spinning as a whole, with little or no mixing accomplished. To prevent

ABOVE: Fig. 10-23 Belt tensioner. *Easily constructed device to slowly engage a high-speed agitator on the top of the distillery tank.*

ABOVE: Fig. 10-24 Chain drive. *A chain drive works much better than pulleys and belts on slow-speed agitators.*

that, weld small stationary vanes to the tank wall between the agitator blade paths to break up the mass. Make sure there's plenty of clearance for the blades to pass the vanes, or you may end up with a jam.

If you're using an electric motor, you should have an automatic thermal overload shutoff built in so that if the mash jams the agitator, the motor shuts off until it cools down and you restart it. Jamming or over-torque most often occurs if you fill the tank with a minimum amount of water and the full amount of starchy feedstock, and then try to turn on the agitator. The time to turn on an agitator, especially if it's underpowered for a particularly thick mash, is when you first start adding the feedstock to the tank to begin cooking. (A mini-bike clutch allows you to start up an agitator in a thick mash.) Cooking techniques also dictate how much horsepower you want (see Chapter 7).

In most mashes, bubbling gas and pump recirculation techniques are not effective in the cooking stage—the exceptions are very watery fruit (e.g., watermelon), molasses, and whey. A mash pump, especially a positive displacement one, may be all the agitation you need for cooking these liquid mashes, simply by pumping out of one end and back in the other. Or, if your mash contains only a slight amount of solids, you can circulate it by drawing liquid from the middle of the tank and blasting it in across the bottom to prevent settling. If you're using the pump in boiling conditions, make sure it can take it—mixtures that are much hotter than a pump is rated for can damage it, or reduce the life of your seals and bearings.

Agitation in Fermentation

Agitating during fermentation is much easier than during cooking. The mash is a lot more fluid than it was, there shouldn't be any chunks, and the action of yeast further thins the mixture with lots of carbon dioxide bubbles.

Agitation in the fermentation step enhances yeast's natural process of absorption and digestion of sugar to alcohol. As yeast progresses through its three-day process, it begins to have a harder and harder time absorbing food. This situation is aggravated if yeast cells become surrounded by a semi-static cloud of excreted alcohol through which their food doesn't penetrate. Mixing keeps fresh food coming into contact with the yeast. Also, yeast in a non-agitated fermentation can get buried

at the bottom of the tank under spent feedstock solids, which makes it difficult for them to carry out any further fermentation.

Tank temperature becomes more homogeneous with agitation. In unagitated tanks, I've recorded over 10° F differences between the mash sitting nearest the tank wall and the hotter mixture at the core. If your probe is attached at the tank wall, as most are, you may think your unagitated mash is cool enough to almost stop fermenting, when in fact the mash core is hot enough to kill your yeast.

In some cases, it's necessary to agitate so as to free carbon dioxide from thick mashes such as beet or Jerusalem artichoke. Otherwise, the pockets of carbon dioxide can swell big enough to "pop" and

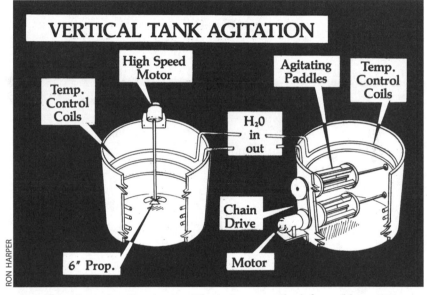

Fig. 10-25 Vertical tank agitation. *The agitator on the left would use a motor running at high speed, and the configuration on the right would use a slow-speed gear-reduced motor.*

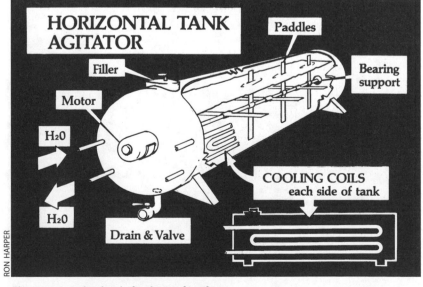

Fig. 10-26 Agitation in horizontal tanks.

Fig. 10-27 Nine-inch-wide auger used as an agitator. The spiral part known as flighting can be ordered to accommodate different diameter shafts and any length.

High-speed agitators are not used in fermenters; although they might speed up conversion, their energy usage far outweighs the benefits. Slow-speed, continuous-paddle-type agitators turning at four to six rpm work just fine with a one-horsepower gear-reduced motor. Agitator paddles should reach above the mash cap, slap it down, and break it up. The blades should pass very close to the bottom of the tank to stir up sediment and keep the yeast suspended. This is easy to do in a horizontal tank and less easy in a vertical tank. In fact, it may be a good idea to have two agitators in a vertical tank, one for the top and one for the bottom—both chain-driven by a single motor.

Thin mashes of fruit juices, whey, flour, finely ground grain, etc., can be agitated in a couple of other ways. One method is by using a pump to recirculate the mash. The inlet will draw liquid from halfway up the tank (where there should be few solids) and return the mash to the bottom fitting (where you usually empty the tank). This will stir the bottom solids and create some pretty good turbulence.

The mash pump method is too energy-intensive to be done continuously, but recirculating the mash for ten minutes three times a day does a pretty good job. This method doesn't work too well in thick mashes, since it's not effective in terms of breaking up the cap.

If you're thinking of putting your pump on a timer, remember that each time you turn on the pump you'll get a lot of foaming from all the carbon dioxide that's suddenly released. It's also important to flush your pump hoses well before using them. It's very easy to lose an entire batch if you hook up a pump and hoses that have a bunch of gross, infected sludge from being used the day before on another fermenter.

A surprisingly good agitation method for relatively thin mashes is to pull warm carbon dioxide from the top of the tank through a compressor and inject it back across the bottom of the tank through a ring of tubing with dozens of tiny holes punched in it. In this method, carbon dioxide is a vehicle for yeast and carries it around on its little gas bubbles, like a mini-balloonist.

spew mash everywhere, including up into your fermentation lock.

Thick mashes tend to float like a cap on top of their liquid. Exposed, the cap dries out a bit, and kills yeast that gets lifted out of the liquid. Agitation will break up and sink the cap in these cases. Agitation also cuts thick feedstock's fermentation time by half.

If you're using plastic tanks, make sure they have heavy-duty walls before installing an agitator.

Fig. 10-28 Direct drive coupler. Directly driving a shaft with a slow-speed motor, a coupler like this one is attached to both the motor shaft and the implement shaft. A hard rubber "spider" is sandwiched between the two mounts and takes some of the shock out of starts and impacts.

In thick mashes, a compressor-recycled CO_2 would only aggravate the cap flotation problem. Tall, narrow cylindrical tanks (especially those with cone bottoms) are more conducive to carbon dioxide compression agitation than squat tanks. Never use your compressor to pump outside air into the tank, as bacteria in the air will infect your mash, and oxygen stimulates yeast to reproduce rather than make alcohol. You can heat or cool your carbon dioxide before returning it to the tank, in order to regulate the fermentation temperature.

Agitation in Distillation

Distillation agitation shortens heat-up time and improves energy efficiency by slowly and constantly circulating the mash being heated. Circulation also reduces surging during distillation, since turbulence homogenizes tank temperatures.

Pumps

Trying to move thousands of gallons of liquid with a bucket gets old fast. Pumps fill tanks with water, move thick mash, and circulate water for condensers, heat exchangers, and cooling towers, and even to mix materials in tanks.

Pumps you intend to use for alcohol should be rated to handle fuel. They're usually referred to as "explosion-proof" pumps, since either the rotor, housing, or both are made of non-sparking materials. Alcohol pumps used indoors, powered by electricity, must be of the "totally enclosed explosion-proof" design to prevent possible ignition of any escaping vapors by electrical motor parts.

Generally, two types of pumps in varying sizes are used in an alcohol plant. The first type is the relatively inexpensive **centrifugal pump**. It works by spinning the solution to be transported from the center of the pump to the outside of the **pump chamber**, where the solution is sent downstream. These pumps depend on the flinging motion (centrifugal force) of spinning spiral chambers to force the material to move (see Figure 10-30). Such a pump **primes** slowly if the chamber is dry to start with. If the liquid source is level or higher than the pump, the pump should prime immediately. Centrifugal pumps work beautifully with water.

The problem with a centrifugal pump is that very thick mixtures tend to plug up the inlet and spiral chambers, even if the pump is rated to handle particle sizes up to an inch. If you use a centrifugal pump for mash, use one with a minimum two-inch inlet and outlet. Even then, it may be necessary to dilute your mash more than usual. If your pump plugs, disconnect it from the mash line and connect it to a water supply. The pressure should push the plug loose as you turn the motor over a few times.

Some pumps come with a fitting on the pump housing to which you can connect a water line and blast water in to free the pump. You can rig a back-flush fitting yourself, by plumbing a reducing T between the pump outlet and the shutoff valve. The T should have a 3/4-inch valve and hose fitting for attaching a garden hose in the event of plugging. Some pumps come with a housing that is set up with quick-release fittings, so you can easily take it off and have direct access to clear the vanes.

It sounds worse than it is: In a micro-plant, the incidence of plugging is infrequent enough to allow you to use a centrifugal pump without much problem.

Or you can use a positive displacement pump. This is what you'll need if you're running a larger plant, especially if you're using a mash heat exchanger. There are many types of positive displacement pumps. They don't depend on centrifugal force, but actually *push* material to be moved. In general, a chamber is filled with the material, and some device pushes it through the chamber down the line. The device can be a piston, a diaphragm, a peristaltic squeezing of a tube, or a spiral-shaped rotor.

Fig. 10-29 Two-inch centrifugal pump. *Note back-flush/drainage plug on the bottom.*

MATT FARRUGGIO

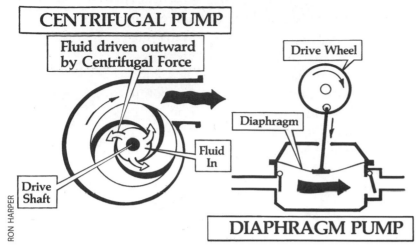

CENTRIFUGAL PUMP

Fluid driven outward by Centrifugal Force

Drive Shaft

Fluid In

Drive Wheel

Diaphragm

DIAPHRAGM PUMP

RON HARPER

Fig. 10-30
Centrifugal and
diaphragm pumps.
These are the two
main types of
pumps that you
will be using in
small plants.

Small-scale alcohol plants most often use the **diaphragm pump**. Diaphragm pumps use a flexible sheet (diaphragm) to change the shape of the chamber. When the piston is pressed down, the diaphragm compresses the space in the chamber and pushes the material out through a one-way valve. When the piston rises, it draws the diaphragm up. The exit valve flaps close when the one-way entrance valve opens, allowing the material to be drawn in by the suction of the expanding chamber. Then the piston lowers, pushing the diaphragm down and forcing the material out (see Figure 10-30).

Fig. 10-31
Three-inch single-
diaphragm pump.
A slow-speed,
unstoppable type
of pump. Note
the pipe plug on
top of the pump
outlet, where a
back-flush fitting
could be installed.

Such a pump's only limiting factor, in terms of particle size, is the diameter of its lines and valves. Some diaphragm pumps are so well built that they can suck up towels, footballs, and other unlikely soft objects. Such a pump rarely plugs up. And each stroke of the piston fills and pumps a more or less identical quantity each time, which means that a displacement pump's rate of delivery is relatively steady—important for regulating the speed of mash moving through a counterflow heat exchanger.

MATT FARRUGGIO

A two-inch pump is adequate for micro-plants, but a three-inch pump is even better for most uses. Three-inch plumbing, pumps, and fittings are more expensive, but the cost is well worth it, since they avoid time lost in clearing blockages, they pump twice as much material in the same amount of time, and they lose less energy to friction with pipe walls (see Figure 10-33). (Also, there are charts in Appendix C for calculating **pressure loss** per foot in pipe and hose.)

Another type of pump you may want to consider for your plant is a progressive cavity pump. Nothing can plug up this baby—in alcohol production, anyway; it can even pump peanut butter. It's been used with great success with thick undiluted Jerusalem artichoke and fodder beet pulp. It works by turning a spiral-shaped rotor in a close-tolerance stator. In essence, it's a three-dimensional Archimedean screw. Its only disadvantage (aside from the expense) is that it can't run dry for any length of time. Even a couple of minutes can cause extensive, costly damage.

If you use a progressive cavity pump to move material from your shredder up into your cooker, place a large hopper above it so it's always supplied with pulp. This pump is really good for pumping mash through a heat exchanger; it's easy to get it to deliver mash at a slow, steady rate, and it can tolerate high temperatures.

A safety shutoff should be set up to stop the pump when it runs out of material to transport. You can use a spring-loaded level switch which, when not buried under feedstock or mash to be pumped, opens a solenoid valve and gives the pump a minimal amount of water to prevent damage.

In a larger plant, you'd use a progressive cavity pump to pump warm mash from the fermenters through the heat exchanger to the still to preheat it, or to pump boiling spent mash back out through the heat exchanger to the mash drying area to extract the heat. In larger plants, separate pumps are needed for transportation and heat exchange, but, for the smaller plant, varying the speed on one good pump suffices. If you're using the same mash pump for both, you'll have to provide a method to slow it down for the heat exchanger and speed it up for transporting mash (see discussion of agitators earlier in this chapter).

If you don't have a municipal water source or a healthy well pump in addition to your mash pump, you may need yet another small pump (generally electric-powered) to send water through your condensers, and through your cooling coils in the still and fermenters. In my co-op's old plant, the well was not sufficient to supply our water needs, so once a day we filled our water tank from the creek with a mash pump, and then used a small electric pump to transfer it from the tank during the day's run. A similar technique can be used with irrigation ditch water.

In a small plant, the pump should be capable of pumping 40+ gallons per minute at 25 feet of height, at 50 psi. Size your pump to be somewhat larger than you need at peak use; pumps last longer if they aren't run for long periods at greater than 80% of their rated capacity. My pump had ³/₄-inch inlets and outlets and was powered by a 3450-rpm ³/₄-horsepower continuous-duty motor.

Take into account possible reductions in effective pumping capacity due to the size of your lines, the distance pumped, and the number of valves and bends through which the water has to travel (see Appendix C for more on pressure loss in common fittings and valves). You'll want reserve pressure in your pump capacity at all times.

The amount of resistance (referred to as either *pressure drop* or *pressure loss*) in your distillery's control coils can affect your ability to produce a consistent proof. If your coils need more water and your pump doesn't have enough reserve pressure to overcome the resistance in the coils when the manual or automatic valve is opened, water flow will not increase sufficiently. This is an extreme example, but your pump should be able to move enough water so that inlet pressure to your distillery cooling system is 40 psi when the cooling coils are on at full flow, while all the other demands are being made on the pump.

Larger plants with several simultaneous needs for cool water often opt for separate small pumps to do the job. Or they use a surge tank or two with

sufficiently oversized main supply lines to reduce pressure losses and to even out demand on the pump's capacity.

Don't overlook pipe diameters and length when installing your plumbing. Too small a diameter pipe to a system that has a very long run or many friction-producing fittings can cause a severe loss of pressure. Rule of thumb: when moving water long distances, figure out how big a diameter pipe you can afford and then buy the next biggest size.

You may need a water pump at your cooling tower (if you have one) or at your heat exchanger. Remember, if you plan on pumping hot water often, you should be sure to use a pump rated for higher temperatures. Solar hot water circulating pumps are often a good choice. Water pumps are almost always of the centrifugal type.

A water pump's ability to handle output demand is greatly enhanced if the water source is placed as high above the pump as possible. Every 2.31 feet of elevation (head) creates one psi pressure. If you're in a hilly location, placing the water tank 20–30 feet above the distilling site is an excellent idea. If your tank is 30 feet above your pump, the pressure at the pump's inlet will be 12.99 psi,

Fig. 10-33 Two-inch hose versus three-inch hose. *A three-inch hose has about twice the volume of a two-inch hose. It is far less likely to clog than the smaller hose.*

Fig. 10-32 2000 Moyno progressive cavity pump. *Cut open to show working parts. The spiral pump rod is spinning within a rubber sleeve.*

Fig. 10-34 Compressed-air-driven pump. *This twin-diaphragm pump is extraordinarily reliable. Although only a two-inch model, it rarely plugs up due to its dual-diaphragm design. The speed of the pump is controlled by the valve, which meters compressed air.*

MATT FARRUGGIO

minus pipe losses during transportation down the hill. A well or water supply capable of filling a tank 150 feet above your distilling site gives you 64.96 psi, minus pipe losses. With pressure like this, it's likely you *won't even need* a pump for cold water.

It is a great idea to have water stored at a high point on your property or in a water tower. This provides a buffer of steady pressure to your entire operation. It also provides a measure of safety in the event of fire, which may knock out electricity but is unlikely to knock out gravity.

How you power your pumps—by alcohol-powered utility engines, electric motors, compressed air, or hydraulic motors—is a matter of choice. I used alcohol-powered engines at my first plant, since I prefer not to be hooked up to the local utility any more than I have to, and alcohol power is cheaper.

If your main water source is down at the distilling level or anywhere below the tank, before you put a water tank way up the hill, make sure your mash pump is powerful enough to fill the tank. It still may not be able to pump water at a rapid enough rate to fill the tank conveniently. Positive displacement pumps are the best at performing this job and are required if you start trying to push water higher than 30 feet above your water source.

How you power your pumps—by alcohol-powered utility engines, electric motors, compressed air, or hydraulic motors—is a matter of choice. I used alcohol-powered engines at my first plant, since I prefer not to be hooked up to the

local utility any more than I have to, and alcohol power is cheaper. Unfortunately, when the pump, agitator, and shredder are all going at once, you can't hear yourself think. Electric motors are a little expensive, but they're quiet. NOPEC, a Bay Area alcohol cooperative in the '80s, had the best approach—using alcohol to run a cogenerator, and powering its pumps and equipment with self-made electricity.

Air and hydraulic motors have the advantage of absolutely explosion-proof design, but they're more expensive. If you have multiple uses for one of these motor types (such as an agitator or pump), their easily adjustable drive rates may save you the expense of having to buy an additional pump.

A five- to seven-horsepower alcohol-powered internal combustion engine (ICE) is good for a three-inch pump, while a three- to four-horsepower alcohol engine is adequate for a two-inch pump. Electric motors should be rated at least two-horsepower for three-inch pumps and 3/4- to 1-1/2-horsepower for two-inch pumps. Ratings similar to electric motor ratings are appropriate for air- or hydraulic-drive motors.

In general, you can assume that one electric horsepower is equal to about three internal combustion engine horsepower. You can use more powerful motors than those rated here, but unless you routinely handle mashes of high viscosity (20,000 **Saybolt Seconds Universal (SSU)** or more), performance won't be much improved. If you're lucky enough to have **three-phase power** (normal electricity is **single-phase**), you may want to splurge on slightly larger electric motors, which are cheaper than similarly sized single-phase motors.

Plumbing

Common-sense plumbing for mash transfer and hot and cold water lines can save you lots of aggravation. The most helpful rule I can offer is to use as little rigid pipe in your plumbing as possible. Let's say you have all this nice two- or three-inch expensive pipe all screwed together. Some fool tosses a half-eaten apple into the fermentation tank; it gets sucked into a line, and plugs it solid.

Maybe you're lucky and the plug is in a spot that will let you redirect water back down the line to back-flush it, and maybe it unplugs. If you're not so lucky, it's either Roto-Rooter time, or time to disassemble all that pipe. This is when you discover it doesn't come apart as easily as it went together.

Even if you use an all-liquid feedstock like whey, the same rule applies. There are hundreds of tiny pockets throughout the pipe and fittings where mash resides between uses, especially in the hard lines coming from the fermenters to the pump.

The bacterial level in piping can infect everything every time you turn on the pump. So if you use a pipe-based system, you need to be fastidious about disinfecting the lines after each use.

Hard-plumbing the entire system requires feed and return pipes to each tank, still, cooker, and to the heat exchanger. At the main pump, a **manifold** with many valves is used to isolate and direct fluids throughout the plant, which includes other pumps and mandatory back-flushing systems.

Of course, metal pipe eventually rusts and corrodes and is difficult to clean. Disassembling such a system to rearrange or move your tanks is a damned nuisance. (My extensive plumbing experience leads me to conclude that plumbing is designed more to wear out and leak than it is to stay together. The fact that nuclear power plants are one big plumber's nightmare doesn't sit well with me, either.)

The best mash plumbing for micro- and small plants, in my opinion, is a mobile pump mounted on a handcart, with a short inlet hose and several lengths of flexible outlet hose, all of which should be fitted with quick-release fittings. The only real disadvantage to a mobile pump is if electricity is your power source, you have to haul around a long extension cord. This same minor disadvantage exists with air-powered or hydraulically powered pumps and their attendant hoses.

If you're using a diaphragm pump, the entire inlet hose should be wire-reinforced to sustain powerful suction without collapsing. The first ten feet of outlet hose should be of the same quality. After ten feet, the outlet hose can be coupled to a softer, more flexible (less expensive) hose material. The sections of outlet hose should also be coupled with quick-release fittings—that way the pump can be moved to where it's needed, the hoses instantly disassembled, if needed, and completely flushed between uses. Plastic-type quick-releases are much less costly than the stainless steel or bronze varieties. If your fermentation tanks' outlets point in the general direction of the still, cooker, and heat exchanger inlets, you won't have to deal with bending a stiff, heaving mash line in an impossible arc. If you're using a centrifugal pump, the hoses can be quite a bit cheaper.

A back-flush system, to clear plugged exit ports on your tanks and to simplify flushing the pump hose, is a good feature to include. Fit a T between the two- or three-inch drain valve and the tank into which a ¾-inch line is attached, isolated by a gate valve. In the event of a jam, open the water gate valve to blast water back into the tank, clearing the plug and allowing the pump to draw again. Water pressure in the back-flush line in these circumstances should be as much as 80 psi. And you can avoid hard plumbing here by inserting a garden faucet instead of a gate valve in the T, and attaching a hose for back-flushing when necessary.

The best mash plumbing, in my opinion, is a mobile pump mounted on a handcart, with a short inlet hose and several lengths of flexible outlet hose, all of which should be fitted with quick-release fittings.

Outlet plugging is better prevented than remedied. The best prevention is keeping your agitator active all during the emptying of a tank. If you don't have an agitator in the tank (as is sometimes the case with fermenters), use your pump to

Fig. 10-35 Pump cart. *Angle iron, a little pipe, and some inexpensive wheels were all it took to make this 95-pound pump and suction hose easy to move. Hose can be stored on the two arms at the top.*

Fig. 10-36 Three-inch to two-inch quick-release reducing coupler. It's made from a bell reducer fitting and two quick-release fittings screwed into either end. Used if you need to go from three-inch hose to two-inch.

vigorously stir the mixture, to suspend as many solids as possible before emptying. Commonly, you will drop your suction line through the top of the tank into the fluid portion, then attach the outlet end of another mobile pump. Pump the fluid back into the tank through a separate back-flush fitting on the bottom, which keeps the mixture agitated.

The hard-plumbing version of the back-flush system is desirable if you frequently process thick mashes. In this case, the back-flush line can also be a convenient water source for rinsing your pump hoses and pump following mash transfers. Hot water is more effective than cold for this, so you can't use cheap plastic water lines. **CPVC**, galvanized pipe, or copper will work.

In areas of the country where winters are cold, main water supply lines for all the water for the distillery, cooker, back-flush system, heat exchangers, etc., should be installed at least 18 inches below the frost line. Nothing is more discouraging than trying to repair a frozen or split line when your mash is ready to go. All above-ground plumbing must be well insulated, since it is most often the weak link in your weatherproofing system. Your heat exchanger should also be well insulated, and water should never be left in the jacket during winter when there's a possibility of freezing. Attaching drain valves at the low points on the heat exchanger allows you to drain the water out when the exchanger isn't in use.

Fig. 10-37 Three-inch versus two-inch quick-release fittings. Three-inch fittings are more than twice as expensive as two-inch fittings because of their much heavier construction.

Although full-flow, bronze, quarter-turn ball valves are nice to have on the outlets of fermenters, cooker, and distillery, most of us have to buy cheapo gate valves. The cost of valves has come down dramatically since the 1980s, now that they are made overseas.

You may be tempted to use plastic three-inch ball valves since they seem so cheap and convenient. Most of them are cheap because the ball is made of plastic or nylon, and the valve bodies are often made of polyethylene, which will all distort the first time you pump 180+°F mash through them. It's leak city from there on out.

In areas of the country where winters are cold, main water supply lines for all the water for the distillery, cooker, back-flush system, heat exchangers, etc., should be installed at least 18 inches below the frost line.

If you use a lot of tanks, as in an operation making 40,000+ gallons per year, and you're set on brass or bronze valves, you'll save several hundred dollars buying used three-inch ones. Remember, these were designed to take quite a bit of pressure, usually up to 125 psi, and you'll hardly be putting any pressure on them at all. As long as the gate or its socket shows no nicks or cracks, you shouldn't have any leaks.

One simple and expensive mistake to make is not filtering the water entering your electric water pump. Even a little pebble can instantly destroy the impellers. This is especially true when water is drawn from a creek or irrigation ditch. A good sediment filter on the front of the pump does a fine job, although keep in mind that most filters cut the pump's capacity a bit.

Cartridge-type filters are common and effective (see Figure 10-40). Change the cartridge often; a clear cartridge housing helps to determine when it's time to change. For added safety, I also recommend a separate line strainer (see Figure 10-41), installed in front of centrifugal water pumps and at the inlet to distillery control coils.

In working on my own plant and doing design for others, I've found that I refer to certain charts and tables of information more frequently than others. I've reproduced them at the end of the book in Appendix C, so you don't have to chase all over to find this information.

HEAT EXCHANGERS AND HEAT RECOVERY

Recovering waste heat can save you part of the cost of alcohol production in several ways. There's the direct savings of energy, but in plants producing 30,000 gallons per year or more, time is even more valuable than energy savings. Recovering waste heat in your plant can cut your cooking time and the time it takes to warm up the still; it can also speed mash cooling and make cleaning easier.

Both micro-plants and small plants produce high- and low-quality waste heat in the course of distillation. High-quality, for purposes of this discussion, is hot water over 160°F. Low-quality is 120–160°F water or hot (200–400°F) exhaust gases from a flue.

High-quality heat can be used for preheating mash from the fermentation tank on its way to the still, adding hot water to some feedstocks, preheating shredded pulp before cooking, radiating heat for drying mash, heating your home, partially disinfecting tanks, and warming fermentation tanks. If you're drying grain on trays, high-grade water heat is preferred, although low-quality heat can be used for drying grain in a rolling drum dryer.

If your hose is rated for it, 180+°F water can go a long way toward disinfecting your tanks, since even short contact with temperatures that high kills the majority of bacteria. Hot water combined with dilute chlorine by way of a mixer bottle attached to the hose nozzle gives you a pretty effective sterilization system. Always wear a protective mask to avoid inhaling this or getting it in your eyes.

If you recover heat from a cooker, both the high-grade heat and any low-grade heat of about 140°F could efficiently be used by a vacuum distillery. There will be no high-grade recoverable heat from mash after vacuum distillation, due to its low temperature.

Low-quality heat is used for diluting feedstocks to cut heating time; as cogenerator or exhaust heat exchanger "coolant" (thereby upgrading the coolant to high-quality water heat); for radiant space heating; for malt drying; for heating of methane digesters, fish tanks, mushroom spawning rooms, or greenhouses; and for general wash-down water, or even irrigation water. You can use it for drying malt, running the water through lots of small coils (5/16 to 1/2 inch), **fin tubing**, or a vehicle radiator

under or in front of drying screens. I would recommend a fan, too, to blow the heat from the coils through the malt.

Water that is at least 140°F is great for washing down your equipment. It could even be run

GRANT HALLER

Fig. 10-38 Chuck Bondi's plant. *Chuck decided to go with hard plumbing instead of hose since his company, Allied Metal, had the pipe available in great quantities. Chuck went with milk coolers for fermenters, and designed his own variation of a solid-fuel fire-tube-type cooker, which heated the mash rapidly. Since Chuck was doing his project in cooperation with a cattle breeding farm, plans called for the use of methane from manure to power the plant and the use of a local cannery's peas and potatoes for feedstock.*

MATT FARRUGGIO

Fig. 10-39 Camlock-type quick-release mash line fittings. *The lever arms press the face of the male fitting against a gasket in the base of the female fitting. Plastic fittings are appropriate for general mash, but for high pressure (40 psi or greater) or continuous high temperature (over 180°F), such as might be found at the heat exchanger, metal fittings are necessary.*

**Fig. 10-40
Cartridge-type
water filter.**
*This should be
installed imme-
diately after the
water pump to
keep fine sedi-
ment out of the
plant's plumbing.
Cartridges should
be changed at the
first sign of
loading up.*

MATT FARRUGGIO

**Fig. 10-41 Line
strainer.** *A line
strainer is a
coarse screen
filter which can
be easily cleaned
by removing the
plug in the upper
branch line. Line
strainers are nec-
essary in front of
centrifugal water
pumps to prevent
small rocks, etc.,
from damaging
pumps. Finer fil-
ters are used after
the pump.*

MATT FARRUGGIO

through hot-water baseboard heaters for space heating. Radiant floor heat typically uses water no warmer than that. Many of your byproducts can make use of heating, especially in winter.

If you use a cogenerator in your plant to provide process electricity and heat, you'll be able to use a lot more low-grade water. Warm water that is 120+°F can be boosted to 200°F high-grade water by circulating it through the cogenerator's cooling jacket. An even spiffier system is to run engine-jacket-heated water through an exhaust gas counterflow exchanger to pull the heat out of 1000°F alcohol exhaust. Thoroughly engineered cogenerators also pull the heat out of 250°F engine oil, before going

on to absorb heat from exhaust and store all this energy as hot water or sometimes hot oil.

Your plant has several sources of high-quality heat, some more accessible than others. The easily accessible ones are the distillery condenser water (generally 170°F), the cogenerator "coolant," (easily 200°F), the exhaust from a stationary engine (1000°F), and processes where gas is compressed and made hot (air compressors, carbon dioxide compressors, or heat pump compressors). The somewhat more difficult sources to extract heat from are the distilled mash (generally 212°F or a little higher), because of the salts in it, and the cooked feedstock, which has to be cooled from 212°F down to 100°F. Being obsessive about recycling heat may reduce your heat input needs by 70%.

If you're designing your own equipment to exchange heat, here are a few things to keep in mind. As you recall, the farther apart the temperatures of the heat source and the material to be heated, the greater the rate and amount of practical heat exchange. Also, liquid-to-liquid heat exchange is much more efficient than liquid-to-gas, or gas-to-liquid. Therefore, trying to heat water with hot air will not be very easy, nor will extracting heat from 300°F exhaust gas.

There are several types of heat exchangers, but the two you're likely to find in a small alcohol plant are immersion coil and counterflow types. Immersion coils cool and heat the fermenters, cool cooked mash in a single-tank system, and sometimes are used in the condenser. They're useful pretty much only for general heating or cooling of extreme temperatures. Cooling alcohol vapor from 180 to 170°F with 60°F water is easy with immersion coils if you're not recovering the heat. To recover the heat as hot water, reduce the amount of water and flow it slowly through a very long coil.

The immersion system has difficulty with bigger jobs, like cooling 212°F mash to 100°F. At the 140°F point, the coil's rate of cooling slows to a crawl; also, trying to get below 120°F might require large additions of coolant to effect any further temperature reduction.

For recovering the most heat possible at the highest temperature, use a counterflow-type exchanger, which is essentially a tube inside another tube. The liquids that are going to exchange their heat are started at opposite ends—one fluid entering the inner tube at one end, the other fluid entering the space between the tubes at the other. As they pass each other, one liquid absorbs the other's released heat. By the time each liquid reaches the end of its respective tube trip, each should closely approach the starting temperature of the other. The length of the tube determines how close the inlet and exit temperatures can become, the material it's made of, the speed at which liquids are sent through the exchanger, and the amount of turbulence in both the jacket (the outer tube) and the inner tube.

Self-built counterflow heat exchangers are most easily made of copper, at least for the inner tube. There are a few superior metals for heat exchange (such as silver or gold), but copper is the only affordable one. Mild steel works reasonably well as an exchanger, but stainless steel is much less conductive.

Around the inner tube, in the narrow space between the two pipes' walls, you should wrap a wire and tack it down with solder to help direct the water in a spiral path.

You can do something similar to the inside of the inner tube, if it's carrying vapor, not liquid. Slide a thin, flat piece of copper or stainless steel sheet metal into the tube, with the opposite protruding end gripped by a vise. Then twist the metal with a pair of vise-grips on your end. What you want is a corkscrew of metal in the pipe to force the hot gases to spiral around and continually run into the tube walls. This creates a lot of turbulence and slows down gas flow, which increases contact time, for more effective heat exchange. Since this **tang** (corkscrewed insert), sometimes called a **turbulator**, increases turbulence and the distance the jacketed water must travel, it allows you to use a shorter exchanger.

If you don't use a tang in your counterflow tubing, and you're working with gases, you can get quite a pronounced **laminar flow**. This is a thin layer of gas that is cooled on the outside, while the majority of the hot gas barrels down the center core. The tang down the center is not as necessary if there is water in the tube, and of course should not be used if the material in the heat exchanger is mash.

Counterflow heat exchangers of this kind make excellent distillery condensers, since alcohol will leave the inner tube cool. Immersion coil exchangers can condense the alcohol to a liquid, but it will still be very hot and have the potential to evaporate

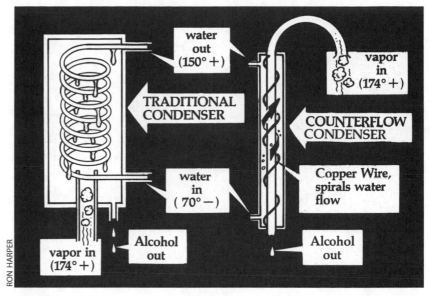
Fig. 10-42

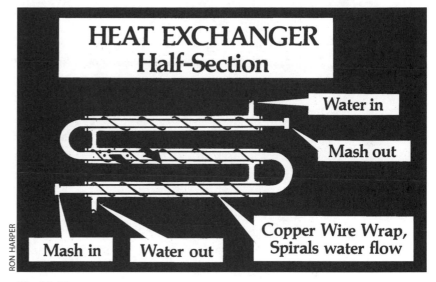
Fig. 10-43

profusely after it leaves the condenser. Extracting heat from the exhaust pipe of a stationary engine (such as one for generating electricity) can be done efficiently using a counterflow heat exchanger.

You'll save a lot of time if you use the counterflow exchanger to preheat fermented mash before distillation. Pump your mash slowly through the center tube (with no tang), while very hot water or oil circulates slowly through the outer jacket.

To get the maximum transfer in an hour's time, for 1000 gallons of medium viscosity mash, you'll need 80 to 120 feet of travel through a 1-1/2-inch tube. The jacket should be two-inch pipe, with a #10 copper wire tacked onto the inner tube to direct the water. The exchanger can be made of ten-foot sections, with hose making the 180-degree turn at the ends connecting the sections. If you have the room, the exchanger will be cheaper to build if

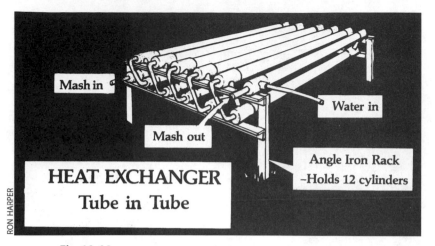

Mash in

Mash out

Water in

Angle Iron Rack
–Holds 12 cylinders

HEAT EXCHANGER
Tube in Tube

Fig. 10-44

Tube-in-tube heat exchanger.

the sections are 20 feet long rather than ten feet. Use either quick-release camlock fittings or **pin-lug fittings** to make the return hoses removable for clearing any plugged part of the heat exchanger (see Figure 10-47).

The process is similar to cooling cooked and/or liquefied mash to fermentation temperatures. This time, instead of providing hot stored water, you are providing cool water and recovering the heat from the mash. Ideally, to save the most time and energy, it would be great to begin preheating the mash for the next batch while cooling the spent mash from the previous batch.

But you can't pump fresh mash into the still while the spent mash is being emptied through the heat exchanger. One solution is to send the preheated mash temporarily to an insulated holding tank. Hot spent mash can then be emptied from the still—either through the heat exchanger to recover its heat in a hot water tank, or have it go directly to

Fig. 10-45

Insulation for hot water lines. You can lose more than a degree from your hot water per foot of copper pipe if you don't insulate it. This effective foam insulation snaps on around the pipe.

screens to recover hot mash solids (as in the case of mushroom growing). Then you would send the hot liquid through the heat exchanger. You can also make room for the new batch by transferring the spent mash to either an empty fermenter or one more tank in order to speed up getting the new mash into the still.

Once you finish pumping preheated mash into the still, the hot spent mash is sent through the heat exchanger's inner tube. With extra holding tanks, two pumps, and two heat exchangers, you could both recover the spent mash's heat and preheat the batch to be distilled at the same time.

To recover high-quality heat, you might think it would be a lot simpler to just run your hot spent mash through the outer jacket, and the mash to be preheated through the inner tube, to accomplish the whole process in less than half the time. But there would be severe problems in cleaning the jacket of mash, especially if it plugged anywhere (remember, this exchanger is 80+ feet long).

Since commercially made mash-to-mash heat exchangers are extremely expensive, the answer on the small scale is to use one heat exchanger to scavenge the waste heat from spent mash and use that hot water to preheat the incoming fresh mash, or store the hot water for later use.

In addition to using counterflow heat exchangers to preheat fermented mash for distillation, both micro-plants and small plants that use separate fermenters can use counterflow heat exchangers for quick-cooling cooked feedstock. Fruit pulp, tuber pulp, grain mash, etc., can go from boiling to fermentation temperature, while you're recovering high-temperature water. And quick-cooling (while avoiding contact with the air) is very helpful in preventing bacterial contamination before yeast is introduced to the mash. It's often the case, particularly with starchy feedstocks, that the water you add to reach an 18–24% sugar concentration is not enough to cool the mash sufficiently.

The counterflow method can also be used to cool mash from 190 to 140°F for the conversion step—giving you more high-quality hot water in the bargain. The wort is pumped from the top of the tank through the heat exchanger, and redeposited through the bottom fitting to reduce mixing of cool and hot worts.

Immersion-type exchangers are often used in micro-plants, especially those with inexpensive combination cooker-fermenter-distillery setups.

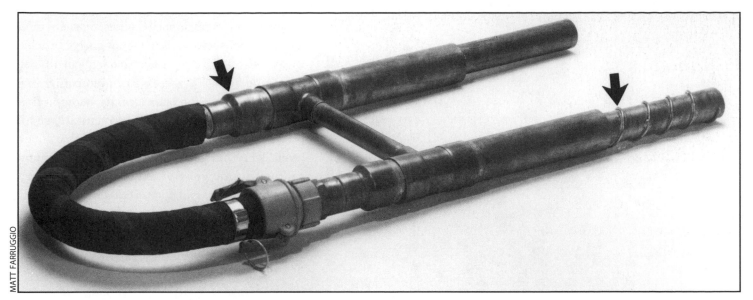

Fig. 10-46 Counterflow heat exchanger end detail. *Tube within tube heat exchangers have been used for mash cooling/heating for decades. This simplified version is easy to build and quite efficient. On the right, you can see that the inner 1½-inch tube is wrapped with #10 wire that is spot-soldered in place. The wire-wrapped tube just slips inside a 2-inch copper pipe. The helical wire forces the coolant water to flow in a spiral path around the central tube. Close to the end of the 2-inch pipe, solder on a 2-inch by ¾-inch T. The end of the 2-inch pipe should have a 2-inch by 1-½-inch reducer soldered on the end, with a 1-½-inch pipe protruding a couple of inches (see left arrow). On the protruding end, solder on a 1-½-inch copper × mip adapter (bottom tube). This adapter will allow you to thread a quick-release fitting or pin-lug fitting from the return hose, as you can see here. Solder a 12-inch piece of ¾-inch pipe between the 2-inch Ts. This will carry jacket coolant from the end of one heat exchange line to the beginning of the other. The return hose does the same with the mash. Five 20-foot sections of this type of heat exchanger are needed to cool 1000 gallons of cooked wort to fermentation temperatures in 60–80 minutes.*

It's difficult to recover much high-quality hot water in these plants with immersion heaters, except that which comes from the condensers and control coil. Coils in the main tank do extract some heat from cooked feedstock in the process of cooling spent mash, but you generally use a great deal more water, and most of it ends up as low quality warm water. That might be just fine if you have needs in other parts of your system for it.

Cooling coils in a 500–2000 gallon single-tank system should be ¾-inch or one-inch diameter and quite long—200 to 300 feet is probably long enough to cool most mashes in 90 minutes or less. A refinement to increase cooling efficiency, albeit with some loss of high-grade heat recovery, is to divide the 300 feet into six separate 50-foot coils, with a common feed line and common returns. This prevents the water from getting too hot in the line and reducing the heat exchange rate. The difficulty in trying to cool mash quickly in a single-tank system often dictates this design.

MASH DRYING

Drying the mash allows you to store it longer for feeding your animals, or to extend its use as a seasonal feedstock. Sometimes, drying it allows you to sell it later for a better price after others sold theirs during the height of production.

The simplest drying technique is to simply strain your feed over a fine mesh (start with #200–#300 and see if it's fine enough for your particular stock), collect the drained liquid for other uses, and dry the feed in the sun on a warm day. The pulpy byproduct remaining on the screen is raked out to a depth of an inch or less, and the sun bakes the water out of it. It helps to rake and turn the feed a few times a day.

Untreated wet feed lasts only two or three days. When you strain off the excess water, you end up with a feed that has 80+% water content; it can be stored for a couple of weeks in a sealed tank or a bin saturated with carbon dioxide from your fermenters. But, in that same carbon-dioxide-saturated bin, mash dried to 65% moisture can store for weeks. The mash is a non-sticky, semi-dry consistency; if you squeeze it, it will not remain in a ball.

Large distilleries often dry their feed to below 20% moisture to store for years in sacks or silos. Medium- and large-scale plants, which of course produce a lot of mash, use rotary drum or

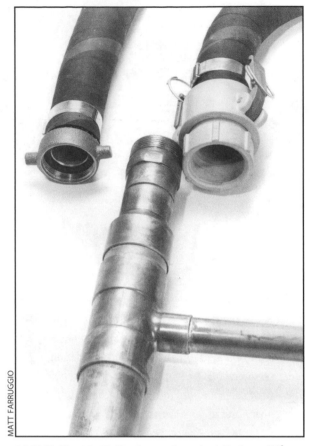

MATT FARRUGGIO

*ABOVE: **Fig. 10-47 Pin-lug and camlock fittings.** When installing the flexible hose on the ends of your heat exchanger, you can use either system. The pin lug set-up is half the price but is a bit more inconvenient to disassemble for cleaning. If you're using thick mashes, camlock fittings are recommended. Also note the 1-½-inch pipe protruding from the 2-inch by 1-½-inch reducer. The 1-½-inch copper-to-male (mip) thread adapter is attached to the 1-½-inch pipe. The water surrounding the 1-½-inch pipe exits the jacket through the ¾-inch line to the right, since the reducer seals the end of the jacket.*

BELOW:
Fig. 10-48
Cooling coil.
Works well for cooling mash in fermenters.

BOB FITCH PHOTO, <WWW.BOBFITCHPHOTO.COM>

rolling-screen-type dryers. The commercial equipment is expensive and uses a lot of energy (usually fossil fuels). Such equipment in a large plant may constitute a large percentage of total equipment costs, and double the plant's energy consumption per gallon of alcohol produced. Luckily, small plants don't have to dry their feed to such extreme levels, since marketing feed is a local matter and short-term storage is all that the small producer needs to smooth out the undulations of supply and demand.

An industrial-strength version of solar drying is planned at a Southern California facility that will be using beets as a feedstock. The hot, dry desert climate has led plant designers to plan to pave several hundred acres with black asphalt. They calculate that pumped beet pulp will dry in about an hour on the surface of the solar-heated **thermal mass** and then can be scooped up by loaders for storage.

Of course, the weather can be counted on to turn rainy the moment you rake out your mash. So on the micro-plant scale, drying the mash on screens in a simple farm-scale drying cabinet shed works well. Pump the mash onto the bottom screen and then up to successively higher screens so the lower ones pick up the finer suspended solids draining from the upper screens. Once they are all loaded with a smooth level of feed, close the shed and draw hot air or hot carbon dioxide through it with a fan (similar to a cooling tower fan) from one side to the other. Drawing the air from one or more squat, long, plastic-covered "greenhouse" tunnels would also supply lots of nearly free drying heat.

There are several places within your plant where waste heat is available as hot air for drying. Flue gases can be directed through multiple fire-tubes across the air-inlet end of your drying shed (it would look like a giant radiator). You could even build a separate small firebox to feed these tubes, which would cross the face of the drying shed and come together in a flue.

Excess hot water can be circulated through a similar "radiator" of copper fin tubing, or even a used truck radiator, and its heat exchanged to the moving air that dries the feed. This is practical in plants where water is limited and a cooling tower is employed to recycle condenser coolant: For micro or very small operations, you could use the same design for drying malt in the drying of feed (see Chapter 7, Figure 7-34). The condenser-

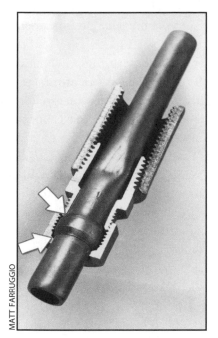

Fig. 10-49 Cutaway through bushing model. *I showed you this assembly in the distillation chapter earlier. It is equally useful for installing cooling coils through the wall of a fermentation tank or cooker. This hardware is used whenever a tube must pass through the wall of a liquid-or gas-filled vessel. The largest threaded fitting is a coupling, which would be welded to the vessel. The middle fitting is the body of the compression through fitting. The compression ring is sealed against the tube by the nut. The sealing surfaces, which prevent escape of vapor or liquid from the vessel, are indicated by arrows.*

heated liquid (maybe water mixed with antifreeze) is run through your dryer radiator, and then sent to the cooling tower. Of course the hot side of a heat pump used to condense alcohol will exchange a lot of heat into air blown over it, and that's how it's designed to work anyway.

Any drying scheme is enhanced by dewatering the feed before placing it on racks. Various presses work fine with corn or barley, but applesauce-consistency mashes are difficult to separate. A popular method of dewatering **particulate** mashes is with an auger press in which a nine-inch auger is turned inside a heavy, perforated **screen cylinder**.

As wet feed is fed into the press, the auger drives it toward a spring-tensioned door at the end of the cylinder. The mash begins to compact, and water is squeezed out through the screen. Auger speed, screen perforation size, and tension on the end door determine how much the feedstock is compacted and dewatered, with variations also from

feedstock to feedstock (see Chapter 27, interview with Floyd Butterfield).

Another dewatering method uses something like a wine press. The wooden staves of a real wine press cylinder aren't fine enough to prevent much of the wet mash from escaping through the cracks. You'll need a cylinder of perforated steel inserted snugly inside the wooden **stave cylinder**. If you're building from scratch, the stave cylinder can have a third as many staves as a regular wine press, since it's really only a reinforcement for the perforated cylinder.

There's a trick to increase the efficiency of such a press that I learned while making cider with friends at the Alpha Community in Oregon. Fill the cylinder about a quarter full of mash and top it with a perforated inch-thick plywood plate. Add mash to the halfway height, and top with another plate. Begin pressing, preferably with a hydraulic ram rated at 30 tons. When spread out over the entire surface of the plates, this pressure is the equivalent of 60–100 pounds per square inch.

Fig. 10-50 Grain dryer. *A few screens like this are all a microplant needs to solar-dry mash byproducts. A small greenhouse over these drying screens would permit drying on days with changeable weather and speed drying on sunny days.*

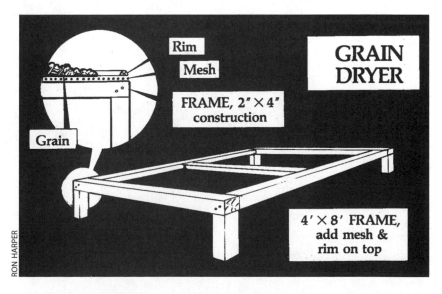

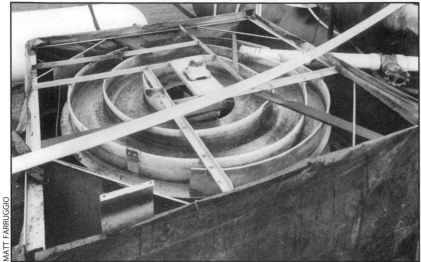

Fig. 10-51 Shaker table. *Wet mash enters a spiral path from the 2-inch pipe on the right. By the time the feed has reached the center of the table, it's been vigorously vibrated and falls into the center. Liquid exits through the perforations in the table surface. Drained mash is much easier to tumble dry.*

Now that the mash has been compressed and dewatered a bit, fill more of the cylinder, top with another plate, add more mash, another plate, and press again.

Intermittent pressure plates allow you to squeeze out the liquid efficiently and quickly. You may find it beneficial to run the once-pressed mash through a second time.

Larger plants use centrifuges, which can really dewater the pulp very well but are out of the price range of small plants, unless you can find them at auction. Once you are up to 25+ gallons per hour, commercial centrifuges start to become affordable, especially used ones. At micro-plants, putting the mash in mesh bags and using an old washing machine to spin them will dewater to about 70%.

After any of the dewatering steps, you proceed to actually drying the material. We discussed the shed method above; both it and the drum method are doubly useful in drying and storing raw feedstock for a period of time before fermentation. Crops like cassava, Jerusalem artichokes, and sweet potatoes can be sliced or coarsely shredded and then dried into chips for storage, with little loss of fermentable carbohydrates.

At least one pair of plant owners I know (see Chapter 27, interview with Kent and Thurly Heintz)

*Fig. 10-52 **Mash dewatering.** A shaker table (not visible in the photo) drains the easily separated water and allows wet solids to drop into the white hopper. Solids are transported from the hopper and compressed in the perforated drum by a tight-fitting auger, and water is squeezed out through the perforations. The amount of compression is determined by how tightly the threaded hooks connected to the long springs holding the exit door are adjusted. DDG drop out the door and into a storage bin. The liquid is collected in a tray beneath the drum to be sent to the fermenters. This system separates the grain before fermentation. The drum is usually covered by a hood to keep things neat.*

Fig. 10-53 Apple washer. *Ken Ratzlaff's self-built device is an important one to study, since its basic construction is appropriate for beet or tuber washing as well as for feed-drying equipment. The electric motor is connected to a white gear-reducing unit with a belt and pulleys, which is then connected by chain drive to the rubber wheel. The wheel turns the drum by friction. The other side of the drum is supported by wheels that are not driven. This washer's drum is salvaged from two commercial clothes dryers, but you can make one yourself with perforated steel. This illustrates very clearly the proper use of chain drive and pulley belt. The belt is used in high-speed application in front of the white gear-reducer. The reducer's output shaft puts out high torque at a low rpm, which makes chain drive the appropriate choice to turn the drum. For washing, Ken installed a perforated (white) plastic pipe inside the drum to spray his apples, and sprayers at the exit where the apples are lifted by elevator to a shredder.*

constructed a cost-effective miniature version of a drum dryer. They inclined a perforated metal drum that rotates on casters via a single drive wheel, then created a jacket around the drum. Hot gases from a large water heater flue are directed through the drum jacket. Hot air is also blown down the center of the drum. The drum is about ten feet long and tumbles the fruit pulp over and over on its way from the upper end of the drum to the lower end. Liquid is collected in a tray under the drum, and the rest of the moisture is driven out of the pulp by heat. Dewatered pulp exits the end of the drum, and the quantity to be stored is dried further, when necessary, with another trip through the drum.

Although the Heintzes use flue gases for drying, the drum method works a lot better if the hot air source is a dry one. Flue gases (especially from natural gas) are full of their own moisture. Drum drying, which requires more heat and air circulation than other methods, is useful for such pulpy feedstocks as Jerusalem artichokes, potatoes, most fruits, almond hulls, and possibly some food processing wastes.

METHANE

Using methane to run your plant achieves the dream: no fossil fuel inputs in producing the fuel. We go into much greater detail on methane's

Fig. 10-54 Feed chute. *This homemade chute directs dried feed, dropping it from the outlet near the top of the tanks adjacent to the motor, alongside the tank, and into the auger lifter exiting to the right.*

MATT FARRUGGIO

impact in our co-products section (see Chapter 11), but let's examine the ins and outs of its production.

The decision on whether to use tanks or ponds for our small-scale methane production boils down to this: Tanks work best when there is little space, and more money. Ponds (deeper than six feet) are best if you've got lots of space but less money.

Making Methane Using Tanks

If you find a nice big tank needing to be hauled away for free, you can turn it into a methane digester. The ideal proportion is a length-to-diameter ratio of about 5:1. Shorter than this, and the tank will not fully digest the material flowing into it.

Baffling the tank to force the fermentation to take a sinuous route also aids in production. An intermittent agitation by a horizontal shaft with paddles will gently circulate the bacteria without swirling up sand or other objects, which act like meteors, smashing our little workers.

The great thing about alcohol plants is that you often have a surplus of hot water. **Methanobacters** (methane-making bacteria) like a toasty environment, even warmer than yeast like. To really crank up gas production, use the low-grade hot water (about 140–150°F) in heat exchange tubing in

the bottom of the tanks or in a jacket to heat the fermenting liquid to 100°F.

It's important to not let the tank lie on the ground and to give it a good eight inches of insulation to help keep the temperature stable. In fact, an even better strategy than using a heat exchange coil is to either use a jacket or set the tank in a bath where the hot water is introduced under the center of the tank. This is a very good system for using water that is not much good for other purposes, being down in the 100–120°F range. The overflow from the bath can be run into a nice hot tub for you and the crew to enjoy, since it will be a toasty 101–104°F.

If you are using tanks instead of a pond, it's easiest to work in batch mode versus continuous mode. This usually means methane production will peak in about 30 days and then drop off to near zero in two months, in a typical bell curve. Heating the tanks can reduce processing time to as little as 20 to 40 days. Using four tanks, starting one per week, will even out your gas production so that it will seem continuous. These residence times are for manures. Residence times for all-liquid stillage can be as little as five days, meaning your digester can be much smaller and produce the same amount of gas.

The disadvantage with tank methods is that you generally end up with a 60/40 mix of methane and carbon dioxide in the gas. If you use a tank system, you'll want to clean the gas of the carbon dioxide (in a countercurrent packed column with water) to more or less purify it—so as to have a better gas to burn, since carbon dioxide is not flammable. The water now containing the carbon dioxide can be used in an algae-growing tank.

The other alternative is to have a tall gas storage tank. The heavier carbon dioxide will naturally settle to the bottom, and you can bleed it off with a valve connected to the bottom of the tank; use a carbon dioxide gauge to tell when you've finished bleeding off the noncombustible gas.

For long-term use in an internal combustion engine, such as a cogenerator, you should scrub the methane, if you use the tank method, to remove hydrogen sulfide gas, which causes engine wear and acid rain when the sulfur is released to the air. The best treatment is prevention. If you are successful at keeping the tank between pH 7.0 to 8.5, not only will you get more methane faster, but you will reduce the production of hydrogen sulfide gas to a level not worth worrying about. But if you smell even a hint of rotten egg odor in your final effluent, the tank gas should be bubbled through a countercurrent column with very alkaline water flowing down and gas going up. Water with baking soda is the best way to remove the trace sulfur in the gas.[2]

Making Methane Using Ponds

In general, though, I favor the pond system for methane production. In a pond deeper than six feet, the lower depths are always anaerobic, which makes them an ideal area in which to produce methane.

A corral built in the bottom of a ten-foot-deep pond will prevent oxygen-laden water from **convecting** to the anaerobic area in the corral. You flow your stillage into the bottom of the corral. As the bacteria begin to devour bits of organic matter and produce methane, the particles start to float up. But before they've gone too many feet, the methane bubble breaks loose and continues up to be collected. The bit of organic matter drops back down to the bottom, as the bacteria continue to eat it. Then it's up for another balloon ride. The gas has to travel though almost ten feet of water, which strips the carbon dioxide pretty well.

This makes methane of nearly 85% purity, which is usable either for alcohol production heat or, better yet, to be run through an internal combustion engine to produce both electricity and hot water.

You collect the methane by suspending a tent of impervious fabric over the corral, fastened to its top. The bubbles will follow the tent to its apex, which is open to an inverted 55-gallon barrel, still underwater, to collect the gas. If the top of the inverted barrel is held under a floating dock exactly 11 inches underwater, the pressure of the gas in the barrel is exactly the right pressure to run natural gas appliances.

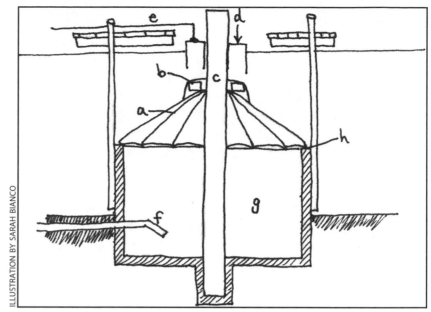

ILLUSTRATION BY SARAH BIANCO

Fig. 10-56 Compact pond-type digester. *This is a compact configuration of a pond-type digester: a) underwater tent to direct methane to be collected; b) floating collar; c)central support column; d) open-bottomed tank to collect the gas; e) gas line leading to shore; f) feedstock input line; g) area where digestion to methane occurs.*

Most likely, however, you will run the methane through a pipe under the bottom of counterbalanced, inverted, open-bottomed tanks in the pond. The tanks will float higher and higher as the methane fills the tank and displaces water. The weight of the heavy tank provides pressure when a line from the top of the tank takes gas to where it is needed. At the house, or plant, a **pressure regulator** reduces the pressure from **line pressure** to 11 inches **water column**, which is about 0.5 psi. (Gas appliances are rated by water column inches; 11 inches water column is equal to the amount of pressure created if a tank were submerged 11 inches under water.)

If you are producing surplus gas, and the floating tank reaches its highest point and gas is about to overflow out of the open bottom of the tank, the gas is then drawn into a two-stage compressor, where it can be stored in a tank on land just like propane. It is then piped to where you need it after going through

Fig. 10-55 Pond-style methane digester. This type of methane digester is less expensive to build than tank-based systems. The trade-off between the two is that pond systems take up more space and are slightly less efficient at producing methane.

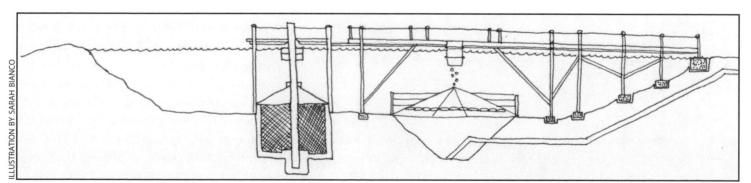

ILLUSTRATION BY SARAH BIANCO

a pressure regulator. Two-stage compressors typically produce about 175-psi gas, and the propane tank is rated for 250-psi pressure. The process of compressing and cooling the gas will remove almost all the water and any water-soluble impurities.

In pond systems, unlike tank systems, hydrogen sulfide gas is not a problem. The ability to keep the pH stable and buffered due to the pond's larger volume keeps the sulfur in solution, where algae use it in metabolism.

SAFETY AND STORAGE

At one time, I preferred storage of alcohol underground. The big advantage, of course, is that your lot size for a station can be much smaller than one using above-ground tanks. But since I first wrote this book, literally hundreds of thousands of single-wall steel gasoline tanks have rusted and leaked, polluting untold millions of cubic yards of soil and water. Moist soil can rapidly corrode its way through steel tanks. This has come to a head with **methyl tertiary butyl ether (MTBE)**, a carcinogenic additive to gasoline that is polluting groundwater all over the country. I used to prefer underground storage, but the irresponsibility of oil companies has resulted in enormous regulation of underground installations.

Environmentally, an alcohol leak would only be as dangerous as the alcohol's **denaturant**. Alcohol itself is readily metabolized by fungi and bacteria,

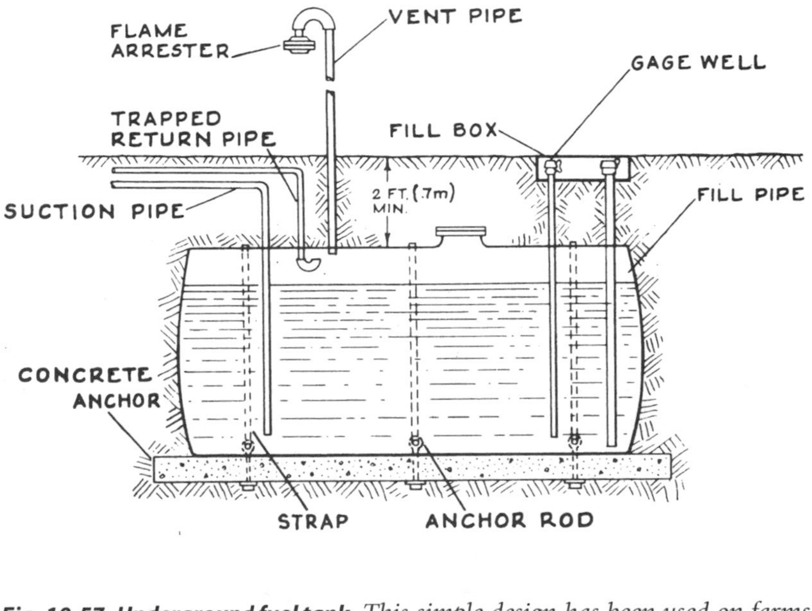

so its life in the soil is between hours and days, not months or years. But if you are using a petroleum denaturant, it could take a long time for the nasty stuff to biodegrade. So underground storage nowadays is a more highly regulated affair, usually requiring a double fiberglass tank so that if a leak happened in the fuel tank, it would be contained by the outer second tank, which would have alarm technology to alert you to the problem. It costs well over a million dollars to install underground tanks in even a small new gas station.

So I am not going to go into the installation of underground tanks, but there are books and codes available to refer to if you plan on doing it.

Alcohol storage issues often revolve around the **flashpoint** of alcohol. Flashpoint is the temperature at which an organic compound gives off enough vapor to ignite in air. This is the lowest temperature at sea level, with alcohol at its leanest air/fuel mixture, at which the fuel will fire with a source of ignition, such as a spark.

Flashpoint varies with alcohol proof and difference in air pressure. It's measured with a closed- or open-cup test. Closed-cup flashpoints are usually lower and are fairly accurate in very dry air in an enclosed space. Open-cup tests are a better indicator of outdoor flashpoints. The closed-cup flashpoint of 180-proof alcohol is about 58°F, while the open-cup flashpoint for the same proof is approximately 73°F. The more humid the air, the higher the flashpoint. The lower the air pressure (for instance, at higher elevations), the lower the flashpoint.

Alcohol has a much higher latent heat of vaporization than gasoline—395 Btu/pound compared to gasoline's dangerously low 147 Btu/pound—so it takes far more energy to make alcohol evaporate. Gasoline is never safe. Its flashpoint is 46°F below zero, and it always gives off a great deal of vapor in storage. Gasoline is ranked as a Class 1A flammable liquid; alcohol is rated as a Class 1C flammable liquid (at 190 proof) because its flashpoint is below 100°F. Alcohol's Class 1 designation unfairly requires that it be regulated in the same manner as gasoline.

People are allowed to smoke and drink in the vicinity of alcohol, pretty good evidence that alcohol is not highly flammable. If you refer to Figure 10-58, you will see that 86-proof alcohol, i.e., your favorite martini, has an open-cup flashpoint of about 95°F. Yet the 1200°F tip of a burning

Fig. 10-57 Underground fuel tank. *This simple design has been used on farms all over the world. Note heavy concrete pad to hold tank in place if water table rises.*

cigarette does not cause tequila shooters to explode. Given that restaurants and bars have no limits to how much whiskey they can store in glass bottles, it is discriminatory for a fire department to prohibit you from storing alcohol for fuel in a properly installed above-ground tank.

Above-ground tanks are much safer than underground tanks, since any leaks are easily detected and repaired. In rural areas, the simplest fuel tanks are called **ladder tanks** and usually range from 100 to 1000 gallons. The frame holds the horizontal tanks six feet off the ground. A fuel truck comes out to fill your tank, and then you fill your vehicle with alcohol, which leaves the bottom of the tank by gravity.

You should always use an automatic shutoff handle similar to what you use at the gas station. I always install a good-quality quarter-turn ball valve on the tank outlet, which I shut off after filling and open just before I fill, so I have double protection against leaks.

If other people use the tank for filling their vehicles and you can't count on them not to screw up, you should install a **breakaway fitting** on the hose.

Flashpoints and Fire Points of Alcohol/Water Mixtures

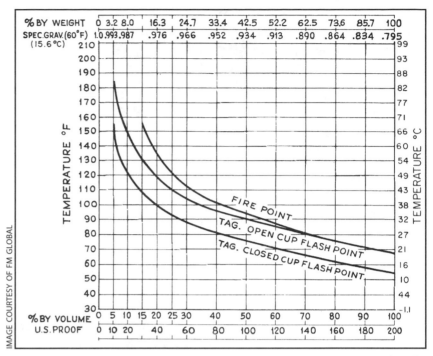

IMAGE COURTESY OF FM GLOBAL

Fig. 10-58 Flashpoints and fire points of alcohol/water mixtures. *Note that the flashpoint of your average martini is not much higher than 190-proof fuel. When's the last time you heard of an exploding martini?*

Above-Ground Fuel Storage—Suction System

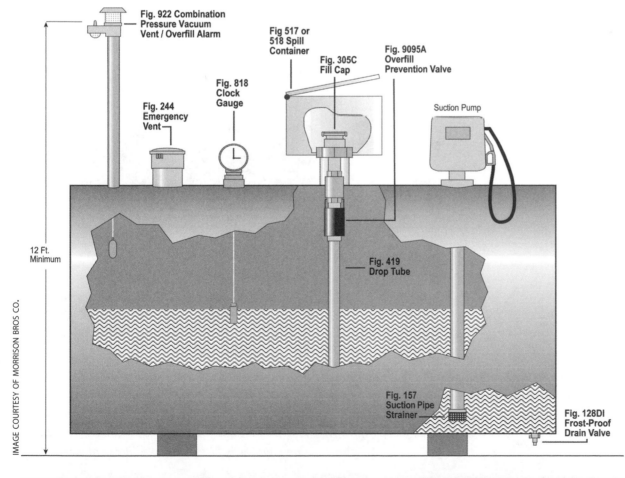

IMAGE COURTESY OF MORRISON BROS CO.

Fig. 10-59 Above-ground fuel storage, suction system. This is a horizontal cylindrical tank with top fill and a top-mounted pump. This above-ground tank has all the bells and whistles. Tanks 1000 gallons and below won't need an emergency vent or the splash-proof filler cap.

Fig. 10-60

For a gravity tank, you will have a stand, usually made of angle iron, that supports the horizontal tank. Do not put the four feet of the stand on plain soil. The weight of a tank full of fuel could push one of the feet right into the ground after it rains, with the tank falling over. The minimum support should be an 18-inch square of 2 × 4 lumber filled with fence-post concrete and soaked well to set in place. It would be even better if you mixed normal concrete for this, but that isn't strictly necessary.

Because I live in earthquake country, I choose not to bolt the tank down to the pads. If we have a shaking, I want it to be able to slide around rather than be rigidly attached, which might cause the stand to buckle. Some jurisdictions might require you to have a concrete slab somewhat larger than the tank footprint, with a cement block wall around the legs to contain the fuel if it somehow started leaking. This is a reasonable request, since

That way if they are idiots and drive away with the nozzle still in the car, they won't pull over the tank and cause a big spill. The breakaway allows them to drive off with the hose hanging out of their car but shuts off any fuel from the tank. Every co-op has at least one part-time idiot, so install one of these—unless you are the only one using the tank, and unless you think it might be necessary to ensure against your own senior moment.

PRESSURE RELIEF PALLET

VACUUM RELIEF PALLET

COARSE SCREEN

BODY

COARSE SCREEN

WELD TO TANK SHELL

ABOVE: **Fig. 10-61** **Typical conservation vent.** *If pressure increases in the tank, the pallet on the left moves upward to let the pressure out. If vacuum forms in the tank, the pallet on the right lifts to permit air in. For smaller tanks, a conservation vent screws into a two-inch threaded coupling on the top of the tank.*

THIS COMPONENT WAS TESTED TO AND MET THE FOLLOWING CRITERIA

PRES. 3 IN. W.C.
VAC. 8 IN. W.C.
.05 CFH @ +2 IN. W.C.
.21 CFH @ -4 IN. W.C.

ABOVE: **Fig. 10-62** **Modern conservation vent.** *This standard conservation vent screws into the 2-inch standard vent opening on tanks (or in this case to the 1-1/2-inch fitting). It delays opening under light pressure and delays sucking in air under light vacuum.*

your denaturant is likely something that you don't want in the soil.

Sometimes tall ladder tanks are not going to work, especially if you are having to be discreet around your neighbors. Horizontal surface tanks should be supported on fire-resistant material, such as reinforced steel. Frequently, the steel ground support for a fuel tank will be made for forklift blades to lift the tanks as necessary. This ground support, which might be four feet by four feet, should be sitting on a slab of fence-post concrete or real mixed concrete. In either event, put concrete reinforcing wire suspended an inch off the ground in the form before putting in either the dry fence-post mix or wet real mix. The wire prevents the slab from cracking.

This tank will use one of its two-inch fittings for you to install the electric fuel pump. The pump should be wired to its own breaker, along with another wire for a light in the pump area so that people can see at night while pumping and recording their fuel use.

On both tall ladder tanks and horizontal surface tanks, there will be at least two other openings. One will be for your vent. The fuel vapor in the tank will expand and contract, depending on temperature. To avoid a tank collapse, you need to be able to let air into the tank, and to avoid the tank expanding and perhaps sprouting a leak, you need to be able to let out a little air/fuel mix.

I use a conservation vent and suggest that you do, too. A simple open vent will allow too much exchange with the outside air and loss of fuel. The conservation vent resists opening until there is sufficient pressure in the tank that some should be released. But if the tank cools before releasing any pressure, you haven't lost any fuel. Also, the tank resists taking in air until a substantial amount of vacuum forms, from cooling and from fuel contracting from vapor to liquid. This helps prevent air that is full of water vapor from coming into the tank unless necessary, helping prevent water absorption by the fuel. Now, a little water is no problem with alcohol, but it is for owners of gasoline or diesel tanks. A vent like this usually costs about $75.

Another two-inch fitting on the top of your fuel tank will be the filling cap, which is usually sealed by a square **bung wrench** and has a cork gasket. This is where the fuel delivery truck will put the fuel in the tank. You want it to seal well so that the

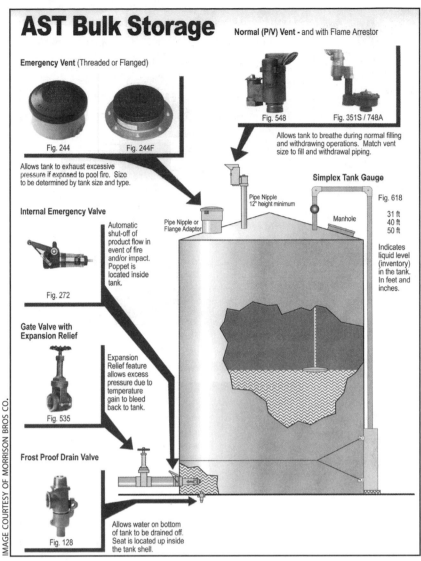

IMAGE COURTESY OF MORRISON BROS CO.

AST Bulk Storage

Emergency Vent (Threaded or Flanged)

Fig. 244   Fig. 244F

Allows tank to exhaust excessive pressure if exposed to pool fire. Size to be determined by tank size and type.

Internal Emergency Valve

Automatic shut-off of product flow in event of fire and/or impact. Poppet is located inside tank.

Fig. 272

Gate Valve with Expansion Relief

Expansion Relief feature allows excess pressure due to temperature gain to bleed back to tank.

Fig. 535

Frost Proof Drain Valve

Allows water on bottom of tank to be drained off. Seat is located up inside the tank shell.

Fig. 128

Normal (P/V) Vent - and with Flame Arrestor

Fig. 548   Fig. 351S / 748A

Allows tank to breathe during normal filling and withdrawing operations. Match vent size to fill and withdrawal piping.

Simplex Tank Gauge

Pipe Nipple 12" height minimum

Pipe Nipple or Flange Adaptor

Manhole

Fig. 618

31 ft
40 ft
50 ft

Indicates liquid level (inventory) in the tank. In feet and inches.

conservation vent works correctly. After a while, you might want to replace the cork gasket with a thin neoprene gasket.

Except in ladder tanks, all fittings should enter from the top. Bottom fittings run the risk of leaking and can be a source of tank failure in a fire.

Don't use gravity to deliver fuel from the tank to a car, unless you're using an automatic shutoff spout as mentioned above.

Always remember to ground the tank and/or filling nozzle to avoid sparks from a buildup of static electricity. Official fuel-filling hose has the ground built in so the filling nozzle is grounded to the tank. On plastic tanks, *you must ground the metal fill pipe.* On a steel tank, the whole thing is on ground, but on plastic tanks, use a fill pipe that goes to within an inch of the bottom of the tank, and has a ground wire attached to it, attached to a known ground, or a ground stake pounded into the ground. It takes only one spark under the right conditions for an early Fourth of July.

Fig. 10-63 Bulk storage of fuel.
For large alcohol fuel tanks, the safety equipment is not too much different than what's used on a small tank.

Grounding also allows you to make use of plastic tote containers, which have a footprint of four feet by four feet (pallet-sized), but are between four to six feet tall, encased in a metal frame. These tanks can be had for free or less than $100 used and can be converted into fuel tanks that hold over 200 gallons of fuel. They are a great alternative to 55-gallon drums. Like any plastic tank, they should have a grounded fill pipe.

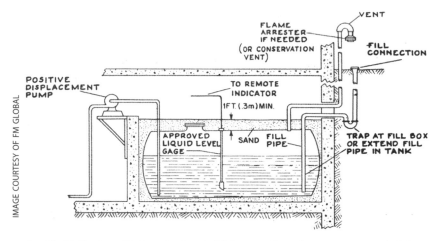

Fig. 10-64 Enclosed indoor fuel tank. This sort of configuration is used if the fuel tank is to be in an enclosed space.

Fig. 10-65 Thousand-gallon ladder tank. This tank is filled by a pump truck delivering alcohol. Fuel flows by way of gravity into your vehicle. A tank like this could serve ten people in a mini community-supported-energy distribution node.

You could build a stand out of 4 × 4s and 2 × 4s to make it higher than your car's filling cap, so as to have a gravity tank, or use it on the ground with an electric fuel pump. Your filling neck can be grounded to the grounding wire of the electric pump for convenience.

One very nice way of storing approximately 275 gallons of alcohol is by using a converted heating fuel oil tank. If you live where heating is done by fuel oil, these tanks often come up used for almost nothing when people install larger tanks. In fact, fuel oil companies will often let you take the tank away if you make arrangements in advance. Fuel oil tanks typically have three two-inch openings on the top. One of the openings is used for filling the tank, the second is used as a vent, and the third can have a gas pump installed in it. There is a bottom fitting where the line that goes to the fuel oil burner gets its fuel.

For decades, diesel vehicle owners have been using their heating fuel oil tanks this way (see Chapter 29, Figure 29-9). That is because heating oil and diesel are virtually identical in terms of fuel quality. So pumping heating fuel into cars avoids paying road taxes normally applied to diesel vehicle fuel.

Obtaining a permit to install a fuel oil tank is easy and acceptable to fire departments. You do the installation of your fuel pump later. You can use the fuel line fitting at the bottom of the tank to run from the tank to your cogenerator, if you want to power your house on alcohol. This ought to be legal almost anywhere.

It's always best to keep the tank out of direct sunlight. A small lean-to or peaked roof made of a noncombustible material is fine protection. Above-ground tanks should have a metal or concrete dike around them to contain all fuel, should the tank rupture. The dike should be low; build it a little wider on the three sides away from the filling side.

Alcohol should never be stored in a garage, where fumes might connect with a water heater, pilot light, or other ignition source and cause a fire. Any indoor tank arrangement must have exceptionally good ventilation for exhaust fumes from spills and so forth. Indoor tanks in buildings other than garages may be encased in a sand-filled concrete bin.

Although you can run into trouble with above-ground storage of alcohol in more densely populated areas, simply point out to authorities that propane or fuel oil can be regularly delivered to people's homes (see Chapter 26).

If you're storing low-proof alcohol, use plastic drums. *Do not* try to use a drum with gravity feed. They don't have tapered threads and will certainly leak.

If storing drums above-ground outside is a little too public, you can store them in a safe but stealthy manner, in a lined, partially underground shed (see Figure 10-67). The shed should have a six-inch depth of gravel over fiber-reinforced cement floor and walls. In areas far from inspectors, an EPDM liner can be used instead of concrete to protect in case of a drum leak. The shed should be fitted with screened windows for good ventilation, a metal roof, and stovepipe-type vent to extract any vapors not blown out through the windows. Install an approved mechanical (arm-powered) or electric fuel pump tight in the two-inch opening on the drum—these pumps have built in **overpressure** and vacuum breakers. Since alcohol fires can easily be put out with water, a fire sprinkler over the top of the area isn't a bad idea, either.

Alcohol dissolves normal pipe joint compounds. Use RectorSeal #7, or a similar alcohol-proof compound for pipe connections. RectorSeal does not guarantee it works with alcohol, but it has served me well. For sealing threaded connections, Teflon tape and Teflon pipe dope work well, and seem to hold up pretty well.

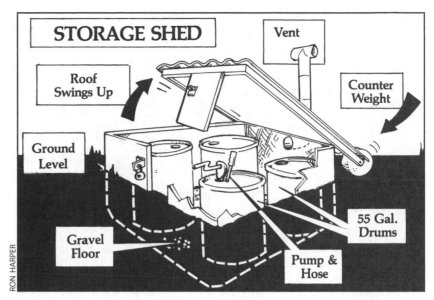

Fig. 10-67
Ventilated fuel storage shed.
This partially underground shed is unobtrusive and can store two to four months' worth of fuel.

Normal gasoline hoses do not seem to have any trouble with alcohol fuel, but some areas choose to enforce methanol rules to ethanol installations. In this case, you are required to make sure your gas-dispensing nozzle is not aluminum but nickel-plated steel, and your hose must be Teflon-coated with stainless steel end fittings. It's ridiculous to force people to meet methanol standards, but sometimes you have to pick your fights with the powers that be; this is one to give in on if it will keep the peace. Try to think of it as chrome-plated exhaust pipes or chromed valve covers—something expensive, unnecessary, but sort of cool.

You should filter your fuel to one micron, so that when you put it in your car, you are not going to

Fig. 10-66 Manual fuel pump. *Designed to be threaded into a drum top, this diaphragm pump will deliver 20 gallons per minute. It has a built-in pressure relief valve/vacuum breaker.*

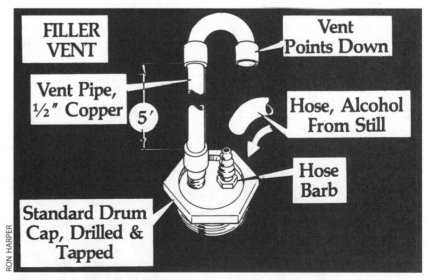

Fig. 10-68 Safety cap for filling 55-gallon drums. *If you are draining alcohol from your condenser into a drum, use this device to let air out of the drum, which may have alcohol vapor, well above where you might have a static electricity spark or any other source of ignition.*

Fig. 10-69 Fire sprinkler assembly detail. *Fire sprinkler heads such as this one are available from any good-sized plumbing supply house. They are simply threaded into ¾-inch pipe fittings. The rosemetal strip (the piece with the temperatures stamped on it) softens and melts at a low temperature. The strip is under compression and holds a plug in the end of the water line. When the metal reaches 165°F, it softens sufficiently for the water pressure to force open the plug. Water striking the blades of the "fan" flings the water in a wide radius.*

plug up the one-micron filter that your automobile uses to protect fuel injectors. You will find that special alcohol-tolerant spin-on filters are usually required. This is because normal fuel filters today are designed to absorb any water in the fuel, and alcohol looks like water to these normal filters. They typically do not spin onto the same hardware as diesel/gasoline filters, so you'll have to get the nickel-plated filter head, too.

Alcohol fires can be extinguished with ABC-type fire extinguishers or by dilution with water. Once alcohol gets down to 100 proof, it has a very

difficult time burning, but it must be diluted to about 30 proof before you can be confident that no hot object will reignite it. Remember that until it's denatured with gasoline or diesel, an alcohol fire is difficult if not impossible to see, particularly on a sunny day. So use lots of water when putting out an alcohol fire.

Normal firefighting foams are ineffective against alcohol. Foams designed for **polar solvents** should be used. The small amount of alcohol in **gasohol** is enough to either delay or make ineffective some firefighting foams. Post a sign in your still and storage tank areas that warns: Alcohol Fuel—In Case of Fire Use Water or Polar Foams.

I hope that you have found the suggestions about setting up your plant useful. I wish I'd had the benefit of others' experience when I got started, so I wouldn't have had to reinvent the wheel so often. Even so, don't let my experience blind you to even better ways of doing things based on the particulars of your site or equipment. I am always interested in hearing about clever solutions and designs. Feel free to write them up and send them, with or without pictures, to me at <info@permaculture.com>, and I'll make the best ideas available on the website.

So far, you've mostly thought of yourself as a moonshiner. But when you start to apply permaculture design to your alcohol operation, you will find that you have a lot of potential surpluses that need to be designed into productive uses and products. In fact, you'll quickly find that the various co-products, and their marketing, will have a major influence on your plant design, and certainly on your profit if you are making alcohol as a business. With good design of all system surplus, you will probably find that the alcohol is the least profitable part of your well-designed plant. Let's look at a sampling of potential co-products and what to do with them.

Endnotes

1. "Ethanol Car Beats Fuel Cells to Win European Eco-Marathon," *Environment News Service*, May 22, 2006 (www.ens-newswire.com/ens/may2006/2006-05-22-03.asp).

2. *Air Pollution*, Fahan School Library, www.fahan.tas.edu.au/libraries/senior/envexpedition/air_pollution.htm (July 14, 2005).

PHOTO COURTESY OF MARK HOFFMAN

Fig. 10-70 Home fuel tank. *Tanks like this in suburban homes may become a standard sight as more people start to make and use alcohol.*

BOOK 3

CO-PRODUCTS FROM MAKING ALCOHOL

By now, you know that I try to never waste anything. I always design any surplus in a system to produce another valuable product. In this section, we develop that idea more fully.

There are literally dozens, perhaps hundreds, of potential co-products from an alcohol plant. I've focused on a relatively small number of them to show how they can complement one another in the design of your own facility.

The most important thing to think about in this section is that the uses for surplus material and energy result in co-products, not byproducts. Terming something a "byproduct" means that you have made a judgment that it is of secondary importance to a "primary" product. Most manufacturers take a raw material, make it into a widget to sell, and throw away the remains (into the nearest river, if they can get away with it). That thrown-away part, that surplus, needs to have a designed use in your plant. Otherwise, it will create waste, pollution, and more work.

One of my engineer friends has told me, "Dave, you need to decide the one thing you're good at and do that. It doesn't work to try to be good at a variety of things because it means all of them would suffer." That's not true, in my experience. It has become vital to consider all the connections between parts of your operation to make the best use of everything. As the plant designer, you need to focus on the relationships between the components. That's what an ecologist does, and it's what successful businesspeople do.

So you will find that the co-products, when properly valued and designed into a productive use, will almost always be worth more money than the alcohol. What's going to be more important for some time to come is that the co-products will not suffer the volatility in price that auto fuel does. Having a well-developed co-product business protects your overall operation against failure.

But the best reason I can give for having a diverse operation is that it is a lot more fun to operate than simply making alcohol and throwing away the remains.

CHAPTER 11

ALCOHOL FUEL IS ONLY THE BEGINNING: TURNING WASTE INTO PROFIT

How many times have you seen some process, shrugged your shoulders, sighed, and muttered, "What a waste!"? It increasingly does not make sense to try to produce a product without looking at how you can use that waste to turn a profit. That's ecology and economy working together—and it's about time. Today's energy shortages and high prices are accomplishing something that ecological protesting could not—the elimination of waste.

In a good permaculture design, the byproducts generated by the production of the primary product are almost always far more valuable than the product itself. After all, a "waste product" of a process is simply a surplus that does not have a system in place to use it.

What follows is a discussion of just a few of the possible products that can be designed into an alcohol/agricultural (fuel/food) system to make profitable use of surpluses. There isn't a fixed recipe because there's no real reason why any two integrated small-scale alcohol operations should be identical, considering the different needs in your community, the potential crops that grow in your area, and the markets for co-products.

DISTILLER'S FEEDS

Distiller's feeds are the de-alcoholized residues that remain after distillation. In the grain fermentation process, all proteins, fats, vitamins, and minerals in the initial feedstock are concentrated threefold due to the removal of the carbohydrates (often starch). Yeast used in fermentation, which multiplies many times over while consuming the carbohydrate, contributes to the high nutritional quality of distiller's feeds. In corn, the process of making alcohol concentrates the protein from 9% to 30%, and fat from 3.5% to about 11%. In addition to the feed benefits, this concentration reduces the space needed to store the now-compact animal feed and the energy needed to ship it by two-thirds.[1]

IT INCREASINGLY DOES NOT MAKE SENSE TO TRY TO PRODUCE A PRODUCT WITHOUT LOOKING AT HOW YOU CAN USE THAT WASTE TO TURN A PROFIT. IN A GOOD PERMACULTURE DESIGN, THE BYPRODUCTS GENERATED BY THE PRODUCTION OF THE PRIMARY PRODUCT ARE ALMOST ALWAYS FAR MORE VALUABLE THAN THE PRODUCT ITSELF.

SINGER (STREDWICK)

COPYRIGHT 1992–2002 ANDREW B. SINGER

Fig. 11-1

The discussion in this chapter is based primarily on grain byproducts, since most research has been done in this area. But the figures and information should be considered as guidelines for the byproducts of other feedstocks, many of which also have a useful animal feed or human food value. As I've noted in Chapter 8, many other crops, from cacti to Jerusalem artichokes, and their byproducts are used as animal feeds, and virtually all are improved by either fermentation or some of the processes described below.

First, some general information and a few definitions; then we'll look at specific animals and how distiller's feeds relate to them. I've condensed a huge amount of information here, and recommend you conduct your own research before committing to radical changes in your livestock-feeding program. Studies on distillery byproducts for animal feed go back more than 100 years—and are continually being updated.

"The material left after the alcohol is extracted is very important to the farmer. It makes excellent cattle food…. Alcohol is nothing but carbon, hydrogen, and oxygen. When these are extracted there still remains all the nitrogenous [elements], phosphates, and other elements. Fed to cows, it enriches the dairy product; later it becomes the best possible fertilizer to go back into the soil to increase and enrich next year's crops Such farming is a perfect cycle of usefulness."

—HENRY FORD, *DETROIT NEWS*, DECEMBER 13, 1916

General Information on Distiller's Feeds

In addition to their known nutrients, distiller's feeds possess **cellulose digestion factors**, unknown growth stimulants, and **urea proteins**. These collectively are often called **grain fermentation factors (GFF)**, especially in older studies. They substantially stimulate growth and proliferation of rumen **microflora** (powerful bacteria which aid digestion in ruminants). The gains in growth exceed the apparent feed value just based on nutrient analysis. Distiller's feeds also aid in generally breaking down cellulose by ruminant animals, presumably by allowing them to use nitrogen much better and derive more energy from other feed ingredients.

There are four main types of dried distiller's feeds from brewing residues: **distiller's dried grains (DDG)**, **distiller's solubles (DS)**, distiller's dried grains with solubles (DDGS), and **condensed distiller's solubles (CDS)**. Each is preceded by the name of the feedstock that it came from.

Distiller's dried grains (DDG) are the solid, granular portion left over from the mash (equaling about 59% of it) after distillation. They're usually screened or centrifuged out of the liquid portion of the mash (stillage), and then dried.

What's left in the liquid after the DDG have been removed typically contains about 10% solids. These solids are too fine to be separated. In addition to the fine solids that are suspended in the stillage, a substantial amount of nutrients are actually dissolved and in solution. Together they form distiller's solubles (DS), or **thin stillage**. Although I'll talk about some uses for DS, they are usually too bulky and perishable to be easily used, due to the large amount of water they are dissolved in.

Large plants use a fair amount of energy to evaporate the water until the stillage becomes a thick syrup. This syrup is called condensed distiller's solubles. Drying syrupy CDS further until it is essentially solid produces **distiller's dried solubles (DDS)**, something the small distiller will probably never see.

Once the water has been largely removed, the CDS can be used as a separate feed or combined with DDG, which is what most beverage industries do. The resulting product is called distiller's dried grains with solubles (DDGS), and this is typically what large alcohol plants use to make an animal feed that can be stored, and to eliminate what would be a waste stream. No water leaves the plant other than by being evaporated from the stillage, so there is no dumping of liquid into a sewage system.

Although the dry matter (suspended solids) of DS is only 10% of the liquid stillage, it comprises about 41% of the total residual corn, while the granular solids of DDG constitute about 59% of the corn residue after distillation. So it's easy to not realize that there is a huge resource in close to half of the residual grain—in liquid form.

To evaporate and concentrate all that liquid is energy-intensive. Some large plants are now changing their processes to avoid the evaporation step. The most practical approach is to design a small-scale operation to avoid having to concentrate it, and instead use it as **wet distiller's grains (WDG)** and DS.

DDG are the primary storable animal feed byproduct for the small distiller who doesn't have the technology available to condense the solubles. To reduce his energy consumption, the small distiller separates his WDG by using fine screens or an auger press, and lets his grain dry in the sun, or uses waste heat from a flue or from some other part of the distillation process. Once dry, the WDG becomes DDG. Later on, we will discuss multiple uses for the solubles in the integrated plant.

DDG (or DDGS) can be stored for a long time if kept cool and dry. For long-term storage, moisture content after drying should be 8 to 12% water, but drying to this level isn't easy. If you do dry it to this level, it must be kept airtight, since it will readily absorb additional H_2O from the air. There are a variety of commercial preservatives available, which claim to be cattle-safe and are relatively cheap, and which store DDG (or DDGS) for up to two years.

If it's not dried to at least 65% moisture, DDG will freeze in very cold temperatures, which makes it difficult to feed. This is dry enough that, when squeezed in your hand, no water will drip from it.

The grains can be stored for several months at 50 to 60% water if saturated with carbon dioxide. Approximately 50 to 60% moisture content is attained with an auger press, or by screening and raking the grain across darkened concrete or much less expensive compacted clay, with exposure to hot sun for one or two days. Drying tunnels, essentially moveable, low, clear plastic "greenhouses" with a fan to exhaust the water-saturated air, are very practical for the small-scale plant.

WDG are essentially DDG without drying and without solubles. The best results in terms of weight gain in livestock are attained using WDG. These are what small-scale producers will primarily produce, since drying is energy-intensive and unnecessary except when you might have some reason to store the distiller's grains.

For the small-scale operation, it's easiest to use the feed wet (WDG) within a couple of days of distillation. The second-best alternative is to sell to or barter with someone nearby with an immediate use for the WDG.

Storage of WDG without preservatives, drying, or ensiling is not possible for more than three to ten consecutive days (depending on how warm it is) because of bacterial invasion. WDG are 80% moisture and can be ensiled while wet in airtight silos. The process of making silage preserves the feed by fermenting it without air until lactic acid

bacteria dominate and drop the pH below what other bacteria can tolerate.

Some of my ex-students report storing WDG for several weeks by keeping the feed saturated with surplus CO_2 from their fermenters. Flooding the storage container with carbon dioxide inhibits most bacteria, and would kill any insects or rodents looking for a free meal.

Using waste heat to dry the grain should work well, but a separate drying operation is expensive, unless your energy is cheap biomass or solar.

Feeding Your Livestock

Ruminants, such as cattle, have a rumen (first stomach) as part of their complex digestive

Typical Nutrient Composition of Co-Products from Corn Dry-Milling Fermentation

NUTRIENT[A]	UNIT	DDGS	DDG	CDS[B,C]
Dry Matter	%	90	90	100
Protein	%	25	27	28.5
Fat	%	8	7.6	9
Fiber	%	9.1	13	4
NDF[B]	%	44	44	23
ADF[B]	%	18	18	7
Ash	%	5	3	7.4
Lysine	%	0.7	0.5	0.9
Methionine	%	0.6	0.2	1.5
Cystine	%	0.3	0.2	0.4
Arginine	%	1.05	0.6	1
Threonine	%	0.93	0.82	0.98
Valine	%	1.63	1.2	1.6
Isoleucine	%	1.52	0.93	1.2
Tryptophan	%	0.2	0.2	0.3
Linoleic Acid	%	3.9	3.6	4.4
By-Pass Protein[B]	%	38	38	N/A
TDN[B]	%	85	77	74
NE-L[B]	mcal/lb.	0.92	0.99	0.89
NE-G[B]	mcal/lb.	0.68	0.67	0.6
NE-M[B]	ncal/lb.	0.99	0.89	0.85
ME Swine	kcal/lb.	1540	1005	1540
ME Poultry	kcal/lb.	1250	940	1250
DE Horses	kcal/lb.	1586	N/A	N/A
Starch[B]	%	3	N/A	N/A
NSC[B]	%	15	N/A	N/A
Phosphorus	%	0.71	0.37	1.3
Potassium	%	0.44	0.16	1.75
Calcium	%	0.15	0.1	0.3
Magnesium	%	0.18	0.07	0.65
Sodium	%	0.57	0.9	0.3
Sulfur	%	0.33	0.43	0.37
Iron	ppm	223	200	600
Copper	ppm	58	44	83
Zinc	ppm	89	N/A	N/A
Manganese	ppm	25	N/A	N/A
Cobalt	ppm	0.18	N/A	N/A

DATA PROVIDED BY JERRY C. WEIGEL, ARCHER DANIELS MIDLAND, DECATUR, ILLINOIS

Fig. 11-2 Corn distiller's feed nutrient composition. *A) Moisture basis; B) DM (dry matter) basis; C) Dry matter varies, so check your actual percentage and adjust accordingly.*

Fig. 11-3 Wheat and wheat DDG. When fermented to alcohol, wheat yields an excellent concentrated protein source for humans or animals. The two piles show the proportion of final DDG to original wheat.

systems, enabling them to digest difficult materials, such as woody cellulose. The rumen is like a biological chemical reduction factory, providing agents to break apart those compounds (like cellulose) that would pass through undigested in most animals.

The stomachs and intestines of ruminants break down cellulose to more easily digestible carbohydrates. The population and biological activity of the bacteria involved can vary greatly, and some feeds stimulate and encourage them, while others inhibit them. The positive stimulation by distiller's grains, in part because of increased **proline** (an amino acid) in the diet, has stumped chemists for over 30 years.

The protein in WDG, DDG, and DDGS is called a protected protein, which means that it escapes microbial digestion in the rumen. The efficiency of a protein's digestion is directly related to the protein's degree of protection. The greater the amount of protein that escapes the rumen, the more there is for the animal to break down in the digestive tract into amino acids it needs for growth.

A great deal (about 70%) of the protein in soybean meal (SBM) is digested in rumen, and that decomposed protein isn't of as much value to an animal. In the case of DDG, only 30% of the protein is degraded in the rumen, and the rest is digested and absorbed by the animal. So on an absorbed-protein basis, DDG is more than twice as valuable as SBM.[2]

Unfortunately, DDG and DDGS are underpriced or even severely discounted because they're often directly tied to SBM, based on crude protein equivalence. David Pimentel at Cornell incorrectly has pegged the value of DDGS to SBM. But the *price* of protein in the two feeds doesn't take into account the protein's *digestibility*; pound for pound, you get more weight gain from the DDGS than from the SBM. Of course, in your own livestock operation, you'll get the full value of the feed in your animals' superior weight gain.

Distiller's feeds cannot be fed straight to livestock. The protein and fiber levels are concentrated enough to kill if they're consumed alone. But dramatic results are achieved by feeding a little less DDG than the amount of the original grain it came from, mixed with less-expensive cellulosic feeds.

Cattle

Beef Cattle

Beef farmers have used DDGS for over 140 years, since they began feeding **whole stillage** to their stock. In the late 1930s and 1940s, research began in earnest to determine the optimal use of DDGS. The results were dramatic. Given hay for roughage along with corn DDS (30.4% protein), animals gained 2.2 pounds a day. Fed soybean meal instead of DDS, the animals gained 1.99 pounds per day.

Modern research indicates that DDGS can replace soybean meal completely, and one study reported "keener appetites, and despite DDS's lower protein content, a slightly higher pound-for-pound replacement value than soybean meal."[3] Of course, DDG and DDGS are both superior to simple DDS. Remember, DDS are just evaporated thin stillage.

Greater weight gain using distiller's feeds has to do with the increase in ability to digest cellulose. The digestive system of cattle and other ruminants is a diverse biological fermentation system. DDGS increases **cellulytic** digestion 26%, DDG increases it 11%, and DDS increases it 13%. So there is something contained in DDGS that favors the development of cellulase-producing bacteria in the rumen. It's not just protein, either, since soybean meal, which is higher in protein, increases the cellulytic activity only 4% when used in place of DDGS on a protein-equivalence basis.[4]

DDGS increases use of urea (a common form of nitrogen in animal feed) in liquid diets as well. An addition of 2.5% DDGS to Purdue's Liquid 64 formula has resulted in weight gains from 2.48 to 2.85 pounds per day, while reducing overall feed rations from 700 pounds of feed per ten pounds weight gain, to only 650 pounds. The percentage of retained and absorbed nitrogen also increases: dietary nitrogen up 63–78%, absorbed nitrogen 72–83%.[5]

Although many older DDGS studies show a 15% faster weight gain with 7 to 10% higher total weight gain, other more recent studies show up to 30% faster weight gain and 14% greater total weight gain, depending on the base diet that the DDGS formula is compared to.

Wet distiller's grains, which contain only 30 to 40% dry matter (as opposed to DDGS at 98%), have been shown to provide the equivalent of 124 to 152% of *dry* corn's weight gain when fed for 17 to 40% of beef cattle's diet. WDG not only were able to replace the protein found in SBM, but also were a primary source of energy. The distiller's grains also provided energy from fiber, fat, and protein.[6] Results strongly suggest that the most dramatic increase in meat production is attainable feeding WDG instead of DDGS.

Feeding up to 30% of the total ration as DDGS will enhance weight gain.[7] But higher levels are still practical economically if other sources of feed become too expensive or unavailable, such as when China bought up just about all U.S. soybeans in 2004.

Dairy Cattle

The wide variety of feed recipes for dairy animals makes it difficult to say how much better DDGS is for them than normal rations, but they seem to derive the most benefit of all livestock. Their milk production can be expected to rise up to 16%, while the milk's fat content may rise 8 to 18%, compared to feeding whole corn. Higher milk fat content is very important, since farmers receive a much higher price for each point of milk fat above average. Dairy cows find DDGS very palatable, so much so that they often balk if DDGS-fortified rations are discontinued.

Calf Starters

It's economically advantageous to use DDG, DDS, and DDGS in calf starter formulas, partially or totally replacing many other expensive ingredients. Twenty percent DDS could replace dried skimmed milk, dried brewer's yeast, and part of the cracked corn and crushed oats in dry starter. DDGS can completely replace linseed meal, brewer's yeast, and dried whey in dairy cattle starters.

Studies have found that feeding 40% DDGS is comparable to a standardized urea-supplemented whole grain diet. Interestingly enough, the weight gain is even better with 40% wet distiller's grains. One study showed that drying, which is energy-intensive, did not improve the feed quality. Some alcohol plants are beginning to take advantage of this, setting up feedlots so that cattle can use the WDG right after distillation.[8]

Sheep

Linseed meal, one of the more expensive items in a sheep's diet, can be replaced with DDG quite

---

CHAPERONING MAD CATTLE

The terrifying malady that goes by the common name of mad cow disease, which is invariably fatal, seems like it eats your brain. More accurately, it causes holes to form until your brain resembles a sponge. It's an awful way to go. But it's even more terrifying when you realize that the disease is caused by something that isn't even alive. It's not bacteria, or even a virus, but goes by the name of **prion**.

Prions are really nothing more than chaperones. When DNA wants your body to respond to genetic instruction, it produces RNA as a messenger, and then the RNA causes the tissue involved to produce proteins that direct cells to do whatever the DNA instructs. Or so we thought. In reality, chaperones take the protein and twist it into the proper shape to do the job at hand. A single protein made by RNA can be made to be the key to many different metabolic functions, depending on the action of the chaperone.

Mad cow disease results because this chaperone starts making the protein in nervous system tissue into slightly different shapes than usual. The shape changes the normal function of the brain tissue. Instead of doing what it's supposed to do, the nervous system cell is hijacked into making many copies of the chaperone itself and sending them out; the cell is eventually consumed in the process of replicating the prion. And, eventually, enough brain protein is converted into prions to make the brain fail.

But the chaperone isn't technically alive, even though it is "reproducing." It's only an assembly of amino acids that tells other proteins what to do, and it is virtually indestructible. Radiation doesn't affect it. High temperatures don't affect it. Incinerated cows, reduced to dust, are being stored in thousands of drums in warehouses because breathing the dust would infect you; if the cows were buried instead, you would get infected from drinking the prion-contaminated water.

But if there is one thing mad cow has taught us, it's that our whole paradigm that DNA is the master control of a rigid hierarchy is completely and fatally false. Mad cow disease violates the central dogma of genetic engineering that claims that our genetic structure resembles a corporation where the CEO (DNA) sends instructions via middle management (RNA) to be carried out by lower management (hormones, enzymes, etc.) to produce the specific proteins. At no time is any level of this hierarchy supposed to be able to reproduce itself or affect the level above it. The existence of mad cow disease has disrupted this theory entirely.

---

satisfactorily. Sheep may make better gains on linseed meal, but overall costs are reduced substantially with DDG. Sheep apparently digest the protein in DDG at about the same efficiency as that of soybean meal, but absorption of crude fiber (cellulose) and gross energy of the ration is much higher with DDG and DDGS. Once again, this is due to stimulation of cellulose digestion.

Revenuers had a sure-fire way of spotting moonshiners in rural areas. They hung out at county fairs and looked for the farmer with the fattest hogs. Almost without exception, the biggest livestock were fed on rich mash from a moonshine still.

Pigs

There are some questions as to the validity of past scientific research on DDGS and swine. In the 1980s, all the studies indicated that the optimal ration was about 10% DDG, but several students, **revenuers**, and old moonshiners I've talked to thought that a 20%, or even a 25–30% mix works fine. Others have told me that they have fed WDG exclusively to their pigs and had excellent results.

Revenuers had a sure-fire way of spotting moonshiners in rural areas. The tax agents would hang out at county fairs, looking for the farmers with the fattest hogs. Almost without exception, the biggest livestock were fed on rich mash from moonshine stills.

Now the college boys have finally come around to the point of view of the moonshiners. Modern data show that nursery pigs tolerate up to 25% of their diet as DDGS. When raising grow-finish pigs, up to 20% DDGS has been successfully tested (although higher levels might make the meat too lean); gestating sows have been successfully tested at up to 50%, lactating sows up to 20%, and boars 50%. On this sort of diet, the piglets are larger at weaning, and the sows often have larger litters.[9]

As swine are not ruminants, they have a more difficult time with high amounts of fiber and excessive protein, so it's good to reduce DDGS's moisture content as much as possible for manure (diarrhea) control.

Since most of the calories are in the oil, DDGS are far more efficient than soybean meal, which they completely replace. Soybean meal must have its oil removed before it is fed to pigs.

DDGS provide a high amount of **nonphytate phosphorus**, a material contained in plants, which swine can only partially digest. When fed on DDGS,

Fig. 11-4

Poultry

DDGS and DDS have been used beneficially as feed for poultry since the 1930s. One of the more expensive ingredients in starter feed is dry skim milk—which DDGS and DDS can replace entirely. Without the milk, and using DDGS or DDS instead, chicks have improved feathering and fleshing, and egg color improves slightly.

Broilers and Chick Starters

Corn/soybean chick feeds are often supplemented with meat and bone scraps. Since the advent of mad cow disease, this is just asking for trouble. Chicks fed 10% DDGS mixed with alfalfa meal and minerals grow as well as, or up to 12.2% better than, those fed meat scraps. With DDS supplementation, broilers have recorded 14.4% increases.

Some older research indicates that 15% is the optimum DDGS mix for corn/soybean feeds; 20% is recommended in corn/soybean/fish diets. More recent studies showed good results at 18% DDGS for broilers.[10] A few studies have shown levels up to 25% suitable for meat birds.[11]

Layers, Breeders, and Turkeys

DDGS and DS have very good effects on hatching and production of layers and breeders. Twenty percent of their diet can be either; if no solubles are used, 15% DDGS is the optimum ration.[12] In diets that include meat scraps, DDGS can replace all the soybean meal and part of the ground barley and wheat bran without any effect on egg production or hatching. Maximum egg production has been attained without any animal protein, using DDS and soybean meal. DDGS and DDS can provide roughly one-third of the protein for layers and breeders.

As for turkeys, when they are fed 10% DDGS with an all-vegetable diet, their growth increase is 17 to 25%. It is thought that DDGS increases poultry's ability to assimilate important nutrients such as phosphorous and methionine.

In many parts of the Third World, a farm without fish production is not a farm.

Fish

In many parts of the Third World, a farm without fish production is not a farm. In the United States, aquaculture is a growing interest for homesteads and ranches. Trout and catfish are among the fish raised commercially in the U.S. Both respond well to DDGS or DDS in their feed. DDGS are preferred over DDS due to the difficulty in handling the pasty-textured DDS in water. Trout diets have been very successfully tested up to about 21% DDGS; catfish diets have been tested up to 15%, and an even higher level may be possible.

A new, very profitable fish crop is **tilapia**. Although it's a tropical fish, it can be raised in temperate regions in greenhouse-covered tanks. Very high levels of DDGS seem to work well with this fish. In fact, Archer Daniels Midland's Decatur, Illinois, plant has a five-acre greenhouse largely devoted to raising tilapia using a mixture of corn and soybean byproducts. Up to 75% DDGS has been used in fish feed.

FERTILIZER, COMPOST, AND MORE FROM STILLAGE

In the process of fermenting carbohydrates, we really remove from a plant only those components which are renewed each year. Everything the soil contributed to the plant is still there in the spent mash. Using stillage as a fertilizer, you can recycle the same soil nutrients into the earth over and over again, removing only solar energy collected by the plants you grew.

DOGS AND DISTILLER'S FEEDS

There hasn't been a whole lot of research done on dogs and distiller's feeds, but the little that has been done is encouraging. A Cornell University study reports that a 7% ration of DDGS improved the reproductive rate of beagles.[13] Diets of up to 39% have been attempted. When using DDS alone, 30% is about the limit before diarrhea occurs.

Several of my students have fed their dogs wet mash. The animals ate it up, quite literally, and with special gusto if the grain was separated before distillation. However, residual alcohol in the feed tends to affect a dog's equilibrium for short periods of time.

Stillage is concentrated stuff with an extremely high biological oxygen demand. If you put too much of it in one area, you can "tie up" nitrogen and oxygen in the soil and "burn" the soil temporarily. If this happens, your soil will be out of production for up to a year, until nature can correct the imbalance.

So, how much stillage should you use as a fertilizer? You can have a chemical analysis done and determine the amount of various nutrients of the solid, liquid, or combined components of the stillage. You would then apply the amount necessary to grow your next crop based on these figures. From a common-sense viewpoint, you wouldn't want to put any more stillage in one spot than a plant growing there would have left if it had died and fallen in the field. In an orchard, for instance, don't put more fruit stillage under a tree than the equivalent amount of fruit that would have dropped there naturally.

In general, thin stillage or spent mash is too acid to be applied as is to your soil. Bring the pH up with lime to the same pH as the soil that it's going to be added to.

In Brazil, the spent sugarcane mash is pretty much all-liquid, and it is generally pumped through irrigation pipes out to sprinklers that apply the fertilizer back to the field. Care is taken not to overfertilize any one area.

In the U.S., large alcohol plants concentrate and dry the thin stillage to a syrup and then mix it with the DDG to make DDGS. However, small plants won't find that practical, so using thin stillage as fertilizer is one way to go.

Although using the liquid as a fertilizer is a sound strategy, we can add value to the thin stillage and improve our soil without risking overfertilization of any one spot. A fail-safe way to return nutrients to your soil is to use stillage as part of compost. In making compost, we start with some sort of organic matter, such as sugarcane bagasse or corn stalks (corn stover). If heaps of this lignocellulosic plant matter are watered, or better yet irrigated, with stillage, all sorts of fungi

Fig. 11-5 Composition of solids and solubles in grain mash.[14]

Mash Constituents

GRAINS FRACTION	AVERAGE %
Dry Matter	34.3
Crude Protein	33.8
Crude Fat	7.7
Crude Fiber	9.1
Ash	3.0
Calcium	0.04
Phosphorus	0.56
SOLUBLES FRACTION	
Dry Matter	27.7
Crude Protein	19.5
Crude Fat	17.4
Crude Fiber	1.4
Ash	8.4
Calcium	0.09
Phosphorus	1.3

BELOW: Fig. 11-6 Stillage transport. *In India, stillage is considered a high-quality fertilizer and is worth trucking to fields that are not adjacent to the alcohol fuel plant. Here, stillage is being mixed with the irrigation water in the ditch.*

ABOVE: Fig. 11-7 Stillage field-spreading. Stillage is being directly sprinkled onto a field in India prior to planting.

and bacteria will rapidly reproduce, decomposing the organic matter. The nutrients in the stillage will initiate rapid reproduction and growth of bacteria and fungi that are necessary for good, fast decomposition.

All this rapidly reproducing biology throws off a lot of heat. The interior of the compost pile will go through a cycle where it will reach 160–190°F, destroying insect larvae, pathogenic organisms, and any pesticides that may be mixed with the stillage or cellulosic material in the pile. Periodically, the compost must be turned or fluffed up to keep enough oxygen available for all the microscopic living creatures that need to breathe in order to eat, reproduce, and poop.

Now, you might be thinking, "Why do all this work if I can just send the stillage out into the field as liquid fertilizer?" The reason is that compost should not be thought of as fertilizer. That would be underestimating its value considerably. Adding fertilizer brings soluble nutrients to feed plants directly. Compost is much more valuable, since it inoculates the soil with uncountable beneficial organisms that provide ongoing fertility by living in the soil.

The topsoil of the Earth is mostly composed of living things, not minerals. Just as with larger visible animals, there is a food chain in the soil; as we saw in Chapter 3, it begins with fungi eating the organic matter, then mites and other microbes eating the fungi, predatory mites and other creatures eating the mites, and finally, springtails, nematodes and visible creatures like earthworms eating, too. Compost piles are a concentrated microcosm of the biology we want to see in our soil. Adding compost adds this biology to the soil, but also adds the food that the biology needs to continue to reproduce once in the soil.

Of course, everything that eats also poops. So, at each meal of the various biological processors, soluble fertilizer in the form of micro-creature poop is created. Plants crave this soluble fertilizer. Adding compost is adding the living biorefineries to make fertility possible and sustainable. Compost feeds the soil life, and the soil life feed the plants.

Compost can be used as a component in a strategy to create deeper topsoil. Below the topsoil is subsoil, which usually contains less than 2% organic matter, and is often compacted with little airspace. If compost is spread on the surface and the soil is loosened to allow the compost to sift

Fig. 11-8 Composting with stillage. *These* **windrows** *are composed of sugarcane bagasse (plant fiber), liquid stillage, and the insoluble "mud" left over after molasses fermentation. This rich fertilizer contains everything the next sugarcane crop needs. More than just minerals, compost contains an enormous concentration of living beneficial bacteria and fungi.*

Fig. 11-9 Spraying compost with stillage. *This is a fail-safe way to recover the liquid byproduct as an easily spread solid, and an excellent way of dewatering the liquid stillage.*

Fig. 11-10 Making windrows. *This large compost operation at Rajshree Sugars in India uses the press mud left over after molasses fermentation and sugarcane bagasse. The compost is turned and sprayed with liquid stillage to make a superior compost.*

down into the subsoil, that subsoil can be converted into topsoil—from two to six inches in the first year or two.

The tool to do this loosening, aerating, and sifting is a **chisel plow**. Once you have created enough topsoil depth by introducing enough organic matter, you can use then stillage in a number of ways to enhance fertility. Once the soil is alive and has oxygen, pH-corrected stillage can be introduced through a special chisel plow, like the **Yeoman plow**, which allows injection of liquids in the furrow formed immediately behind the plow as it is dragged through the field. Provided there is enough organic matter in the soil, the stillage will feed the organisms in the soil, and their nutrients will be converted into forms more easily stored by soil humus particles.

A variation on this concept is to use stillage in the making of compost tea. Compost tea is usually a solution made of actively reproducing compost, diluted in water and aerated with bubblers. The compost tea can then be used to inoculate soil with the solution of microorganisms in a powerful variation on the dry compost sifting technique I described above.

But this works only if there is already enough organic matter in the soil. So it's most likely to be of use later in a soil-building program. Stillage can first be fermented into a compost tea with a wide range of organisms, often dominated by lactic-acid-forming bacteria. Mixing a few shovelfuls of good-smelling topsoil from a forest, or from an active but cool compost pile, will provide the initial organisms; you then need to provide lots of aeration. The compost tea can then be introduced to the field behind the chisel plow or even added in the cool of the evening, mixed with irrigation water.

Stillage can also be used very profitably with or without other organic matter in a methane digester. The digester converts nonfermentable carbohydrates, fats, and components into methane and now-soluble nutrients which would be considered fertilizer. The solubles in thin stillage can actually produce enough methane to power the entire alcohol plant, and even generate a surplus of methane for other uses (see Chapter 10). This may be one of their best uses in a small-scale plant.

Alternatively, stillage can be used to raise algae for feeding to fish, which we will discuss later. It can be safely run though a **constructed wetlands**, both to use up the nutrients in the stillage (preferably by conversion to cattails, which can be harvested as an energy crop) and to produce micro-crustaceans for feeding fish (see Aquaculture later in this chapter).

MUSHROOM PRODUCTION

While working in Mexico on a project producing mushrooms on coffee pulp, I saw a saying on the wall, "To help your kids grow healthy and strong, make sure they eat their mushrooms." This couldn't have been more accurate. In fact, in China, mushrooms are considered a fifth critical food group in addition to the traditional four food groups touted in the United States. People in Mexico have been known to eat more than 200 species of mushrooms, unlike the relatively fungophobic U.S. or England, although the market for mushrooms other than the standard white button mushroom has been consistently growing in the U.S. at about 10% a year since World War II. Americans are acquiring a taste for mushrooms. In cities, it is now commonplace to find shiitake and oyster mushrooms year round.

Mushrooms are not plants; they have very different metabolisms from plants. They are also much higher in the amino acids methionine and lysine. They also produce vitamin B_{12}, which plants contain only if they are grown organically (vegetarians can get this vitamin from eggs or dairy).

Since alcohol feedstocks are plant-based, they tend to be low in methionine and lysine. For protein to be properly absorbed by and available to animals (including us), all essential amino acids have to be in proper balance. If one or two are

Fig. 11-11 Oyster mushrooms. These mushrooms are growing on brewer's grains, a close relative of distiller's dried grains. Note how thoroughly the fungus has grown through the grain substrate.

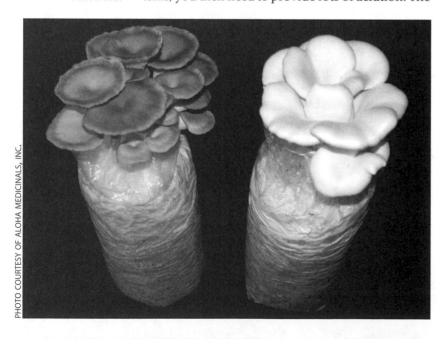

PHOTO COURTESY OF ALOHA MEDICINALS, INC.

deficient, then we cannot absorb all the protein in the food. The deficiencies become limiting factors. Grazing or browsing animals get the proper balance of amino acids by eating a wide variety of plants, plus mushrooms and the occasional earthworm or insect. But when plant materials are used to create animal feed, the low levels of critical amino acids cause an overall low efficiency in availability of protein.

Unlike plants, which have all sorts of differentiated cells to do different jobs, all the cells in a fungus are the same except for the spores. So, where plants have certain kinds of cells that make leaves and others that make roots, mushroom mycelium (the cottony fiber) and the mushroom fruits are all made of the same cells. The mycelium essentially winds itself up into the shape of a mushroom, then produces spores which complete its reproductive duties.

Mushroom cultivation in a wide range of agricultural waste has been practiced at many scales over the last few decades. Materials such as cotton waste, orange peels, coffee pulp, textile mill waste, olive press paste, etc., have been used as the substrate (material mushrooms are grown on) for both the oyster mushroom *Pleurotus* species and the shiitake *Lentinus* species. In Nature, these mushrooms are found growing out of dead trees and are sometimes referred to as "white rot fungi," since their

mycelium grows throughout the wood, digesting it. These mushrooms are valuable human foods. In fact, extracts of shiitake are used to fight cancer in Japan. Both shiitake and oyster mushrooms act as antiviral, and to some extent antibacterial, agents in the human body.[15]

Oyster and shiitake mushrooms, and, to a lesser degree, other edible mushrooms, cause significant changes in the substrate over the period of cultivation. To produce mushrooms, the fungus must first grow throughout the substrate, digesting it to generate stores of energy and dissolved nutrients, which it uses to produce the edible mushrooms.

Meanwhile … the alcohol fermentation and distillation process results in byproducts in which the easily fermentable carbohydrates have been removed. The compositions of the byproducts vary with the feedstock initially used to produce the alcohol, but they share some common attributes. The remaining carbohydrates (cellulose and hemicellulose) are sometimes bound to lignin or pectins. Lignin is the resinous material in wood and is a rich source of nutrients for fungi, making up about 20% of the straw of wheat, barley, rice, or oats.[16] Pectin is the gel-like material in fruits.

When used as a mushroom substrate, the carbohydrates in the alcohol byproduct become more

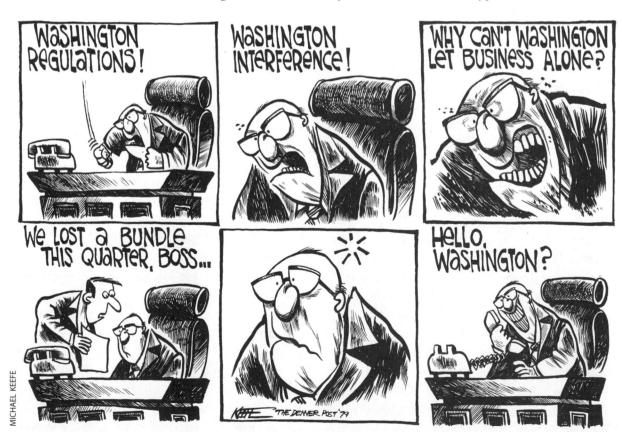

MICHAEL KEEFE

Fig. 11-12

available as lignin is preferentially degraded and eaten by the mushrooms. (The mushrooms also eat the pectins.) As the mushroom mycelium consumes the substrate, the proportion of amino acids changes to more closely resemble the protein of meat.

Much of the material that becomes the new amino acids comes from the digested lignin. By the time the process of cultivation has run its course, the substrate is almost entirely made up of mushroom mycelium in many ways identical to the mushrooms harvested from it.

A pound of DDGS, which commands a price of only three to five cents per pound dry weight, will yield up to one pound of oyster mushrooms, wet weight. The value of the oyster mushrooms is currently around $3 per pound wholesale; shiitake is worth $5 to $7 per pound.

One of the overall effects is to bring the substrate's proportion of methionine and lysine levels up, making the substrate a much more valuable animal feed, from a protein point of view. The removal of the lignin also frees residual hemicellulose and cellulose for efficient digestion by ruminants. For both of these reasons, cellulosic feeds, like wheat straw and cassava pulp, are far more digestible when used as a mushroom substrate than when used in their raw form.

An excellent study using another of the "white rot" fungi, *Trametes versicolor*, a.k.a. **turkey tails**, a group of wood-digesting fungi, demonstrated results not only in the lab, but also in actual animals' digestive tracts.[17] The fungi were allowed to grow through straw for 14 days and in that time reduced the lignin up to 36% with minimal loss of carbohydrates.

Then the mushroom-spent substrate was put in mesh bags attached to fishing line. The bags were allowed to reach the rumen of the cow's four-stomach system and after four days were retrieved from the cow's gut, using the fishing line. In this way, it was possible to see in real life what the difference in digestibility was between plain straw and straw partially digested by fungus.

The researchers found that dry matter digestibility was improved by 64%, turning the low-value straw into as good a feed as hay, which has 70% dry matter digestibility.[18] (Hay is the whole grain crop with seeds, and straw is just the stalks without

seed.) What's more, the cattle liked eating the fungus-digested straw better than the plain straw.

Some highly cellulosic distillery byproducts, such as cassava pulp, must usually be supplemented with expensive animal protein sources, such as fishmeal, to make them sufficiently valuable for sale as animal feed. This amino acid supplementation is not necessary if byproducts are first used as a substrate to produce mushrooms. The improvement in amino acid structure would be especially valuable in earthworm and aquaculture crops, such as fish and shrimp. Spent oyster mushroom substrate has proven itself as a fine feed for tilapia in Thailand[19] (see next section).

DDGS, spent cassava pulp, sugarcane bagasse, and beet pulp are among the substrates substantially improved by cultivation of fungi in the substrate, whether or not they result in saleable fruiting bodies (mushrooms), valuable products that they are. A pound of DDGS, which commands a price of only three to five cents per pound dry weight, will yield up to one pound of oyster mushrooms, wet weight. The value of the oyster mushrooms is currently around $3 per pound wholesale; shiitake is worth $5 to $7 per pound.

The byproduct substrate will be unlikely to lose value during cultivation, since it retains higher protein and higher digestibility, even though some of it is consumed in the process of making fruiting bodies. Only about 3% of the cellulose would be consumed, while the amount of lignin removed would be about three and one-third times the cellulose removed.[20]

This process thus produces both a high-value primary product in the form of edible mushrooms, and a moderate-value byproduct in the form of improved animal feed.

A further benefit of using distillery byproducts for mushroom production is the greatly reduced rate of microbial contamination. This is due to the lack of easily fermentable carbohydrates, which are the primary food source of most of the common microbial contaminants of mushroom cultivation. Since the alcohol fermentation process removes starch and sugar to produce the alcohol, they do not remain in significant quantities in the spent pulp.

In preparing a substrate for mushroom cultivation, it is necessary to heat it to pasteurize or sterilize it prior to inoculation with mycelium. This is done to give the fungus a big head start over contaminants, which hopefully will be overwhelmed by the fast growth of the fungus in the semi-sterile substrate.

Preparation of the substrate can be done readily in an alcohol plant, since the substrate leaves the alcohol distillation process in a hot, fully sterile state ready for immediate inoculation with fungi upon cooling. This removes the need to use energy or special sterilizing equipment in mushroom cultivation. The costs of labor and materials in handling the substrate through the sterilization process are a major part of the cost of production in stand-alone mushroom farms.

Since most alcohol byproduct mashes will be made of small granules of material, it can be difficult to get sufficient air circulation through the substrate when the fungus is growing. Mixing the substrate with a material such as straw—both to help aeration and as food for the fungus—is common. This material, too, has to be pasteurized. Using hot distiller's solubles from a small plant is a great way to heat and pasteurize the straw, and to add another jolt of soluble nutrients that the straw will absorb.

Mushroom substrate is usually packed into bags which have a sanitary air exchange device mounted in the bag opening to let air but not microbes in while the fungus is growing. Like animals, fungi use mostly oxygen and release mostly carbon dioxide while growing. In the case of oyster mushrooms, cultivation usually results in harvestable crops in three weeks to a month and

continues for two or three more weeks, when the spent substrate can be fed to livestock.

AQUACULTURE

As farmer Joel Salatin, a master at designing good livestock-raising systems, puts it, "The smaller the livestock, the more profit you make." So the least profitable are cattle, and when you get as small as, say, earthworms, the profit is the greatest.

Smaller livestock convert feed to flesh at a much more efficient rate. Where it might take ten or more pounds of feed to make a pound of beef, it takes only three pounds of feed to make a pound of chicken, only 1.6 pounds to make a pound of fish, and, better still, one pound of feed to produce a pound of shrimp or earthworms. In the case of fish or shellfish, and to some degree, earthworms, not having to maintain a resistance to gravity dramatically drops the food energy costs.

Tilapia

There are many fish that can be raised using DDGS or similar products, both saltwater and freshwater, but the most study has been done with tilapia. These fish vary a lot genetically, from carnivorous cultivars to full-time vegetarian varieties. They also vary quite a bit in their resistance to cold, with some that will live only in balmy tropical water and some, like Israeli varieties, that will survive temperatures as cold as 56°F.

Almost all commercially sold tilapia are male; fish buyers buy only large male tilapia. It's not an issue of flavor but of established habit. Males can grow up to a pound and more than a foot long

Fig. 11-13 Mature tilapia. These vegetarian fish are a hybrid between Egyptian and Israeli varieties. These fish are ready to go to market.

FISH-FARMING TRAILBLAZERS

Neris González, after being tortured for teaching permaculture in El Salvador, now teaches Chicago ghetto dwellers to raise their own tilapia in 55-gallon drums in their basements.

Martin Schreibman, a professor at Brooklyn College, has helped Israel set up fish farms in the desert, and has pushed sustainable fish farming in the urban community.

In Vietnam, permaculturists reap huge yields from fish raised in floating bamboo cages in nutrient-rich rivers. Food is literally constantly flowing through the cages. Some fish farms keep the males in large floating cages in the pond to facilitate easy harvesting.

DAVID BLUME

pretty easily, while females generally slow down or stop growing at only eight to nine inches, weighing far less than a pound. This is because tilapia are mouth breeders; the females protect their young in their mouths until they are big enough to live on their own. During this time, the females can't eat larger-sized food without endangering their young, but they can eat algae safely.

So, if you feed your fish algae, you could sell both your males and females as food. But right now, selling female fish isn't generally done, and education of fish buyers would be necessary.

RIGHT: Fig. 11-14 Small fry. These tiny tilapia are being raised under shade, since they are sensitive to ultraviolet light.

Fig. 11-15 Duckweed. *This floating aquatic plant is being eaten by small fry. It has nitrogen-fixing bacteria on its roots dangling in the water. The nitrogen is a key nutrient in producing protein, and the bacteria harvest it for free from air dissolved in the water.*

Tilapia breed about every two weeks, which means you get large numbers of fry, or young fish. Whenever you are overstocked for the size of your pond, the average size of your fish drops dramatically, even though you have the same pounds of fish per cubic foot of water. To get commercial-sized fish, you have to constantly cull out most of the females and excess fry. Fry can be sold, if male, for 50 cents to $1 apiece to large fish farms that choose not to raise their own fry.

Tilapia will live in a wide variety of environments and can be produced on a wide range of scales. They eat many types of food and have even been raised on a simple diet of animal manure. After all, manure is largely bacteria and could be thought of as a high-protein, single-cell-protein source.

Marketing of fresh tilapia increases in profitability the farther from the coast you get. Let's face it, you don't get a lot of fresh fish in a place like Kansas City. In colder climates, you generally have to protect the fish from the weather by raising them in ponds in greenhouses. Over 90% of the cost of raising fish, averaged across several species, comes down to two things: the feed, and the energy to aerate the ponds. So a good design would tackle both of these big-ticket items.

Large-Scale Production

Let's first look at bigger-scale production. Some aquaculture operations, like ADM's in Decatur, use a mixture of DDGS and soybean byproducts along with some other nutrients to produce a pelleted fish food for their vegetarian tilapia. The DDGS costs maybe three cents per pound. It takes less than two pounds of DDGS to produce a pound of tilapia worth at least $3, gutted, on ice, as human food. But it's possible to get as much as $7 to $12 per pound for live fish, depending on where you are delivering.

As retired Senior Vice President Martin Andreas of ADM explains, "You have to remember, in the wintertime in the Midwest, shrimp tastes like you are chewing rubber balls. They have to be shipped a long way, and they are frozen a long time. The chances of getting fresh shrimp in the Midwest are very slim. So getting fresh tilapia and shrimp is a real treat for people." You can see that raising tilapia close to where they are going to be consumed is the best design. Delivery of live fish to restaurant chefs is at the top of the profit chain.

A study done in 1997 used a variety of diets for tilapia and found that one of the best was 77.75% DDGS and 15% soy flour (with other minor ingredients). It performed very well. The feed conversion rate was 1.62, with a 1920% increase in weight in 56 days. The feed only had 28% protein. The control was over 32% protein and contained expensive fishmeal, blood meal, etc. It had a conversion rate of 1.25 and a weight increase of 2890%. So although the yield was lower on the vegetarian diet, the cost per pound of fish produced using DDGS was within a penny per pound of the highest-protein diet.[21] The high-protein diet used fewer pounds of feed to get the same weight of fish, but the DDGS feed was so cheap that the extra quantity needed still came out costing much less.

In its five-acre aquaculture greenhouse, ADM produces more than 50,000 pounds a month of tilapia year round on an all-vegetarian diet. The company raises the fish from eggs to maturity all in-house. The fish are raised in stages, using 180 10,000-gallon tanks.

ADM starts by raising its own fry from eggs. At first, the fry were not surviving. Like a permaculturist, ADM looked at its failures and at how the fish are raised in nature to solve the problems. Observing that tilapia are mouth breeders led to the understanding that ultraviolet light was killing the tiny unprotected fish, so the problem was solved by building a fry propagator protected from sunlight.

ADM begins the commercial process by putting 100,000 fry in one tank covered with duckweed (*Azolla sp.*), a common tiny floating plant. When the fry have finished eating the duckweed, it's time to move the fish up to the next tank.

Pellets drop into the tanks automatically, while acidity and concentration of fish excreta are controlled by both a filter at the end of each tank and a chive bed. Since chives are heavy nitrogen users, a large bed of them is used as a **biofilter** to draw a large amount of nitrates (from fish poop) out of the water, making the water clean enough that it can be recycled to the fish tanks.

The chives could be a commercial crop, but they serve a more important function for the next-door greenhouse that's growing lettuce and cucumbers. When there is an unusually large outbreak of pests that beneficial insects can't control fast enough, the gardeners make an emulsion of chive concentrate

Perhaps the simplest high-value product to make from these surplus fish is fish emulsion, a favorite fertilizer of organic farmers and gardeners. It has high amounts of nitrogen, with good quantities of phosphorus and potassium. Fish emulsion from culls can sometimes actually sell for much more than fish for eating, based on the volume of fry produced.

to use as a human-safe, biodegradable, organic spot insecticide.

Dan Helfrich, the production manager of ADM's five acres of fish tanks, told me that the company produces enough nitrogen in the fish water to fertilize more than 100 acres of intensive crops. Early in its research, ADM found that the quantity of fish water that was produced totally dwarfed the nutrient needs of its other five acres of intensive hydroponic lettuce production. In considering whether to use the fish water as a nutrient source for the lettuce, Dan pointed out, "All of our customers are upscale markets, so we have a very stringent safety program. We didn't think our customers would be too excited about our using the fish pond water."[22] People can be so fussy, can't they?

ADM delivers its fish live to markets all over the Eastern U.S. in its own trucks, powered by its own biodiesel. The trucks drive right into the greenhouse to collect the fish from a tank, taking about 10,000 pounds at a time, essentially all the fish from one tank. They can be in Chicago in three hours or New York City in 19 hours, with the fish alive and healthy on delivery. In New York, some buyers pay a premium, since the fish are certified kosher. With Wal-Mart committing in January 2006 to selling

Fig. 11-16 Chive beds. ADM uses chives as a biofilter for its tilapia tanks, and then harvests the chives to make a natural insecticide from the juices. Chives can also be sold as an herb, or the flower stalks can be sold for dried flower arrangements.

DAVID BLUME

100% sustainable fish, the market for this kind of fish production can only skyrocket.

So, the ADM model is a pretty simple and profitable way to add value to the DDGS by converting it into fish. But we don't have to stop there. Let's look at what happens to all the fry and females that have to be culled. One answer might be to feed them to omnivores (e.g., hogs), a potentially important part of a small-scale integrated fish system.

Algae can reach photosynthetic efficiencies in the 10% range, compared to land plants' 1–3% efficiency.

Perhaps the simplest high-value product to make from these surplus fish is fish emulsion, a favorite fertilizer of organic farmers and gardeners. It has high amounts of nitrogen, with good quantities of phosphorus and potassium. Fish emulsion from culls can sometimes actually sell for much more than fish for eating, based on the volume of fry produced. Algae researcher and fish farmer Marc Cardoso estimates that his operation culls quite a bit more tonnage of fry than it sells of finished fish.[23]

It takes about a pound of fry to make a gallon of standard fish emulsion. There are several ways to make it, but the way I've done it is to take the expired fish or fish scrap left over from processing fish for wholesale frozen fillets, and put it in drums with a minimum amount of water to cover it. Then add and mix enough concentrated phosphoric acid to bring the pH down below 3.0 (see Chapter 7). In a matter of days, the fish will dissolve totally; then add lime to bring the pH back up to at least 6.0 before diluting the emulsion to use in your fields or garden.

The process goes faster at a higher temperature, so I have buried drums in hot compost windrows and let the continuing composting process keep the drums at about 140°F. After it's done, I filter the fish emulsion and pack it in drums or whatever container I plan to store or sell it in. Fish emulsion sells for about $300 per 55-gallon drum, or double to triple that price ($8 per pint) in smaller containers. So over the same six- to eight-month time period, what you make from adult fish gutted and sold on ice is doubled selling fish emulsion.

Small-Scale Production

Since small-scale producers will have a lot of stillage rich in distiller's solubles, we can employ a different strategy to eliminate the two major costs, aeration and feed, in the raising of fish. Tilapia can live in a pond with algae (for instance spirulina) so dense that a Sacchi disk (white test disk) lowered into the water will become invisible at a half-inch deep; by contrast, a normal farm bluegill/bass fish pond often is clear enough to see the disc at three feet deep.

The algae are both food and the source of oxygen for the fish. Fish need dissolved oxygen to breathe while they release dissolved carbon dioxide into the water. In the growth process, algae produce oxygen and absorb carbon dioxide, as well as the nutrients from either thin stillage or methane digester liquids. Algae can reach photosynthetic efficiencies in the 10% range, compared to land plants' 1–3% efficiency.

Spirulina has the advantage over other algae in that its corkscrew shape clumps and is easier to harvest. Since it is technically not a true algae but a **cyanobacteria** that also photosynthesizes, *it fixes nitrogen from the air.* It has been proven to make a great bioconversion agent of thin stillage. This nitrogen-fixing should not be underestimated, since it is a key component of the high protein found in the "algae." In fact, researchers in India have concluded that raising spirulina for its nitrogen-fixing capacity, then using it directly as a fertilizer, is cost-competitive with fertilizer made from natural gas.

Researchers such as Marc Cardoso have created algae-dense ponds to raise their fish. Using a pond divided by a fine screen, Cardoso raises massive quantities of algae on one side. The pond, which is only four to six feet deep, has a soaker hose on the algae-raising half, winding back and forth at the bottom. Carbon dioxide from the fermenters is bubbled from the soaker hose up through the algae-breeding side.[24] As the algae are constantly agitated by the bubbling carbon dioxide, they all get their turn at the sunlight. Spirulina is so efficient at photosynthesis that it photosaturates at only 10% of full sunlight, so it's important to keep it agitated.

Meanwhile, the algae pond is kept at the proper tropical temperature using waste hot water heat from the distilling process. The hot water can be run through inexpensive radiant heating tubing at the bottom of the pond in a serpentine pattern alongside the CO_2 soaker hose. At the end of the pond, the algae are permitted to flow into the fish tank through the screen. The algae have all the nutrients, carbon dioxide, heat, and sunlight they need in order to produce large amounts of biomass.

Research has confirmed freshwater algae yields of nearly 15 tons per acre per year without all the enriching strategies employed above.[25] This many algae would produce as much as ten tons per acre per year of combined fish and fish emulsion.

If there's algae left over after fish production, methane production is an excellent use. Raw methane contains a high percentage of carbon dioxide, which can be stripped out by bubbling it through water. The carbon dioxide becomes trapped in the water, and the methane bubbles out, now clean, and is collected. This CO_2-rich water can also be added to the algae pond. The elevated carbon dioxide level makes the algae reproduce right up to the solar saturation point.

Thus, excess carbon dioxide, thin stillage nutrients, and nitrogen from the air raise food for the fish.

Cardoso first runs most of the DDGS through his hogs to get the first yield of animal protein. (Although he uses pigs, you could substitute chickens or ducks.) Then he uses the manure in a methane digester to produce his second yield, methane, which he uses to run the alcohol plant (plus, he gets the carbon dioxide). The fully digested liquid animal manure is flowed at a measured rate into the algae pond to provide the nutrients needed to keep up with the reproduction rate of the carbon-dioxide-crazed algae.

The algae slowly diffuse through the fine screen into the fish side of the pond, where the fish eat them. The algae, now in the fish side of the pond, are turning the carbon dioxide into oxygen, which the fish need to breathe. The fish breathe out carbon dioxide, which keeps the algae growing. The algae release so much oxygen that it bubbles out of the water. Cardoso has noticed that the excess oxygen in the greenhouse atmosphere gives visitors the sort of high that Japanese businessmen get at Tokyo oxygen bars. Now that's what I call a stimulating system input!

The fish eat the algae; some of it turns into fish flesh, and the rest they poop out, making high-nitrogen fish manure soup out of their pond, which can be drawn off and used to irrigate the same fields to grow the next crop. That starts the process in motion once again.

One of the benefits of using the algae system with thin stillage, rather than just feeding the DDGS directly to the fish, is that algae-fed fish can be certified organic, while fish fed with GMO-tainted DDGS grain cannot. Also, fish farmed in an alcohol fuel loop contain no mercury or PCBs.

So how much fish can you produce in a system like this? Both Cardoso's and ADM's operations produce about a pound of fish per cubic foot of water every six months. But look at what the integrated algae-based system using thin stillage (or methane digester effluent) produces in addition to fish: a premium price due to organic certification, other animal protein (since the DDG can be directed to other livestock production), fish emulsion, oxygen (aeration), and fertilizer for the next field crop—all powered using surplus heat and carbon dioxide from the alcohol plant and nitrogen from the air.

ADM makes a good profit, too, but it has to mix and pelletize the food, filter the water to get rid of fish poop, and aerate the tanks. It does sell its CO_2. ADM's system essentially uses 180 separate tanks

Fig. 11-17

WHY DON'T YOU STOP KILLING PEOPLE WITH YOUR SECOND-HAND SMOKE...YOU INCONSIDERATE JERK

COPYRIGHT 1992–2002 ANDREW B. SINGER

SINGER

which, in reality, are many small systems, so the economy of scale scarcely applies here.

I have only scratched the surface of what is possible when combining alcohol production with aquaculture. It's just enough to get your own imagination going about what you could do to make sure that every output of the system is an input for some other part of the system. The challenge is to design permaculturally so that there is as little leakage of valuable resources or energy from the system as possible, and so that there are surpluses that generate the necessary capital to keep everything going.

MARICULTURE

The next frontier for Dan Helfrich at ADM is using DDGS and soybean co-products to raise Pacific white tiger shrimp. A saltwater variety, these can be acclimated to very low salt levels—five to ten parts per trillion (ppt), compared to 40 ppt in natural seawater. With their attractive 1:1 feed conversion ratio and fast growth rate (around 150 days), they are potentially a good aquaculture product.

Since shrimp are filter feeders, meaning that in Nature they scavenge micro-life from the ocean, they concentrate the toxins that the micro-life had absorbed. Helfrich thus sees farmed shrimp as a superior product: "We can guarantee that these fish and shrimp have never been exposed to heavy metals, mercury, or PCBs and all of that."[26] "Prawns go for $13.50 a pound," says ADM VP Andreas. "So there's a little bit of incentive there to work this out."[27]

Shrimp molt several times as they grow, leaving behind their outgrown exoskeleton. As an organic farmer, I know the value of "shrimp shell," both as an organic fertilizer and as a partial cure for a soil-borne pest known as symphylans. The shells are made of **chitin**, and when they are added to soil, organisms which like to eat chitin have a population explosion. Since the microscropic, lobster-like symphylans have a chitinous exoskeleton, they, too, are attacked by the chitin eaters, and their population is reduced. So shrimp cultivation will produce another valuable co-product.

Of course, taking a saltwater shrimp and acclimating it to freshwater wouldn't be necessary if you lived near a saltwater shore. But the problem with land mariculture of saltwater species is usually that the wastewater is released back into the environment, causing too high a concentration of nutrients and deoxygenated water, not unlike heavy sewage.

Good "three species profitable" ecosystems that address this problem have been studied, using sea bream, seaweed (*Ulva lactuca L.*), and abalone.[28] The fresh seawater is pumped in, and the fish eat pelleted grain feed (DDGS). Eighty percent of the nitrogen from protein digestion is excreted into the water as ammonia. That water can be pumped into the seaweed tank, where the seaweed is able to harvest up to 80% of the nitrogen from the ammonia. Ten percent is lost to evaporation. The seaweed is then cut and fed to abalone, reducing the "lost" nitrogen to 10%.

But, of course, much of that 10% could be harvested by micro-algae and microcrustaceans, which would use the abalone poop and feed the Pacific

---

SHRIMP OR TSUNAMIS?

Around the world, mangrove swamps, which line the coasts of tropical areas, are being ripped out at an astounding rate. These ecosystems have been hailed as some of the most diverse and least understood. At a minimum, mangrove swamps are often the safe breeding ground for fish that later inhabit coral reefs. And they have been credited as huge bioremediation filters when they line the outflow of rivers loaded with sewage. Nipa and sago palms co-exist in bands within the mangrove zone and provide an untapped wealth of sugar and starch for alcohol. People living in these coastal areas have long made diverse types of livings from the mangrove ecosystem.

A major reason these diverse ecosystems are being destroyed is to make room for shrimp farms. But, since we can easily raise all the shrimp the world needs in conjunction with alcohol production, there is no need to destroy this delicate and intricate marine ecology.

At least 150,000 people died in the 2004 tsunami in Southeast Asia. Organizations like the M.S. Swaminathan Research Foundation of Chennai, India, have been warning for years that coastal mangrove forests are "the first line of defense against devastating tidal waves on the eastern coastline." Dr. Swaminathan pointed out that, "Our anticipatory research work to preserve mangrove ecosystems, started 14 years ago … has proved very relevant today.… The dense mangrove forests stood like a wall to save coastal communities living behind them,"[29] he said. All along the Orissa coast where mangrove forests have been regenerated, the damage to people and property was minor.

Perhaps all the corporate shrimp farms should be required to house their boards of directors at their seaside farms as a condition of ripping out the mangroves.

tiger shrimp (adding a fourth species to the system) coexisting with the abalone, reducing the nitrogen loss to 5% or less. This dilute fertilizer solution would then flow through a cattail bed (see Chapter 8) before returning as clean water to the ocean. Of course, the cattails would be periodically harvested for alcohol production.

A simpler variation of this system was operated for two years in a study with just the sea bream and *Ulva lactuca* seaweed. The seaweed absorbed the fish waste, and then was used to feed the fish. There was a lot of surplus seaweed. The most important finding was that the balance between the fish and seaweed was stable and practical.[30]

EARTHWORMS

Earthworms make excellent livestock to raise at an alcohol plant. The commercial worms that I'll be discussing are known by a number of names: red wigglers, hybrid red worms (a misnomer), angleworms, or compost worms. Worm breeders actually end up raising two visually similar species: The majority are *Eisenia foetida*, and a smaller percentage are *Lumbricus rubellus*.[31]

Since they do not have to resist gravity like cattle and other standing livestock, earthworms do not have skeletons. Their digestive system runs nearly the full length of the body and can break down a huge range of materials. These traits contribute to

earthworms being among the most efficient converters of feed to flesh.

Depending on temperature, moisture, and the continuity of a high-quality food supply, earthworms reproduce every three weeks to 30 days.[32] Now, livestock that more than doubles its herd size every month or two is pretty … well … sexy. In fact, in six months, six worms can become 1500 worms, since each time they reproduce they can generate 20 new worms, although not all survive.[33] Since worms are hermaphroditic, both partners can be pregnant at the same time!

Once you have a large enough standing herd and a well-controlled environment, you can harvest nearly half of the worms every month and never slow down worm production. Commercial worm farmers generally use ground grain or coffee grounds (which are simply ground seed) to maximize worm reproduction. The worms seem to relish the added aphrodisiac zing of caffeine, as well.

When raising worms, you have choices about end products and marketing. You can sell worms wholesale to bait companies and a few worm farm brokers. If you have a concession at a fishing area, you can sell worms retail for around $2 for 50 worms, or about $40 per pound.[34]

Worm castings (worm poop) are a significant product, as well. While they're in the growth/

Fig. 11-18

reproduction phase, worms can convert feed to flesh at a ratio of nearly 1:1. So a pound of dry grain will result in about a pound of worms, plus a lot of semi-moist castings. Worm castings sold at retail prices locally to organic gardeners or others in need of high-quality potting mix will typically bring in about $15 to $22 per cubic foot,[35] which weighs about ten pounds.[36] Under good conditions, you will reap up to four pounds of worms from a cubic foot of bed when you're ready to sell the castings.

But if you focus on selling castings instead of selling worms, you will eventually reach a livestock-saturation point of more than 4000 mature worms per square foot. When the earthworms sense that the quantity of food is balanced by their population, they slow, then nearly stop reproduction. If they run short of food, they can end up turning into cannibals, until their population dwindles sufficiently to live in balance with their food supply.

Like any animal, when an earthworm reaches maturity, and stops putting on weight, most of the food it eats ends up as manure. A little food energy easily maintains worms' life functions. They can eat about half their weight per day at maturity or up to their full weight per day while in a growth phase under controlled conditions. So if you have 200 pounds of worms in a large bin with screening on the bottom suspended above the ground, you could feed them 100 pounds of DDGS per day, and all of it would disappear. With a little shaking of the screen-bottomed bin, you'd harvest a pretty large mountain of moist castings each day, which could be sold by either the cubic foot or pound.

Fig. 11-19

The worms in essence become little bioreactors that process the DDGS that come from the alcohol plant. They eat the bedding and DDGS. They mix it with a veritable soup of enzymes using powerful muscles. Digestion results in amino acids, sugars, plus a phenomenal amount of varied organic molecules. Inside the guts of earthworms live fungi, bacteria, protozoa, nematodes, and a variety of other organisms that all play a part in digestion.[37] The earthworm digestive system is really a microcosm of the universe of beneficial life found in the soil.

From a pound of dry DDGS, which sells for about three to five cents per pound in bulk,[38] you can produce about a pound of slightly moist castings, which retails for $1 to $1.50.[39] It is possible to produce 2.5 gallons of alcohol from a bushel of corn, worth maybe $5 ($2 per gallon). But the same bushel of corn will concurrently produce 18 pounds of DDGS, which, when fed to worms, will result in about 20 pounds of castings worth $20 to $30.[40] Intelligent use of the "byproduct" of a system is often more valuable than the so-called primary product.

To give you perspective on the potential profitability of worm farming, consider a Pennsylvania State University study done under controlled conditions in a 42-square-foot box.[41] Based on a conservative harvest of four pounds of worms per square foot per year (about a third of what other literature says is possible), the researchers calculated a net profit of $132 per year (in 1994 dollars). No profit was attributed to the castings. Even so, the total revenue from worm sales alone comes to more than $3 per square foot, compared to corn's one-tenth of one cent per square foot.

David Hoffman, an earthworm experimenter extraordinaire in California's Marin County, collaborated with me in 2005 on a trial of feeding DDGS to earthworms. The first thing he did was test for any agricultural toxins or residues in the DDGS. He was surprised to find no residues at all. I pointed out to him that the process of fermentation, which requires cooking at pretty high temperatures, followed by enzymatic breakdown and consumption by yeast, probably evaporated and broke down most volatile chemicals. Whatever might be left was vaporized in distillation.

Hoffman first tried feeding the worms by adding a large quantity of DDGS to the worm bin. The resulting biological frenzy of bacteria and fungi

THE LAST PERSON IN, SWITCH OFF THE LIGHTS!

ENERGY POLICY

ENRON

ANY MORE BRIGHT IDEAS?

caused a great deal of heat to be produced, due to DDGS's high nitrogen content, similar to what happens in a compost pile. The worms in this first test were literally cooked.

After several trials, Hoffman ended up doing what commercial worm growers do when feeding grain to stimulate reproduction or fatten up their worms; he added a thin surface layer of DDGS each day instead of large quantities.[42] "The worms went crazy over it. I've never seen them react to any other food more enthusiastically," he said.[43] This outcome compared favorably to the trials I did in the 1980s using distiller's waste made from nongrain materials.

Hoffman provided me with the castings produced from the DDGS, and I had them analyzed both as a saturated paste, using a process similar to compost analysis, and via acid hydrolysis prior to characterization, using a process similar to a soil analysis. The castings came out at 4% nitrogen, 4.38% combined phosphorus, and 1.89% potassium;[44] the fertilizer value of the castings is superior to even fish emulsion. Due to the long interval between collection and testing, a large portion of the volatile nitrogen was likely lost in evaporation, and I would speculate that the nitrogen level would have been over 8% when fresh. The finished castings were also high in most trace elements, except manganese and iron.

METHANE

In the U.S., thin stillage is evaporated and the solubles concentrated to a syrup, then added back to the DDG, making DDGS. This is very energy-intensive, but it avoids releasing water with a very high biological oxygen demand.

If stillage isn't evaporated, concentrated, and added back to the DDG, it has to be disposed of, which is difficult and costly. So, in a way, making DDG into DDGS is necessary to avoid alcohol plants having a big liquid disposal problem.

Methane production is a way of using up all the high-BOD nutrients profitably without having to waste all that energy to evaporate the water out of the stillage. When the thin stillage is anaerobically fermented by bacteria instead of yeast, you get methane, essentially natural gas.

Depending on the site, the choice for renewable process energy is generally between firewood (if you are growing it for heat energy or can get it cheap) and self-produced methane. Setting up

for methane production is less work over time, but a bit more cost up front. The methane produced will reduce to zero the energy required to make alcohol, using direct heating and cogeneration for electricity. (And remember, the byproducts of its combustion are carbon dioxide and water, so you can duct both into your greenhouse.)

Pimentel and Patzek (see Appendix A), in their 2005 paper denigrating alcohol, made a big production of the fact that for every gallon of alcohol produced, there is essentially more than ten gallons of sewage produced. This sewage, of course, is the spent mash. Dr. Pimentel and Patzek then dutifully assign a very substantial energy penalty for treating the spent mash as sewage in their flawed analysis of the energy balance of ethanol and imply that in some way this mash is discharged into waterways.

It's important to note that the largest alcohol plant in the country, the 200-million-gallon-per-year plant in Decatur, owned by Archer Daniels Midland, has no sewer. The only way that water leaves the plant is as evaporation in the cooling towers. They do have a small sewage treatment facility on-site to deal with the human waste produced by its 6000 employees, but the plant itself is sewer-free.

So where does the liquid mash (distiller's solubles) go? In most plants, it is dewatered to a syrup and then added back to the dried distiller's grains to make DDGS. This avoids the environmentally unacceptable issue of mash becoming sewage. But it takes a lot of energy to evaporate off all that excess water. On the small scale, we are not even going to begin to try to evaporate and concentrate the thin stillage we have left over.

Fig. 11-20 Methane digester in India. Most Indian alcohol plants use large, tank-style digesters like these. They conserve on space but are more capital-intensive to build than pond systems.

PAUL GAYLON

Although making DDGS is one clever way to turn a surplus and potential water pollutant back into a product and profit, let's look at what other parts of the world do with their mash. Remember the permaculture maxim, "The problem is the solution."

In India, in 285 of 290 distilleries, we can see the power of integrating methane production with small- and medium-scale alcohol production. Unlike the 50- to 200-million-gallon-per-year behemoths of the U.S., distilleries in India average 2.5 million gallons per year.[45] As of 2002, 130 distilleries already had methane plants in India, with another 50 either being built or planned.[46]

The methane potential in India from these plants is 35% more than is needed to produce all the heat and electricity at 40,000 Btu per gallon (the average efficiency of the Indian distilleries).[47] So no fossil fuels are required to convert the crops to alcohol. This energy comes just from the stillage.

Sugarcane alcohol is a massively energy-positive system that uses almost no fossil fuels at all.

In the past, the thin stillage was poorly handled and was either dumped on too little land or even dumped into Indian rivers because the plant operators thought of it as waste. No longer. Now stillage has been recognized for the resource that it is. All the soil nutrients are still in the thin stillage, as well as some substantial carbohydrates that could not be fermented in the alcohol-making process.

The noncarbohydrate portion of the stillage— the nitrogen, phosphorus, potassium, etc.—is still there in the liquid byproduct of the methane plant, as well as all the dead bacteria that did the work. This means that the fertilizer percentage per gallon of stillage is much higher, and more easily available in an organic but soluble form, for crops and soil organisms. The liquid outflow from the digester can be returned to the fields as excellent fertilizer.

For example, in processing sugarcane, an "average-sized" plant of 2.5 million gallons would generate (dry weight) 96,491 pounds of nitrogen, 16,211 pounds of phosphorus, and 1,891,228 pounds of potassium. At 71 pounds per acre yield, the amount of nitrogen in the digester fluid is just about perfect enough to grow the next sugarcane crop.[48] Over-fertilization with nitrogen has many negative effects on cane. The phosphorus levels of 12 pounds per acre from the effluent are much greater than is needed to fulfill the needs of the next crop. Phosphorus availability is really tied to activity of soil microbiology, so having a high organic matter content is key to keeping phosphorus available. Potassium is the really key element with sugarcane. Recommendations are often 800 pounds per acre or more, and potassium tends to be expensive. The 1401 pounds per acre in the digester effluent is even more than is needed to grow the next 7000 liters per acre of ethanol.[49]

The excess nutrients available for the next crops should come as no surprise. Sugarcane roots exude huge amounts of sugar to support a huge diversity of fungi and bacteria responsible for making minerals soluble and useful to the plant. Sugarcane roots can go as deep as 18 feet, where potassium is abundant[50] (see Chapter 8). In this characteristic, both sugar beets and cane are top harvesters of deep potassium.

Another exciting use for the digester effluent is to feed algae for fish production (see earlier in this chapter). The strategy of anaerobically digesting organic matter and then feeding the nutrient solution to algae to remove the nutrients from the water has been proven effective,[51] as has the recovery of nutrients in cattail marshes (see Chapter 8). Although it takes more space, this kind of anaerobic stillage treatment entails a construction cost of 1% of standard energy-intensive aerobic sewage digesters, which produce little methane, use enormous amounts of electrical energy for aeration, and produce substantial sludge which needs disposal.

Also, you can feed the DDG to cattle and then use their manure for methane production along with the thin stillage. This will produce far more methane than you can use at the plant. Remember, we were getting plenty of gas just from the solubles. At least one large U.S. plant is following this strategy and will certainly sell a lot of energy to the grid, as do Brazilian alcohol plants that use bagasse to fire their boilers.[52] The amusing thing is that this U.S. plant considers this a new idea, when feeding out stillage to cattle and using manure as a methane source have been done sporadically in Brazil for years.

Now that cellulose processing is becoming practical, there's a new reason to produce methane.

Half the sugars produced in the cellulose process are not fermentable by normal yeast. (As I discuss elsewhere, the use of GMO yeast might be an undesirable direction for cellulose ethanol to go.) One of the easiest uses of those unfermentable sugars is to produce methane. These sugars would spend only two to five days in the digester, and would certainly generate more energy than is needed to process the fermentable sugars to fuel.

If some of the sugarcane bagasse were also used in the digesters, the output of gas would be exponentially greater, since the cellulose would be converted to methane; the alcohol plants would have to install large cogeneration systems to feed methane-derived electricity into the grid. In fact, an alcohol plant with more extensive methane production could even be a local source of cooking gas and reduce deforestation further (see Chapter 13). So, as you can see, sugarcane alcohol is a massively energy-positive system that uses almost no fossil fuels at all.

When using corn as a feedstock, 41 percent of the residual corn is in liquid form, with the rest being DDG. This rich soup of nutrients has much more methane-producing compounds than the wash from sugarcane fermentation. In particular, the fat content of the DS, about 17% of the dry matter, is a major contributor to the amount of methane that can be produced.

CARBON DIOXIDE

Carbon dioxide (CO_2) is an odorless, colorless, and noncombustible gas. All animal life and some simpler life forms inhale oxygen and exhale CO_2. Plants, on the other hand, do just the opposite, breathing in CO_2 and releasing the oxygen part, while using the carbon.

Carbon dioxide has gotten a bad reputation, since it's the main gas implicated in greenhouse effects of global warming and it is also not something that humans want to breathe. It's not poisonous, but in a closed environment, being heavier than air, CO_2 displaces what we do need to breathe.

In the course of fermentation, yeast converts half of the feedstock's starch and sugar into carbon dioxide, which is often just released into the air. Ecologists (and smart business people) are rightly annoyed by that kind of waste. As a surplus product of alcohol production, CO_2 has many uses: for carbonating beverages, for preserving food by displacing oxygen atmosphere in a container, for pressurizing spray cans (avoiding

Fig. 11-21 Farm-scale carbon dioxide compressor. This size compressor would do the job of compressing CO_2 gas generated at night so it can be stored until the next day, when plants can use it.

THE WITTEMANN COMPANY, LLC

CO$_2$ SAFETY CONCERNS

If you go into a greenhouse with a carbon dioxide level above 5000 parts per million (ppm), you will begin to experience symptoms of oxygen deprivation, such as hyperventilating, ringing in your ears, dizziness, a splitting headache, or maybe even falling down in convulsions—in other words, you'll probably realize something's wrong. Get out, pronto! Breathing normal air will revive you.

To avoid unexpected incidents, install a carbon dioxide meter that can be read from outside the greenhouse door. If the CO_2 level is too high, use a fan to draw air into the greenhouse, diluting the CO_2 so it's safe to work inside. This is rarely an issue in normal, single-wall greenhouses whose leaks often give rise to a complete replacement of air in an hour. But double-wall, plastic greenhouses may turn over only a quarter of the air in an hour, so gas buildup is a possibility.

Fig. 11-22 English cucumbers. *These are producing in the middle of an Illinois winter. Note the high density of fruits, large leaves, and lack of cucumber beetle scarring in these cucumbers, grown in a greenhouse using carbon dioxide, and using beneficial insects for pest control.*

DAVID BLUME

Fig. 11-23 Greenhouse butter lettuce. *These nearly perfect heads of lettuce are near maturity at less than 30 days from seed due to carbon dioxide enrichment and optimum climate control. No insecticides were used to raise this crop.*

DAVID BLUME

the use of ozone-depleting CFCs), enriching the atmosphere of greenhouses, enriching outdoor field crops, filling welding tanks and fire extinguishers, etc.

Although, by weight, CO_2 represents about half the original carbohydrates fermented by yeast, it contains only 10% of the chemical energy that was stored in the carbohydrate. So about 90% of the chemical energy of the sugar is retained in the alcohol. Another way of looking at it is that each unit of carbon dioxide used by plants has the potential to trap nine units of chemical energy as carbohydrates as a result of photosynthesis. At any

rate, half your product stream by weight is a surplus—for which we permaculturists can certainly find uses.[53]

CO_2 is a growth stimulant for plants and a necessary component of their metabolism. Plants use the carbon dioxide to produce carbohydrates like sugars, starches, and cellulose. Once the carbon atoms are harvested by photosynthesis, much of the oxygen is vented out of the leaves or roots.

Our atmosphere today has roughly 340 parts per millions (ppm) of carbon dioxide. Although this is a lot higher than 100 years ago, CO_2 is still often the limiting factor in plant growth, when there is enough sun, water, and nutrients. Fed extraordinary amounts of CO_2, plants grow healthier and larger, flower earlier, produce higher fruit yields, reduce flower bud abortion in some plants (for instance, roses), increase stem length in flowers, and augment flower size.[54]

Using CO_2, many plants may be brought to maturity in 30 to 60% of their usual time and at a higher weight per plant. Some crops' photosynthetic rate, and therefore biomass, is increased by up to 300% with carbon dioxide.[55] How quickly plants mature with increased CO_2 depends on a matrix of conditions.

For instance, in the case of popular and profitable greenhouse crops like tomato, cucumber, and peppers, the bumper crops are due to a combination of more fruits per plant and faster flowering at CO_2 levels above 1300 ppm.[56] The ADM greenhouse gets almost 50 one-pound English cucumbers per plant in about four months, compared to perhaps 15 cukes without the extra gas.[57] Those big cucumber leaves really suck up the carbon dioxide and set flowers like there's no tomorrow.

With energy conservation in mind, modern greenhouses are being built a lot tighter than they used to be, with far less outside air circulation so as to prevent heat loss. This makes getting even a minimum amount of carbon dioxide into the greenhouse from outside air difficult. Modern greenhouse operations report that their plants consume naturally occurring CO_2 in the greenhouse by 10:00 a.m. This radically slows their growth, since once the carbon dioxide level drops to 200 ppm, photosynthesis is inhibited; growth is cut by 50%.[58] Eventually, additional carbon dioxide will be a required ingredient in all energy-efficient greenhouses, not just those looking for exceptionally high yields.

Seedlings and plants harvested for their vegetation fare better at CO_2 levels between 800 and 1000 ppm.[59] Although leafy greens grow best at a lower concentration of CO_2, they actually can consume greater quantities per hour than flowering or fruiting plants. This is due to the vegetative phase of plant growth being much more rapid than the flowering or fruiting phase. The input of CO_2 can reduce the growth time until harvesting time for leafy greens by half. The ADM greenhouse produces a mature lettuce head in 30 days from transplant,[60] compared to the 60 days I was used to at my California farm.

For most ways of producing vegetables or flowers, you want to maintain a level of CO_2 somewhere between 1000 to 1500 ppm.[61] In general, you'll find most crops will increase their photosynthesis about 50% when you maintain a 1000-ppm level.

Since plants use carbon dioxide only during the day, you would generally store the gas collected at night for use the following day. You can compress the gas and store it in a tank like compressed air or use the floating tank method as is done in methane storage. Of course, if you are extending day length with high-pressure sodium lights, you'd need to supplement the greenhouse with CO_2 during the additional lighted hours as well.

So, how much CO_2 should you use? Well, of course, it varies with light intensity from day to day and week to week, but to simplify a complicated calculation, we can average things out. The following example analyzes the CO_2 requirements in a greenhouse eight feet tall at the sides and twice that high in the middle. This calculation assumes nearly mature plants, which are using gas at a higher rate than little sprouts.

The first addition of CO_2 would usually occur an hour before sunrise. On an "average" light intensity day, to increase the CO_2 from 300 ppm to 1300 ppm would take about 1.7 pounds of CO_2 for every 120 square yards of bed space.[62] However, on a sunny day, in a pretty tight double-wall polyethylene greenhouse (where one-third of the atmosphere is replaced per hour), you'd expect to add an additional 1.3 pounds of CO_2 per hour per 120 square yards of bed space. For a standard 30-foot by 100-foot greenhouse (111 square yards), that would come to 144 pounds of CO_2 per hour of daylight.[63]

Since the study quoted above was conducted in Ontario, Canada, you can imagine that the brightness of sunlight varies widely from winter to summer. The "average" day cited above would require 4754 pounds of CO_2 per acre per hour in the month of January, and, due to the short day length in January in Canada, the CO_2 would only be added for 82 hours during the month. In July, an average day would require an enrichment rate of 16,408 pounds of CO_2 per acre per hour, and the longer days of summer mean that we would need to add this CO_2 for 283 hours over the course of the month.[64]

Over the entire year, you'd need about 143 tons of CO_2 per acre.[65] To produce this much carbon dioxide, you would need to ferment about 360 tons of corn. This would mean that you would be producing approximately 44,000 gallons of alcohol (143 tons divided by 6.5 pounds per gallon). Of course, you would end up with about 120 tons of DDGS, as well.

Among greenhouses' multiple uses is the nighttime heating of residences or other buildings. A greenhouse will almost definitely pay for itself in a couple of years if you use it for heating your home. Greenhouses can store the day's heat in some sort of thermal mass—in as simple a device as a concrete block wall painted black or 55-gallon water drums painted black and stacked in the "back" of the greenhouse, on the side away from the sun. During the day, sun will heat the mass, and at night the mass will radiate the heated greenhouse air, which can then be blown to where you need it.

Plants would rather not breathe CO_2 at night, preferring oxygen when the sun goes down. An hour or so before sunset, use exhaust fans at the

Fig. 11-24 Carbon dioxide storage. *This tank holds several tons of carbon dioxide in liquid form.*

IMAGE COURTESY OF TOMCO EQUIPMENT COMPANY

bottom of your greenhouse to remove all the carbon dioxide from the greenhouse. Allow the greenhouse temperature to stabilize after your thermal mass heats the new air you've drawn in. Then circulate the warm air into your home at night. In very cold climates, it's possible to use an air-to-air heat exchanger to use the carbon-dioxide-rich warm air being blown out to heat the fresh air coming in. As mentioned above, carbon dioxide should be compressed and stored until the following sunrise.

Carbon dioxide can be fed to some outdoor crops, but is most commonly used in a greenhouse environment where it's not as easily dissipated. If you are growing row crops that eventually have a closed canopy, carbon dioxide in the field can bring excellent results. This has also been proven true in tree crops. Since the stomata that absorb the gas are on the bottom of most leaves, keeping the carbon dioxide distribution tubing (which can double as drip irrigation tubing) at ground level is the best strategy with row crops.

Experimental plots of field-grown rice exposed to higher concentrations of CO_2 than usual have developed up to and in excess of 50% higher yields.[66] Several studies have shown that enrichment with CO_2 increases sucrose, starch, and total nonstructural carbohydrate (TNC) concentrations in rice vegetative tissues.[67] Also, rice can be ratooned like sugarcane, where it is cut about three feet above the ground and allowed to produce a second crop. When grown with a doubled level of CO_2, the second crop of rice had double the normal production for a ratooned crop.[68] This was not unexpected but was still a dramatic result.

Ominous response to high temperatures has been a real problem in studies on rice, which has serious implications for rice production in areas being heated by global warming. Unlike with many other crops, CO_2 enrichment of rice did not increase high-temperature tolerance or yield. At lower temperatures ($82.4°F$), and with the CO_2 at about twice the normal level, the carbon dioxide increased open-air yields of U.S. Southern rice varieties from 46 to 71%, depending on variety. But the benefits of CO_2 were less noticeable at higher temperatures, the yield was lower at $104°F$, and the rice died just as it did at lower CO_2 levels.

The longest-running open-air CO_2 enrichment study has been done on sour orange trees. At low light levels of 5% of full sunlight, in normal air, photosynthesis ceases. But with 300 ppm extra CO_2, photosynthesis continues all the way down to 0.5% sunlight. Thus, lower leaves and trees with an overcanopy would continue to produce carbohydrates during low-light periods that were not productive otherwise.[71]

GLOBAL WARMING—IS THE PROBLEM THE SOLUTION?

Lots of wild schemes have been posed about what to do with fossil fuel carbon dioxide. Many have to do with somehow disposing of it as if it's toxic waste rather than a surplus to exploit. "Solutions" talk of injecting carbon dioxide deep in the earth to sequester it and prevent its release to the atmosphere.[69]

In the case of cars, it's clearly better to burn alcohol, so as not to put net carbon dioxide into the atmosphere. But once that problem is solved, what about all the utility companies and factories that burn natural gas, coal, or oil to make process energy for products and electricity?

Permaculturally, all this fossil carbon dioxide is a resource that is not currently properly designed into a system for use. If all that CO_2 were scrubbed of toxic chemicals, it could be pipelined to where we want to increase photosynthesis. It seems like common sense to sequester the carbon dioxide in plants that breathe it and make useful carbohydrates out of it while generating oxygen, rather than just hiding the CO_2 underground.

If CO_2 were piped to agricultural fields, it would at least double the yields. If we bubbled it through coastal ocean kelp beds, we'd recapture the gas in the algae carbohydrates, while soaking up the ocean-heating sunlight, oxygenating the water to the benefit of krill and sea life, and more than doubling the ability of kelp to absorb coastal sewage.

Maybe the most urgent use would be to pipe it to the tundra, seasonally, where we need to rapidly grow snow-trapping shrubs or trees; these would help keep the tundra cold, so it doesn't kill us with methane releases.

Twenty-two percent of the world's forests are in Northern Russia, and there is talk of clear-cutting them. These are just about as important to the Earth's breathing as the Amazon; cutting there should never be greater than biomass produced. Massive amounts of European carbon dioxide could be put to profitable use by being piped in, substantially increasing the forests' growth; this alone might make it possible to economically sustain a reasonable harvest of wood products without clear-cutting.[70] Pipelines carrying carbon dioxide could be cheap, made of plastics made from carbohydrates, since they would operate under modest pressures.

Perhaps when it comes to global warming, we can harness the problem to create the solution.

In the high-stress situation of high temperatures, the sour orange tree, unlike rice, did well. At 107°F, with increased CO_2, it increased its photosynthetic rate by 200%. High temperatures like this stop growth in normal trees at normal levels of carbon dioxide, but the additional CO_2 increased the temperature range of photosynthesis to 15 degrees higher than controls.

At a more ordinary 95°F, the increase in photosynthesis was 100% above normal. At this temperature, leaf weight (which is one way of measuring photosynthetic surface, the size of the leaves) increased by 40%.[72] And more green "solar panel" area equals more carbohydrate production. The yield of oranges was triple that of trees not getting the extra CO_2.[73] The study also found that the density of the wood of orange trees that received extra CO_2 was 5.8% higher. (That result has implications for fuel wood production, as well.)

A caveat: Not all plants will increase photosynthesis in high temperatures. Full-canopy crops use their thick green foliage as protection against wind, which would dissipate the carbon dioxide gas.

Crops such as Jerusalem artichokes, sweet sorghum, sugarcane, lettuce, broccoli, leafy greens (like kale and collards), and perhaps squashes like buffalo gourd, would all benefit from additional carbon dioxide diffusion. In arid areas where high-value crops are grown with drip irrigation, it is possible to use the drip system to deliver carbon dioxide when it's not delivering water.

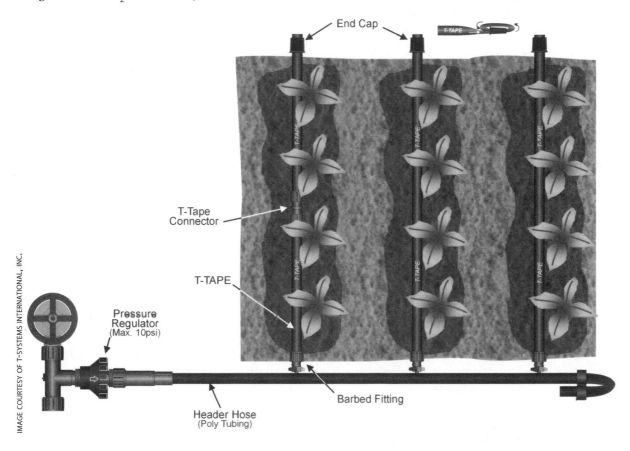

IMAGE COURTESY OF T-SYSTEMS INTERNATIONAL, INC.

Fig. 11-25 T-tape drip irrigation. Drip irrigation conserves up to 75% of water compared to sprinklers and permits accurate delivery of liquid fertilizer, such as filtered fishpond water.

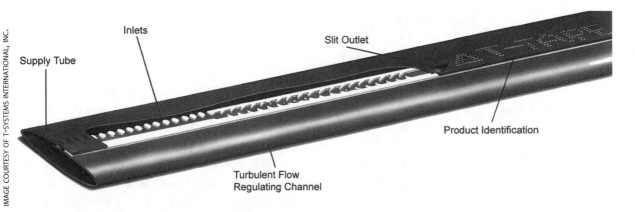

IMAGE COURTESY OF T-SYSTEMS INTERNATIONAL, INC.

Fig. 11-26 T-tape interior. T-tape design prevents clogging and can be used for delivering low-pressure carbon dioxide to the field instead of water.

Carbon dioxide from fermentation is often superior to that produced by flue gases from natural gas, kerosene, and propane boilers, or exhaust from generator engines, since the CO_2 does not contain toxins like sulfur, carbon monoxide, or oxides of nitrogen. However, the fermentation yeast often make some trace organic chemicals that can affect plant growth. These organic chemicals evaporate with the CO_2 and will be carried into your greenhouse or fields along with it unless you do some minimal processing of the CO_2.

So it's best to at least wash the CO_2 gas in a countercurrent water column filled with pall rings (see Chapter 9). The water column could be as small as ten inches in diameter by six feet tall; this size can process 50 to 100 pounds of CO_2 per hour. The water captures most of the organic impurities as it flows downward, while the CO_2 gas flows upward, bubbling through the water.

After the CO_2 has been washed, you could get more sophisticated and run it though another column with activated charcoal to remove even the smallest traces of organic vapors, making it saleable for purposes other than greenhouse use. For a greenhouse, though, the activated charcoal step isn't generally necessary.

If you are not going to use the carbon dioxide yourself, it may pay to scrub, dry, and compress it, and sell it as a liquid. You can retail it to welders or soft drink or fire extinguisher companies.

Your ability to market carbon dioxide profitably depends on how close you are to sources of natural-gas-produced carbon dioxide. In a medium-sized operation making 40,000+ gallons of ethanol a year, this byproduct can be a significant enough source of revenue to warrant marketing to retail users of compressed CO_2. The 143 tons discussed above for the Ontario farm could be worth $71,500 if cleaned to soda pop standards and sold for 25 cents a pound! You'd make more money by using it for lettuce or algae to feed tilapia, but if there's no market for those, selling it industrially may be your only option. You could use the proceeds from selling to industry to finance the building of your greenhouses.

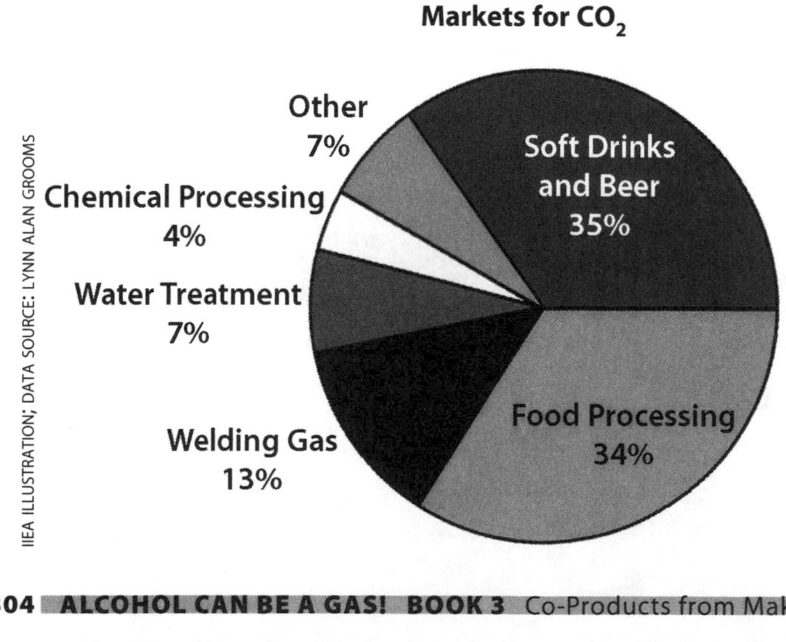

SINGLE-CELL PROTEIN

Single-cell protein (SCP) is a general term for bacteria, fungi, or yeast that are grown in a nutrient solution and then harvested in some way for food. Yeast, bacteria, or cyanobacteria, which are almost 100% digestible because of their relatively simple structure, can be excellent food for animals and, in many cases, humans. SCP has the potential to replace many more expensive ingredients in animal feeds, such as skim milk. It's an excellent ruminant feed, as well.

In general, though, Americans reject plant or bacterial protein in favor of beef and other animal proteins. For this reason, the most important studies in bacteria farming are being carried out in other countries. The field is growing fast, and discoveries of valuable new food sources appear regularly.

An important paper at the Fifth International Alcohol Fuel Technology Symposium indicates great promise for the small-scale alcohol producer.[74] The bacterium under consideration was *Geotrichum candidum* (CBS 187.67).[75] It has the potential for providing an affordable basis for stillage treatment, and a valuable byproduct for the producer. The stillage in the study had been used as backslop several times to concentrate the solubles. Reusing spent mash reduces the amount of water your plant will use, and the amount of stillage for disposal.

In most cases, liquid stillage is a mediocre animal feed, coming from feedstocks like sugarcane or fruit, with minimal nutritional value, and is a nuisance to dispose of except by composting or methane production. Although thin stillage doesn't have the proper nutrients for large animals, bacteria find it quite tasty.

Molds are almost the only other life that can coexist with *Geotrichum*. Introduced to your stillage, *Geotrichum* rapidly forms a surface layer of

Fig. 11-27
Markets for CO_2.
From natural sources and as a byproduct of industrial processes. Food processing and carbonation of drinks are the big markets.[76]

Markets for CO_2

- Other 7%
- Soft Drinks and Beer 35%
- Chemical Processing 4%
- Water Treatment 7%
- Welding Gas 13%
- Food Processing 34%

IIEA ILLUSTRATION; DATA SOURCE: LYNN ALAN GROOMS

mycelium (fiber-like strands), like a crust that limits the kinds of organisms that can invade and contaminate the liquid. This fibrous growth characteristic makes it easy to harvest compared to other single-cell proteins, like yeast, which need centrifuges or other ways to separate them from the residual liquid.

As the bacteria grow, several things happen. The thin stillage is dewatered up to 90% by the evaporative effect of the bacteria. The oxygen demand level in the stillage drops 50 to 60%. The fibrous bacteria yield about a pound (dry weight) per ten gallons of stillage. Dry solids content of the fresh bacteria varies between 10 to 30%, of which 50% is protein. But there is more value in this than protein. Many minerals, vitamins, and metabolites that aren't in soybean meal are in single-cell protein.

Two thousand gallons of mash would produce enough stillage to yield about 200 pounds of bacteria (dry weight), or 100 pounds of pure protein. These figures apply to bacteria grown on sugarcane stillage, while the yield from thin grain stillage will most likely be quite a bit higher. SCP produced on the more nutrient-dense stillage from corn should at least double the yield from sugarcane.

If you are feeding animals, SCP production would be a much more energy-efficient way to dewater the solubles from grain alcohol production, rather than evaporating the liquid from the solubles, which is responsible for a lot of the energy used in producing DDGS. In this case, we get the bacteria to do it with almost no energy demand.

Since *Geotrichum* is a bacterium and not a plant, the proportions of amino acids are rearranged, more like what an animal needs. In one test of a single-cell protein's suitability as feed, the bacteria were cooked with steam for ten minutes and fed to tropical carp in various percentages, including 100%. The fish showed no **histological** damage, even at 100%. The best growth rates were at levels of approximately 15%. But the researchers noted that the highest growth rate was not significantly higher than the other growth rates, so if single-cell protein were the cheapest feed source around, it would definitely make sense to use it exclusively and suffer a slightly lower growth rate.

There have been many other trials with single-cell protein using a wide variety of organisms. In fact, the yeast that ferments lactose in cheese whey has been thoroughly investigated as an SCP organism, as early as 1980.[77]

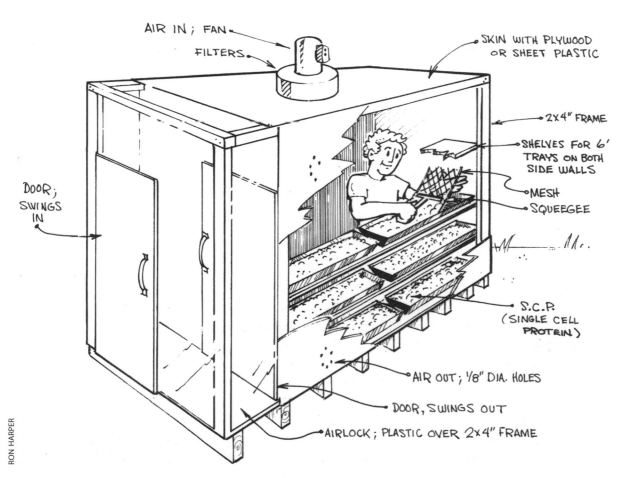

AIR IN; FAN

FILTERS

SKIN WITH PLYWOOD OR SHEET PLASTIC

2x4" FRAME

SHELVES FOR 6' TRAYS ON BOTH SIDE WALLS

MESH

SQUEEGEE

DOOR; SWINGS IN

S.C.P. (SINGLE CELL PROTEIN)

AIR OUT; 1/8" DIA. HOLES

DOOR, SWINGS OUT

AIRLOCK; PLASTIC OVER 2x4" FRAME

RON HARPER

Fig. 11-28 Single-cell protein farm. A grow room like this would work for a small-scale operation. Ocean-going shipping containers would work well for this purpose.

Since we feed corn to cattle primarily for its protein content (in part for its fat content), it becomes obvious that with a large enough alcohol industry, yeast as a high-quality animal feed becomes quite feasible.

Using SCP to raise tilapia might very well be one of the best ways to turn this protein source into a high-value human food, rather than a low-cost cattle feed. Using thin stillage to produce SCP as fish food should yield roughly .4-.5 pounds of fish for every gallon of alcohol produced. This amount of fish can be worth as much as $5, if sold live to local restaurants. *Geotrichum* is very easy to propagate. A used shipping container makes a good, cheap, airtight grow room. No sunlight is necessary since these are bacteria, not plants. To prevent contamination by wild molds or bacteria, a fan blows air into the grow room through a HEPA (high-efficiency particle air) filter. This ensures that there's always positive pressure of reasonably antiseptic air in the room. So when you open the door, air rushes out instead of carrying microbes into the room.

Pump hot, sterile stillage from your still through a large, disinfected, solids-separation screen to remove solids large enough for animal feed, mushroom cultivation, or other uses, and into plastic trays about five inches deep. Stack the trays bunk-bed style in a simple greenhouse-type structure (see Figure 11-28). Add the *Geotrichum* bacteria to each tray and allow them to grow for about a week. At the end of that time, *Geotrichum*'s fibrous crust is skimmed off the top of the remaining inch or so of liquid.

If you plan to store the bacteria, dry it at temperatures below 200°F. Drying desiccates the bacteria, without denaturing the protein. If you plan to use the bacteria fresh, you may or may not want to briefly cook it to make it more palatable to fish. Try it wet first to see how much they like it before bothering with a cooking step.

THE NEW MOONSHINERS

Back in the 1980s, one of my students from Ukiah, in Mendocino County, California, had an interesting use for carbon dioxide. His crop (which is also the county's largest cash crop, even though it's not strictly legal) used to take him about seven months to grow. Pumping CO_2 from his fermenters in among his plants via perforated plastic pipes allowed his crop to mature in five months—with particularly potent characteristics.

All the liquid stillage left over after SCP cultivation has a much-reduced oxygen demand and is ideal for mixing into a compost pile for use as fertilizer.

Although there's no established futures market for SCP at this time, we can make some general conclusions as to its resale value. Based on the current price of soybean meal (SBM) of $221 per ton at 45–48% protein,[78] SCP should be worth at least $225 per ton of dry bacteria. That comes out to only 11 cents per gallon of alcohol produced.

But SCP is much closer to yeast in composition than SBM is, having B_{12} and other B vitamins, a more ideal amino acid balance, and other vitamins and minerals. It would therefore be much more likely to be comparable to dried brewer's yeast. A ton of dry yeast, also considered an SCP, goes for $600, or 30 cents per gallon of alcohol produced, as animal feed supplement.[79]

It's entirely possible that an efficient SCP production with a good market could actually offset most, if not all, of the cost of the corn going into the alcohol plant. But, obviously, processing it through small livestock, such as fish, is far more profitable than selling it to others.

YEAST

In large alcohol plants, yeast can be separated from spent mash by centrifuge. In small plants, it becomes part of the distiller's solubles, which can be used to produce other products, or else it is partially reused as nutrients in backslop to feed the new living yeast in the next batch of mash.

As a source of single-cell protein, yeast is a very nutritious product. Australians certainly know all about Vegemite sandwiches, made from a nearly black yeast concentrate that has a flavor that grows on you. Lots of snack food products use yeast for flavoring (e.g., "smokehouse" flavoring in sauces or on almonds).

Like many sources of single-cell protein, yeast contains a lot of protein (45%), fat, and carbohydrates.[80] It is currently harvested in some of the larger alcohol plants as a high-value feed additive. While taking a tour of the old Williams ethanol plant in Pekin, Illinois, I saw a very efficient collection and packing system that made yeast sales one of the most profitable divisions of the plant.

Since we feed corn to cattle primarily for its protein content (in part for its fat content), it becomes obvious that with a large enough alcohol industry, yeast as a high-quality animal feed becomes quite feasible. Biologist Barry Commoner thought about this when doing projections for what alcohol production might mean for the nation and concluded that we could replace essentially all the corn we use in cattle feed with yeast coming from sugar crop and cellulose fermentation.

But Commoner did not anticipate the strange path we've taken to get to cellulose fermentation with GMO enzymes and yeasts. In the process, there would be a primary fermentation of cellulosic, C6 sugars first with standard yeast, followed by separation of that yeast as a feed product. Then would come a second fermentation of the stillage with the GMO C5-sugar-consuming yeasts. This stillage would then go back into the soil as compost from the spent mash, in theory fully cooked during distillation, where such yeast are unlikely to survive.

The GMO question certainly complicates what looked to be a simple solution to the animal feed equation. But if this process can be adequately worked out, and all our animal feed protein can come from yeast as a byproduct of cellulose conversion to alcohol, the land used for 87% of corn production (i.e., animal feed) can be freed up for other crops.

WASTE HEAT

Many alcohol plants have a great deal of excess low-grade heat in the form of 120 to 150°F water. Water-cooled distillery condensers produce water this temperature. If you use a heat exchanger in the cooling of mash before inoculation by yeast, you will produce more of this low-grade warm water. The water can be stored in an insulated tank for home use, and can be made hotter for space heating by circulating it through solar panels or a biomass-fired heat exchanger.

Water at 130–150°F can be used very effectively for baseboard heaters or especially for radiant floor heat, where plastic tubing is run in serpentine patterns before the concrete of the floor is poured. Running the hot water through the tubing heats the concrete, radiating heat into the room. It also puts the heat where it is needed most, at the bottom of the space. Heated air usually immediately rises to the top of a room, while the cool air stays down where the users are.

Radiant heat can be very valuable anywhere that people need to live when the weather is cold, and also in mushroom-raising facilities, at the bottom of fish tanks, under chicken or duck coops,

PHOTO COURTESY OF CALIFORNIA SOLAR

Fig. 11-29
Radiant heat tubing. This tubing is attached to the concrete reinforcing wire and suspended above the ground. When the concrete is poured, the tubing will be encased where it can be used to heat the floor.

for keeping a methane digester at optimum temperature, or even for warming the ground below greenhouse crops. In all these locations, the surplus hot water can replace expensive fossil fuel and be used to create income.

I favor storing hot water in tanks that minimize their surface area to volume. Generally, the larger the tank, the lower the ratio of surface area to volume is, which limits heat loss. Tanks should be insulated with one, or preferably two, layers of straw bales, with a decent roof to protect the bales. Bales can be clay- or lime-plastered to protect them from very occasional windblown rain. Otherwise, siding should protect the bales from direct rainfall.

Well-insulated tanks should lose only a few degrees a week. Once you have this hot water "battery," any heat source can add heat to it—solar panels, cogenerator waste heat, even heat exchangers embedded in black asphalt on a hot day.

BIOMASS FUEL

After fermentation, most solid and liquid waste is used as DDG and liquid feed, depending on your feedstock. But some fruits, like apples, yield no real valuable feed—just a bit of pulp and the skins. If you aren't going to compost or make methane out of these materials, you might consider making them into dry fuel chips.

Sometimes it is possible to use screw-press animal feed pelletizers to take less valuable byproducts, dewater them, and make them into pellets that either have a feed value or can be used as pelletized fuel in stoves or heating. Most fruit and cane-type feedstocks can be good sources of energy if you sun-dry them, then use them to fuel your boiler.

Woody biomass contains roughly 25% lignin, the energy-dense resin that holds the wood together. Most cellulose-to-alcohol processes can pretty easily separate out lignin as a byproduct which air-dries quite nicely and would contain just about the right amount of energy to fire the boilers needed to process the wood sugars to alcohol.

Just as you can't separate a tree from its mycorrhizae and expect it to flourish, you can't separate alcohol production from the intelligent use of its surplus products and expect to make a living. The co-products are even more important to your plant's bottom line than the alcohol itself. Poor design—treating the co-products as waste—would actually increase work, potentially become pollution, and, instead of creating profit, end up costing money to dispose of.

The world has more problems than just getting enough fuel. Designing uses for all your plant surplus, in the permaculture way, ensures a clean facility that is environmentally beneficial, steals markets from fossil fuels, and generates an abundance of food.

The best thing about this sort of design is that it does not follow the usual economy of scale. It benefits the smaller producer more than the giant plants. Ultimately, if you're anything like me, orchestrating an elegant design in which all the components balance each other and make a profit is where all the fun resides in doing this work. So, let's look at some scenarios for integrating alcohol with its co-products.

Endnotes

1. *Distillers Grains By-Products in Livestock and Poultry Feeds, Nutrient Profiles*, University of Minnesota, St. Paul, Dept. of Animal Science, www.ddgs.umn.edu/profiles.htm (July 27, 2005).

2. T. Klopfenstein, "Distillers Grains for Beef Cattle," National Corn Growers Association Ethanol Co-Products Workshop (Lincoln, NE), November 7, 2001, Table 5.

3. W.P. Garrigus, "Digestibility Studies with Distillers Grains with Solubles," *Kentucky Agricultural Experiment Station Bulletin* (June 1951), 564–579.

4. W.M. Beeson, et al., "Nutritional Factors Affecting the Utilization of Dry and Liquid High-Urea Supplements," *Purdue Agr. Exp. Sta. Cattle Feeders Day Report* (1969), in *Distillers Feeds* (Distillers Feed Research Council), 46.

5. Beeson.

6. Klopfenstein.

7. A. DiCostanzo, "Use of "New Generation" DDGS in Ruminant Diets," U.S. Grains Council Spanish DDGS Users Meeting, November 17–18, 2003.

8. A. Trenkle, "Evaluation of Wet and Dry Distillers Grains with Solubles for Finishing Holstein Steers," *Iowa State University Animal Industry Report* (November 2003).

9. Jerry Shurson and Mindy Spiehs, *Feeding Recommendations and Example Diets Containing Minnesota-South Dakota Produced DDGS for Swine*, for the University of Minnesota, St. Paul, Dept. of Animal Science.

10. B.S. Lumpkins, A.B. Batal, and N.M. Dale, "Evaluation of Distillers Dried Grains with Solubles as a Feed Ingredient for Broilers," Southern Poultry Science Meeting, January 2003.

11. L.M. Potter, "Studies with Distillers Feeds in Turkey Rations," *Proceedings Distillers Feed Research Council Conference 1966*, 47–51, www.ddgs.umn.edu/diets-swine.htm.

12. B.S. Lumpkins, A.B. Batal, and N.M. Dale, "The Use of Distillers Dried Grains Plus Solubles (DDGS) for Laying Hens," Southern Poultry Science Meeting, January 2003.

13. C.M. McCay, et al., "The Place of Corn Distillers Dried Solubles in Dog Feeds," *Proc. Eighteenth Distillers Feed Conference (1963)*, in *Distillers Feeds* (Distillers Feed Research Council), 72.

14. Jeff Knott, Jerry Shurson, and John Goihl, *Effects of the Nutrient Variability of Distiller's Solubles and Grains within Ethanol Plants and the Amount of Distiller's Solubles Blended with Distiller's Grains on Fat, Protein, and Phosphorus Content of DDGS*, for the University of Minnesota, www.ddgs.umn.edu/research-quality.htm (December 12, 2005).

15. R. Chang, "Functional Properties of Edible Mushrooms," *Nutrition Reviews* 54:11 (November 1996), S91–S93.

16. Chang.

17. N. Abdullah, A.D. Khan, and N. Ejaz, "Influence of Nutrients Carbon and Nitrogen Supplementation on Biodegradation of Wheat Straw by *Trametes versicolor*," *Micologia Aplicada International* 16:1 (2004), 7.

18. Abdullah, Khan, and Ejaz, 7–12.

19. John Holliday, Director of Research, Aloha Mushrooms, communication with author.

20. Abdullah, Khan, and Ejaz, 7–12.

21. Y. Victor Wu, Ronald R. Rosati, and Paul B. Brown, "Use of Corn-Derived Ethanol Coproducts and Synthetic Lysine and Tryptophan for Growth of Tilapia (*Oreochromis niloticus*) Fry," *J. Agric. Food Chem.* 45:6 (1977), 2174–77, Table 4.

22. Dan Helfrich, Greenhouse Manager, Archer Daniels Midland, communication with author, September 2003.

23. Marc Cardoso, CEO, EcoGenics, communication with author, May 28, 2005.

24. Cardoso.

25. Robert Hodam, *Energy Farming* (California Energy Commission, 1978), 10.

26. Helfrich.

27. Martin Andreas, Vice President, ADM Corporation, communication with author, 2003.

28. Andreas.

29. G. Venkataramani, "Mangroves Can Act as Shield against Tsunami," *The Hindu* [India], December 28, 2004 (www.hinduonnet.com/2004/12/28/stories/2004122805191300.htm).

30. Amir Neori, et al., "Seaweed Biofilters as Regulators of Water Quality in Integrated Fish-Seaweed Culture Units," *Aquaculture* 141 (1996), 183–199.

31. Alice Beetz, "Baitworm Production," *ATTRA National Sustainable Agriculture Information Service*, http://attra.ncat.org/attra-pub/PDF/baitworm.pdf (May 18, 2005).

32. J.P. Martin, J.H. Black, and R.M. Hawthorne, *Earthworm Biology and Production*, for University of Florida Institute of Food and Agricultural Sciences, (2005), 4.

33. Martin, Black, and Hawthorne.

34. Telephone survey of San Francisco Bay Area bait stores by Molly Graber, April 2005.

35. Survey of San Francisco Bay Area garden stores by David Blume, ongoing.

36. Survey of San Francisco Bay Area garden stores.

37. Martin, Black, and Hawthorne, 3.

38. "DDGS Market Update," *Ethanol Today* (November 2004), 13.

39. *Common Sense Organic Products*, www.commonsensecare.com/vermicompost.html (May 17, 2005).

40. *Common Sense Organic Products*.

41. Beetz.

42. Martin, Black, and Hawthorne, 7.

43. David Hoffman, communication with author, April 2005.

44. *Organic Amendment Report*, no. 05-109-033 (A & L Western Agricultural Laboratories).

45. C. Senthil, "Treatment and Bio-Energy Potential from Distillery Effluents," *Environmental Pollution Control Journal* 5:3 (March–April 2002), 2, Table 1.

46. Senthil, 3.

47. Senthil, 2–3, Table 1.

48. Cornelis van Dillewijn, *The Botany of Sugarcane* (Waltham: The Chronica Botanica Company, 1952), 234.

49. van Dillewijn, 245.

50. van Dillewijn, 150.

51. F.B. Green, et al., "Methane Fermentation, Submerged Gas Collection, and the Fate of Carbon in Advanced Integrated Wastewater Pond Systems," *Wat. Sci. Tech.* 31:12 (1995), 55–65.

52. Chris Clayton, "Ethanol Plant Would Run on Methane Produced from Cattle Manure," *Omaha World-Herald*, November 19, 2004, www.enn.com/today.html?id=410 (July 2005).

53. G.A. Deluga, et al., "Renewable Hydrogen from Ethanol by Autothermal Reforming," *Science* 303:13 (2004), 993–97.

54. T.J. Blom, et al., *Carbon Dioxide in Greenhouses* (Ontario, Canada: Ministry of Agriculture and Food, 2002), 1.

55. S.B. Idso, B.A. Kimball, and S.G. Allen, "CO_2 Enrichment of Sour Orange Trees: 2.5 Years into a Long-Term Experiment," *Plant Cell & Environment* 14 (1991), 351–353.

56. Blom.

57. Helfrich.

58. Blom.

59. Blom.

60. Helfrich.

61. Blom.

62. Blom.

63. Author's calculation.

64. Blom, Table 1.7.

65. Blom, Table 1.7.

66. Jeffrey T. Baker, "Yield Responses of Southern U.S. Rice Cultivars to CO_2 and Temperature," *Agricultural and Forest Meteorology* 122:3–4, (April 20, 2004), 129–137.

67. Baker, 135.

68. Baker, 134, Table 4.

69. Terry Macalister, "Buried at Sea: Shell's Plan for Greenhouse Gases," *The Guardian* [UK], March 9, 2006.

70. Patrick Perner, "The Business of Russia's Forest," *BISNIS Bulletin*, April 2004, www.bisnis.doc.gov/bisnis/BULLETIN/apr04bull5.

71. Idso, Kimball, and Allen.

72. Idso, Kimball, and Allen.

73. Idso, Kimball, and Allen, 351–353.

74. W. Tentscher, "Ethanol Production by Direct Fermentation of Sugarcane in Small-Scale Plants With Respect to Appropriate Technology," *Proceedings Fifth International Alcohol Fuel Technology Symposium* (1982), 301–07.

75. Tentscher.

76. Lynn Grooms, "Sparking CO_2 Interest," *Biofuels Journal* 2 (2005), 8.

77. Deborah S. Simpson, *Production of Yeast-Whey Biomass for Single Cell Protein* (Ontario, Canada: University of Guelph, 1980).

78. *Soybean Report*, USDA, www.ams.usda.gov/mnreports/GX_GR211.txt (July 15, 2005).

79. Mark Scott, of Diamond V Yeast, communication with author, July 15, 2005.

80. *Pekin Brewers Dried Yeast Data Sheet*, 43-P, no. A65MK501 (Pekin Brewers, October 2000).

CHAPTER 12

MICRO-DISTILLERY MODEL FARM

One afternoon, I sat down with my calculator and turned on the permaculture designer part of my mind. What might a farm produce with a micro alcohol fuel plant as its central component? Bear in mind, this is only a snapshot of a simple idea of integrating some of the co-products into a production system. It is not a fixed recipe, and you could find endless variations, based on the markets and climate of your area.

In permaculture, we have a tradition of failing small and often to learn lessons we can apply on the larger scale, so I am starting this experimental design on a very small scale. To be very conservative,

BEAR IN MIND, THIS IS ONLY A SNAPSHOT OF A SIMPLE IDEA OF INTEGRATING SOME OF THE CO-PRODUCTS INTO A PRODUCTION SYSTEM. IT IS NOT A FIXED RECIPE, AND YOU COULD FIND ENDLESS VARIATIONS, BASED ON THE MARKETS AND CLIMATE OF YOUR AREA. I'LL USE FISH, MUSHROOMS, AND WORMS AS AN EXAMPLE.

Fig. 12-1 Greenhouse cucumbers, grown under optimal conditions of elevated carbon dioxide and warm temperatures. *Archer Daniels Midland can produce three times the yield per square foot of outdoor farms year round.*

it is based on a micro-plant operating on purchased grain (at least initially) and running off a batch every four days. To make calculating easy, I am assuming it will process 1.15 tons of corn and will produce 100 gallons of alcohol per batch.

We are not using the stalks (corn stover) for cellulosic alcohol production in this scenario, but are instead discing them back into the ground to improve fertility. We will use a small fraction of the stover, about 11 tons, in mushroom cultivation.

Initially, this setup will produce a little more than 9000 gallons of 196- to 200-proof alcohol per year—enough to fuel a mini community-supported energy (CSE) co-op of about 18 vehicles traveling 10,000 miles a year at 20 mpg.

In this micro-model, the distillery also serves as a cooker and fermenter, which simplifies the equipment needs. We will also be using an external heat exchanger and a straw-bale-insulated hot water storage tank as part of the scheme. The flue of the distillery/cooker will have a heat exchanger in it to capture waste heat to be stored as hot water. The condenser will use a heat pump both to chill the alcohol and recapture that heat for a little more hot water. The same heat pump will be used to control the distillery column. The electricity will come from an on-site cogenerator that also heats water. If the plant owner drives a hybrid, this system will allow him to convert to plug-in hybrid electric, using surplus electricity from the alcohol-powered generator.

Process heat energy initially will come from waste wood products, such as orchard prunings and broken pallets, but eventually will come from a coppiced woodlot planted to produce firewood. We'll burn the wood waste or firewood in a cob rocket stove to keep the skill level and welding to a minimum. I'll briefly mention how an optional methane digester might be worked into this model, as well.

The distillery model is based on an eight-inch-diameter column, producing 180- to 190-proof alcohol at 16 gallons per hour. The alcohol could simultaneously be dried to 196–200 proof in a pressure swing corn grit water extraction system for dehydration of the alcohol. Although it's not strictly necessary to dry the alcohol to this level, the energy efficiency of this two-step approach is significant.

So there are some things we can observe right away. We will operate the still about 91 times a year, so we'll buy or grow 105 tons (91 × 1.15 tons per batch, or 3750 bushels) of grain per year to operate the plant. Assuming a yield of 160 bushels per acre, that means we are looking at approximately 23 acres of corn. Over time, this acreage will go down, and the yield will go up, as we apply what we learned from the experiments in Chapter 3.

Initially, this setup will produce a little more than 9000 gallons of 196- to 200-proof alcohol per year—enough to fuel a mini community-supported energy (CSE) co-op of about 18 vehicles traveling 10,000 miles a year at 20 mpg. We will sell the alcohol at $2.50 per gallon to CSE members and pass the federal tax **VEETC (Volumetric Ethanol Excise Tax Credit)** of 51 cents per gallon on to the drivers. The driver's net cost will then be $1.99 per gallon plus sales and road taxes. We'll keep the producer's tax credit of ten cents per gallon.

This plant will produce about 33 tons dry weight of wet distiller's grains (WDG), and another three tons of nutrients in the thin stillage, a.k.a. distiller's solubles (DS). Although it's possible to use solar energy to dry the WDG to distiller's dried grains (DDG) for storage, here we are using the WDG as it's produced. Unlike large alcohol plants, we will not be evaporating the water and condensing the solubles to be mixed with DDG.

The plant will also produce 50 tons of carbon dioxide, which we plan to use entirely on the farm. If we could find a local market, the CO_2 might sell for greenhouse use at $5000–$8000.

From the permaculture vantage point, making alcohol and then selling the feed and CO_2 would be letting most of the valuable resources leave the property, without extracting all the various yields from these surpluses. If we simply sold the 33 tons of DDG wet as feed to a local dairy, we might get only $1500 for it.

Our use of the DS is a key factor in how we proceed to co-product development. For this scenario, I have chosen to integrate the DS in both mushroom and fish-raising components. At a minimum, we could use the grain products in our own livestock operation. We could choose, for example, to convert WDG at a dry feed/product ratio of 10:1 in beef, a 3:1 ratio in chicken, 1.6:1 in fish, or close to 1:1 for shrimp or earthworms.

I'll use fish, mushrooms, and worms in this example. Even though I am using wet distiller's grains,

I give you what the dry weight of those grains would be in order to let you compare apples to apples.

Let's divide our annual production of 33 tons dry weight of WDG into three 11-ton parts. The first third of the WDG will directly become fish food. At least half of the food that tilapia eat can be WDG. The other half must include expensive lysine and methionine amino acid feeds to balance the amino acids in WDG—or we can figure out how to produce the needed amino acids on our farm. As you'll see in a minute, a byproduct of what we do with the second 11-ton portion of DDG will take care of our fish needs nicely.

The second 11 tons of the WDG will be used for oyster and/or shiitake mushroom production. The first step will be to take the WDG immediately from the distillery tank at the end of distillation when it is boiling hot and fully sterile. We will separate the granular WDG from the liquid DS in a modified washing machine (modified by changing pulley sizes on the motor and washing drum) on spin cycle. At 250 pounds per run (90 runs per year), using three mesh bags per spin cycle, this would require six five- to ten-minute batches. This process dewaters the WDG and leaves us with separate hot DS.

We will use some of these 900 gallons of boiling-hot DS to pasteurize and enrich shredded dry corn

stover with the soluble nutrients. We will use about the same dry weight of stover as the dry weight of the second third of the WDG (11 tons) or, to put it in terms of a single batch, a little less than 500 pounds of combined stover and WDG. That's about three three-string bales or ten two-string bales. The stover is needed to provide good aeration for the mushrooms' growth. We'll add some lime to balance pH. Remember, we have removed most of the carbohydrates from the grain already by fermentation. The carbohydrates that the mushrooms will eat will be mostly the cellulose in the stover and the small amount of cellulose in the WDG. Protein, fats, and minerals are provided by the WDG and the DS.

Mushroom-growing takes two small buildings, which can be made from ocean shipping containers or, better yet, straw-bale construction plastered with lime. The first building is a clean room (where all the air has been filtered to remove

Fig. 12-2

bacteria, etc.), and the second building is a spawn-running room.

The DS-pasteurized stover and the hot DDG are mixed on a table by hand in the clean room until the mixture cools to below 90°F. We then inoculate the stover/DDG mixture by mixing about ten pounds of oyster or shiitake mushroom spawn into the mass as it is packed into plastic bags. The spawn is laboratory-raised fungus on grain that is grown through with mycelium. The bags are then closed with a breathing device that lets gases exchange from the bag, but keeps bacteria out.

Earthworms are more efficient by far than chickens or fish. There is close to a one-to-one ratio of dry weight of food eaten to weight of worms produced. But once the worms are mature, most of the feed goes to produce castings.

Fig. 12-3 Agrocybe aegerita. *The swordbelt mushroom, growing on brewery grain. This is a prime edible mushroom, commanding a high price. Many say that it tastes like bacon.*

The bags then go on racks in the spawn-running room for the next two weeks. This room can be kept warm by running the warm water recovered from the alcohol plant through radiant heating tubes in the floor. The two weeks in the spawn-running room allow the mycelia of the mushrooms to branch out from the spawn particles to digest much of the DDG/stover substrate. The content of the bags will take on the appearance of dirty cotton.

After the mycelium has fully grown through the mix, the bags are moved to a cooler, more humid, fruiting room. The bags are slit to let out the CO_2 and to let the oxygen in. These new conditions imitate the onset of late fall, which induces fruiting of

the mushrooms through the slits made in the bags. The mushrooms are then harvested over about three weeks' time.

The 22 dry tons of mushroom substrate (combined stover and DDG) will produce about 22 tons of fresh oyster mushrooms (wet weight) annually, or a little less than 500 pounds per week. At a wholesale price of $2.50 per pound, the gross income for the 22 tons of mushrooms would be $110,000. Sure beats selling the WDG as cattle feed. Any mushrooms not meeting the cosmetic standards of the market can be fed to either the fish or earthworms.

Roughly half of the 22 tons of dry matter of the stover/DDG that we started with has now been used in producing the mushrooms. The spent mushroom substrate, now about 11 tons dry weight (although it is in a wet form), is essentially almost fully converted to mushroom mycelia. Whereas the original plant matter was low in the amino acids methionine and lysine, the process of growing the mushrooms rearranges the ratios of amino acids, and the substrate will now be an excellent feed, balanced in amino acids. We will return to using this surplus product in a moment.

So the first 11 tons of WDG became fish food. The second 11 tons of WDG were mixed with stover and turned into mushrooms. The last 11 tons of the WDG will go into specially built "livestock" pens.

As a permaculturist, I am allergic to *extra* work, since it usually indicates poor design. I often tell my students who want to raise livestock that, with a few exceptions, you are never going to have a vacation again, because those animals depend on you every day to keep them in line and alive. An exception to this rule is earthworms (see Chapter 11).

Part of the plan for this small plant is to build special bins that functionally optimize the production of worm castings (worm poop) over worm production. Although it's entirely possible to include worm production for worm sales, we will keep things simple in this example by leaving out that particular yield and focusing on the castings.

Earthworms are more efficient by far than chickens or fish. There is close to a one-to-one ratio of dry weight of food eaten to weight of worms produced. But once the worms are mature, most of the feed goes to produce castings. Conversion of the 11 tons of WDG dry weight at maximum saturation of worms (see Chapter 11) would yield about 22 tons of castings at their normal finished moisture content of 50%. It is not desirable to dry worm castings much below this, since it would kill much of the microlife that makes castings so valuable.

Remember the leftover 11 tons of mushroom substrate, after we harvested the mushroom? If we take half of the spent mushroom substrate (5.5 tons) and feed this to the worms, they will convert this fungal delicacy to another 11 tons of castings for a total of 33 tons of castings. (Or we can use all this spent substrate elsewhere. Bear with me here.)

Although worm castings are often sold by volume, e.g., cubic foot or cubic yard, a growing number of producers are selling by the pound; prices range between $1 and $1.90. So 22 tons of worm castings should be worth, retail, $44,000 to $83,600. If we include the castings from conversion of spent mushroom substrate, 33 tons would be worth $66,000 to $125,400.[1] You can expect to sell your castings at half this price wholesale, if you sell by the ton or cubic yard to a nursery supply.

In the states where medical marijuana is now legal, your best retail customers for organic castings could be the growers of this medicine. They are a very good retail market, especially if you also supply fish emulsion, carbon dioxide, and kelp solution. And they pay cash, too!

So far we've converted the WDG into worm castings, fish, and mushrooms. We still have a quarter of the spent mushroom substrate to consider. We will use that material to feed our livestock: tilapia.

Our fish's diet can be half DDG, but, for the other half, we need to add the nutrients the fish need in addition to the DDG. We can eliminate the cost of this supplemental feed (typically soybean meal) by feeding the other half of the spent mushroom substrate to them instead. So now the fish diet is 11 tons of DDG and about 5.5 tons dry weight of mushroom mycelium made of WDG/stover.

This 16.5 tons of feed at a conversion rate of 1.6 to 1 yields about 20,000 pounds of live fish biomass. About 45% of fish sales will be the 9000 pounds of full-size male fish for food, delivered in an oxygenated tank truck to nearby city restaurants. We can reasonably expect to get $7 to $10 per pound for this 4.5 tons of fish. That would mean the live fish income stream would yield a gross of $63,000 to $90,000.

The other 55% of the tilapia will be close to 11,000 pounds. These are females and culled fry, which currently don't have a market. These can be converted to fish emulsion. It takes about a pound of fish to make a gallon of fish emulsion. So let's be conservative and say these 55% will yield about 10,000 gallons of fish emulsion. This would be worth almost $120,000 if sold in five-gallon pails.[2] But sold in one-gallon bottles at $17.50,[3] it would be worth $175,000. If you sold it in quarts, the price would go up to $6.95, for a whopping $278,000 retail.[4] You could wholesale these products at half the retail price, if you don't want to market them yourself—although you could easily sell all of your production at a booth at a farmers' market, along with your vegetables.

How much space would fish-raising take? This quantity of fish could easily be raised in two 32,000-gallon tanks or, better yet, ponds inside a standard 30 × 100-foot greenhouse with two crops of fish a year. You could add another tank or two of this size to take advantage of algae and duckweed production as additional or alternate fish feed.

Let's take the model micro-distillery farm idea a little further, make it a little more complex and interesting. Remember the liquid DS we still have unaccounted for? We used some of it right away when we soaked the stover for mushroom production in it, but we still have most of it left. Since it is already liquid, we might as well figure out a way to use it with our fish.

As we said in Chapter 11, one way to feed fish without even using DDG is by raising massive quantities of algae. The limiting factors on algae growth are similar to those for land plants. We need to supply nutrients, sunlight, and carbon dioxide for a healthy algae population explosion, and if any one of the three is in short supply, then that will limit production. We can't do a lot about increasing the sunlight, but sunlight is not usually

---

MADE IN CHINA

As hungry humans in the U.S. and across the globe are demanding more organic food, a massive and profitable market niche has opened, and Chinese farmers are ready to fill it. While many U.S. farmers still seem hypnotized by the PR blitz of pesticide and genetically engineered seed companies like Monsanto, Chinese family farmers are happily enjoying the 20–50% higher annual profits they are receiving by going organic. "They earn more money. They don't have to worry about sales. They don't have to worry about storage. There's no reason why they shouldn't go for organic farming," said Wang Tingshuang, an organic farmer in China's northeastern agricultural province of Jilin.[5]

limiting. If we use DS as a nutrient source in the algae pond, we're okay as far as nutrients go. But even though the algae would have all the food and sunlight they'd need, they would quickly use up all the carbon dioxide in the water, so that would limit production. So we need more CO_2.

The carbon dioxide can come from three sources. One source, of course, is the fishpond. The fish have been breathing in the oxygen in the water and "exhaling" the CO_2 into the water. So their pond has a surplus of carbon dioxide in the water that the algae would lust for (if algae have such thoughts). Circulating the fishpond water into the algae pond would do double duty, adding not only CO_2, but also removing and converting surplus nutrients (fish poop).

But even if we recovered all the CO_2 that the fish make, we'd still be quite short. That's where the second source of carbon dioxide, fermenting alcohol, comes into play. The CO_2 pouring out through the fermentation locks on the still/fermenter can be bubbled into the bottom of the algae tank. If we do this, together with adding all the DS and some of the pond water, the reproduction rate of the algae will go through the roof.

The algae growth oxygenates the water, producing a massive oxygen surplus, and eliminates the oxygenating costs of normal fish farms, where they usually consume huge amounts of energy running various sorts of bubblers. The green, algae-dense, oxygen-rich water is slowly pumped over to the

fishpond, oxygenating it and feeding the fish even more tasty food and carbohydrates.

This process uses the nutrients in the DS, via carbon dioxide and sunlight, to produce several more tons of fish food, which would give you lots more fish to sell. This photosynthesis only happens during the daylight hours. At night, we would capture the CO_2 from nighttime fermentation, clean it, and compress it (see Chapter 11) for use the next day.

If we were to add methane production to the simple plant, it would be our third possible source of carbon dioxide. Instead of using the surplus DS to produce more fish food, we might first run the DS through the methane digester to extract carbohydrate energy as natural gas, and then use the outflow of the methane digester to fertilize the algae. (The conversion of DS to methane is standard practice in India.) The gas from the digester is 60% methane and 40% carbon dioxide. The process of cleaning methane in a water column dissolves the CO_2 into the water, which can be used as makeup water for the algae pond. This would make a lot of sense if you didn't want to use wood for process heat in your alcohol plant. The DS can produce enough methane to do all the heating and even produce electricity as a generator fuel for the plant.

Let's say you don't want to do something so intricate. You just want to feed the DDG to the fish and harvest the fish and not get all involved in this algae/carbon dioxide complexity. After all, the extra cost may not be worth the hassle if you don't have local markets where you can sell the fish or fish emulsion.

Let's look at the alternative. If we feed all this tonnage of high-protein DDG feed to the fish, they are going to eat it and then poop out a very high percentage of it. As much as 70–90% of the nitrogen is excreted, depending on where they are in their growth cycle. Fish water is loaded with concentrated nutrients and ammonia. In fact, the fish would die if the water wasn't changed or processed.

One use of the fish water is to pump it and the DS into low-value corn or other energy crop fields, where it would provide all the fertilizer for the next crop needed for next year's fuel. This assumes you turned the cornstalks or bean trash back into the soil at the end of the season. The combination of organic matter and fish water actually increases soil fertility and soil depth each year.

If you choose not to raise surplus algae, all of the 50 tons of surplus carbon dioxide could be piped into an attached greenhouse to rapidly grow high-value, off-season fruit or vegetables, such as strawberries, cucumbers, or tomatoes. The quantity of CO_2 from this micro-sized project could be used intensively in three to four standard 30 × 100-foot greenhouses (see Chapter 11).

So let's assume we are combining some fishpond soup, a small quantity of the leftover earthworm castings, and the carbon dioxide in greenhouses filled with raised beds. It would be *simple* to produce organic vegetable yields of three pounds per square foot, per plant cycle, of densely planted complementary crops. Furthermore, bottom heating of the soil via radiant heating tubing can profitably use lower-temperature surplus hot water from the plant. Three crop cycles can be easily completed in 11 months.

So that's nine pounds per square foot per year of raised-bed space. In three standard greenhouses, there would be 7200 square feet of crop area out of the 9000 total square feet (taking into account 20% used for paths between raised beds, and a 200-square-foot staging area). With a focus on high-value organic salad mix, tomatoes, cucumbers, or strawberries, the annual yield could easily be 65,000 pounds.

At retail prices of $2 to $4 per pound, you are looking at $120,000 to $240,000 of organic vegetable sales (on top of your fish sales). The farther away from California, the higher the average price will be.

Sound too good to be true? One of the most productive farms in the United States made far more money than this on organic salad mix, in a city lot in Berkeley, California, without all the luxurious rich and controlled conditions we are outlining in this system.[6] In a similar-sized greenhouse in New England, Anna Edey and her integrated livestock and greenhouse projected at full capacity

earnings of $180,000 in eight months producing salad alone, without most of the advantages of an alcohol-derived greenhouse.[7] (She did some CO_2 enrichment by housing animals in the same air space as the vegetable operation.)

While I am conservatively projecting three crop cycles, it should be noted that the Archer Daniels Midland greenhouse in Decatur, Illinois, gets more than ten crops of lettuce per year, as do some coastal California lettuce farms without any greenhouses. Pushed to the limit in a highly organized CO_2-enriched greenhouse, you could turn over a salad mix or head lettuce crop every 20–25 days.

Still sound like too much work? Then let's look at a really basic setup. Using inexpensive drip irrigation tubing, you can pump filtered fish water *and* carbon dioxide into your fuel-wood forest and energy crop fields, which would not only increase the yield of wood for fueling your distillery and eliminate the cost of fertilizer, but also increase the soil fertility and energy crop yield over time (see Chapter 3). So to keep operating the system using your own corn or wheat (wheat provides handy straw instead of stover), you'd end up having to farm less land in order to get the same yield, and to also make alcohol, and to produce value-added crops of mushrooms, fish products, high-value veggies, and earthworms.

To cut costs with our micro-distillery, there are a lot of ways to provide process heat for cooking and distilling by making use of the low-quality warm water that comes out of the plant. We mentioned above that the DS (along with other surpluses)

Fig. 12-4 Organic insect control. Between the author and Marty Andreas of ADM is what looks like a potted plant with a shade cover. These are hung throughout the greenhouse and are beneficial insect habitat. The beneficial bugs reproduce in these tiny thickets and then fan out to control the cucumber pests.

Fig. 12-5
***Greenhouse
lettuce at Archer
Daniels Midland.***
*Although ADM
employs soilless
growing tech-
niques, similar
yields and quality
are attained
with soil-based
methods in green-
houses, too.*

can first go to a methane digester, heated with hot-water coils. If, for instance, you are raising chick-ens on the mushroom waste and surplus worms, the chicken manure and bedding can go into the digester, too. Local surplus deep-fryer grease makes huge amounts of methane when mixed with the DS. Culled tilapia can be put in the digester as well, even though it is more profitable to make them into fish emulsion (see Chapter 11).

In any event, you should be able to produce more than enough methane to do all your distill-ing, cooking, and electricity production. Surplus warm water is useful for heating greenhouses, tila-pia/spirulina ponds, mushroom houses, methane digesters, and buildings.

We need to realize just how obscenely wasteful it is to put ten to 20 Btu of fossil fuel into each Btu of grain and then take ten Btu of grain to feed cattle in feedlots to get only one Btu of meat, only to poop almost all of that energy away in mountains of ground-poisoning manure.

If you want to fire your plant with wood instead of methane, and if you have the space, you can use part of the solubles and/or fish water to irrigate a fuel-wood forest. Providing firewood for process heat to make 9000 gallons of alcohol would take about 270 million Btu from wood. This amount of heat can be reasonably obtained from seven cords of wood (each stacked 4' × 4' × 8'), weighing some-where in the neighborhood of two tons per cord. Raising the wood sustainably with renewal pruning (coppicing) would take about 1½ acres of woodlot.[8] Of course, the understory of the 1.5 acres of wood-lot could produce crops of strawberries, blueberries,

hazelnuts, currants, flowers, raspberries, greens, etc., if you chose to expand into outdoor crops.

Here's how to use coppicing to grow high-quality firewood: Establish four major branches, or main leaders. Cut the oldest of the four branches each year. The four-year-old branch should be perfect stove-wood diameter and not need any splitting (extra work to be avoided). When multiple sprouts shoot out from the cut in the spring, pinch back all but one to replace the branch that was cut. In four years, you will have another full-sized branch to cut from that shoot. Each year, cut the oldest leader.

Coppiced woodlots work well with dense, hard, fast-growing, nitrogen-fixing leguminous trees or even fruit trees. According to Bob Cannard, master farmer and orchardist in California, using this cop-picing technique has resulted in apple trees pro-ducing for 600 years in Siberia and olive trees still producing after 1000 years in Greece.[9] He's also able to prune 25 acres of apple trees in a few days with this efficient method.

All of this is a simple example of permaculture thinking on the micro-plant scale. We can do much more complex stacked designs. Just think about what it would be like if we added a few fermenta-tion tanks, a cooker, and a slightly larger still, so we were running one shift each day, making 60,000 gallons a year instead of 9000 gallons.

The major point of this short exercise is to help you realize that alcohol production should be part of an overall diversified farm design. Furthermore, we need to realize just how obscenely wasteful it is to put ten to 20 Btu of fossil fuel into each Btu of grain and then take ten Btu of grain to feed cattle in feedlots to get only one Btu of meat, only to poop almost all of that energy away in mountains of ground-poisoning manure.

Since basic variations on the model farm could be accomplished, depending on complexity and climate, with perhaps five to 40 acres, the average Midwest farmer who controls hundreds, some-times thousands of acres is going to have a lot of extra land on his hands. In most places, the model becomes even more efficient when we start designing with crops *other than corn*.

Permacultural wisdom totally changes the modern-day question of how many square miles one person can farm, growing monoculture crops; the new question is how many good jobs for how many people can we produce per acre with energy/food polycultures? You can see we'd have

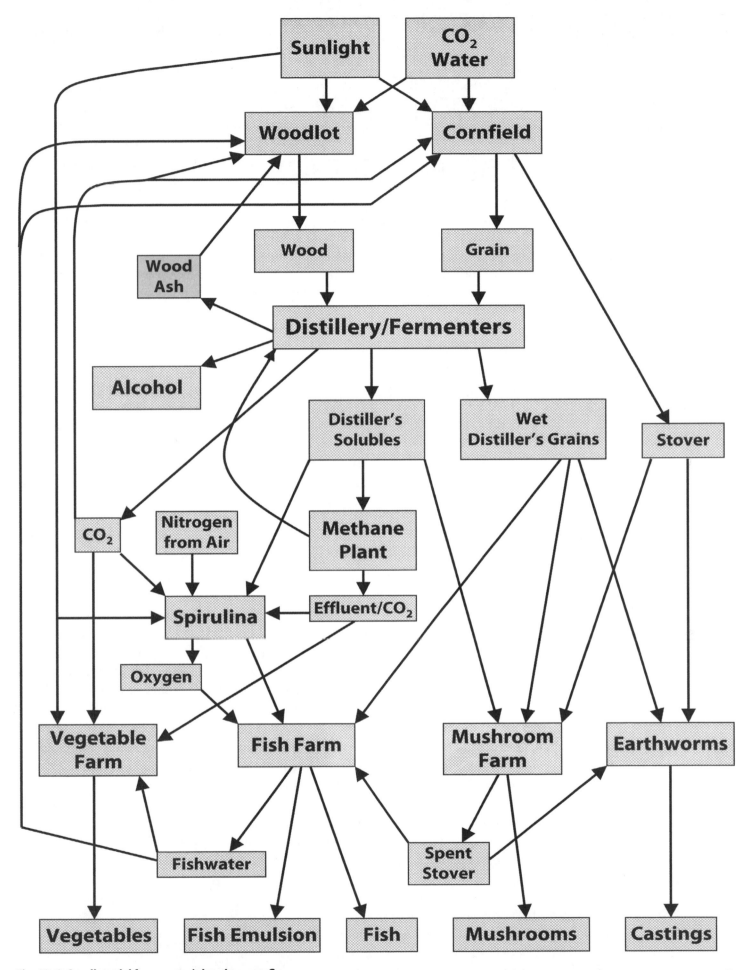

Fig. 12-6 Small model farm material and energy flow.

Micro-Scale Alcohol Plant Income

CROP	QUANTITY	UNIT PRICE	GROSS VALUE
Alcohol	9000 gal.	$ 2.50/gal.	$ 22,500
Mushrooms	44,000 lbs.	2.50/lb.	110,000
Worm Castings	33 tons	1.00/lb. whlse.	66,000
Live Fish	9000 lbs.	7.00/lb.	63,000
Fish Emulsion	10,000 gal.	6.00/gal. whlse.	60,000
Greenhouse Veggies	65,000 lbs.	2.50/lb.	162,500
TOTAL			$ 484,000

Fig. 12-7 Micro-scale alcohol plant income. Gross income from simplified model micro-farm biorefinery. Note that the alcohol is the least valuable product in this simplified example.

a big crew making middle-class wages if we had a complex permaculture design, making sure we got every yield possible out of whatever feedstock we chose to make alcohol from.

So what do the numbers add up to in a model micro-farm? In this simplified version, our main products are alcohol, fish, fish emulsion, mushrooms, earthworm castings, and high-value greenhouse vegetable crops. Surplus DS would go out as fertilizer along with surplus fish water, to provide all the fertilizer we need for the next year's energy crop and wood for process heat. The estimated gross receipts of $484,000 are detailed in Figure 12-7.

These are the gross figures, of course. A lot of labor would be involved in handling this much production, and that's a good thing. We can generously assume the nonlabor costs of production and marketing are 25% of the gross. That's $363,000 dollars available for labor and profit. Ten good rural incomes would be met by this amount, even if the laborers only worked about 25 hours a week.

And this is a simple permaculture design. There are much more complex and remunerative layers of production that can be built onto this simple model. A somewhat larger plant, producing 200–300 gallons 325 days per year (65,000–97,500 gallons) would take very little additional equipment, but the number of opportunities and complexity of products would multiply. What's more, *the density of jobs per acre would go up.*

So is it more efficient to produce alcohol in 50-million-gallon-per-year plants that have only one or two low-priced co-products? It depends on what you call efficiency. If you are trying to make the most alcohol with the least number of people, then *maybe* a big plant would be more efficient. If efficiency is defined as producing the most money from a given ton of corn, *maybe* wet mill corn biorefinery plants can do better.

But when it comes to how much money you can produce from an acre of energy crop, being small enough to get full retail from your value-added products sold locally makes the small scale far and away more efficient than large plants.

A most important yield is the number of jobs per acre you can create. If you look at the maximum number of farmer-owners who can make a good living on such an integrated farm, there is no comparison; large plants barely produce one direct alcohol plant job for every million gallons of alcohol and increase the price of a bushel of corn only about 20 cents. Yes, alcohol plants do circulate a lot of money into the local economy, and that creates jobs. But our small model farm can produce several good jobs for every 9000 gallons of alcohol produced, and circulate even more money locally.

In the long run, large plants cannot compete with smaller, multiple-yield, multiple-feedstock, permaculturally designed plants when it comes to benefits to a state and its people. Right now, there are places for all sorts of production scales in a national system of combined energy/food production. As time passes and the concern over energy is no longer price, but whether it is even available—decentralized, local production of fuel and food becomes vitally important. We should fill all the niches at each level of fuel production until we are energy-independent, self-sufficient in food, and fully employed.

Now that your head is spinning with the possibilities of all the products that can issue from your plant, it's time to refocus on alcohol. After all, that's what got us into the whole discussion to start with. Although it may not turn out to be the most profitable part of your enterprise, alcohol is the core product, and understanding how to use it is essential, whether you're selling it or making it part of your own energy supply. We'll start by hitting some of the surprising high points and bust a few myths in the process.

WHAT'S A PATH GOOD FOR?

Here's how I use an excavated path next to a raised garden bed:

One: It's a path, a way to get to my garden beds without compacting the soil.

Two: It's a swale if I lay out the path, and therefore the bed, on contour; during rainstorms it fills with water evenly, like a long skinny reservoir. The water then sinks into the soil in all directions instead of running off.

Three: I can improve the usefulness of the path and maintain its porosity to water if I fill it with partially rotted springy compost or wood chips. No more muddy feet, and I don't have to periodically loosen the path soil to get it to drain.

Four: The path provides rooting space. Now that it is filled with decomposing compost, roots can grow into it where before there was only air. Plants at the edge of the bed now get bigger than plants in the middle of the bed, since they have twice the space to root in and a rich nutrient source to boot.

Five: Seeding the paths with manure worms makes my paths into earthworm compost bins and produces more worms.

Six: When the path fills with water, it now becomes compost tea, fertilizing the beds. To avoid drowning, the worms burrow into the bed, with compost tea flowing in behind them through their fresh tunnels.

Seven: The path provides aeration. When the water in the swale recedes, the worms turn around and head back for the paths and food. Their returning tunnels provide aeration for the bed.

Eight: So what's a byproduct? If you take a one-foot slice across the four-foot-wide bed and path, you have four square feet of yield of vegetables from the bed and a one-cubic-foot yield of worms and castings from the path. At $1.50 per pound, the vegetables are worth $6. Harvesting half the worms in the cubic foot of path (one pound) brings in $22, and the castings are worth about $15. So which is the primary product and which is the byproduct?

Endnotes

1. Author's calculation.

2. Peaceful Valley Farm and Garden Supply, www.groworganic.com (June 24, 2005).

3. *Maxicrop Fish*, www.biconet.com/soil/MaxicropFish.html (June 24, 2005).

4. *Maxicrop Fish*.

5. Nao Nakanishi, *Organic Farming Set to Boom in China: Chinese Farmers Cash In on Need for Organic Food*, www.organicconsumers.org/organic/china102703.cfm (December 4, 2006). [Original Source: Reuters (October 25, 2003).]

6. Michael Olson, *MetroFarm: The Guide to Growing for Big Profit on a Small Parcel of Land* (TS Books, 1994).

7. Anna Edey, *Solviva: How to Grow $500,000 on One Acre and Peace on Earth* (Trailblazer Press, 1998), 158–159.

8. Stephen T. Cockerham, "Irrigation and Planting Density Affect River Red Gum Growth," *California Agriculture* 58:1, 43.

9. Bob Cannard, Master Farmer and Orchardist, communication with author, June 2000.

WHY MEGAOILRON IS OBSESSED WITH CONTROLLING THE PRESS

What follows is a story about Will Rogers, told by Utah Phillips.

I was thinking lately about old Hank Penny. Hank Penny was an old man. I don't know if he's still extant. We shared a park bench together in San Diego some years ago. Well, Hank Penny was an early Grandpa Jones. He played the banjo and the fiddle and the guitar and told tales and sang old songs.

He told me on that park bench that in the early days of radio, he had been on the first coast-to-coast radio broadcast in the history of the known universe. I questioned him about that and he said, "Yes, it emanated from Camden, New Jersey. It was sponsored by Standard Oil of New Jersey." He said he was in the studio tuning his instruments and he understood that there were people out on the poles all the way across the country physically holding the wires together so they wouldn't blow down, because this would be the first time anybody was going to actually speak to millions of people all at the same time. That's extraordinary. "Well the Master of Ceremonies," he told me, "was the great Will Rogers."

Hank Penny said he overheard Will Rogers talking to the president of Standard Oil who was there for that historic occasion. And Will Rogers said, "You know, I'm about to talk to millions of people for the first time. Now, Caesar couldn't do it, Alexander couldn't do it, but I'm about to do it, so it probably better be important. What do you think I oughta talk about?"

The president of Standard Oil said, "Well, I think you should remind the people out there that Standard Oil of New Jersey is a service company." Will Rogers didn't understand what he meant. He walked around and thought about it for a while, and then they put him in front of the microphone and threw the switch, and there he was alive in front of all of America and parts of Mexico and Canada for the first time, millions of people.

He said, "Ladies and Gentlemen, I've been asked to offer a few remarks about Standard Oil of New Jersey, and especially as regards their policy of service, and I didn't know what that meant. I cast my mind back to my daddy's ole ranch in Oklahoma years ago and remembered that we had a big stud bull …"—he's talking to millions of people, ya know—"… and the neighbors would bring their cows by for service. I didn't understand what that meant. I asked my daddy 'Daddy, what is this service we've been offering,' and my daddy said, 'You're too young to know about that, son. Nature will provide that information in due course, and by the time you got it, you'll be so glad you got it, you won't care where it came from.'" They always tell little kids that.

Fig. 12-8 Utah Phillips.

"Well, I was disappointed, but then my daddy went away on a selling trip; he sold lightning rods part time for a firm out of Des Moines, Iowa. The hired hands were out stretching wire, and I was up at the big house alone, and a neighbor came by, leading a cow, asking for service. I thought this was my opportunity to find out what that might be. I run him into the pen, closed the gate, clambered up on the top rung, sat down there to watch.

"My neighbor said, "Get down off of there, son, you're too young to know about that," and he sent me packin'. Well, I was sort of disappointed, but I knew around back of that pen there was a high board fence, and in that fence were some knotholes. I snuck around quiet, rolled some cordwood up against the fence, clambered up on it, put my eye up to one of them knotholes, and I peered through. And, ladies and gentlemen, I saw right there in front of me exactly what it is Standard Oil has been doing to the people all these years."

BOOK 4

USING ALCOHOL AS FUEL

This volume has so far made the case for what must be, to some of you, some pretty startling points and proposals, and has included much information that will be eye-opening to many curious about alcohol fuel and its potential. We talked about how producing alcohol can reverse the greenhouse effect, how we have more than enough land (and sea) to grow starch, sugar, and cellulose crops for ethanol, and how we can clean up our waterways by harvesting treated sewage through cattails. Not to mention, we stated that we can replace all petroleum fuel use with ethanol!

In the next part of the book, which is about how to use alcohol in vehicles and machinery, I will go over a lot of stuff for motorheads. Many of you reading this book aren't mechanics and will be relying on those who are to help you with engine conversion. However, if you've ever been curious about how your car works, you'll be pleased to note that, for the most part, this section has been written at a level that will permit you to learn about and understand what's going on under the hood. You may not get all of it, but you will pick up quite a bit. Anyone who has experience working on vehicles should understand everything I discuss.

There is some exciting stuff about using alcohol as fuel that everyone should appreciate, so I've put some of these things right up front so you nonmechanics don't have to wade through all the technical discussions just to get to these gems. These are the items in Chapter 13. They merit special attention because they are as startling as some of my earlier arguments. For instance, we'll show how using alcohol for cooking can stop much of the deforestation in the world. We'll show how properly designed alcohol engines get better mileage than gasoline engines.

Later in this section, I'll spend a chapter discussing myths and facts about alcohol's fuel properties, and then devote several chapters to the nuts and bolts of conversion. To accommodate the differences between alcohol and gasoline, several basic changes need to be made in an engine's carburetion, fuel delivery, air supply, ignition timing, and starting on cold mornings. Most of these changes can be accomplished in a few hours. The first time you convert a car, though, some experimentation will be necessary to get it running the way you like it.

In 2005, some colleagues and I did a couple of car conversions to test some ideas about what it would take to make modern cars run on alcohol. Appendix B provides the details; it profiles two pretty simple conversions using different technologies and different fuel injection systems to give you some idea of how you could experiment to get your vehicle running well on alcohol.

The final chapters in this "Alcohol as Fuel" section discuss flexible-fuel vehicles, other types of alcohol that can be used for fuel, cogeneration, and using alcohol with diesel engines.

There is some technical language in this section, but don't let that stop you. Those of you who are just learning about the technical aspects of using your alcohol may want to come back and reread Chapter 13 after you have read the chapters following it. You will find that you understand even more about why Chapter 13 is exciting.

CHAPTER 13

SURPRISE! ETHANOL IS THE PERFECT FUEL

To begin this section on using alcohol as fuel, I've brought together a few items that are so exciting that I felt that everyone, including those who don't consider themselves mechanically inclined, would want to hear about them. So if you are planning to just hand this book to your mechanic when it comes time to convert your car, you will still want to read this chapter.

You'll learn herein how to one-up your friends who are already running partially green on biodiesel. You'll learn how to deny the gas and electric companies their monthly check. And how we can reverse deforestation using alcohol. How about junking your nuclear-electricity-powered stove and cooking with something more environmentally friendly? Read on!

Here are some reasons why ethanol is the perfect fuel:

WE CAN PUT E-85 (85% ALCOHOL/15% GASOLINE) IN OUR CARS NOW! REALLY!

For some time, I had been hearing these rumors about a bunch of crazy farmers in South Dakota. When I'd ask about them among the upper echelons of the alcohol industry, there would be a lot of rolling of eyes and embarrassment—so I knew they were upsetting some serious apple carts. I knew I had to seek these guys out and see what they were doing with their cars. Little did I know I'd soon be doing my own versions of their experiments, with spectacular results.

One December, I flew into Sioux Falls and paid a call on a tall, rangy fellow in shiny cowboy boots and a baseball cap that proclaimed he was a member of FOIL (Foreign Oil Independence League). Orrie Swayze is the past president of the South Dakota Corn Growers Association, a former Air Force pilot, active in the Veterans of Foreign Wars, a farmer himself, and a man with a mission.

IF YOU ARE PLANNING TO JUST HAND THIS BOOK TO YOUR MECHANIC WHEN IT COMES TIME TO CONVERT YOUR CAR, YOU'LL STILL WANT TO READ THIS CHAPTER.

YOU'LL LEARN HEREIN HOW TO ONE-UP YOUR FRIENDS WHO ARE ALREADY RUNNING PARTIALLY GREEN ON BIODIESEL. YOU'LL LEARN HOW TO DENY THE GAS AND ELECTRIC COMPANIES THEIR MONTHLY CHECK. AND HOW WE CAN REVERSE DEFORESTATION USING ALCOHOL.

HOW ABOUT JUNKING YOUR NUCLEAR-ELECTRICITY-POWERED STOVE AND COOKING WITH SOMETHING MORE ENVIRONMENTALLY FRIENDLY? READ ON!

Fig. 13-1 *"Pa, feller down there says he's [California Governor] Jerry Brown … says he's lookin' into alternative fuels."*

Fig. 13-2 Orrie Swayze. *Pilot, patriot, veteran, and long-time alcohol fuel activist.*

ORRIE SWAYZE

Orrie Swayze was running a non-flexible-fuel vehicle on E-85 (which is actually about 70% alcohol in the cold months). He's been doing it in this vehicle for over 20,000 miles, and he has other vehicles that have more than 150,000 miles on them altogether, running on E-85, with no conversion.

On the way to his town, two hours from the airport, we cruised in his 1993 Cadillac running on E-85, while he regaled me with story after story of his relationship to alcohol fuel, dating back to 1965. The amazing thing about his Cadillac is that he hadn't done a darned thing to convert it to alcohol. Yet, there we were, roaring up the highway at 90 mph, doing something that almost everyone says is impossible. Orrie was running a non-flexible-fuel vehicle on E-85 (which is actually about 70% alcohol in the cold months). He'd been doing it in this vehicle for over 20,000 miles, and he had other vehicles, with more than 150,000 miles on them altogether, running on E-85, with no conversion.

And that was his point. He explained to us that there are a lot of vehicles that can run high percentages of alcohol without any of the modifications usually done to convert an engine. He noted a Minnesota State University study where participants were running a wide variety of cars on 30% alcohol, and every car they tested ran on it. No auto manufacturer notifies you that you can run more than 10% ethanol in your car, unless you have a flexible-fuel vehicle.

Also notable is that there was a wide variation on the miles per gallon of various makes and models. The 1992 Ford Taurus (pre-flex-fuel model) lost only 1.28% mpg, compared to the 1996 Oldsmobile Achieva, which lost 14.66% mpg. The 1994 Buick Regal lost only 2.96%.[1] No writer or pundit who buys into the "heating value equals mileage" argument would have predicted that the majority of unmodified cars running on 30% alcohol would have less than 10% loss in miles per gallon.[2]

We had heard since the 1980s that Brazilian vehicles routinely ran at least 30% alcohol in their fuel-injected engines, so I wasn't too surprised by this study, except for the very low mileage drop in some of the best examples. But Orrie wants the study done again, this time pushing each car to its limit on E-85, to see what the maximum is that each model can run. He tells us that no one he knows who has tried using alcohol has had to go any less than 40%, and some people have gone all the way up to using E-85 like he has—all without modification. (Based on the work of these pioneers and the Minnesota State University study, the governor of Minnesota has passed a mandated minimum of E-20 for the state.)

Orrie pointed out that if all of our fuel-injected cars—and that's almost all of them since the late 1980s—ran on just 40% alcohol, we probably wouldn't actually need any foreign oil.

Orrie pointed out that if all of our fuel-injected cars—and that's almost all of them since the late 1980s—ran on just 40% alcohol, we probably wouldn't actually need any foreign oil. That is, we can be energy-independent with current technology—and simultaneously be slashing CO_2 output—if we just start producing the alcohol. No massive multibillion-dollar research project would be necessary just to get the cars running, as in the case of hydrogen. Just make the alcohol, put it in the cars we have on the road right now, and stop the cause for much of the conflict between the U.S. and the Middle East. Period. End of discussion. This was one of the most revolutionary statements on alcohol fuel I had ever heard.

Orrie talks to anyone who will listen about this alternate reality, which only requires that people have enough courage to go to the pump and try it. He's been successful, too. There are now a lot of people up in his part of South Dakota who have taken the challenge and tried running at least half E-85 and half gasoline. He meets a lot of resistance, due to the propaganda over the years about how "experts" say that even 10% alcohol will ruin your vehicle. He quotes Albert Einstein, "Unthinking respect for authority is the greatest enemy of truth."

Orrie is incensed that almost all of the alcohol produced in South Dakota leaves the state. "In South Dakota, we use about 400,000 gallons of gasoline, and we produce 400,000 gallons of alcohol. If we used what we produced right here, all that money leaving our state would stay here, making everyone prosperous."

Orrie recently had discovered that the E-85 made in South Dakota is denatured with "natural gasoline," which is made up of the very volatile, hard-to-dispose-of, natural gas liquid condensates which have been banned in other places, such as California. This overly volatile stuff evaporates way too easily and causes vapor lock in cars using E-85. This means, of course, that when you can't start your car because the gasoline has turned to vapor instead of liquid—vapor-locking the engine— you'll turn your back on the fuel. As Orrie puts it, "The consumer only has to have one bad experience with E-85 before they will give up on it forever." The fault is either with the alcohol plants buying cheap "gas" to denature the alcohol, or a deliberate blending of evaporative trash by the oil companies to sour E-85's reputation.

Orrie and his compadre, Al Kasperson, have been analyzing fuel from different parts of the state to get to the bottom of this affair. Lenient about its evaporative emission standards, South Dakota gasoline is allowed to have a **Reid vapor pressure (RVP)** of over ten pounds, while in California nothing more than a smidge over six pounds is allowed.

Al turned out to be a pioneer 20 years ago when E-10 (10% alcohol) was considered controversial. He remembered how oil companies predicted dire results and put up people to come to hearings to slam their chainsaw up on the podium and say, "That gasohol wrecked my chainsaw."

Orrie and Al set out to prove them wrong. They received a small grant and ran a variety of small equipment, from chainsaws to leaf blowers, on alcohol and did the same experiment on gasoline. All the engines running on E-10 were far cleaner, with less wear than the gasoline models. Their photographs and data went a long way to clearing the way for the mandate of E-10 in South Dakota. Nowadays, Al is working on mixtures of alcohol, diesel, and biodiesel for tractors and heavy equipment in his spare time.

Orrie is about as patriotic as they come, but he worries that the country is going into a decline in its ability to innovate, and that people are only too willing to give up their freedom of thought in the face of fear.

Orrie talked with me at some length during our drives between interviews. He lamented the lack of curiosity in people; he said he sees this often when he tries to get folks to just add a few gallons of E-85. He is about as patriotic as they come, but he worries that the country is going into a decline in its ability to innovate, and that people are only too willing to give up their freedom of thought in the face of fear.

Orrie feels strongly that we should not fight wars for resources. After he returned from the service, his wife, angry with the war protests, asked him to tell her that those protesters were wrong. Orrie replied, "I can't do that." To avoid going to war for oil is one of his strongest motivations in advocating use of alcohol.

A Vietnam vet, he lost a brother in that war. He told me stories of his being a pilot, and how it feels to see the enemy on a low pass, puffs of smoke from their guns shooting at our troops and to drop napalm "deciding which five or six hundred people would die."

Another South Dakota pioneer, Jim Behnken, came out to talk about the use of E-85 in regular cars and planes. Although Jim has focused most of his attention on aircraft (we'll come to that later) and on combustion engineering, he has studied alcohol use in vehicles, as well.

Jim confirmed that the fuel injection computer has a fairly wide range of response to air/fuel mixtures. Although some engines have big enough fuel injectors to accommodate the instructions from the computer to deliver more fuel in the case of alcohol, others do not. Sometimes the limitation is in the software of the computer, but more commonly it's in the injector size. Jim explained that "compared to gasoline, pumping alcohol through the injector is like pumping maple syrup. It's more viscous, which limits flow, too." These injector problems can be overcome with minor modifications in most cases (see Chapter 16).

Jim has developed a method of predicting the approximate amount of alcohol an unmodified car can use. Using a diagnostic scanner to gain access to the vehicle's **electronic control unit (ECU)**, he looks for a parameter called the **long-term fuel trim**, which deals with the engine when it appears to be running too lean. Running a lot of alcohol in an unmodified car can trigger a "check engine" message on your dash if you exceed the preset value of the long-term fuel trim.

When the check engine light comes on, it means that the emissions system thinks that the engine conditions won't let it work properly. What you have are so few pollutants coming out of the car that the **catalytic converter** doesn't have enough crud to burn and stay hot. Jim points out that this is most likely to occur momentarily on moderate to rapid acceleration, and also during engine warm-up.

Fig. 13-3

EVERY TIME YOU FILL YOUR GAS TANK, YOU'RE SUPPORTING DICTATORSHIP *AND* TERRORISM.

SINGER

UNITED WE STAND

THE SAUDI MONARCHY, THE BIN LADEN FAMILY, NIGERIA'S MILITARY, COLUMBIA'S PARAMILITARIES AND RUSSIAN ORGANIZED CRIME ALL THANK YOU FOR YOUR SUPPORT!

COPYRIGHT 1992–2002 ANDREW B. SINGER

As a general rule, subject to how lead-footed you tend to be on acceleration or the altitude where you are driving, you can use twice as much alcohol as a percentage of your fuel as the long-term fuel trim number. So, if you have a Silverado pickup truck with a 40% allowable long-term fuel trim, you could reasonably expect to use around 80% alcohol without tripping the check engine light or having drivability problems. In the case of some vehicles, an advanced scanner can be used to reprogram "flash" memory in the ECU if your car is so equipped. This can allow a mechanic to reset the long-term fuel trim to a higher number.

Hybrid owners could immediately reap the benefits of E-85. Although a lot of hoopla is made of hybrid cars such as the Prius, I can't say they thrill me or anyone I know in the automotive industry. Their main advantage is that they recover energy that is usually lost to braking. Much of the fuel savings of the vehicle comes from lightweight construction, and most of that advantage is taken away by the weight of batteries.

But people have latched onto them as something positive to do in the face of Peak Oil and global warming, since they burn cleaner and ostensibly get better mileage. On the highway, their mileage isn't improved much at all; it improves mostly in city driving. The problems with hybrid cars will become apparent once they are a few years old and drivers are stuck with the repair costs of keeping two completely separate drive systems maintained just to save maybe 10% mpg. Hybridization makes much more sense on big trucks or city buses, since the heavy braking use and much larger fuel consumption makes that 10+% difference count.

Paul Sullivan, a teacher at Minnesota State University in Mankato, supervised a student project that produced some highly controversial results. The goal was to see how much E-85 an unmodified Toyota Prius could run on. They slowly ramped up the percentage of ethanol until they reached all the way to E-85, and the car ran well. It was a little sluggish during the first 15 minutes while it warmed up, but, otherwise ran even better than on gasoline. Once warmed, the car went from 64 horsepower at 4500 rpm to 77 horsepower at 4700 rpm on E-85.[3] Anyone who has driven a Prius can appreciate that extra horsepower.

The only problem they had was that after about 150 miles, the check engine light would come on—which meant that the engine was running lean, due to the adjustments the car made for running alcohol. In fact, there was no actual problem.

When it came to emissions, the Prius dropped nearly 90% of its already low hydrocarbons and was almost unmeasurably low. NOx was so low as to actually be unmeasurable. Carbon monoxide was higher during the initial start-up phase, and then, once the engine warmed, it settled down to a drop of between 39.2 to 48.8%. The start-up of the car in cold conditions initially caused a rise in CO. As Paul pointed out, some modification for heating the fuel or air at start-up would have eliminated this anomaly. He also noted that nothing was done to optimize the engine for use of E-85, which did result in lost mileage, although the car gained both horsepower and dramatic reductions in emissions.

Our Tests

A few months after visiting South Dakota, I decided to personally test Orrie and Jim's claims of unmodified cars running on at least 50% alcohol. Although I wanted to do this experiment at the level of a peer-reviewed scientific paper that would be published in a respectable journal, the least expensive quote for doing testing with all the bells and whistles was $40,000. Not having time to raise that money, I decided on a practical approach with three questions in mind:

1. If you fueled eight unmodified average vehicles with 50/50 alcohol and gasoline, would they still be drivable?

2. Would 50/50 alcohol substantially reduce their emissions?

3. If people were suddenly able to run 50/50 alcohol, would the average car, in its average neglected condition, violate smog laws? (Normally, people tune up their car before getting their smog test to ensure that they pass; this is not the day-to-day state of their engine most of the year.)

I found eight volunteers with average vehicles and asked them to bring their cars to an official dynamometer emissions testing station certified by the California Air Resources Board. These stations have the cars run on a set of rollers, imitating a standardized test track while measuring emissions.

All the volunteers received an emission test on their completely unmodified cars, even if there were obvious vacuum leaks, items not working, or other kinds of wear and tear issues that would affect emissions while still on gasoline. This is more representative of what's on the road every day and very different from when you receive a smog test to certify your car for registration. All those vacuum leaks or non-operating emissions equipment would have to be fixed before you could legally be tested.

All the vehicles' tanks were at least 50/50 alcohol/gas. I asked each driver to go out and try to make his car misbehave.

Once the cars had their emissions tested on gasoline, I emptied the tanks and refilled them with a 65% alcohol/35% gasoline mixture, since I was certain that most of the cars retained a gallon or so of gasoline, even after draining. This ensured that all the vehicles' tanks were at least at 50/50.

I asked each driver to go out and try to make the car misbehave. I sent them on a route that would take them through lots of stop signs, on level freeway stretches, and on a long hard climb on a freeway. I asked them to challenge the car—to slow down periodically on the hard climb and then floor it, and to try as many combinations of acceleration and deceleration as they could think of. I told them to be on the lookout for coughing, or sputtering, or hesitation, or anything unusual at all.

I found that every single car performed well on the mix. In only two cases, there were minor hesitations at stop signs during the first five minutes of driving, but those symptoms disappeared during the test. This is due to modern **electronic fuel injection (EFI)** systems going through an adaptive learning phase, which they regularly do for seasonal gasoline formula changes or for aging of the engine. After a few minutes, the EFI systems "learned" the new mix, and the slight hesitations in those two cases went away.

The most common comment was that the drivers could not really tell the difference

Emissions Change in Unmodified Vehicles Using 50% Ethanol Compared to Those Using 5.7% Ethanol Blended with California Reformulated Gasoline; Testing Protocol of California Air Resources Board ASM Emission Test

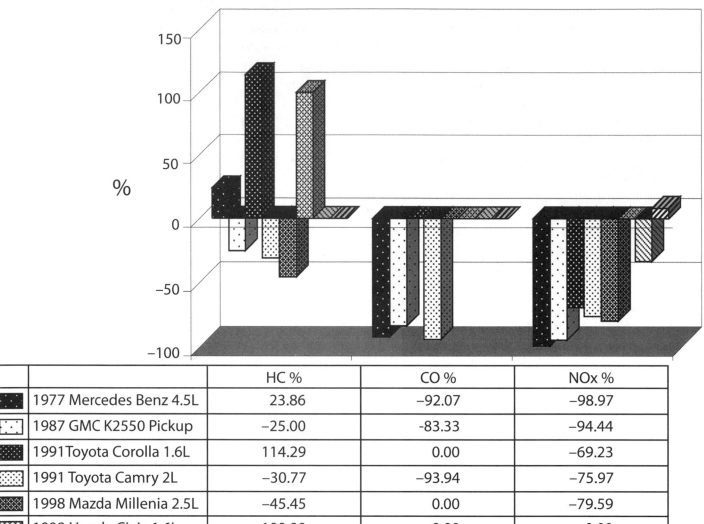

		HC %	CO %	NOx %
■	1977 Mercedes Benz 4.5L	23.86	−92.07	−98.97
▦	1987 GMC K2550 Pickup	−25.00	-83.33	−94.44
■	1991 Toyota Corolla 1.6L	114.29	0.00	−69.23
▦	1991 Toyota Camry 2L	−30.77	−93.94	−75.97
▦	1998 Mazda Millenia 2.5L	−45.45	0.00	−79.59
▦	1998 Honda Civic 1.6L	100.00	0.00	0.00
▨	2000 Toyota Tacoma 2.4L	0.00	0.00	−33.33
▨	2004 Toyota Tacoma 3.4L	0.00	0.00	8.33

Fig. 13-4 E-50 emissions test in unmodified vehicles. The ASM test is done on a dynamometer at both 15 and 25 mph. In this chart, all vehicle data is based on the 25-mph data stream, except in two cases (1991 Camry & 2004 Tacoma), where we relied on the 15-mph data stream due to these vehicles having large vacuum leaks at the higher speeds. These leaks would have had to be repaired before certification in a normal smog test.

In constructing this chart, we calculated the change in emissions between California pump gas, which has 5.7% ethanol added, and a 50% minimum ethanol/gasoline blend. To avoid visual distortions, if an emission changed less than 3.5% from its legal limit, it was considered insignificant, and read as zero. As you can see, most of the emissions are drastically reduced or remain virtually unchanged.

The very clean Honda Civic showed an apparent 100% increase in HC, but in reality, this means the Civic went up from 8 ppm to 16 ppm. That isn't much, but it is 100%. To put it in perspective though, 16 ppm is one-third of the Civic's legal limit of 47 ppm of HC. The other "high" reading was from the 1991 Corolla, which went up from 14 ppm to 30 ppm. This increase was most likely an anomaly, due to any number of vehicle equipment faults. The HC legal maximum for this vehicle is 105 ppm. This means that the higher HC reading was still just one-third of the level permitted by law. The 1977 Mercedes, which did have a slightly elevated HC reading, actually failed the overall smog test when on gasoline, but passed when using the 50% ethanol blend, due to its dramatic reductions in CO and NOx. These eight cars were completely unmodified and untouched vehicles. It would be possible to lower emissions even more, if modifications and standard repairs were made. The net result is that unmodified cars running on 50% alcohol ran fine and had an overall dramatic reduction in pollution.

between gas and the mix. Several participants said the car was much quieter on alcohol and there was less vibration. They also remarked how much smoother their acceleration was under load. Nothing really remarkable happened, and there were no failures. But they were all mighty proud to be using American-made fuel!

Volunteers then drove back to the testing station and had their emissions tested on alcohol. There were a variety of results. In most cases, hydrocarbons and CO either stayed similar if they were already really low or dropped substantially in the cases where they started off higher. In the case of NOx, one reading was higher on the alcohol mix than on gas. This is consistent with there being vacuum leaks between the **mass airflow sensor** and the engine. Excess oxygen was recorded in the trial that had higher than expected NOx, confirming this hypothesis. But even in the case where NOx increased, that vehicle was still well within legal limits.

So, the conclusions of the study were that overall emissions were generally dramatically lower, drivability was either unremarkable or improved, and cars that didn't pass the first gasoline smog test did pass on the alcohol mix. So having alcohol at the pump and drivers blending their own fuel would have a definitely positive effect on emissions.

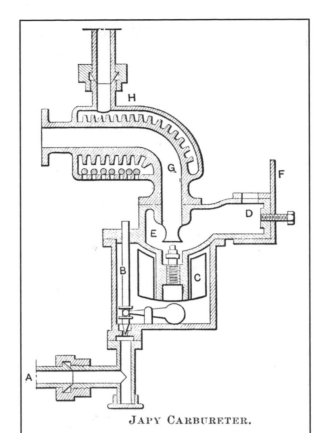

Fig. 13-5 Japy vaporizing alcohol carburetor.

JAPY CARBURETER.

A, liquid fuel supply pipe. — *B*, needle valve regulating fuel supply.— *C*, float which maintains constant level. — *D*, air entrance. — *E*, mixing chamber for air and liquid.— *F*, valve which regulates supply of air. — *G*, passage leading to admission valve. — *H*, surrounding jacket in which a part of the exhaust gases or of the cylinder cooling water passes.

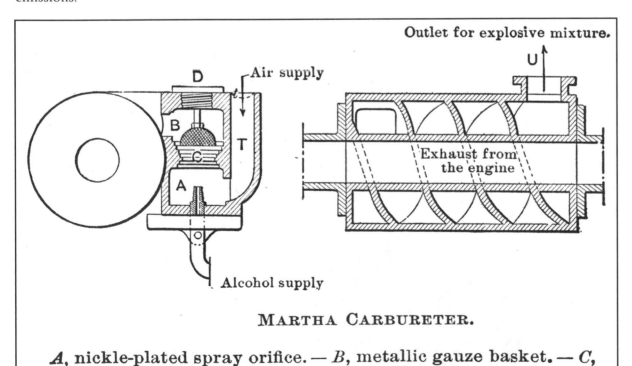

MARTHA CARBURETER.

A, nickle-plated spray orifice. — *B*, metallic gauze basket. — *C*, steps. — *D*, plug for inspection. — *t*, metal gauze strainer. — *T*, air passage. — *U*, outlet tube for explosive mixture.

Fig. 13-6 Martha vaporizing alcohol carburetor.

WE CAN GET HUGE MILEAGE INCREASES WITH VAPORIZING ALCOHOL

In the early part of the last century, alcohol was considered the best vaporization fuel, and whole books have been written on **vaporized alcohol engines**. There are wonderful old illustrations of them being used with tractors, autos, and other internal combustion engines.

Over 250 patents have been filed for one or another technique to vaporize liquid fuel to a gas to power an ICE. Some alcohol-vaporizing carburetors bubbled their fuel with hot air; some dripped it onto hot surfaces. Others boiled it in exhaust pipe heat exchangers. Wick systems were even used in some engines.

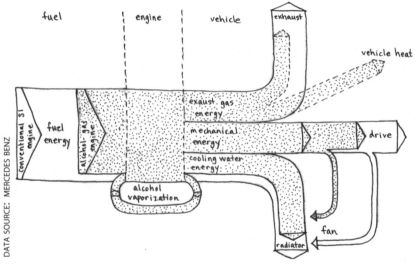

Fig. 13-7 Sankey diagram of a vapor alcohol engine. *This compares the uses of fuel energy between a conventional spark ignition engine and vapor alcohol engine. Less energy is wasted in a vaporized alcohol engine.*

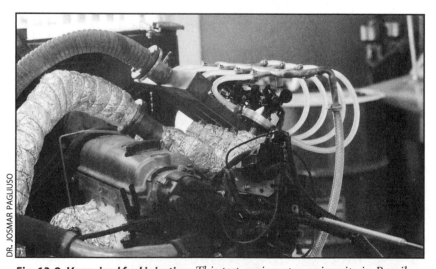

Fig. 13-8 Vaporized fuel injection. *This test engine at a university in Brazil runs on exhaust-vaporized alcohol.*

Some of these carburetion systems have been proven to increase gasoline mileage to 200+ mpg, and many simpler systems have squeezed 80+ miles per gallon out of gas. Even NASA has developed a vaporization system for planes. To my knowledge, not one of these patents has ever been developed into a commonly marketed product.

The history of the vaporizing carburetor reads like a grade-B thriller, full of intrigue, industrial sabotage and espionage, threats, patent suppressions, and suspected murder. In the recent past, some of the most promising patents were purchased and owned by major automakers or oil companies. Many entrepreneurs attempting to sell patents on this to anyone other than major corporations have allegedly been the victims of physical or legal harassment, until they were forced to close their businesses. After all, less than 30 years ago, automobiles commonly averaged about ten miles per gallon. What powerful energy company would support an easily available vaporizing system that worked to increase mileage to 50–80 miles per gallon? How far would power companies go to suppress the information that a simple vapor-fueled home generator could beat the price of their electricity?

A vaporizing system creates an efficient mixing of air and fuel for near-perfect combustion. Vaporization ideally turns a combustible liquid into a vapor of individual molecules of fuel for extremely efficient combustion, converting a lot more fuel energy to work.

Normal carburetion, and to some extent fuel injection, use engine vacuum to atomize liquid fuel to a spray, which is ignited in the cylinders. The surface of the droplet of fuel is essentially all that fully combusts, the only thing that creates work (miles per gallon). Only a portion of the droplet burns, from the outside toward the core. The rest degenerates before it's pushed out of the cylinder as exhaust and heat.

Modern gasoline may contain more than 400 chemicals that boil at a whole variety of temperatures, from 80 to 450°F.[4] In the past, 95% of the chemicals in gasoline would boil below 200°F. Almost all the older vaporizing systems were designed to use fuel with lower consistent boiling points. If used on today's gasoline, they would distill lower-boiling components into vapor, but leave behind oily residues that would quickly render the device useless.

Alcohol boils at a fixed point of 173°F, which makes engineering a vaporizing system quite manageable. The vaporizer is essentially a heat exchanger or boiler. The methods of vaporization are many: radiator coolant heat, exhaust heat, electric heat, engine oil, microwave cells, electrostatic generators, and injection of exhaust-generated steam.

Historically, vaporizing carburetors vaporized fuel either before carburetion or downstream from the carburetor's **venturi**. The most successful designs separated the functions of producing vapor and mixing vapor with gas into two units. This is also the general approach being used by modern developers working with fuel injection.

The **shell and tube heat exchanger** is a common configuration in many patents. Some inventors prefer to use the shell for heating, while others run their heat source through the tubes. In either case, the goal is to have fully vaporized fuel available to deliver to the air/fuel mixing device, be it the carburetor, or a special gaseous fuel injection system.

Recently, in Brazil, scientists have had partial success using exhaust to vaporize alcohol at the exit of a normally liquid fuel injection system. Although the control system still needs work, their research showed impressive fuel efficiency, and more importantly, dramatic reductions in emissions.

For modern inventors to have a chance to succeed in development of vaporizing systems, the single most difficult problem is that alcohol is mixed with 15% gasoline in the U.S. and other countries like Sweden. This large amount of gasoline dramatically complicates engineering, an impediment that does not exist in Brazil where straight alcohol is sold at the pump. Those who produce their own alcohol, however, do not have to add the 15% gasoline—and so can take advantage of partial or full vaporization techniques. Modern research into vaporized fuel goes by the name of **homogeneous charge compression-ignition (HCCI)**. We'll talk more about this in the next section to prepare you for what the future of alcohol engines holds.

THE FUTURE: ALCOHOL-ONLY ENGINES! NO GAS!

Although pundits often declare that mileage always suffers when using alcohol, it turns out to be not quite accurate. Certainly a simple conversion of a standard gasoline engine to alcohol

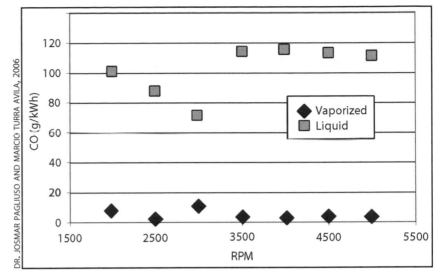

CO Emissions of Vaporized versus Liquid Ethanol

DR. JOSMAR PAGLIUSO AND MARCIO TURRA AVILA, 2006

Fig. 13-9 CO emissions of vaporized versus liquid alcohol. *Note the dramatic drop below the already low liquid alcohol carbon monoxide levels. This test measured exhaust straight from the engine without the use of a catalytic converter.*

generally results in a small drop in mileage. After all, it would be expecting too much for an engine with 80 years of engineering optimizing it for gasoline to work as well with alcohol. But what if we put our engineering genius to work optimizing an engine for alcohol? What would a dedicated alcohol engine look like?

Over the last 20 years, the limited attempts at building a dedicated alcohol engine have concluded that such an engine would be very different from a gasoline engine. Generally, researchers have discovered that a dedicated alcohol engine would combine aspects of both spark-ignition and compression-ignition diesel engines. In general, a spark or ignition point is used to ignite the fuel when the engine is running at idle or slow speed. The spark ignition system turns off when the engine is up to cruise or under load, during which time the fuel is ignited solely by compressing it until it explodes (like diesel engines). Successful

"I also pointed out in the article that as governmental laboratories had developed from 40 to 55% efficiency in alcohol engines as against 20% in gasoline machines, the use of alcohol at double the cost of gasoline for power purposes, was cheaper for motor[s] than gasoline in common use today."

—C.M. CHESTER, REAR ADMIRAL OF THE U.S. NAVY, WRITING TO HENRY FORD ON DECEMBER 15, 1916

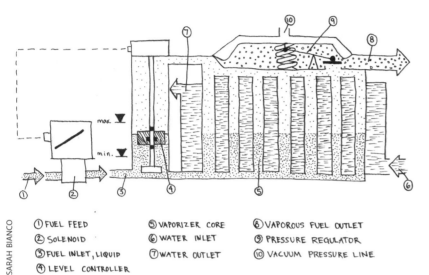

Fig. 13-10 Mercedes Benz alcohol vaporizer. *Used to run "diesel" city buses in New Zealand on straight alcohol in the 1980s.*

① FUEL FEED
② SOLENOID
③ FUEL INLET, LIQUID
④ LEVEL CONTROLLER
⑤ VAPORIZER CORE
⑥ WATER INLET
⑦ WATER OUTLET
⑧ VAPOROUS FUEL OUTLET
⑨ PRESSURE REGULATOR
⑩ VACUUM PRESSURE LINE

SARAH BIANCO

experiments with such a system were made in the 1980s with modified Volkswagen Rabbit diesel engines running on methanol.

More recently, a very serious improvement in this idea has been carried out on an EPA contract using a Volkswagen TDI diesel engine. The goal was to build a high-efficiency small engine for hybrid electric use, using only alcohol as the fuel. The study looked at both methanol and ethanol, and the results were revolutionary.

Most cars use an inexpensive, low-pressure pump to inject fuel into the engine's manifold in the vicinity of the valves, where the mixture is sucked into the engine. The TDI diesel engine normally uses an expensive, very high-pressure pump to directly inject diesel fuel into the engine cylinder. Researchers abandoned the expensive pump and installed a standard gasoline-style, low-pressure fuel injection system that delivers fuel into the manifold. The compression ratio, in this case, had to be reduced to 19.5:1 from the stock 22:1 when using methanol, and to about 18:1 when using ethanol. The **glow plugs** normally used in starting a diesel engine were replaced with electronically controlled **spark plugs**. The vehicle used a variable-geometry turbocharger.

Researchers also recirculated a large proportion of exhaust gas back into the manifold. The best fuel consumption was achieved at about 50% **exhaust gas recirculation (EGR)**. The high EGR also helps prevent any **preignition (pinging)**, which can occur with alcohol above an 18:1 compression ratio. As described above, the modified engine operated as a spark-ignited engine at idle and low speeds, but as a compression-ignition engine at higher speeds.

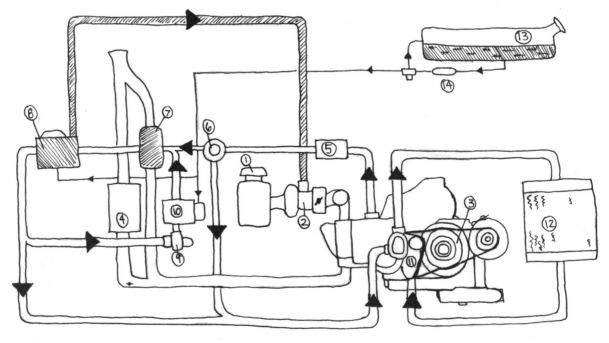

Fig. 13-11 Mercedes Benz vapor alcohol engine flow diagram. *This design uses heat from multiple sources to fully vaporize alcohol for diesel engine use.*

SARAH BIANCO

① AIR CLEANER
② GAS MIXER WITH THROTTLE VALVE
③ ALCOHOL-GAS ENGINE
④ MUFFLER
⑤ TRANSMISSION OIL COOLER
⑥ THERMOSTATIC
⑦ EXHAUST GAS HEAT EXCHANGER
⑧ ALCOHOL VAPORIZER
⑨ ELECTRIC WATER PUMP
⑩ HEATER
⑪ WATER PUMP WITH THERMOSTAT
⑫ RADIATOR
⑬ ALCOHOL TANK
⑭ FUEL PUMP WITH PRESSURE CONTROL

EGR helps slow the **flame propagation** endemic to pinging, and substantially interferes with the formation of NOx. The EGR was cooled before introduction into the turbocharged airstream, using a water **intercooler** (a small radiator in the exhaust gas duct to absorb heat).

The fuel mileage was up to *22% higher than on diesel,* and, of course, the engine put out more horsepower. The **thermal efficiency** was measured at 43%, which approaches the efficiency of hydrogen fuel cell technology, but at a fraction of the cost. Normal gasoline and diesel vehicles achieve thermal efficiencies in the vicinity of only 20%.

Although final emissions data were not made publicly available, I had a chance to discuss them with one of the scientists involved. The preliminary emission results, using methanol, were exceptionally low, and these should be quite comparable to ethanol readings. NOx was at 0.1–0.2 grams per kilowatt-hour, and hydrocarbons (HC) were at less than 0.2 g/kWh, as were CO emissions. Emission readings like these are very low in comparison to gasoline or diesel emissions.

Bear in mind that this was a converted diesel engine. There remain further improvements that would result in even higher thermal efficiencies, making the efficiency difference between an alcohol internal combustion engine and a mythical hydrogen fuel cell engine indistinguishable, although they are dramatically different in price.[5]

Another EPA study looked at the differences between the **torque curves** for alcohol and gasoline and how they interact with the transmission. As

---

PROPANE CARBURETORS

Let's touch on propane carburetors and their vaporizing capabilities. Tinkerers in the 1980s found them to be ideal for mixing vaporized fuel with air for delivery to the engine. The propane carburetor is already designed for gaseous fuels, so simple changes in diaphragm materials and regulator spring tensions to allow for alcohol's different air/fuel ratio are usually all that's needed to make it compatible for alcohol. The regulator is designed to hold a certain amount of slightly pressurized fuel vapor in reserve for when additional power is needed (for acceleration, primarily). A lot of farmers who used propane carburetors on older gasoline tractors and on their cars during the '80s were getting mileages in heavy old Cadillacs that a Prius owner would be proud of.

Jim Pufahl, a South Dakota farmer, used such a system to fuel his International Harvester tractor, with dramatic reductions in fuel consumption. (Tractors run at a steady rate most of the time and don't often have to contend with the rapid accelerations and decelerations of the automobile.) Jim burned gaseous alcohol at a 16:1 air/fuel ratio—much leaner than normally carbureted alcohol's 9:1 mixture or gasoline's 14:1 ratio.

Jim used an older-style stock propane vaporizer, which runs much hotter than the newer types. He found that several modern propane vaporizing chambers (which use engine coolant to heat the alcohol) in a series work well, but they are too bulky to fit under the hood of most cars. On his tractor, there was more than enough room. But there are good, small, compact heat exchangers available within the refrigeration industry that could do a better job than the stock propane fuel heaters. I use one of these on my liquid-fueled Volvo, for instance.

This combination of vaporizer, regulator, and a air/fuel mixing device, with some sophisticated additions, was used in the 1980s by Mercedes-Benz in converted diesel buses in Germany and New Zealand. After only a few years, even though they were still in the field-testing phase, these alcohol-vapor-fueled buses were approaching the efficiency of the same engines on diesel fuel. The buses were reliable, and vaporization seemed to be a major answer to retrofitting diesels to run on pure alcohol. Adding heat to the alcohol to vaporize it seemed to solve alcohol's difficult ignition problem (low cetane rating) and eliminated other problems like lubrication of high-pressure diesel pumps.

One of the hottest new research areas in diesel engine technology is called homogeneous charge compression-ignition (HCCI) engines, a fancy name for vaporized alcohol in a diesel engine (see Chapter 25). So after all this time, vaporization is being given the credibility it deserves.

Newer developments in propane injection may make it an ideal delivery technology for vaporizing systems in both spark-ignited engines and homogeneous charge hybrids that have the aspects of both spark-ignited and diesel engines. Normal liquid fuel injectors and their fuel supply **rails** are more efficient when the fuel is heated, but once the fuel turns to gas in the fuel line, there can be multiple problems in getting even delivery. Propane/natural gas injectors use a gaseous fuel already, so no problems there.

you'll soon learn, alcohol develops its peak power at seven degrees after **top dead center**, much sooner than gasoline. This translates to more power (torque) at lower speeds than gasoline. This characteristic can be put to use to extend mileage, rather than just for thrills in acceleration. Generically, this is called **downspeeding**—changing the gearing of the transmission so it matches the

Fig. 13-12 Gordon Cooper and Bill Paynter, March 15, 1981. *Departing from Sacramento, California, for Washington, DC. This was the first transcontinental flight powered by alcohol fuel, a world record set May 15 through 23, 1981.*

Fig. 13-13 Plane running on E-100. *This Brazilian plane demonstrated the effectiveness of straight alcohol as a reliable lead-free general aviation fuel.*

higher power output of alcohol at the right point in the torque curve. The net result is that the car will accelerate in a manner similar to a gasoline vehicle, instead of the vroom-like effect experienced with a converted alcohol vehicle.

Downspeeding turns that excess acceleration into mileage. In the EPA test, downspeeding the transmission by 23% doubled the increase in miles per gallon, due to high compression alone—for a total of 24% better mileage. This test was done with a compression ratio of only 11.19:1 in a GM 4-cylinder 2.2-liter engine.[6]

Combining downspeeding with the high-compression/high-EGR strategy discussed above would likely increase mileage quite a bit more. Taken further, vaporizing the fuel would make an HCCI engine that might not even need spark at idle.

So, there are frontiers left to explore in the quest for the dedicated alcohol engine. What's certain is that gasoline is certainly less efficient than alcohol, and that dedicated alcohol engines will get better mileage than gasoline or diesel, with emissions so low they may not even register on the current smog testers.

Saab has taken the lead in announcing in 2006 its development of the first post-petroleum, dedicated alcohol sports car.[7] I can't begin to tell you how excited I was to hear this sounding of the eventual death knell of gasoline.

WE CAN PUT ETHANOL IN PLANES OF ALL KINDS!

In 1971, the U.S. airline industry spent $10.6 billion on operating expenses, personnel, amortization of aircraft, fuel, lubricants, maintenance, and advertising. In 1981, it spent $11 billion for fuel alone—and flew fewer total miles, with more fuel-efficient engines. Fuel is the industry's major economic problem. Based on April 2006 prices, the 19.9 billion gallons of jet fuel we are likely to use this year will cost $41.85 billion dollars.[8]

General aviation, which comprises the smaller planes that frequent the smaller airports and rural areas of the nation, is feeling the crunch, as well. High-performance aircraft engines require high-quality, high-octane, leaded fuel, and oil refiners aren't crazy about maintaining the expensive equipment needed to produce it for only about 1% of the fuel market. Aviation gasoline is very expensive, since its formulation is regulated and the oil companies can't just dump their waste into it. Alcohol-based fuels are far less expensive, have the high octane needed, and eliminate the use of lead. Orrie Swayze thinks this might result in a renaissance of sport aviation, where average people could afford to take up flying again.

Making a switch to E-85 is no big deal in general aviation planes. Unlike automobiles, planes experience constant changes in elevation, and therefore changes in available oxygen; smaller craft have thus traditionally relied on the pilot to check and adjust the air/fuel mixtures. You have to get into planes worth over a quarter-million dollars before computerized control with fuel injection takes the place of pilot skill in mixing fuel with air.

A modern pioneer in ethanol aviation is Jim Behnken, whom we mentioned earlier. He has been instrumental in having a particular blend of E-85 certified for general aviation use. As a fuel engineer and president of Great Planes Fuel Development, he has burrowed into the minutiae of the qualities of alcohol.

His solution to cold-starting problems, instead of trying to get a cold-start device certified, was to streamline the certification by the FAA of an 85% ethanol/15% petroleum mixture that would start in any weather. He was successful in his long quest to get E-85 partially certified for general aviation, where it is called AGE-85. Using primarily **isopentane** instead of generic gasoline for the 15% of the fuel that isn't ethanol ensures that the plane will start in any weather. Part of the formula includes about 1% biodiesel to provide superior lubrication in place of **tetraethyl lead** that is otherwise still added to general aviation fuel. Ethanol's high octane is actually superior to the octane level in petroleum-based aviation fuel.

Although Jim has jumped through all the hoops the FAA has put in his path, they continue to refuse to release AGE-85 for public use. It seems that the FAA is an agency that is virtually paralyzed in terms of decision-making. If you pilot a small plane, you need to write to the FAA and tell them you want this fuel certified and available.

In Brazil, they have taken a different tack to cold-starting. Brazilian aircraft company Embraer uses a cold-starter in its plane, like in Brazilian cars, because it runs on a fuel mixture of 96% alcohol with 4% water. Basically, what they've done is segment the in-wing fuel tanks so that there is a small tank for cold-starting fuel. The rest of the wing has

Fig. 13-14 Dedicated alcohol high-horsepower airplane. *This powerful low-flying crop duster made by Embraer in Brazil runs on unleaded HE-100 (96% alcohol and 4% water).*

the main fuel supply. The pilot can switch from one tank to the other while in the cabin.

When I visited Embraer's plant and airstrip, I had the privilege of seeing its first four dedicated, high-compression alcohol aircraft being assembled for sale. This was a historic occasion, and everyone, including me, beamed with pride. This aircraft is primarily used in agriculture for spraying chemicals, which is a heavy payload for a small plane. The engine must be very powerful and absolutely reliable, since the plane often flies just a few feet above the ground, and has to rapidly climb and turn at the end of fields. We watched as Embraer put the aircraft through aerobatics for me and my film crew.

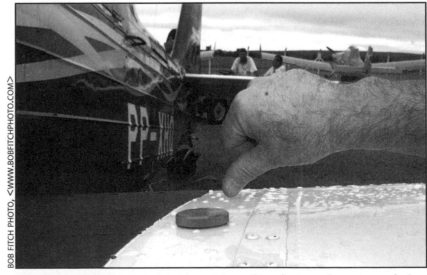

Fig. 13-15 Cold-start tank for planes. *Many planes' wings also serve as fuel tanks. The first foot of the wing is a separate tank with a separate cap for cold-starting fuel to get the engine running before switching to straight alcohol.*

The company told us that it is responding to demand. The farmers want to use alcohol, so Embraer is not even offering a gasoline option for this aircraft any longer. Work is now in progress on redesigning some of its larger aircraft for dedicated alcohol use.

The late Bill Paynter and his partner, astronaut Gordon Cooper, who is also now deceased, were working on a multimillion-dollar venture to perfect a high-mileage alcohol-vaporizing system for aircraft and to certify with the FAA several alcohol-powered engines, including jet engines. Cooper and Paynter performed the first American transcontinental flight on alcohol fuel in 1981.

"The private automobile has been the single most important source of capitalist profit for nearly 100 years. It has sustained automobile manufacturing—as well as steel, plastic, rubber, glass, and upholstery manufacturers, advertising, insurance, credit companies, and, of course, the petroleum industry—from exploration, refining, distribution, and service stations."

—ALLAN ENGLER

Paynter had been in aviation since World War II, serving as Ronald Reagan's personal pilot for many years prior to Reagan's presidency, and owned Union Flights of Sacramento, a chartered air service. Much of his and Cooper's research went into methanol, but they did some flying on ethanol, as well. Their success with their Piper Cub, a common piston-engine small plane, was the result of lots of trial and error, and is offered only as a guide to the course that your own experimentation might take.

Aware of what alcohol fuel could do on the road, Paynter and Cooper decided to test its practicality in the air. Their record-making cross-country flight gives us a lot of encouragement. For one thing, their fuel consumption using methanol, which has comparatively less heating value than ethanol, was drastically lower than had been predicted.

Based on its heating value, methanol should deliver less than half of the mileage of aviation fuel. But over the 25,000 miles they flew, the fully analysis-instrumented, methanol-converted Piper Cub repeatedly measured an unusual fuel consumption curve. On takeoff, the fuel flow monitor indicated that the engine was using 1.5 times as much methanol as it normally would use gasoline.

That fit their expectations and calculations, but they didn't expect the 10% more horsepower on takeoff.

While climbing or cruising below 8000 feet, Paynter and Cooper used 1.2 to 1.3 times as much fuel as they would have on gas. Between 8000–12,000 feet, fuel consumption was about the same as their usual gasoline consumption. But once they reached the higher elevations, 12,500–16,000 feet, consumption was only 0.7–0.9 times that of gasoline![9]

They recorded the same results many times under many different air conditions on subsequent cross-country flights. The results were due to alcohol's oxygen content, and the fact that in the upper elevations, gasoline engines are very inefficient, whereas alcohol's efficiency is maintained.

Paynter talked about several major differences in engine operating conditions on gasoline and alcohol. The most striking was alcohol's extremely low exhaust temperatures. Running under power (with a **rich mixture**) during a climb or on takeoff, the Piper's exhaust temperature was 50 to 75°F cooler than it had been on gasoline. When the engine was properly **leaned out** for cruising, the temperature was so cold that it went completely off the cold side of the thermometer (the exhaust temperature probe is 1-1/2 inches from the exhaust port). There wasn't enough heat in the alcohol exhaust to run the cabin heater (which usually exchanges heat from the exhaust), so they were freezing cold at upper elevations.

The oil temperature dropped, too, at cruising, to as low as 125°F, cool enough to start becoming too viscous. They remedied that problem by covering a portion of their oil radiator. When using alcohol, oil consumption dropped dramatically: The Piper normally used a quart of oil every 25 hours, but while burning alcohol it used one quart every 60 hours.

After 25,000 miles traveled on alcohol, Paynter and Cooper used a **boroscope** (a fiber optic device that lets you see the inside of anything the boroscope's cable can thread through) to see that the engine was exceptionally clean—and there was no sign of wear.

The Piper Cub conversion was based on what Paynter and Cooper knew about alcohol automobile conversion. So as not to alarm the FAA, they used exclusively off-the-shelf parts for enriching their mixture. They replaced the fuel injectors with the next-larger-sized injector. The major fuel metering experimentation, though, was done at the injector pump. For each fuel line feeding the new injectors, they drilled out the bronze, jet-like parts that meter fuel from the pump. These jets are replaceable, so you don't have to panic if you over-drill them. Paynter recommends starting lean and working toward a richer mixture during experimentation. Running too rich on one occasion, they burnt a cylinder.

Working slowly, opening up the jet 10% at a time, they finally got the right power output. The best indicator for a more or less proper mixture is exhaust gas temperature. Paynter also used an **exhaust gas analyzer** for checking the exhaust's oxygen content to see if the fuel was being burned efficiently.

Paynter and Cooper's Piper conversion was fairly basic, and they expected that mileage performance would increase significantly once they turbocharged the engine. They were also developing a microwave vaporizer that promised to significantly decrease their already impressively low fuel consumption.

Paynter noted that ethanol, which they'd only used for a short time below 8000 feet, was more efficient than methanol and that the operating cost per hour was about the same for both fuels (at the time, Paynter was quoting a much higher industrial cost figure for ethanol!).

I asked Paynter if it's possible to use alcohol in jet (turbine-powered) aircraft, and he told me an interesting story about Cooper's experience:[10]

"Gordon learned firsthand about alcohol as a potential jet fuel years ago. He was an active astronaut then, and trained regularly with a T38 trainer jet [the Northrop version of the F5 supersonic fighter]. Gordon was at Indianapolis working on a racing car he owned with Jim Rathman on this occasion, and he was on standby as a pilot for Apollo 10. When the first-string pilot got sick, he was called in as a replacement—when NASA calls, you go.

"It was a weekend, and the local airport didn't have any jet fuel. Knowing that a jet turbine engine will run on almost anything because of its inherent flexibility, he decided to try the T38 on methanol that he could have quickly delivered by tank truck. He assumed he'd have much less range because he knew methanol had a low heating value compared to jet fuel. So although he thought the methanol would burn fast, he figured he could just make it to another airport to refuel.

"Gordon told me himself that he tested his engines' power all the way to the takeoff end of the airstrip. Satisfied with their performance, he took off under full power and, in his climb, hit the afterburners. Gordon had hoped to get up to 30,000 feet. At 40,000 feet, the jet was running better on methanol. He made it all the way to Houston. He did burn more on takeoff, but once he reached altitude, it ran great.

"When he landed and told the engineering officer that he had flown in on alcohol, the officer got all upset, saying he had probably ruined the jet's engine, but there was no indication of any damage. "There's been other military testing and flying with methanol, but the records of those tests seem to have disappeared."

Worldwide, 2.4 billion people cook their food with biomass fires. Indoor air pollution ranks as the fourth leading cause of premature death in the developing world. Most of the people who are exposed to daily inhalation hazards and eye damage from wood smoke are women and children. When wood is scarce or expensive, water boiling is sacrificed in favor of food cooking.

COOKING WITH ALCOHOL: THE GLOBAL IMPACT

In most of the world, cooking is done over a fire indoors, without a flue. Worldwide, 2.4 billion people cook their food with biomass fires. Indoor air pollution ranks as the fourth leading cause of premature death in the developing world.

Most of the people who are exposed to daily inhalation hazards and eye damage from wood smoke are women and children. Women in developing countries typically spend three to seven hours per day by the fire, exposed to smoke, often with young children nearby. Smoke in the home is one of the world's leading child killers, claiming nearly one million lives each year.[11]

Dr. E. Bates sums it up, "If people do not have fuel for lighting, they must sit in the dark; if they do not have fuel for cooking, quite simply, they starve."

Of course, suggested solutions include better ventilation, but, more importantly, "the use of cleaner-burning fuels."[12] Solar cookers, which use reflected light of the sun, are a good daytime answer, but for night and early morning cooking, alcohol makes a lot of sense.

RIGHT: Fig. 13-16
Henrietta Obueh
of Nigeria, cook-
ing with alcohol.

COURTESY OF DOMETIC AB

*She is using an
Origo Clean
Cook made by
Dometic. This
simple alcohol
stove is safe to
use indoors and
replaces hundreds
of hours of work
gathering wood
miles away from
home.*

Cooking with alcohol need not be done with an expensive stove. Very effective stoves have been made using nothing more than three soda cans, some epoxy, a heavy needle, and a few used bike spokes. These stoves do emit carbon dioxide, but emit almost none of the other toxic compounds that biomass or petroleum products do.

Go to my website at <www.permaculture.com> for current links to the latest soda can stove designs. Although one of these little stoves will boil a quart of water in a few minutes, it's possible to arrange a ring of six or seven of them to heat a large pot of water very quickly. When traveling in developing countries, all you need to bring with you are a few tubes of J.B. Weld high-temperature epoxy, some sewing needles, and a pair of scissors or shears, and you can start a revolution using soda cans.

Right behind smoke-related deaths, a major cause of childhood death below the age of five in developing countries is diarrhea, which is largely caused by the inability to boil water to sterilize it before drinking it. When wood is scarce or expensive, water boiling is sacrificed in favor of food cooking. In some parts of the world, half the children die before the age of five from avoidable dysentery. Mothers then have to produce approximately ten offspring to have enough of them (two per parent) to survive to support the parents in old age.

Developing nations would relieve enormous burdens on their citizens by distributing denatured alcohol and cheap stoves like these. In some countries, people can spend up to six hours walking,

one way, to where there is still wood to cut, to gather a supply to last a family a week for cooking food. Offering free workshops to help people make their own alcohol stoves from cans may be the best strategy to prevent young tree seedlings from being harvested as stick wood for cooking.

Clearly, fueling stoves rather than SUVs is the most important use of fuel in Ethiopia. In light of this, Swedish manufacturer Dometic AB's Project Gaia "is predicated on the belief that the household fuel market is the highest and best use market for Ethiopian ethanol."[13] Project Gaia does not yet have a small-scale alcohol production component, relying instead on alcohol made by larger regional distilleries. If small-scale alcohol fuel does not develop, but cooking fuel demand increases, the project expects that methanol from unsellable natural gas will fire the stoves.

Providing villages with small-scale biomass-fired distilleries would provide fuel for both heating and lighting, while providing employment for people to grow energy crops for the village distillery. At a minimum, distillery byproducts could be used to provide dairy cattle with feed, with the milk being distributed to needy families.

Eventually, the distilling operation could evolve to fuel a small cogenerator to provide hot water

*Fig. 13-17
Alcohol marine
range. Alcohol
is by far the saf-
est fuel aboard
boats. Marine
ranges have been
made for decades.
The galley crew
always appreciate
using an alco-
hol stove, since
kerosene stove
emissions are a
frequent cause of
seasickness!*

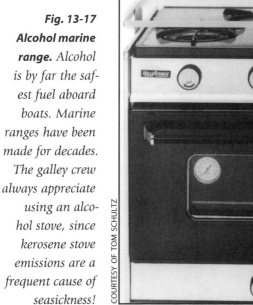

COURTESY OF TOM SCHULTZ

and electricity for refrigeration and equipment of a health clinic. Alternatively, heat absorption refrigerators run by alcohol are a practical alternative to electrical refrigerators at clinics, due to the often erratic supply of rural electricity in developing countries. Dometic AB makes alcohol-powered refrigerators of this type, and Brazilian concerns are also beginning to produce them.

Lighting in developing countries is often done by lanterns that use kerosene or animal fat. These release large amounts of particulate matter, and when the fuel is made from petroleum, toxic chemicals are released which are both carcinogenic and mutagenic. A good solution in areas without electricity is to use alcohol for lighting via low-cost lanterns designed to vaporize the alcohol before firing a **mantle**. The use of mantles can be controversial, since they can contain a small amount of naturally occurring radioactive material. Prior to mantle use, before kerosene existed, alcohol mixed with a little turpentine provided clean, bright light in American cabins via simple lamps with wicks for decades. Turpentine can be collected from pine trees in a method not too different from maple sugar tapping.

See, I told you this chapter would be pretty cool. Now that you've had your horizons widened as to what's possible with alcohol fuel, let's dig in to the details. We'll start with the nitty-gritty: the differences between gasoline and alcohol as fuels.

Endnotes

1. Grant Bonnema, Gregory Guse, et al., *Use of Mid-Range Ethanol/Gasoline Blends in Unmodified Passenger Cars and Light Duty Trucks* (Mankato, MN: Minnesota Center for Automotive Research, January 1999), 6.

2. Bonnema, et al.

3. Andy Tan, Ahmed Shebe, et al., *The Creed Project: 2003 Toyota Prius Hybrid Conversion to Ethanol* (Mankato, MN: Minnesota State University Automotive Engineering Technology, April 21, 2004).

4. *Material Safety Data Sheet for Gasoline*, www.brownoil.com/msdsgasoline.htm (June 24, 2005).

5. Matthew Brusstar, et al., *High Efficiency and Low Emissions from a Port-Injected Engine with Neat Alcohol Fuels* (Washington, DC: U.S. Environmental Protection Agency, Society of Automotive Engineers, 2002).

6. David B. Gardiner, Robert W. Mallory, et al., *Improving the Fuel Efficiency of Light-Duty Ethanol Vehicles—An Engine Dynamometer Study of Dedicated Engine Strategies* (Washington, DC: Society of Automotive Engineers Technical Paper no. 1999-01-3568), 1–10.

7. TCC Team, *Waiting in the Wings: Saab Aero × Concept Bows*, http://autos.aol.com/article/future/v2/_a/waiting-in-the-wings/20060303153909990001 (March 9, 2006).

8. John Heimlich, Vice President and Chief Economist, Air Transport Association, www.airlines.org/news/d.aspx?nid=9194 [By June 2006, average annual price for jet fuel already averaged $88.33 per barrel; annual projection is based on 2005 total jet fuel use of 19.9 billion gallons].

9. Gordon Cooper and Bill Paynter, communication with author, 1982.

10. Bill Paynter, communication with author, 1982.

11. "Indoor Cooking Smoke Quietly Killing Millions," *Solar Cookers International: Solar Cooker Review* 10:2 (November 2004).

12. "Indoor Cooking Smoke Quietly Killing Millions."

13. Harry Stokes, Project Gaia, *Commercialization of a New Stove and Fuel System for Household Energy in Ethiopia Using Ethanol from Sugarcane Residues and Methanol from Natural Gas*, presented to the Ethiopian Society of Chemical Engineers at its forum on "Alcohol as an Alternative Energy Resource for Household Use" (October 30, 2004), 2.

Fig. 13-18
James McMurtry

PHOTO COURTESY OF JAMES MCMURTRY

"We Can't Make it Here"
Music and Lyrics © 2004 by James McMurtry

Vietnam Vet with a cardboard sign
Sitting there by the left turn line
Flag on the wheelchair flapping in the breeze
One leg missing, both hands free
No one's paying much mind to him
The V.A. budget's stretched so thin
And there's more comin' home from the Mideast war
We can't make it here anymore

That big ol' building was the textile mill
It fed our kids and it paid our bills
But they turned us out and they closed the doors
We can't make it here anymore

See all those pallets piled up on the loading dock
They're just gonna set there till they rot
'Cause there's nothing to ship, nothing to pack
Just busted concrete and rusted tracks
Empty storefronts around the square
There's a needle in the gutter and glass everywhere
You don't come down here 'less you're looking to score
We can't make it here anymore

The bar's still open but man it's slow
The tip jar's light and the register's low
The bartender don't have much to say
The regular crowd gets thinner each day

Some have maxed out all their credit cards
Some are working two jobs and living in cars
Minimum wage won't pay for a roof, won't pay for a drink

If you gotta have proof just try it yourself Mr. CEO
See how far 5.15 an hour will go
Take a part-time job at one of your stores
Bet you can't make it here anymore

High school girl with a bourgeois dream
Just like the pictures in the magazine
She found on the floor of the laundromat
A woman with kids can forget all that
If she comes up pregnant what'll she do
Forget the career, forget about school
Can she live on faith? live on hope?
High on Jesus or hooked on dope
When it's way too late to just say no
You can't make it here anymore

Now I'm stocking shirts in the Wal-Mart store
Just like the ones we made before
'Cept this one came from Singapore
I guess we can't make it here anymore

Should I hate a people for the shade of their skin
Or the shape of their eyes or the shape I'm in
Should I hate 'em for having our jobs today
No I hate the men sent the jobs away
I can see them all now, they haunt my dreams
All lily white and squeaky clean
They've never known want, they'll never know need
Their shit don't stink and their kids won't bleed
Their kids won't bleed in the damn little war
And we can't make it here anymore

We'll work for food
We'll die for oil
We'll kill for power and to us the spoils
The billionaires get to pay less tax
The working poor get to fall through the cracks
Let 'em eat jellybeans let 'em eat cake
Let 'em eat shit, whatever it takes
They can join the Air Force, or join the Corps
If they can't make it here anymore

And that's how it is
That's what we got
If the president wants to admit it or not
You can read it in the paper
Read it on the wall
Hear it on the wind
If you're listening at all
Get out of that limo
Look us in the eye
Call us on the cell phone
Tell us all why

In Dayton, Ohio
Or Portland, Maine
Or a cotton gin out on the great high plains
That's done closed down along with the school
And the hospital and the swimming pool
Dust devils dance in the noonday heat
There's rats in the alley
And trash in the street
Gang graffiti on a boxcar door
We can't make it here anymore

CHAPTER 14

ALCOHOL VERSUS GASOLINE IN YOUR ENGINE

In the late 1970s and early 1980s, there was a lot of funding all over the globe to study the uses of alcohol fuels. The oil shortages during this period spurred governments to make this a priority, and a great deal of solid data resulted.

In the past 25 years, very little research has been done on ethanol as a fuel in this country. The one exception was the National Renewable Energy Laboratory (formerly the Solar Energy Research Institute, presently going through cutbacks), in conjunction with the U.S. Department of Energy, which continued working on ethanol, notably on engine use.

At first, I was surprised to see so little research. But as I delved into the subject, a familiar pattern formed. The change from government funding of scientific research to corporate funding has caused scientists to self-censor which research they pursue. I found this to be true over and over again in talking with people who had done studies on ethanol in the 1980s and who now do completely unrelated work.

Most studies since then have been done with methanol, which is largely a product of natural gas or coal, which is a field of study acceptable to energy corporations. Many of the benefits that were demonstrated using methanol as a fuel would have been even greater with ethanol.

It has been hardest to find new data on vehicle emissions using **neat ethanol** fuel—this is nearly pure ethanol, with only trace amounts of other materials used as denaturants. Since we were unable to find *any* recent directly comparable internal combustion engine (ICE) emission data for neat ethanol, we decided to perform our own studies.

Some data comparing emissions of liquid and vaporized alcohol (both dramatically low) appeared in the previous chapter, and emissions data for liquid alcohol appear in the detailed vehicle conversions in Appendix B. Hopefully, this will help to counter the extensive propaganda against alcohol

THE CHANGE FROM GOVERNMENT FUNDING OF SCIENTIFIC RESEARCH TO CORPORATE FUNDING HAS CAUSED SCIENTISTS TO SELF-CENSOR WHICH RESEARCH THEY PURSUE.

MATT FARRUGGIO

Fig. 14-1 Alcohol and gasoline fires. *A tablespoon of each fuel was ignited in each bowl. The difference in combustion qualities is apparent. Alcohol burns clean and was easily extinguished with a plant mister. The gasoline fire on the right thoroughly blackened the bowl with a tarry, carbonaceous deposit and could not be extinguished with the mister.*

fuel in the field of emissions. But even more pervasive than false emission data are highly promoted myths about the use of alcohol fuel in general.

"The crisis is very, very near. World War III has started. It has already affected every single citizen of the Middle East. Soon it will spill over to affect every single citizen of the world."

—ALI SAMSAM BAKHTIARI, VICE PRESIDENT OF THE NATIONAL IRANIAN OIL COMPANY (NIOC)

MYTHS ABOUT PUTTING ETHANOL IN YOUR CAR

In Chapter 2, we touched on some of the bigger myths surrounding ethanol production, involving energy return and food versus fuel. Now let's look at some of the myths associated with putting ethanol in your car. Some of them are: Alcohol burns much hotter than gasoline and will wreck a car's valves; its emissions add to air pollution; it gets half the mileage of gasoline; it is corrosive, ruining most of the various metals in a car's fuel system and engine, not to mention gaskets, seals, rubber, and plastics; the high amount of water in its exhaust will rust out a car's exhaust system; crankcase oil is severely compromised by alcohol contamination which wrecks bearings, etc.; it washes oil from cylinder walls, causing rapid bore and ring wear; and it won't start on cold mornings. Sheesh, after hearing all this, who would ever try the stuff?

Burning Hotter

Let's first take the myth about alcohol burning hotter and destroying the engine. Each engine is a little different, but, in general, alcohol's exhaust temperature is in the vicinity of 1000°F, while gasoline's is around 1400°F (with a maximum of 1650°F). Under ideal research engine conditions, where everything is tuned perfectly, the difference between alcohol and gasoline exhaust can be as

Fig. 14-2 Energy losses from gasoline. *Energy losses and available power from a gasoline engine. As you can see, a gasoline engine makes a pretty good heater but doesn't get a lot of work done. Alcohol engines have attained more than 40% efficiency (energy converted to work).*

Energy Losses from Gasoline

Energy Lost from:	%
Exhaust	20–45
Coolant (Radiator Heat)	15–30
Friction	7-38
Incomplete Combustion	2–5
Radiation	1–5
Transmission	1–5
Available as Road Power:	7–15

little as 130°F, but those conditions are not found in most cars on the road. The difference is far more pronounced on older vehicles without sophisticated electronic control units (ECUs, the engine's computer "brain").

So, alcohol burns much cooler than gasoline and extends engine life by double to triple over gasoline combustion with its higher temperatures. Although the exhaust temperature is much lower, the radiator water temperature doesn't change on your in-dash gauge, but the thermostatic valve that releases water from the engine to the radiator for cooling opens far less often in an alcohol-run car than in one using gasoline.

Alcohol's high latent heat of vaporization cools air entering the car's cylinders upon its induction stroke.[1] Gasoline, with its higher-temperature combustion, requires much more cooling. Because gasoline's average latent heat is very low, it can absorb very little heat. The higher the temperature, the more fuel is lost to cylinder heat exchange.

Cooling is done by the cooling jacket. But during acceleration or when the vehicle is under load, gasoline's exhaust temperature and total waste heat climb. Gasoline engine design often reduces this peak heating by using excess fuel to cool the **combustion chamber** by evaporation. This is very wasteful.

In a properly converted engine, burning alcohol in the cylinder is a much more complete process than burning gas. There is much lower temperature and far less unburned or partially burned fuel, making it more efficient, less wasteful, and less dangerous for the environment. Also, alcohol's cool combustion results in lubricating oil temperatures of 50°F or more lower than with gasoline. This makes a big difference in the lubricant's performance.

Furthermore, the combustion of alcohol generates a larger volume of combustion gases than gasoline, which automatically develops higher peak cylinder pressures.[2] After all, the purpose of an ICE is to drive pistons to do work, not generate heat. Energy losses from gasoline in an ICE are generally as seen in Figure 14-2.[3]

So where does the myth come from? In auto racing, where methanol is used, very low air/fuel ratios are used to get the maximum horsepower. When you dump that much extra fuel into the engine, higher temperatures are reached and can cause engine damage. This is especially true with corrosive methanol.

Since much less alcohol fuel is wasted as heat, you can get away with things you might not want to try with gasoline. Back in the early '80s when I was first converting vehicles, vehicles had cooling fans driven from the engine by belts. These fans drew as much as 17 horsepower, which was work not being delivered to the wheels and which therefore reduced mileage. In some cases, while running on alcohol, I was able to completely dispense with a cooling fan, which by itself gave a 10–15% increase in miles per gallon.

I bring this up to show how cool the engines ran. Today's vehicles have largely eliminated the energy lost to fans by using low-power electric fans that go on only in relation to a rise in temperature. Belt-driven fans used today have a clutch

Properties of Gasoline, Ethanol, and Methanol

CHEMICAL PROPERTIES	GASOLINE	ETHANOL	METHANOL
Formula	C_4-C_{12}	Ch_3Ch_2OH	Ch_3OH
Molecular Weight	wide range	46.07	32.04
Carbon % (Weight)	85–88	52.14	37.48
Hydrogen % (Weight)	12–15	13.12	12.58
Oxygen % (Weight)	0	34.74	49.94
Carbon/Hydrogen Ratio	5.6–7.4	4.0	3.0
Optimum Air/Fuel Ratio	14.7:1	9.0:1	6.5:1
Tolerable Air/Fuel Ratio	13.2–15.1:1	5.3–23.3:1	2.8–13.6:1
PHYSICAL PROPERTIES			
Specific Gravity @ 60°F	0.70–0.78	0.794	0.796
Vapor Pressure @ 100°F	9.0–12.5 RVP	2.5 RVP	4.6 RVP
Boiling Point °F	80–440	173.3	148.1
Freezing Point °F	−70	−174.6	−144.2
Solubility in Water	240 ppm	miscible	miscible
Viscosity @ 68°F (centipoise)	0.288	1.17	0.547
Weight per Gallon (lbs.)	5.8	6.59	6.64
THERMAL PROPERTIES			
Low Heating Value Btu/lb.	18,900	11,550	8,600
Btu/gal.	110,960	76,230	57,760
High Heating Value Btu/lb.	20,250	12,800	9,776
Btu/gal.	118,250	84,348	64,521
Octane Rating (Research)	90	106	106
Octane Pump (R+M/2)	86	98	98
Maximum Compression Ratio (Spark-Ignited Engine)	8.5:1	18.5:1	19:1
Auto-Ignition Temperature °F	430–500	685	878
Flashpoint (Closed Cup) alcohol @ 200 proof in °F	−50	56	53
Flashpoint (Open Cup) alcohol @ 185 proof in °F	−45	70	69
Latent Heat of Vaporization @ 77° F Btu/lb.	147	395	503
Btu/gal.	852	2603	3320

Fig. 14-3
Properties of gasoline, ethanol, and methanol. *A useful collection of characteristics to refer to in comparing fuels.*

Fig. 14-4 The four-stroke cycle. *This is how your engine works. In the intake stroke the piston goes down, sucking in air/fuel mixture. In the compression stroke the piston travels up, squashing the fuel into a small space. In the power stroke, you see the spark plug firing, and the exploding mixture driving the piston down. (In reality the spark plug fires just before the end of the compression stroke.) After delivering its power to the drive train, the piston travels back up to push out most of the exhaust.*

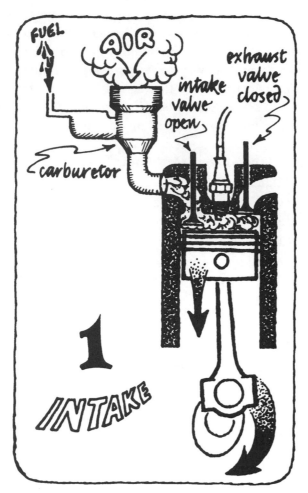

FUEL

AIR

carburetor

intake valve open

exhaust valve closed

1 INTAKE

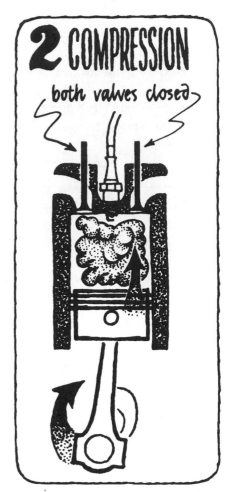

2 COMPRESSION

both valves closed

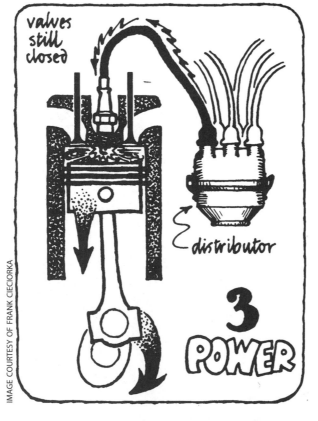

valves still closed

distributor

3 POWER

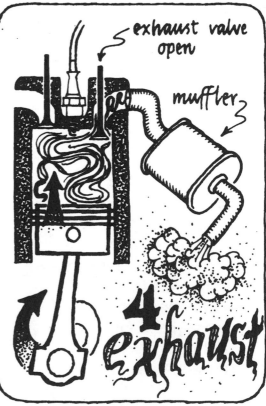

exhaust valve open

muffler

4 exhaust

that lets the fan spin only when you are idling or traveling too slowly for simple air movement through the radiator to be sufficient to cool the engine.

Emissions

The cooler, more efficient combustion of alcohol has significant implications for emissions. The major emissions from automobiles are carbon dioxide, carbon monoxide, nitrogen oxides, and hydrocarbons.

Alcohol itself burns nearly as cleanly as hydrogen. Unlike gasoline, which is a mixture of straightforward hydrogen/carbon compounds, alcohol is an oxygenated hydrocarbon. Alcohol's extra oxygen, along with its faster flame propagation, gives it a substantial advantage over gasoline in emissions.

Clean burning is important for three reasons, the first and most obvious one being that the environment doesn't need any more pollutants. Second, a clean-burning fuel means less carbon residue to wear down engine cylinders and rings, contaminate oil, and, in some vehicles, block exhaust mufflers. And third, the less unburned or partially burned fuel you leave behind, the more work you'll get out of your fuel, and the less you will waste.

Statistics from older emission studies of alcohol's exhaust vary widely. Research done independently of energy corporations, including some studies by the U.S. Departments of Energy and Transportation, has shown dramatic reductions of regulated pollutants, especially carbon monoxide, hydrocarbons, and nitrogen oxides. Several energy corporation studies indicate that alcohol exhaust contains high levels of hydrocarbons (HC) and nitrogen oxides (NOx), with reductions in carbon monoxide (CO). However, the validity of the techniques that oil-company-sponsored organizations have used to measure HC and NOx is disputed in the scientific community.

Results that are representative of many other independent tests were reached in an independent study done in 1981 by *Mother Earth News* in cooperation with the New York City Environmental Protection Agency. New York City pollution standards were considered the strictest in the nation. Their test vehicle was a 1978 Chevrolet taxi with full pollution control gear, tuned to factory specifications. The *gasoline* results were 1.5% CO and 200 ppm HC, which is *much* cleaner than the average American car of the era.

The researchers proceeded to remove all of the pollution-control equipment from the engine, except for the **positive crankcase ventilation (PCV)**. They drained the tank and filled it with alcohol, and found that there was an incredible reduction in emissions: CO dropped to 0.08% and HC to 25 ppm. That's 95% less CO and 87.5% less HC in alcohol exhaust than gasoline exhaust,[4] but without any pollution control equipment!

Even though standards have been substantially tightened since this study, the increased engine control in modern engines permits proportionally even greater reductions in emissions and in some instances actually reduces them to zero (see Appendix B, Tundra Conversion).

Most of the CO resulting from any alcohol test, and much of the HC, comes in the first minute or so of operation, due to issues related to cold-starting an alcohol vehicle, and the two minutes or so until the engine warms enough for complete computer control.

According to one methanol study,[5] in order to reduce unburned methanol emissions at start-up (at 25°C with the M100 (100% methanol) engine), flame propagation must be completed early enough to avoid flame quenching (partial burning). Advanced spark timing, which starts the flame earlier, and igniter modifications (hotter spark plugs) that decrease flame propagation time were shown to be effective.[6] (See more on these sorts of changes in Chapter 18.)

Fig. 14-5 Chainsaw combustion chamber. *After a few hours of running on an alcohol/biodiesel mixture, the interior of the chainsaw engine is absolutely carbon-free.*

In the average low-compression conversion to alcohol, emissions of nitrogen oxides (the brown agent in **photochemical smog**) are 50 to 99% less than gasoline's emissions. It takes very high cylinder temperatures to form NOx—cooler-burning alcohol doesn't deliver intense enough temperatures to form large amounts. In the case of extremely lean-burning mixtures or high-compression conversions, the reduction of NOx may only be 20 to 50% less than gasoline emissions—although by increasing the amount of exhaust gas recirculation, many researchers have been able to essentially eliminate NOx as an emission in high-compression engines.

Using alcohol at 190 proof also radically drops NOx readings, since the small amount of water absorbs a disproportionate amount of heat, which makes NOx formation difficult.[7]

Although it is good to limit NOx as much as possible, it should be remembered that hydrocarbons, also known as **particulate emissions**, and carbon monoxide have been shown to be a far more serious health threat. The purpose of emission controls is to provide conditions that keep hydrocarbons low, by making sure there's plenty of oxygen, without increasing NOx by overdoing the oxygen proportion in the combustion.

Contained within alcohol's very low hydrocarbon emissions is buried a slight difference between the composition of alcohol and gasoline hydrocarbons—in older engines, alcohol does give off three times as much **acetaldehyde** as gasoline. This sounds like a lot, but three times a tiny amount of acetaldehyde is still very small. Acetaldehyde is thought to be a possible human carcinogen, since in very large quantities it causes cancer in rats. And it is slightly irritating to the eyes in larger dosages. But it is not considered a readily reactive component of photochemical smog, which forms in the air on exposure to sunlight.

Acetaldehyde certainly does not pose anywhere near the same carcinogenic threat as do benzene, **toluene**, or **xylene (BTX)**, which are major components of gasoline. In fact, when humans drink liquor, the liver turns some of the alcohol into acetaldehyde, part of the reason we get tipsy. When you compare the toxicity of gasoline hydrocarbons to alcohol's, you find that instead of acetaldehyde there are far more dangerous emissions stemming from the BTX used in gasoline. Alcohol emissions carry none of the heavy metals and substantial sulfuric acid that gasoline and diesel exhausts do.

Still, who wants to add any new pollutants to the air? As it happens, modern three-way catalytic converters strip all traces of acetaldehyde from the exhaust of a properly tuned alcohol vehicle.

Another small category of pollution is evaporative emissions of raw fuel into the air. These are what escape from the fuel tank when some of the fuel becomes vapor. Straight alcohol's evaporative emissions are dramatically less than gasoline's, and are not any more toxic than what you'd find in the air of your local bar.

UNRELIABLE TESTS

Studies by certain energy corporations have indicated high hydrocarbon *and* aldehyde emission rates for alcohol fuels, but the Department of Scientific and Industrial Research of New Zealand has evaluated almost all of these studies as unreliable. Most tests on aldehyde emissions from methanol and ethanol engines are performed by entities related to oil companies and use the MBTH method, or a modification of this method. It turns out that MBTH is just about the most fickle collection method around, accurate only in very special circumstances. The New Zealand report definitively states, "The assumption that this [MBTH] technique can be used as a general tool for the measurement of aldehyde levels in exhaust gas is erroneous."

CATALYTIC CONVERTERS IN BRAZIL

In the '80s, before the use of catalytic converters, it was possible to notice the odor of acetaldehyde on bad air days in São Paulo, Brazil, even though the overall pollution level in that city had dropped dramatically with the increase in alcohol vehicles. Ironically, the pollution level in São Paulo, an enormous sprawling city, had dropped so much during the era in which alcohol dominated the fuel market that the implementation of catalytic converters on cars was postponed, since there didn't appear to be a need for them. After the drop in alcohol-powered cars in favor of gasoline-powered cars there in the 1990s, which resulted in a dramatic rise in pollution, it became necessary for auto manufacturers to install catalytic converters on new cars.

Fig. 14-6 Los Angeles, a day with clean air.

Fig. 14-7 Los Angeles, a day with not-so-clean air.

The same cannot be said for the evaporative emissions of benzene, toluene, and other carcinogenic or mutagenic **aromatic hydrocarbons** found in modern gasoline. These garbage byproducts are added to gasoline by oil companies in greater amounts in winter, when their high vapor pressures are easier to mask, than in the summer, when the excessive evaporation would violate air quality laws—and cause vapor lock in the vehicle's fuel lines.

Because it is a **polar liquid**, when alcohol is mixed in small quantities with a large amount of certain volatile hydrocarbons (especially cyclobenzene, cyclopentanes, and cyclobutanes, garbage byproducts condensed out of natural gas), a chemical reaction occurs that causes additional vapor pressure. Reid vapor pressure (RVP) measures how evaporative a fuel is. The higher the vapor pressure, the more evaporation occurs.

The increase in evaporative emissions when 5-10% alcohol is added to gasoline is usually less than one pound RVP, depending on the season and what state regulations permit. Gasoline has six to 14 pounds RVP; if the level of alcohol is near 4%, nine-pound RVP gasoline would become ten-pound RVP in terms of evaporative emissions.

The vapor pressure reaction is most notable in gas containing about 4% alcohol, but still occurs to a lesser degree in the range of 2 to 20% alcohol. Outside of this range, alcohol reduces vapor pressure. An independent lab analysis I had performed on E-65 (65% alcohol) showed reductions in vapor pressure of the remaining gasoline from over six pounds RVP to 4.38 pounds, which equals quite a dramatic reduction in evaporation. The RVP of **E-98** is between two to three pounds.

Most of the evaporative emission problems were solved years ago when automakers were required to have sealed fuel systems so that fuel vapors couldn't escape from gas caps. Nowadays, Mega-Oilron-sponsored studies focus on permeation emissions, the amount of evaporative emissions that get through flexible fuel lines. The increase in the amount of emissions from this source at 5.7% alcohol is less than one gram per day! This amounts to less than three-tenths of one percent of the total emissions from the exhaust of a gasoline car on a normal day's commute. So only in the case of a small amount of alcohol added to a lot of gasoline is there a tiny increase in evaporative emissions.

But adding the 5.7% alcohol to the gasoline dramatically dropped the vehicle's tailpipe emissions of CO—100 times more than it increased its permeation emissions. Bear in mind that 90+% of the tailpipe emissions are eliminated by alcohol use, and you can understand how permeation emissions are simply a propaganda tool. If you were serious about eliminating this one gram per day, you could: 1) eliminate the addition of the garbage to the gasoline in the first place; or 2) use impermeable fuel lines; or 3) legislate that all gasoline needs to have more than 20% alcohol added so as to actually reduce the already small amount of gasoline's permeation emissions.[10] When you see discussion of permeation emissions, credit American Petroleum Institute propaganda.

So, to sum up the emissions situation, running neat ethanol would result in a 90% drop in emissions over gasoline in even a simple conversion. A well-designed dedicated alcohol engine would actually emit cleaner air than the air entering the engine.

GASOLINE IS GARBAGE

The entire sale of gasoline is, in effect, the largest waste disposal system on the planet. It is the proverbial elephant in the middle of the room that no one talks about and pretends isn't there. Just as in Rockefeller's day, when dangerous volatiles (gasoline) were flushed into rivers at night, the oil companies are disposing of their waste, this time charging us for it as auto fuel.

The real money from oil is made on industrial chemicals and products[8] that, even 25 years ago, totaled more than $25,000 per barrel of oil,[9] while the gasoline from a barrel of oil only amounted to $106. The spread is wider today. A small percentage of oil produces most of the profit, and all the rest is waste.

Gasoline might contain over 400 different toxic waste products on any given day at a refinery. Their boiling points will range from 80°F to over 400°F. It won't be the same mix two days in a row, since it all depends on just what high-value products are being made that day, and what is left over as waste.

Now that cyclobutane, cyclohexane, cyclobenzene, and other natural gas condensates (NGC) are considered "conventional" fuel, gasoline contains the garbage from the natural gas business, as well. Natural gas condensates actually constitute nearly 10% of the fuel from the Middle East, and perhaps even more from the United States. If you subtract NGC from gasoline, the figures would show that *the world peaked in oil supply years ago.*

So, by changing the "accounting" and including NGC (and tar from tar sands) as conventional oil, MegaOilron can temporarily dodge the panic that will ensue when Peak Oil is finally officially admitted.

Given the dramatic benefits to the environment, you would think that you'd be given incentives to convert your vehicle to alcohol. Instead, the legal requirements for pollution control devices on alcohol-driven vehicles are still somewhat ambiguous. Air pollution control laws are actually designed to enforce the placement of smog devices on vehicles, regardless of lower emissions. What's more, most things you might add to the engine to enable clean running of alcohol would also be illegal. Any alteration is considered "tampering" by the government.

In the 1980s, court victories in California and other states challenged those laws, and, in those cases, vehicles that could document emissions within legal limits were allowed to dispense with emission control devices. Nowadays, most smog agencies on both state and federal levels ignore these court decisions and refuse to recognize the legitimacy of converting gasoline vehicles to alcohol *unless they are certified.* Flexible-fuel vehicles are certified, and thus are legal to use alcohol. In theory, you could have your conversion certified, but the process for being certified generally costs more than the purchase price of your car (see Chapter 26 for more on this subject).

Mileage

The myth that alcohol fuel gets only half or two-thirds the mileage of gasoline is often repeated, but no one ever cites any road tests. A properly designed alcohol engine will generate more miles per gallon than a gasoline engine. Alcohol burns at a much faster speed than gasoline, turning more of its energy into work faster, and more of its chemical energy into work rather than waste heat.

The mileage myth assumes a simple but spurious comparison of the two fuels' heating value (expressed as Btu/gallon) as the basis for low mileage figures for alcohol. The oft-cited David Pimentel, Ph.D., at Cornell University, has declared that "because of the relatively low energy content of ethanol, 1.5 gallon[s] of ethanol have the energy equivalent of one gallon of gasoline." Let's examine this erroneous assumption.

Heating value is the measure of the amount of Btu generated by the perfect combustion of the material with an ideal amount of oxygen, as in burning wood for a fire. The heating value of one and a quarter pounds of wood, a cup of gasoline, and four pounds of dynamite are about the same

(7000 Btu).[11] But the work each of these materials generates is vastly different.

In an engine, what we care about is how well the fuel drives a piston. The issue is work, not heat. To measure the energy of work, the standard unit is the **joule**. That's about the amount of energy it takes to toss an apple vertically a foot. Simply defined, it's the force against the apple (or a piston) multiplied by the distance.

Power is the speed at which a joule takes place. When you generate one joule per second, that's a **watt**, a term you should be familiar with in rating the power of motors or power tools, which do work. Showing how irrelevant heating value is to work, the wood generates 30 kilowatts (thousands of watts), the gasoline, 3.2 gigawatts (billions of watts), and the dynamite, 2000 gigawatts.

In general, since there is a limited amount of time to complete the combustion of fuel in an engine, the speed at which the fuel burns in an engine determines both how efficiently the engine converts the fuel into power (watts). The gasoline in the example above will explode in about 2.5 milliseconds (thousandths of a second), while the wood takes several minutes to release its heat. The dynamite delivers its energy in four microseconds (millionths of a second).

Pundits will argue that it's not fair to use wood or dynamite in a comparison because they are not liquid fuels. So, let's compare liquid fuels.

The reason you don't use heating value to compare auto fuels is that it is not a valid way to compare the work of liquid fuels in an engine. Liquid fuels have other characteristics that are more important than heating value. If heating value were

Fig. 14-8
Pinging. In pinging, the fuel explodes due to auto-ignition when the temperature in the cylinder gets too high during compression. This is caused by low octane rating of fuel for the compression ratio of the engine. Pinging spikes temperature and causes engine-damaging stress due to mistiming of the combustion.

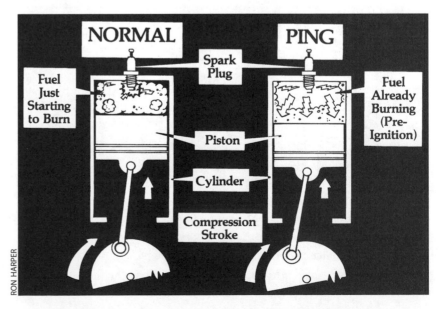

RON HARPER

the valid way to compare fuels, why wouldn't we run our gasoline engines on diesel fuel or melted candle wax (an excellent rocket fuel), which have far more heating value than gasoline? The answer is simple: Other characteristics are far more important. Octane, efficiency in combining oxygen with fuel, volume of the combustion products, and speed of the **flame front** are all more important than heating value.

A miles-per-gallon comparison between gasoline and alcohol, considering alcohol's higher efficiency, is much closer than the difference in heating values would seem to indicate. Field tests by my students and other independent studies commonly

indicate reductions in mileage of 10 to 15% in low-compression conversions of engines designed for gasoline.[12] These are the largest reductions to be expected, except in very simple conversions.

In 1982, *Mother Earth News* performed some mileage studies on a 1970 Chevrolet 250-cubic-inch inline (carbureted) six-cylinder engine, with some very good results. The truck lost 5% mpg while unloaded, but, when loaded with 2200 pounds, the comparison swung in alcohol's favor—16% better than gasoline.[13] With a load, alcohol maintained its mileage, while gasoline's mileage dropped.

In doing its study, *Mother Earth News* altered the vehicle's carburetor to deliver more fuel to match the fixed amount of incoming air. The main jet determines the amount of fuel to be mixed with the incoming air. Assuming heating value was the proper measure for fuel mixing, their first tests used a carburetor metering jet drilled out 40% larger. But instead of losing 40% mpg, they showed a loss of only 12% mpg. With further experimentation, they found that the optimum increase of jet size was only 19%. Losses are often less, or nonexistent, in high-compression vehicles (which I'll discuss below) or in vehicles under load.

At that time, several inventors, some major auto companies (e.g., Mercedes), and NASA recorded mileages way above alcohol's heating-value-predicted mileage by completely vaporizing the fuel and using a propane-type gaseous fuel carburetion system. The highest scientifically verifiable figures for liquid fuel efficiency are more than double the amount of work per Btu compared to gasoline (see Chapter 13 on alcohol engines). But during the 1980s, I personally witnessed several vaporized alcohol engines getting almost double the mileage of gasoline.

In Brazil in the 1980s, government-licensed alcohol conversion shops were required by law to deliver mileages on alcohol no more than 25% lower than what the same car would get on gasoline. Over a hundred shops in Brazil's capital routinely exceeded government minimums, but all shops had to at least meet this standard. Even this modest standard should have been impossible, if heating value corresponded to mileage.

When comparing fuels, it's important to compare Btu per mile and the efficiency with which a fuel is converted to work. Although alcohol starts off with fewer heating value Btu, it burns with a much higher efficiency than gasoline, so that the miles

Energy Content Versus Efficiency

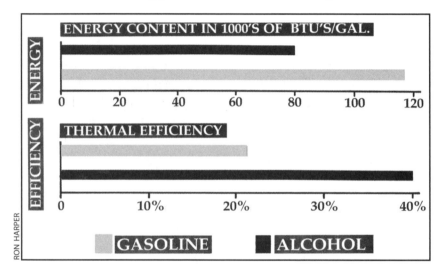

Consumer Automotive Buying Preferences
in Years 1980, 1981, 1983, 1985, 1987, 1996, 1998, 2000, 2001

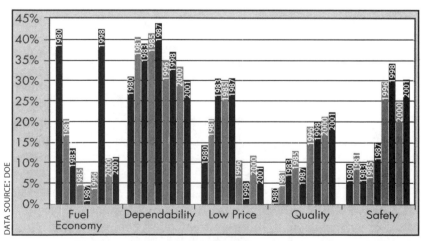

Fig. 14-10 Consumer automotive buying preferences. Although we've been discussing mileage as if it is important, consumers care less about mileage than nearly any other characteristic about their vehicle. The exception in 1980 reflects the sharp spike in prices during the Arab oil embargo. But note the almost immediate reversal of concern about mileage after the embargo ended.

per gallon are either very close (as in a gasoline engine converted to alcohol) or, when an engine is optimized for alcohol, actually higher than gas or diesel due to the increased efficiency. Alcohol has achieved 48% work efficiency in the lab, and 43+% efficiency on the road.[14] Gasoline generally hovers between 15–20%. Alcohol's higher level of efficiency had been measured in the early 1900s by the U.S. Navy and has therefore been known as fact ever since.

And today we need to think about other qualities of a fuel, too. As Dr. Josmar Puglioso elegantly states, "What really matters is cost per mile, pollutants per mile, and, most important, CO_2 per mile."[15]

Standard gasoline engines converted to using alcohol will get much better mileage than the heating value would indicate, although generally a little less than gasoline. Engines designed to take advantage of ethanol's fuel characteristics will generate more miles per gallon, at a lower cost per mile, with less than 1% of gasoline's pollutants, and with negative CO_2 emissions per mile when its whole life cycle is taken into account.

Corrosion

Methanol, a fuel used by high-performance race vehicles, is severely corrosive and will attack **terneplate**, aluminum (some alloys are immune), zinc, and magnesium. Methanol will also loosen rust from steel. Most "rubber" fuel line materials are compromised by methanol, which is so nasty that, even used as a 10% octane booster with gasoline, it will strip acrylic lacquer paint from cars. (Acrylic *enamel* paint is almost unaffected, though.)

Ethanol at 190 proof or above will have none of these effects on virtually all fuel system components in vehicles made after 1983. Since then, alcohol has been blended into much of the fuel in

At 160 proof, 80% alcohol is quite capable of causing significant corrosion. When water content of alcohol fuel is high, there can be corrosion of metals containing aluminum (with less than 12% silicon) or zinc. Water sets up an electrolysis that draws metals most easily dissolved into solution. However, rusting or corrosion does not readily occur at levels of 185+ proof.

the U.S. at from 5.7 to more than 10% without any of these corrosive effects.

At 160 proof, 80% alcohol is quite capable of causing significant corrosion. When water content of alcohol fuel is high, there can be corrosion of metals containing aluminum (with less than 12% silicon) or zinc. Water sets up an electrolysis that draws metals most easily dissolved into solution.

However, rusting or corrosion does not readily occur at levels of 185+ proof. Industrial 190-proof alcohol is sold in 55-gallon steel drums every day. Commercially sold alcohol in the U.S. has no water in it at all. In Brazil, 96% alcohol with 4% remaining water has been sold at the pump for more than 25 years without mishap.

If you insist on experimenting with low-proof alcohol, adding alcohol-soluble synthetic oil retards corrosion. Corrosion inhibitors can be used in low-proof alcohol fuel to eliminate any corrosion of metals in the fuel system. The Brazilian inhibitor I used in the 1980s was mixed at a ratio of 1 to 4000 with alcohol, and added less than one cent per gallon to the cost. Less than 0.5% biodiesel also is an effective corrosion inhibitor.

There are aberrations. In bad fermentation batches, invading bacteria may produce a high amount of acetic acid, which has a low enough boiling point to be distilled with the last part of the alcohol run. High acetic acid content and low proof together are corrosive, and if your alcohol's

"The bias aroused by the use of alcohol as a motor fuel has produced results in different parts of the world that are incompatible with each other. In general, we can detect two schools of thought with regard to the use of alcohol as a motor fuel. Countries with considerable oil deposits (such as the United States) or which control the oil deposits of other lands (such as Holland) tend to produce reports antithetical to the use of fuels alternative to petrol; countries with little or no indigenous oil tend to produce favorable reports.... The contrast between the two cases presented is most marked; one can scarcely avoid the conclusion that the results arrived at are those best suited to the political or economic aims of the country concerned, or of the industry which has sponsored the research. We deplore this partisan use of science, while regretfully admitting its existence, even in the present writer."

—J.W. PLEATH, CHAPMAN HALL, LONDON, 1949

pH is below 6.0, it is high enough in acid to cause problems. To avoid these problems, it is standard procedure to neutralize your mash to pH 7.0 before distillation by adding lime.

The various esters found in fusel oil can also enter your fuel at the end of a run. Rather than cause corrosion, esters have a tendency to deposit a little gum or carbon on valves. This has never been known to cause a problem.

Many of today's cars come with an alcohol-proof plastic gas tank. If you have an older car or a tank that is made of metal, you may have a few concerns. But I've only seen gas tank coatings corroded on very old vehicles that were using low-proof, highly acid-contaminated fuel.

Electrolysis nickel coating on a gas tank to corrosion-proof it generally costs over $100. Some Teflon solutions can create a corrosion-proof coating. A good preparatory step for Teflon coatings is to have the tank boiled out and "pickled" at a plating shop. Check your J.C. Whitney catalog or truck parts dealer for sources.

Carburetors made of an inexpensive alloy called Zamack can be surface-corroded when used with low-proof or acidic alcohol; Zamack is composed of zinc and some other metals susceptible to electrolysis. Corrosion appears as a white powder that, after a few months, plugs fuel orifices in the carburetor. Corrosion inhibitors or synthetic oil reduce or nullify that effect, or you may choose to take your disassembled carburetor to a plating shop and have it plated with electrolysis nickel. The springs in the power valve, **accelerator pump**, and any other non-stainless or non-brass parts should be plated at the same time. (Of course, you wouldn't plate rubber, leather, plastic, or foam parts.) But remember this is only necessary if you are using alcohol below 185–190 proof.

As far as rubber parts, plastics, seals, gaskets, fuel pump diaphragms, and floats in the carburetor and fuel tank, there are some materials to avoid and some to replace in pre-1983 vehicles. Ethanol is not harmful to seals, gaskets, normal neoprene fuel lines, or the newer high-pressure fuel injection hoses.

In the United States, most fuel pumps have been alcohol-safe since the introduction of gasohol. However, some cars made in the early 1980s were fitted with particularly cheap-looking clear fuel lines. These clear lines can be softened by alcohol and will eventually collapse, cutting the fuel off to the pump or carburetor. Replace them with standard neoprene fuel lines. It's been noted that

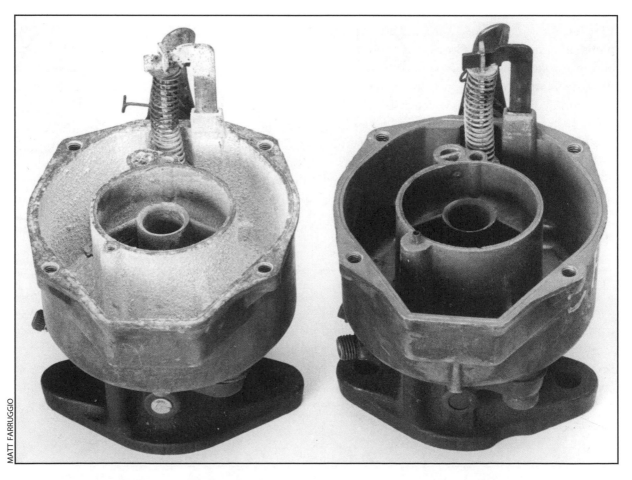

Fig. 14-11 Carburetor corrosion. The carburetor on the left was left sitting cold, filled with untreated 160-proof high-acid content alcohol for three months. The carburetor on the right used only 185-plus-proof hot alcohol with inhibitors for a longer period of time. Only slight surface corrosion of the accelerator pump spring was noted.

MATT FARRUGIO

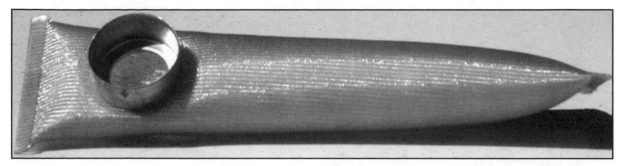

Fig. 14-12 Fuel pump intake filter. *This sock-style filter protects the small pump in the gas tank from sucking up chunks of crud.*

neoprene swells slightly with hot alcohol fuel, but in my experience this does not reduce the line's life, and it certainly hasn't caused me any trouble. Fuel lines based on urethane should be replaced, but these are more common on older planes than in cars.

The swelling of fuel system materials is far more pronounced when you mix alcohol at *low* dilutions with gasoline. In fact, you can't design a much worse mix than today's gasohol, which has less than 10% alcohol. For instance, 100% alcohol causes a 2% swelling in a common Viton **fluorocarbon elastomer**, and 100% gasoline causes about 1% swelling, but 10% gasohol causes 6% swelling. Neoprene does not swell at all.

Nitrile elastomers will swell 11% on 100% alcohol, still within usable design specifications. But when exposed to 10% gasohol, they will swell 68%, due to the reaction between alcohol and the volatiles in gasoline.[16] These elastomers are being phased out as fuel system material. Most good-quality fuel pumps made today use Viton, or a similar **fluoroelastomer**. In the few cases where a nitrile elastomer was used in the fuel pump, if it fails, then make sure its replacement does use Viton.

In general, pre-1983 fuel pumps may have some rubber parts that could fail. But considering the amount of use they've already given you by this time, it isn't much of a hardship to replace them with modern materials. They have already exceeded their design life.

On very old vehicles, 1950s and earlier, fuel lines were made of **butyl rubber**, which doesn't stand up well to alcohol. These lines are only found on collector's vehicles kept in original condition—you're not likely to run into them often.

Owners of fuel pumps from the 1950s will run into a unique problem. The diaphragms used in old pumps, invented before modern elastomer technology, were made of cloth and varnish—easily dissolved by alcohol. This was the case on my old 1952 fire truck. Cut out a thin neoprene or silicone elastomer diaphragm and replace the cloth one. Similarly, the floats in old carburetors were made of cork coated with varnish. Remove the float and coat it with a thin layer of epoxy (see Chapter 15 for a more detailed description of floats).

Some common clear plastic fuel filters use paper filters, held in place by an alcohol-soluble glue. If the filter paper comes loose, it can plug the filter

---

FOOL ME THREE TIMES?

These words are from Senator Tom Harkin (D-Iowa):

The oil companies first started putting lead into gasoline—as an octane enhancer. This went on for years and years [75 years] until finally we found out that lead was poisoning our kids and poisoning the atmosphere.

So we told the oil companies, "You've got to take the lead out of gasoline." They said, "Well, we've got to have octane enhancers." And they came up with what they called the "VOCs," the volatile organic compounds—xylene, toluene and benzene—and they put those in there and kept the octane up.

Well, guess what we found then? After a few years of this, we found out VOCs are highly carcinogenic, so we said to the oil companies, "Hey, you've got to do something about this. You've got to get rid of that."

…Now the oil companies are saying, "Get rid of the oxygen standard." They were happy with two percent when they could use MTBE. Now they're saying, "Get rid of this. Trust us. We'll come up with something else. We've got alkylates. We're going to come up with some other kind of witches' brew here that we're going to put into gasoline that will keep the octane up and will keep our air quality standards high."

Fooled once, fooled twice, fooled three times. Are we going to be fooled another time by the oil companies while we've got something [ethanol] that will both enhance the octane and at the same time clean up the air and won't pollute the water?

Senator Harkin failed in keeping the oxygenate standard, which was repealed in 2005 at the demand of Republican lawmakers.

Fig. 14-13 Plugged sock filter. *When I first converted my 1982 Volvo to alcohol, I did so suddenly. All the accumulated crud inside the tank washed off the walls and encrusted the sock filter. This caused the pump to fail. This can be avoided by increasing the alcohol percentage in the tank slowly over several fill-ups, dissolving the crud gradually.*

and prevent your carburetor from getting its fuel. Replace with metal-bodied fuel filters or, even better, small can-type cartridge fuel filters.

The slightly higher water content in alcohol's exhaust *does* slightly increase basic rust in the exhaust system more than gasoline exhaust. But sulfuric acid, not rust, is the major culprit in exhaust system failure. Alcohol's exhaust *does not* contain sulfuric acid, which causes most of the damage to many cars' exhaust systems. In an International Fuel Alcohol Symposium presentation, several studies concluded that there is almost no difference in exhaust system longevity when comparing alcohol fuel with gasoline.

One myth that pops up from time to time suggests that **blow-by** (vapors getting past the rings and down into the oil supply) from alcohol fuel contaminates oil and forms highly acidic sludge in a crankcase. This would reduce lubrication, and bearings would be damaged, or at least wear more quickly. During the Fifth International Alcohol Fuel Technology Symposium, several papers referred to major problems in lubricating *methanol*-fueled vehicles. However, the studies done on *ethanol* indicated that ethanol-fueled vehicles enjoy better lubrication qualities than gas-fueled cars.

During the symposium, Shell Oil's paper on lubrication made the following conclusions: "The bench and field trial results suggest that lubricants meeting API SE/CC and CCMC [industry's standards for automotive oil] performance requirements satisfy the demands of vehicles running on HE 100 [96% ethanol/4% water] fuel. Using this fuel, the engines were on the whole exceptionally clean, particularly in respect of piston deposits."[17] They went on to say, "Other aspects of lubricant performance, such as wear protection, prevention of oil thickening, sludge and rust formation, were similar with HE 100 and with gasoline."[18]

When you first start burning alcohol, you may need to replace your car's **oxygen sensor**, since some states permit the use of **MMT** (a manganese-based chemical compound) as an additive in gasoline. Over time, this substance coats valves and almost everything else in the system with manganese, similar to the way that leaded gas leaves deposits on surfaces. When alcohol starts cleaning the MMT out of the engine, the oxygen sensor can plug up with manganese.

I believe the final word on the safety of normal, acid-free alcohol in fuel systems is to be found in the 1996 publication, *Changes in Gasoline III, The Auto Technician's Gasoline Quality Guide.* In tests, autos were operated for extended periods of time on alcohol blends, and then their fuel tanks and fuel system components were removed, cut open, and examined. The conclusions showed that alcohol generally does not increase corrosion in normal, everyday operation.[19]

Blending

There is a myth that anything less than 200-proof alcohol will separate from gasoline due to the small amount of water in the alcohol. Gasoline, alcohol, and water are miscible (stay dissolved in one another), depending on temperature and on water and alcohol content. In fact the bottled additive to combat water in your tank, generically known as "Dry Gas," is nothing more than 200-proof alcohol, which causes the water to blend with the gasoline.

In Brazil, they pump alcohol that contains about 4% water. In warm climates there is absolutely no problem in mixing wet alcohol with gasoline, but all of Brazil is not warm and balmy. When I visited there, a General Motors engineer showed me a study that accurately outlined the physical limits of mixing water, alcohol, and gasoline. According to the paper, published by the Society of Automotive Engineers, at about 68°F, alcohol with as much as 45% water will mix with gasoline and not separate. At 4% water, alcohol will form a stable mix with gasoline down to about minus 22°F![20] This means that those of you who live in milder climates don't have to go through the extra step of producing dry 200-proof alcohol to get it to mix properly with gasoline. And if you do live in minus 22°F, you would generally only have to use 200-proof during the winter and only if you were going back and forth between alcohol and gasoline in a non-flexible-fuel vehicle. Flexible-fuel vehicles will simply adjust to phase-separated fuel.

ALCOHOL AND OTHER OCTANE ENHANCERS

The oil we are extracting nowadays contains little of the natural high-octane components of past, high-quality oil. We are beginning to scrape the bottom of the oil barrel. This means that base gasoline needs to be fortified with octane enhancers to reach the 87- to 92-octane range of regular to premium fuel.

Alcohol's **blending octane value** (alcohol mixed with gasoline) is 112.5[21]; gasoline basestock, straight out of the refinery, is about 67 octane.[22] Alcohol's **running octane value** (alcohol used by itself) is at least 105, but can reach 112.5 when used as an octane booster for gasoline. When you add alcohol to gasoline, you get a disproportionate increase in octane rating. So adding a gallon of 105-octane alcohol to a gallon of 67-octane gas does not end up giving you two gallons of 86-octane fuel—you would get two gallons of 89.75-octane fuel.

Tetraethyl lead, the original toxic octane booster, is now banned in most developed countries, except in airplane fuel. In the U.S., the carcinogenic MTBE (methyl-tertiary butyl ether), which had been used as an octane booster, has now been phased out. More commonly, reformulated gasoline is used as an octane booster—portions of oil are "severely reformed" to produce benzene and toluene, plus xylene (BTX). These octane enhancers are largely what chemists call **aromatics**. In some states, up to 25% of 87-octane regular gasoline is allowed to be made of these carcinogens.

In general, each one of these chemicals can raise octane about two points for every 10% content in the fuel. So, 10% benzene added to 67-octane fuel equals 69-octane fuel. Today's "regular" gasoline

Fuel Compatibility Among Water, Ethanol, and Gasoline

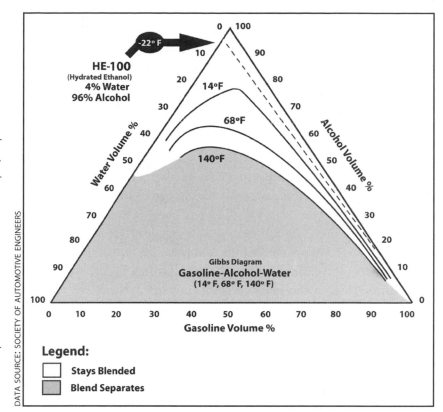

DATA SOURCE: SOCIETY OF AUTOMOTIVE ENGINEERS

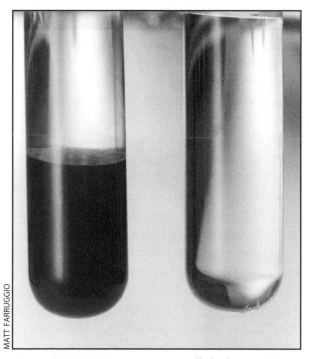

MATT FARRUGGIO

Fig. 14-15 Phase separation. *Initially both test tubes held gasohol (E-10). A few drops of water and water-soluble blue dye were added to the tube on the left. This would be the equivalent of adding gallons of water to your tank. The alcohol and water promptly separated out and settled at the bottom. The less dense gasoline floats on top.*

ABOVE: Fig. 14-14 Fuel compatibility among water, ethanol, and gasoline. E-100 (which is 96% ethanol and 4% water) is shown to stay mixed with gas all the way down to minus 22°F. Also, note that alcohol containing as much as 45% water will stay mixed with gasoline at 68°F. It's clearly not necessary to use anhydrous alcohol except during winter in the coldest areas.

contains about 28% BTX in addition to other more expensive boosters, but "premium" contains over 40%.

These chemicals are not naturally available in crude oil in sufficient proportions to raise gasoline to the octane rating at which it needs to be sold, so high-energy processes are used to convert other components of oil to BTX. Refiners keep the actual information secret on how much energy and volume of product is lost in converting low-octane crude products into octane boosters. But from looking at prices on the pump and comparing them to wholesale prices, we can get a close approximation of the oil and refinery energy needed.

In a recent month, the spread between the pump price of regular and premium gasoline was about 30 cents. This difference represents the addition of the aromatic octane boosters, and reflects a 17.5%

higher price. Since we know that an increase in volume of 12% aromatics is needed to go from 87 to 92 octane, we can see that to boost octane by five points costs oil companies at least 17.5% more for energy.

On the other hand, if you add 10% of alcohol to basestock gasoline, in Winter Reformulated Gasoline in California, you get a whopping 11.25 point increase in octane[25]—2.7 times the octane enhancement of the carcinogenic BTX.

So, if ethanol replaces the need for severely reformed products by 2.7 to 1, and those products require 17.5% more energy, this means alcohol as an octane booster deserves an energy credit of approximately 47% of the oil refinery energy it replaces—stretching existing petroleum supplies. This is never accounted for in the studies that claim alcohol has a barely positive energy balance.

Alcohol's efficiency as an octane booster is also why oil companies won't buy it to blend with gasoline unless regulated to do so—the volume of the fuel would be less. Adding a smaller amount of alcohol to do the job of a larger amount of aromatic poison means that the oil companies have fewer gallons to sell. One of their biggest complaints about taking MTBE out of the fuel was that 5.7% alcohol did the job of 15% MTBE.

ALCOHOL'S OCTANE RATING AND COMPRESSION RATIOS

In the simplest sense, octane is a measurement that indicates the point at which fuel will ping in an engine. The **octane number** refers to the ping resistance of a fuel, as compared to pure chemical octane (rated at 100).

As a fuel mixture is compressed, the fuel's heat also becomes compressed; compressed gas sharply increases the temperature of the mixture. So, as the piston comes up, the mixture approaches a temperature at which the fuel will explode by itself with no ignition from a spark plug. The plug should fire shortly before the temperature reaches the **auto-ignition point**. The auto-ignition point of unleaded regular gasoline can be as low as 430°F. Alcohol's is 685°F.

Your engine's compression ratio dictates the degree of compression on the mixture and thus how hot the fuel will get. A fuel with too low an octane rating for your engine's compression ratio causes preignition, or pinging. The stress put on the cylinder, along with the incredibly hot temperatures of a pinging mixture, shorten engine life considerably.

LEAD STILL USED AROUND THE WORLD

Lead as a fuel additive has been banned in the U.S. for a long time now. Soil levels of lead are starting to finally drop, slowly but surely, as lead becomes **chelated** by the soil into salts, and it can be carried down below the root zones of most plants.

Lead was added to gasoline for over 75 years as an inexpensive octane booster, over the strenuous objections of leagues of scientists. The battle to get lead out of U.S. gasoline reads just like the history of recognizing the dangers of tobacco, or the current battle over global warming. A few industry scientists and industry-funded think tanks, along with lobbyist money, tie up regulatory machinery with sufficient doubt that it takes decades or more to get action. In the case of lead, it took 75 years.

We now know that there is no known minimum threshold of neurological damage from lead in children. The tiniest amount of lead we can measure is enough to cause damage. So, we should all literally breathe a sigh of relief that lead is gone, right? Wrong. Whereas lead is banned in most developed countries, products containing lead are still used extensively around the world in less-developed countries. Tetraethyl lead is poisonous to breathe, eat, or even touch.[23]

Even though the auto, oil, and chemical companies knew this, they persisted in telling the public that there was no danger, while frankly admitting it in journals.[24] They funded studies like a 1925 work by the U.S. Bureau of Mines to say that lead in fuel was innocuous. Men at lead plants went insane or were killed. Workers in Dupont and Standard Oil lead plants called the facilities the "House of Butterflies," since workers would hallucinate insects crawling on their skin. More than four dozen researchers were killed in the development of the stuff.

Pinging is a problem for those who still drive late 1960s and early 1970s vehicles. During this period, in which the U.S. had access to huge quantities of high-quality sweet crude oil, high-octane fuel was produced as a matter of course. "Regular" gas was 94 octane, and most cars had at least an 8:1 or 9:1 compression ratio. In order to take advantage of premium fuel's 104 octane, many cars came off the assembly line with 10.5:1 and even 12:1 compression ratios. Nowadays, "premium" has a lower octane rating than old-fashioned regular, and modern regular is a disgustingly low 85 to 87 octane.

Alcohol is able to tolerate a compression ratio of 15:1 in general and up to 18:1 in specially designed engines. This means that cars with high compression can take advantage of alcohol's octane rating, obtaining more horsepower and mileage. As mentioned above, mileages as high as 22% above diesel's have been recorded in studies with high-compression engines.

If you get a compression ratio as high as 12.5:1 in a normally designed gasoline engine, that's generally about all you need. Above this, the increased efficiency in compression ratios usually isn't worth the trouble of increasing the strength of most of the related engine components. Really high-compression conversions (such as in the 19:1 conversion in Chapter 20) start with a stout diesel engine and *reduce* the compression ratio to the ideal level. Those engines are designed with the strength needed to tolerate high compression.

Unfortunately, automobile companies have reduced the strength of engine parts because of the low-compression-ratio designs of modern engines, which exist to accommodate the low grade of modern fuel. An engine that doesn't have to stand up to the stress of a high-compression application and quality fuel can be built much more cheaply, for higher profits.

In the '80s, the manufacturers found that in higher-compression carbureted engines, some of the cylinders farthest from the carburetor would get insufficient fuel, burn lean, and therefore pre-ignite. So they built engines to accommodate the lowest common denominator among the cylinders. Nowadays, with fuel injection, the problem of uneven fueling has gone away, and some engines are being produced with modestly higher compression ratios.

Low compression penalizes the driver in terms of mileage and handling. It may not be possible to raise the compression ratio of some vehicles to the optimal range for alcohol, but you can take some advantage of alcohol fuel's high octane in a low-compression vehicle by advancing the ignition timing (see Chapter 18).

Alcohol burns at a faster flame propagation rate than gasoline; ethanol's flame speed is 0.4 meters per second, compared to gasoline's 0.33. Although alcohol's flame speed is faster, you would actually start the ignition early. This would seem counterintuitive, but alcohol's wide range of combustible air/fuel mixtures, compared to the very narrow range of gasoline, allows us to start combustion early. Part of the reason alcohol gets higher efficiency is this combination of faster flame propagation and wider range of usable air/fuel mixtures.

So, instead of a sharp peak in explosion temperature and force, alcohol's explosion is a wide

Fig. 14-16 Preignition. *This is subtly different from pinging. Instead of the fuel simply exploding because compression makes it auto-ignite, preignition is usually caused by hot carbon deposits. Preignition and pinging both sound the same, but preignition is more damaging. 1) Hot spot preignites fuel; 2) and 3) piston compresses, and spark plug creates regular ignition; 4) flame fronts collide, causing pinging noise.*

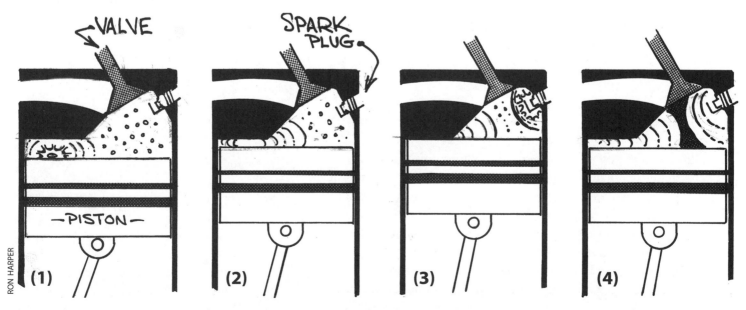

VALVE SPARK PLUG

PISTON

RON HARPER

(1)   (2)   (3)   (4)

plateau. This is why we advance the time the spark is fired for alcohol, versus gasoline. This dramatically reduces vibration, as well as peak temperature—both key factors in engine life.

Because alcohol's peak pressure happens when the piston is higher in the cylinder (ideally at about seven degrees after top dead center, versus about 12 to 15 degrees for gasoline), it develops a lot more torque faster than gasoline. This translates to more "horsepower," especially when passing. If the transmission gearing were matched to this torque profile, then it would result in better mileage, too. An ideal flexible-fuel vehicle would have a variable range in the transmission in order to take advantage of alcohol's higher torque.

Alcohol burns cooler, cleaner, and with less vibration than gasoline. It extends engine life, delivers more horsepower, and is ideal for hardworking engines.

COLD-STARTING ALCOHOL-FUEL VEHICLES

The one perfectly valid criticism of pure alcohol as fuel is that a car won't start without help in weather below 30°F, and at times, 50°F (see sidebar in Chapter 17 on why some cars cold-start better than others). Those qualities that make alcohol so safe—high latent heat and flashpoint—make it difficult to vaporize enough alcohol to start the engine on cold days, although once they're slightly warmed, cars will run fine.

Up to a point, gasoline has the same starting problems as alcohol. Gasoline sellers often add butane, isopentane, or natural gas condensates to winter gasoline so that it will start in cold weather. Some of these additives are so volatile that they can cause vapor lock (fuel turning to vapor before it gets to the fuel pump), normally a summertime driving problem.

The usual cold-starting solution for alcohol fuel is to turn the engine over for even a few revolutions on something more volatile than alcohol. Alternately, you could heat up the fuel going into the engine enough to start without using a separate starting fuel.

In the United States, flexible-fuel vehicles do not need either of these two systems! The alcohol fuel sold at the pump in the U.S., otherwise known as E-85, contains at least 15% gasoline. This addition of gasoline raises the RVP of alcohol high enough to start at low temperatures.

The 85/15 split has to do with the deal that got automakers in the U.S. to produce flexible-fuel vehicles. To get manufacturers to produce alcohol-capable cars, companies were offered credits under the corporate average fuel economy rules. Getting credits for making flex-fuel cars made it possible to sell more fuel-guzzling SUVs and still meet their mandated fleet average.

But in developing the program for flex fuels, manufacturers balked at the $50 or so that it would cost to provide cold-start devices. They demanded and got an E-85 standard that mixed 15% gasoline with alcohol so that no cold-start devices were necessary. Now we have the E-85 albatross around our necks, when most of us would prefer to use no petroleum at all.

Those of us who want to run on E-98 (the ethanol with the highest percentage of alcohol that is commonly available in bulk), or alcohol made at a local co-op, need to deal with cold-starting systems. I'll detail some of them in Chapter 17.

To summarize: Alcohol burns cooler, cleaner, and with less vibration than gasoline. It extends engine life, delivers more horsepower, and is ideal for hard-working engines. With all that in its favor, you're probably interested in knowing something about conversion.

Endnotes

1. Keith Owen, Trevor Coley, and Christopher S. Weaver, *Automotive Fuels Reference Book* (Warrendale, PA: Society of Automotive Engineers International, 1995), 590–591.

2. Owen, Coley, and Weaver, 591.

3. Owen, Coley, and Weaver, 84.

4. Mother's Alcohol Fuel Seminar, "Making Alcohol Fuel: Alcohol as an Engine Fuel," *Mother Earth News*, 1980, http://journeytoforever.org/biofuel_library/ethanol_motherearth/me1.html (July 16, 2005).

5. David P. Gardiner, et al., *Vehicle Implementation and Cold Start Calibration of a Port Injected M100 Engine Using Plasma Jet Ignition and Prompt EGR* (Warrendale, PA: Society of Automotive Engineers International, 1995).

6. Gardiner, et al.

7. Ben Harder, "No Deep Breathing: Air Pollution Impedes Lung Development," *Science News* 166:11 (September 11, 2004), 163.

8. *Petroleum (Oil)—A Fossil Fuel*, Energy Information Administration, www.eia.doe.gov/kids/energyfacts/sources/non-renewable/oil.html#how%20is (January 2006).

9. "The Magic Oil Barrel," Dow Canadian Insight Edition, 18.

10. See <www.permaculture.com> for more extensive studies on permeation.

11. Jim Lux, *Comparison of Relative Energies and Powers*, http://home.earthlink.net/~jimlux/energies.htm (December 7, 2006).

12. Matthew Brusstar, et al., *High Efficiency and Low Emissions from a Port-Injected Engine with Neat Alcohol Fuels*, for U.S. Environmental Protection Agency, SAE Paper 2002-01-2743, 2002.

13. Mother's Alcohol Fuel Seminar.

14. Brusstar.

15. Josmar Pagliuso, communication with author, March 2006.

16. Ismat A. Abu-Isa, *Effects of Mixtures of Gasoline with Methanol and with Ethanol on Automotive Elastomers*, for General Motors Research Laboratories, October 31, 1979, Table VIII, 12.

17. H. Krumm, et al., "Lubrication of Spark Ignition Engines Running on Alcohol Containing Fuels," in *Fifth International Alcohol Fuel Technology Symposium Proceedings* II (Shell Research Limited/West Germany, May 13–18, 1982), 434.

18. Krumm.

19. *Changes in Gasoline III, The Auto Technician's Gasoline Quality Guide* (Downstream Alternatives, Bremen, Indiana,1996), www.ethanolrfa.org/objects/pdf/AboutRFA/Gasoline.pdf (April 2006).

20. A.C. Castro, C.H. Koster, and E.K. Franleck, *Flexible Ethanol Otto Engine Management System* 942400 (Warrendale, PA: Society of Automotive Engineers International, 1994).

21. Robert E. Reynolds, *Replacing the Volume and Octane Loss of Removing MTBE from Reformulated Gasoline Ethanol RFG versus All Hydrocarbon RFG* (South Bend, IN: Downstream Alternatives, 2004), 13.

22. Robert E. Reynolds, President, Downstream Alternatives, communication with author, August 2005.

23. Graham Edgar, "The Manufacture and Use of Tetraethyl Lead," *Industrial and Engineering Chemistry* 31 (December 1939), 1439–46.

24. Jamie Lincoln Kitman, "The Secret History of Lead: Special Report," *The Nation*, March 20, 2000 (www.thenation.com/doc/20000320/kitman).

25. Reynolds, *Replacing the Volume*, 13.

Fig. 14-17

DRIVE TO WORK/ WORK TO DRIVE

Even if your vehicle has fuel injection, I still recommend that you read this chapter. The design of fuel injection systems will be much easier to understand once you learn about all the various parts of the carburetor and how they compensate for different conditions in the driving cycle, from **idle** to **wide-open throttle (WOT)**. The programming of electronic fuel injection systems largely mimics the changes made in the various systems in a carburetor. It's kind of like facing the workings of a digital watch for the first time, versus understanding the gear works of a clock.

When I wrote the first version of this book in 1982, most vehicles had carburetors. Fuel injection was still an exotic system on expensive European cars. Nowadays, carburetors are virtually a thing of the past; since the late 1980s, virtually all car manufacturers have made the switch to fuel injection systems.

However, in many parts of the world, cars are kept on the road for 30 or more years, and, even in the U.S., there are millions of cars with carburetors. In 2004, actress Daryl Hannah asked me to convert her T-top Trans Am (the car she drove in the movie *Kill Bill*) to alcohol fuel. It certainly didn't have fuel injection.

Also, there are a lot of stationary industrial engines (e.g., generators) and smaller utility engines (e.g., lawnmowers) that still use carburetors and that are likely to be in service for a long time. Of course, there are whole classes of auto, motorcycle, and boat racers who still use carburetors, too.

In some vehicles, the main and idle jets are all that need to be adjusted for reasonably good performance. Most vehicles, though, require some further carburetor adjustments. This is definitely the case with working trucks, semi-high-performance cars, and engines with poor manifold design.

So what is a carburetor? It's a mixing device responsible for keeping air and fuel at the proper level for optimal burning. Both gasoline and

THE DESIGN OF FUEL INJECTION SYSTEMS WILL BE MUCH EASIER TO UNDERSTAND ONCE YOU LEARN ABOUT ALL THE VARIOUS PARTS OF THE CARBURETOR. ALSO, IN MANY PARTS OF THE WORLD, CARS ARE KEPT ON THE ROAD FOR 30 OR MORE YEARS. EVEN IN THE U.S., THERE ARE MILLIONS OF CARS WITH CARBURETORS.

MATT FARRUGGIO

Fig. 15-1 Cutaway carburetor. *Shows power valve piston assembly (with spring on top) and float needle assembly (to the right).*

The carburetor mixes gas and air.

Fig. 15-2

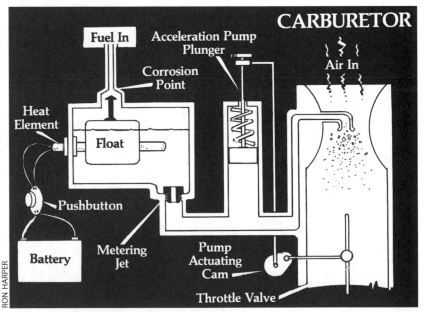

Fig. 15-3 Carburetor components. *The heating elements shown in this illustration are not stock equipment. They represent a method of cold-starting your vehicle on chilly mornings. The corrosion point is the* **float valve** *assembly. It will plug up with bits of corroded material, or flakes of tar/varnish left behind by gasoline, if the fuel filtering is not sufficient. This will sometimes cause sticking of the float valve in a partially open position. The venturi is the narrow neck of the carburetor below the air intake.*

alcohol have to be mixed with air before they'll burn properly inside an automobile engine. Gasoline needs 14.7 parts of air to one part gasoline. Alcohol needs less air: nine parts of air to one part alcohol. These are the **stoichiometric ratios** that theoretically produce the most complete combustion of fuel. In reality, some departure from the ideal is needed at different times.

The throat of the carburetor is a narrowing of the chamber through which air passes, called the venturi (see Figure 15-3). When air passes through this narrow passage, the same volume of air passes from one side to the other, which forces the air to move more swiftly through the narrowed venturi. The faster airflow translates to increased engine vacuum, pulling fuel into the airstream through small holes around the venturi. These small holes are fed through a small internal line to the fuel bowl. Unless it's regulated, the vacuum will suck as much fuel as it can from the fuel reservoir through the passages.

At the end of the fuel line in the fuel bowl is a part known as the **main metering jet**, a key part of the system.

MAIN METERING SYSTEM

The main metering system is what determines how much fuel is mixed with the incoming air. It primarily controls this mixture when the vehicle is moving, providing a minimum background amount of fuel during idle and acceleration.

The metering jet determines the air/fuel mix at normal speeds; it restricts the opening from the fuel reservoir, limiting the amount of fuel that can get to the venturi. This is the first part that must be modified to run on alcohol.

You should make adjustments to the **idle circuit** (see Idle Circuit section, below) at the same time.

Begin your alterations with the car engine cold. First, remove the four to six screws that hold down the top of the carburetor. Release the **throttle linkage** that attaches to the foot pedal at the carburetor. (If your carburetor top is independent of the **linkage**, you don't have to release it.) Some carburetors may require that you remove the fuel and/or vacuum lines for easy access. At the end of the fuel passage, from the venturi back into the fuel bowl, you'll find the main metering jet(s).

In most cases, the main jets are threaded into the **float bowl** floor, underneath the **carburetor floats**. (Old single-barrel carburetors use a hollow post that dips down into the bowl, with a jet threaded into the end of the post.) Carburetor floats are foam or metallic "balls" connected to arms that float in the fuel. The jet is nothing more than a small bolt with a tiny hole bored through it. It has a screwdriver slot on the top. Unscrew it.

Once the jet is removed, you'll enlarge the size of the hole by approximately 15–35% in diameter. (The richness allowed in a 35% increase is almost always more than what a cruising engine requires for a good, rich-burning mixture.) Some auto manufacturers stamp jet size in thousandths of an inch right on the part. If yours isn't conveniently coded, measure the diameter yourself with a set of numbered metal drill bits by slipping them one by one into the jet. When you find the bit that fits snugly, that'll be the size you'll work from. (Numbered drills measure in thousandth-inch increments, as opposed to common wood drill bits, which read in eighths or sixteenths of an inch (see Figure 15-11).)

In practice, no matter how you measure, you will have to use trial and error to determine how much to drill, so it doesn't much matter how you calculate it. You can start experimenting for proper fuel mixture from either the lean (too much air) or the rich (too much fuel) end of the scale. Back in the 1980s, spare jets were cheap and easy to find, so it made sense to start rich and work your way back to lean. But today, jets for carburetors can be hard to find. So you'll probably want to start lean by minimally drilling out your existing jet, and increasing in size until you get good running.

To begin our conversion, we will guess at what we think will be the lowest (leanest) practical air/fuel ratio. A standard gasoline jet for my old '53 Chevy truck is 0.056 inches in diameter. A 15% increase in the diameter of the jet is the right place to start. To calculate this figure, multiply 0.056 inches times 0.15, which gives you 0.0084 inches. Add that figure to the original jet size and you get 0.0644 inches. A #52 drill bit is the equivalent of 0.0635 inches; use that bit to drill out the first test jet.

It is essential that you remove the jet from the carburetor before you start drilling; otherwise, you'll have brass chips in your manifold and then in your cylinders, which will cause extensive damage when you start the engine.

Another warning: Use a pin vise (drill) to enlarge the jet holes—a power drill's lowest-speed setting is too fast for this job and will cut a very rough hole. A drill press may be okay at its lowest speed, but it's safer to stick with a pin vise.

You may want to have an extra jet, since you may drill the original out too large during experimentation. If you don't have an extra jet, you can solder over the hole in the existing one and then drill the correct size through the solder.

It is essential that you remove the jet from the carburetor before you start drilling; otherwise, you'll have brass chips in your manifold and then in your cylinders, which will cause extensive damage when you start the engine.

After drilling the main metering jet, take your car out and try it on a reasonably steep freeway grade. Keep experimentation times short and do whatever you can to avoid running the engine on too rich a mixture for very long, as this may cause engine damage.

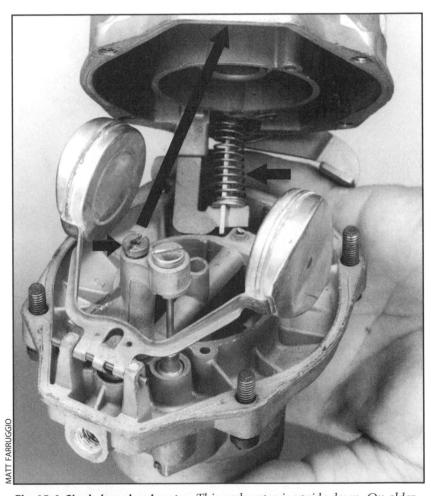

MATT FARRUGGIO

Fig. 15-4 Single-barrel carburetor. *This carburetor is upside down. On older-style single-barrel carburetors, the main jet (short arrow) is threaded into the end of a post that dips into the float bowl (long arrow). Adjacent to the main jet is the power valve assembly. The spring-loaded assembly in the rear of the carburetor (medium arrow) is the accelerator pump assembly.*

If you can't maintain cruising speed at normal levels (55, 60, 70 mph) without loss of power or misfiring, then take your car back down to the shop and drill out the jet one size larger. A #51 drill bit will measure 0.067 inches in diameter, about 19.6% larger than the standard 0.056-inch jet.

Test-drive again. You may have to go up two or three more drill sizes before you stop experiencing the power loss or misfiring at high speeds that tells you your main metering mixture is too lean. Use the jet one step larger than the slightly too-lean size.

My old truck ran without misfiring at 0.067 inches, a #51 bit. A #52 drill bit gave me too lean a mixture; although it increased my mileage a little,

Fig. 15-5 Main jet removed. The main jet has been removed from the carburetor and is being checked for size before drilling.

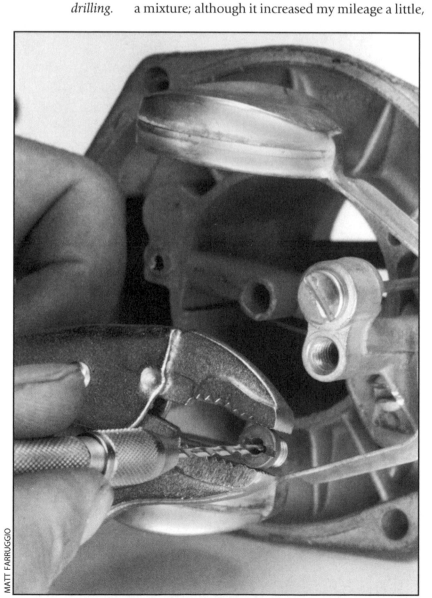

Fig. 15-6 Pin vise and number drills. The pin vise can hold drill bits as large as these, and down to very fine bits.

I lost a little power. With a #51 bit I lost about 7% of my mpg, unloaded, in comparison to gasoline. A #51 bit will measure 0.067 inches in diameter, around 19.6% larger in diameter, and roughly 40% larger in area, than the standard 0.056-inch-diameter jet. Vehicles with good manifold design, along with heated air and fuel, may be able to get down to a 30% area increase or less.

Changing jets with the season used to be a standard home mechanic chore up until the '70s. In winter, I would run a little richer to accommodate winter's colder, denser air, using a #50 hole in a replacement jet.

There are more accurate ways to tell if your air/fuel mixture is correct than by simply test-driving. The safest way is to do this work where you have access to an exhaust gas analyzer. You'll be checking the exhaust with each jet until you first reach the lowest carbon monoxide reading as the goal to shoot for. This will indicate both low emissions and the most efficient conversion of fuel to work. I like to shoot for a carbon monoxide reading that is either zero or close to zero. Start a bit rich, and keep reducing the amount of fuel until you first hit zero or whatever is your lowest CO reading. If you keep reducing fuel after this point, you most likely will start elevating the NOx readings.

Perhaps the best way to tune an air/fuel mixture without an exhaust gas analyzer is to install an **exhaust gas temperature (EGT) gauge** and monitor the exhaust gas temperature as it leaves your cylinder and enters the manifold. (Install the gauge while you're still on gas, to get gas readings before you change over to alcohol.) As you drill jets, you record the temperatures, and at the point

Decimal Equivalent Sizes for Drill Bits

DRILL LTR.	DEC. EQUIV.
A	.234
B	.238
C	.242
D	.246
E	.250
F	.257
G	.261
H	.266
I	.272
J	.277
K	.281
L	.290
M	.295
N	.302
O	.316
P	.323
Q	.332
R	.339
S	.348
T	.358
U	.368
V	.377
W	.386
X	.397
Y	.404
Z	.413

ABOVE COLUMN:

Fig. 15-7 Decimal equivalent size of the letter drills.

DRILL NO.	DEC. EQUIV.	DRILL NO.	DEC. EQUIV.
80	.0135	41	.0960
79	.0145	40	.0980
78	.0160	39	.0995
77	.0180	38	.1015
76	.0200	37	.1040
75	.0210	36	.1065
74	.0225	35	.1100
73	.0240	34	.1110
72	.0250	33	.1130
71	.0260	32	.1160
70	.0280	31	.1200
69	.0292	30	.1285
68	.0310	29	.1360
67	.0320	28	.1405
66	.0330	27	.1440
65	.0350	26	.1470
64	.0360	25	.1495
63	.0370	24	.1520
62	.0380	23	.1540
61	.0390	22	.1570
60	.0400	21	.1590
59	.0410	20	.1610
58	.0420	19	.1660
57	.0430	18	.1695
56	.0465	17	.1730
55	.0520	16	.1770
54	.0550	15	.1800
53	.0595	14	.1820
52	.0635	13	.1850
51	.0670	12	.1890
50	.0700	11	.1910
51	.0670	10	.1935
50	.0700	9	.1960
49	.0730	8	.1990
48	.0760	7	.2010
47	.0785	6	.2040
46	.0810	5	.2055
45	.0820	4	.2090
44	.0860	3	.2130
43	.0890	2	.2210
42	.0935	1	.2280

ABOVE TWO COLUMNS:

Fig. 15-8 Decimal equivalent size of the number drills.

DAVID BLUME

Fig. 15-9 Exhaust gas temperature gauge. This gauge is useful both for determining when the exhaust is hot enough to operate the oxygen sensor and also to measure the highest exhaust gas temperature while attempting to tune for the leanest mixture.

Fig. 15-10 Exhaust gas temperature sensor. The sensor is installed as close to the cylinders as possible to get the most accurate reading. In this case, the exhaust gas manifold was removed and taken to a machine shop to have a hole drilled and tapped.

at which you first reach the highest temperature, you'll be pretty close to ideal.

By drilling jets, you increase the fuel supply. The temperature keeps going up, and then it plateaus. If you keep drilling out the jet and increasing the amount of fuel, at first there is no change, and the temperature remains the same—but you are wasting fuel. So you're looking for the leanest mix (lowest air/fuel ratio) at the highest temperature.

As long as the alcohol exhaust temperature is a little lower than what it would be if you were burning gasoline, usually below 1350°F, you should be all right. You can measure the exhaust gas temperature at the junction of the manifold and exhaust pipe. Although this point is a bit cooler than the temperature at the exhaust valves, it represents a mix of several or all of the cylinders, rather than one or two.

Another alternative for tuning your air/fuel mixture is to measure manifold vacuum. A manifold **vacuum gauge** measures the amount of vacuum created by the engine in the manifold (air

DAVID BLUME

Diameter versus Area

DIAMETER (INCHES)	AREA (SQUARE INCHES)	DIAMETER (INCHES)	AREA (SQUARE INCHES)	DIAMETER (INCHES)	AREA (SQUARE INCHES)
.020	.000314	.078	.004478	.136	.014527
.021	.000346	.079	.004902	.137	.014741
.022	.000380	.080	.005027	.138	.014957
.023	.000415	.081	.005133	.139	.015175
.024	.000452	.082	.005281	.140	.015394
.025	.000491	.083	.005411	.141	.015615
.026	.000531	.084	.005542	.142	.015837
.027	.000573	.085	.005675	.143	.016061
.028	.000616	.086	.005809	.144	.016277
.029	.000661	.087	.005945	.145	.016513
.030	.000707	.088	.006082	.146	.016742
.031	.000755	.089	.006221	.147	.016972
.032	.000804	.090	.006362	.148	.017203
.033	.000855	.091	.006504	.149	.017437
.034	.000908	.092	.006648	.150	.017672
.035	.000962	.093	.006793	.151	.017908
.036	.001018	.094	.006940	.152	.018146
.037	.001075	.095	.007088	.153	.018385
.038	.001134	.096	.007238	.154	.018627
.039	.001195	.097	.007390	.155	.018869
.040	.001257	.098	.007543	.156	.019113
.041	.001320	.099	.007698	.157	.019359
.042	.001385	.100	.007854	.158	.019607
.043	.001452	.101	.008012	.159	.019856
.044	.001521	.102	.008171	.160	.020160
.045	.001590	.103	.008332	.161	.020358
.046	.001662	.104	.008495	.162	.020612
.047	.001735	.105	.008659	.163	.020867
.048	.001810	.106	.008825	.164	.021124
.049	.001886	.107	.008992	.165	.021383
.050	.001964	.108	.009161	.166	.021904
.051	.002043	.109	.009331	.167	.021904
.052	.002124	.110	.009503	.168	.022167
.053	.002206	.111	.009677	.169	.022432
.054	.002290	.112	.009852	.170	.022698
.055	.002376	.113	.010029	.171	.022966
.056	.002463	.114	.010207	.172	.023235
.057	.002552	.115	.010387	.173	.023506
.058	.002642	.116	.010568	.174	.023779
.059	.002734	.117	.010751	.175	.024053
.060	.002827	.118	.010936	.176	.024329
.061	.002922	.119	.011122	.177	.024606
.062	.003019	.120	.011310	.178	.024885
.063	.003117	.121	.011449	.179	.025165
.064	.003217	.122	.011690	.180	.025447
.065	.003318	.123	.011882	.181	.025730
.066	.003421	.124	.012076	.182	.026015
.067	.003526	.125	.012272	.183	.026302
.068	.003632	.126	.012469	.184	.026590
.069	.003739	.127	.012668	.185	.026880
.070	.003848	.128	.012868	.186	.027172
.071	.003959	.129	.013070	.187	.027464
.072	.004072	.130	.013273	.188	.027759
.073	.004185	.131	.013478	.189	.028055
.074	.004301	.132	.013685	.190	.028352
.075	.004418	.133	.013893	.191	.028652
.076	.004536	.134	.014103		
.077	.004657	.135	.014314		

Fig. 15-11 Diameter versus area.

tubing leading from the carburetor to the engine). Changing jets until you first get the highest manifold vacuum will put you close to ideal. You'll see the vacuum increase until it plateaus, not changing as you continue to increase fuel. So the leanest setting is right when you increase the amount of fuel no more than is necessary to reach the highest manifold vacuum setting.

IDLE CIRCUIT

The idle circuit is a simple bypass system that allows the engine to get a somewhat richer mixture when idling. The air/fuel mixture at idle is adjusted by a needle-type jet, which restricts how much fuel the engine can suck through the idle jet opening in the venturi wall. Adjust the idle circuit by retreating the idle air mixture screw, almost always located at the base of the carburetor. Rotate the screw out, counterclockwise. The more you turn it out, the more fuel for a given amount of air can go through the carburetor at idle. (Do this while you're adjusting your main metering jet, described above.)

You'll need to experiment to know how far out to turn the idle screw. While you're still running on gasoline, start turning the screw in half a turn at a time until you feel it gently bottom out. Keep track of the number of half-turns you take, so you can adjust back to the gasoline setting easily.

After you've changed the main metering jet the first time, return the idle air mixture screw to the gasoline setting, start your engine, and begin turning the screw out, counterclockwise, counting each half-turn. You'll notice that the car idles quite roughly in the beginning. As the mixture at idle begins to enrich, the engine starts to progressively sound smoother. This is best measured when idling the engine somewhere around 1500 rpm. You can adjust the idle by ear—when it sounds smooth, stop turning the screw out.

If you can't use an exhaust gas analyzer to tune for lowest CO, you can use a **tachometer (rpm gauge)**. The tachometer tells you how fast the engine is turning in revolutions per minute (rpm). As you turn the idle air mixture screw out, the idle will begin to smooth, and the engine will begin to run faster. Over the course of this adjustment, engine speed can rise as much as 200 rpm. As your readings rise, keep an eye on the tachometer. There's a point within a couple of turns at which the rpms will not go higher. Past those couple of turns, engine rpm drops, and the engine sounds rough. Too much fuel is being allowed in, and the fuel mixture is too rich. The point you're looking for is when the rpms first peak.

A manifold vacuum gauge works similarly to a tachometer, and is probably the clearest, most accurate measurement of a good air/fuel mixture. Instead of looking for the highest rpms, you'll keep backing out your screw until you get to the first point of highest manifold vacuum.

Sometimes further adjustments are necessary. On some carburetors, for instance, backing the screw out still doesn't let enough fuel in. In such a case, turn the screw all the way out, remove it, and you'll see a fixed jet hole where fuel enters the venturi. Drill this jet out around 30% in area. (As I mentioned earlier, drill with the carburetor off the car to prevent any shavings from dropping into the engine.) Since the mixture screw threads in the carburetor body are uncomfortably exposed while you're drilling, wrap the drill bit with tape to avoid damaging them. This surgery shouldn't prevent you from going back to gasoline for short periods; just turn the screw back in to achieve a leaner idle mixture for gasoline.

In some carburetors, the passages leading from the fuel bowl to the idle circuit are so small that they limit the fuel getting to the needle area. Enlarging those passages with a pin vise takes a bit of time and ingenuity. Since every carburetor is different, you may want to get a detailed schematic of your carburetor from a local carburetor rebuild shop; or, better yet, enlist the help of the shop in your project. The alternative is to switch to **throttle body fuel injection (TBI)** (we're coming to that).

A caveat: Beginning in the early 1980s, manufacturers (and the EPA) didn't trust you to leave your idle mixture where the factory set it, so they covered the idle screw with a plastic limiting cap. Fortunately, these caps are easily broken off "accidentally." More diabolical is the super-smog-engineered carburetor with no idle adjust at all. If you've got one of those, your best bet is to replace it with an older carburetor or a throttle body fuel injection kit with the same basic engine size, bolt pattern, and linkage assembly.

ACCELERATOR PUMP

It's relatively easy to get an alcohol-powered vehicle to cruise and idle well, but getting the acceleration cycle just right is a little more involved. Symptoms of acceleration circuit problems are coughing

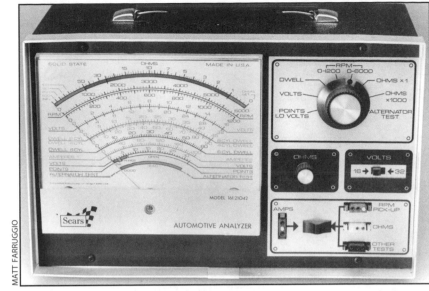

Fig. 15-12 Automotive analyzer. *Common, good-quality tune-up analyzer. The tachometer is built in and is read as the second scale (rpm). Do your testing for highest rpm reading in the 1500 rpm range. The* **dwell meter,** *which is used in setting the* **points** *in old-fashioned distributors, doubles as a serviceable* **duty cycle meter** *in testing fuel injection* **pulse width.**

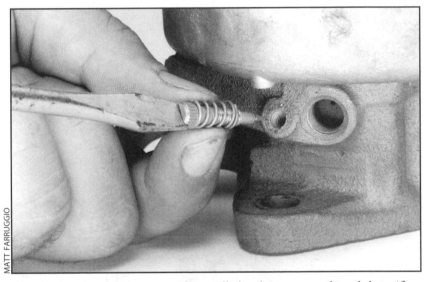

Fig. 15-13 Idle mixture needle. *The needle has been removed, and the orifice, that may have to be drilled out is now accessible. The large threaded opening next to the smaller idle mixture passage is where the* **vacuum advance** *hose barb is attached to the carburetor.*

and sputtering; dying on acceleration from a stop; and/or slow, unresponsive acceleration when passing at high speeds. Be sure these are carburetion problems, though; they could also be symptoms of vacuum leaks and/or electrical malfunctions.

Carbureted vehicles are more sensitive during acceleration because as you accelerate, the manifold vacuum drops. That drop in vacuum reduces the amount of fuel spray that stays mixed with air in the manifold on the way to the engine. Much of the spray will drop to the manifold floor or walls and not get to the cylinder in a form adequate for burning. It then inefficiently evaporates from where it has puddled on the surface of the manifold. The net effect is to make the mixture that arrives at the cylinder lean, causing stalling or loss of power.

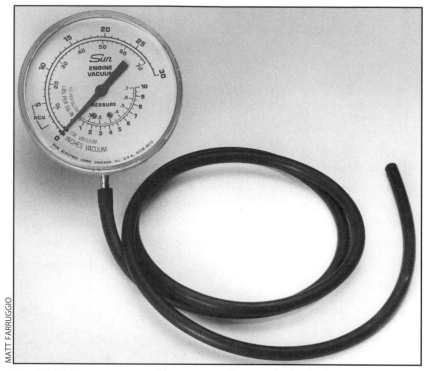

Fig. 15-14 Vacuum gauge. *The neoprene line can be hooked up to any constant vacuum line on your engine. Such gauges are relatively inexpensive and come with a variety of adaptors to allow you to tap into several locations on the engine.*

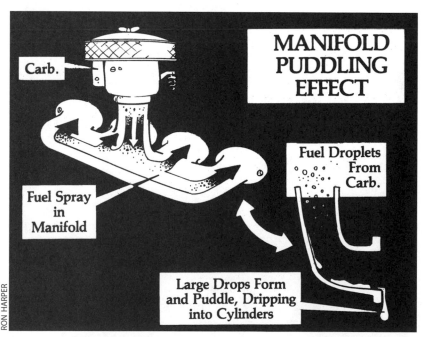

Fig. 15-15 Manifold puddling. *Fuel spray drops out of air and forms a film on the manifold walls. The problem is most pronounced with distance from the carburetor, sharpness of turns, and smoothness of the inner surface finishes.*

The carburetor accelerator pump tries to make up for the drop in manifold vacuum by delivering excess fuel in hopes that enough of it will get to the cylinders as a spray to deliver the power needed for acceleration. The pump kicks in, spraying raw fuel into the venturi in a squirt gun effect, which can cause your air to fuel ratio to drop to almost 4:1.

While this is a crude, if generally effective system, the greater amount of coarse spray causes evaporative manifold cooling, which further aggravates **puddling**—the condensing of fuel out of the airstream onto the cool manifold walls. When the air/fuel mixture becomes this rich, there are substantial increases in noxious emissions.

When adjusting for alcohol, you may need to supply more fuel for a longer period of time during acceleration. Many older carburetors (e.g., Ford, Autolite, Weber, and Holley) have adjustable linkage (a control rod that pulls the trigger in the squirt gun). With these, you often need do nothing more than remove the spring clip, then move the rod up out of its hole on the lever arm to the hole giving the longest stroke, and replace the clip. Careful with that clip—it likes to fly off and get lost.

There are a couple of ways to set accelerator pumps without adjustable linkage. Since the pump forces liquid through a jet, increasing the jet size will increase flow. Drilling out the jet gives you more juice, but for a shorter period. Don't drill out more than 25%, or you could end up with too little back-pressure, in which case you'll get a dribble down the wall of the throat instead of a spray.

Drilling this very tiny jet in the throat of the carburetor is a delicate operation, since there's not a lot of room in there to work. It's easier to drill the jet from inside the float bowl or pump chamber than from inside the carburetor throat (see Figure 15-15). On many carburetors, the unit containing much of the idle and accelerator circuitry is easily removable, making the whole project much easier.

Some Holleys and a few other high-performance carburetors have replacement accelerator pump assemblies available, with larger-volume chambers that give you more fuel over a longer period of time. These assemblies are preferable because of their accelerator squirt's longer duration. If the duration is too short, you'll have to adjust your **power valve** (see next section) more than you'd like. Don't use the highest accelerator setting instead of a lower-but-adequate setting; it gives

Fig. 15-16
Two-barrel
carburetor accel-
erator pump jets.
The tiny orifices,
one of which
is indicated by
the end of the
drill bit, can be
easily drilled out
with very fine
drills. The entire
accelerator pump
assembly can be
removed with a
screwdriver to
make the job easy.

Carburetor Troubleshooting Chart

Fig. 15-17
Carburetor
troubleshooting
chart.

PROBLEM	CHECK FOR THE FOLLOWING
Rough idle	1, 3, 4, 5, 6, 7, 9, 10, 11, 12, 15, 16, 17
Flooding	1, 8
Flat Spot	1, 3, 4, 5, 6, 7, 8, 9, 10, 11, 12, 15, 16, 17
Stalling	1, 2, 4, 5, 6, 7, 9, 10, 11, 12, 15, 16, 17
Whistling	3, 4, 16
Hard Starting	1, 3, 4, 5, 6, 7, 10, 11, 12, 13, 16, 17, 18
Leaking	1, 8, 14
Poor Gas Mileage or Performance	1, 5, 6, 7, 8, 9, 10, 11, 12, 13, 17, 19

1. Dirt in carburetor (lodged between needle and seat). Lightly tapping carburetor adjacent to gas inlet will often cure this problem.
2. Blocked fuel filter.
3. Cross-threaded vacuum line fittings, or vacuum lines not connected, or leaking.
4. Old flange jacket not completely removed from mounting surface, or new gasket incorrectly installed.
5. Fouled spark plugs, faulty ignition wires.
6. Improper point setting or timing.
7. Defective coil, condenser, distributor cap, rotor, or vacuum advance.
8. Faulty fuel pressure.
9. Defective heat riser valve.
10. Dirty air cleaner filter.
11. Poor engine compression.
12. Choke plate is completely open when engine is fully warmed up and running. If the choke plate is not completely open, automatic choke may require adjustment.
13. Defective or blocked choke tube.
14. Worn fuel line fittings.
15. Improper setting or idle adjustment screws.
16. Uneven tightening of mounting studs.
17. Defective crankcase ventilating valve.
18. When engine is cold, choke should be closed. If not, choke may require adjustment.
19. All items under rough idle, blocked gas cap or tank vent, dragging brakes, low tire pressure, auto transmission malfunction, faulty thermostat, restricted exhaust.

you lots of extra power but affects mileage and possibly emissions, just as it would with gasoline.

On some special vehicles, such as motorcycles, and for stationary motors, the accelerator pump may be better off providing less of a blast. The increased manifold puddling you get from cooling and condensing excess fuel can actually decrease the net fuel spray reaching your engine. *Mother Earth News* found this to be the case in its Harley-Davidson conversion, and

I've seen the same results on several stationary engines. Using air or fuel heat can provide better results.

POWER VALVE ADJUSTMENT

If you're losing power or misfiring in the middle of a hard acceleration or when going up a long grade with a load, the problem may be your power valve, a spring-loaded, vacuum-drop-activated valve. Like the accelerator pump, it dumps extra fuel into your carburetor when you need it, but it does not pump it in.

The manifold vacuum drops during acceleration, and in a long acceleration the vacuum may not catch up during the time that the accelerator pump is spraying. If the manifold vacuum drops even further, the power valve senses it and comes to your aid.

There's a spring holding the power valve closed until vacuum pressure drops to around six to eight inches Hg from its normal 15–20 inches **mercury vacuum scale (Hg)**. At that point, the spring overcomes the vacuum that's holding the power valve piston closed, which allows a jet to open, which provides a very coarse spray of fuel to be sucked

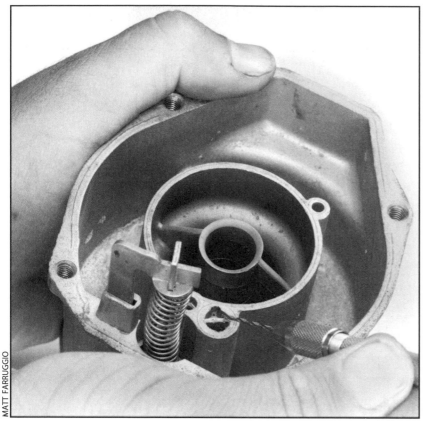

ABOVE: *Fig. 15-18 Single-barrel carburetor accelerator circuit jet. The tiny orifice indicated by the tip of the drill bit leads to the carburetor venturi at a steep angle down and away from you. Fuel from the spring-loaded plunger adjacent is traveling up the larger-diameter passage directly beneath the drill bit. The fuel is forced to go down through the tiny orifice when the carburetor is assembled, since the larger passage is capped off with a carburetor gasket. To enrich the accelerator circuit, you need to drill out the orifice about 25% larger.*

RIGHT: *Fig. 15-19 One-barrel carburetor power valve assembly. The top of the carburetor has been inverted and the entire jetting assembly removed. At the top is a spring-loaded check valve, which is usually immersed in fuel. The stainless steel, weakly sprung ball keeps fuel from entering the main fuel passage adjacent, unless it's depressed. Just below the check valve is the power valve piston, which is usually retracted from the check ball by a vacuum being drawn from the side of the passage that the piston has risen above. Still in the passage is the power valve spring, which tries to overcome the vacuum to force the piston to open the check valve. Removing the piston completely and adding some washers underneath the spring will help tension the spring and let the check valve open at a higher vacuum setting.*

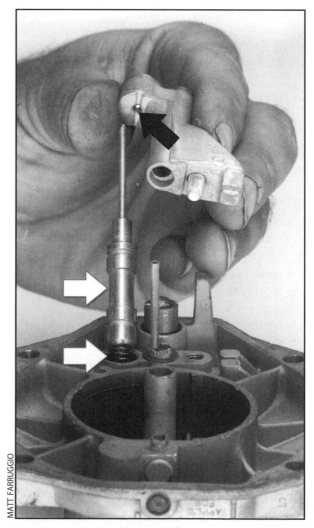

into the venturi by the weak vacuum. When this happens, you can almost see your fuel gauge take a nosedive (and your week's paycheck going down the throat of your carburetor). But you will get a jolt of acceleration.

If you own a good semi-high-performance carburetor, your power valve may be replaceable. Ask your local speed shop for a power valve that opens on a higher vacuum setting. Either ten or 12 inches Hg should be fine. This valve will open earlier, picking up or overlapping the accelerator pump's fuel delivery, and providing power.

If you can't get a replacement power valve for your car, increase the spring tension by adding washers, or solder a few wraps of wire on the plunger to act as a washer (see Figure 15-20). Increased spring tension allows your power valve to open earlier than it otherwise would. Some of my students have run into isolated cases in which the power valve is set up in an opposite configuration, and weakening the spring allows earlier opening. Consult an exploded diagram of your carburetor, and the configuration should be obvious.

Carburetors are pretty crude devices. So, neither power valve nor accelerator pump alterations have to be reversed in the event you want to temporarily return to that other fuel. You'll get pretty poor mileage in town, you'll have high emissions, but you'll have a lot of power. If you intend to remain on gas

MATT FARRUGGIO

for several thousand miles of city driving without changing back the carburetor, change the oil twice as often as you usually do, to guard against excessive oil degradation from the extra-rich mixture's blow-by. Contaminated oil (oil diluted by gasoline) creates a situation where oil with unburned fuel is washed more easily from the cylinder walls, and the diluted oil provides inadequate bearing lubrication.

CARBURETOR FLOATS

Carburetor floats control the amount of fuel entering the carburetor by shutting fuel off at the needle valve (the float valve) where it enters the carburetor. As the float rises, it pushes a soft-tipped

*Fig. 15-20 **Power valve actuator.** Several wraps of steel wire were wrapped around the base of the rod to stiffen the spring (acting like added washers). Once you've determined the proper number of wraps (by experimenting), solder them in place.*

*Fig. 15-21 **Twin main jets and power valve.** The two jets in the bottom of this two-barrel carburetor float bowl have their size stamped on their faces. The pin protruding from the unit in front of the jets is part of the power valve. When the carburetor is assembled and the manifold vacuum drops, a spring-loaded rod depresses the pin, allowing fuel to flow around it into the passage leading to the venturi.*

MATT FARRUGGIO

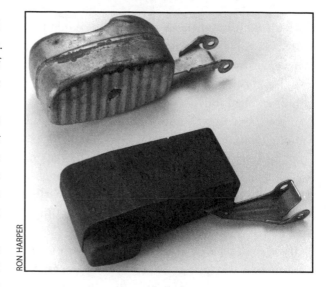

RON HARPER

needle into the fuel entry jet, shutting off the fuel supply. Lever arms amplify the float's lifting force. It works like the float ball in a toilet that shuts off the supply of water to the flush reservoir (see Figure 15-23).

The setting of the float height determines the amount of fuel in the carburetor's fuel reservoir bowl. The reservoir level is critical to an engine's proper operation. At too low a level, you can run out of fuel on a long acceleration; levels higher than factory specifications can cause flooding, leaks at the center gasket, and a big drop in fuel economy.

Some carburetors actually have windows in the side of the float bowl so you can see the fuel level with the carburetor fully assembled and running. In this case, you are simply setting the float height to bring the fuel in line with the mark on the window.

Alcohol has a different density than gasoline, which changes the buoyancy of the floats. First, make sure they are adjusted properly for gasoline and mark the fuel level in the bowl. Then

Fig. 15-23 Carburetor float assembly. It works kind of like your toilet where the float rises and cuts off incoming liquid when it reaches the right height.

adjust the floats for alcohol by simply bending the float arms until you've reached the mark you made inside the float bowl, while it is filled with alcohol.

If your carburetor floats are made out of foam plastic, you may want to replace them with brass. Although most foam floats aren't affected by alcohol, there is a general fear that they'll somehow get "waterlogged" and float too low. If you're concerned about this unfounded myth, and are unable to get replacement brass floats, you can paint the originals with two thin layers of methanol-resistant, model airplane paint.

Or you can dip the floats in epoxy to prevent supposed corrosion or waterlog. Such a coating makes the floats slightly heavier, which means they ride a bit lower in the fuel than before. This needs to be taken into account when setting float height. If you plan to experiment with heated low-proof alcohol, it will affect the solder that seals the floats, so I recommend going a step further and painting them with epoxy or epoxy resin.

ELECTRONIC CARBURETORS

In the evolution of carburetors to fuel injection, there was a brief era of electronically controlled carburetors. At that point, it wasn't clear that fuel injection would be that much of an advantage, and carburetor manufacturers were trying to make sure that their technology wouldn't be made obsolete. Over time, though, it became clear that this hybrid technology was very prone to failure, and more expensive. As fuel injection came down in price, it no longer made economic sense nor emissions sense to stick with electronic carburetors.

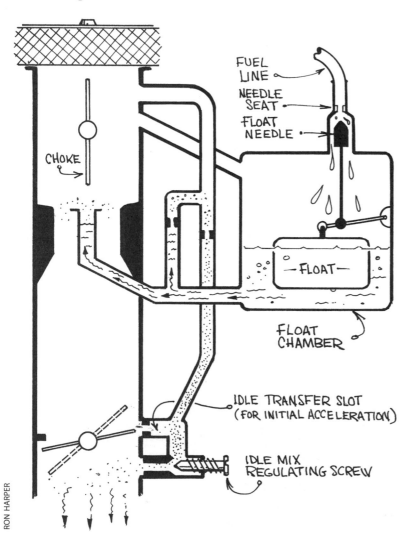

FUEL LINE
NEEDLE SEAT
FLOAT NEEDLE

CHOKE

—FLOAT—

FLOAT CHAMBER

IDLE TRANSFER SLOT (FOR INITIAL ACCELERATION)

IDLE MIX REGULATING SCREW

RON HARPER

But a lot of late '80s and early '90s cars have these carburetors. The basic idea was that the carburetors received rudimentary feedback from electronic sensors and could adjust their air/fuel mixture more readily than simple mechanical carburetors. The primary sensor was an oxygen sensor in the exhaust (for details, see Chapter 16). This permitted the carburetor to change the mixture based on actual engine conditions. By sensing the oxygen in the exhaust, these systems could adjust the fuel flow in the carburetor to be closer to optimum for whichever fuel was in the tank.

Fixed mechanical carburetors on gasoline settings could tolerate no more than 25–30% alcohol added to gas before they could no longer properly mix the fuel with air. (That's why gasoline in Brazil is mixed with no more than 25% alcohol, to accommodate older carburetors.) Although not as accurate or precise as fuel injection, electronic carburetors are not fixed and can crudely adjust to changing engine conditions. This lets them meet emission standards, more or less.

Sometimes crudeness is an advantage. Many vehicles' very rudimentary controls permit them to accommodate any percentage of alcohol without any changes! I have met farmers with early '90s big eight-cylinder cars, filling up on E-85 with impunity. In general, though, I feel that it's better to replace these electronic carburetors with a programmable aftermarket electronic throttle body fuel injection system, as we'll see in the next chapter, on fuel injection.

Endnotes

1. Energy Information Administration, *Petroleum (Oil)—A Fossil Fuel*, www.eia.doe.gov/kids/energyfacts/sources/non-renewable/oil.html#how%20is (January 2006).

2. "The Magic Oil Barrel," *Dow Canadian Insight Edition*, 18.

"It's a 4,000 gallon tank but I only keep 2,000 in it . . . can't stand hoarders."

Fig. 15-24

KEN ALEXANDER

Fig. 15-25

"As production falls off this cliff, prices won't simply increase; they will fly. If our oil dependence hasn't lessened drastically by then, the global economy is likely to slip into a recession so severe that the Great Depression will look like a dress rehearsal. Oil will cease to be viable as a fuel—hardly an encouraging scenario in a world where oil currently provides 40% of all energy and nearly 90% of all transportation fuel."

—*LOS ANGELES TIMES* EDITORIAL

I approached the subject of fuel injection with some trepidation, as would many shade-tree mechanics my age. We all grew up with carburetors, so this newfangled fuel injection stuff was pretty foreign. But once I pierced the mystery about these systems and started working with them, I found that they were, in general, much easier to convert to alcohol than carburetors, and that it was possible to conserve a lot more fuel.

HISTORY

Historically, carburetors' crudeness at responding to the needs of the engine at different points in the engine cycle meant a general overuse of fuel and a correspondingly high level of emissions. Over a ten-year period, carburetor manufacturers tried to make adjustments—and the price soared. So, driving fuel-injected cars became more attractive to consumers.

Fuel injection did not magically appear in the 1970s. Diesel vehicles began using high-pressure mechanical direct fuel injection in the late 1920s. In the early 1930s, some European two-stroke gasoline engine manufacturers were using timed mechanical fuel injection on cars in an attempt to reduce the loss of around one-fourth of the fuel—the intake and exhaust valves were both open at the same time, and the raw air/fuel mixture was going out the exhaust without ever being burned.

The first well-known four-stroke fuel-injected engines were the 1952 Mercedes M196 Formula One racer and the 1954 300SL gull-wing sports car,[1] both developed by Bosch. These early systems, although better than carburetion, shared many of its limitations. They were also solely air/fuel systems, and had nothing to do with other parts of engine control, like spark advance. The early systems operated best at wide-open throttle (WOT) or at idle, but had many idiosyncrasies at the partial throttle settings and during acceleration.

FUEL INJECTION SYSTEMS ARE GENERALLY ALL CARBON STEEL WITH GENERALLY ALCOHOL-PROOF MATERIALS. THEY REQUIRE AT LEAST 185+-PROOF FUEL, AND, IF POSSIBLE, 192+-PROOF. LOW-PROOF, ACID-CONTAMINATED ALCOHOL COULD DAMAGE FUEL INJECTION SYSTEMS BY PULLING LEAD OR TIN FROM SOLDER, OR BY CAUSING RUST. IF YOU ARE USING HOMEMADE ALCOHOL OF SUSPICIOUS ORIGIN, USE A CORROSION INHIBITOR—ONE-HALF OF ONE PERCENT OF A SYNTHETIC, ALCOHOL-SOLUBLE, TWO-STROKE OIL; OR ABOUT THE SAME PERCENTAGE OF BIODIESEL—TO LENGTHEN PUMP AND INJECTOR LIFE.

DAVID BLUME

Fig. 16-1 Mass air tuner. *The four screwdriver-adjustable knobs adjust the varied conditions an engine encounters, from idle to wide-open throttle. The knob on the left is for idle, the next one's for initial acceleration, the third for high-end acceleration, and the last on the right for WOT. This unit can also be plugged into a laptop for more precise adjustment.*

Innovators quickly figured out that direct injection and timed, high-pressure delivery of fuel weren't required in gasoline engines! After all, carburetors delivered fuel full-time outside the cylinder into the engine manifold without exotic high-pressure injector timing. Early racers came close to matching fuel injection's even fueling of each cylinder by having manifolds that sported a carburetor for each cylinder.

So, for gasoline, indirect low-pressure systems were created that either delivered fuel on top of the manifold where the carburetor normally would be, or sprayed it directly on the outside of the valves where the head and intake manifold meet. Although many of these early systems did pulse the fuel on and off, it became unclear whether pulsing fuel was necessary. Fuel delivered outside the cylinder would be sucked into the engine each time the intake valve opened, dozens of times per second. Some companies even dispensed with pulsing altogether and instead continuously injected fuel.

But acceleration required increasing the amount of fuel periodically. The increase in fuel for different levels of acceleration came from systems that would either increase the pressure in the fuel line feeding the injectors, or increase the rate of flow to a fuel distributor that had individual lines running to an injector pointing at each set of intake valves.

Fuel injection, even in these early systems, successfully cured many problems that occurred in carbureted engines. First, delivering fuel to the top of each cylinder, instead of sending it down a long manifold, all but eliminated puddling. The fuel atomized better, and there was no unequal feeding of cylinders, which could create a very lean

condition in some of them. These lean mixtures would tend to ping, so engines had to be designed with low compression ratios, since high compression would work great for the cylinders with enough fuel, but cause even more destructive pinging in lean cylinders.

Other benefits of fuel injection became apparent. Since you didn't have to contend with manifold puddling, then you didn't have to use exhaust heat to raise the temperature of the manifold. The engine could "breathe" better, using cooler, denser air. And, as with carburetors, there tended to be extra fuel delivered in many partial acceleration throttle situations to ensure smooth running.

The one drawback was that some unburned mixture would still be lost when both exhaust and intake valves were open. More precise systems were still to come.

The next big advance, which formed the basis of modern fuel injection, happened in the early 1950s. The "Electrojector" was developed by the Bendix Aviation Corporation. Instead of continuous injection with mechanical controls to increase fuel delivery, an electric valve known as a solenoid was installed on each injector. This valve pulsed open each time electricity coursed through the valve and closed when electricity was withdrawn. The amount of fuel delivered depended on how long the valve was kept open during each pulse (pulse width). So the valve would open a fixed number of times per minute, but the time it was open would vary. During idle, the solenoid valve would open very briefly during each pulse. Under sustained acceleration, the time open would greatly increase until it was almost half or more of each pulse cycle.

The pulse width was controlled by the first electronic "brain box," forerunner of the modern electronic control unit (ECU). Long before computer chips, the earliest version calculated the needs of the engine using vacuum tubes.[2] Sensors measured manifold pressure, engine speed, atmospheric pressure, and air and coolant temperatures; this information was routed to the brain box for calculation of fuel pulses. This was the first truly electronic fuel injection (EFI).

Although the Electrojector was tried on a few models of cars, it wasn't ready for full production until 1965, at which point the manufacture was turned over to Bosch. It then appeared for the first time in the transistorized (no vacuum tubes) 1967 Volkswagen 1600.

Fig. 16-2

In the 1980s, continuous fuel injection systems gave way to pulsed systems: Instead of a fuel distributor sending fuel via separate fuel lines to injectors, there would be one common fuel line, with each injector branching off of it. Instead of always being on, each injector would have an electric on/off solenoid valve.

In today's engine, the ECU takes in information about the airflow into the engine, the oxygen in the exhaust, the temperatures of the coolant and the air, and many other things, to determine just how long to open the fuel injector. The injectors receive a signal from the ECU, and open and close rapidly—often 50 or more times per second, per injector. This way, the proper quantity of fuel is delivered by varying the length of time of each pulse.

Today, fully computerized engine control handles far more than just the air/fuel mix. The ECU now gathers information on elevation, air, engine water, exhaust temperature, ignition timing, engine knocking, and more, with oxygen sensors in the exhaust both before and sometimes after passing through the catalytic converter. The ECM has taken the place or taken control of other formerly unconnected engine parts (like distributors, coils, warmup dampers, chokes, etc.).

For the most part, I am going to focus on the various models of EFI developed by the Bosch Corporation. This history provides a strong foundation to understand how modern engine systems can be easily converted to alcohol. To this day, the Bosch folks are the pioneers, and, with few exceptions, other systems can be said to be at least heavily based on Bosch—if not outright knock-offs. Learning what works on Bosch systems will apply to other manufacturers in most cases.

GENERAL ISSUES REGARDING ALCOHOL AND FUEL INJECTION

Let's take a quick look at a couple of general issues pertaining to alcohol and fuel injection. Then we'll briefly examine sensor systems and catalytic converters before I get into the actual conversion information.

Fuel injection systems are generally all carbon steel with generally alcohol-proof materials and require at least 185-proof fuel, and, if possible, 192-proof or more. Low-proof, acid-contaminated alcohol could damage fuel injection systems by pulling lead or tin from solder, or by causing rust.

If you are using homemade alcohol of suspicious origin, use a corrosion inhibitor—one-half of one percent of a synthetic, alcohol-soluble, two-stroke oil; or about the same percentage of biodiesel—to lengthen pump and injector life.

Always use extremely clean containers for your fuel. Fuel injection filters are often as fine as one micron (sometimes smaller), as compared to typical fuel filters of ten microns used at fuel stations. One-micron filters will plug up readily if they run into a large quantity of crud.

"The small producer of alcohol need never fear competition from the big producers, such as the Whiskey Trust. In the first place, the supply of raw material is unlimited. Not until someone learns how to control the sun and its light can there be a monopoly in the raw material for alcohol manufacture."

—HENRY FORD, *DETROIT NEWS*, DECEMBER 13, 1916

OXYGEN SENSORS AND YOUR CATALYTIC CONVERTER

It's important to understand why the oxygen sensor system is part of virtually all fuel injection systems. The raw truth is that it's there to protect your catalytic converter. In theory, catalytic converters reduce the temperature at which exhaust will burn—so, exhaust below 1000° in the presence of the catalyst will ignite at much lower temperatures.

In reality, with gasoline, a catalytic converter is a compromise between cost and environmental protection. Due to the almost daily variation in the composition of that toxic waste known as gasoline—its vast number of hydrocarbon partial-combustion products, the sheer volume of pollutants—and the modest size of catalytic converters in today's cars, it's a miracle how much pollution they do prevent.

But, to keep the pollution control system cheap and small, it became necessary for the auto companies to find ways of lowering the amount of pollution produced. It was cheaper to solve the pollution problem part of the way upstream, so to speak, than to have to process all the pollution downstream in a catalytic system. If all industries understood this basic economic law, our planet would be a lot cleaner!

In the case of gasoline, the engine runs to a stoichiometric ratio of 14.7 to 1, air to fuel, a theoretically perfect combustion mixture that should result in a minimum of pollutants being exhausted to the dinky catalytic converter. If the ratio varies too much from this number, the car won't run, as gasoline has only a limited range of air/fuel mixtures combustible in an engine.

Stoichiometric Air/Fuel Ratio of Gasoline

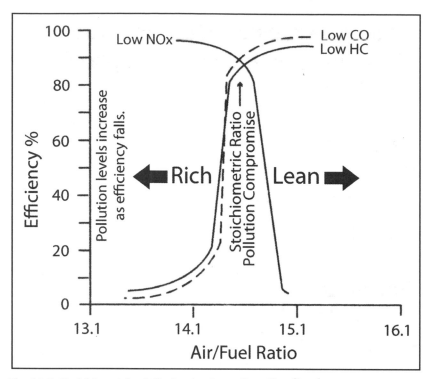

Fig. 16-3 Stoichiometric air/fuel ratio of gasoline. Gasoline has a very narrow range of acceptable air/fuel mixtures at which it will combust. The 14.7:1 ratio is not actually the most efficient level, but the point that is the best compromise between NOx on the one hand and HC/CO on the other hand.

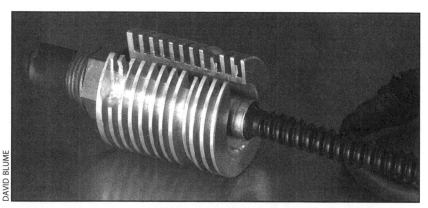

Fig. 16-4 Oxygen sensor with fins. On the left, you can see the sensor bulb and the threads (the same sort as on a spark plug) that screw into the exhaust pipe. Dan Fodge made a sleeve with fins, which fits snugly around the part of the sensor housing that extends outside of the exhaust pipe. It carries excess heat away from the oxygen sensor, which extends the life of the sensor tremendously. This modification has been done to a sensor that is used temporarily with a handheld, air/fuel exhaust analyzer. The sensor on your vehicle wouldn't have these fins.

But between the ideal and the impossible, there exist a range of air/fuel ratios that, if allowed, could burn up or plug the catalytic converter. So, to protect the converter, there is a delicate balance to be maintained.

If the car ran on gasoline with a slight excess of oxygen, then less pollution would be emitted; hydrocarbons and CO would be the first to drop. But if the oxygen level became too high, nitrogen oxides would go up. With alcohol, burning lean without increasing pollution is much more attainable.

The oxygen sensor (or, in some cars, a **wideband sensor**) is installed very near the engine in the exhaust manifold. If it detects too much oxygen, it signals to the ECU (or electronic carburetor) that the engine is not getting enough fuel to use the available oxygen. If, on the other hand, the sensor smells too little oxygen, it sends a signal that too much fuel is being delivered. In each case, the engine instantly calculates the next pulse width to make the mix as close as possible to stoichiometric (typically, within 1%).

When the car is operating under the influence of the oxygen sensor feedback loop, it is said to be operating in **closed loop**. If, for some reason, the car goes outside of the feedback loop's ability to keep up with engine changes (such as when you suddenly push the pedal to the floor to pass someone at WOT), it is said to be operating in **open loop**, and relies on stored generic data to keep operating. Over time, the kinds of incidents that escape closed-loop running have been whittled away, until there's little you can do on a late-model car that can't be fully controlled by the computer.

An ECU today often can calculate the near-perfect pulse width, not just for every revolution of the engine, but for each pulse of fuel, for each cylinder, hundreds of times a minute, using the information from all its sensors, but with the oxygen sensor dominant. The one exception—and that is being quickly closed down—is when you start the vehicle. Oxygen sensors typically don't start operating accurately until the sensor has warmed to 600°F, and then the closed loop kicks in. In older cars, that can take up to two or three minutes. Some oxygen sensors made today have as many as three or four heating wires built in to speed up heating the sensor, even before the exhaust has reached 600°F. The start-up and, to a lesser degree, the cool-down of an engine are when it is most polluting.

TWO INTERMITTENT EFI SYSTEMS

Intermittent (noncontinuous) EFI systems can be divided into two types: throttle body fuel injection (TBI), which introduces fuel on top of the manifold, where a carburetor used to be found; and the more common, **multi-port fuel injection**, which has multiple injectors in the manifold, pointing at the valves that open into each cylinder.

Throttle Body Fuel Injection (TBI) Systems

In throttle body fuel injection systems, one or two injectors sitting in the **collar** send fuel through to the engine via the manifold. This means that the manifold is still wet, and many of the supposed gains of injecting the fuel at the valves would seem to be lost.

But since the fuel is much more effectively atomized by the fuel injector, far less of the mixture drops out and puddles in the manifold. Unfortunately, many TBI systems have the **throttle plate** underneath the injectors, so the well-atomized fuel runs right into the plate, condenses, and drips into the engine, or at least turns into a coarser spray.

TBI is of great interest to owners of older cars with carburetors, since fully programmable electronic fuel injection can be had by bolting on aftermarket replacement TBI units in place of the original carburetor. As you will see later, many programmable ECUs work just fine on TBI units, so you can more easily go dual-fuel than you could with a carburetor.

Because of the ability of TBIs to reduce emissions, they are now frequently used to replace carburetors on forklifts, wood chippers, and other medium-sized industrial equipment, where people have to work around the exhaust. In enclosed or close-in applications, carburetors running on gasoline don't have a snowball's chance in a catalytic converter of meeting workplace air safety standards.

TBI systems, and other fuel injection systems, operate under much higher fuel pressure than your average carburetor does. Fuel injection systems typically deliver fuel to the injectors at 28 to 70 psi, depending on the system. Atomization increases when the high-pressure fuel is sprayed through the injector.

Multi-Port EFI Systems

The second type of intermittent EFI system is the multi-port system, which is divided into two types. The first opens one or both banks of injectors simultaneously on each pulse, and the other opens injectors in a sequence at the optimal time for each cylinder—this allows the ECU to make changes to fuel delivery at each pulse of each cylinder, rather than making across-the-board changes that apply to whole banks of injectors.

The biggest difference with port fuel injection, over TBI, is that the air/fuel mixture does not have to travel down the manifold to get to the engine, thereby avoiding a lot of the problems of TBI or carburetors (see Chapter 19).

CONVERTING OLDER FUEL INJECTION SYSTEMS

Let's look first at how to convert older fuel-injected systems. We'll discuss Bosch's D-, L-, and K-Jetronic systems, since they allow us to make various mechanical adjustments or workarounds to alter the air/fuel ratio for running on alcohol. Millions of cars with these older systems are still on the road today.

Older Volkswagen vans (circa 1968), many American cars built from 1979 to 1983, and some Jaguars use a "primitive" version of the modern electronic fuel injection system, called the D-Jetronic. This system has a fuel pump feeding a common fuel line, with branches to each injector, with excess fuel returning to the gas tank.

Fig. 16-5 Basic fuel injection system. As you can see, fuel is pumped to the right through fuel rails. The individual fuel injectors branch off the common fuel supply. A pressure regulator installed in the fuel rail after the area of branching injectors restricts fuel flow, which determines how much fuel goes back to the tank and what pressure is in the common fuel line feeding the injectors.

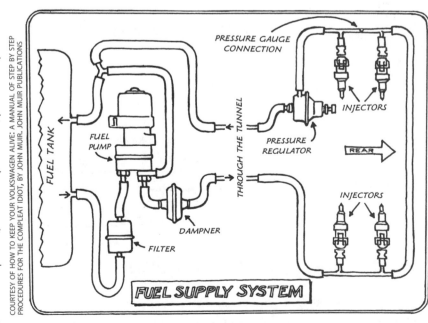

COURTESY OF HOW TO KEEP YOUR VOLKSWAGEN ALIVE: A MANUAL OF STEP BY STEP PROCEDURES FOR THE COMPLEAT IDIOT, BY JOHN MUIR, JOHN MUIR PUBLICATIONS

Each time the ECU sends a pulse to the solenoid valve to open, a precise amount of fuel squirts into the engine, based on an assumed fuel pressure. A simple way to enrich the mixture for alcohol is to alter this assumption.

Fuel pressure in a D-Jetronic gasoline-fueled car is usually around 28 psi. If we increase the pressure in the fuel line, more fuel will be delivered during the same pulse width. The fuel pump is capable of 90 psi, but a practical maximum based on other parts in the system is around 72 psi—and, for alcohol fuel, there is no need to go even that high.

A fuel pressure regulator determines how much back-pressure is in the fuel lines leading to the injectors, and how much fuel passes through the regulator on its way back to the gas tank. So, a fuel pressure regulator acts to restrict the fuel line leading back to the tank, causing back-pressure in the fuel lines leading to the injectors.

With this system, the fuel pressure regulator is adjustable. We can use a wrench to alter the pressure regulator to a higher setting for use with alcohol. The shade-tree measure of when you've increased the pressure enough is the point at which you first reach the highest manifold vacuum reading.

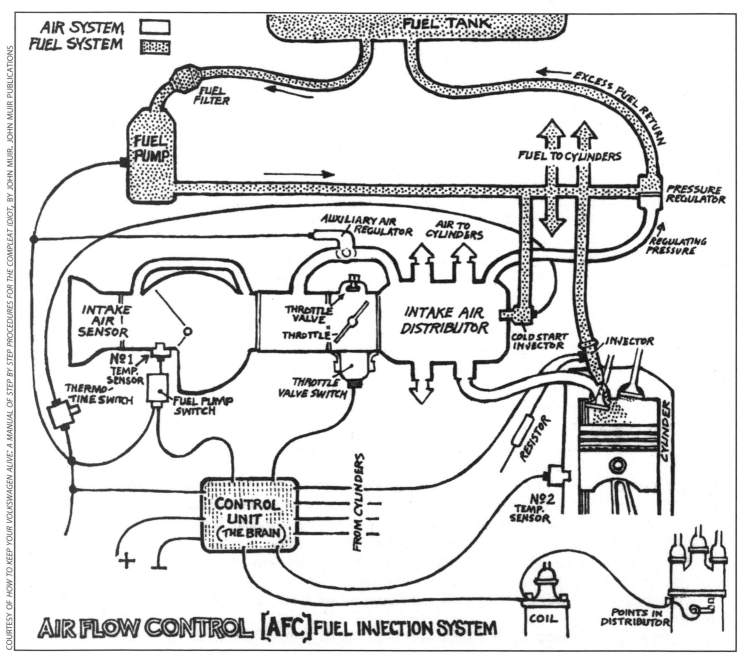

Fig. 16-6 L-Jetronic fuel injection system. *This system is still part mechanical and part electronic. Note the cold-start injector, which puts extra fuel into the air supply when starting on cold mornings. Using a resistance heater on this fuel line makes for a good alcohol cold-start system.*

DAVID BLUME

An even more accurate measure is the highest exhaust gas temperature. If you have access to a carbon monoxide tester, you can increase the pressure until a substantial amount of CO is detected, then back off pressure until it disappears or hits its lowest level.

The brain box will operate using the oxygen sensor information to keep the air/fuel mixture zeroed in at the stoichiometric ratio on alcohol, since we have brought the fuel delivery pressure to the right range that permits full control by the ECU.

Two other types of older electronic fuel injection systems are found on a variety of vehicles: the L-Jetronic and the K-Jetronic. The Bosch L-Jetronic system (also called the Air Flow Control System) uses a spring-loaded, bias-closed damper in the air supply to sense engine needs on several small Japanese cars, 1975–76 VW vans, 1975–79 VW Beetle Convertibles, and the BMW 530i, among others. The damper is connected to an electrical contact traveling across a **potentiometer** (a sort of dimmer switch), which is read by the brain box to control the duration of the fuel injection pulse delivered to the cylinders.

The potentiometer's range can be adjusted to enrich the mixture by loosening the center set screw and rotating the scale until you get the highest manifold vacuum—or, if you install one, the highest reading on your exhaust temperature gauge—indicating proper air/fuel mixture. Setting for lowest CO reading is a good way to make sure you have the cleanest emissions.

Like the L-Jetronic, the K-Jetronic is a semi-mechanical electronic fuel injection system. It was used in many European and some higher-performance Japanese makes in the mid-'70s to mid-'80s. It is also called a **constant injection system (CIS)**, since it constantly injects fuel, as opposed to pulsing fuel delivery. There are an awful lot of

these systems still on the road today because the system is so solid and trouble-free.

The **metering rod** in the fuel distributor housing of the Bosch K-Jetronic receives pressurized fuel for delivery through tapered slots (which I understand were cut to shape with a laser!) to the injector lines, which continuously release this fuel through very simple injectors.

The CIS system has two fuel pressures: the control and the line pressure. The **control pressure** responds to changes in the air inflow and controls the fuel distributor. The line pressure goes though the fuel distributor and then to each line on its way to an injector. Excess fuel goes back through a pressure regulator in early models and in later models through a **frequency valve** in the fuel return line to the fuel tank. The frequency valve works like an adjustable fuel pressure regulator.

One way to increase fuel flow to the injectors is to fool the sensing system into thinking it needs to deliver more fuel. The large damper, which rises as air rushes into the engine, pushes the fuel distributor rod upward (against control fuel pressure) as the needs of the engine increase. The housing around the damper is engineered to model fuel delivery needs based on how much air gets past the damper.

One way to increase fuel flow to the injectors is to fool the sensing system into thinking it needs to deliver more fuel.

You can change the air/fuel mixture pretty precisely using a long, three-millimeter Allen wrench to adjust the range of the damper. The adjustment screw is located between the fuel distributor body and the air funnel of the damper arm (see Figure 16-8). Using the measurement methods already discussed, you can adjust the screw to the optimal point.

You do have to remove the anti-tamper device, a ball that blocks your access to the adjustment screw. After taking the damper housing off and flipping it over, you can easily drive the ball out with a punch and hammer. When you reassemble the housing, you'll have access to the adjustment screw. Cover the hole with a piece of duct tape instead of a new anti-tamper pellet. Removal of the anti-tamper device and adjustment of fuel is nicely detailed in the owner's manual in the case of early '80s Volvos. Some of the Volkswagen and Mercedes engines using this system had a nice

LEFT: *Fig. 16-7 Fuel pressure regulator. This regulator determines how much fuel goes back to the tank and how much fuel pressure is in the common fuel rail. It's adjusted by the "bolt" on the right and fixed at a particular setting by its lock nut.*

Of course, it violates several laws if you drive without the anti-tamper device reinstalled. Most mechanics not only forget to reinstall one but also forget to cover the hole with tape. But it is more important to clean up the emissions than to be overly concerned about laws on tampering which, although well-meaning, preclude the possibility of owners improving upon existing products.

Fig. 16-8 Constant injection system. *The fuel distributor is shown disassembled from the air-sensing base. The piston protruding from the fuel distributor usually goes through the hole in the base and rests on a lever arm, which is attached to the damper on the left. The more vacuum the engine creates, the higher the damper's lever arm raises the fuel distributor piston. As the piston goes up, more fuel is delivered to the injectors. To enrich the mixture, the lever arm is adjusted by an Allen wrench through the small hole in the center of the base. One full turn will take you through the entire range of too-rich to too-lean.*

rubber plug on a wire handle that sealed the hole but was easily removable with just a tug when you needed to get at the adjustment screw.

Of course, it violates several laws if you drive without the anti-tamper device reinstalled. Most mechanics not only forget to reinstall one but also forget to cover the hole with tape. But it is more important to clean up the emissions than to be overly concerned about laws on tampering which, although well-meaning, preclude the possibility of owners improving upon existing products.

Of course, you can replace the anti-tamper ball after you have made your adjustments, but that makes switching back to gasoline a hassle, should you need to do it.

Later versions of this system let you be more precise in making adjustments, since they have a frequency valve that operates in concert with an oxygen sensor. A frequency valve is a solenoid valve that regulates the fuel pressure by controlling the amount of fuel allowed to return to the tank.

To adjust the vehicle for use with alcohol, use a dwell meter (this doubles as a pulse width or duty cycle meter) connected to the test lead for the oxygen sensor/frequency valve. Adjust the air/fuel mixture on alcohol with your 3mm Allen wrench, until the dwell meter reads approximately 45° at the engine speed specified in your vehicle's manual. The 45° setting is one-half of 90°, and therefore corresponds to the 50% duty cycle (how long the solenoid valve is open) that is normal on gasoline.

What you've done is recalibrate the fuel delivery pressure to make the vehicle's ECU think that you are running on gasoline. It makes it much easier for the oxygen sensor and controlling frequency valve to precisely zero in on the stoichiometric air/fuel ratio and to thus minimize emissions.

MATT FARRUGGIO

Fig. 16-9 CIS line pressure regulator. *If you know you aren't going back to gasoline, you can increase the line pressure by adding shims (calibrated washers) like the one draped over the spring.*

CONVERTING NEWER FUEL INJECTION SYSTEMS

You have several main routes to follow when converting newer, fully electronic, fuel injection systems to alcohol:

1. The crudest way is to use larger injectors or an adjustable pressure regulator to increase the fuel pressure, which will deliver more fuel at the same pulse width that the ECU normally uses.

2. The second route is to use **fooler technology** to modify the information that the ECU is getting or sending, so it is fooled into running on alcohol while it thinks it's running on gasoline.

3. The third and most powerful and accurate way is to use an aftermarket ECU, which allows you to actually program in different values than the factory settings in order to get the engine running the way you want.

Increasing Fuel Pressure

Let's look at the first choice. If you have a stock injector like the one on a late-model Toyota Tundra, it is rated at 19 pounds of fuel per hour. (Of course, it rarely puts out that much, unless you are going flat out.) If you want about 30% more capacity to account for peak alcohol demand, then you would look for another injector rated at close to 24 pounds per hour. As it turns out, there is a 25-pound-per-hour injector available for the Tundra. Once you've installed this, every time the injector opens, it will deliver more fuel than it did on gasoline. After all, the computer only measures how long the injector should open, not how much fuel is delivered.

This crude conversion works, but on six- or eight-cylinder vehicles it can take a full day to disassemble your intake manifold to get at the injectors. It is usually extremely easy, however, on four-cylinder engines.

If this is all you do—simply change the injectors—mileage will probably drop as much as 25%. Your horsepower will go up very noticeably. Atomization of cold fuel will not be great, since you will be sticking with stock fuel pressure. Most likely, you would have to change the injectors back to stock in order to revert to gasoline, since the larger injectors would have a hard time opening at very short intervals to deliver gasoline properly.

The larger injectors get you started, and then the oxygen sensor system fine-tunes the fuel flow by testing how much oxygen is in the final exhaust. It tunes the engine to use just the right amount of fuel to be within a percentage point of ideal air/fuel mixtures (stoichiometric combustion).

Instead of using larger injectors, you could use an aftermarket, adjustable pressure regulator and heat the fuel.

Instead of using larger injectors, you could use an aftermarket, adjustable pressure regulator and heat the fuel (see Chapter 19). This regulator determines how much fuel is allowed to return to the fuel tank, and how much pressure remains in the common injector line (also called the **rail**). An increase of the rail pressure simulates a larger injector size, since under higher pressure, more fuel will flow during a given pulse width. Heating reduces fuel viscosity, and that increases fuel delivery from stock injectors quite nicely.

As with the D-Jetronic system discussed above, the injector delivery rate is based on the assumption that the fuel pressure is what is specified for the stock engine. So, if you use an adjustable pressure regulator, and go from 30 psi up to 60 or 80 psi, you are going to get a much bigger delivery of fuel, even though you haven't increased the pulse width. You will also get much better atomization of the fuel with higher pressure, which significantly improves efficiency. This also allows you to run on alcohol without changing injectors. But if you need to go back to gasoline, you simply get out your wrench and adjust the pressure back—a five-minute job.

Ford flexible-fuel engines apparently do this to some degree, since their specifications for fuel pressure show a range of 47 to 63 psi. What's more, the fuel pressure regulators from 1998 and later Ford 3.0-liter engines are not on the rail where they normally are found, but all the way back in the fuel tank, as part of the fuel pump assembly.

Fooler Technologies

A second way to convert your fuel injection system is to use one of the electronic fooler technologies. These fool the ECM into thinking it's running on gasoline, when it's not. Foolers let you do a much more precise and fuel-thrifty conversion, or refine a pressure increase conversion.

In the 1980s, IPD, a Volvo high-performance aftermarket product company in Portland, Oregon, came up with a simple but cunning adaptation of Bosch's electronic brain box as an early fooler technology. They rebuilt the box with electronic parts of their own, and wired it to a dash-mounted dial, which allowed enriching or leaning the fuel mixture using percentages designated on the dial. It intercepted the signal going from the ECM to the injectors and modified it to change the pulse width of the injector.

After you have determined the proper settings (using the measurement methods described earlier), IPD's method lets you switch from alcohol to gasoline and back again, more or less at the turn of a knob. This early device only allowed you to increase the air/fuel mixture across the board; refinements for various phases of acceleration or cruising were not taken into account. So, in many ways, it's not much of an improvement over just putting in larger injectors. But you could run the engine at a percentage other than stoichiometric, and, as we will see, that's a potential advantage with alcohol. Devices like this are now sold as a

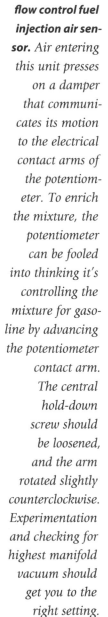

Fig. 16-10 Air flow control fuel injection air sensor. *Air entering this unit presses on a damper that communicates its motion to the electrical contact arms of the potentiometer. To enrich the mixture, the potentiometer can be fooled into thinking it's controlling the mixture for gasoline by advancing the potentiometer contact arm. The central hold-down screw should be loosened, and the arm rotated slightly counterclockwise. Experimentation and checking for highest manifold vacuum should get you to the right setting.*

MATT FARRUGGIO

universal unit that will work on most fuel-injected engines—elegant, simple, and, under current air pollution laws, quasi-legal in most states.

Another kind of fooler is an airflow engine tuner, sold primarily in car magazines or over the Internet, which fools the computer into thinking the engine has far more air coming in than it actually does. Most cars nowadays are equipped with a mass airflow (MAF) sensor, or **manifold absolute pressure (MAP) sensor**, which typically gives an accurate signal to the ECU of how much air is coming into the engine, even at varying altitudes or temperatures. The fooler takes the signal being generated by this sensor and runs it through a processor before sending it on to the ECU. When the ECU takes the signal from the air tuner and makes its calculation for pulse width, it uses data generated by the tuner to come to its conclusion.

This is actually a pretty good approach, since by using alcohol, we are in reality introducing more oxygen into the system. The oxygen is coming in by way of the chemical structure of the fuel, instead of through the air duct. So, using air tuners can give you excellent control over switching back and forth between alcohol and gasoline.

Unlike the old IPD single-knob devices, most of these tuners give you several real knobs or computer "virtual" knobs for idle, acceleration, cruising, and wide-open throttle (see Figure 16-1). This allows you to get high efficiency at each engine phase, rather than being limited to the across-the-board approach of earlier electronics. For maximum mileage on alcohol, you can reduce the amount of fuel during acceleration more than would be possible with gas, since alcohol's cooler burning does away with the need for excess fuel to cool cylinders during acceleration. Best of all, these tuners tend to be cheap.

Since you are cutting into only a single wire from the MAF or MAP sensors, going back to gasoline is as simple as installing a switch to send the signal straight to the ECU instead of through the tuner. This can be quite invisible under the hood, and will not upset state emission testers' peace of mind.

One advantage of fooler technology is that late-model cars use the data from the MAF meter to partially figure out ignition advance. When the new signal is sent to the ECU, the vehicle also advances the timing, which is a boon to improving power and mileage. So you get some of the benefit of an

aftermarket ignition system just by using a MAF tuner.

Make sure that the air tuner you purchase is capable of increasing the fuel delivery by 40%. At hard acceleration or WOT, you may briefly need this much fuel if you are planning on dual-fueling and not changing fuel pressure. Some tuners only go as far as 25%, often not enough to switch back and forth between gasoline and alcohol, unless you also provide a way of varying the rail pressure. As you'll see in a minute, combining a 25% MAF tuner with an adjustable fuel pressure regulator or an oxygen sensor fooler will give you dual-fuel capability. If you are going to run only alcohol, a combination of fuel heat and a 25% tuner ought to work fine.

In Brazil, the most popular aftermarket conversion to alcohol is a fooler technology that is very easy to install. It intercepts the signal going to each fuel injector, processes the signal coming from the ECU, compares it to another fuel map stored in its memory that has the proper alcohol pulse width, and sends a new signal to the injector. In essence, it's like one of IPD's dial-injector signal

Fig. 16-11 Brazilian fooler. *There are two sockets for each of four cylinders. You plug the socket coming from the vehicle ECU into one of the pair, and the other you snap onto the fuel injector. You then connect the one ground wire to the negative on your battery.*

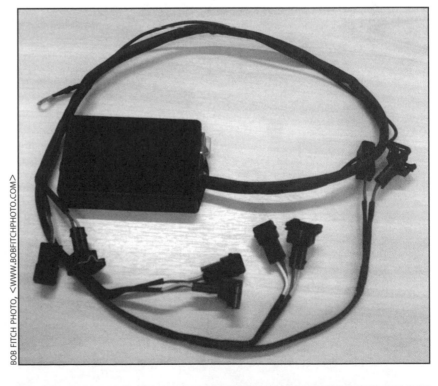

BOB FITCH PHOTO, <WWW.BOBFITCHPHOTO.COM>

processors from the past, but already programmed to go between alcohol and gasoline at the flip of a switch.

To install this kit is pretty simple. You detach the electrical plug from each injector and plug them into receptacles that lead to the fooler's **wiring harness**. Now the signals for each injector are routed to the new brain box. You then snap another plug from the wiring harness onto the fuel injector. This plug carries the new reprocessed signal from the box to the injector. The device is grounded to the battery. It's a simple matter at that point to flip a switch on the box to have it run on gasoline (turning off the box functions), or to switch it to allow the vehicle to run on alcohol, even though the ECU thinks it's running on gasoline.

Using this kind of dual-fuel device works best if you are below a quarter-tank before switching to the other fuel, since it does not have the infinite adjustment for every possible mixture of alcohol and gasoline that a factory flexible-fuel vehicle can handle. It works best when it is closest to a high percentage of one fuel or the other.

Newer, more advanced foolers imitate the GM Brazil flex-fuel systems, noting when you are filling the tank and then automatically adjusting to whatever fuel mix is in the tank. These foolers usually require you to start with approximately half alcohol and half gasoline to get the initial calibration.

Brazilian-style foolers only work on air/fuel mixtures and don't address ignition advance. But in many late-model cars, the ignition advance is now automatically controlled, and needs no monitoring or alteration.

Fooler technologies are rarely advanced enough to do what we really need, which is to let an engine run leaner than stoichiometric when burning alcohol. Most foolers cannot override the oxygen sensor authority to control the mixture to stoichiometric without setting off the "check engine" light on your dash.

So with most of these fooler technologies, it's very beneficial to also use an oxygen sensor fooler. Every oxygen sensor uses a standard output range of 0–1 volts. The output voltage tells the engine's ECU how close to stoichiometric the air/fuel ratio is running by sniffing the exhaust. Stoichiometric burning, regardless of the type of fuel, should read 500 millivolts (one-half a volt.) So, even if you use one of the other fooler technologies to get the air/fuel mix to be in the proper range for alcohol burning, you can't run leaner than stoichiometric, due to the priority given to oxygen sensor data.

While this is not important to running on gasoline, it makes a lot of difference in mileage for an alcohol-fueled vehicle. Since alcohol has a wider range of acceptable air/fuel ratios, being able to fool the computer into permitting lean operation will result in better mileage. It will also virtually guarantee zero CO and HC emissions.

The most important aspect of this technology is in cars that can otherwise use E-85 with no modifications except for tripping the check engine light. Using an oxygen sensor fooler will permit you to exceed the long-term fuel trim setting and use a higher percentage of alcohol or perhaps even straight alcohol in an otherwise unmodified car without tripping the light.

Oxygen sensor foolers generally run under $100. This fooler is also easily concealed from smog test station examiners by installing the fooler in your glove box and running the oxygen sensor wire first to the fooler and then back out through the same hole in the firewall. By wrapping the two oxygen sensor wires in a plastic wire harness, the break in the wire becomes invisible.

Oxygen Sensor Voltage Curve

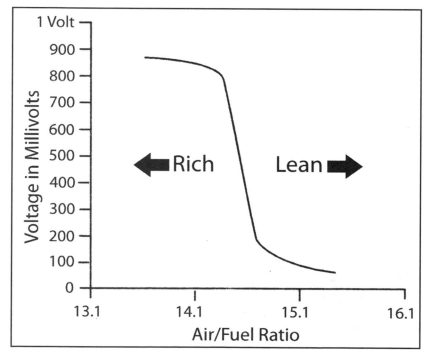

Fig. 16-12 Oxygen sensor voltage curve. The full range of oxygen sensing is spread over one volt. It is split into one thousand parts, with 500 millivolts representing stoichiometric combustion. An oxygen sensor fooler can minutely adjust the mixture to be approximately 15–20% lean while running on alcohol, while the ECU thinks the engine is running at stoichiometric.

The simple foolers have a knob that allows you to turn them to simulate either more or less oxygen than what is really in the exhaust. For alcohol running, you want the fooler to indicate that there is less oxygen in the exhaust than there really is. This one change can increase your alcohol mileage 15% or more.

We can expect a proliferation of these and other fooler technologies to be available in the near future as E-85 stations start to be more common. Check <www.permaculture.com> for what's available.

Aftermarket Programmable Electronic Control Units

I've had to approach this section three times since I began working on it in 2002. At that time, there were very few aftermarket programmable ECUs; most were focused on American cars and were designed with racers in mind. The main contenders were the Accel/DFI system and the F.A.S.T. system. Instead of using fooler technology, these ECUs either replaced, or piggybacked onto, the factory ECU. They permitted direct modification of engine conditions using a variety of sensors on the car.

In doing R & D for this book, we converted an Acura Integra, using a relative newcomer to the aftermarket ECU arena, the AEM ECU. Although it came later into the game than others, it is by no

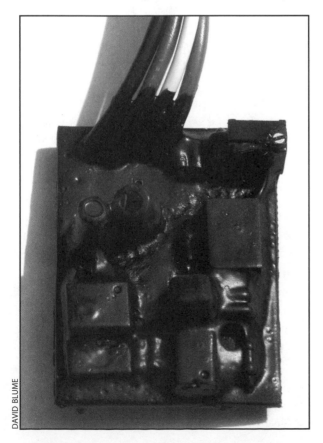

DAVID BLUME

means an underdeveloped computer. It is primarily compatible with Asian engines.

Programmable ECUs allow virtually unlimited control of the engine in every stage of acceleration, or environmental condition. So, you don't really need any type of fooler technology, since you now are the ECU and can directly tell the system how you want it to run. Typically, you can program the system while driving the vehicle, using a laptop plugged into its ECU. Of course, that makes it hard to pay attention to what going on in front of you on the road, so usually one person programs while another drives (see Appendix B).

Most of the major companies that make these computers have Internet user groups that load data from their vehicles onto the Web. So, what used to be the hardest part of using an aftermarket ECU—getting the car to start with guessed-at data—is pretty much a thing of the past. You can download a pretty close data map for your car and engine and then start modifying from there.

Almost all the aftermarket computers use a wide-band sensor instead of an oxygen sensor, for precision. This is even better than a good CO sensor, since you are getting readings while you are on the road instead of just in a shop. We've found that shooting for an air/fuel mixture at about 13:1 when starting off is often a good base figure, once the oxygen sensor warms up and takes over control from the default open loop map. The actual air/fuel ratio is not 13:1 but a little lower. Air/fuel meters are calibrated for gasoline and don't read quite accurately on alcohol, so you end up working with the numbers you get on the read-out.

Each manufacturer is somewhat different in how the programmable ECU screens are set up, but most will allow you to have an exhaust gas temperature window, manifold vacuum display, your fuel flow window (mpg), air/fuel mixture window, duty cycle (pulse width) window, and a spark advance window, all open at the same time. So you can see exactly what happens as you, say, bump up the pulse width a little, or advance the timing, while at wide-open throttle. For a gear-head, it's a dream come true. Using the computer to test the car at idle, off idle, mid-range acceleration, cruising, and wide-open throttle, you can create a map that you can save.

So, you save an alcohol high-mileage lean-burning map for highway use, an all-around good mileage map for mixed conditions of city and highway, maybe a high-performance rich-burning hotshot

LEFT: Fig. 16-13 Oxygen sensor fooler. This device installs under you dash. Intercepting the signal from the oxygen sensor, it sends a modified signal to the ECU permitting you to run a leaner air/fuel mixture than stoichiometric. It is adjusted by a small screwdriver turning the potentiometer at the upper right hand corner.

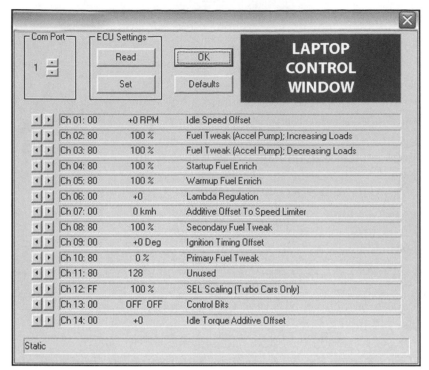

Fig. 16-14

Software control panel for hacking a car's computer.

map, and a ho-hum gasoline-burning map. When it comes to multiple maps and the ease of switching among them, different computers have different capabilities. They may require you to plug your laptop back in, or let you use a personal digital assistant (PDA), or even let you use an old-fashioned switch.

The drawback to having all this digital power at your fingertips is the cost. When I started researching this years ago, you could spend $1800 to $2000. Gulp! But in 2006, aftermarket ECU companies are springing up like mushrooms. There must be more than a dozen now, with many of them pricing their units at $500 and under. Some are for specific types of engines, and some are more general. They vary in how much control you have.

Most of what we really need for alcohol is good, multiple-phase air/fuel control, good spark advance control at different engine phases, and a good cold-start cycle that can be set to work with whatever system you want to install. The more advanced issue is whether you can burn leaner than stoichiometric, to get the best mileage from alcohol start-up, and not need to use an oxygen sensor fooler—which a fully programmable ECU will let you do.

There is at least one Canadian company that is claiming it will be coming out with a line of chips that will inexpensively (probably for less than $200) replace your original ECU master chip with a flexible-fuel version that is essentially indistinguishable from factory flexible-fuel vehicles. Keep an eye out.

HACKING YOUR CAR'S COMPUTER

Today's motorheads are into more than taller pistons and headers. Some of them are computer geeks who love to get in and mess with the factory programming of your car. You don't necessarily have to buy a full aftermarket computer to have control of your engine parameters. There are various hacker approaches that will work on various families of computer/fuel systems. The online user's groups for your car are a good place to start looking for these resources..

Here's an example of a panel that is part of software that taps into the VW and Audi computers after 2001. As you can see, you can easily increase fuel at start-up, which could eliminate the need for cold-start devices in all but the coldest climates. You can also see where you can control the amount of acceleration fuel to get better mileage or more horsepower. You can adjust ignition timing to advance it for alcohol. The primary fuel tweak controls the long-term fuel trim, and increasing it about 25% puts the air/fuel mixture in the range where the O_2 sensor can control the car.

Using this software, you could program your car to run on alcohol with no material costs whatsoever, as my student Patrik Jonsson did. You can also switch back to gasoline within a minute or two by plugging in your laptop and restoring the factory settings. If you find one of these hacker programs for your car and like how it works, please let us know at <info@permaculture.com>.

Now you know how to convert your engines to breathe properly and purr like tigers. But even the most powerful cat can have trouble getting out of bed when the weather is less than tropical. The same is true of your alcohol engine. So next, we'll address starting up when it's cold.

Endnotes

1. Forbes Aird, *Bosch Fuel Injection Systems* (New York: HP Books, 2001).

2. Aird, 29.

In Chapter 14, I touched on criticisms that alcohol can't start engines easily at lower temperatures. I promised more detail about cold-start systems with E-98 and other higher fuels, so here we go.

Cars running alcohol fuel need help to start in cold weather. You can solve the problem by adding volatiles to your fuel—either gasoline or **diethyl ether** made from alcohol (see sidebar on adding ether). Even better is to use a cold-start device.

By its very nature, fuel injection will start in colder weather than carburetion will—as low as

CARS RUNNING ALCOHOL FUEL NEED HELP TO START IN COLD WEATHER. YOU CAN SOLVE THE PROBLEM BY ADDING VOLATILES TO YOUR FUEL—EITHER GASOLINE OR DIETHYL ETHER MADE FROM ALCOHOL. EVEN BETTER IS TO USE A COLD-START DEVICE.

BOB FITCH PHOTO, <WWW.BOBFITCHPHOTO.COM>

Fig. 17-1 General Motors Brazilian cold-start system. *My left hand is pointing at the injector screwed into the intake manifold. My right hand indicates a small, cube-shaped electric fuel pump, which is supplied from the plastic reservoir with the "red" cap. If you look carefully, just above my right finger, you'll see the injector fuel line and clip where it attaches to the top of the fuel pump. In Brazil, drivers fill this reservoir with half gasoline and half alcohol.*

20 to 40°F without any aid. Lower temperatures will require a cold-start device unless you are using E-85 or alcohol with ether added to make it more volatile in cold weather. The winter blend of E-85 actually contains as little as 70% alcohol in cold Midwestern states, the rest being made up of gasoline or other petroleum volatiles. Since gasoline and/or natural gas condensates have a very low flashpoint, they will evaporate and explode when sparked at temperatures of 45°F below zero. When added to alcohol, these substances permit the engine to get enough vaporized fuel to start in most weather conditions without the use of a cold-start device.

Many of these substances are considered toxic waste by the oil or natural gas companies, and they are only too glad to dump them into the fuel during the winter. Sometimes, however, the oil companies go a bit overboard. In 2004, in South Dakota, they were disposing of so many volatiles in the E-85 blends that widespread vapor-locking was happening throughout the state *in the winter!*

If you don't care to use petroleum additives full time in your fuel just to provide for cold-starting help (and who could blame you), there are several external methods for cold-starting an engine.

ADDING ETHER OR OTHER FUELS

Truckers have used pure diethyl ether to start diesel engines for years, and their system will work on alcohol-fueled engines, as well. The ether comes in a metal bottle, similar to a small propane bottle, that can be screwed into an electric metering or solenoid valve wired to the vehicle key, or to a switch mounted on the dashboard.

In cold weather, you simply push a button that triggers the solenoid valve to give the engine a measured amount of ether. Be precise with the amount of ether, as too much can cause engine damage. Two cubic centimeters—about a thimbleful—seems right for four- and six-cylinder engines. Four cubic centimeters is plenty for eight-cylinder engines.

When ether leaves the metal bottle, it travels to your engine through a 1/8-inch tube connected to the manifold. The instructions in commercial kits recommend that you tap a hole into your manifold and thread in the injector so that the ether can be atomized directly into the manifold. The atomized ether is then sucked into the engine by vacuum. (If you do have to drill and tap a hole, remember to do this with the manifold off the engine, so that drill shavings can't get sucked into the engine.)

Fig. 17-2 Ether or propane cold-start. In either case, a small bottle of compressed starting fuel is mounted on the firewall. The electric solenoid valve opens when you turn your key to start, and the thermostatic switch senses the temperature of the engine is low. You get a shot of volatile starting fuel, and that starts the engine. Although this diagram shows fuel entering through a constant vacuum fitting on the carburetor, any constant vacuum fitting on the manifold will serve in fuel-injected systems.

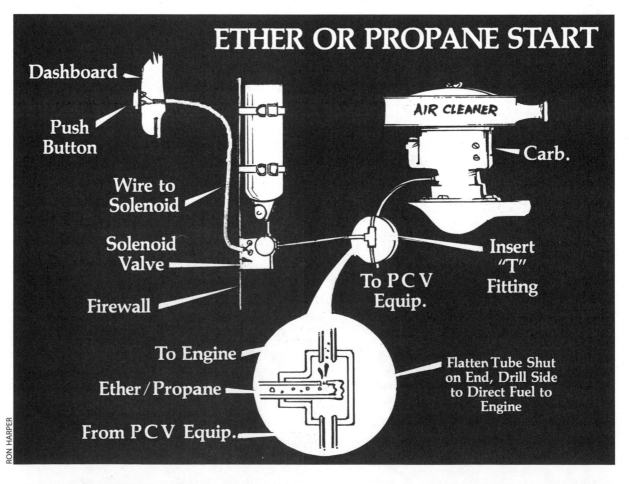

ETHER OR PROPANE START

Dashboard

Push Button

Wire to Solenoid

Solenoid Valve

Firewall

AIR CLEANER

Carb.

To PCV Equip.

Insert "T" Fitting

To Engine

Ether / Propane

From PCV Equip.

Flatten Tube Shut on End, Drill Side to Direct Fuel to Engine

RON HARPER

An easier way to get the ether to the manifold is to insert a T directly into a vacuum line. Your positive crankcase ventilation (PCV) line is an excellent line to tap for this purpose. This line is designed to continuously suck oil vapors from your valve cover or other crankcase location and shoot it back into your engine, where it's supposed to re-burn. (In much older engines, these oily vapors used to be vented to the air.)

Use hose barbs on the through line of the T, with a compression fitting on the T's branch. Insert your injector—a piece of tubing with one closed end and a tiny hole drilled in the other end—and seal it into the branch by tightening the compression fitting. Place the T as close to the carburetor as possible.

The kits come with a sophisticated atomizer that spreads the dose of ether over several seconds. This atomizer comes with threads, so you can screw it into either the manifold or a threaded T fitting.

If you want to spend more money and get a totally automated cold-start system, you can purchase a thermostatic switch. When the engine block water is above some preset temperature, usually 90°F, the thermostatic sensor switch stays open

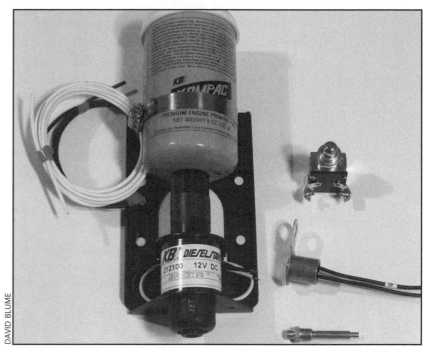

DAVID BLUME

Fig. 17-4 Commercial ether start system. Although commercial systems are more expensive, they deliver a measured amount of ether, which more or less idiot-proofs the system against flooding the engine with explosive ether. On the right (from top to bottom) is a momentary contact switch for manually starting, a thermostatic switch to be attached under any bolt on the block, and a very nice atomizer to be threaded into a drilled and tapped hole in the manifold. The roll of tubing on the left is used to connect the valve to the atomizer.

MATT FARRUGGIO

Fig. 17-3 Cutaway model of PCV-type cold-start fuel injector. This model is intentionally oversized to illustrate its construction. The tube should be ⅛ inch with the end soldered and drilled with a tiny hole (the same way this ⁵⁄₁₆-inch line has been constructed). The hole would actually point to the left, in the direction of the airflow, and the compression fitting would be a ³⁄₈-inch mip × ⅛-inch compression adapter. Hose barbs should be ³⁄₈ inch for most PCV lines.

and won't allow electricity to pass though and open the ether solenoid valve. When you turn the car key and the engine block water is below 90°F, the circuit is allowed to complete, and ether runs into the manifold.

This system is not manual. The thermostatic sensor is simply screwed into your block or water system and wired to both your key and the temperature sensor on the way to the ether cold-starter's solenoid valve.

Ether bottles are good for about 600 shots. On a very cold morning, your engine may require one or

Fig. 17-5 Propane cold-start system model. The normal propane torch nozzle is removed, and a solenoid valve is attached in its place and wired to the dashboard. A push of the button releases liquefied gas into the manifold to start the vehicle. As the pressure of the fluid drops inside the manifold, the propane should almost completely vaporize. Make sure the valve is a lock-off type that doesn't let propane seep when closed.

more shots before the engine starts, but ether is the surest of engine starters. Even in cold climates, you

ADDING ETHER TO DENATURE ALCOHOL

If we follow the route of adding a volatile to alcohol, for cold-starting or denaturing, there are much better solutions than gasoline or isopentane. Diethyl ether is even more effective, ounce for ounce, than isopentane, if added at low levels; 6% ether has been found to be as effective as 10% isopentane. The nice thing about ether is that it's often produced from alcohol.

In the ASTM specifications for E-85, ethyl ether is allowed, so technically you should be able to meet these specifications without using gasoline. Older, high-quality chemistry books will tell you how to make it. Ether is extremely explosive, and you must be extremely careful about its production. So this tends to be a "don't try this at home, kids" sort of thing.

Low percentages of ether will raise your combustion temperature slightly but still below gasoline's combustion temperature. Because of its high oxygen content, ether also has a slight emission-reducing effect on ethanol's already tiny tailpipe emissions (whereas isopentane increases them).

Although you are unlikely to actually work with ether, it is usually stored in five-gallon steel drums, in a well-ventilated place, preferably a shady spot outdoors. Ether should be pumped at a cool, even cold, temperature—using a totally enclosed explosion-proof pump—into cool alcohol. Never pour it into the alcohol through the open air. Inhaling ether can anesthetize you before you know it. And long exposure to ether vapors can kill you.

Pumping liquid ether into liquid alcohol, however, should eliminate the hazards related to vapor releases. Ether fires will burn fiercely but, like alcohol fires, can be extinguished with water.

As the lawyers say, "Tell them they should leave this to professionals." Consider yourself warned.

may have to cold-start only once in the morning, since the engine will often be warm enough between errands. One bottle should last four to six months. Replacement cost for a bottle of ether is no more than a few dollars; keep a spare bottle in your trunk. Bottles can be stored easily and safely as long as you keep them away from heat above 250°F.

Retail prices for basic ether cold-starting systems can run around $200. Do-it-yourselfers can save half this price by buying all the parts separately.

Propane bottles also fit on electric ether cold-starters, but they cause the seals to swell, jamming most common solenoid valves. Manual valve kits operated from the dashboard by means of a cable and lever accept either propane or ether bottles, but propane will wear the valve out sooner, and isn't as effective as ether for starting.

Electric solenoid valves used on propane fork-lifts are suitable, since they can be relied on to fully close when off. Normal solenoid valves tend to allow propane to seep.

Using ether can mean avoiding petroleum alto-gether—diethyl ether can be made from ethanol! (See sidebar.)

Old fuel injection systems had an extra fuel injector mounted right in the manifold to deliver extra fuel during start-up. You can hijack this sys-tem by installing your own fuel tank and electric fuel pump, and wiring it to either a button or the ignition. You can then mix 50/50 gasoline and alcohol, or alcohol with 15% ether, and start in almost any weather.

You can also use an electric pump with momen-tary contact (an electrical switch that works only when it's pressed), instead of a mechanical pump, to transport starting fuel to the injector. You can use an electric thermostatic valve like the one described for the ether bottle method with this sys-tem, as well. It's important to use a fuel pump that closes completely when not actually on; otherwise, the starter fuel reservoir could be sucked dry by the manifold vacuum.

If you think this cold-start system is some kind of flaky hippie thing, think again. This is exactly what car companies do today on the assembly line in Brazil. They use a plastic reservoir similar to your windshield washer container, a tiny one-inch cube-shaped fuel pump, and a line right into the manifold.

Another excellent and inexpensive method for cold-starting is to attach a small tank from a lawn mower under the hood. Run a line from the tank to a mechanical fuel pump bolted to your firewall, with the pump lever protruding through the wall into the passenger compartment, somewhere near your brake pedal. Then, run a line from the pump outlet to a simple sprayer, similar to the injector described for the ether line above.

The fuel gets sprayed into the PCV line, or mani-fold fitting, to be sucked into the manifold. Sim-ply stepping on the lever a couple of times will squirt the solution into the manifold and start the engine. If you drill the hole good and small, and use ⅛-inch tubing, the fuel will spray several sec-onds as the pump unloads its fuel into the line (see Figure 17-7). A fuel filter should be interposed in the line between the pump and the injector to

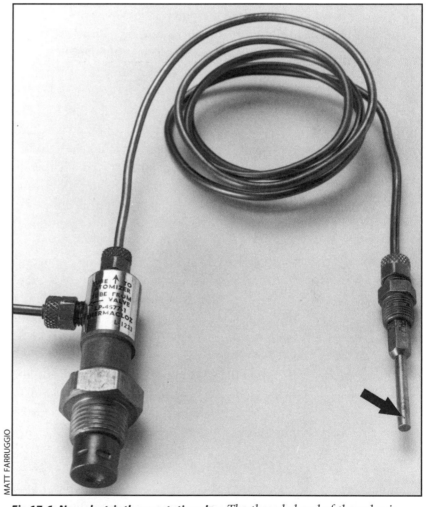

MATT FARRUGGIO

Fig 17-6 Non-electric thermostatic valve. *The threaded end of the valve is inserted into any convenient fitting on the engine cooling jacket, or at the point in the thermostat housing where water temperature is sensed for your dash thermometer (idiot light). The valve remains open until water tem-perature reaches 90°F. Cold-start fuel in a cold engine will pass through the valve to the injector. If your engine is warm, no cold-start solution passes on to the injector. This is a really nice way to go if you are using the mechanical fuel pump. Note the very small size of the injector hole (see arrow).*

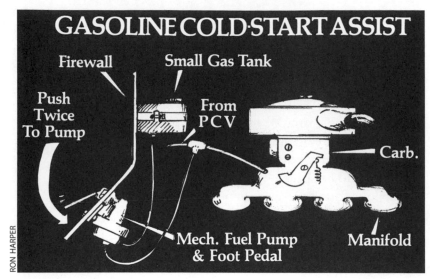

GASOLINE COLD-START ASSIST

Firewall   Small Gas Tank

Push Twice To Pump

From PCV

Carb.

Mech. Fuel Pump & Foot Pedal   Manifold

Fig. 17-7 Liquid fuel cold-start system. Seen from a side view, the mechanical fuel pump system is simplicity itself.

RIGHT: Fig. 17-8 Liquid fuel cold-start with a mechanical fuel pump. The cheapest of all methods, the mechanical fuel pump costs about $10, as does a used lawnmower tank. A piece of rod is attached to the pump, which extends through the firewall near your brake pedal. Fuel is then pumped from the tank above through the 1/8-inch white plastic line, to an injector in the manifold. One or two pumps with your foot and off you go.

prevent plugging. This was the system I used on all my vehicles back in the 1980s.

Never attach a sprayer to the top of the air cleaner pointing at the carburetor, or, in the case of fuel injection, into the air intake. Although this modification is recommended by many older alcohol fuel books and pamphlets, it can cause a terrible fire if the car backfires while you're squirting fuel into the airstream, at its furthest from the engine. So, make sure it is being sucked or injected right into the manifold. (Besides, if the car is cold, either the choke, MAF sensor, or throttle plate will obstruct the fuel spray from going down the airstream.)

Thurly Heintz (see Chapter 27) came up with a cold-start innovation that works well with his motorcycle and with cars, and requires no mechanical or electric fuel pump at all. He took a tank from a Coleman lantern, which has a hand pump built into the tank housing. Filling the tank with cold-start fluid, he pumped it up with about 50 strokes. Then he opened the fuel valve, and the fuel went through a line to the manifold, courtesy of the tank's air pressure. The tankful lasted him six months on his motorcycle, and needed no further pumping.

OTHER METHODS

Another cold-start system uses a diesel glow plug as a heater. A glow plug generally looks like a bolt with a wire coming out of it. The end of the "bolt" is the heater, which is heated internally by electricity. The threads of the "bolt" are screwed into the side of the carburetor float bowl through a drilled and tapped hole (see Chapter 15, Figure 15-3). Insert the glow plug in a place that won't interfere with the carburetor float(s). The

glow plug is essentially an electric heater and, as such, can bring the alcohol in the float bowl up to 160+°F. Alcohol this hot atomizes readily in the venturi, and should stay hot enough to get to the cylinders.

This method requires 60 to 100 seconds from the time you turn on the glow plug switch until everything is hot enough to start the engine successfully (which means it's not recommended for early morning bank robberies requiring quick getaways.) The nice thing about it is that you don't need to use any gasoline or external systems to get started.

Mounting glow plugs in your manifold where a cold-start injector can spray straight alcohol on it permits you to start without using any gasoline.

Another cold-starting system is to wrap your fuel injection rail with a heating strip like the ones used in well houses for preventing pipes from freezing. The trick here is to find one rated for use with 12 volts, or you'll probably have to install a 12-120 volt inverter.

Some of my students use fuel line heaters (from tractor or diesel truck supply stores) to heat the fuel electrically as it enters the carburetor or

fuel injection rail. One student told me that the Cummins Company heater from the 1980s works particularly well. A line heater is not as effective as a glow plug for cold-starting, and is best used in conjunction with the standard dipstick heater that is common in the Midwest and Northeast for exceptionally cold mornings.

Synthetic oil that is rated at 5w–40w also aids in cold-starting, since initial 5w viscosity hardly impedes the engine as it turns over.

A new, promising approach to cold-starting is the heating of the fuel injectors themselves. A patent application currently held by Chrysler seems like an elegant solution to the cold-starting issue. Basically, a layer of semiconductor material is built into the fuel injector. When given electricity, semiconductors can generate intense heat—think how much heat the cooling fans on your computer have to carry away. Before the fuel leaves the injector, the semiconductor both filters it and nearly instantly heats it. The semiconductor has many tiny channels that maximize the surface area and can heat the fuel in ten to 20 seconds by as much as 100°F. What is even more beneficial with this

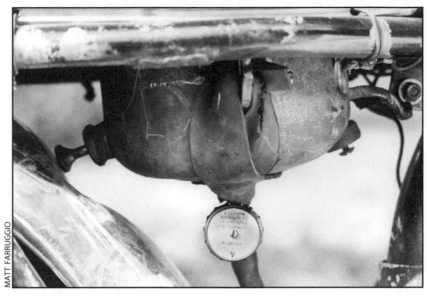

MATT FARRUGGIO

Fig. 17-9 Motorcycle cold-start system. *Thurly Heintz's ingenious adaptation of a Coleman lantern tank allows him to simply open the valve to deliver a pressurized stream of fuel to his manifold. The tank is flexibly mounted under the seat with inner-tube strips. The tank is removed from the straps to fill.*

approach is that the semiconductor material can tell the computer how hot the fuel is, working like a sensor for precise control.

We have asked Daimler/Chrysler if it plans to go into production of this injector, but, as of this writing, we haven't heard back. This would be a very hot idea, indeed.

Send a photo and description of your ingenious cold-start idea to me at <info@permaculture.com>, and I'll publish it online.

Although you could consider your conversion finished—now that you've optimized the air/fuel mixture and can start your vehicle in any weather—there are more tricks of the trade to learn. Since you are working with a race-quality fuel, you'll want to learn to tune your engine to take advantage of alcohol's high octane. Next, we'll see just how your engine's ignition system can be adjusted to get the most from your fuel.

WHY SOME CARS COLD-START BETTER THAN OTHERS

For a long time, I couldn't figure out why some fuel-injected engines start at really low temperatures and some start only at higher temperatures. Dr. Josmar Pagliuso, a veteran Brazilian alcohol researcher, solved the mystery for me.

He told me that he and a driver of a car of a similar year were parked side by side one cold morning, and one car started easily with no cold-starting fluid, and the other wouldn't. This remarkable comparison led him to look further into the operation of the two engines.

The one that started most easily on cold mornings had fuel injector timing that sprayed fuel past the open valve into the cylinder. The other car had a design that sprayed the fuel against the back of the valve, which, after the engine is running, is a very hot surface. But when cold, it literally blocked most of the fuel spray from making it into the cylinder.

Fig. 17-10

Cars used to have distributors to control the times at which spark plugs fire, and to direct electricity from the high-energy coil to the plugs. Like carburetors and fuel injection systems, distributors evolved from mechanical to electronic, and now have almost completely disappeared, with their functions being absorbed into the vehicle's computer. Over the next 20 years, new vehicles will do away with spark ignition altogether as they go to vaporized fuel in designs of homogeneous charge compression-ignition engines. But, as with carburetors, there are still all sorts of spark distribution systems on millions of vehicles on the road today.

HISTORY

Twenty years ago, a distributor was a mechanical device. In many ways, it was only a little more complicated than the same device found on the Model A. As we saw in Chapter 1, that original distributor counted on you, the driver, paying attention to power and the sound of the engine to digitally (using the five digits on your left hand) adjust spark timing. Over time, distributors have evolved to take care of these adjustments automatically.

Until recently, distributors were powered directly from internal gearing of the engine and spun as fast as the engine. Points, which were essentially switches for opening and closing the electrical circuit, were once strictly a mechanical affair.

Changes in the spark timing with different engine speeds and conditions were accomplished by a diaphragm and lever arrangement, which varied with engine vacuum, and a set of whirling weights that changed the setting depending on how much centrifugal force was created by their spinning.

All of this was done based on abstract projections worked out in advance, which were averages of safe but not necessarily efficient settings. With the advent of electronic distributors, engine sensors could provide information to more precisely

WHAT WE WANT TO DO WHEN CONVERTING TO ALCOHOL IS TO START THE SPARK FIRING MUCH EARLIER THAN WE WOULD WITH GASOLINE. THIS HAS MANY ADVANTAGES.

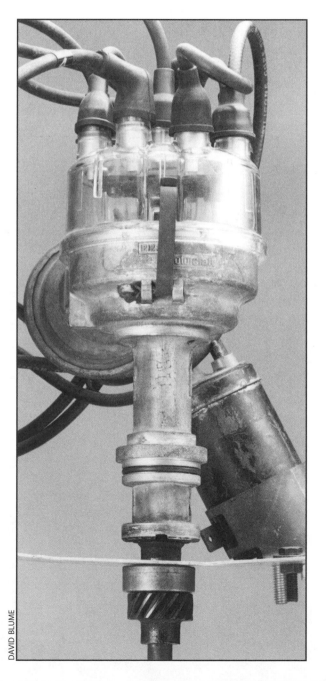

DAVID BLUME

Fig. 18-1
Distributor. *The spark plug wires are connected around the rim of the distributor cap, while the coil wire enters in the center. You can just see the rotor inside the clear cap. The high-energy spark coil is behind on the right. At the bottom is the helical gear that drives the rotor.*

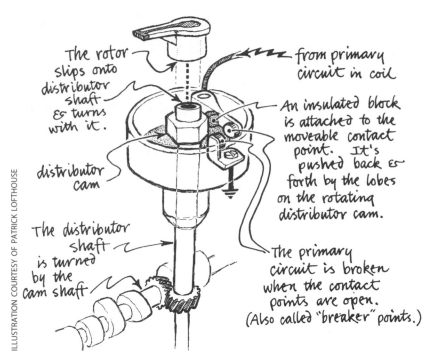

ILLUSTRATION COURTESY OF PATRICK LOFTHOUSE

The rotor slips onto distributor shaft & turns with it.

from primary circuit in coil

An insulated block is attached to the moveable contact point. It's pushed back & forth by the lobes on the rotating distributor cam.

distributor cam

The distributor shaft is turned by the cam shaft

The primary circuit is broken when the contact points are open. (Also called "breaker" points.)

Fig. 18-2 Basic distributor. *The switch that turns the spark on and off is known as the* **breaker points** *(as in circuit breaker). When the points touch, low-voltage (12V) electricity crosses and signals the release of high-energy electricity from the coil. The coil electricity enters the top center of the distributor to the top of the rotor. Spark then jumps across, from the rotor to a contact for a spark plug on the rotor cap (not shown). In this drawing, the distributor is sending spark to a six-cylinder engine; that's why the distributor* **cam** *has six sides.*

and constantly adjust timing. Now distributors and their functions have largely been taken over by the vehicle computer, using sophisticated sensors and rapid feedback on many actual engine conditions.

MAKING TIMING CHANGES

I am going to start by showing you how to handle the basic timing changes with mechanical distributors, and then follow up with more modern methods of ignition timing and their modifications. Even the most basic system can be upgraded with new aftermarket ignition products.

In general, what we want to do when converting to alcohol is to start the spark firing much earlier than we would with gasoline. This has many advantages. Gasoline burns in a very narrow range of air/fuel mixtures that, of course, change rapidly as fuel is consumed during the explosion in the engine. This means that, in most cases, gasoline must be ignited just as the piston reaches the top of the cylinder, or just a very short time before. If the spark is fired too early, gasoline will combust abnormally and cause engine-damaging pinging, as well as waste a lot of its energy as partially combusted fuel.

Fig. 18-3 Model A "digital" distributor. *The distributor is the black device on top of the center of the engine. Look closely, and you'll see the copper straps lead from it to the top of each spark plug. The rod going down and to the right (see arrow), through the firewall, was attached by a linkage to the spark advance level on the steering column. Digital? Yes, you used the digits on your left hand to adjust the timing.*

MATT FARRUGGIO

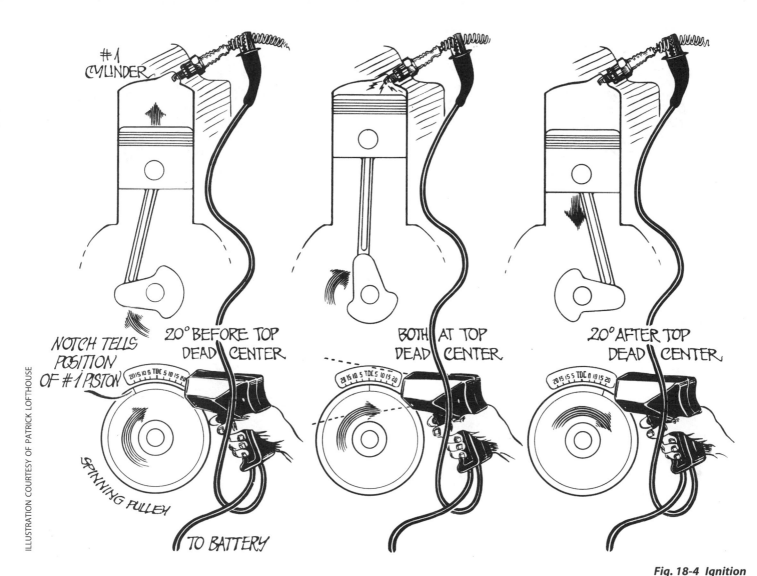

#1 CYLINDER

NOTCH TELLS POSITION OF #1 PISTON

20° BEFORE TOP DEAD CENTER

SPINNING PULLEY

TO BATTERY

BOTH AT TOP DEAD CENTER

20° AFTER TOP DEAD CENTER

Mechanical Systems

When the piston is at the top of its compression stroke, that is considered 0° or top dead center, so the piston and connecting rod are pointing straight up. The connecting rod angles one way before it reaches the top and angles the other way on its way back down. Its angle is measured in degrees from top dead center.

Older gasoline-powered cars had a very narrow range of acceptable timing settings—0 to 8° **before top dead center (BTDC)**. This produced peak cylinder pressures at about 12–14° after top dead center.

With gasoline, you can't deviate from ideal specifications by more than one to three degrees; if the timing varies any more than that, a gasoline engine will run terribly, if at all. Alcohol is much more forgiving. If you aren't on optimal timing for alcohol, you may notice only a slight drop in mileage or perhaps a little trouble starting. In many cases, alcohol operates at a range of initial timing settings from 0 to 25° BTDC at idle.

Ignition timing controls the point in time when spark plugs are ignited to fire your mixture in the cylinder. Since it isn't possible to see the plug fire in the cylinder, you need a **timing light**, which flashes like a strobe every time your #1 spark plug fires. Shine this strobe light on the fan belt pulley or the **harmonic balancer** in the area adjacent to the timing plate. The timing plate is numbered a few degrees on either side of zero. The strobe light illuminates the spinning fan pulley every time that first plug fires, so you'll be able to see where the timing mark on the pulley aligns with the marks on the block-mounted plate.

To change the position of the timing mark to a specified number for your tune-up, loosen the hold-down bolt at the distributor's base and turn the whole distributor slightly, either clockwise or counterclockwise (see Figure 18-5). One direction advances the timing (firing earlier), and the other retards it (firing later). Once you've lined up the mark where you want it, tighten the hold-down bolt.

Fig. 18-4 Ignition timing. *Using a timing light— which shines on a scale each time the #1 sparkplug fires—tells you when in the cycle the spark ignition is timed. When the piston is at the very top of its stroke, it's at top dead center. Fired while the piston is still traveling up is before top dead center. Fired after the piston is on its way back down is after top dead center.*

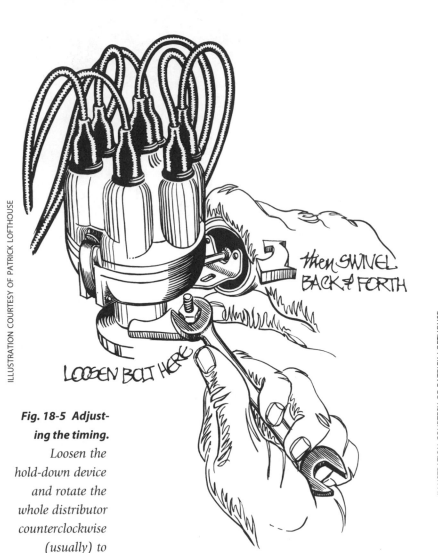

then SWIVEL
BACK & FORTH

LOOSEN BOLT HERE

Fig. 18-5 Adjusting the timing.
Loosen the hold-down device and rotate the whole distributor counterclockwise (usually) to advance, clockwise to retard, the timing.

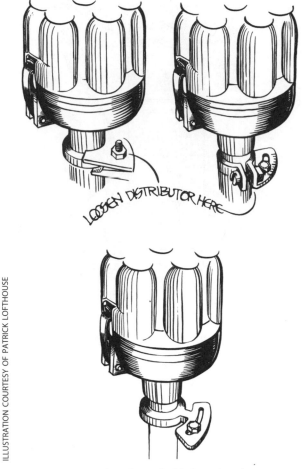

LOOSEN DISTRIBUTOR HERE

Fig. 18-6 Various distributor hold-down brackets.
Different manufacturers use different methods to hold the distributor in place once the timing is set.

MARK POSITION OF NOTCH *in* RELATION TO DISTRIBUTOR RIM

MARK POSITION OF DISTRIBUTOR BODY IN RELATION *to* ENGINE BLOCK

OR MARK POSITION OF VACUUM ADVANCE *in* RELATION *to* ENGINE BLOCK

Fig. 18-7 Keeping things straight.
Use any of these marking methods to indicate the proper setting for gasoline and for alcohol.

DAVID BLUME

Your timing plate may not have enough marks to measure how much you want to advance your timing for alcohol. A **timing tape** (available at speed shops) may be appropriate for anticipating correct timing. It's a metallic or paper tape with markings up to 360 degrees, calibrated for the diameter of the balancer or pulley on your engine. You attach it with adhesive to the balancer or pulley. Read the tape against the zero mark on your timing plate.

If you lack a timing tape, you can set your present timing mark at point zero and, using the timing gauge on the engine as a degree ruler, paint a new mark on the pulley at 10° BTDC. Rotate the pulley until the new 10° mark is on zero, and mark off another 10° from which you can work. You can use either mark to get you into the ballpark, and then use the marks on the timing plate for more exact readings between 10 and 30°.

To determine the proper alcohol timing exactly, start by gradually advancing the timing to 15° at idle without the vacuum advance. Reconect and test-drive at normal speeds. To disconnect the vacuum advance, remove the vacuum hose from the distributor and plug the hose with a pencil or bolt.

If your vehicle misfires, pings, or loses power when it gets up to 65 or 70 mph, you've advanced the timing too far. The idea is to keep advancing the timing (at idle) a degree or two at a time, until you hit the exact point at which you begin to lose power, or misfire while accelerating to high speeds. When this happens, back up the timing a minimum of two degrees from the point at which you first heard the misfire (inaudible misfiring occurs a degree or two before you can sense it).

There are electronic gauges that use a **knock sensor** to listen to the engine and signal you with sound and/or lights when you begin to have inaudible

pinging. This sensor is a microphone that can hear inaudible knocking (preignition). If you are part of a co-op, this should be one of the community tools to lend when first converting to alcohol.

The higher your vehicle's compression, the less ignition advance is required. This is because flame propagation is more rapid when the fuel mixture is compressed and turbulence is enhanced. For example, cars with 8.5 to 1 compression ratios typically require 15 to 25° initial timing advance. You may go as high as 50 to 60° overall timing advance during the accelerating cycle, due to vacuum advance. Vehicles with turbocharged or supercharged engines, high-energy ignition, multi-spark ignition, or some types of newer precombustion spark plugs, need even less overall advance.

On some cars, in order to advance timing 15 to 20°, you have to loosen the distributor hold-down bolt, and remove it along with the hold-down plate to free up the entire distributor. A slight counterclockwise twisting motion allows you to lift it out and replace it one drive-tooth back, counter-rotational from the direction the rotor spins (most rotors spin clockwise). This gives you an automatic adjustment of 10° or so, depending on the number of teeth on your distributor drive gear (see Figure 18-1). Make sure you mark the distributor plate and engine block so that you know

Fig. 18-8 Aftermarket knock detector. This meter and the microphone below it can be used to determine the best settings for an aftermarket ignition-timing device. Once you've got your best average settings for gasoline and alcohol, you don't need this any more and can lend it to someone else in your fuel co-op.

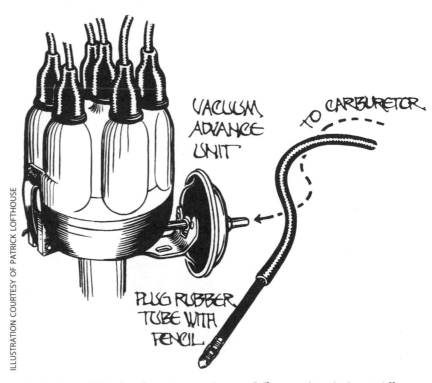

ILLUSTRATION COURTESY OF PATRICK LOFTHOUSE

Fig. 18-9 Deactivating the vacuum advance. When setting timing at idle, you must deactivate the vacuum advance to get an accurate reading. Be sure to plug the hose, or you'll be creating a giant vacuum leak.

Fig. 18-10 Electronic timing advance. *This electronic device or ones like it can be used with mechanical distributors and electronic distributors to permit the driver to have control over ignition timing. You can advance timing as much as 15° without having to make any under-the-hood adjustments. It lets you go back and forth between alcohol and gas, and have optimum timing for both.*

the original position of the distributor, in case you need to return to original timing (some distributors are not constructed to permit this).

A much easier way to get an additional 15° range without changing the position of the distributor is to use an aftermarket ignition timing control. A knob on the dash will give you about 15° of control over spark advance, and will allow for more easily converting back to gasoline if your car has a non-computerized distributor (see Figure 18-10). Tow trucks sometimes use aftermarket ignition timing controls to increase horsepower under load. These devices work on mechanical distributors and also on many electronic ones.

High-energy ignition systems designed to break the spark into smaller sparks don't require much adjustment. They're excellent systems for alcohol fuel because they speed flame propagation, so you'll need less spark advance.

If you anticipate switching back and forth from alcohol to gasoline, you might choose to leave your timing at the gasoline standard and suffer mileage loss when running on alcohol. Or you can advance the timing somewhat, but use a good water injection system while you're running on gasoline to hold down pinging. Another possibility is to mark

the distributor position for using gasoline on your adjustment plate, and then move the distributor to the proper location for using alcohol (which you've also marked).

If you expect to use different percentages of alcohol (a car can run on 175 proof or higher), I recommend setting the ignition timing for the highest proof. Since low proofs contain more water than higher proofs, their burn time is lengthened, and they thus benefit from advancing the timing. But timing that is advanced for lower proofs will cause misfiring (pinging) when used with higher-proof alcohol.

All this is a bit bothersome compared to the various adjustable electronic ignition systems that have come along since the '80s. But if you are working on a 1960 Chevy half-ton pickup truck 190 kilometers from Managua, Nicaragua—well, you might just have to put up with a little bother.

Electronic Systems

Once you get beyond the purely mechanical distributor systems, a wide array of electronic systems makes it easy to tune ignition timing for almost any conditions. I'd divide these newer systems into two rough categories: those that use knock sensors, and those that don't.

Soon after electronic ignition became popular, various programmable systems were developed and adopted by people who like to race. With the advent of electronic components, it became possible to advance or retard ignition timing so as to optimize power at various phases of acceleration. This capability meant that engines could dramatically advance ignition timing, a boon to alcohol users.

Early systems, such as the simple MSD Ping Control, actually allowed you to dial in your advance manually. Using a knob on your dashboard, you could choose up to a 15° advance over stock timing. Similar products are on the market today. The idea is that you can manually advance your timing until you hear some pinging, or see it displayed on a knock sensor, and then back it off a couple of degrees on the fly.

Feedback systems also started to become available early in the 1980s. Since inaudible pinging occurs a couple of degrees before you can actually hear it in the engine, feedback systems do automatically what your ears and brain tried to do manually. They use what is essentially a special microphone to sense the frequencies of the first

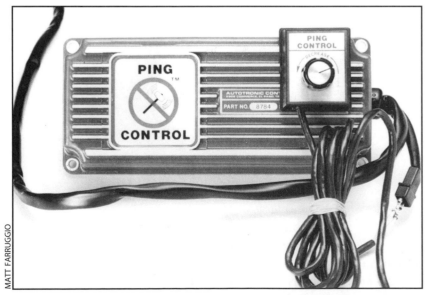

MATT FARRUGGIO

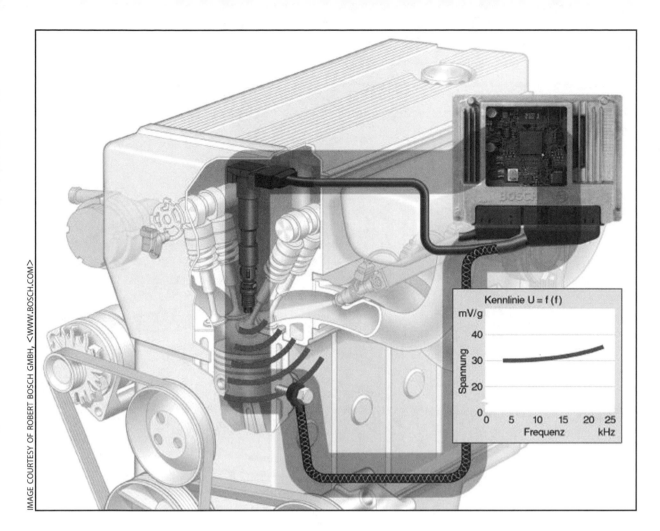

Fig. 18-11 Knock sensor. *A microphone tuned to listen for engine detonation advances ignition timing until inaudible knocking is detected. The information is sent to the vehicle's computer, which then retards spark ignition and/or turbocharge boost.*

incidents of preignition, and they back the timing off 2–20° from there. These systems are constantly advancing the timing, testing the limit, and backing off—keeping the engine always at the optimal spark timing.

Over time, feedback systems began to be built into electronic ignition systems right from the factory. Now, most modern ECUs incorporate this capability into the software that runs the car.

For older cars, automatic aftermarket ignition systems with knock sensors are available. Although they tend to be a little pricey ($500 or so in 2006), the savings in fuel economy, the lower emissions, and the convenient automatic operation are easily worth it, if you plan to keep the car.

In choosing an aftermarket timing control system, get the specification on what range of advance is possible. Some of these units don't have a wide enough range (look for at least 15°, and preferably more) to adjust for alcohol and gas. Although you don't need to have this full range if you are only running on alcohol, it's obviously important if you want the system to automatically accommodate a switch back to gasoline. This range is standard on flexible-fuel vehicles. For instance, my Ford Ranger,

under acceleration, uses 40° of advance on gasoline, but 60° on alcohol. There's a nearly 15° difference at idle.

If you go for a higher-priced, programmable aftermarket ignition system, you can develop maps for differing levels of alcohol, and for different performance. Maximum advance gives you the best emissions and often the best mileage. But alcohol reaches its peak torque earlier than gasoline. So, these ignition maps can be made more precise than simply making an across-the-board increase of, say, 20 degrees. Given how inexpensive some of the new programmable ECUs are getting to be, it starts to look cost-effective on later-model vehicles to use one to do both your air/fuel adjustments and your ignition timing, rather than buying separate devices for fuel and ignition.

We've now covered the big three issues in conversion—air/fuel mixture, spark timing, and cold-starting. Although these alone are enough to get you running on alcohol, there's more you can do to take advantage of alcohol's superior qualities.

Fig. 18-12

CHAPTER 19

ASSORTED ADJUSTMENTS

This chapter describes several tricks of the trade to help your vehicle take advantage of alcohol's characteristics. These modifications can even be done (for the most part) on flexible-fuel vehicles, to squeeze more mileage out of your fuel. On older carbureted or throttle body fuel injection (TBI) vehicles, some of these techniques can make the difference between a smooth, reliable conversion or one that is balky. So the details do matter.

PREHEATING FUEL

The major disappointment in the design of the internal combustion engine is that it doesn't burn "large" droplets of fuel efficiently. Only the drop's surface actually burns, leaving a great volume of fuel unburned or partially burned. This results in waste heat and increased emissions. For greater efficiency, smaller droplets are required. Theoretically, the best results possible would be obtained by burning individual molecules, such as in a vapor. We discussed vaporization in Chapter 13 and will have more to say on it later.

But simply preheating, even without vaporizing, is helpful. Preheating reduces the surface tension of large drops, letting them break up into smaller droplets. Fuel is usually heated through the exchange of heat from hot radiator water, exhaust, or engine oil.

Carburetors

When fuel is heated, it leaves the carburetor or TBI better atomized and at a higher temperature to withstand evaporative cooling in the manifold. When vehicles have carburetors or TBI, the fuel's trip through the manifold tends to cool the fine spray on its way to the engine. It does so by absorbing the spray's heat and reforming the fog into large drops, which are often deposited on the manifold surface and never reach the engine as a burnable spray.

THIS CHAPTER DESCRIBES SEVERAL TRICKS OF THE TRADE TO HELP YOUR VEHICLE TAKE ADVANTAGE OF ALCOHOL'S CHARACTERISTICS. THESE MODIFICATIONS CAN EVEN BE DONE (FOR THE MOST PART) ON FLEXIBLE-FUEL VEHICLES, TO SQUEEZE MORE MILEAGE OUT OF YOUR FUEL. ON OLDER CARBURETED OR TBI VEHICLES, SOME OF THESE TECHNIQUES CAN MAKE THE DIFFERENCE BETWEEN A SMOOTH, RELIABLE CONVERSION OR ONE THAT IS BALKY. SO THE DETAILS DO MATTER.

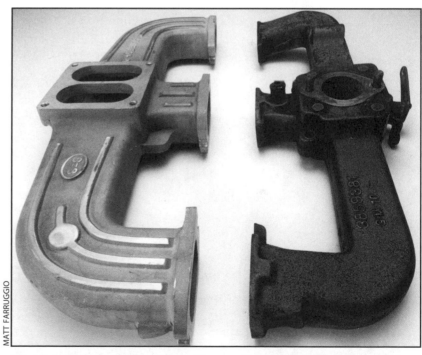

MATT FARRUGGIO

Fig. 19-1 Equal-feeding manifolds. *The manifold on the left has been engineered so that the cylinders on the end get the same amount of air/fuel mixture as the cylinders in the center. You'll also note that the turns are smooth, which can make a world of difference on a carbureted or TBI vehicle.*

When the fuel spray falls out of the air and onto the walls of the manifold as a liquid, it doesn't just stay there. The rushing air heading for the engine causes the liquid to evaporate to a vapor (not an atomized spray). If the engine were operating at a steady speed, the amount of fuel evaporating would equal the amount of fuel dropping out of spray to liquid—i.e., equilibrium. But, in reality, engines don't run at a steady speed. So when you accelerate, a lot more fuel drops out of the spray, and when you are idling and cruising, less fuel drops out of the spray. Overall, equilibrium happens, but there are times when more fuel is dropping out than evaporating and other times when the evaporation catches up to the fuel puddling (see Chapter 15, Figure 15-15).

Today's port fuel injection systems have largely eliminated most of the issues of manifolds, since they carry only air and not fuel. Since the fuel does not travel through the manifold, the atomization enhancement by heating the fuel is more likely to be maintained all the way into the engine's cylinders.

Gasoline can't be heated for better fuel efficiency. It is made up of a combination of all kinds of gums, tars, varnishes, solvents, etc., with boiling points ranging from 80 to 400°F. Heating gasoline with a standard carburetor would cause vapor lock in the fuel lines or pump, sudsing, a loss of power, engine overheating, and carbonization.

But because alcohol boils at 173°F, you can usually safely heat it up to about 160°F before it enters the carburetor. My earliest fuel-heating experiments employed a style of fuel heater that

was common during the fuel-rationed era of World War II, when inventive citizens discovered that heated kerosene could run gasoline engines (kerosene—essentially diesel—wasn't being rationed). The top (hottest) radiator hose was sliced open and a couple of inches of hose removed, leaving a gap. A piece of copper pipe with the same outside diameter as the hose's inside diameter was cut about five inches long. A copper fuel line was wrapped around the center of the pipe several times, and affixed by solder. The pipe was then inserted into the radiator hose and clamped at each end to prevent leaks. Radiator water ran through the pipe into the radiator and heated the fuel line attached to it.

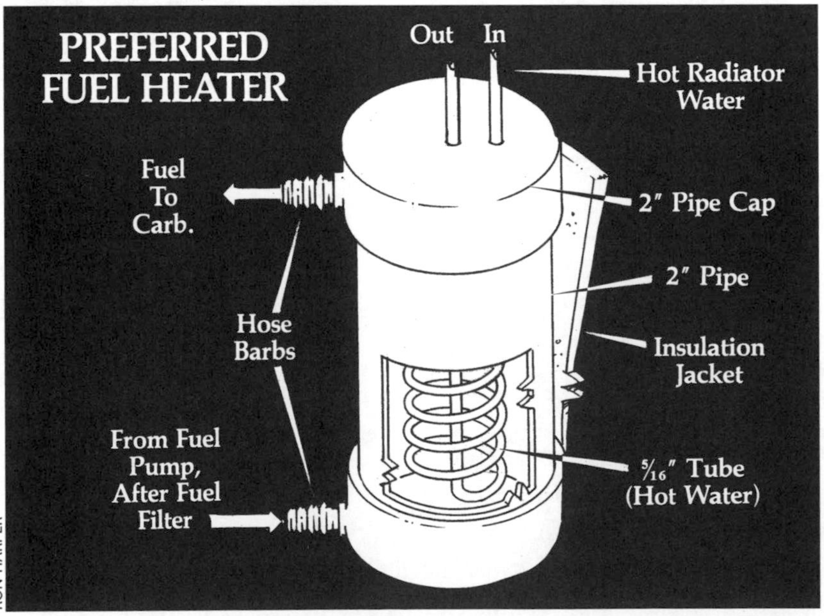

This procedure made the thick kerosene hot enough to atomize and burn after a vehicle was warmed up using rationed gasoline. It works well with alcohol, too. Wrapping the fuel line three times around the pipe is sufficient to raise alcohol's temperature to 150+°F.

A warning: When you're using a World-War-II-style fuel heater and you turn the engine off, heat continues to permeate the radiator hose. In fact, heat often goes way up in the minutes following engine shutdown, actually boiling the fuel. That pressure can force the fuel past the float needle(s) to flood the carburetor. If that happens, it's entirely possible that overflowing fuel will run down the cylinder walls and contaminate your oil.

To get around that potential, I redesigned the fuel heater so that a small flow of water ran through a coil in a large reservoir of fuel (see Figure 19-2). While the car is running, the fuel maintains heat because of its long contact time in the reservoir. When the car stops, the coil doesn't transfer enough heat to boil the fuel, so the float valve isn't forced open. The redesigned heater has the additional advantage of employing engine coolant from the heater hoses, which have a continual source of recirculating block coolant. This is much more efficient than a radiator hose, which is subject to the intermittent action of a thermostatic valve releasing water from the block.

If you have an air-cooled engine, like a Volkswagen, its risers are a good heat source. You'll have to experiment to find the number of coils to expose to this source. When the engine is fully warmed up, shut it off, pull off the fuel line, take an alcohol fuel sample, and check the temperature. It should be about 160°F. Generally, two or

Fig. 19-2
Preferred fuel heater for carburetors.
This design, heating a large amount of alcohol with a small amount of engine coolant, avoids carburetor flooding.

PREFERRED FUEL HEATER

Out In

Hot Radiator Water

Fuel To Carb.

2" Pipe Cap

2" Pipe

Hose Barbs

Insulation Jacket

From Fuel Pump, After Fuel Filter

5/16" Tube (Hot Water)

RON HARPER

three wraps of copper line around the risers are sufficient. Or wrap and solder your fuel line with perforated copper plumber's strap, bolting the other end of the strap to a hot part of the engine so the strap conducts heat to the line—a crude-looking, but quite effective, system for any air-cooled engine. Air-cooled engines can also heat their fuel using engine oil.

Fuel heating is generally a very safe procedure. The only real safety consideration is that the copper fuel line not extend as one piece from the fuel pump through the heater to the carburetor—copper is susceptible to fatigue from vibration. Always connect the fuel lines at the fuel heater with short pieces of neoprene fuel lines in front of and behind the heated portion, to build in some flexibility and shock absorption.

Fuel Injection

Let's divert ourselves from carburetors and talk about heating fuel in fuel-injected engines—a tricky, though rewarding, process. Because alcohol is several times as viscous as gasoline, it takes more pressure to get alcohol through the same size hole or line compared to gasoline. So, one of the benefits of heating ethanol is that it becomes more fluid, and the delivery of properly heated fuel through your injectors increases by as much as 25%. Reducing viscosity may mean avoiding having to switch to larger injectors or having to use an aftermarket fuel pressure regulator.

The high pressure of fuel injection allows you to pack more heat into the fuel without it boiling. So, on the surface, a simple counterflow heat exchanger seems as if it would work to heat fuel-injected engines. In this system, hot engine water (or engine oil) would enter the center tube at one end, and fuel would enter the jacket at the other. In theory, by the time the fuel reached the end of the exchanger, it would be as hot as the incoming fluid.

But, unlike carburetors, most fuel injection systems circulate the fuel in a loop back to the tank. This means your hot engine water has to heat both the fuel going into the engine via the fuel injectors, and the portion of the fuel continuously returning to the tank. This risks heating the entire tank. Volvo gas tanks used to use a tall-profile design in order to help radiate heat that returning fuel would naturally pick up from being around hot engine parts. An easier but wasteful way to deal with excess heat is to dispose of it by running the

DAVID BLUME

return fuel through a transmission-oil-type radiator to air-cool it before it returns to the tank.

The other problem with using a counterflow heat exchanger in this setup is that you would need a bigger heat exchanger to get your fuel hot enough to use, since not all of the heat is going into the injected fuel.

Since I'm a permaculture designer, it really bothers me to see this surplus heat energy not being used, or being intentionally thrown away. So, what we need here is a second fuel-to-fuel heat exchanger. Warmed fuel heading back to the tank flows down the center tube of this new exchanger, while new, cold fuel coming from the tank begins circulating at the opposite end in the jacket. This preheats your cold fuel before it reaches the main heat exchanger, and scavenges the heat from the return line.

Fig. 19-3 Oil filter adapter. *This device "sandwiches" between the oil filter and the block. It gives you an easy outlet from the oil pump and returns to the oil sump when using engine oil for heating air or fuel.*

"A terrorist is someone who has a bomb but doesn't have an air force."

—WILLIAM BLUM, AUTHOR, ROGUE STATE

Since the fuel injection system is a high-pressure system, it is possible to pack a lot more Btu into the fuel than you could with a low-pressure system; alcohol under high pressure will not boil until over 250°F. This high pressure will increase the vaporization and atomization of the fuel. When the fuel leaves the injector, the pressure is released by injecting the fuel into the relative vacuum of the manifold. The boiling point of the fuel plummets past the elevated fuel temperature, causing some of the fuel to immediately boil.

Fig. 19-4 Counterflow fuel heater. *This type of heat exchanger can be used to heat fuel and then recover waste heat from the excess fuel returning to the fuel tank in fuel-injected engines.*

The amount of the fuel that boils depends on how many Btu per pound of fuel you transferred into the fuel at the higher temperature. At 100 psi, you can pack an extra 250 Btu per pound into the alcohol without boiling it. That's more than half of the energy needed to fully vaporize the fuel at atmospheric pressure. Once released onto the backs of hot valves in the manifold, much more energy will be absorbed by the fuel, and a great deal of the fuel can be further vaporized.

This is an enormous advantage over carburetion when it comes to increasing mileage on alcohol. It's still not a full vaporizing system, which would generally require the use of more waste heat leaving the engine, but it is a viable way to increase mileage in a standard fuel-injected engine.

Do not heat gasoline in this way. The lighter components of the fuel will boil very easily, and, if your fuel rail is full of vapor instead of liquid, you will generally not get enough fuel. You also risk leaks, which could lead to fires if the pressure in the line skyrockets due to boiling volatiles.

HEATING INTAKE AIR

Most motor vehicles sold in the past decade have air preheaters. When the engine is cold, heated air is introduced into the carburetor or manifold to help vaporize gasoline. Once the engine is warmed up, the preheaters introduce cold air. This switching function is accomplished by a temperature-controlled damper in the air cleaner housing. Using a sheet-metal screw, you should be able to fix the flap in place so that it always directs heated air into the carburetor when running on alcohol. With some heaters, you may be able to connect the damper's vacuum line to a constant vacuum fitting to hold the flap closed.

Heating the air helps to atomize fuel and reduce the evaporative effect, improving mileage and overall drivability of a converted alcohol engine. Although heated air usually causes a drop in power in a gasoline engine, the oxygen contained in ethanol complements the lower oxygen content of the heated air. In running straight alcohol, your goal should be to have the air/fuel mixture enter the engine at 200 to 400°F.

If your car doesn't have a preheater, you can make one: Fit a section of flexible exhaust tubing—about double the diameter of the present exhaust tube and 12 to 18 inches long, space permitting—over the standard exhaust tubing. Seal the opening on the forward end by welding a doughnut-shaped piece of metal in place. On top of the flexible tubing, at the forward end, cut a hole in the tubing and weld a stub of pipe over it for a place to attach a metallic air heater hose to run to the air cleaner. The other end of the flexible exhaust tube remains open to pull in air (see Figure 19-5).

The exhaust pipe temperature will be high enough to heat incoming air. Air temperatures entering the system should be 250 to 400°F. You can experiment with somewhat higher temperatures, but if you go too high, you may get pinging and higher NOx readings.

With gasoline, reduced oxygen in warmer air decreases horsepower. When you heat the fuel, you increase the mileage due to better atomization but reduce the increased horsepower you'd otherwise get with alcohol. However, horsepower reduction with alcohol is much less of an issue, since an engine configured for gasoline inducts

Fig. 19-5 Air heating for carburetors. *This method works well for six cylinder vehicles with air filters on top of carburetors.*

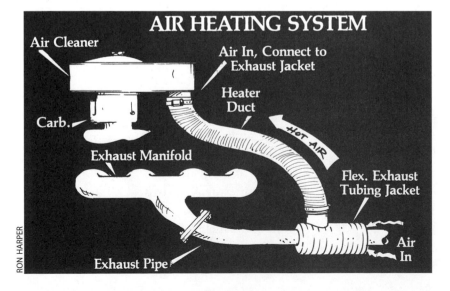

AIR HEATING SYSTEM

Air Cleaner

Carb.

Exhaust Manifold

Exhaust Pipe

Air In, Connect to Exhaust Jacket

Heater Duct

HOT AIR

Flex. Exhaust Tubing Jacket

Air In

far more air than alcohol requires. So, in the converted alcohol vehicle, heated air does reduce your otherwise increased horsepower but rewards you with higher mileage. Thanks to better atomization, you may get 5 to 10% more mileage and smoother accelerations than you would with unheated air.

In fuel injection systems, preheating fuel and air results in shorter pulse widths. In carburetors, it allows you to use a smaller metering jet, since fuel will flow more freely and atomize better, and the air will contain less oxygen. Heating the air thins it some and helps enrich the mixture, since hotter air is less dense. But heating air in front of the air sensor on a fuel-injected system can cause a shake-up in the air temperature sensor department, even though it does enrich the mixture. So, heating the air is likely to cause fewer problems if you heat it after it passes the sensor and before it reaches the manifold. Since there is a long, flexible duct leading from the air sensor to the manifold, inserting a heating device in the duct is the way to go.

MANUAL AND AUTOMATIC CHOKES

Older vehicles used a choke at startup to restrict air coming into the carburetor, in order to richen the mixture. A choke is a flap that restricts the amount of air admitted to the carburetor or TBI system to make the air/fuel ratio richer. Automatic chokes designed for gasoline carburetors often sense the temperature of the intake manifold or engine block. When the temperature sensor indicates the engine

has warmed, the choke is slowly released, admitting more and more air until it is fully opened.

When you run on alcohol, the engine block actually takes longer to warm up than the manifold temperature would indicate. An automatic choke that reads a manifold temperature rather than a block temperature will unload (open) and allow too much air to enter the carburetor too early, causing the engine to stall.

Some old automatic chokes can be bent around so that they open later, and some are electrically heated and adjustable, but I would still prefer a manual choke. I would just leave the choke partially closed until the engine warmed up and then open it all the way.

A manual choke is also a good diagnostic aid for approximating correct air/fuel mixture. If the mixture is too lean, then choking the engine should simulate a richer mixture and make the car sound smoother. But because a manual choke, if not properly set by the owner, could increase emissions, it has not been permitted on vehicles since the '80s. Rudimentary automatic and manual chokes can still be found on industrial equipment and small motors.

Fuel-injected cars typically program a greater delivery of fuel at startup rather than utilize a choke.

SPARK PLUGS

While it's not strictly necessary to change your spark plugs in an alcohol-fueled car, many people have found that hotter plugs aid in cold-starting, and help alcohol ignite when the timing hasn't

Fig. 19-6 Air heater. Use this configuration for carbureted eight-cylinder engines.

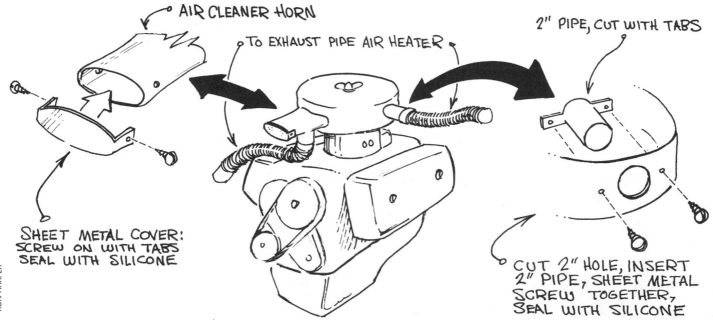

AIR CLEANER HORN

TO EXHAUST PIPE AIR HEATER

2" PIPE, CUT WITH TABS

SHEET METAL COVER: SCREW ON WITH TABS SEAL WITH SILICONE

CUT 2" HOLE, INSERT 2" PIPE, SHEET METAL SCREW TOGETHER, SEAL WITH SILICONE

RON HARPER

been advanced. Also, in some tests, hotter plugs have been shown to increase mileage a little. They do have to be changed back if you go back to using gasoline.

For the committed alcohol user, hotter plugs may be worth it, especially since spark plugs can last well over 150,000 miles on alcohol, compared to 10,000 miles on gasoline. The spark plug gap should be checked once a year, and reinstalled with new compression rings, if necessary.

Some aftermarket ignition control systems specify spark plugs that are compatible with their hardware. This assumes you are using a programmable electronic ignition and plan to go back and forth between alcohol and gasoline.

MANIFOLD AND CARBURETOR GASKETS

This modification is only helpful with carbureted or TBI engines. Between a car's cool intake manifold and the very hot head to which the manifold is bolted, sits a gasket made of fiber or heavy paper; these materials act as insulators to keep heat **Fig. 19-7** out of the manifold. If you allow your normally cool intake manifold to get hot, you can make use of that heat to keep your fuel reasonably well atomized on its way to the engine. So, you want to replace the gasket with one made of copper or aluminum to allow the heat to transfer.

There's another gasket between the manifold and the carburetor or TBI. In cars with no air heating, the carburetor or TBI can get so cold (from the freezing effect of fuel rushing and evaporating through it) that ice will form on the outside of it! Since all your fuel is evaporated through this narrow passage, replacing this gasket helps maintain heat. (Some carbureted engines had the exhaust manifold integrated with intake manifold(s) so exhaust could be directed under the carburetor when the engine was cold.)

A replacement copper gasket should be about 0.035 inches thick, **annealed** in an oven at 250°F for 20 minutes (annealing softens the metal and makes it a more effective gasket), and allowed to cool slowly. Burying it in warm sand is a great way to slow the cooling. An aluminum replacement gasket is fine, as it is. Copper-spraying a paper gasket won't work.

IN THE FUTURE we can expect spot gasoline shortages...

UNFORTUNATELY, IT'S THIS SPOT.

MICHAEL KEEFE

KEEFE THE DENVER POST '79

Replacing either of these gaskets will result in better fuel atomization, letting you use a slightly smaller jet or pulse width. But the changed gasket(s) will make it very difficult to go back to gasoline.

MANIFOLD SHROUDING

Manifold issues apply to TBI or carbureted systems. On straight six-cylinder, and some straight four-cylinder engines, manifold problems are compounded by uneven cooling, particularly in the most forward cylinders. This can cause fuel to become liquid and to cling to the walls and floor of the manifold, instead of being delivered evenly to the cylinders.

By shrouding (encasing) the manifold with sheet-metal housing, you stabilize air around the manifold, and (on four- and six-cylinder engines) trap some of the exhaust manifold's heat to warm up the intake manifold. During the 1980s in Brazil, some companies used a one-piece intake/exhaust manifold to provide even heat. Other companies there put a jacket of hot water around the intake manifold.

Shrouding is a crude method compared to direct manifold heating, but under the right circumstances, it's something the homegrown enthusiast can do himself at little expense. On a new car, shrouding the manifold can be a headache, since there's so much "plumbing" to thread in and out of anything that might cover the manifold. On older vehicles, which have little or no smog equipment, it's a piece of cake. More and more new engines come with engine covers, which do a fine job of preventing air movement.

Feel the intake manifold carefully with your hand while the engine is running. (Don't touch the scalding exhaust manifold!) If the leading edge is cooler than the rear edge of the intake manifold, or if the entire manifold is still cool even after you've heated the fuel and air—you should shroud it. Straight six-cylinder engines are most likely to need this alteration. As an added precaution, you might also line the inside of the sheet metal with some insulation. Use mineral wool or, better yet, a refractory wool, designed to handle high temperatures.

The farther the fuel has to travel, the cooler it will get, and the more fuel will drop out of the airstream and coat the sides of the manifold. This is particularly true in poor manifold design, when there are sharp turns in the manifold. The corners of these turns feel especially cold, since much of the heavier fuel vapor runs into the manifold wall and becomes liquid again, rather than making the turns.

As a general rule, manifolds don't come "finished" inside. Since they're most often made of cast iron or cast aluminum, the inside looks like the surface of the moon. All those bumps and craters cause turbulence, which further interferes with vapor flow. **Porting** (where a machine shop smoothes the intake manifold's inner surfaces) is very helpful in reducing turbulence, although it increases the volume of the manifold.

Aftermarket manifolds, usually made from smooth aluminum, perform well with any fuel, and can be designed to be equal-feeding—engineered so the mixture moving to each cylinder travels at an equal rate. Efforts are made to reduce differences between the distance to each cylinder by controlling the turns and volume of each branch of the manifold.

In older vehicles, some stock manifolds are so crudely designed that they have a difficult time distributing fuel equally to each cylinder. It is possible to retrofit these vehicles with equal-feeding manifolds, which will substantially sustain atomization all the way to the furthest cylinders when using TBI or carburetors.

Once you've made the leap to a dedicated alcohol vehicle, you can see how making small changes that take advantages of alcohol's unique characteristics can result in greater efficiency. Needless to say, the sooner we get rid of E-85 and move to E-100, the better off we'll all be when it comes to fine-tuning dedicated alcohol engines. Please feel free to send details and photos of your successful experiments to me at <info@permaculture.com>, and I'll put the best ones on the website.

You've now learned a lot of tricks to take advantage of many of alcohol's superior characteristics as a fuel. To fully get the most power and mileage from an internal combustion engine running on alcohol, you need to take full advantage of alcohol's high-octane characteristics. In the next chapter, we'll look at different ways to convert your gasoline engine to simulate an engine fully optimized for alcohol.

Fig. 19-8

"There are, of course, still those who deny that any warming is taking place, or who maintain that it can be explained by natural phenomena. But few of them are climatologists; fewer still are climatologists who do not receive funding from the fossil fuel industry. Their credibility among professionals is now little higher than that of the people who claim that there is no link between smoking and cancer. Yet the prominence the media give them reflects not only the demands of the car advertisers. We want to believe them...."

—GEORGE MONBIOT, AUGUST 12, 2003

CHAPTER 20

CONVERTING TO HIGH COMPRESSION

You can convert your car more closely to an ideal alcohol engine by increasing its compression ratio to take advantage of alcohol's high octane rating. With its practical rating of 105, you should be able to run alcohol in an engine with a compression ratio from 12:1 to 18:1, and sometimes a little higher, without experiencing pinging or preignition. Your present engine's compression is probably somewhere between 7:1 and 9:1—higher ratios cause pinging using modern gasoline.

To some degree, the upper limit of the compression ratio depends on the design and shape of the combustion chamber. The practical limit on how high you can go depends on how strong the connecting rods, bearings, and crankshaft are. Sometimes, you are limited by the number of bolts that hold down the head; some cheap engines have too few bolts and can't stay sealed, no matter how tightly you turn the bolts.

An increase in compression ratio commonly raises horsepower from 15 to 40%. Increases of up to 30% don't reduce engine life at all, while increases of 30 to 40% may reduce it slightly, in certain cases. Added compression also helps speed the initial flame front, letting alcohol deliver more power before the exhaust valve opens.

A higher compression ratio can also increase alcohol mileage by up to 22%, depending on several engine design factors. In general, higher-compression engines are more efficient than lower-compression ones. The highest efficiency is attained by the difference in the highest and lowest pressure in an engine cycle. High-compression alcohol engines and diesel engines turn more of the fuel into work than gasoline engines.

Synthetic oil is a necessity in high-compression conversion, since petroleum oils aren't rated for engines with a decent amount of compression. The **film strength** and flashpoint of synthetic oil are much higher than petroleum's. This contributes to

YOU CAN CONVERT YOUR CAR MORE CLOSELY TO AN IDEAL ALCOHOL ENGINE BY INCREASING ITS COMPRESSION RATIO TO TAKE ADVANTAGE OF ALCOHOL'S HIGH OCTANE RATING.

ILLUSTRATION COURTESY OF FRANK CIECIORKA

Fig. 20-1 A piston. *Burning fuel expands, pushing the piston down.*

Fig. 20-2
Compression
ratio of 8 to 1.

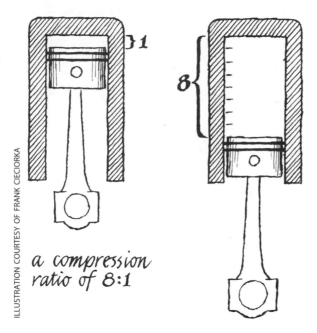

a compression
ratio of 8:1

an even longer healthy engine life in an alcohol-run, high-compression car.

RAISING ENGINE COMPRESSION MECHANICALLY

The major advantage of raising compression mechanically, with different parts in the engine, is that it costs very little extra to do this on the assembly line. Ideally, you design the car on the assembly line to have stronger parts and a taller piston that compresses the fuel into a smaller space before firing.

But after you own the car, rebuilding the whole engine is expensive. So you are limited in your mechanical choices.

Shaving heads is practical, and using longer pistons is expensive but possible. Neither of these methods usually permit you to raise the compression to the ideal level for alcohol. That would only

generally be possible with a proper design of all parts, assembled at the factory.

In converting your current engine, the most inexpensive and straightforward mechanical way to raise engine compression is to shave the head or even plane the block to reduce the volume of the cylinder. The piston, which still travels its same distance (stroke), will now compress the mixture into a smaller space. This technique can raise your compression by two or three points.

Having the heads shaved doesn't require that you completely rebuild your engine. In fact, you don't even have to remove the engine to remove the heads. Shaving the head on my 1982 Volvo is going to cost me $40 once I take the head off and take it to a machine shop. The cost of parts should be less than $200 with new valves (which I might as well do with the head already off). There will be about three or four hours of work involved in taking it out and putting it back in.

On some engines ('60s and '70s straight-six Chevy, slant-six-Chrysler, and some eight-cylinder American cars, for instance), you can replace the head with one from a smaller **displacement** engine. This has the same effect as shaving a head, but instantly gives you more compression. Check the bolt patterns in your owner's manual and double-check with your local machine shop to make sure that the heads are interchangeable.

While the head is off, there's another move to consider. Reworking the head is an opportunity to change the **camshaft** of overhead cam vehicles. Higher-lift cams can fill the cylinder earlier and hold the mixture later, allowing you to advance the ignition timing quite a bit and to get more complete combustion and more power/mileage. The disadvantages are that there is more overlap between intake and exhaust, and that it is possible to lose some fuel, unburned, out of the exhaust. So, choose a cam that will give you more torque earlier in the stroke. These are sometimes called **RV cams**, since pickup trucks that haul recreational vehicles sometimes specify them.

Another way to raise compression mechanically is to pull out the engine and rebuild it with domed pistons or with longer pistons than the ones you were using before. The pistons accomplish what a shaved head does by reducing your cylinder's volume so the mixture compresses into a smaller space. High-performance pistons for most engines are available at speed shops. A good set for an

---

"I converted my VW Passat Variant to ethanol. Within four months and 10,000 miles, I have saved US$600 in purchasing US$1700 worth of fuel. This data is extracted from 60 consecutive measurements made every time I refuel the car, and not just mileage, but also the costs of both gas and ethanol are recorded. No problem with the car, and average consumption is 7.05 km/L [about 17 mpg]. I think I could get more once the car is turbocharged, and possibly it is working with a too large advance in ignition. I will try to change that. The cost of conversion, just the substitution of the chip, was less than US$200. For less sophisticated cars, it is about half of that."

—DR. JOSMAR PAGLIUSO

eight-cylinder engine will probably also include stronger connecting rods and bearings to take the higher pressures. The advantage here is that you can increase your compression more than you could by shaving the head.

Labor is the major cost for either mechanical method of raising engine compression. Both procedures are time-consuming, and thus, at prevailing mechanics' fees, expensive. Rebuilding with new pistons, etc., can end up costing as much as $3500 if you're paying for someone to remove, rebuild, and reinstall your engine. Doing it all yourself will save money, but will take a few days of careful work. For weekend mechanics, changing the head should be within your abilities, whereas rebuilding the engine is a much bigger job. Check to see if your state provides tax deductions for conversion to alcohol. It might make enough of a difference to make it worthwhile.

Different engines have different limits in how far the compression can be raised. Going above 10:1 in most cars requires evaluating connecting rods and perhaps other engine parts. You'll find that the higher you go, the more parts you'll begin to consider replacing with heavier-duty units. Connecting arms are often too light to take a ratio of 12.5 to 1. Many new cars have lightweight main bearings, which should be replaced with heavier main bearings, and crankshafts if possible. Most engines built before 1980 should be able to tolerate a compression ratio between 10.5 to 1 and 12.5 to 1. Some new engines that require premium fuel are already a bit higher in compression, and are made with stronger parts.

I still remember, with a good deal of humor, the researchers at the University of Santa Clara converting a Ford Pinto to a compression of over 14 to 1.[1] The car's lower-end parts worked fine. The pistons, connecting rods, etc., all held up, but the engine kept stretching the bolts that held the heads to the engine. The bolts were replaced with a very high-grade variety, usually found at nuclear power plants, and they stretched, too. Since recasting the block and heads with more bolts didn't seem to be a practical solution, the researchers reduced compression to a little over 13:1. The engine showed no further stress-related problems.

When you've converted to a high-compression configuration, you'll find it takes much more energy to compress the air/fuel mixture than it did before, and your starter is going to have to work

High-compression engines will need less ignition advance, less air heat, and much less cold-start help. However, excessive air and fuel heat may cause pinging. You can avoid this and tolerate an even higher compression by using heads or pistons with a swirl (also called a yin/yang) pattern built into their surfaces.

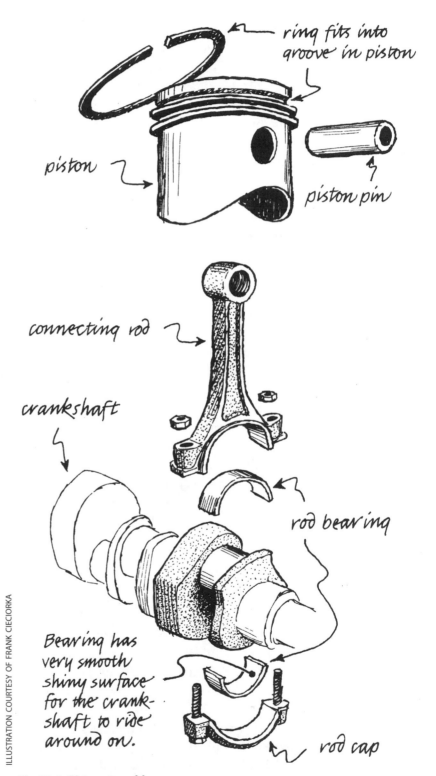

ring fits into groove in piston

piston

piston pin

connecting rod

crankshaft

rod bearing

Bearing has very smooth shiny surface for the crankshaft to ride around on.

rod cap

ILLUSTRATION COURTESY OF FRANK CIECIORKA

Fig. 20-3 Piston assembly.

much harder to turn over the engine. I recommend installing a heavy-duty one. Check with your local speed shop or a J.C. Whitney catalog.

High-compression engines will need less ignition advance, less air heat, and much less cold-start help. However, excessive air and fuel heat may cause pinging. You can avoid this and tolerate an even higher compression by using heads or pistons with a swirl (also called a yin/yang) pattern built into their surfaces. The swirl creates a turbulent atmosphere for your air/fuel mixture, which resists pinging. Mercedes-Benz reported in the 1980s that its heated (vaporized) alcohol engine would run as high as 12:1 with a swirl pattern; without it, a safe operation level was 10:1. United Parcel Service did something similar in its early 1980s trials, using Chevy 292 straight–six engines in its delivery fleet. They found the yin/yang pattern beneficial.

Turbochargers can be made dual-fuel by using an adjustable wastegate, which controls how much of the exhaust gases power the compressor.

Depending on manifold and engine design, carbureted high-compression engines can start at temperatures as low as 30°F without cold-start assistance. Colder than that, and the usual built-in, factory cold-start systems fire up the engine quite rapidly. Fuel-injected, high-compression engines may start as low as 15°F.

If you blend ether with your alcohol for cold-starting , a high-compression engine will need about half the recommended dose. Also, don't run a special extra-hot spark plug as you would with a lower-compression engine—try one or two heat ranges below stock.

Start with a small diesel engine if you want to convert a high-compression diesel engine into a high-mileage alcohol one. Volkswagen's TDI has been a favorite test engine for this sort of thing for a while now. In the case of the TDI, you need to decrease the compression ratio from over 20:1 down to about 18:1, and increase pre-cooled exhaust gas recirculation to about 50% to prevent pinging.[2] You would also turn off the direct fuel injection and install a standard multi-port fuel injection system in the manifold or a TBI system on the manifold. Alternatively, you could keep the stock injection system but add biodiesel to the fuel to provide the lubrication needed. Spark plugs and electronic ignition would replace glow plugs. You may find that full-time glow plugs are just as effective as spark plugs. This is a lot of work, but you will end up with a very powerful alcohol engine, which will get better mileage than it would on diesel or gasoline. Since it was built to sustain the stress of diesel operation, it should last darn near forever on alcohol.

RAISING ENGINE COMPRESSION NONMECHANICALLY

The nonmechanical technique for raising compression is by means of a turbocharger or **supercharger** that pumps compressed air into the engine for a higher pressure at the end of the piston stroke. A turbocharger is a compressor driven by the engine's exhaust gases. Superchargers are usually powered by the engine through large drive belts, although some are driven by their own electrical motor.

One of the big advantages of a nonmechanical charger system is the great air acceleration that is delivered to the engine. With higher-velocity compressed air racing through the manifold, fuel atomization is greatly increased. Air leaving the compressor for the manifold will be quite warm, too.

Turbochargers generally start giving your engine compressed air, both as you accelerate and at higher cruising speeds. Superchargers can be set to start giving you pressure right at idle. Turbochargers can be made dual-fuel by using an adjustable **wastegate**, which controls how much of the exhaust gases power the compressor. You can get wastegate adjustment controls mounted on the dashboard, but they're more expensive, compared to under-the-hood, wrench-adjustable ones.

Electric superchargers can just be turned off when you switch back to gasoline. Automatic pressure controls determined by a knock sensor are also available, which makes dual-fueling sensible. This is already offered stock on engines like the older Volvo 850 turbo and the modern Swedish Saab 9-5 flex fuel engine.

Another advantage of the charger systems over mechanical systems for raising compression is that you can use blowers to take advantage of your engine's low-compression characteristics, like easy starting and smooth idling. Blowers are pretty easy to install by yourself, even if you aren't a practiced mechanic. You can buy them in a kit for most engines, and even for a widening variety of motor-cycles. They run anywhere from $700–$2000 and up, depending on their quality and application.

Setting your blower for alcohol requires some experimenting. **Boost pressure** will be a little less than double that recommended for gasoline. I suggest starting at the recommended gasoline boost, and working your way up from there to the point just before you start pinging. Pinging at high compression damages your engine much faster than pinging at normal compression.

As you can see, I've given you an outline for experimenting with high-compression conversions. The especially tempting conversion of diesels to alcohol is virgin territory. But the best part is that those of you who experiment with this will be in a great position to embarrass the heck out of gasoline or diesel users when you fully tap the potential for converting fuel to work. This is a frontier that takes an especially good mechanic to blaze the way for others. So if you'd like to share your successes and failures in converting diesel engines or making other high-compression conversions, write me at <info@permaculture.com>.

Now that we've taken a look at what are usually considered industrial engines, let's change tack to look at the little guys. These are the countless engines that power things smaller than your car. There are an awful lot of them in our lives, and, in some ways, it's more important to convert them than our cars.

Endnotes

1. L.H. Browning, "Thermokinetic Combustion Process Modeling," *Characterization and Research Investigation of Alcohol Fuels in Automobile Engine: Final Report* (1981), 53–80.

2. Matthew Brusstar, et al., *High Efficiency and Low Emissions from a Port-Injected Engine with Neat Alcohol Fuels* (Washington, DC: U.S. Environmental Protection Agency, Society of Automotive Engineers, 2002).

Fig. 20-4

"When we burn a seven-pound gallon of gasoline we create about 21 pounds of CO_2. This means that two-thirds of fuel we are burning is [atmospheric] oxygen. In the first six months of life, a baby requires about 400 cubic feet of oxygen to survive. Burning one gallon of gasoline consumes about 400 cubic feet of oxygen."

—STEVE HECKEROTH

Fig. 20-5

CHAPTER 21

SMALLER ENGINES

Most people think of engines as synonymous with cars or trucks, and the discussion of air pollution is largely dominated by these same engines. But there are many more small engines in the world than there are car and truck engines. In many cases, they have a huge impact on economies and the environment.

Yard work has been re-termed "mow, blow, and go," since the traditional reel mower and rake have been replaced by small motor-driven lawnmowers and leaf blowers. Scythes, sickles, and hoes have been replaced by weed eaters. Trees are felled and bucked with chainsaws, not axes and bow saws. In many parts of the world, and most ominously in China, bicycles have given way to both two-stroke and four-stroke motorcycles.

For better or for worse, a huge amount of human labor has been replaced by small engines and fuel. Are we willing to do these jobs with labor once again when Peak Oil drives gasoline prices over $10 a gallon? Not likely. These motors are here to stay.

The hidden danger is that none of these small motors have any sort of emission controls. A seven-horsepower, two-stroke chainsaw puts out more dangerous pollution than a giant SUV. So, far from being a little problem easily dismissed, the issue of what to use to power our small engines is quite significant.

MOTORCYCLES

When running on alcohol, your bike will run cooler with a great deal less vibration and noise than it ever did on gasoline, and there'll be no exhaust pipe carbonization to deal with. (It isn't clear how much noise and vibration reduction would apply in high-compression conversions.) With some minor differences, four-stroke motorcycle engines are similar to automobile engines. Older cycles use air-cooled engines for the most part. Newer bikes are becoming more like miniature cars all the time—fitted with electric starters,

FOR BETTER OR FOR WORSE, A HUGE AMOUNT OF HUMAN LABOR HAS BEEN REPLACED BY SMALL ENGINES AND FUEL. THE HIDDEN DANGER IS THAT NONE OF THESE SMALL MOTORS HAVE ANY SORT OF EMISSION CONTROLS. A SEVEN-HORSEPOWER, TWO-STROKE CHAINSAW PUTS OUT MORE DANGEROUS POLLUTION THAN A GIANT SUV. SO THE ISSUE OF WHAT TO USE TO POWER OUR SMALL ENGINES IS QUITE SIGNIFICANT.

MATT FARRUGGIO

Fig. 21-1 Vertical shaft vaporized alcohol engine. This eight-horsepower engine powered my distillery's high-speed agitator for some time. The vaporizer cut our fuel use many times over. Stock exhaust pipe and intake manifold were removed, and a new manifold was constructed with a box around it. Exhaust surrounds the manifold and exits through a small car muffler. Estimated temperature of alcohol vapor entering the head is 300°F.

Fig. 21-2 Motorcycle air heater and cold-start setup. *A short length of flexible exhaust tubing is placed around the motorcycle exhaust pipe, and a hose adapter is constructed as shown here on my moped. Hot air draws from the front of the exhaust pipe through the adapter and into the carburetor through the hose. Note the drilled hose clamp on the air intake hose (see arrow)—it's snug but can be rotated on the hose by hand. To cold-start the bike, line up the hole in the clamp and the hole in the hose in order to spray ether or butane into them. When the bike starts, the clamp is rotated so that the hole in the hose is covered and dirt can't get in.*

radiators, liquid cooling, electric fuel pumps instead of gravity, and even fuel injection instead of carburetors.

Carburetors on older motorcycles are usually **side-draft** rather than downdraft types, and are often much easier to adjust than automobile carburetors. After 1990, carburetors evolved toward downdraft configuration. The diameter of the jets controls fuel mixture at cruising speeds just like with autos, but acceleration, low speeds, and sometimes mid-range speeds are controlled by the position of a needle, or needles, in jets.

The needles are usually adjusted by a screw, or by "stepping" (snapping into positions on a retaining ring). Use a slightly larger carburetor jet than you normally would, and raise the needle as much as necessary, so there's no stumbling or coughing at idle.

A more fuel-efficient method, which also eliminates stumbling on acceleration, is to alter the needle's taper. To enrich the mixture for idling

and acceleration, the needle should be reduced in diameter near the top and feathered out to the original taper toward the tip end. These adjustments give you a richer mixture, coinciding with the needle's initial lifting (as you twist the accelerator grip).

Motorcycles often idle rough until they warm up. Always replace the manifold-to-head gasket with copper or aluminum to encourage heat transfer. Heating the manifold can make a great difference in performance, permitting you to use less fuel while accelerating than you did with gasoline. On an older bike, if there's enough room, the manifold can be replaced by an exhaust system vaporizer. A water-heated jacket can be constructed for some newer bikes. I recommend wrapping your manifold(s) with insulation so that moving air doesn't cool them further.

Alcohol racers (dirt track racers, particularly) have been known to remove most or all of their heat fins, since they never idle during a race, and air moving over the bare engine provides more than enough cooling. Heat fins don't get all that hot with alcohol. For efficient running in street cycles that run too cool at high speeds, cover the leading edge of the fins to deflect most of the air and retain some heat in the block and head. Heating the air helps, but requires the cyclist to fabricate his own air heater with heat around the exhaust pipe (see Figure 21-2). One of my students enclosed his heat fins and ducted the heat into his carburetor with good results.

Cold-starting a motorcycle is relatively easy. One way is to use the hose clamp and hole method. On mopeds, motorcycles, or even old tractors, you can make use of the hose between the carburetor and head as a site for a quick cold-start. Drill a ⅛-inch hole in the hose and cover it with a hose clamp that also has a similar-sized hole drilled in it. To start the engine, rotate the almost snug clamp by hand to line up the two holes. Then use a butane cartridge (used for filling lighters) or canned ether to shoot a two-second blast into the hose. Turn the clamp to cover the hole and start the engine. On really cold mornings, I've had to do this two or three times for my moped. (Of course, if I had any sense, I wouldn't ride my moped on cold days.)

You can also cold-start your motorcycle by installing a starting fuel reservoir made from a small utility tank engine, white gas lantern, or stove tank (see Chapter 17, Figure 17-9). Install a selector valve to allow you to send fuel from either your main tank or starting fuel reservoir. Since motorcycles are often a problem to kick-start on cold days, I recommend getting in the habit of switching over to the starting fuel each time you prepare to turn off the bike. This primes your carburetor with a full bowl of starting fuel. Kick-starting isn't a problem if you use the lantern or stove tank to make your starting fuel reservoir—they can be pressurized from a built-in pump to deliver fuel directly to the manifold.

High-temperature fuel softens normal clear plastic tubing, and aldehydes tend to make it brittle. If you want to keep those clear fuel lines so you can see if your fuel is getting to the carburetor, use tubing rated for high-temperature alcohol, such as Tygon. I've personally given up on clear plastic tubing; I use time-tested neoprene, with a short length of thick Pyrex glass tubing inserted into the line to see the fuel flow.

Fuel heating is not strictly necessary if you have good air heat and some heating on the manifold. You can heat fuel for a bike, but it's tricky. Most bikes feed fuel to the carburetor by gravity without the aid of a fuel pump. The fuel lines tend to be large-diameter, which slows fuel flows. This long residence in the vicinity of your heat source can cause small amounts of the fuel to boil, and the bubbles may stop flow altogether. For this reason, you wouldn't want to wrap your line around the exhaust pipe.

Mother Earth News suggested a good way to gently heat fuel, which I've successfully used. Slip a perforated copper ¾-inch plumbing strap under a bolt on the hot engine head, and wrap and solder the other end to the fuel line. The strap conducts heat to the copper fuel line.

An increasing number of high-performance bikes are now using either TBI or port fuel injection. So follow the guidelines for these systems in Chapter 16.

SMALL TO MEDIUM UTILITY ENGINES

Lawn mowers, rototillers, generators, pumps, heat pumps, and other utilitarian devices too numerous to mention are powered by small- to medium-sized utility engines. These engines are generally quite simple, since they're designed to run at a relatively steady rpm, and are not expected

Fig. 21-3

MIKE PETERS

GOOD GOD.... IT'S GENE KELLY...

to go through the various rapid acceleration cycles of an automobile engine. They are also allowed to remain simple, since they don't currently have to meet emission standards.

The air/fuel mixture in these smaller machines is usually controlled by one screw in one jet. These engines are totally adjustable. On most engines under ten-horsepower, the only necessary alteration for alcohol is backing off the air/fuel mixture screw, maybe one to one and a half turns. Some engines will also have an idle jet to adjust.

Cold-starting is still necessary, since utility engines aren't very high-compression. On stationary engines, a really convenient cold-start assist is to install a second small tank with starting fluid in it. In the fuel line, in front of the fuel pump (or in front of the carburetor on gravity-fed systems), insert a selector valve so you can control which fuel to draw from. When you are about to shut off your engine, switch over to starting fuel for 30 to 60 seconds to prime your carburetor and/or fuel pump. Place the selector valve as close to the fuel pump or carburetor inlet as is convenient.

On stationary engines with fuel pumps, it's helpful to heat your fuel as you would for a regular automobile engine. But heating the fuel doesn't

work well on a gravity-fed system, such as on lawnmowers or many motorcycles. If it overheats even a little, there's no way to stop hot fuel vapor bubbles from going toward the tank, rather than toward the carburetor. You could end up with a fuel-starved, vapor-locked carburetor while waiting for a light to change at an intersection. In gravity-fed systems, heating the air and manifold are your best bets.

ABOVE: Fig. 21-5 Counterflow vaporizer. *Positioned to replace the present exhaust and intake manifold, this vaporizer is simple to build. Exhaust travels through the center tube of the unit, while the fuel mixture travels around the jacket. The mixture exits the short manifold at the base of the exchanger and is inducted into the head. The carburetor is attached at the end of the vaporizer, which necessitates lengthening the fuel line and linkages. Several lathe chips are placed in the exhaust tube to create turbulence, which improves the heat exchange.*

Fig. 21-4 Stock utility engine exhaust, intake manifold, and carburetor. *The carburetor throat is the dark circle in the center to the right of the filling cap. The exhaust is on the right, threaded into the head with ¹/₂-inch pipe thread..*

To heat a utility engine manifold, you may as well vaporize the fuel completely. Because you won't have acceleration problems with a stationary engine, vaporization can be accomplished at a steady rate. A counterflow heat exchanger, in which exhaust runs through the center tube and alcohol spray from the carburetor is circulated in the jacket, is an excellent way to vaporize fuel. You can experiment with the length—a foot-long vaporizer works well on my pump (see Figure 21-4).

In the counterflow heat exchanger, a couple of long, coarse lathe chips in the exhaust tube create turbulence. A coarse, expanded metal screen, welded to the end of the exhaust tube, secures the lathe chips, and acts as a spark arrester. For safety's sake, install a weighted, hinged cap on an outlet cut in the vapor jacket (not shown). The cap should seal well, so that you don't get any vacuum leaks. In case of a backfire, the cap will fly open and then shut, releasing pressure, so there's no damage to your engine or vaporizer.

A second method to vaporize fuel for a stationary engine is to build an exhaust box around the intake manifold (see Figure 21-1).

With either heat exchanger, vapor should enter the head at 250 to 300°F. Fuel consumption is similar in both heat exchange methods, but the counterflow type may make starting a bit difficult because of the distance the vapor has to travel to the engine. Turn the carburetor needle in to reduce fuel flow. You'll experience much better fuel consumption than you did on gasoline.

Utility engines seem to work better with proofs above 180. I've used lower proofs, but there always seemed to be a bit of surging at idle—probably due to manifold puddling. So I recommend using 190+ proof.

Lubrication in small engines is not as efficient as in larger engines. I like to mix alcohol-soluble two-stroke oil with alcohol (even though utility engines are four-stroke) in very dilute quantities, about 1 to 200 or even more dilute, to extend engine life. When using these additives, you should always use either 192-proof or 200-proof alcohol to guarantee good mixing. Biodiesel can also be used, at this level of dilution, as a lubricant and anti-corrosive. It would probably be almost as effective as using the alcohol-soluble two-stroke oil.

It's best to use the exhaust-box-type vaporizer when you're adding a lubricant, since synthetic oil and biodiesel boil at a very high temperature, and otherwise may not vaporize completely. In a counterflow-type exchanger/vaporizer, oil will often be left behind, causing problems, so it shouldn't be used with this type of vaporizer.

In some utility engine carburetors, a float needle and seat may both be made of Viton. If only one or the other was Viton, there'd be no problem. But Viton swells slightly in long contact with hot alcohol, and the swelling causes slight deformities in the sealing surface between the needle tip and the seat, which may cause leaking and a flooded carburetor. To avoid this, I would take your rubber-tipped needle to a machine shop and have a new stainless steel one-piece needle made (like they used in the old days). It shouldn't cost more than $5. This problem is almost unheard of nowadays, since Viton rarely swells in modern formulations.

A chainsaw can be far more polluting than several SUVs. It is frightening to see the number of two-stroke motorcycles coming into regular use in the developing world.

TWO-STROKE ENGINES

After checking with my colleagues in Brazil and searching the literature here in North America, I was floored to realize that little or no published research had been done on two-stroke engines running on alcohol fuel. These engines appear everywhere maximum horsepower and minimum weight are considerations.

The problem with two-stroke engines, compared to the heavier four-stroke engines like those in cars, is that there is a brief time when both the exhaust valve and intake valve are partially open, and some of the unused fuel goes directly out the exhaust. What's more, the lubrication system of a two-stroke engine requires that lubricant be included with the fuel or injected into the engine along with the fuel. So the lubricant goes through partial combustion and adds its own load of toxins to the exhaust.

A chainsaw can be far more polluting than several SUVs. It is frightening to see the number of two-stroke motorcycles coming into regular use in the developing world. And in North America, there are enormous amounts of weed whackers, leaf blowers, chainsaws, motorcycles, snowmobiles, and other equipment powered by these particularly filthy engines.

In California, and increasingly everywhere, the problem is even more grim. Reformulated gasoline in two-stroke engines raises octane by including a large amount of the proven carcinogens benzene, toluene, and xylene. Since these smaller devices do not have catalytic converters, some of the unburned mixture flows out the exhaust, and users are exposed to highly dangerous chemicals. These chemicals, especially benzene, have been banned from most other industrial uses in California and much of the world.

In the '80s, I used a number of different substances to lubricate alcohol-fueled two-stroke engines. Castor bean oil is an old standard; I also used synthetic two-stroke oils made for alcohol-powered motorcycle racing. Blendsol and Klotz have been used by methanol-powered auto racers for some time; there is also Red Line Alcohol Pre-Mix, which is, in my view, a newer and superior product. There are also new lubricants made from vegetable oils that successfully mix with alcohol or gasoline for use in two-stroke engines.

Using a synthetic oil/alcohol mix makes a two-stroke engine very happy. The plugs never seem to foul. There is little noticeable smoke from the exhaust (other than a slight acetaldehyde odor while warming up), and hardly any oily residue to drip from the exhaust. Lack of fouled plugs is often a problem when chainsawing (see Chapter 14, Figure 14-5), since it deprives the woodsman of his best excuse for taking a break. The reduction of carbon in the exhaust pipes of two-stroke engines will save you from poor performance, maintenance trips to the shop, and messes to clean up.

The degree to which you should dilute your fuel with synthetic oil varies with the product, but it's always much more diluted than a gasoline two-stroke mix. Depending on the manufacturer, 60 parts (or even 100 parts) alcohol mixed with one part oil is the ratio.

Although it's possible to get these synthetics to mix with low-proof alcohol, they work much better with high-proof. Fuel below 190-proof can cause cylinder or bore wear in two-stroke engines. My moped let me know, with quite a clatter, never to feed it 151-proof rum again. Oil injection engines, in which oil is injected into the cylinder rather than mixed with the fuel, are more accommodating than most. But even with these, use of alcohol fuel below 190-proof is not recommended.

A Two-Stroke Alcohol/ Biodiesel Experiment

In the 1980s, I ran a variety of two-stroke equipment (and my moped) using alcohol-soluble two-stroke oil for the lubricant. But recently, I decided to set my sights a bit higher. Since I knew that Brazil used castor oil mixed with alcohol in diesel engines, I reasoned that biodiesel mixed with alcohol might work in North America, on two-stroke engines. After all, biodiesel's lubricity has been well documented.

I picked up a used Husqvarna 51 chainsaw and had it disassembled and inspected to see that it was up to specs, with no unusual wear. After reassembly, I took the saw up to T. Gray Shaw, a master arborist in Berkeley, California. He runs his own vehicles on biodiesel and veggie oil, so he was keen to participate.

We mixed up a gallon of E-95 and added biodiesel in the same proportion that gasoline is mixed with two-stroke oil, about 40:1. The saw had two mixture-control jets with adjustable needles in them, one for idling and one for main operation. I said, "Let's see how many quarter-turns we have to go in, on both the low- and high-range mixture screws, to hit bottom. That will tell us where to start." So we tried to turn in the low-range screw, and after a quarter-turn, it stopped. Odd. We tried to turn it out, and we only got about a half-turn and it stopped again. So did the high-range adjustment screw.

We pulled off the cover so we could see the carburetor. What we found were two anti-tamper plastic pieces that fit over the mixture screws, with ears sticking out so the screws could only go about a quarter-turn each way. Obviously, a shop mechanic had pressed these on after tuning the saw. Well, I'm sure he meant well, but they were kind of flimsy, and sort of fell off when we tugged on them with pliers.

We backed them out a few turns to start rich. On the low range, we started turning the screw in, clockwise, until the chain stopped rolling, and we kept turning in until it started up again. We then turned the screw back out, enriching the mixture until the chain once again stopped rolling. When you go too lean on a two-stroke, it actually speeds up the motor until it damages itself. Watching the chain helped us find the point at which the mixture was just rich enough to not send the chain into motion at idle, avoiding lean-burning conditions.

We adjusted the high range by making sure the saw had enough power to cut smoothly through an old cherry log. We began leaning it out until it hesitated, and then reversed the screw counterclockwise, enriching it until the power came back to normal.

The saw ran for about half an hour the first time. If we had been running on gasoline without two-stroke oil, the saw would have been junked. So it

MATT FARRUGGIO

appeared that biodiesel was doing the job of lubrication, and we were both pretty encouraged. There was no nasty, carcinogenic, petroleum two-stroke exhaust to breathe.

When Gray took the saw to work and ran it at the same 40:1 alcohol/biodiesel mix, the saw was sufficiently lubricated, but there was a smell that he found quite disagreeable. After a day of working with it, he was really sick of the smell.

I took over testing and successfully ran the Husqvarna saw on a 75:1 dilution of alcohol to biodiesel, but at 150:1, there was over-speeding and piston ring wear on the cylinder wall. When I took it into the shop for disassembly, the mechanic discovered that one of the air cleaner housing mounting screws was stripped, admitting much more air than normal. So I couldn't tell if the wear was from lean burning or from the mixture. More testing and analysis of the exhaust is needed to determine the correct dilution of biodiesel and alcohol. I don't think this exhaust could be anywhere near as dangerous as the carcinogenic BTX content of gasoline, which escapes unburned into the exhaust on two-stroke engines.

A few final tips: When the spark advance is adjustable on a two-stroke engine, you should increase it less than you would for a four-stroke engine. Cold-starting is about the same as for a four-stroke, but if you're blending ether into your fuel, use no more than 5%.

Some saw repair shops provide a service to raise the compression ratio of your chainsaw to greatly increase torque and overall power. People who have

LEFT: Fig. 21-6 Alcohol-soluble two-stroke oil. Although the bottle is marked for methanol only, the oil is equally soluble in ethanol.

TWO-STROKE LUBRICANTS

One two-stroke test run by a friend ran into a novel problem. A modern chainsaw that had seen a lot of hours initially ran great on alcohol with a plant-based synthetic lubricant; it had plenty of power, and the crew loved the lack of foul exhaust. But as the carbon buildup on the cylinder head dissolved away, it caused an unexpected problem. The reduction in carbon caused the volume of the cylinder to increase enough that the mixture became lean, and the saw started to over-speed. The larger volume of the cylinder exceeded the ability of the carburetor to deliver enough alcohol to get a safe mixture!

The solution would have been to take the carburetor to a machinist and have the very tiny jet in the carb drilled out slightly so it would be adjustable with the carburetor needle. Older chainsaws with more leeway shouldn't run into this problem.

done this use premium gas with an octane booster; they find that it's worth this hassle, since they get so much more power out of their saw. Alcohol's high octane gets you the benefits without the hassle.

SYNTHETIC LUBRICANTS

Synthetic oils are essentially plastics, and are superior to common petroleum lubricants. They are often composed of **diesters** or other extremely long-chain molecules, and their chemical structure gives them incredible film strength. This is a measure of how difficult it is to squeeze the oil from between two metal surfaces—if metal isn't touching metal, there's little or no wear on the parts. Synthetic oil's high film strength and its resistance to viscosity loss at high temperatures reduce friction up to 45% in some studies. This can increase mileage slightly. And less friction results in oil temperatures up to 50°F cooler than in an engine running with typical motor oil.

Synthetic oils usually have a wider range of viscosity than petroleum lubricants. Several are rated at 5w–40w, a few at 5w–50w. When starting a cold engine, a low 5w viscosity is beneficial, since the oil's distribution through the engine is faster, facilitating engine cranking. As the engine heats up, viscosity increases to provide additional protection to the parts.

Many synthetic oils are formulated to flow and stick to hotter areas of the engine and are not affected by exposure to engine heat. Synthetic oils have been run in many different engines (including long-haul diesels) for over 500,000 miles without requiring oil changes, except for the normal replacement quart when filters are changed every 12,000 miles or so. Synthetic oil's very high flashpoint keeps the cylinder more or less continuously lubricated—not the case in a petroleum-oiled engine.

Most **polyol ester** lubricants can be made from alcohol and the fatty acids of plants, but since petroleum is still cheaper, they are currently made from petroleum basestocks. Synthetic oil is priced three times higher than high-quality petroleum oil, but it can really end up costing much less per mile, considering oil changes will be about eight times less frequent. Increased engine life is a bonus.

In a two-stroke engine, you can't add a normal alcohol-based synthetic crankcase lubricant (engine oil) to your alcohol, because it's not soluble. Two-stroke engines require that the lubrication oil be mixed in with the fuel, whereas four-stroke engines have a separate pressurized lubrication system, and the oil is stored in the crankcase, i.e., your car.

A four-stroke gasoline engine can't tolerate the addition of a lubricant to its fuel. But because of alcohol's very low carbon production during combustion, an alcohol-run four-stroke can tolerate in-fuel lubricants by adding $\frac{1}{2}$ of 1%, or even $\frac{1}{4}$ of 1% alcohol-soluble two-stroke oil. This slight addition can extend piston ring and valve life. Biodiesel does a great job at these dilutions.

Synthetic lubricants are available for transmissions and differentials, as synthetic grease for drivelines, wheel bearings, and suspension lubrication, for automatic transmissions, and even as brake and power steering fluids. Not all companies carry these products. Most synthetic lubricants are available through local speed shops, motorcycle shops, some auto parts stores, and by mail order.

Start writing to these manufacturers; tell them you want to know when they switch from using petroleum and go to alcohol and fatty acids. The price of the raw materials to make the lubricants is pretty small, and it would make little difference to the manufacturers' bottom lines to make them from renewable ingredients. They need to know there's a market for those who don't want to give any of their money to oil companies!

Now it's time to think about cars again. Until we have enough small alcohol plants and community stations in every town to fuel our alcohol-converted cars, vehicles that can switch easily between gasoline and alcohol, or burn any mix of the two, will be helpful. Next, we'll look at current and future flexible-fueled vehicles, and learn some tricks to use on those that are out there today.

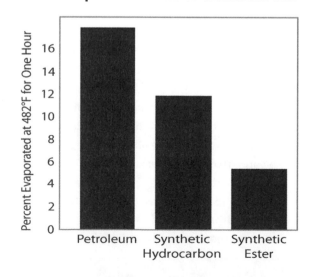

Fig. 21-7
Evaporation rates of crankcase oils.
Synthetic ester lubricants, such as those that can be made from alcohol, are highly resistant to degradation under high heat, unlike petroleum lubricants.

Evaporation Rates of Crankcase Oils

CHAPTER 22

FLEXIBLE-FUEL AND DUAL-FUEL SYSTEMS

People think that flexible-fuel engines are a new technological triumph, but nothing could be further from the truth. The Model A and Model T Fords were the first flexible-fuel vehicles (FFVs), able to run on any mixture of alcohol and gas. Their computer was your brain and ears, and you made the adjustments yourself. It's taken us more than 75 years to duplicate the function of these early vehicles with modern technology.

Most people are led to believe that FFVs can only run on 85% alcohol (E-85) or less. But by law, E-85 can actually be anywhere from 70% to 98% alcohol (E-100 has 2% gasoline as a denaturant). I have run my flex-fuel 2000 Ford Ranger for more than 50,000 miles on E-100. The only thing a FFV lacks when running on E-100 is a cold-start device, which you can easily install yourself. Other than cold-starting, there is no technical reason for E-85 to exist.

ORIGINS OF THE FFV AND E-85

Since 1994, a limited number of FFVs have been made in the United States and, more recently, in other countries, such as Brazil and Sweden. In Brazil, almost all new GM and Ford, and all new Volkswagen vehicles, are FFVs.[1] Today, there are about six million FFVs on the road in the U.S. Yet most owners here don't know they're driving one, and the car dealers haven't been aware they've been selling them. This is especially amazing since until 2005 there were substantial federal tax credits for buying FFVs and juicy state credits, too.

So why would a car company sell a technologically advanced vehicle that gave the driver a choice over what to put in the tank and not brag about it? The answer is to be found in the conundrum auto companies had over the CAFE and SUVs. No, I'm not talking about soccer moms getting together for a latte after dropping the kids off at school. The corporate average fuel economy (CAFE) has been the bane of American car companies for decades now. Each company must hit a target average mileage

AN FFV ENGINE CAN USE ANY PROPORTION OF ALCOHOL AND GASOLINE IN THE SAME GAS TANK. THE CAR RUNS ON ALCOHOL IN TOWN (95% OF MOST PEOPLE'S DRIVING), WHILE RETAINING THE ABILITY TO USE GASOLINE FOR TRIPS THAT MIGHT TAKE YOU OUT OF RANGE OF YOUR ALCOHOL SOURCE. BUT IT'S NOT THE IDEAL WAY TO RUN AN ENGINE ON ALCOHOL, SINCE IT HAS TO COMPROMISE ON NEARLY EVERY FRONT TO ACCOMMODATE BURNING MEGAOILRON'S TOXIC WASTE. FFVS ARE A NEAR-TERM BRIDGE TO A FUTURE WHERE WE END UP HAVING ALCOHOL STATIONS ALWAYS WITHIN RANGE.

BOB FITCH PHOTO, <WWW.BOBFITCHPHOTO.COM>

Fig. 22-1 GM tri-fuel. *This flexible-fuel vehicle made in Brazil runs on alcohol with 4% water in it, gasoline (which contains 25% alcohol in Brazil), and natural gas. The compression ratio is over 11:1, which is a compromise limited by the octane rating of the gasoline/alcohol blend sold at the pump. Both the straight alcohol and natural gas could tolerate a much higher compression ratio.*

for all the cars it sells, or pay hefty fees for going over the average. So the rising popularity of SUVs that dragged down the CAFE of the American companies cost them plenty, since they couldn't meet their goals.

"Democracy is two wolves and a lamb voting on what to have for lunch. Liberty is a well-armed lamb contesting the vote."

—BENJAMIN FRANKLIN

Now enter the chicken or the egg problem of alcohol vehicles and alcohol pumps. In the 1980s, oil companies complained that they shouldn't be required to sell alcohol at the pump if there weren't cars to run on it. Car companies said they couldn't make cars run on alcohol if there was nowhere to buy it at the pump. To break this impasse, two pieces of legislation were passed in the early 1990s. The first required government fleets to buy vehicles that could run on renewable fuels; the second gave car companies CAFE credits if they made cars that could run on alcohol.

So auto companies got their credits, which allowed them to build more highly profitable SUVs, and the government got to buy cars that could be powered by renewable ethanol. But there was no requirement in either of these laws that mandated that these vehicles had to actually run on alcohol to earn their benefits. So FFVs were the perfect political solution. In other words, they fulfilled the letter of the law without upsetting any apple carts at the oil company boardrooms or in Detroit, while state and local governments received funding to buy flexible-fuel vehicles. Of course, nothing much changed at all. The bottleneck of lack of distribution did not change—even though there were now millions of FFVs on the road—because they could still run on gasoline.

Part of the devil's bargain in this deal came in the demand of the auto companies to make E-85 the legal definition of alcohol fuel. The car companies said they'd make the FFVs, but not if they had to pay for installing cold-start devices. That would cost them $50 per car. Now you might say, "What's $50 on a $30,000 SUV?" Well, on a million SUVs it's $50 million of Detroit's favorite dollars.

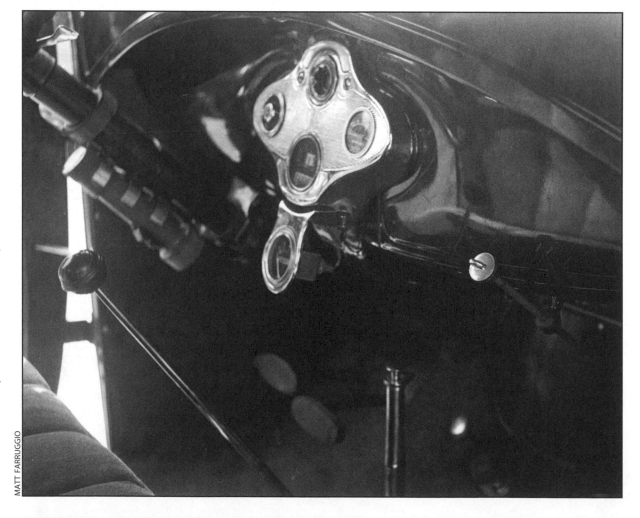

Fig. 22-2 Model A dual-fuel carburetor control. *The shiny knob on the right is for inserting or retracting the needle in a Model A carburetor's single jet. Depending on the setting, either gasoline or alcohol could be used.*

MATT FARRUGGIO

By permitting 15% gasoline to be added to alcohol and still call it alcohol fuel, vehicles can start in any weather without the inexpensive cold-start devices. So to allow the car companies to avoid this small expense, we are now saddled with E-85 and all the problems associated with trying to deal with toxic waste added to our otherwise clean alcohol fuel.

The other thing E-85 accomplished was the invisibility of FFVs. Most people who own them don't even know it. If they had to fill up their cold-start tank under the hood every six months, they would be quite aware that they owned an alcohol vehicle, something MegaOilron strenuously wanted to avoid.

In 2006, however, both Ford and GM changed their approach on FFVs, actively advertising their availability and even participating in a modest way in helping increase the number of alcohol pumps. We can all hope that this is the start of a duplication of Brazil, where in a few short years gasoline-only cars are becoming a thing of the past.

BASICS

An FFV engine can use any proportion of alcohol and gasoline in the same gas tank. In a way, it is what we always wanted in a car. The car runs on alcohol in town (95% of most people's driving), while retaining the ability to use gasoline for trips that might take you out of range of your alcohol source.

This is not the ideal way to run an engine on alcohol, since compromise is required on nearly every front, in order to accommodate burning the Oilygarchy's toxic waste. FFVs are a near-term bridge to a future where we end up having alcohol stations always within range. Our real visions will take the form of dedicated E-100 vehicles in the not-too-distant future.

The early line of flexible-fuel vehicles worked by having sophisticated software and a special fuel line sensor to tell the computer how much alcohol was in the mixture. Newer FFVs have done away with the fuel line sensor and just interpret data from the other standard oxygen sensors to provide the information needed by the ECU during and after each fill-up.

The computer, which has a fuel "map" for each increment of alcohol, relays how much alteration is needed for the particular mix of the day—to the car's fuel injection timing, ignition timing, and, in some cases, turbocharger. Using information from

REFORMING

Scientists at the U.S. National Research Energy Laboratory (NREL, formerly the Solar Energy Research Institute) have rediscovered a school of "vaporization" that was popular decades ago. This process is called "reforming." If you heat alcohol (or other fuels) in the absence of air to temperatures far above their boiling point, and duct them through certain catalytic metals, they will decompose into a variety of combustible gases. Different metals catalyze with different alcohols in different ways, and there are a variety of flammable gas byproducts, including hydrogen, carbon monoxide, methane, acetylene, and various aldehydes.

When this mixture is combusted, the thermal efficiency rivals that of hydrogen fuel cells, but without all the cost. Reforming can be done today using alcohol without a complete reworking of our fuel system's infrastructure; reformers could be manufactured to run home- or farm-scale cogenerators, getting a lot of people off the grid.

NREL's tests in the early 1980s on a methanol-fueled engine, which had a compression ratio of 14:1 and automatic spark advance, reached essentially double the mileage of alcohol's liquid mileage at 2000 rpm. This, of course, means that methanol was getting better mileage than gasoline. The difference with ethanol would be even greater. (More work has been done on methanol than ethanol due to research funding preferences.) The NREL engine fit underneath the hood of a Chevy Citation, and advancing technology has since meant smaller and more efficient engines.

Reforming is once again being worked on in a big way as part of the research for the front-end of fuel cell technology. Onboard reformers are considered the first step on the hydrogen fuel extravaganza. And, while most of the catalysts needed to reform gasoline to get hydrogen depend on pricey, monopoly-controlled platinum, several good catalysts for alcohol (nickel compounds, rhodium, and cerium) are cheap and widely available.

So, how does reforming take place? In a nutshell, alcohol is preheated to a boil by radiator water in a countercurrent heat exchanger. Then it's superheated by exhaust to very high temperatures and sent to a metal catalyst reactor. The issuing gases are cooled to a temperature (about 200°F) that doesn't tend to pre-ignite in the very high-compression engine that can be used. Radiator water is used as a coolant this time, rather than as a heating medium.

Depending on the catalyst, reactions occur at specific temperatures, and vapors inducted at different temperatures result in different byproduct breakdowns. Most of the catalytic metals require alcohol to be vaporized in the neighborhood of 600°C, but a few metals, such as nickel, yield excellent results—an incredibly clean combustion, high thermal efficiency, and very attractive mileage—at about half this temperature.

the oxygen sensor and the appropriate map, the fuel injectors open the optimum length of time to deliver the right amount of fuel to satisfy the oxygen sensor's limits. On alcohol, the injector would stay open approximately 80% of the time; on gasoline, it would open for just a very short period of time.

The map's main impact occurs before the oxygen sensor warms up, right after you start the vehicle. Until the oxygen sensor takes control of the engine in "closed loop operation," it is in "open loop" and needs data to substitute for the signal it would normally get from the oxygen sensor. FFVs have a different open loop data map at start-up than a normal car, to prevent them from running rough or too lean while the car is cold and to assist in starting on cool mornings.

"It has been the policy of every American president since Harry Truman, that as long as our energy resources are dependent on that part of the world (i.e., Middle East), we are going to be there in force."

—SENATOR JOHN MCCAIN (R-ARIZONA)

Although I don't have confirmation of this, I think some FFV vehicles also have a variable fuel pressure regulator to increase the fuel pressure on alcohol. My Ford Ranger specifies a 50-pound range of acceptable pressures for the fuel pump in the service manual. Also, the fuel pressure regulator, instead of being in its normal place on the fuel rail, is back in the fuel tank at the end of the return line. I speculate that it logistically makes sense to alter fuel pressure, so as to avoid the need for special fuel injectors that open for very brief times when running on gasoline and much longer times on alcohol. After all, if we are smart enough to figure out how to fool the ECU, the bright guys at Ford could have made it standard on the assembly line.

MODIFICATIONS IN FFVS

To squeeze better mileage out of an FFV running on E-100 (which you can produce or buy), heat the fuel, install an aftermarket fuel pressure regulator to increase fuel pressure for increased heating efficiency, and add a cold-start device. Adding an oxygen sensor fooler will add to your mileage as well. You will have to have a way of turning off the fuel heater and fooler if you start putting gasoline in the tank.

Overall, FFVs work okay. They are primarily designed to run on gasoline and are optimized for that, even though they will run on alcohol. The main problem is with the fuel line sensors, which don't seem to last and have to be replaced. These fuel line sensors add at least $50 to the cost of manufacturing the car (although you'll pay hundreds of dollars for a replacement).

Fuel lines have had to be replaced in some FFV vehicles made in countries where inexpensive urethane-based fuel lines were being used instead of the fuel lines normally used in the U.S. Also, in some early FFVs, a rubber part in the fuel pump was made of a cheap material that would wear out a little more quickly. Its failure rate was often just within the warranty period, so manufacturers had a motivation to use a material that would last a lot longer. Most now make all their pumps with Viton to avoid any potential problems.

In addition to fuel system modifications, FFVs also have a modified ignition system. Spark advance—the timing of when the spark plug fires—is altered to fire much earlier, depending on how much alcohol is in the tank. In a Ford Ranger, the gasoline spark timing is 40° before top dead center; on E-100, it is 60° before top dead center at one point in the acceleration cycle. This was simply a software change and didn't require any new hardware for the car companies. No real alterations were needed, except to extend the ignition maps in the ECU to work with alcohol. The ignition control is automatic when using a knock sensor, to advance the timing as much as possible.

BRAZIL

Flexible-fuel is the standard in Brazil. During the big spike in oil prices as the 21st century kicked off, Brazilians started to really see the difference in fuel prices. Gasoline was at least twice the cost of alcohol—and their gasoline already had 25% alcohol in it, which kept the gasoline price down compared to other parts of the world. People with gasoline cars started experimenting at the pump, mixing alcohol with the gasoline, calling the mixture "rabo de galo" after the Brazilian cocktail where two dissimilar liquors are mixed together.

As more and more Brazilians found out that the modern fuel injection systems in their gasoline cars would tolerate a high proportion of alcohol, it briefly looked like there might be a shortage. It's like when you have a potluck and some people bring vegetarian dishes, but the meat-eaters decide the veggie stuff looks better than what they are being served, and so they use up the meat-free food before the vegetarians can get their share.

So, several companies saw a clear market opportunity in selling FFVs in Brazil, the Volkswagen Gol being the first. About 30,000 FFVs were sold there in 2003, the first year they were introduced. In 2004, 330,000 FFVs were sold in Brazil, almost a third of the total automobile sales. By the third year, four out of five new cars sold were FFVs. In 2006, Volkswagen and Fiat announced they would no longer offer gasoline-only models. As several people there told us, when either alcohol producers or oil producers raise their prices, the driver can choose the other fuel. People felt that it would make both kinds of producers behave themselves, since neither will have a monopoly. By 2008, in Brazil, vehicles manufactured to run solely on toxic dinosaur fuel will probably follow those creatures into extinction.

One big difference between U.S. and Brazilian FFVs is the compression ratio of the engine. The compression ratio in GM cars in Brazil is between 10.5:1 and 11.2:1, depending on the model. This is the highest compression ratio that those cars can have and still use gasoline that has 25% alcohol. Any higher and you'd have pinging. American FFVs don't have elevated compression ratios, so they can't take advantage of alcohol's higher octane like the Brazilian FFVs do.

Since all the gasoline in Brazil contains 25% alcohol, and since vehicles manufactured there have higher compression to match the higher-octane fuel, FFV engineering is easier, since there is never an all-gasoline condition to deal with.

The engineers figured that with good information from the exhaust oxygen sensor, they could work out an algorithm to give the computer the information it needed without an expensive extra sensor. They switched to a more accurate heated oxygen sensor than was formerly being used in Brazil and which is more or less standard in the U.S., and figured out a clever strategy. Whenever the tank is filled, the float sensor sends a blip to the computer to alert it that the fuel mix will be changing. So, for a little while, exhaust sampling occurs more often in order to minutely monitor the changes in fuel combustion while the fuels are mixing. It takes maybe ten minutes of driving to fully mix two fuels together. When the changes

As more and more Brazilians found out that the modern fuel injection systems in their gasoline cars would tolerate a high proportion of alcohol, it briefly looked like there might be a shortage. It's like when you have a potluck and some people bring vegetarian dishes, but the meat-eaters decide the veggie stuff looks better than what they are being served, and so they use up the meat-free food before the vegetarians can get their share.

stabilize, the computer goes back to its normal rate of sampling the exhaust.

This technology is the basis of the GM flexible-fuel cars in Brazil today. In the U.S., GM is using the Brazilian technology in several 2006 models with FFV as a standard feature. There are no real hardware costs, just better software. I expect all American cars to be officially FFV in a few years, if any of the numerous bills requiring it ever get out of congressional committee. So far, most legislators are under the impression that making cars FFV is a big burden to the car companies and would increase consumer prices of vehicles. What a laugh!

Brazil has both old and new facilities making alcohol, and some of the older, less advanced continuous distilleries pass tiny plant particles all the way through distillation to the fuel storage tanks. GM found it necessary to add two inexpensive filters to protect their expensive fuel injector filter from tiny bits of organic matter. This is not necessary in the U.S.

VARIABLE-COMPRESSION ENGINES IN FFVS

No matter how you look at it, making a car a flexible-fuel vehicle means you have to make compromises with design to accommodate gasoline. High compression seems to be the most expensive characteristic to include in a flexible-fuel design.

Many researchers have approached the idea of having a variable compression ratio. An intriguing engine design was developed and built 25 years ago by researchers at San Jose State University in California. They broke the normal connecting rod from the piston into two pieces that a third rod connected to. This third rod, rotating adjacent to the main rods, would vary the length of the two-piece piston rod depending on engine need. It was a clever mechanical system that would be very easy to install on the assembly line for a flexible-fuel vehicle.

SAAB 2005 EQUALS VOLVO 1994

When I first heard about the Saab 9-5 turbo flexible-fuel engine, I was pretty excited. I thought this was a revolutionary new way to do flexible-fuel. What makes it work for both fuels is that the amount of turbocharger boost is maximized based on the knock sensor. The turbocharger can provide all the benefits of high compression for alcohol fuel, while dropping back to low compression when running on MegaOilron brew. On alcohol, the engine would tolerate a lot more boost before signaling the knock sensor than on gasoline.

I figured I'd never own a Saab like that, but I talked with the guys at IPD, a high-performance parts company in Portland, Oregon, and said, "It sure would be nice if Volvo would come out with something like that." They starting laughing and told me that Volvo came out with that exact system in 1994 on their 850 turbo.

A few weeks later at a contra dance, I was talking to Tom Fischer about alcohol and mentioned this recent development. He said, "You're telling me this since you know I have an 850 turbo, right?"

Well, one thing led to another, and he brought his car out to my place with a gallon of gas in it. We put about three gallons of alcohol in it and did everything we could to make it cough or sputter, and all it did was run more smoothly. We topped it up, so it was in effect running E-95, and went for a hard ride with steep hill climbing. The only thing Tom could notice was that it shifted more quietly and with less "clunk" between gears in his automatic transmission—alcohol's peak pressure better matched the torque curve of the transmission than gasoline's did. We took it up to 90 mph, up a steep highway, without a moment's hesitation or any flat spots. It definitely didn't have any of the symptoms of overly lean burning.

The Volvo 850 Turbo is an advanced flexible-fuel vehicle right out of the box.

So, it turns out that the Volvo 850 Turbo is an advanced flexible-fuel vehicle right out of the box. We did trip the "check engine" light since the engine was burning too clean, but since the car was pre-1996, the computer didn't put the car in limp mode. After his mechanic reset the check engine light, it never came on again. The adaptive learning function of the computer had "learned" the new fuel and was managing it properly.

Unlike the new Saab, the Volvo still lost a little mileage over gasoline. At press time, we were planning to see how to adjust the maximum turbocharge boost to get better mileage. And there are tricks with fuel heating that could make a difference, too. Adding an airflow engine tuner would allow us to fool the computer on the 850 into accepting less than stoichiometric air/fuel ratios, which is almost certainly what the Saab computer is capable of.

Other companies (e.g., Saab) have been experimenting with heads that are essentially hinged, with all moving parts for the valves incorporated into the head, so the entire head can move and increase the volume of the cylinder to lower pressure when necessary. Although it is not in production as of yet, this would make an excellent FFV design.

Sweden, which implemented an extensive E-85 distribution system virtually overnight, is starting to revolutionize the flexible-fuel market. It will probably be the first country to have high-performance dedicated alcohol vehicles rolling off the assembly line.

A very exciting FFV engine now in production there is the Saab (GM) 9-5. It has a variable compression ratio that permits the car to take full advantage of the high octane of whatever mix of alcohol is in the tank. Its adjustable turbocharger wastegate is controlled by the ECU, which determines the optimum compression via a knock sensor. As soon as a little pinging happens, the engine control opens the wastegate a little and drops the boost pressure! We should demand this be done in the U.S., too; then we would see equal or better mileage on alcohol versus gasoline.

You can find this setup on other older cars, too. Starting in 1993, Volvo front-wheel drive vehicles like the 850 have a vacuum-operated wastegate that adjusts the turbocharger pressure based on a knock sensor. The only thing you'd need to add to a system like this would be an airflow engine tuner or oxygen sensor fooler in order to run leaner than stoichiometric air/fuel mixture.

Saab exploited something very interesting in developing the 9-5 FFV engine, and they were able to get 15% better mileage than gasoline in the mid- to high-load ranges. They discovered that, unlike with gasoline, engines running alcohol needed no extra fuel for cooling the combustion chamber while climbing a hill or accelerating. This helped even out the small loss of mileage when cruising on alcohol, so miles per gallon with gasoline and alcohol were quite close, or even indistinguishable, depending on the kind of driving being done. There was some loss on highway driving (due to infrequent acceleration), but in city driving the difference could go either way.[2]

Variable-compression FFVs may be a new trend with the continuing expansion of natural gas fueling stations. Natural gas and propane have an even higher octane rating than ethanol, so either would seem a better match for alcohol in a dual-fueling arrangement than gasoline.

Some new cars are leaving the factory with stock, natural gas, sequential-port-injection systems. The Honda GX NGV has a high-compression engine (12.5:1) and multi-point-injected propane or natural gas fuel systems.

I haven't tried it yet, but I believe the GX would easily handle vaporized alcohol through its gaseous fuel injectors. I just know that you gear-heads out there could make this a natural gas/vaporized alcohol dual-fuel car if you put a couple of days' work into figuring it out (see next section).

General Motors already makes tri-fuel vehicles, primarily used by taxi drivers in Brazil, that run on alcohol, gasoline, and natural gas. In 2006, Volvo demonstrated a multi-fuel car that purportedly was able to optimize for each fuel, including natural gas and biogas.

PROPANE/ALCOHOL DUAL-FUEL SYSTEMS

Alcohol and propane are compatible fuels with very high octane ratings. Using propane with alcohol, rather than dual-fueling gasoline with alcohol, allows you to advance the timing dramatically, making both fuels more efficient. Advanced timing, which isn't possible with gasoline, also helps alleviate loss of power, a common beef about propane. Propane is available enough to be a practical long-distance fuel, while alcohol is at present more convenient when used locally.

Both fuels are extremely clean, extend engine life, and keep air pollution down (although propane doesn't really reduce greenhouse gas emissions). The only problem with propane is that it's a little hard on crankcase oil, since it burns much hotter than alcohol, or even gasoline.

A propane/alcohol dual-fuel system lets you start your engine on propane and switch over to alcohol. You won't need cold-start help in such a system, unless you happen to run out of propane on a cold morning. On older cars, the easiest way to accommodate a propane/alcohol dual-fuel system is to modify the propane/gasoline system—specifically, the gasoline side through which you'll be running your alcohol. Dual-fuel systems are designed so you can use your old carburetor or fuel injection system for the liquid fuel, and attach the propane carburetor right on top of it. Converting the pro-

pane carburetor is done the same way you would convert any gasoline carburetor.

Carbureted propane won't work very well on most electronic fuel-injected cars built after 1995, due to the **on-board diagnostics (OBDII)** protocol. Your check engine light will come on if you try to use carbureted propane. There are already a few independent conversion systems on the market that use propane fuel injection, either alone or as a dual-fuel system, that do work with modern engine computers. Aftermarket multi-port propane/natural gas dual-injection systems are now on the market that interface with OBDII and allow you to run both gaseous fuels and liquid fuels.

For liquid alcohol dual-fueling, as with the Honda GX NGV mentioned in the previous section, a TBI could just be dropped into the airstream for liquid alcohol dual-fueling, if vaporizing is something you don't want to take on. But since the car has a continuously variable transmission, it would automatically adapt to alcohol fuel's higher low-end torque characteristics. This ability of the transmission to downspeed should be quite beneficial in terms of mileage.

The GX engine comes close to being an ideal liquid-fueled stock alcohol engine, and deserves some real attention by engine buffs looking for an experimental vehicle. With a little fooling around for alcohol, it could easily be the greenest vehicle on the road. These cars have been sold for a few years, so it will soon be possible to buy used GX NGV engines from Japan and install them in a Civic body.

So, the upshot of our discussion on FFVs is that they're useful in the short run. While you can use some of the alterations we've talked about to squeeze more mileage out of your present-day FFV,

"A corporation, essentially, is a pile of money to which a number of persons have sold their moral allegiance. As such, unlike a person, a corporation does not age. It does not arrive, as most persons finally do, at a realization of the shortness and smallness of human lives; it does not come to see the future as the lifetime of the children and grandchildren of anybody in particular. It can experience no personal hope or remorse, no change of heart. It cannot humble itself. It goes about its business as if it were immortal, with the single purpose of becoming a bigger pile of money."

—WENDELL BERRY

dedicated alcohol engines are going to be more efficient and get better mileage. When there are enough alcohol fueling stations, the need to dual-fuel switch to gasoline will be as obsolete as the need for buggy whips.

But we need to be watchful on the road to a renewable future. Along the way, we will be offered a variety of *alternative* alcohols to ethanol. Other alcohols can be used as fuels, and you should know enough about them to stand up as an intelligent citizen when Presidents or legislators offer us alternative wolves in renewable sheep's clothing.

Endnotes

1. Dr. Josmar Pagliuso, communication with author, March 2006.

2. *New Ethanol FFV for Saab 9–5 Range*, General Motors, www.gm.com/company/gmability/adv_tech/100_news/saab_e85_092004.html (June 29, 2005).

Fig. 22-3

Fig. 22-4

There are hundreds of alcohols, ranging in form from liquids to greases to waxes. In addition to ethanol, several other alcohols—methanol, **butanol**, and, to a lesser degree, **propanol**—can be used as auto fuels.

Like most other organic chemicals, alcohols are named by the number of carbons that are in the molecular structure typical of all the chemicals in the same group (in this case, the alcohols group). Methanol has a single carbon, ethanol has two carbons, propanol has three carbons, and butanol is a four-carbon alcohol. (Beeswax is a 13-carbon alcohol.)

These other alcohols are a little different than ethanol. Each has advantages and disadvantages.

We won't say much about propanol here, since it is not generally considered a good candidate for auto fuel. Yeast make a tiny amount of it in the process of producing ethanol. Propanol is hard on fuel system materials. Commercially, it is made from petroleum products or propane, and its emissions are quite a bit higher than ethanol's.

METHANOL

At one time, methanol (**wood alcohol**) was commercially produced by the **destructive distillation** of wood. Destructive distillation is essentially how you make charcoal, by heating wood in the absence of oxygen. Instead of the wood burning, volatile components react and gas off, leaving behind the almost pure carbon of charcoal. Much of what gases off is methanol.

If methanol were made by destructive distillation from biomass, it would actually be a renewable fuel. But that's not the case right now. Today almost all methanol is made from natural gas (methane)—the cheapest source for methanol on the large scale. MegaOilron has periodically proposed that methanol could be made from coal, as well.

Small-scale production of methanol is generally neither economical nor desirable. But in the case

IN ADDITION TO ETHANOL, SEVERAL OTHER ALCOHOLS—METHANOL, BUTANOL, AND, TO A LESSER DEGREE, PROPANOL—CAN BE USED AS AUTO FUELS. THESE OTHER ALCOHOLS ARE A LITTLE DIFFERENT THAN ETHANOL, AND EACH HAS ADVANTAGES AND DISADVANTAGES.

Fig. 23-1

where people make charcoal, for instance in Haiti, arranging to condense the issuing vapors instead of venting them to the air would provide methanol as a useful byproduct for cooking. This more efficient use of all the wood would help reduce deforestation.

The biggest drawback to using methanol as fuel is its toxicity. Ingestion of small amounts of methanol is lethal—partially due to its decomposition into formaldehyde in your liver (methanol can literally pickle it, since formaldehyde is used to toxify and preserve tissues). Ingesting as little as 30 milliliters can kill you. And methanol does not have to be ingested to be fatal. Methanol can sicken or kill if it is in contact with skin for prolonged periods. Barrier creams, commonly used to prevent skin absorption of toxic chemicals, actually encourage the skin's absorption of methanol. M-85 (85% methanol/15% gasoline) is a blend proposed by natural gas and coal companies, and it tests many times more toxic than pure methanol! Combining methanol with gasoline helps them both pass right through your skin.

Methanol's emissions are cleaner than gasoline's, and can be nearly as clean as ethanol's, when it comes to the big three: CO, NOx and hydrocarbons. The real problems in methanol emissions come from chemical emissions that are relatively unique and endemic only to methanol. In addition to tailpipe emissions of poisonous formaldehyde and **formic acid** (bee sting venom), methanol emits methyl nitrite gas. Methyl nitrite is born out of a spontaneous reaction between unburned methanol and NO_2, after they leave a hot tailpipe for cooler air.

Methyl nitrite may form immediately behind an exhaust pipe in the cooling exhaust, or later in the chemical soup of smog. It's so poisonous that

---

A FEW DRUNKEN FISH

If an ocean tanker full of ethanol were to sink or spill, the alcohol would dilute almost immediately with seawater to below 100 proof, which would prevent any fiery spectacle. The biggest tragedy would be a few drunken fish. And sea life shows an amazing tolerance for alcohol of all kinds. Even methanol, an extremely toxic alcohol, could cause only localized damage to marine life. In fact, methanol is less poisonous to many forms of sea life than to humans. Some sea creatures even manufacture small quantities of methanol on their own.

METHANOL VERSUS SYNTHETIC GAS FROM METHANOL

ExxonMobil has patented a high-tech process to turn methanol into synthetic unleaded gas. Methanol is 105 octane, but ExxonMobil's synthetic fuel is only 90 octane. The process takes four gallons of methanol and makes about one and a half gallons of synthetic gasoline. In the words of early alcohol activists, that's like taking four pounds of steak and making it into a pound and a half of hamburger. A ton of coal converted to methanol is good for 2160 miles. Synthetic gasoline from the same ton of coal yields only 1296 miles. This scheme has been resurrected today in the pursuit of **dimethyl ether (DME)** as another wasteful process to make synthetic gasoline from coal or natural gas.

---

exposure to 250 ppm has killed laboratory rabbits within four hours. Levels from 20 to 3000 ppm have been recorded at the tailpipes of vehicles using methanol and methanol blends for fuel. The worst levels were recorded with a 20% methanol mix with gasoline.

Methanol has several problems when it comes to its use as a fuel. When you hear propaganda that "alcohol" is corrosive, what is cleverly left unsaid is which alcohol is being referred to. Methanol is corrosive to many metals and has a pronounced effect on aluminum, magnesium, zinc, brass, and solder (which last two both contain zinc), as well as pot metal (a mixture of metals used to make carburetors), and the plating in the metal fuel tanks on older vehicles. Methanol also has destructive effects on many plastics, and many other "rubber" products found in fuel systems. It harms standard lubricating oils and even has a slightly negative effect on neoprene or silicone elastomers. Auto racers who use a lot of methanol go to a lot of expense to use fuel system materials to counter methanol's corrosive nature, e.g., stainless steel fuel lines.

Methanol has only 40% of the heating value of gasoline, but burns with close to the efficiency per Btu of ethanol. Typically, it would lower the mileage per gallon by 30% or more compared to gasoline. Since methanol and ethanol have about the same octane rating, some of methanol's mileage loss is mitigated if the engine compression ratio is raised.

Methanol is going to be cheaper to make from fossil fuels for some time to come, but it will not reduce greenhouse emissions of carbon dioxide. Methanol's use in biodiesel production unfortunately makes "*bio*diesel" a misnomer.

BUTANOL

Butanol can be used as an auto fuel. Although it is currently produced from petrochemicals, it can be produced by fermentation of carbohydrates, and some fats. During World War II, butanol produced this way was used for making explosives, and many butanol plants sprang up around the country. One of the largest wartime plants was based in Peoria, Illinois. Many of these facilities were converted to making ethanol after the war.

The organism that converts carbohydrates to butanol is not a yeast, but a member of the *Clostridium* group of bacteria. Once the bacteria ferment the carbohydrates, you can then boil and condense the vapors out of the mash to 180 proof without a fractional distilling column. You end up with a mixture of approximately 30% acetone, 60% butanol, and 10% ethanol (with some water). This mixture is generally referred to as **ABE**.

The butanol and acetone tend to float on top of the ethanol and water. The three solvents tend to separate into three distinct layers after cooling. Adding various mineral salts to the mixture makes the three- phase separation even more distinct. You can literally (carefully) pour off, or drain off, each layer by itself. A **fractional distillation**, such as in our packed column designs, will distill acetone and ethanol first, leaving a relatively pure butanol to follow.

Sounds easy, right? But there are a number of problems and issues to consider.

One big problem with butanol fermentation is that it is toxic to the very bacteria that produce it. So in traditional butanol fermentation, once the concentration reaches 1.3%, it kills the bacteria that make it. (Yeast can tolerate more than 13% ethanol, or ten times as much ethanol as the bacteria can tolerate of butanol.) This very dilute mixture means that a lot of energy is used to boil the bacterial mash in order to distill off the relatively tiny amount of ABE.

Newer butanol fermentations exceed this limit a little bit by using two separate species of *Clostridium*. When *C. tyrobutyricum* and *C. acetobutylicum* are mixed, one of them turns carbohydrates into hydrogen and butyric acid, and the other turns the butyric acid to butanol. So in this fermentation, you end up with very little ethanol and acetone. Done this way, it's claimed that after distillation, you will get as much straight butanol from a bushel of corn as you would get ethanol, plus some hydrogen to boot. It's important to note that this process has not been used outside of laboratory scale, though.[1]

"The Stone Age did not end for lack of stone, and the Oil Age will end long before the world runs out of oil."

—SHEIK ZAKI YAMANI OF SAUDI ARABIA

Butanol fermentation is pretty smelly. Odors of rotting meat or gangrene are a result of other *Clostridium* species, to give you an idea of how bad the smell can be. It's not something you'd want to do in your garage in any large quantity without good odor control. And disposing of the smelly spent mash is an overlooked issue in butanol production.

Butanol is also quite toxic, but not nearly as deadly as methanol. Workers who routinely inhale butanol vapor are prone to lose their hearing, and suffer damage to their liver and kidneys. Absorption through skin has a similar effect to ingestion or inhalation. Less serious symptoms of butanol poisoning are neurological, e.g., blurred vision, nausea, and dizziness. Very good plant design is necessary to avoid exposure to butanol.

Butanol is often cited as being corrosive. Supposedly, it corrodes everything that methanol does, plus almost every other common rubber material used in automobiles. Modern fluoroelastomers may be immune. Very little modern work has been done on emissions, but at least one report claims a high level of cleanliness compared to gasoline—but still quite a bit dirtier than ethanol.[2]

Even with the drawbacks discussed above, butanol has some distinct advantages and potential uses which you may find valuable, depending on your overall plant design. First of all, butanol is a valuable commercial solvent selling for about $6 a gallon. Acetone is also priced higher than auto fuel.

These prices will rise along with the price of oil. So until the butanol and/or acetone markets are saturated, it makes more sense to sell butanol into the industrial chemical market instead of burning it as fuel.

One of these industrial markets is biodiesel production. In making biodiesel, alcohols are typically used as part of the process. Butanol is virtually insoluble in water. Making biodiesel with butanol, as compared to methanol or ethanol, is far more reliable (less water-sensitive). It also produces higher-mileage butyl esters, rather than methyl or ethyl esters. The value of butanol in the production of biodiesel would trump its use as a straight fuel.

A useful plus with butanol is that the bacteria that produce butanol can use the glycerin leftovers from biodiesel production to make ABE. The glycerin market is already oversaturated by biodiesel production. In typical permaculture fashion, making butanol from the glycerin waste would eliminate most of the cost of the butanol used in making the biodiesel. This would also produce biodiesel *without the use of fossil fuel* (methanol), and therefore make truly biological diesel, unlike the current product based on methyl esters.

Another advantage is that butanol is unregulated by the U.S. Treasury Department, since it's not a beverage alcohol. Its heating value is only about 15% less than gasoline's; with a slight enriching of the mixture, most vehicles will run it with little or no modification to the air/fuel system, other than replacing materials incompatible with it.

My favorite reason for incorporating butanol fermentation into plant design has to do with using cellulose feedstocks to make ethanol. *Clostridium* can make butanol using the five-carbon sugars that are not yeast-fermentable to ethanol—but which make up about half the sugars produced in the cellulose process. This gets around the temptation to use GMO yeast to turn the five-carbon sugars into ethanol. All you would have to do is take the sterile spent mash after distillation, adjust pH, and inoculate it with *Clostridium*.

In this way, butanol production offers an alternative to digesting the unfermentable sugars (or glycerin from biodiesel production) to methane. If the solvent market becomes saturated, dropping its price to the industrial market, butanol can be used as a relatively clean replacement for fuel oil for boiler fuel in the alcohol plant. And

yes, it could be another biofuel, used alone with converted fuel system components or harmlessly added to gasoline (while vehicles using that fuel still exist).

To make ethanol, methanol, or butanol, you have to use both heat energy and electricity. In a well-designed alcohol plant, you would use your surplus carbohydrates and/or lignin to satisfy your needs for these forms of energy.

But what about your home or business place? These need heat and electricity, too. In the next chapter, we'll discuss cogeneration as the way to provide these forms of energy—demonstrating that there's more to alcohol than just running your vehicles and chainsaws.

Endnotes

1. David Ramey, "Advances in Biofuels," *Acres U.S.A.* 34:10 (November 2004), 20.

2. Ramey.

CHAPTER 24

COGENERATION AND OTHER SYSTEMS TO PROVIDE ENERGY FROM ALCOHOL

Once you've freed yourself of gasoline addiction, there's an itch to escape the clutches of other monopolies. I can offer you some advice on loosening your local utility company's grip, either with an alcohol plant or simply as an alcohol consumer.

COGENERATION

You can, of course, use alcohol as a source for electricity by fueling an engine which, in turn, runs a generator. But there's more to be gained than this. Cogeneration is the harvesting of the heat

ONCE YOU'VE FREED YOURSELF OF GASOLINE

ADDICTION, THERE'S AN ITCH TO ESCAPE THE

CLUTCHES OF OTHER MONOPOLIES. I CAN OFFER YOU

SOME ADVICE ON LOOSENING YOUR LOCAL UTILITY

COMPANY'S GRIP, EITHER WITH AN ALCOHOL PLANT

OR SIMPLY AS AN ALCOHOL CONSUMER.

Fig. 24-1

produced when a fuel is used to generate electricity or mechanical energy. Commercially, this concept is called cogeneration or, more recently, **combined heat and power**.

An internal combustion engine gives you only about a fifth of the fuel's energy as work, in this case gasoline. The rest of the fuel's energy becomes waste (oops, I mean surplus) heat, sent to the radiator, and hot exhaust. So, your ICE is really a dandy heater that also produces some work as a byproduct. Normally, this heat is wasted when running a car or generator. Since heat is the major product of your fuel, it is a major permaculture design error to fail to put it to use.

Collecting heat from the engine jacket, the exhaust, and the oil will recover just about all of the surplus heat from an engine. Most of the heat is captured by engine coolant pumped through a heat exchange jacket around the cylinders. The heated fluid continues on to the radiator, which is a series of very flat tubes soldered to fins. The fluid flows through these tubes; the vehicle's speed and the radiator fan result in a fast flow of cooler air over the tubes and fins. This moving air absorbs the heat from the engine coolant. The fluid leaves the radiator cooled, and returns to the engine to absorb more heat.

Instead of pumping the engine's hot coolant to a radiator, you can use the engine's water pump to propel the heated water through a heat exchanger installed into a well-insulated hot water storage tank. Hot water leaving the engine will be around 200+°F—what we refer to as high-quality heat, useful in the alcohol plant but typically too hot for safe home use.

You can use the heat exchanger to bring a larger quantity of cold water up to 140°F, producing "low-quality" warm water. Water at this temperature would be useful for household hot water or heating through hot-water baseboard heaters, or for radiant floor heating in buildings, under greenhouse vegetable beds, or at the bottoms of fish tanks or methane digesters.

In an alcohol plant, we generate a lot of low-quality heat but use a lot of high-quality, very hot water. A cogeneration system lets you convert the low-quality warm water to high-quality heat (200+°F), which can be used for preheating mash, distilling alcohol, or cooking starchy feedstocks. You can save two-thirds of your process energy and produce all your electricity this way.

Remember, too, that much of the electricity used in an alcohol plant is for agitation and pumping. Both of these uses transfer much of the electricity to the mash as heat due to friction. You've probably experienced this yourself when using a blender, when the heat from the spinning blades warms up whatever you are blending.

HOT SANDWICH

If you really want to squeeze every last drop of potential heat from your engine, another source is engine oil. Engine oil contains the heat generated by friction of moving engine parts. Sandwich adapters bolt on under your oil filters to permit easily plumbed fittings to route your engine's hot oil to a separate heat exchange coil installed in your hot water storage tank (see Chapter 19, Figure 19-3 for a look at a **sandwich adapter**).

PLUG IN YOUR HYBRID?

One of the worst black-ideas-painted-green is the proposition of all-electric cars or plug-in hybrids. The proponents talk about how we could use little or no oil if we just plugged our cars in at night and charged their batteries at home. If you don't drive too far, the gasoline engine in the hybrid might never be needed during the day, and think about how clean the air would be.

But the electricity doesn't appear by magic. It's made from dirty coal or nuclear power for the most part. So when you charge batteries with grid electricity, the toxins are moved from the tailpipe to the power plant. What's worse is that nighttime charging makes a market for otherwise unmarketable, expensive, nighttime coal and nuclear power.

The only exception to this otherwise awful scenario is when you generate your own power cleanly from alcohol and use it to charge your electric- or alcohol-powered hybrid car batteries and heat your home at the same time. Solar electric panels are useless for charging your car at night, and even if you put solar electricity into the grid during the day, it wouldn't change the fact that nighttime plug-ins would make a market for dirty power after dark.

You can build your own cogeneration system relatively inexpensively. A used straight four- or six-cylinder engine mounted on a frame is the heart of the system. It can drive the generator either via the fan belt pulleys or by a direct coupling to the driveshaft. Put the frame in a shelter out of the weather or in your basement.

The best values are complete, used Japanese car engines. In Japan, to stimulate new car sales, vehicle registration fees go up each year. This results in a lot of two-year-old vehicles being traded in. These vehicles are then cut up into parts, which are then sold used around the world, since there are rarely tariffs on used parts. So, in effect, the car companies in Japan sell the vehicle twice. A complete 100-horsepower engine from the companies who deal them comes truly complete, with starter, alternator, water pump, distributor, etc., and costs a few hundred dollars.

In addition to capturing the heat normally wasted by the radiator, you can harvest surplus engine heat from other engine systems. Exhaust heat, which can be 1000°F, can be extracted in much the same way a distillery condenser extracts heat from alcohol vapor. Using a separate counterflow heat exchanger (see Chapter 10, Figure 10-43), exhaust gases flow down a center tube while water is run in the jacket around the center tube in a spiral path in the opposite direction.

You can run the water leaving the engine cooling jacket through the exhaust's heat exchanger jacket to bring the temperature up to well above the boiling point of plain water. The water entering the heat exchanger should be cooler, about 200°F, and will rise to over 250°F after absorbing heat from the exhaust. The engine coolant doesn't boil at 212°F because it contains antifreeze, which prevents boiling below about 275°F. This makes engine coolant with antifreeze a very high-quality heat source that can either heat the distillery or raise the temperature of a much larger quantity of cooler water.

To determine just how much heat energy a cogeneration system can provide your particular operation or home, consider your engine's

MICROTURBINE COGENERATORS

New players in the small-scale cogeneration field are small turbine-driven generators like the Capstone microturbine (below). Instead of an internal combustion engine, a small turbine engine, kind of like a mini jet engine, drives the generator. The big advantages of a turbine over an ICE are that there are so few moving parts, and that all the waste heat is recoverable in the exhaust rather than via the radiator/exhaust/oil recoveries needed for ICEs. Currently, turbines run on a variety of liquid or gaseous fuels and should soon be certified for use with ethanol.

Fig. 24-2 Capstone microturbine. Any one of these compact units, just running a few hours a week, could easily power your home and provide you with all the hot water you need.

fuel consumption at 1500 rpm, or at whatever rpm you're planning to run it. Don't run it at more than half the rpms you use for cruising down the highway. Let's say you're running a 7.5-kilowatt generator powered by a straight six-cylinder engine that uses seven gallons of alcohol per eight-hour day, generating for the operation of your still and your home. There are about 85,000 Btu per gallon (the so-called higher heating value) of alcohol, since you are recovering the waste heat. So total energy available for both electricity and heat is 595,000 Btu (seven gallons × 85,000 Btu).

Let's assume about 180,000 Btu go to produce nearly 53 kilowatt-hours of electricity per day (far more than your small alcohol plant will need). For discussion's sake, let's conservatively assume that you're 80% efficient in extracting the remaining 415,000 Btu as heat from the cooling jacket, oil, and exhaust, so you'll be recovering 334,400 Btu. This is quite a substantial sum of energy, especially when used as a booster to turn warm water into hot water. This many Btu should bring 3344 gallons of 110°F water up to 210°F.

Even more efficiency would be gained for heat capture if any surplus electricity ran a heat pump to add heat to stored water.

For home use, you'd have to run the cogenerator only once per week, or maybe twice per week in the winter. The electricity you produce would last you a week in most cases, stored either in batteries or in grid credit. Your hot water tank should be highly insulated. With R-50 insulation (one straw bale thick), you should lose only a few degrees from the tank over the week.

If you have a tank of alcohol at home to fuel your car, it can make sense to use some of the alcohol to heat and electrify your home. This would be not too different from having fuel oil or propane delivered to your home and using these fuels to run a cogenerator.

On the flip side, using alcohol as a generator fuel is a good way to get a permit to have an alcohol tank installed on your property. After all the permits and inspections are done, you can add your pump and filling nozzle to fuel your vehicle, too. In an alcohol plant, however, it would be a better design to use self-produced methane to power the

CAN YOU POWER YOUR HOUSE WITH YOUR CAR?

Although combined heat and electricity units (cogenerators) are available in sizes that would power small businesses and homes, it might make more sense to use the cogenerator you already own: your car. It has a small belt-driven alternator that produces around 50 amps of 12-volt electrical power for your vehicle's electrical system and charging the battery. But any offshore boat owner knows that 110- or 220-volt units that are only slightly larger than your car's alternator produce the voltage that's required to operate home appliances, as they do on cruising boats.

Back in the '70s at *Mother Earth News,* we built a custom bracket to mount a homemade 110-volt alternator, which we ran from a separate belt on the engine of our venerable Chevy pickup truck. We were able to use it to charge a battery bank at the Eco-Village that would keep the building lit and run construction tools. We also rigged bypasses on the radiator hoses to shunt the hot water through a heat exchanger installed in the building's insulated solar hot water tank. That old Chevy made enough heat for a lot of showers.

So the day might come that you arrive home from work and plug the house electrical and hot water system into your car for a few hours, while compressing the carbon dioxide exhaust into a storage tank to feed your organic greenhouse plants the next day. Too far out? In 2004, Dodge debuted a concept hybrid electric pickup truck for contractors that would generate all the electricity for the worksite.

So, Ford, GM, and Chrysler—I double dare you to come out with flex-fuel, factory-manufactured hookup systems like this for people's homes. (In the meantime, I guess we'll just have to do it ourselves.)

JACK DAILEY, COGENERATION PIONEER

While doing historical research at the Ford Museum, I found a fascinating story about how Henry Ford had Jack Dailey build him a generator for the Ford farm using a large Ford 8-cylinder engine. The generator powered the shop—and the house quite some distance away, which was connected by a service tunnel. After the generator was running, Ford had Dailey plumb the engine to send the hot water to the main house. They also recovered heat from the exhaust as part of making the generator quiet. This is the earliest example of an internal combustion engine cogenerator that I've found.

cogenerator, saving the more valuable alcohol for its primary purpose in running cars.

While there are many attractive methods for making electricity—windmills, hydroelectric, and photoelectric cells each have their place in particular situations—the problem, and biggest cost factor, with any of them is storage. Storing energy in batteries ensures that you save excess power production, and allows you power at night (in the case of photovoltaic cells) and on windless or cloudy days.

But you can expand your definition of what a battery is, thinking of it as any way to store energy. While your generator is running, you can store its heat in an insulated hot water "battery." You pump water uphill into a tank, or up to the top of your apartment house; this creates a "gravity battery," so you have pressurized water long after the cogenerator has been turned off.

You can have an air-conditioning battery by storing excess chilling capacity as ice, made by a heat pump driven directly by your engine or from surplus electricity production. That ice can cool your home or chill water to condense your alcohol vapor to liquid. You might use the ice to cool water that you then pump into a tank jacket to cool overheated fermentation tanks.

If you don't want to build your own cogenerator, several companies have made them since the 1970s. Fiat makes a variety of smaller cogeneration units that can run on both liquid and gaseous fuels. In the 1980s, Fiat's units had the flexibility

to run on gasoline, alcohol, methanol, methane, and natural gas. The product is still called the Fiat TOTEM, and it is still sold principally in developing countries. Nowadays, you'd have to convert it to use alcohol. Look for a proliferation of smaller cogenerators as utility prices climb. If it hasn't already occurred to you, your car can be used as a cogenerator when you come home from work (see sidebar "Can You Power Your House with Your Car?" on previous page).

HOOKING INTO THE GRID

In hooking your electricity production to the grid, you need to make sure your power matches theirs (if your power is out of phase with theirs, it's lost as heat in the transmission lines) and that no lineman working on a job gets zapped because there's power in a line he thought was shut off.

Once you're on-line, so to speak, your meter will turn in the usual direction when you're using utility company power, but when you're transmitting your excess power to them, your meter will run backward. At the end of the month, you find out who owes whom! Check with your state's public utility commission or similar agency for rules on power buyback. Then talk to your legislators to change the rules to favor small producers like yourself.

States vary in how they deal with buying or giving credit for power. In states that allow **net metering**, you may do your alcohol production seasonally, but earn electrical credit for an entire year's worth of energy from cogeneration. In some states, if your operation produces more than 50 or 100 kilowatts per hour, the utility is obligated to buy back your electricity for money.

Some states will allow you to credit photovoltaic electricity put into the grid—but not from a cogenerator, a clear form of discrimination against alcohol. In California, you get credit for what you put into the grid, which you can draw against later. But if you put in more than you get back in a year, the utility gets to keep and sell the surplus you gave them. This gives you much more incentive to store your energy as gravity, ice, or even hot water, rather than give your surplus electricity to the utility company for free.

"The farmer ultimately will not only drive his automobile and tractor and threshing machine with alcohol, but he will light his house with it, and his wife will have an alcohol stove on which she will do all her cooking."

—HENRY FORD, *DETROIT NEWS*, DECEMBER 13, 1916

HEATING, LIGHTING, AND COOLING WITH ALCOHOL

As we pointed out in Chapter 13, there are significant environmental and health benefits to cooking with alcohol. But alcohol can also be an ideal fuel to perform such other necessary functions of human habitation as heating, cooling, and lighting. In some cases, you might even argue that these uses preempt use of alcohol for vehicles. In much of the world, where there might be only one car per 50 to 100 people, the use of alcohol for vehicles is not relevant, but an alcohol-powered refrigerator may mean the difference between life and death.

Heating

Although oil is used primarily for transportation fuel in the United States, a significant amount is burned for heat. This is especially true on the East Coast and in the Midwest. On the West Coast, and in other scattered areas throughout the country, natural gas, which is less polluting than oil, is the primary heating fuel. But the peak of natural gas production is likely to be a short number of years after the peak of oil, so it really isn't a long-term solution.

Heating with fuel oil, and to a lesser degree with natural gas, is an expensive affair. The monthly cost of heating a home in New England in winter can often exceed the mortgage payments. Although insulation and solar heating are probably the best long-term strategies for homeowners trying to cut costs, if you're going to run your vehicle on alcohol, it can make sense to use alcohol for heating your home, as well. And if you're running an alcohol plant, the low wines that are full of fusel oil and other contaminants can be burned cleanly in your home heater.

Fuel oil burners are a crude but fairly efficient heating method. Fuel is injected as a spray through a nozzle into a combustion chamber, and burning gas then travels through a heat exchanger to warm your house. While you'll have to use almost a third more alcohol than you would fuel oil to heat your home, alcohol efficiency is a little higher—since its combustion results in exhaust gas with a higher steam content, and high-moisture gas transmits heat better than fuel oil exhaust.

The real savings in heat transfer and maintenance using alcohol is due to its clean-burning properties. Carbon and creosote that build up in the firetubes of fuel oil burners act as insulators, reducing heat transfer and running up bills every so often for cleaning. A burner run on alcohol will hardly ever need cleaning.

Converting a home fuel oil burner is quite a bit simpler than converting an automobile. There are two main adjustments to be made: First, using a crescent wrench or socket wrench, remove your present nozzle and replace it with a larger nozzle. The area of the new nozzle's orifice should be 35 to 40% larger than the original. This adjustment allows you to get the same amount of heat from your unit as you did on fuel oil. Second, since alcohol has its own oxygen, you should set your air control to shut draft down to a bare minimum. Allow just enough draft so your fire doesn't starve. Also, since fuel oil pumps rely on fuel oil for lubrication, you should add 1% biodiesel (or several percent of kerosene) to your alcohol to achieve the same effect.

As of this writing, it's more expensive to heat with alcohol than with natural gas. (But even with its lower heating value, alcohol is considerably less expensive than heating with fuel oil, and so much less disgusting.) This situation is not expected to last long. Natural gas prices will reach parity with oil soon. As that time draws near, you may want to consider heating with alcohol.

Lighting

Alcohol was once the most common illuminant in America, and it may come in handy again. You currently may have the luxury of electric lights at the flip of the switch, but a very large portion of the world does not. For much of the world, when the sun goes down, light goes away. It's hard for someone who lives with electricity to imagine what a change in lifestyle alcohol lamps can make to those who don't have lighting.

Alcohol mixed with a little turpentine (renewable wood sap) makes a very nice illuminant in wick-type lamps. Pure alcohol burns with a barely visible flame, so the addition of turpentine makes a nice warm yellow flame. Since alcohol burns extremely clean and carbon-free, it doesn't cause the mess and health issues of kerosene used for lamp fuel. It is also safer in the event of an accident, since an alcohol fire can be extinguished with water.

Cooling

Alcohol can also run a refrigerator. Although young people find it hard to believe, only a couple of generations ago people kept their food cold with ice that was delivered to the house once a week. And long before electricity came to rural America, people operated actual refrigerators on a flame produced by alcohol. Some RV refrigerators today still run on a flame, usually from propane.

The Icyball and other refrigerators took advantage of the low boiling point of ammonia. If you remember from the energy primer in Chapter 9, when a liquid expands to a gas, it cools and has the ability to absorb heat. The ammonia refrigerators used a small flame, from alcohol or other fuels, to cause the liquid ammonia to boil to a gas, which then went to chill the cabinet via heat exchange tubing. The warmed ammonia then circulated outside the cabinet and condensed to liquid, shedding its heat—no expensive heat pump or electricity was needed.

The disadvantage in the old refrigerators was that if a leak developed, ammonia, which is toxic to inhale, would be released. Using modern materials, this flaw is no longer an issue. Companies in Sweden and Brazil are now making these sorts of refrigerators again. In rural areas, in much of the world, these refrigerators would inexpensively permit medical clinics to keep perishable medicines cool, once again making an enormous difference. And off-the-grid homes in developed countries could also make use of this simple, reliable technology.

When new homes are built these days, buyers tend to demand "all-electric," with heating, lighting, cooling, and cooking all powered by electricity. If you own or buy one of these generally wasteful homes, you can certainly power it cleanly with your own cogenerator. But if we were to imagine the home of the future, wouldn't it to be nice to think of contractors building "all-alcohol"-powered homes, like the Grangers of the early 1900s envisioned for every farm in America?

So now we've learned how to run all our gasoline cars, trucks, chainsaws, and generators on alcohol. We even know we can cook and heat with it. That leaves us with just one more category of big users of fossil fuels—diesel engines. Biodiesel, made from vegetable oil, is the simple, renewable alternative to diesel fuel. It is very positive in regards to climate change gases, but is still about as dirty as gasoline when it comes to smog emissions. And it is tempting owners of tropical rainforests to cut them down to make palm oil. You'll be glad to know that we can largely replace diesel fuel with alcohol. Let's take a look at how to do it.

SKYSCRAPER ICE MOUNTAINS

In New York City, apartment building or condominium associations, often known as co-ops, are leaders in using cogeneration to provide inexpensive electricity, heating, and cooling for their buildings. New York also has a number of large apartment buildings that hold mountains of ice in their basements. Most of the ice is made during winter, when bitter cold air is circulated though the basement, which is being filled with water, to make an enormous block of ice. Excess electricity from cogeneration runs heat pumps to make more ice, even in the summer.

The ice block is fitted with heat exchangers through which nearly freezing water, or ice-chilled refrigerant, is pumped. In hot weather, the water travels up to the apartments, where it flows through chilling coils that absorb the heat from the rooms. The water then returns to the basement and is again chilled by the ice. This avoids most of the cost of expensive air conditioning.

Fig. 24-3
Iris DeMent.

"Wasteland of the Free"
Music and Lyrics © 1996 by Iris DeMent

Living in the wasteland of the free…

We got preachers dealing in politics and diamond mines
and their speech is growing increasingly unkind
They say they are Christ's disciples
but they don't look like Jesus to me
and it feels like I am living in the wasteland of the free

We got politicians running races on corporate cash
Now don't tell me they don't turn around and kiss them peoples' ass
You may call me old-fashioned
but that don't fit my picture of a true democracy
and it feels like I am living in the wasteland of the free

We got CEO's making two hundred times the workers' pay
but they'll fight like hell against raising the minimum wage

and If you don't like it, mister, they'll ship your job
to some third-world country 'cross the sea
and it feels like I am living in the wasteland of the free

Living in the wasteland of the free
where the poor have now become the enemy
Let's blame our troubles on the weak ones
Sounds like some kind of Hitler remedy
Living in the wasteland of the free

We got little kids with guns fighting inner city wars
So what do we do, we put these little kids behind prison doors
and we call ourselves the advanced civilization
that sounds like crap to me
and it feels like I am living in the wasteland of the free

We got high-school kids running 'round in Calvin Klein and Guess
who cannot pass a sixth-grade reading test
but if you ask them, they can tell you
the name of every crotch on MTV
and it feels like I am living in the wasteland of the free

We kill for oil, then we throw a party when we win
Some guy refuses to fight, and we call that the sin
but he's standing up for what he believes in
and that seems pretty damned American to me
and it feels like I am living in the wasteland of the free

Living in the wasteland of the free
where the poor have now become the enemy
Let's blame our troubles on the weak ones
Sounds like some kind of Hitler remedy
Living in the wasteland of the free

While we sit gloating in our greatness
justice is sinking to the bottom of the sea
Living in the wasteland of the free
Living in the wasteland of the free
Living in the wasteland of the free

CHAPTER 25

HOW DIESEL ENGINES CAN RUN ON ALCOHOL

Learning how to run diesel engines on alcohol gives us a peek at what alcohol engines of the future may be capable of doing. Diesel engines are used in heavy equipment, generators, cars, ships—almost anything requiring reliable, efficient, industrial power. They are noted for very high thermal efficiency, and the price of diesel fuel is often lower than gasoline. Diesel engines are increasingly being found in European cars, which get higher and higher mileage rates as improvements are made over the years.

Gasoline engines are spark-ignited, meaning they fire their fuel with spark plugs. Diesel engines

THE VERY THING THAT MAKES ALCOHOL AN IDEAL FUEL FOR GASOLINE ENGINES—ITS HIGH RESISTANCE TO PINGING, DUE TO ITS HIGH OCTANE RATING—WORKS AGAINST ITS EASY USE IN DIESEL ENGINES. LEARNING HOW TO RUN DIESEL ENGINES ON ALCOHOL GIVES US A PEEK AT WHAT ALCOHOL ENGINES OF THE FUTURE MAY BE CAPABLE OF DOING.

Fig. 25-1

MICHAEL KEEFE

are compression-ignition engines. This means that they compress the air/fuel mixture until it gets hot enough to explode by itself (auto-ignition) without the need for a spark. They work by intentional pinging (kind of). But since diesel engines are made to withstand the stresses of auto-ignition, it isn't dangerous.

Compression ratios in diesel engines are as high as 18:1 in tractor-trailers, and 23:1 in cars like the Volkswagen TDI. Very high compression gets more of the easily extractable energy from a fuel.

The very thing that makes alcohol an ideal fuel for a spark-ignited engine—its high resistance to pinging, due to its high octane rating—works against its easy use in diesel engines. Diesel fuel is given a cetane rating, which measures how easily it will "ping." It's a rating system that's the opposite of octane rating. Diesel fuel itself has a cetane rating of about 45, while ethanol has only 8 (and methanol a pitiful 3). This rating expresses itself as the auto-ignition temperatures of each fuel. Ethanol will self-ignite at 395°C; diesel at 245°C. So, diesel wants to explode more easily at a lower temperature. When pure alcohol finally does ignite, it does so at the wrong time in a diesel engine and causes damage.

Using alcohol in a diesel engine presents immediate general problems. Presently, very high-pressure diesel injection pumps and injectors require lubrication, which diesel fuel normally supplies. (Diesel fuel does not lubricate the engine, contrary to what many people think.) Alcohol doesn't provide enough lubrication by itself. Also, if water in the system gets past the fuel line water trap, it will damage the fuel pump. The most significant obstacle is the low cetane rating. But depending on which approach you use, anywhere from 50 to 85 to 100% of diesel fuel can be replaced with alcohol and/or a combination of biofuels.

Sulfur, heavy metals, HC, NOx, CO, dangerous chemicals, and carbon dioxide could all be reduced drastically under alcohol's influence. Research done at the Institute for Technological Research in Brazil has shown that even a 3% addition of ethanol mixed into diesel will significantly lower particulate (HC) emissions. In fact, CO emissions from diesel engines run on E-95 alcohol can even be slightly less than from gasoline engines running on alcohol.[1] New European demands for cleaner diesels are leading the way for better versions of these noxious polluters.

RIGHT: Fig. 25-2 Al Kasperson, renewable fuel pioneer, blending fuel. Al has just shaken three layers—diesel, biodiesel, and alcohol—which have blended into a homogeneous mixture.

BLENDING ALCOHOL AND DIESEL

Researchers are interested in finding ways to make alcohol and petroleum diesel mix. So far, most petroleum-based emulsifiers that have been proposed are too expensive to be worthwhile; these permit blending of 10% alcohol with petroleum diesel. Even this small addition dramatically drops some of diesel's emissions, but more can be done.

South Dakota farmers have discovered that biodiesel makes blending of dry alcohol and normal diesel possible in almost any proportion. Biodiesel mixes well with alcohol and, in theory, should provide the same versatility as castor oil (discussed below), at a lower price. What's more, biodiesel has the fuel pump lubrication that alcohol lacks. As mentioned above, even a 1% addition of biodiesel mixed with alcohol provides effective lubrication.

Most of the farmers' tests used 50% alcohol, and 25% each of normal diesel and biodiesel. Petroleum diesel was added primarily because it was cheaper than biodiesel at that time—clearly 50/50 alcohol and biodiesel should work equally well. The basic idea is that the biodiesel acts as the cetane improver and causes ignition at the right time, and the alcohol then goes right along with it.

More testing needs to be done to determine just how little biodiesel needs to be mixed with

DAVID BLUME

alcohol to burn properly without any modification to the diesel engine. I'm going to guess that it will probably be about 20% biodiesel with 80% alcohol, since that would closely imitate the castor oil mixes of the 1980s. But it might be lower than that. Any experimentation on your part should be done in a shop with a dynamometer and a knock meter to detect problems before they happen. Given that biodiesel will never amount to more than a few percent of the demand for diesel fuel, this would maximize its use, while alcohol would provide the bulk of the energy.

USING ALCOHOL IN THE NORMAL DIESEL FUEL INJECTION SYSTEM

There are a number of ways to use alcohol in the normal diesel fuel injection system, and they all require overcoming alcohol's low cetane rating. There are several sugarcane- or alcohol-based chemicals that, added in very small amounts, can raise alcohol's cetane number to acceptable levels. Some of these cetane improvers are isoamyl nitrate, diethyl nitrate, butyl nitrate, and ethylene glycol nitrate.

Triethylene glycol dinitrate (TEGDN) and **tetrahydrofurfural nitrate** were the most widespread cetane improvers used in Brazil in the 1980s. TEGDN is primarily derived from alcohol and is added at 5% as a cetane-improver—a huge improvement over previous petroleum-based types, which required a 12% additive ratio. Tetrahydrofurfural nitrate is based on sugarcane bagasse or any number of other biomass sources, such as rice hulls, corncobs, straw, etc. These biomass materials easily produce furfural (which is also found in fusel oils or can be an output of cellulosic alcohol processes). Furfural is then converted to tetrahydrofurfuralic alcohol, and then to its nitrate form of tetrahydrofurfural nitrate. You will need a little more tetrahydrofurfural nitrate than TEGDN in your mixture; experiment for best results.

Diethyl ether made from ethanol has been used at 15% or more to raise cetane ratings to acceptable levels, but there have been fewer experimental trials with this method, even though it is potentially attractive.

In any of the cetane improvement methods, it's necessary to address the problem of lubrication. During the 1980s, Mercedes-Benz produced and ran diesel buses running on straight alcohol. Mercedes-Benz mechanics routed an extra lubrication oil line from the normal engine oil pump into

---

SLIPPERY STUFF

Biodiesel is an incredible lubricant and has proven itself in ball scratch tests. This test uses a press and a ball on the end of a shaft pressing against a plate with some of the lubricant on it. At some point the pressure causes the ball to break free of the surface and scratch the plate.

The length of the scratch at a given pressure is a good measure of lubricity. Bigger test rigs have recorded pressures of over 6000 psi in standardized tests, before the biodiesel fails and a small scratch can be made. Engine pressures are far, far below this figure, which explains the life-extending characteristics of biodiesel.

---

the fuel injector pump, and a return line for excess oil to the crankcase. The engine oil passes through a filter and then goes directly into the injection pump housing. Since diesel crankcases are very large and oil changes are frequent, you don't have to worry that oil lost to the injector pump will endanger lubrication of the engine—oil ends up injected with alcohol into the engine, providing the pump and injectors more than enough lubrication. There's only a tiny amount of oil that finds its way into the fuel, about 0.1%.

With any cetane additive, you can eliminate the need for sending oil into the injector pump by adding 1% biodiesel to provide necessary lubrication for the injection pump.

About 50% more fuel is allowed through the control jet, so it takes more alcohol to run a standard diesel engine than diesel fuel. But alcohol does raise the horsepower and torque ratings about 20%!—still within allowable limits for a diesel engine without danger of runaway. When the conversion is limited to these simple changes, a small mileage loss often results.

Extensive tests on such fuel systems uncovered no unusual corrosion or wear. All zinc-coated parts had been stripped and plated with tin, since it was anticipated that wet alcohol would probably have some effect on the zinc.

DUAL-INJECTION CONVERSION

One effective conversion may be to use both fuels where they're best suited in the power cycle. For instance, there's more possibility you'll experience knocking during idle, so diesel fuel alone would be appropriate there. The increased power of alcohol is best used for acceleration, cruising, and heavy loads.

You can deal with lubrication problems by introducing your alcohol through a separate system from the high-pressure fuel injector pump. The Volvo company is a pioneer of the dual-injection approach. Volvo is able to use 80% alcohol injected by one injector and 20% diesel injected by another, with up to 50% water in the alcohol, and lose only 9.5% power. The trick seems to be injecting diesel fuel first and establishing a good hot combustion to ignite the alcohol introduced through separate injectors. This gets around the cetane issues nicely. At low loads and at idle, only diesel is injected.

This system may be the best one yet developed, but it is far too expensive to attempt retrofitting an engine.

It is similar to systems now used to run natural gas/diesel in compression-ignition engines, since natural gas is also a high-octane/low-cetane fuel. This system is rapidly spreading through city bus companies.

A diesel engine that is dual-fueled in this manner has nearly smokeless exhaust and amazingly clean emissions compared to diesel alone. This is a relief, as diesel emissions historically have been practically unregulated, yet have put out about 50 times the particulate pollution of a gasoline vehicle. Diesel emissions also contain major carcinogens, mutagens, and tetragens (chemicals that damage fetuses.) Although the emissions of pollutants are lower, natural gas/diesel engines still put out enormous amounts of carbon dioxide, since both fuels are fossil fuels.

CASTOR OIL CONVERSION

Although it's not common today, diesel conversion has been done in Brazil and some less-developed countries by mixing castor oil with wet alcohol in a regular fuel delivery system. Castor oil—added very precisely—is one of the few oils that will mix with alcohol and water. Too much castor oil and you'll get an unacceptably high carbon (HC) output. Too little castor oil and you'll get knocking, injector plunger seizing, and unacceptably high horsepower and torque readings. A mixture of 20% castor oil is just right for most engines.[2] Optimum compression ratio with castor oil seems to be 23 to 1.

Both castor oil and alcohol are less expensive than diesel, so the mixture should be quite cost-effective.[3] Castor oil looks even better when you realize that it permits the blending of hydrated ethanol (4% water). This small amount of water can dramatically drop the characteristically high NOx of diesel or biodiesel by cooling peak combustion temperatures.

FUMIGATION CONVERSION

Another conversion technique, **fumigation**, quickly became a best-selling kit in the 1980s in several U.S. farming communities. A carburetor is placed on the air intake, matched to the load by

Fig 25-3 Diesel injection/fumigation kit. In this popular 1980s kit, up to 80% of the diesel fuel was replaced with alcohol sprayed into the airstream. The device in the upper right-hand corner is the alcohol injector, which was installed downstream of the turbocharger.

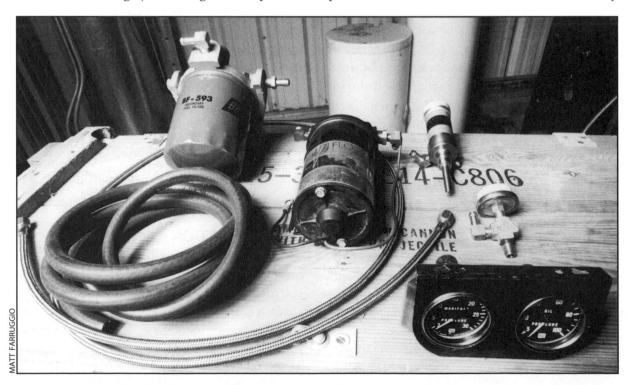

MATT FARRUGGIO

a throttle control. It's designed to introduce alcohol to the engine by way of the airstream, bypassing the present fuel injection system entirely. The diesel injection system is adjusted to deliver about half the usual amount of fuel.

This same system is commonly used now on Brazilian and Chinese tractors. Almost no alcohol is used at idle, but, under a load, up to 50% diesel can be replaced by alcohol of about 140 proof. You have to be sure not to exceed the engine's maximum torque and horsepower ratings. If they get too high, the engine can "run away" (see Chapter 21, section on Two-Stroke Engines) and be severely damaged.

It's recommended that you use a dynamometer to check your power output when converting by using carbureted (fumigated) alcohol. Fumigation kits are most practical for tractors and other high-torque, relatively steady-speed applications (e.g. generators). I've had reports that there may be increased engine wear to turbocharger vanes when using this system.

Antonio Moreira dos Santos of Brazil and others have done research, injecting (instead of carbureting) alcohol downstream of the turbocharger, which safely resulted in increases in power and decreases in particulate matter. Advanced injection timing can result in up to 80% diesel fuel replacement while doing post-turbo-injected fumigation. This is essentially an aftermarket method of achieving the dual-injection method used on city buses, mentioned earlier, with a similar level of fuel replacement. A 1980s U.S. system, which no longer seems to be on the market, used a fuel injector inserted after the turbocharger. This seems like a superior design, since fuel delivery was matched to engine demand by sensing turbocharging pressure, and the system was usable by over-the-road, diesel semi-tractor trucks.

HYBRID SPARK-IGNITED/ DIESEL ENGINES

A more technologically advanced diesel conversion might be called a high-compression spark-assisted diesel, or an indirect-injection stratified-charge alcohol engine. Translation: a hybridization of spark-ignited and diesel-type engines. By using a hot spot, like a spark plug or a glow plug, alcohol ignition can be controlled and fired at the right time in the diesel engine.

This system works on small diesel automobile engines, where many other approaches are too expensive to consider. In many cases, an engine's compression ratio has to be lowered. This is

Fig. 25-4 Injection/ fumigation system diagram.

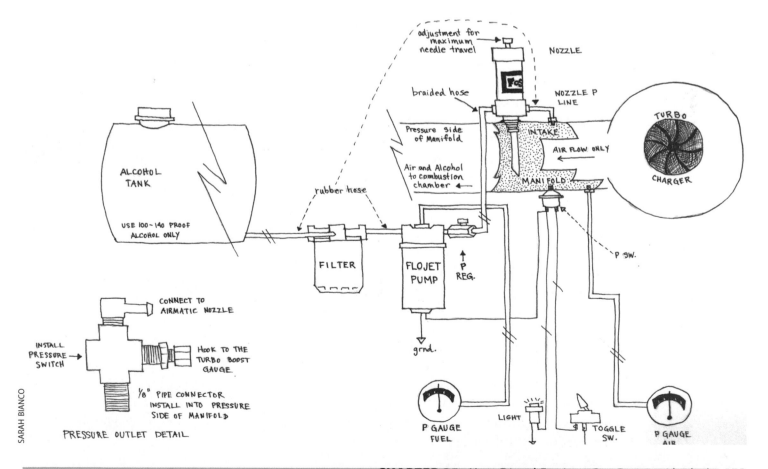

accomplished by removing some of the piston top, or adding a thicker head gasket or spacer between the head and block.

Santa Clara University researcher H. G. Adelman added two head gaskets to a Volkswagen Rabbit engine to decompress the engine to 16.5:1, and added a distributor timed from the camshaft. He replaced the glow plugs with spark plugs and used Redline alcohol-soluble oil mixed 1–3% with the alcohol. At high speeds, the engine operated like a diesel engine, without using the spark plugs—it wasn't necessary to disconnect them; just turning them off was enough. At low speeds and idle, the spark ignition was needed to fire the fuel at the right time. Thermal efficiency was about 5 to 15% less than diesel—but then, no effort was made to optimize efficiency. As discussed in Chapter 13, a more advanced version of this concept, using a VW TDI engine, actually achieved 22% better mileage on alcohol than on diesel.

Archer Daniels Midland did a series of tests on part of its fleet of over-the-road semi-tractor trucks using E-95 (5% gasoline). The primary technique was to leave the glow plugs, which are normally used only on cold mornings, operating full time. This provided a hot spot in the cylinder that would ignite the E-95 at the proper time. ADM's tests were successful but not as efficient as more extensive conversions, since alcohol mileage suffered. But it was a very inexpensive way to convert common, moderately large-capacity diesel engines.

VAPORIZED ALCOHOL ENGINES

The new frontier in diesel engine research right now involves the use of vaporized instead of liquid alcohol in what are termed homogeneous charge compression-ignition (HCCI) engines. These engines, with the heat energy of vaporization, will work with straight alcohol at any speed and eliminate the need for spark at idle. Experimenters at the University of California's Lawrence Berkeley National Laboratory were able to use low-proof alcohol effectively when fully vaporized.

What I see in the near future for dedicated alcohol-fueled engines is a convergence of the technologies of older vaporized alcohol engines and high-compression diesel-like engines, with or without spark-assist, using easily distilled and energy-efficient 180-proof alcohol that completely replaces petroleum diesel and gasoline.

With that, we conclude the Alcohol as Fuel section of this book. I invite you to re-read Chapter 13 (Surprise! Ethanol Is the Perfect Fuel!) with your now deeper understanding of why what's written there is truly revolutionary.

Endnotes

1. Dr. Josmar Pagliuso, communication with author, March 2006.

2. G.W. Phillips, *Series 53 Engine GMC*, for Detroit Diesel Allison, division of GMC.

3. Author's calculation.

BOOK 5

THE BUSINESS OF ALCOHOL: HANDS-ON ADVICE

Now that you know how to produce and use alcohol, some of you will want to go beyond making it. The pioneers among you will want to put what you've learned into practice and go up against MegaOilron, beating them at the game of business. You CAN do it. You have the better product, at a lower price. And with some clever business structures, you can even make use of the tax credits the oil companies normally harvest when they buy alcohol to mix with their toxic brew.

If you are new to business, I have a couple of essential references for you to pick up. The first is *Small Time Operator,* by Bernard Kamoroff, the classic easy-to-understand book on how to take care of your bookkeeping and business paperwork. The other one is *The Seven Laws of Money,* by Michael Phillips and Salli Rasberry. A better book on guerilla marketing has never been written. So go to it, and beat them at their own game.

CHAPTER 26

ECONOMIC AND LEGAL CONSIDERATIONS

So now you know how to make your alcohol and run your car and equipment on it. Next comes the hard part: making a living doing it and keeping the bureaucrats fed and happy at the same time. Although we've all been steeped in the Horatio Alger view of America, where all you have to do is work hard and your labor will be rewarded, the reality is that 95% of small businesses fail within five years.

I was lucky that in my youth I was a member of the Briarpatch Network in the San Francisco Bay Area. Long before anyone had a concept of green business, the Briarpatch formed among people who felt that their businesses should reflect their personal and even spiritual values. We believed that honesty in all our dealings was key, and we maintained open books. After all, our competitors knew what was in them, since their books were probably just about the same.

While employees had a pretty good idea what was going on financially, they tended to believe our businesses were doing better than they actually were. So who were we keeping secrets from? Our customers—the people we depend on for our living? Bad idea.

So, most Briarpatch businesses would publish abbreviated financial statements, and periodically post them at their shops or in their newsletters. This almost always had unexpected positive results because employees figured out where the company could make more money, or save on costs. It also engendered trust from customers, some of whom might be the next lender to the company during a time of expansion.

The keys to staying out of trouble in business rest on a few basic things. Keep good records; without them, you can't really know what's going on. Don't let things get behind. Ask everyone you do business with, from employees to vendors to customers, how things might be done better.

One of the best things about Briarpatch was that when one of us felt stuck, we could reach out to

SO NOW YOU KNOW HOW TO MAKE YOUR ALCOHOL AND RUN YOUR CAR AND EQUIPMENT ON IT. NOW COMES THE HARD PART: MAKING A LIVING DOING IT AND KEEPING THE BUREAUCRATS FED AND HAPPY AT THE SAME TIME. THE REALITY IS THAT 95% OF SMALL BUSINESSES FAIL WITHIN FIVE YEARS.

COPYRIGHT 1992–2002 ANDREW B. SINGER

Fig. 26-1

the membership and ask for a consultation. A team that thought they had the expertise to help would assemble for lunch, and we'd work out the problem together. I encourage you to look for opportunities like this in running your own part of the alcohol movement/business.

At a micro scale, it is relatively easy to use waste products for either energy or feedstock, and it is easier to market the primary co-products at either retail or a good solid wholesale price, since the volume is not enormous.

ECONOMICS OF ALCOHOL FUEL PRODUCTION

Typically, alcohol is produced in plants making 50 million gallons per year, with purchased corn and fossil fuels, at below a dollar a gallon. A micro-system is often the most expensive way to produce alcohol, if you ignore the co-products, since small manufacturers can't get the same price breaks for what they need and also have higher labor costs.

But the many profitable opportunities easily make up for the disadvantages. Ingenuity and creative thinking can be rewarded in the micro-plant. At this scale, it is relatively easy to use waste products for either energy or feedstock, and it is easier to market the primary co-products at either retail or a good solid wholesale price, since the volume is not enormous.

In painting a picture of what the economics of micro-scale production could look like, we are conservatively presenting a worst-case scenario. To allow you to compare different possibilities, we'll be looking at three simple examples—two using starch-based feedstocks and one using a sugar-based feedstock.

Fig. 26-2

All three scenarios assume some level of multiple product outcomes from the micro-plant. There will be primary products that undergo little or no post-alcohol-production processing. Primary products are those that don't need to be changed into something else before you market them, e.g., hot water, carbon dioxide, or wet distiller's grains as an animal feed. There will also be three secondary products that all use the primary co-products to add value.

All the energy costs involved in alcohol fuel production are attributed in these examples. We're using wood biomass as the fuel, simply to narrow down the variables. Reuse of the energy input is described, either within the ongoing alcohol process or in creating the co-products, especially the secondary ones. The calculations are based on what it takes to make a gallon of alcohol at the rate of 100 gallons a day, as in Chapter 12. For simplicity's sake, no scenario using methane as the energy source is included.

Most of these calculations for alcohol production are based on my own experience with the Liberator 925, a combination cooker/fermenter/distillery that I designed back in the 1980s. It is close to the model in Chapter 12, and is capable of producing 9000 gallons of 190-proof ethanol fuel annually, running every fourth day.

A larger, but still "small," alcohol plant able to run every day would use three separate fermenters and a separate cooker and would certainly produce alcohol for less cost per gallon, recycling energy more efficiently at lower labor per gallon, with a lower capital investment. Quite a substantial business would be possible producing 60,000 gallons per year with a modest increase in capital costs. Lessons learned in the micro scale will transfer to this small scale.

To keep things simple, I have drawn a line between the plant and the farm it could (or should) be part of to allow us to focus on the alcohol-specific components. So, while some of the benefits and products aren't discussed as giving value, they would clearly make a difference in a farmer's bottom line. For instance, fish water/manure is a value to the farmer that we don't look at here; if we raised shrimp, the harvested shrimp shell would be useful to control symphylans, a soil pest. So, we're describing only a selection of co-products, not an exhaustive treatment of the value of all the potential stuff we could get out of these things from the farmer's point of view.

For some of the co-products, I have chosen retail marketing, and for others I have specified wholesale marketing. I do this cautiously so as not to paint an overly rosy picture of potential income. Overproducing a co-product may force you to sell it wholesale in order to reach a larger market to absorb your production. Once again, my choices are not all written in stone; they are fully adjustable based on the local market and on what you find enjoyable to do.

Figure 26-3 shows a few of the potential co-products from these three feedstock processing scenarios, to give you a grasp of where the value-added markets might be.

The calculations in this chapter take into account the VEETC (Volumetric Ethanol Excise Tax Credit) and deduct it from the production price, but it is also possible to sell the alcohol for 51 cents/gallon more and pass the tax credit on to the drivers. Also included is the ten-cent-per-gallon **small producer tax credit** that is also being passed through to the community-supported energy (CSE) **limited liability corporation (LLC)**. Although many states have tax credits, **tax deductions**, or even cash payments for producers, I am not including any of these in the calculations, since they vary so much.

The price of alcohol fuel is most heavily influenced by two things: the feedstock cost, and the use and marketing of co-products (both primary and secondary). Obviously, bought feedstock is more expensive than free-to-haul feedstock, but co-products could make purchase worthwhile.

I've attempted to list my assumptions regarding cost, maintenance, and so forth, so you can make revisions specific to your own situation. For the high-cost example, I am using purchased corn; for the medium example, I am using waste

doughnuts (a feedstock on which I have a lot of data); and for the cheapest, I am using seasonal fruit waste.

The price of alcohol fuel is most heavily influenced by two things: the feedstock cost, and the use and marketing of co-products (both primary and secondary).

Ways to Save More Money

With a little additional capital investment, plant operators with an entrepreneurial sense can cut many of the costs. For instance, since heat is a large part of the final cost of your alcohol, locating a waste source for wood can substantially reduce production costs. Using waste materials from sources near your plant site dramatically reduces collection costs. You might consider setting up near a company with big refrigerated storage and collect heat from its refrigeration system, as the Heintz brothers demonstrate in Chapter 27.

For a reasonable fee, you can often subcontract to haul out waste wood from tree-trimming companies, hauling directly from where they are cutting, so they can avoid the labor and dump costs associated with disposal of the wood. But if your local landfill charges a lot per cubic yard, the companies may be willing to go out of their way to dump the wood for cheap or free at your place.

Thin stillage is usually enough to produce all the methane you need to fire the plant. If it is rich in fats, it puts out a lot more gas, something to consider if wood waste or orchard prunings are not available for process heat. If you live anywhere near chicken "farms" or feedlots, you may find you can have as much manure and bedding delivered as you need to produce methane. Your best bet may be to deal directly with the hauling company charged to dump it, rather than the company that produced it.

A primary co-product of alcohol production is hot water. This represents recovery of about 70% of the process heat, which can be used to offset the cost of energy in the next batch or, more efficiently, in the generation of co-products.

Barter may be an effective means to market co-products—in exchange for restaurant dinners, credit against your grocery bill at a market, or in direct trade with someone who produces something you don't. If you raise fish next to a corn

Expenses and Credits for Three Feedstocks

INPUT COSTS	FRUIT CULL	DONUTS	CORN
Feedstock	$ (.15)[1]	.10[2]	.96[3]
Yeast[4]	.01	.01	.01
Enzymes	.10[5]	.05[6]	.06[7]
Miscellaneous Consumable[8]	.03	.03	.03
Energy[9]	.07[10]	.00[11]	.10[12]
Electricity[13]	.035	.035	.035
Cost of Distillery[14]	.083	.083	.083
Maintenance	.02	.02	.02
Labor[15]	.60	.60	.60
Total Input Costs to Produce One Gallon	$ .80	.93	1.90
TAX CREDITS, REIMBURSABLE AS CASH			
Federal Producer's Small Plant Credit	$ (.10)	(.10)	(.10)
VEETC	(.51)	(.51)	(.51)
State Producer's Credit (Counted as Zero)	(.00 to .30)	(.00 to .30)	(.00 to .30)
Gross Production Cost per Gallon after Credits[16]	$ .19	.32	1.29
CREDITS FOR DIRECT BYPRODUCTS			
Hot Water Over 180°, Stored[17]	$ (.05)	(.07)	(.07)
Wet Mash Sold for Dairy or Cattle	(.20)[18]	.00[19]	(.87)[20]
Carbon Dioxide, 6.5 Pounds.[21]	(.65)	(.65)	(.65)
Net Cost per Gallon after Direct Byproducts[22]	$ (.71)	(.40)	(.30)
SECONDARY PRODUCT POTENTIAL[23]			
Type of Mash Byproduct	Skins/Pulp	Liquid	WDG[24]
Mash Byproducts per Gallon (Approximate)	8 lbs.	10 gals.	7 lbs.
Potential for Mushrooms	$ 40.00[25]	25.00[26]	42.50[27]
Potential for Worm Castings[28]	16.00[29]	.00	16.00[30]
Potential for Fish[31]	50.00[32]	60.00[33]	43.75[34]

DAVID BLUME

1. Hauling costs, charging $5 per yard to haul away, in season.
2. Hauling costs @ no charge to source.
3. $2.50 per bushel of corn, @ 2.6 gallons per bushel.
4. Purchase price 1/3 of a cent, but there are materials costs to breed up for inoculation.
5. Pectinase dissolves pectin to sugars and makes fermentable sugars more available.
6. Standard starch-reducing enzymes; less needed due to fine grind of flour.
7. Starch-reducing enzymes.
8. Acids, alkalis, testing materials, yeast nutrients.
9. Distilling to 180 proof and then putting vapor through a pressure swing corn grit dryer to bring final proof up to 200. Assume wood at $160 per cord and starting from room temperature.
10. Wood heat—less than corn due to simpler process.
11. Energy for heat provided by donut fat.
12. Same as #9, but uses more energy for starch process.
13. For pumps and fans; 0.035 kilowatts at $.10/kWh.
14. Stainless steel tank, cob or brick firebox, automatic reflux control, hot water storage tank, heat pump condenser, diaphragm pump, alcohol receiving/distributing tank, non-vacuum 1500-gallon tank, Lightnin mixer, used, with 20-year life. $5,000 if self-build; $15,000 if contract welding, etc. Figure is based on $15,000 at 9000 gallons per year with no depreciation credited.

15. Assume near-automation on distillery, so it doesn't need babysitting. Labor is to load and heat, add yeast, and later distill for three hours. Marketing delivery of products takes three hours @ $10 per hour.
16. This averages $.60 per gallon across the three examples.
17. Figured as Btu wood-equivalent cost. Assume recovery of 70% as 150°F hot water. Can be used for byproduct industries or for next alcohol run.
18. Dewatered fruit pulp price on a dry-weight basis.
19. The thin stillage could be converted to single-cell protein, two lbs. per gallon or $.22, but not included here.
20. $80.00 per ton equivalent to dry weight.
21. $.10 per lb., sold to local greenhouse growers or welders, although it would be worth much more in the growing of high-value vegetables on-site. Cost of compressing equipment is not included here, since it would be less than one cent per gallon of alcohol produced.
22. These totals would be for sale off-farm. If secondary materials are produced, then the primary co-products would not be sold but would instead be used in making secondary products.
23. Gross income only. No tertiary byproducts by using waste of secondary products, except as noted. Value per gallon of alcohol produced.
24. Wet distiller's grains.
25. Eight lbs. of dry pulp with eight lbs. of additional straw soaked in DS, making 16 pounds of mushrooms, at $2.50 per lb.
26. Eight lbs. of straw soaked in DS, yielding ten lbs. of mushrooms.
27. Seven pounds of DDG plus eight lbs. of straw soaked in DS, making 17 lbs. of mushrooms.
28. $1 per lb. wholesale castings, at 50% moisture.
29. Fruit pulp and 50% straw produce 16 lbs. of worm castings, sold at 50% moisture content @ $1 per lb.
30. DDG plus straw yields 16 lbs. castings @ $1.
31. Simplified figure of $10/lb. of fish as combination of male fish live retail and balance as fish emulsion. Fish is an alternative to worm castings.
32 Assumes pulp mixed with an equal weight of straw to make mushrooms first. Byproduct of mushroom production feed to fish. Five lbs. of fish biomass @ conversion rate of 1.6:1.
33. Estimated 6 pounds of fish from using 10 gallons DS liquid and carbon dioxide to produce spirulina to feed fish.
34. WDG converted to fish at 1.6 lbs. per lb. of fish.

Fig. 26-3
Expenses and credits for three feedstocks.

farmer, you ought to be able to work out a deal where you get corn at the end of the season and he gets thousands of gallons of your fish water. If he has to borrow less money for inputs (fertilizer), his risk of failure goes way down when he can pay you in corn at the end of the season.

Tax Incentives

Tax incentives consist of tax credits or tax deductions. Tax credits are subtracted from your total amount of tax due. Tax deductions are subtracted from total gross income before you calculate your taxes.

For many years, the U.S. tax credit for alcohol fuel was a complicated affair that was actually a credit against the federal excise tax on all fuel sold for public highway use. If the alcohol was used for any other purpose, such as for powering machinery or off-road equipment like tractors, you didn't get the credit. This did not provide much incentive for farmers to run their tractors on ethanol. Over time, the tax credit evolved to become even more difficult to claim, and only oil companies buying alcohol to blend with gasoline were able to get the credit.

But, in 2005, a law created the Volumetric Ethanol Excise Tax Credit (VEETC), which can be claimed by anyone in the sales chain of the fuel right down to the retailer. As long as you have a resale terminal, you can also be a distributor, jobber, refiner, or even producer. This is so much better than the former law, since the fuel can be sold without regard for the tax status of the buyer, or what the buyer uses the fuel for. Now you can get the credit for using the fuel to power your generator or off-road equipment, or for heating and lighting your home.

The catch: In order to be eligible for VEETC for alcohol production or use, you need to be a business. (In the final chapter, we talk about a pretty painless way that consumers can get the credit without having to become a stand-alone business.) Happily, being a business is almost always to your advantage. If you're not a business now, you should become one. Hop on down to your county courthouse and fill out the registration form. You'll probably have to go through logbooks of business names to make sure the name you've chosen hasn't already been taken ("Microsoft," for example). In most cases, after you've checked the logbooks, you turn in the application with a check to the city and/or county, then publish the new name of your business in a local paper a certain number of times. Voilà! You're a business.

One of the questions on the application asks for the "purpose" of your business. "Research and development of fuel alcohol processes" is a good

Fig. 26-4

answer. For tax purposes, you are doing business as (dba) a sole proprietor.

You can have a job and still be a business. Any income you earn (in a job, for instance) or loss you sustain (from your new sideline business, for instance) is considered part of your total income or loss as far as the IRS is concerned. Lots of things you aren't able to write off as a citizen, you can write off or deduct from your business income, such as all of your health costs and some or all of your trip to Mexico (since you toured the Corona beer plant "to study fermentation methods"). Of course, I'm not giving you tax advice; that is for your CPA to do.

Federal Tax Benefits

For small-scale producers, defined as under 60 million gallons per year, there is a federal ten-cent-per-gallon producer's tax credit. The producer can choose to keep it or pass it on to retailers or buyers of the fuel. Between the VEETC and the small producer credit, you are looking at a total tax credit of 61 cents per gallon off your federal taxes. So your business can submit a bill for 61 cents per gallon for tax credits, which is refundable as cash!

If you form a fuel co-op as a limited liability corporation (LLC), the tax credits are passed down to the co-op owners/members to write off on their personal taxes (for a simple way to do that, see Chapter 29).

If your business sells alcohol to co-op members, but their intended use is for off-road consumption, you still get the 61 cents per gallon refund, but you don't pay the road tax! (For off-road use, the alcohol is dyed a different color, like farm diesel.) If the user decides to put the alcohol in a cogenerator instead of a tractor, it is still exempt from the road tax, but still qualifies for the tax credit.

In the 2005 energy bill, a couple of new things came into existence. The costs of putting in fuel-dispensing equipment or stations for renewable fuels now qualify for a tax credit. The current credit is for 30% of the cost of establishing the refueling site, up to a maximum of $30,000. If you are a corporation or even a sole proprietorship, you can include most or all of the costs to calculate the credit. That doesn't include any permit fees, staff costs to obtain use permits or building permits, etc. This credit doesn't apply just to big public stations; it also applies to your home (business) fuel dispensing equipment.

Also buried in the bill is a system of renewable energy "credits" that are not tax credits. These credits accrue if you make your alcohol from cellulosic material or if your plant is powered more than 90% by renewable energy. You can earn 2.5 credits for each, for a total of five credits. So what do they mean? Since the federal government has established a Renewable Fuel Standard (RFS) for auto fuel, oil companies must buy a certain amount of renewables to mix with gasoline. But instead of actually buying the fuel, they can buy credits to meet their obligations under RFS. However, the amount of alcohol being produced by all the big plants coming on-line will surpass the RFS. So the credits may end up being worthless.

The bigger issue is that as long as the oil companies have a near-monopoly on distribution of auto fuel, they will manipulate the market. Let's take the summer of 2006 as an example and then discuss how the credits might be used by a savvy producer as a hedge against market manipulation.

Oil companies first bid the price of alcohol up, claiming that the elimination of MTBE from auto fuel was to blame. The American Petroleum Institute portrayed the increase in alcohol's price as a free-market problem, demand outstripping supply. In reality, since oil companies are the only customer for 99% of the alcohol produced (only 1% is sold as E-85) what oil companies will pay for alcohol is totally under their control.

Running up the price had two effects. The oil companies could avoid the embarrassment of having E-85 at the pump for a dollar less per gallon than gasoline (which was $3 per gallon in summer 2006); and the higher price provided fodder for a massive PR campaign casting alcohol as too expensive for the consumer.

Immediately following a devastating article in *Consumer Reports* that dutifully related the oil company position that E-85 was a bad deal for the consumer, the oil companies, with their disinformation campaign in place, boycotted purchasing alcohol in excess of what was required by the RFS, crashing the alcohol futures market by 40% nearly overnight. So, once the disinformation went as far as they could make it go, they dropped their purchases and crashed the market.

Producing your fuel renewably or from cellulose provides a hedge against manipulation, since oil companies would prefer to buy your credits rather than buy alcohol fuel when they are in a "crush the competition" cycle. They have historically boycotted buying alcohol to disrupt the alcohol business and create a lack of confidence, which makes

getting capital difficult. So, at those times, they would rather buy credits than fuel—and if you have credits to sell, that will help protect you from losses when the price drops through the floor as a result of their boycott. In such a case, all the producers, desperate to sell fuel, drop their prices. Producers using renewable energy will be less subject to the artificial volatility of manipulated pricing.

State Tax Benefits

At last count, 36 states have ethanol-related tax incentives: 22 states have incentives to support ethanol production, 32 states have incentives to support use of ethanol, and 18 states have both production-side and application-side incentives. There are also a small but growing number of states that are providing tax credits for establishing E-85 stations, in addition to the federal credits.

Some states give relief from property or other taxes on the cost of building plants, but the primary measures used by states to support expansion of ethanol production are given to producers. Production-based tax credits, currently ranging from 7.5 to 30 cents per gallon of ethanol, either reduce liability for state income tax or transfer a reduced tax liability to the marketer on the ultimate sales of motor fuel. These credits are independent of the federal benefits. You will have to check with your state to see if there are tax credits for production. Check <www.permaculture.com> for periodic updates on state tax credits.

Some states still have deductions, such as no income or sales taxes on the first so many millions of gallons produced. These tax incentives are designed to help attract alcohol distilleries to be built in the state, and should apply to your small plant, as well.

Most state ethanol production incentive programs have legislative and/or administrative regulations defining eligibility, maximum amounts claimable (per facility and/or in total), effective time period, and other terms and conditions. For instance, Minnesota's ethanol producer payment program, enacted in 1986, offered a direct producer payment of 20 cents per gallon for up to 15 million gallons per year of ethanol production per facility over a ten-year operating period. As of 2002, the state was budgeting approximately $34 million per year in producer payments to 14 producers for about 170 million gallons per year of ethanol production. A year later, in the face of severe state budget cuts, the Minnesota legislature reduced the producer payment to 13 cents per gallon, with provisions to reimburse producers the lost seven cents per gallon in future years. The program is scheduled to end payments to producers in 2010.

Some states employ one or more forms of inducement to marketing ethanol-blended gasoline, the purchase of flexible-fuel vehicles, the installation of E-85 fueling facilities, and/or the marketing or purchase of E-85 fuel.

More than half of U.S. states have passed incentives related to various alternative fuels.[1] These vary from executive orders and state agency requirements to more aggressive incentives that provide tax credits for installing fueling equipment and procuring vehicles. The incentives tend to reflect the level of local support for alternative energy sources.

Several states have adopted a tax rate based on a fuel's energy content. In 1998, the state of Minnesota was selected as a national E-85 pilot market by the U.S. Department of Energy Clean Cities Program and other public and private partners.[2] Early on, representatives to this coalition recognized the inequality of the state motor fuel tax. Through legislative action, the Minnesota motor fuel tax was corrected to represent "energy parity" for the fuels.

They figured that since E-85 fuel does contain 29% fewer Btu per gallon based on heating value compared to gasoline, the state motor fuels tax on E-85 should be 29% less than the tax on gasoline. (MegaOilron's propaganda about alcohol's heating value being an indication of mileage came back to haunt them here.) So, the state of Minnesota implemented a $0.142 per gallon rate for alcohol fuel, to equalize the tax an E-85 user pays on a Btu-per-mile basis; gasoline users pay a tax of $0.20 per gallon. This revised policy applies to the use of gasoline, E-85, natural gas, propane, and other fuels in the state of Minnesota.[3]

In January 2005, the fuel tax on ethanol in North Dakota was reduced from 21 cents to one cent.[4] North Dakotan government officials believed that such a significant tax reduction could likely reduce the price of ethanol available at the pump and increase the interest and use of ethanol in the state. Some North Dakotans fear that since the legislation does not guarantee a lower price for motorists,

Costs and Credits for E-85

	NEW INSTALLATION	CONVERSION OF PETROLEUM SYSTEM
Cost	$ 40,000	$ 5000
Income Tax Credit (50%)	20,000	2500
Renewable/Retail Multiplier (× .125)	5000	625
Total Value of Credit	$ 25,000	$ 3125

YOU CAN THANK MEGAOILRON

It used to be that you had to have a tax liability to get any benefit from alcohol fuel tax credits. MegaOilron buys 99% of all the alcohol produced for fuel in the U.S., and, as we are all painfully aware, it's a master of tax avoidance. So Big Oil demanded—and got—the tax credits for alcohol fuel to be *reimbursable as cash!*

This also applies when we claim the tax credits from our production, or when we buy alcohol for our driver co-ops with the tax credits intact. Currently, the forms are filled out once a month, showing how much fuel was purchased, or produced if you are a producer and keeping the credit. You get a check back from Uncle Sam in 90–120 days. So co-op members would get a 61-cent-per-gallon rebate check a few months after they buy their fuel.

Who knows if this great arrangement will continue, but enjoy it while you can. You could find yourself in the position of making doughnut alcohol for 30 cents a gallon in material and energy costs, and then getting a check for twice that amount for doing your patriotic duty of getting America off of transnational trash. Kind of boggles the mind, doesn't it?

fuel sellers could just pocket the tax savings. But even if that were to happen, farmer/producers would still make and spend their additional profit, generating taxes for the state.

Colorado and Kansas have incentive programs for the building of alternative fuel stations, which provide excellent templates for other states aiming to advance the use of renewables-based fuel technologies.[5] For tax years prior to January 1, 2011, Colorado income tax credits are available for the construction, reconstruction, or acquisition of an alternative fuel refueling facility, directly used for the storage, compression, charging, or dispensing of alternative fuels to motor vehicles. The percentage of the income tax credit varies according to the tax year in which the costs are incurred. In 2005, the tax credit is 50% of the cost of establishing the station.

In the Colorado program (similar to the one in Kansas), E-85 is eligible for an increased income tax credit. The percentage of the credit is multiplied by 1.25 if either of two standards is met: 70% or more of the alternative fuel dispensed each year by the refueling facility must be derived from a renewable energy source (e.g., E-85) for ten years (certification is required); and/or the fueling facility is generally accessible for use by persons other than the person claiming the income tax credit. So, for example, a 50% credit becomes 62.5% in the case of a retail E-85 service station facility installed in 2005. A caveat: Because the tax credit rate varies by year, if it goes down in future years and you don't keep the station renewable for ten years, you'll owe all the back taxes.

Consider the examples for E-85 equipment placed into service before January 1, 2006 (see Figure 26-5).

With a small band of advocates lobbying for it, the Colorado law could probably be passed in every state in the U.S. There have been numerous attempts to have these station-building credits made a national law, but each time it's been killed in committee. So for now, I think it's up to each of us to make an appointment with our state representatives and push for these benefits locally. Don't waste energy on Congress.

The bottleneck right now in getting alcohol fuel to the public is the distribution system. With the improved federal credits and Colorado-style state credits, community-supported-energy (CSE) projects (driver-owned stations) would be able to

easily attract private capital to build this alternative system. The funders, as members of the limited liability corporation, would get the generous station credits in exchange for lending the money to the LLC at a very low interest rate.

Go to our website, <www.permaculture.com>, for tools to pass alternative fuel incentive programs in your state, and to join our legislative alert email list so you can be notified when important bills are pending.

Just as George Washington used the government to tax his competing small distillers out of existence, big business can use the regulatory system to make it hard for small entities to enter or compete.

LEGAL CONSIDERATIONS

When it comes to legalities, it's important to know how the U.S. legal system works, as compared to say Europe or other places. It's a main reason why people want to live in the U.S. In many parts of the world, the basis for the legal system is that, "If it isn't specifically permitted, it's prohibited." But in the U.S., sort of the Wild West of business law, the rule has always been, "If it isn't specifically prohibited, it's permitted."

This has both good and bad aspects to it. On the one hand, for those starting something new, it is oftentimes far easier to just go do it if there's nothing on the books about it. On the other hand, this legal attitude leaves the door open to abuses by corporations that simply say, "Well, there's no law against it."

And, just as George Washington used the government to tax his competing small distillers out of existence, big business can use the regulatory system to make it hard for small entities to enter or compete. This has regularly been true when it comes to alcohol as a fuel. Thus, at many times, those making and using alcohol did so in acts of civil disobedience.

That's all well and good for pioneers and the courageous, but most of us would rather not have to face the might of those in power. So much of what I discuss in this section is how to weave your way legally through the red tape thickets and how to use loopholes to legally get your way when the bureaucracy appears to be a solid unscalable wall. I also cover ways to have a good, fun time fighting and winning against city hall when you need to.

Legal Structures for a Fuel Co-Op

The primary reason to set up any sort of corporation is to protect all the people involved in the activities of the alcohol venture from having any personal liability. Someone suing a corporation, whether it's a nonprofit, LLC, or a standard C or S corporation, can realistically sue only for the corporation's assets. Assets of any members or directors are usually at risk only if there is flagrant criminal negligence on the part of any of them. In the litigious United States, you must protect yourself to keep the lawyers from eating you alive.

A lot of people who get fired up over my community-supported-energy idea ask me if they should be a nonprofit corporation. You might be surprised to hear that the answer is usually no. Normally, the use of a nonprofit corporate structure is called for if you know you can receive a grant to establish your alcohol plant or distribution company. These are hard to come by.

Most of the tax benefits that accrue to alcohol fuel producers and distributors are irrelevant to nonprofit corporations, since they don't pay any taxes.

So if you don't care about the tax credits and you do expect to get grants, you could set up as a nonprofit. The new VEETC law does make it somewhat more possible for a nonprofit to pass along credits, but it is much easier for a for-profit entity.

My general recommendation from the point of view of cost, least amount of paperwork, and the most flexibility in allocating tax credits is to form a limited liability corporation (LLC). This gives you the liability protection of a corporation, but without a lot of the paperwork and overhead costs. You can set up your by-laws in a very flexible way. An LLC can take advantage of creative ways of using tax credits. It can have nonprofit members (e.g., school districts or tax-exempt environmental, youth, or senior groups), and their unused tax credit can be given to other members who can use it. The nonprofits can receive fuel at a discount.

In forming your CSE, it should be relatively easy to find a backer who will put up the money at, say, 5% interest for you to establish your station. He would get the juicy 30-92.5% total tax credits in return for financing you. Payback of his capital and

Fig. 26-6

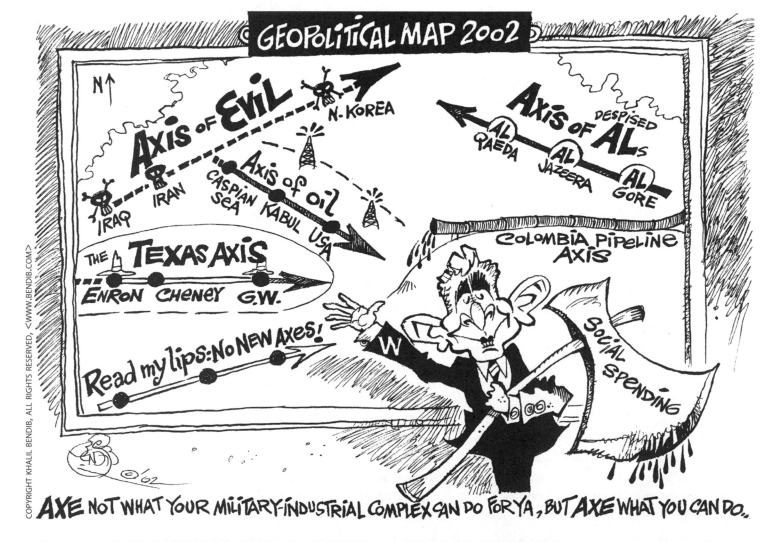

interest could be done by the LLC taxing itself ten cents per gallon until the loan is paid off.

But if 100 people put up $500 to $1000 each to establish the station, they would get the tax credits instead of an outside investor. If you're a natural-born organizer, that could be the way to go.

If the project becomes larger, holding substantial assets (e.g., a farm to produce the CSE fuel, or real estate for distribution stations), you will probably want to set up a separate LLC simply to hold the assets and lease them back to the LLC that does day-to-day operations. If the LLC distribution station is only leasing the land and equipment from the LLC holding company (for, say, one-tenth of a cent per gallon), then there are no assets to be taken if the distribution station loses a lawsuit. If the operating LLC is sued, this structure helps put the legal fire out right away, as soon as the attorney bringing the suit realizes he would be suing a company with no assets. The two LLCs can even have the same board members, but they are two separate entities in the law, and you can't sue one for actions of the other. They really do have to be separate entities, though, with two insurance policies, etc.

This is the usual way corporations do it, and so should you, after you get advice from your attorney. Far be it from me to give you legal advice. That's not legal.

Legalities of Car Conversion

This is another example of how regulation is designed to benefit corporations and keep the little guy out of the market. Even back in the 1980s, during the last big upsurge in interest in alcohol fuel, it was technically illegal to run your car on alcohol—and it still is. The Environmental Protection Agency (EPA) has prohibitive rules for automobile manufacture.

So in order to protect both the environment and the car engine's catalytic converter, which is quite delicate, you are not technically allowed to alter anything on your fuel system unless the modification is certified with the EPA or, in my home state, the far more onerous California Air Resources Board (CARB).

This no-tamper law makes some sense. It does keep people from souping up their cars for high performance and making a lot of pollution. Also, early smog-prevention equipment robbed power or mileage from vehicles, so people would remove it. But nowadays, smog controls do not lower mile-age or reduce power, so most of the motivation for the law has evaporated.

None of these regulations should be applied to discourage alcohol fuel usage. Logically, if we pass our smog test with our modifications, then we are legal—and if we don't, we have to get our vehicles in compliance. Well, nowadays, part of passing the smog test in many states, such as California, is demonstrating you haven't tampered with your car or added any equipment that isn't tested and certified legal by CARB. Thus, for now, the only cars that are legally able to run on alcohol in states with CARB-style regulation are certified flexible-fuel vehicles sold by major auto manufacturers. At this point, the EPA is fully in favor of alcohol fuel, even though the old prohibitions are still on the books.

Now, it turns out that CARB is partially made up of people formerly employed by MegaOilron, who hate alcohol with a passion. There are also those in CARB who think otherwise, so there is an internal battle going on which may shake up the long-standing posture against alcohol conversion.

Fortunately, most states currently take a more relaxed attitude about alcohol fuel and only require you to pass the smog test to get your car registered. And some states don't even require smog tests. They don't feel obligated to enforce the EPA rules in their state. (In reality, regional offices of the EPA support alcohol and haven't busted anyone who converted their car, to the best of my knowledge. But, technically, they could.)

So why does this matter to you? Because most states don't have the budget to do their own air quality studies, so they simply adopt California's standards.

Most states have an experimental permit, which is usually easy to get. In California there isn't even a form, just a letter will do. Obviously, there had to be a way to legally run test vehicles for the certification process. This permit's purpose is for you to experiment with a conversion system that you intend to someday get certified. You could renew it every few years, claiming you are testing a newer generation that you aren't ready to take to certification yet. But this loophole could get tighter at any time.

One useful thing to realize is that enforcement is pretty loose. I mean, no one is stopping cars to see if they are running too clean on alcohol. They nail you when you go for your smog test. So, depending on the type of conversion you have made, you might need to take fuel heaters or other visible alterations off the car for the day of the test, and then put them back in place after the test. Think of it as a tune-up, only it's a de-tune-up.

I have no problem breaking this sort of law, because my actions in running on alcohol are upholding the spirit of the Clean Air Act by dramatically reducing my vehicles' emissions. Could you imagine this case in front of a jury? I can just see it now; the attorney for CARB would say, "Well, the defendant here modified his car, which lowered his emissions by 95%. It's your job as the jury to convict him of breaking the clean air laws."

Also, take a look at what your state sets as the cut-off date for smog testing. In California, it's 1974, so anything before that is unregulated. I sure miss my 1953 Chevy 1-1/2-ton fire truck. It ran great on alcohol and cost me about $50 to convert.

The other legality you have to think about is your car's warranty. If you own a flexible-fuel vehicle, you can run on E-98 without voiding the warranty; legally, E-85 is defined as a blend of alcohol and gasoline which is at least 70% alcohol to as much as 100% alcohol. So E-98 is E-85 in the eyes of warranties and laws.

If, on the other hand, you run your new Prius on E-98 even though it isn't a certified flexible-fuel vehicle, simply because you know it will work, you run a small risk of problems with your warranty. Since cars supposedly have been fully alcohol-compatible since the early 1980s, there shouldn't be any part of the fuel system that is attacked by alcohol.

If you are worried that a dealer mechanic might raise a stink, you add the stink first. Simply adding a little gasoline before you go to the shop should clear the air, so to speak. It is not within the experience of most shops to question the composition of the fuel when doing repairs. As long as you put some gas in the alcohol, no one will know the difference. Well, they might comment on how clean everything is in your cylinders if they take a look, or how your spark plugs look just like new, or how, when they take it on a test drive, it sure has a lot of pep. Running on alcohol, especially with synthetic polyol ester motor oil, should triple your engine

Fig. 26-7

life, so you aren't likely to run into a warranty issue on your engine anyway.

Bear in mind that if you do tell them you used more alcohol than is specified in the warranty, some shops will use this as a very convenient excuse to make you pay full fare on the repairs. I've already heard of an outrageous case in California where a shop did just this with no evidence that the part failure was due to higher than average alcohol fuel content.

But if you are running an unmodified post-1996 car on "too much" alcohol, you could be running so clean and lean that the darn "check engine" light could come on. That light means that the engine is running outside of the limits the computer expects to see in order to protect the catalytic converter. Even though you are not endangering your catalytic converter, the light will go on if it is not getting enough pollutants to burn. So it indicates you need to dirty up your exhaust a little with gasoline, since the computer doesn't quite know what to do when the unmodified engine is too clean-burning. Some mechanics are now beginning to learn to reprogram the memory in the OBDII system so that the check engine light won't come on when you run too "lean" too long.

Legalities of Production

The right to distill one's own alcohol has been an issue in every free country. The rulers of virtually every country tax alcohol in some way, and the United States is no exception. After Prohibition, alcohol was heavily taxed and regulated, and the shadows are still with us today. In the old days, producers of alcohol for industrial use were strictly regulated in order to discourage the diversion of industrial alcohol to the illicit beverage market. Regulation in general favors large entities that can deal with the burden of compliance, while small entities go out of business.

When I first started making alcohol in the '70s, it was still illegal to produce it. Microbreweries were not permitted, and making your own beer and wine, other than a small amount for home consumption, was illegal. I still vividly remember that in the early '80s the tax on beverage alcohol was $21 per gallon, and it cost the former **Bureau of Alcohol, Tobacco, and Firearms (BATF**, a division of the "Treasury Department") $20 per gallon to enforce the law. I doubt the net is much higher than this today.

"As the President said today, there's not enough oil out there to meet the demands we have. Honestly, we have got to reduce our demands so that we need to have a surplus of energy to invest in the renewables. We should've started 25 years ago when we absolutely knew that Hubbert was right. So we have, in a very real sense, blown 25 years. Putting it off is going to make it just more and more painful and more expensive."

—REPRESENTATIVE ROSCOE BARTLETT (R-MARYLAND), APRIL 27, 2005

But this ensured that only large beverage distilleries stayed afloat, since small companies could not bear the regulatory burden. Only corporations big enough to support the huge expense of meeting the regulations and paying a minimum $10,000 bond could manage. There are companies today whose sole purpose is to navigate and file the paperwork that goes along with operating beverage distilleries.

Along came the energy crisis of the '70s. Those few people who knew that alcohol was a cheap fuel began to look for legal ways to produce what they needed for their vehicles. The BATF responded by confiscating fuel stills and arresting and fining the owners.

Then in the late '70s, some of us used our brains and delved into the arcane regulations that oversaw beverage production. It was discovered that there was a provision for a simplified license for experimental distilleries. Although it was intended to help big companies do research and development for new distillation techniques without having to go through a lot of red tape, it was enough of a loophole to theoretically give small distillers a way to legally produce.

The first small distillers to apply were rebuffed, but then it was time to bring democracy to bear on this fiefdom. *Mother Earth News* and other magazines wrote about alcohol, the immoral busts, and the experimental permit, and encouraged people to apply for the license. Although no one knows the exact count for sure, best estimates are that over a quarter-million people sent in requests—and there were only two BATF clerks assigned to process permits.

When the experimental permit requests were answered, the applicants were told to post a $10,000 bond. The permit, which required frequent renewal, didn't allow you to sell or give away your fuel and required poisoning the alcohol with chemicals that would actually damage your engine.

When people got this news back from BATF, most folks took their permit and bond applications and burned them under their stills.

Spurred by public outcry, several members of Congress got into the act, insisting that the BATF cooperate with the budding fuel producers. Several changes took place: The government lowered the permit fee to $100 and then finally did away with it altogether. The bond fee for small plants was dropped, and bonds on medium plants were made quite reasonable.

It's also legal now to sell or give away your alcohol, as long as it's denatured. The present denaturing formula requires two gallons of gasoline or diesel to be added to each 100 gallons of alcohol. Some people prefer to use synthetic alcohol-soluble oil as a denaturant. In several instances, producers have been allowed to use as little as 2% synthetic lubricants.

The BATF/Treasury Department insists you keep three types of records to fulfill permit requirements. Production records should show the quantities of feedstock you use, with all your other data such as sugar concentration, pH adjustment chemicals, yeast added, fermentation times, etc. Denaturing records should enumerate each batch of denatured alcohol and the formula used for each case. Disposal records should indicate alcohol proofs, and account for where all of your alcohol went: how much you used in your cars, heater, for cooking, how much you sold, or anything else you did with it. Every gallon must be accounted for.

Fig. 26-8

Record-keeping is not only necessary for the Treasury Department, but also for the IRS, who may want to see your records after you claim an alcohol credit on your tax return. No records, no tax credit.

Like the feds, most states changed their alcohol laws in the 1980s. Many simply require you to have a federal permit on file with the state's alcohol beverage agency; others have a minimal permit fee. Check with your state beverage-taxing agency to determine the state policy.

Filling Out the Federal Alcohol Plant Permit

Reprinted on the next pages is the U.S. federal permit form required for legal alcohol fuel production, along with the government's directions for filling it out. It's a fairly straightforward document that you can obtain pretty quickly by contacting your regional Bureau of Alcohol Tobacco and Firearms at 877-882-3277. To obtain the form online, go to <www.ttb.gov/forms/5000. shtml#alcohol>, then scroll down and click on form 5110.74.

The permit is free to small plants; larger plants must post a bond in addition to registering. But you don't have to put up the full amount of cash yourself. In most cases, a bonding company will guarantee your bond for a fee (usually between 10–20% of the bond amount).

I've included a few suggestions as a guide to filling out the permit form. These are numbered to coincide with the appropriate permit question.

1. Check "small." Even if you plan to grow, your first venture is likely to be limited, and you won't have to pay the bond you'd have to pay once you go over 10,000 **proof gallons**. (Proof gallons are calculated by taking the proof of the spirits multiplied by the **wine gallons** (a standard American gallon) and dividing by 100.)

2. Your first effort as an alcohol business doesn't require you to fill out this box. As your business grows or changes, though, the permit will have to be updated to let the Treasury Department know what's going on.

5. They don't tell you until later, but you don't have to submit your social security number, and, being the ornery cuss that I am, I never give out this number unless required. If you are a company and hire employees, then you fill in your Employer Identification Number.

6. If you are not a corporation, your date of birth is now required. I would suppose this has something to do with the Patriot Act. It is not optional. (I don't think the Treasury Department is planning on sending you a birthday card, though.)

7. Location refers to the plant's location, not where you live—unless the plant is at your home.

SAVING TIME

An excerpt from the King County, Washington, Fire Marshal's Office Newsletter, April 27, 1981:

Alcohol Stills: The Fire Marshal's office receives numerous inquiries each week relative to installing alcohol stills for the manufacture of fuel alcohol. The vast majority of these are from individual citizens desiring to install a still in their home or back yard. Unfortunately, many of these people are encouraged to pursue this by several federal and State agencies involved in fuel conservation. Little or no attention is paid to the potential fire and life safety hazards involved…. Permits will only be issued for sites in M-H zones [heavy manufacturing] which comply with all flammable liquids provisions of the Uniform Fire Code. [Author's note: As of April 1981, there were no regulations relating to distilleries in the Uniform Fire Code.] If [the still site] is not an M-H, advise them it is illegal. This saves a lot of time.

Here's a letter from "Mark," May 6, 1981:

Dear Dave: I attended your class at Seattle Pacific University. I'm a paid firefighter for one of the rural fire districts here in King County. I was discussing alcohol production with my chief, and he showed me a copy of the King County Fire Marshal's Office Weekly Newsletter [above]. Can you imagine what would happen if they tried to enforce this?

I think, as time passes … attempts to stop alcohol production will be just as you told us in class. They will (1) flood the county and state court systems with honest citizens just trying to be independent of OPEC; (2) look like idiots; or most likely (3) change the laws.

I have a lot of friends, most with very good reputations in the community, who, like myself, are going to continue right ahead with our stills and not be slowed down by this type of bureaucratic nonsense! When I showed the other stillmakers the K.C.F.M.O. Newsletter, we all decided to invest a few bucks in tall shrubs and put our stills "out of the public eye" until the laws change.

I called the A.T.F. yesterday for my permit applications, and the people there were surprisingly helpful and interested in my project. It seems strange to talk to a government agency where the people actually listen and are interested.

OMB No. 1513-0051 (3/31/2009)

DEPARTMENT OF THE TREASURY
ALCOHOL AND TOBACCO TAX AND TRADE BUREAU (TTB)
APPLICATION FOR AN ALCOHOL FUEL PRODUCER UNDER 26 U.S.C. 5181

INSTRUCTION SHEET FOR TTB FORM 5110.74

COMPLETE THIS FORM IN TRIPLICATE. SIGN ALL COPIES IN INK.
PLEASE READ CAREFULLY. AN INCOMPLETE OR INCORRECT APPLICATION WILL DELAY YOUR ALCOHOL FUEL PRODUCER'S PERMIT.

1. PURPOSE. The application is completed by a person (applicant) who would like to establish a plant to produce, process, and store, and use or distribute distilled spirits to be used exclusively for fuel purposes under 26 U.S.C. 5181. Distilled spirits means only ethanol or ethyl alcohol. The production of methanol does not require a permit from the Alcohol and Tobacco Tax and Trade Bureau. The production of distilled spirits from petroleum, natural gas, or coal is not allowed by the Alcohol Fuel Producer's Permit.

2. GENERAL PREPARATION. Prepare this form and any attachments in triplicate. Use separate sheets of approximately the same size as this form when necessary or as required. Identify these separate sheets with your name and attach to this form.

3. WHERE TO FILE. Submit application to Director, National Revenue Center, 550 Main St, Ste 8002, Cincinnati, OH 45202-5215. If required by your state, submit the designated copy of your approved application to the alcohol beverage agency or other State agency.

4. INFORMATION ABOUT APPLICANT CURRENTLY ON FILE WITH TTB NEED NOT BE RESUBMITTED. State in item requesting such information the type and the number of the license or permit for which the information was filed.

5. TYPE OF PLANT (ITEM 1). This item need only be completed on an original application or when the level of operation changes. Determine the type of plant on the basis of how many proof gallons of distilled spirits you intend to produce and receive during one calendar year. Proof gallons are calculated by taking the proof of the spirits multiplied by the wine gallons (a standard American gallon) and dividing by 100.

    Example:
    50 gallons of 190° proof spirits =
    190 times 50 divided by 100 = 95 proof gallons

6. AMENDED PERMIT (ITEM 2). Complete this item when changing the terms and conditions of an existing permit. Fill in only those sections being amended. (Refer to 27 CFR 19.919-19.930.)

7. CAPACITY OF STILLS (ITEM 11 (d)). The capacity of your still(s) in proof gallons equals the greatest number of proof gallons of spirits that could be distilled in a 24-hour period. The capacity of a column still may be shown by giving the diameter of the base and the number of plates or packing material. The capacity of a pot or kettle still may be shown by giving the volumetric (wine gallon) capacity of the pot or kettle.

8. SAMPLE OF DIAGRAM OF PREMISES (ITEM 14). The diagram of your plant premises may be drawn by hand and does not have to be drawn to scale. Below is a sample of such a diagram.

9. SIGNATURE OF/FOR APPLICANT (ITEM 18).

    a. Individual owners sign for themselves.

    b. Partnerships have all partners sign, or have one partner who has submitted an authorization to act on behalf of all the partners sign.

    c. Corporations have an officer, director, or other person who is specifically authorized by the corporate documents signs.

    d. Any other person who signs on behalf of the applicant must submit TTB F 5000.8, Power of Attorney, or other evidence of their authority.

10. ADDITIONAL INSTRUCTIONS FOR SMALL ALCOHOL FUEL PLANT APPLICANTS. Complete items 1-16 on the application form. Be sure that you sign and date the form in items 18 and 20, respectively. SKIP ITEM 17. NO ADDITIONAL INFORMATION IS REQUIRED. Prepare any attachments in accordance with instruction #2.

11. ADDITIONAL INSTRUCTIONS AND REQUIRED INFORMATION/FORMS FOR MEDIUM AND LARGE ALCOHOL FUEL PLANT APPLICANTS. Complete all items on the application form. Be sure to sign and date the form in items 18 and 20, respectively. Prepare all attachments in accordance with instruction #2. SUBMIT ADDITIONAL INFORMATION AND FORMS (ITEM 17) AS STATED BELOW:

    a. Show the following information for an individual proprietor, each partner, or each officer and director of a corporation or similar entity who will have responsibilities in connection with the operations covered by the permit. In addition, large alcohol fuel plant applicants must show the same information for each interested person who is listed as an individual in the statement of interest required by 27 CFR 19.916(b):

        (1) Full name including middle name;
        (2) Title in connection with applicant's business;
        (3) Social security number;
        (4) Date of birth;
        (5) Place of birth; and
        (6) Address of residence.

    b. A statement as to whether the applicant or any person required to be listed by the instructions above has been previously arrested or charged with, or convicted of a felony or misdemeanor under Federal or State laws (other than minor traffic violations).

    c. A statement of the maximum quantity of distilled spirits to be produced and received during a calendar year.

    d. A Distilled Spirits Bond, TTB Form 5110.56, as required by 27 CFR 19.955.

    e. Statement of the amount of funds invested in the business and the source of those funds.

    f. Any other information required by the Director, National Revenue Center after examination of this application.

SPECIAL INSTRUCTIONS FOR ALL APPLICANTS

12. OPERATIONS BEFORE ISSUANCE OF PERMIT. Unless otherwise specifically authorized by law or regulations, an applicant for an alcohol fuel producer's permit may not engage in operations until a permit has been issued by the Director, National Revenue Center.

13. STATE AND LOCAL LAWS. This permit does not allow you to operate in violation of state or local laws. Applicants should check with the appropriate state and local authorities before engaging in alcohol fuel plant operations.

14. TTB FORMS AND REGULATIONS. TTB forms and regulations pertaining to alcohol fuel plants may be ordered by contacting the National Revenue Center at 1-877-882-3277 or from the TTB Web site at www.ttb.gov.

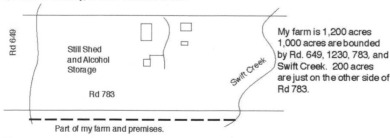

TTB F 5110.74 (12/2006)

Fig. 26-9 Application for an alcohol fuel producer permit.

DEPARTMENT OF THE TREASURY
ALCOHOL AND TOBACCO TAX AND TRADE BUREAU (TTB)
APPLICATION FOR AN ALCOHOL FUEL PRODUCER PERMIT
(Prepare in Triplicate. See Instructions)

FOR TTB USE ONLY	
DATE RECEIVED	DATE RETURNED AFTER CORRECTIONS
PERMIT NUMBER	EFFECTIVE DATE

1. TYPE OF PLANT *(Check applicable box)*

(Complete for Original Application or when level of operation changes)

☐ SMALL - 10,000 Proof Gallons or Less*

☐ MEDIUM - More than 10,000 Proof Gallons but not more than 500,000*

☐ LARGE - More than 500,000 Proof Gallons*

*Proof Gallons to be produced and received during one calendar year *(See Instruction 5.)*

2. AMENDED PERMIT *(Check applicable box(es))*

(Change In)
☐ LEVEL OF OPERATIONS *(Increased operations by small and medium plants only.)*

From _____

To _____

(Change In)
☐ NAME OF PROPRIETOR

☐ LOCATION OF PLANT

☐ OTHER *(Explain)*

PERMIT NO.	STATE
AFP-	

3. NAME OF OWNER *(If partnership, include name of each partner)*

4. DAYTIME TELEPHONE NUMBER *(Include area code and extension)*

5. EIN (If no SSN)	6. DATE OF BIRTH (Sole/Each Partner)

7. LOCATION *(If no street address show rural route)*

8. MAILING ADDRESS *(If different from plant location) (RFD or Street No., City, State, ZIP Code)*

9. PREMISES FOR ALCOHOL FUEL PLANT ARE *(Check applicable box)*

☐ OWNED BY THE APPLICANT *(Skip Item 10, go to Item 11)*

☐ NOT OWNED BY THE APPLICANT *(Complete Item 10)*

10. Officers of the Alcohol and Tobacco Tax and Trade Bureau, and state and local officers, are granted access to the premises described by this application for an Alcohol Fuel Producer's Permit.

NAME AND ADDRESS OF PROPERTY OWNER

SIGNATURE OF/FOR PROPERTY OWNER	DATE

11. STILLS FOR FUEL PRODUCTION ON PLANT PREMISES

STILL MANUFACTURER (If owner is the manufacturer write "Owner") (a)	SERIAL NUMBER OF STILL (b)	KIND OF STILL (Charge, Chamber, Continuous Still, or other (Specify)) (c)	CAPACITY (Proof Gallons) (See Instruction 7) (d)

12. BASIC MATERIALS *(Other than yeasts or enzymes)* TO BE USED IN PRODUCTION OF SPIRITS *(Check applicable box(es))*

☐ GRAIN *(Corn, Wheat, Sorghum, Barley, etc.)* OR STARCH PRODUCTS *(Potatoes, Sweet Potatoes, etc.)*

☐ SUGAR BASED CROPS OR PRODUCTS *(Cane Sugar, Sugar Beets, Molasses, Sweet Sorghum, Beet Fodder, etc.)*

☐ FRUITS OR FRUIT PRODUCTS *(Grapes, Peaches, Apples, etc.)*

☐ FORAGE CROPS *(Alfalfa, Sudan Grass, Forage Sorghum, etc.)*

☐ CROP RESIDUE *(Garbage or other refuse)*

☐ OTHER *(Specify)* _____

13. DESCRIPTION OF SECURITY MEASURES *(Such as use of locks, fences, building alarms, etc.)* TO PROTECT PREMISES, CONTAINER(S), STILL(S), AND BUILDING(S) WHERE SPIRITS ARE STORED

TTB F 5110.74 (12/2006)

Fig. 26-9 (continued) Application for an alcohol fuel producer permit.

14. DIAGRAM OF PLANT PREMISES *(In the space provided or by attached map or diagram, show the area to be included for the alcohol fuel plant. Identify roads, streams, lakes, railroads, buildings, and other structures or topographical features on the diagram. Show location(s) where alcohol fuel plant operations will occur. The diagram should be in sufficient detail to locate your operations and premises.) (See instruction 8 for sample diagram.)*

15. I WILL COMPLY WITH THE CLEAN WATER ACT *(33 U.S.C. 1341(a)). (Will not discharge into navigable waters of the U.S.)*

☐ YES   ☐ NO

16. IF THIS APPLICATION IS APPROVED AND THE PERMIT IS ISSUED, I CONSENT TO THE DISCLOSURE OF THE NAME AND ADDRESS SHOWN ON THE APPLICATION IN A TTB PUBLICATION, "ALCOHOL FUEL PRODUCERS," WHICH MAY BE DISTRIBUTED ON REQUEST TO THE GENERAL PUBLIC *(including media, business, civic, government agencies, and others)*. UNDER 26 U.S.C. 6103 YOU HAVE A LEGAL RIGHT NOT TO GIVE THIS RELEASE.

☐ YES   ☐ NO *(A no response will have no effect on the consideration given this application)*

17. MEDIUM AND LARGE ALCOHOL FUEL PLANT APPLICANTS MUST PREPARE AND ATTACH THE ADDITIONAL INFORMATION SPECIFIED IN INSTRUCTION 11.

Under the penalties of perjury, I declare that I have examined this application, including the documents submitted in support thereof or incorporated therein by reference, and, to the best of my knowledge and belief, it is true, correct, and complete.

18. SIGNATURE OF/FOR APPLICANT	19. TITLE *(Owner, Partner, Corporate Officer)*	20. DATE

TTB F 5110.74 (12/2006)

FOR QUESTIONS CONCERNING YOUR APPLICATION CONTACT THE TTB OFFICE BELOW:

DIRECTOR, NATIONAL REVENUE CENTER
ALCOHOL AND TOBACCO TAX AND TRADE BUREAU
550 MAIN ST, STE 8002
CINCINNATI, OH 45202-5215
TOLL-FREE 1-877-882-3277

PRIVACY ACT INFORMATION

The following information is provided pursuant to Section 3 of the Privacy Act of 1974 (5 U.S.C. 552(a)(e)(3)):

1. AUTHORITY. Solicitation of this information is made pursuant to 26 U.S.C. 5181. Disclosure of this information by the applicant is mandatory if the applicant wishes to obtain an Alcohol Fuel Producer's Permit.

2. PURPOSE. To determine the eligibility of the applicant to obtain an Alcohol Fuel Producer's Permit, to determine location and extent of the premises, and to determine whether the operations will be in conformity with law and regulations.

3. ROUTINE USES. The information will be used by TTB to make determinations set forth in paragraph 2. In addition, the information may be disclosed to other Federal, State, foreign, and local law enforcement, and regulatory agency personnel to verify information on the application where such disclosure is not prohibited by law. The information may be further disclosed to the Justice Department if it appears that the furnishing of false information may constitute a violation of Federal law. Finally, the information may be disclosed to members of the public in order to verify the information on the application where such disclosure is not prohibited by law.

4. EFFECTS OF NOT SUPPLYING REQUESTED INFORMATION. Failure to supply complete information will delay processing and may result in denial of the application.

The following information is provided pursuant to Section 7(b) of the Privacy Act of 1974:

Disclosure of the individual's social security number is voluntary. Pursuant to the statutes above, TTB is authorized to solicit this information. The number may be used to verify the individual's identity.

PAPERWORK REDUCTION ACT NOTICE

This request is in accordance with the Paperwork Reduction Act of 1995. The information is required to obtain a permit under 26 U.S.C. 5181.

The estimated average burden associated with this collection is 1 hour and 48 minutes per respondent or recordkeeper, depending on individual circumstances. Comments concerning the accuracy of this burden estimate and suggestions for reducing this burden should be directed to the Reports Management Officer, Regulations and Rulings Division, Alcohol and Tobacco Tax and Trade Bureau, Washington, D.C., 20220.

An agency may not conduct or sponsor, and a person is not required to respond to, a collection of information unless it displays a current, valid OMB control number.

Fig. 26-9 (continued) Application for an alcohol fuel producer permit.

8. Mailing address is your home, business, or post office box.

9. If you rent space from a friend who leases it from a landlord who is still paying the bank for the property, who owns it—the bank, landlord, your friend, or you? The best answer is your friend. He has immediate control of the land, and he sublets it to you. If you lease the premises directly from a landlord, he is considered the owner.

10. The property owner's signature is necessary so he cannot charge the Treasury Department with trespassing, if it ever shows up for a surprise inspection. The owner's signature in no way makes him liable for your actions. Your lease makes you, not your landlord, responsible for what happens on the leased property. Don't worry about your privacy, though. Unless you're moonshining, you shouldn't be disturbed by the Treasury Department.

11b. The revenuers want to know your still's serial number. If you're going to purchase a distillery, you should be able to get the number long in advance of delivery so as not to slow up the permit process. If you are building your own still, make up your serial number—it's your still.

11c. The kind of still most of you will be building is a "batch-type, packed column still."

11d. Enter the diameter of your column(s) and the type of packing material you use, as well as the size of your tank. For instance, "six-inch column, pall rings, 1000-gallon tank." The revenuers use this information to calculate how much alcohol you can make in a 24-hour period. If you're ever caught moonshining, the government can determine that you're liable for liquor taxes on every gallon you could have produced, irrespective of what you actually did produce. That's a lot of tax liability. If it entered your head to make a little 'shine on the side, reconsider.

12. You are asked what you'll make your alcohol from. Check all the boxes. In the space marked OTHER, write, "Whatever I can get my hands on." Never limit yourself on a government permit.

13. I think it's important to have a lock on an alcohol storage tank, but a fence is going a bit far. If your property is already fenced, you certainly don't need another one around your distilling/storage area, but if it makes the officials feel better, tell them you have a fence. A fence can be as simple as four stakes in the ground with a piece of wire stretched between them.

16. The Treasury Department makes available a list of everyone with a permit who has answered yes to this question. If you want alcohol industry junk mail and don't mind everyone having access to your name and address, answer "yes." A "no" answer doesn't affect your eligibility. On the other hand, if you don't mind being available to help a fellow distiller fill his tank when he is in your state, you could answer "yes" (at least until there is a national nongovernmental network of fuel distributors).

Alcohol Fuel Production and the Local Powers That Be

Oddly, local regulations can be the worst ones for alcohol producers to deal with, far more repressive and backward than state or federal government's. I lived in Woodside, California, which must've held the national record for the number of citizen complaints to the building department per capita. This was a town where everyone expected that it was their right to be in your business, since all the houses were worth millions. You even needed to have a hearing to get the color of paint approved for your fence!

Yet I operated a successful farm there for almost nine years with no neighbor complaints. The secret to doing this sort of thing is buying off your neighbors. For instance, if you want to have chickens, raise the first flock indoors where their effects won't be noticeable. Then when they start producing, you visit your neighbors for a few weeks and bring them boxes of eggs. Then when you build your poultry palace, I mean, chicken coop, and let the chickens run around the yard, there won't be a

---

I THINK, THEREFORE I'M GUILTY?

A young chemist was arrested and tried for owning a distillery, although no alcohol had been found at his still. The chemist protested that he had not made any alcoholic beverages with his still and therefore, had not really committed any crime. The judge said that having the equipment in his possession meant that the chemist must have intended to make alcohol, and that made him as guilty as if he had actually done it. Hearing this, the young man confessed to a far more serious crime, specifically, rape. The shocked judge asked him for the identity of his victim, and the chemist responded, "You don't understand, your Honor, I haven't actually raped anyone, but I do have the equipment in my possession."

If you have to defend your fuel production to the fire department, point out the differences between gasoline and alcohol (see Chapter 13). Alcohol has the higher flashpoint and much higher latent heat of vaporization, which makes it much more explosion-resistant than gasoline. Unlike gasoline, alcohol fires are readily extinguished with plain water.

peep, so to speak, from your neighbors (as long as you don't have any roosters).

You could do the same thing with fuel. Get the neighbors all excited about alcohol and put some in their tank. Once they hit the accelerator and get the extra punch, they'll be hooked. When you want to build a bigger still, there won't be any complaints, and you'll have loyal discount customers right next door.

If you live on land zoned agricultural or agricultural/residential, there's little problem in building and running a still, and local government often requires no permit at all. In certain residential zones, the situation can be quite different. Your community may have no process to evaluate the use of a distillery on your property. If the "buy your neighbors" strategy doesn't work and you have to get a permit, you can perhaps learn from my students' varied practical experiences (see Chapter 27).

When a governmental body is ignorant of a subject, its general reaction is to ban it and hope you will go away. If you persist, they will then ban it until it can be studied. Ideas and projects often die during the study period, and the community changes little, if at all. Many of my students, finding no laws on the books concerning alcohol fuel production, have just gone ahead without troubling local bureaucrats.

Also, if you ask for a permit that doesn't exist, the powers that be may feel compelled to create a permit process. This always takes lots of study and lots of time. I've had discussions with several planning commissioners who admit privately that, for just that reason, they would rather not be contacted.

If you do choose to contact your local government regarding fuel production, you'll be dealing with the planning commission (city and/or county), fire departments (city and county), plumbing inspectors, electrical inspectors, health and safety departments, environmental and/or air pollution boards, possibly the permit appeals

board (if one or more of the above agencies decides to ban your project for whatever reason), and if the permit appeals board defends the agency that turned down your proposal, a board of supervisors, city council, etc.

It's entirely possible that you will have no trouble with any of these offices. But it's more likely you'll have a problem with one of them. The fire department, for instance, is paid to be paranoid. And they may be justified in not trusting ordinary citizens to handle flammable liquids, although we have been filling our own cars 11 billion times per year with very few incidents. Still, their concern is understandable. Supportive fire marshals have shown me photos of raids they made on garages during the Arab oil embargoes against the United States in the 1970s. There were people storing hundreds of gallons of gasoline in garbage cans. As a result, there are strict laws now against storing fuel at your home or business.

In the '80s, I knew of several fire marshals supportive of the alcohol movement who would make a point of helping potential fuel producers, sharing their expertise in fuel handling, and who qualify as enlightened public servants.

If you have to defend your fuel production to the fire department, point out the differences between gasoline and alcohol (see Chapter 13). Alcohol has the higher flashpoint and much higher latent heat of vaporization, which makes it much more explosion-resistant than gasoline. Unlike gasoline, alcohol fires are readily extinguished with plain water.

I've been unable to find, in any part of the country (except in "dry" counties), any regulation prohibiting a citizen from storing an unlimited quantity of 151-proof rum in his home. Alcohol fuel is essentially the same as higher-proof rum. The difference is that alcohol is stored in a safe metal drum. If you can get your fire department to look at your alcohol as a familiar common beverage rather than a fuel, they shouldn't object to your operation.

To the best of my knowledge, there are no laws regulating distilleries in the national fire protection codes. This is the set of laws your fire marshal will first refer to in trying to figure out what to do with you. Objections from fire departments are almost always about storing your fuel. Storage guidelines in this book are in accord with the requirements of several major industrial insurance companies—which are much stricter than most fire departments.

However, if your fire department is unmoved by any of that, you have a couple of choices. You can move your plant site to commercially zoned areas where there aren't any objections to distillation and storage of fuel. Unfortunately, rents in most industrial areas are not cheap, and you'll probably have to form a cooperative to make enough alcohol to justify the monthly rental. From a strategic point of view, that may be the best approach in an urban area. You can also partner up with an out-of-town farmer to make your alcohol out there and just have it delivered to you.

You can also choose to fight back against irrational laws. According to the MSDS (Material Safety Data Sheet), 50% alcohol (a stiff martini) has a flashpoint of 75°F (24°C).[6] That means that (except in California, where indoor smoking in bars is illegal), every bar and most restaurants in the country would have to be either closed down or air-conditioned to below 75°F to prevent cigarettes (1200°F) from igniting drinks. When is the last time you saw someone's drink explode? Demand equal treatment!

The point is that just because the theoretical flashpoint (below 100°F) makes a liquid a 1B flammable substance on paper (at 200 proof), it doesn't mean that you should be subjected to the same regulations as something that blows up at 45° *below zero*, like Class 1A gasoline. If bars, state or private liquor stores, or restaurants can store alcohol, so can you.

In most states, local fire departments have no jurisdiction over vehicles. So here's a suggestion. Go to your local junkyard and pick up a couple of cheap car axles. Put these under your fuel tank and strap the tank down to the axles. You now have a tank trailer, which puts your tank beyond the jurisdiction of the local fire department. The highway patrol, under whose jurisdiction the trailer does fall, can't do anything about the vehicle until you take it out on the road (which, of course, you wouldn't do), where they could cite you for not having lights, turn signals, towing hitch, etc.

On a trailer, your fuel should be unregulated. For a larger co-op, actually buying a 4,000- to 6,000-gallon fuel tank trailer may be the best way to break a recalcitrant fire department into the idea that alcohol is here to stay and they

One of my students dug a hole with his bulldozer and made a pond and put his still on a platform floating a few inches above the bottom on 55-gallon drums full of air held between the floor joists. As a "boat," neither the building inspector nor the fire marshal had jurisdiction over his operation, except for the land side of his electrical and water connections.

should work with you. But use these confrontational approaches only as a last resort, since the fire department may retaliate by challenging the still itself, which would require you to put it on a "trailer," too. One of my students dug a hole with his bulldozer and made a pond and put his still on a platform floating a few inches above the bottom on 55-gallon drums full of air held between the floor joists. As a "boat," neither the building inspector nor the fire marshal had jurisdiction over his operation, except for the land side of his electrical and water connections.

Quite often, a planning commission is reluctant to allow you a permit for a plant operation in a residential or semi-residential area. You should be able to demonstrate that a small plant is not offensive, and that the safety record of small-scale permit holders is extremely good, and get your neighbors to sign a petition supporting you. If you still lose, you can go to the permit appeals board to make your case. If you're unsuccessful there, the last line of defense is your board of supervisors or city council. Opposing energy independence is not a popular position for an elected U.S. official to take, especially now.

When you get on the board's agenda, you have two to six weeks to prepare for the meeting. Try to find supporters to join you. There's strength in numbers. In a large city, 40 to 80 people at a council meeting have a large impact. Remember that each one of you has the right to make a statement, often around three minutes, but the primary person appealing the decision has much more time and can call expert witnesses, too. City councils and planning commissions have been known to drop their opposition to such issues when it looks as if the time it would take to hear everyone might screw up their business agenda. Jamming the system is part of our democratic process—in Congress, it's called filibustering.

If you have a date with the council, make sure—and this is critical—that radio, newspapers, and television stations are notified. They may send news teams. I've found that alcohol safety demonstrations just outside city hall are highly effective. Make cherries flambé (a dessert which uses flaming high-proof rum) to demonstrate common uses of alcohol (and to feed the media). You might run a small motorcycle on Everclear, or any other high-proof liquor, to show that there's no significant difference between high-proof beverages and fuel. Let a reporter pour the booze into the gas tank and take a ride on-camera. Burn a teaspoon of gasoline in one metal dish and alcohol in another. The gas dish will smoke and blacken; the alcohol will burn clear. Then extinguish the alcohol fire with a plant mister. Try to do the same with the gas. The flame will flare and not go out easily. (Rehearse all this once or twice at home to avoid embarrassing foul-ups in front of a camera.) Point out that it is legal to deliver propane to exposed tanks, so why can't you have alcohol delivered to your storage tank? Fuel oil gets delivered, and natural gas comes right into the house by pipes, so why should alcohol be discriminated against?

This kind of information and demonstration not only focuses the press (who have some clout with city supervisors), it's also a positive influence on legislators themselves. My ex-students and I are happy to report unanimous approval by elected representatives whenever such demonstrations have been necessary. When you're forced to fight for self-sufficiency, it's serious business. Use every resource at your disposal. You can fight city hall, and you should, when city hall is wrong.

Pioneers have never had it easy, though, and you may find it less difficult to go ahead and make your alcohol without notifying anyone, provided there are no prohibitions in your community against storing liquor, or specifically against distilling alcohol. It's been said that it is easier to ask for forgiveness than permission.

Write up your story on your experience mud-wrestling with the bureaucracy, and I'll publish it on my website.

Sometimes your local authorities would really appreciate some help in drafting sensible codes. I've been asked to to do this a number of times. In particular, a community might be willing to let you have a still but they don't want to open the door to a 100-million-gallon-per-year behemoth to move in. If you need help, I can consult to your local government and help to tailor code language to protect the community while defending the rights of individuals. Have your public servant contact me at <info@permaculture.com>. I am also available to testify on behalf of projects finding themselves at the mercy of their local authorities.

Dealing with rules and regulations is everyone's least favorite part of producing alcohol. So, in the next chapter, it's with great pleasure that I take you for a bit of fresh air and traveling. I'll be introducing you to former students and other alcohol producers who can tell you themselves if it was all worthwhile.

Endnotes

1. *E-85 Excise Tax Equivalency*, National Ethanol Vehicle Coalition, www.e85fuel.com/members/legislative.htm (February 11, 2005).

2. *E-85 Excise Tax Equivalency*.

3. *E-85 Excise Tax Equivalency*.

4. James Warden, "Bill Cuts Tax on E-85 Fuel," *Associated Press*, www.grandforks.com/mld/grandforks/news/state/10731904.htm (January 31, 2005).

5. *Proposed Infrastructure Incentives*, National Ethanol Vehicle Coalition, www.e85fuel.com/members/legislative.htm (February 11, 2005).

6. *Ethanol (Anhydrous)*, Veggiepower [biodiesel], www.veggiepower.org.uk/pge204a.htm (June 22, 2005).

CHAPTER 27

PRACTICAL EXPERIENCES WITH ALCOHOL PRODUCTION

In the fall of 1982, I traveled 1600 miles through California and Oregon visiting alcohol fuel producers. My companion and chronicler was Matt Farruggio, a skilled technical photographer. Our team also included a film crew from KQED-TV to record our trip for one of the segments of my television series.

This chapter is comprised of interviews we conducted with students of my workshops and with others. These men welcomed us to their plants and generously shared their experience to benefit those who will follow.

The conditions these pioneers in modern fuel alcohol production faced were much tougher than those today. To begin with, there was just the sheer difficulty in finding good advice. Some did it before the government legalized alcohol production. There was no Internet to consult for information, and oil companies were in a full-court press to libel alcohol in the media. One thing that shines through in all these interviews is the persistence and ingenuity brought to all parts of the alcohol fuel process. We all need to remember that we possess these qualities, if not individually, then certainly collectively.

When my television series was canceled after its first airing, the greatest loss was the first-person stories on tape of those we visited (destroyed by KQED). All of them carried a thread of hope and confidence that proved we could provide solutions without government help. It's difficult to capture what's exhibited on tape in a book. I've done my best to give you a good picture of a few of the people and alcohol plants we visited.

This is a slice of history, so I would ask that you go back 25 years and pretend you're leanin' on the side of my truck, listening in.

FLOYD BUTTERFIELD

Floyd's farm is way off the highway and just above the floodplain of a nearby river. Matt and I headed

WHEN MY TELEVISION SERIES WAS CANCELED AFTER ITS FIRST AIRING, THE GREATEST LOSS WAS THE FIRST-PERSON STORIES ON TAPE OF THOSE WE VISITED (DESTROYED BY KQED). IN THIS CHAPTER, I'VE DONE MY BEST TO GIVE YOU A GOOD PICTURE OF A FEW OF THE PEOPLE AND ALCOHOL PLANTS WE VISITED.

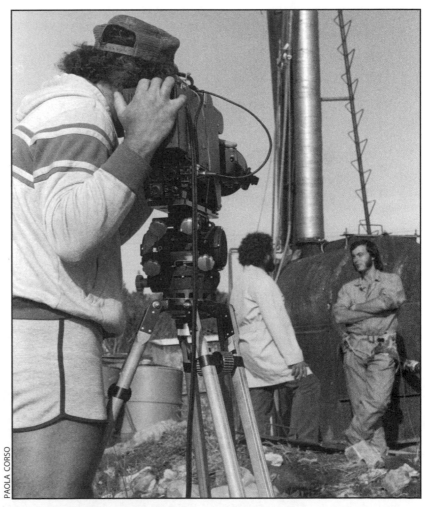

Fig. 27-1 NOPEC alcohol fuel facility. *KQED filming Ian Crawford at the NOPEC alcohol fuel cooperative.*

Floyd Butterfield may think he's a farmer; I believe he's a born engineer. With no formal technical background, he designed his own continuous columns to produce about 13 gallons an hour, and has a verified reflux ratio of only 2.2 to 1!

toward what was obviously a boiler stack poking up out of a metal barn. Inside, an air-powered pump was apparently cooling mash from the cooker through a countercurrent heat exchanger. We hopped a three-foot fence (which kept poultry out of the distilling area) and looked around.

The distillery was running, the mash was cooling, the heat exchangers were cooling the condenser water—and there was no one around. The whole plant was automated. Matt started setting up for some shots, and soon Floyd came down from the house to show us around. His plant is a beauty, and so well organized that it has won awards for its efficient design.

Floyd Butterfield may think he's a farmer; I believe he's a born engineer. With no formal technical background, he designed his own continuous columns to produce about 13 gallons an hour, and has a verified reflux ratio of only 2.2 to 1! Many large beverage stills at the time had reflux ratios

as high as 6:1; the average was 4:1. Floyd's energy costs (all the energy the plant uses) total about 26,000–30,000 Btu per gallon depending on feedstock. This isn't supposed to be possible when you ask the experts. Tell Floyd. This efficiency is better than the 34,000 Btu per gallon that 100-million-gallon-per-year plants boast about today.

He learned how to do all his steam engineering from textbooks. He used a standard distillation engineering text to make his plate design calculations. Unfortunately, the books never mentioned minimum reflux ratio, the critical factor that you need to know in order to figure perforation size, the area of the plate to be perforated, spacing, and other plate factors. Floyd searched for the answer, calling large distillery builders, distilling engineers, and experts. No one would share this one piece of information. How did he finally calculate it himself? "I took a flying guess. Turns out I must have been pretty close."

Floyd's plant is designed to put out around 75,000 gallons a year. He considers this the smallest size a plant should be if it produces alcohol for sale, unless you use a large amount of fuel yourself. At 75,000 gallons, he feels labor costs are reasonable. (Floyd was producing alcohol when the price of gasoline was only $1.50 per gallon, so a smaller

Fig. 27-2 Floyd Butterfield. The author interviewing Floyd Butterfield in 1982.

MATT FARRUGGIO

plant can make sense today.) He points out, "Automation's fine, but you still need someone around for when things go wrong."

Not much goes wrong with Floyd's plant. He and his brother have invested about $100,000 in time and materials over two and a half years. Their project is underwritten by a friend who wants to be assured of an independent fuel supply in the event of a national calamity.

At the time I talked to Floyd, in 1982, he was having little trouble selling his alcohol for about $2 a gallon to businesses that needed alcohol solvents. He was just beginning to sell alcohol as fuel.

Although Floyd has used unusual feedstocks, his plant's primary feedstocks are wheat, barley, and occasionally other grain mill screenings. He weighs his stock on a beautiful old scale he picked up at a brewery auction. Four-inch-diameter augers transport the grain from a bin to the hammermill, and then to the agitated steam-fed cooker.

After cooking and conversion, the wort is cooled by an external counterflow heat exchanger, and then pumped up to a Sweco shaker table which separates most of the solids dropping from the auger into a holding bin for sale later as DDG. He has no difficulty selling distiller's grain feed to feedlot owners, although he's not yet getting full market value. But as Floyd puts it, "Once they learn how good the stuff is … the price will go up some."

Floyd prefers separating solids before fermentation, even though it reduces yield a bit, for more convenience in handling wort and mash. An almost entirely liquid mash means simplified heat exchanger design, and easy tank cleaning after fermentation. The wet remainder drops into an auger press, which squeezes most of the liquid from the grain. The press has to be operated at different speeds for different grains or grain feedstocks. Floyd

adjusts his speed by changing drive gears on the auger's chain drive. Leftover liquid is remixed with the solid-separated wort and fermented backslop.

Floyd agitates his liquid mash a couple of times a day with an air-powered diaphragm pump. Agitating this often allows him to ferment the mash almost completely (residual sugar less than 1%) in 48 hours.

He vents his fermenters' carbon dioxide into his storage bin. He can successfully store feed for a month this way, but demand is so high he hasn't had to hold it for more than three days. He hasn't found any need for preservatives in his operation.

A five-horsepower compressor with a 30-gallon tank operates all his pneumatic controls and his pump. He prefers air power to electric power here, since he can alter pumping speed using a simple valve to control air supply.

Floyd would like to see farmers in his area get together to put up a still a few times larger than his own to serve all of them. He notes that if he doesn't operate his plant on a continuous basis, his costs are twice the price to produce his fuel. A larger plant would be less subject to irregular distilling schedules, since it would be run as a separate business, rather than as a farmer's sideline.

THURLY AND KENT HEINTZ

Thurly and Kent Heintz studied whatever written material on alcohol fuel they could get their hands on for two years while they worked as professional carpenters. In 1982, they set up a distillery at the Brandt Farms (fruit growers) packing shed. They built two 20-gallon-per-hour batch vacuum distilleries, and installed all the accessory equipment, in a phenomenally short period of time.

They had been scrounging for equipment for several months, and began building in February.

Fig. 27-3

The Heintz plant produces 40 gallons per hour and cost them $10,000 in materials to build.

Within two months, they had their fermenters, distilleries, a huge fruit hopper, elevator, a shredder, "boiler," and vacuum system all plumbed together. By May, they were almost ready to begin.

Then disaster struck. An irrigation line broke loose and thoroughly flooded the field where all their equipment sat. They had buried their fermentation tanks to keep them insulated from the California Central Valley's 100°+F summer temperatures; and they had used the buried tanks as the foundation for their fruit shredder, giant hopper, and stillage/pulp holding tank. When the field flooded, their fermentation tanks acted like beached ships when the tide comes in, floating right out of the ground; the attached top-heavy equipment toppled; the plumbing twisted like spaghetti. "When we came down and saw what happened to all our work, we didn't know whether to cry, get drunk, or get to work fixing it."

I don't know in what order they proceeded, but they were lucky to have a crane available, and they lifted their tanks, bailed out the water, and were basically recovered a month later. They began shakedown runs in June, and by the end of summer, fruit processing, fermentation, and distillation systems were "running pretty slick."

The Heintz plant produces 40 gallons per hour and cost them $10,000 in materials to build. Their distilling system is a simple batch vacuum system. They don't use any sort of pump for moving fruit pulp or mash between their shredder, fermentation tanks, distillery, or stillage tank—whole fruit culls are transported by elevator, essentially an inclined conveyor belt with cleats, up 50+ feet into a huge hopper. Since the fruit leaves the packing shed at 42°F, they let it sit for a day to warm up to air temperature (80–90°F). At the bottom of the hopper, a chute opens into a large hammermill; the fruit is pulverized and flowed into the fermentation tanks.

Twenty-four hours later, a hard vacuum is drawn on the still tanks, which are connected to the fermentation tanks by pipe. Once the still is evacuated and the valves on the fermentation tank are opened, the vacuum in the still draws fermented mash quite rapidly. When a still full of spent mash has to be emptied into the stillage storage tank (several feet above the ground), the vacuum is drawn on the stillage tank, which pulls spent mash up into it, leaving a weak vacuum in the still. The still is then evacuated, and the process continues.

Kent and Thurly have two distilleries consisting of 1000-gallon vacuum-proof tanks and eight-inch columns. Their condensers are oversized six-inch-diameter tubes jacketed with eight-inch plastic pipe. They've made an interesting innovation using Calder Couplers to seal the jacket at each end of the six-inch condenser. These couplers are

Fig. 27-4 Kent and Thurly Heintz.

MATT FARRUGGIO

rubber collars normally used to connect clay pipe for underground applications. They look very much like large no-hub couplers, often used with smaller-diameter cast-iron pipe in buildings.

Hot water from their heater circulates through 150 feet of 3/4-inch copper tubing heat exchangers in the distillery tank. It takes a little over 20 minutes to bring the temperature up to boil. Reflux control is accomplished with a cold-water coil in the top of the column.

Their still uses 3/8-inch-diameter copper tubing for control of the column using cool water. A single, tightly wound, long coil in the top of the column allows them to control the proof manually—they had no problem getting the still to settle right down at 190+ proof (typical of a distillation under vacuum).

The mash in the still is heated with hot water through a heat exchanger in the tank. With the kind of water heater the Heintzes use, running on

Fig. 27-5 Rear view of the Heintz plant. *Twin distillery tanks with feed dryer mounted between them. Note the chute (see arrow) where dried feed drops from the drum to the truck loading auger, which lifts up and away from the tank.*

MATT FARRUGGIO

Fig. 27-6 The Heintz alcohol plant. *Working from right to left, fruit goes up the elevator into the large hopper and is transported a few feet to the hammermill (see arrow). Shredded fruit drops from the hammermill into half-buried fermentation tanks. The vacuum pump behind the distilleries draws a vacuum in the two 20-gallon-per-hour, eight-inch column stills, which subsequently pulls the ripe mash in from the fermentation tanks. After distillation, spent mash is drawn up into the white tank. From there, it flows into the separator (far left), and then to the tumble dryer on top of the stills. A feed auger (upper far left) takes the dry feed and drops it into a truck when the plant is running. Dave, Kent, and Thurly are standing around the small **flash heater** that runs the distilleries and feed dryer.*

MATT FARRUGGIO

natural gas, a lot of waste heat issues from the flue—Kent guesses about 60,000 Btu per hour. So they've ducted this heat beneath an inclined, rotating, perforated drum, and they use it to dry their animal feed byproduct. Their roller drum is mounted on top of and in between the two distillery tanks, so it picks up a bit of the heat radiating from the tanks, as well. This inexpensive version of extremely expensive rotary drum dryers used by the grain industry works especially well, since high-speed drying isn't necessary in an alcohol plant of this size.

The sharing of information on alcohol production was really important to the Heintzes. Thurly told us, "When we went looking for fruit shredding equipment, we were told by three different hammermill factory salesmen that anything over 40% moisture wouldn't work and would load up on the hammers. Finally, we talked to a guy with some real experience, and he told us that stuff with 80% moisture, like our fruit, is self-washing and would run right through. He was right—it works like a champ."

When the brothers found a used hammermill, they bought a 40-horsepower electrical shredder. "It cost us $3500—a real good price for a shredder that size, and we were really surprised at how fast it grinds fruit. We can go through more than 20 tons in two hours." They know now that their hammermill is about twice as powerful as it needs to be for their type of feedstock.

Kent and Thurly don't cook their fruit at all. Since the already washed fruit comes from a sanitary refrigerated storage area in the cannery and

is fermented quickly with a strong inoculation of yeast, there aren't significant problems with bacterial contamination. Their pulp floats a bit, while solids such as pits drop to the bottom of the tank. (Peach pit remainders may become a saleable byproduct as an attractive red "gravel" for gardens, yards, and dirt roads—the brothers have already taken orders from some of their neighbors.)

Thurly is a strong supporter of the alcohol fuel movement. (He did a simple conversion on his motorcycle, a Triumph Bonneville.) He warns, "Too many guys who make alcohol think that patenting some little new process or device is going to make them rich. They hoard information that could be helpful to a lot of other folks who are struggling to just become self-sufficient. If anything is going to damage the alcohol movement, it won't be the oil companies—it'll be that secretive attitude on the part of people in the movement."

THE GOODMANS

As Matt and I traveled north through the Central Valley, farmers were working at a furious pace to get their cotton harvested and under cover ahead of the predicted early rains. Paul Goodman, his brother Ron, and father W.L. took the time to talk to us about making and using alcohol fuel.

Paul uses a batch vacuum still that he purchased from a Midwest company. He gets about eight gallons an hour out of it, although it was advertised to produce ten gallons an hour. The still came equipped with electrical heating elements for boiling the mash.

BELOW: Fig. 27-8 24-hour analog temperature recorder. *Attached to Paul Goodman's fermentation tank.*

Fig. 27-7 The Goodman alcohol plant. *This simple setup consists of two agitated fermenters (salvaged milk coolers), a vacuum distillery (in back), and a modified steam cleaner (not shown). The Goodmans' feedstocks are all liquid: melon juice in the summer, molasses in the winter.*

"When we ran the thing on electricity, our energy cost was over 30 cents a gallon. Our electric meter spun so fast that it attracted the attention of Pacific Gas and Electric [PG&E]. They came out and put us on a special meter that charges industrial rates.

"Now we get the mash quite hot, over 125°F, in its fermentation tank. We use one of those portable steam cleaners that run on liquid fuel—it produces more than enough energy without us having to buy a full-scale boiler. We pump the hot mash into the still, and, as we draw the vacuum down, distillation starts on its own."

With some pride, W.L. told us that the fancy industrial electric meter hardly turns at all now. He said they intended to run the steam generator on alcohol the next winter, and they were looking into using their fuel to replace all the electrical energy they were presently buying from PG&E.

The plant is equipped with used milk coolers for fermenters, purchased at a local dairy. The Goodmans drained the freon from cooling coils in the coolers' jackets, and they control the temperature by running either cold water or hot steam through the coils. "Steam at 40 to 50 psi gives us an unbelievable heat exchange," W.L. said. "It comes out of the end of the coils as hardly warm water."

The Goodmans use the vertical stainless steel agitator that came with their cooler. They've installed a temperature log (see Figure 27-7) to record the variation in temperature over a several-day period, and plan to use this information to optimize fermentation techniques for some new feedstocks they're thinking of trying.

W.L. Goodman, a lifelong farmer, is a vocal advocate of alternative energy, and more in touch than most urban dwellers with the Earth's sensitivity to chemicals

---

The feedstocks that they've worked with have been liquid. Since molasses is fairly easy to work with and the price per ton was low when we visited in the fall, molasses was their choice for winter processing.

In summer, the Goodmans have an amazing resource available to them. Melon growers take their culls to a collection area, not far from the Goodman farm, where pulp, rinds, and seeds are separated to be fed to cattle. The melon juice is pumped into an enormous lake, where it's allowed to go through a bacterial decomposition.

Paul's experiments with juice taken from the lake have been disappointing, since the extremely high bacteria level so overwhelmed his yeast that fermentation was stopped. He's negotiating to salvage some of that juice before it's dumped.

The Goodmans run two old tractors on their alcohol fuel. The old-style carburetors on their tractors utilize a single jet, and simply enriching the mixture has been enough to achieve smooth running.

W.L., a lifelong farmer, is a vocal advocate of alternative energy, and more in touch than most urban dwellers with the Earth's sensitivity to chemicals. "For years they told us DDT was safe, and I've been soaked in it plenty of times. When the truth came out, we all stopped using it. No one around here has dropped dead from the stuff, but we know we may not live as long."

W.L. said that "nuclear waste is another thing altogether. A meltdown up there [at Rancho Seco near Sacramento] would be the end of everything around here." In fact, Rancho Seco has since had such a dismal safety record that it had to shut down its reactor and switch to generating electricity with fossil fuels.

RIGHT:
Fig. 27-9
Jim Hall.

Jim Hall's a grain farmer. He became interested in renewable energy and ways of dealing with grain for multiple markets back in 1978.

JIM HALL

It was close to sunset when we arrived at Jim Hall's farm and alcohol plant. His place is in one of the most beautiful areas of California, Mount Shasta's backyard. A storm the day before had made the mountain look like Mt. Fuji. The snowpack is what makes this otherwise arid region of the state an agricultural wonder. Most of the soil here is volcanic, alluvial, and therefore extremely rich. We drove into the farm and up to a barn with a boiler flue rising out of the north wall.

Jim's a grain farmer. He became interested in renewable energy and ways of dealing with grain for multiple markets back in 1978. "Since then, grain prices haven't gotten any better," he told us. In fact, that week, wheat had just hit a five-year record low—a result of the oversupply of two bumper years.

"Alcohol seems to be the way to get our fuel and still market the grain. I have these local dairies just

dying for me to get off my butt and get this plant operating full-time. They know the feed value in it … I won't have a price for them until I see what kind of moisture content I get. I'm shooting for 65%. At 65%, when you feel it, you'd think there was a lot less moisture in it than that. It won't stick together."

The plant is mostly automated. "When I've finally insulated my columns and the building so the temperature in here is set, I'll feel confident that I can walk away from it and still get a high proof."

Jim's setup uses a standard workhorse 12-inch-diameter, twin-column, continuous still. The stripper column has half-inch holes covering about 8% of the perforated plate area. The rectifier plates use 0.764-inch holes. Jim went with round downcomers rather than weir dams (as in Floyd

Fig. 27-10 Jim Hall's self-built 12-inch continuous distillery. *Cooker and grain auger are to the left. Fermentation tanks are shown in the background.*

MATT FARRUGGIO

Butterfield's still). He uses larger than normal downcomers because he decided he didn't want to be limited to grain as a feedstock. "I planned on using potatoes, sugar beets, and that kind of thing, and I knew I'd have a lot of solids. I have a packing shed nearby. Last year, it dumped 77,000 sacks of potatoes. They started sprouting in storage, and the market was really bad."

Because his still is a continuous perforated plate, Jim has to use steam as his heat source, and he's set up his cooker with a steam jacket to do the grain cooking. He told us that when he first started making alcohol, the boiler tried his patience. "It's a good little boiler, but I fought it for a long time. I was using propane for fuel and didn't realize a natural gas boiler won't run on propane. It started out fine, building up temperature, and then after about three minutes, it would shut down. My first run, the column would just start to settle and I'd be getting 155–160 proof, and then all of a sudden the boiler

Fig. 27-11 Propane boiler. Jim Hall's boiler is typical of what you'd need to make 25–30 gallons of alcohol per hour. It puts out roughly 500,000 Btu per hour. Such boilers cost about $6,000 new. Similar-sized boilers that run on wood or vegetable oil are available for a little more.

MATT FARRUGGIO

would kick off and the temperature would go all crazy in the column. By the time I could restart the boiler, I was only getting 100 proof. When I finally found out it was the propane, I changed the orifices in the burner, and we've been on good terms ever since."

His fermenters are coated steel. Jim plans to build a platform across his tall barn above them and add another set of fermenters made of lightweight plastic or fiberglass. His mash plumbing is a simple hose with quick-releases to minimize contamination and to allow for design changes in the plant. Using hose allows him to use a single positive displacement pump for all his mash moving.

His cooker uses an auger agitator—about nine inches in diameter, if I remember correctly. It's unique in that he's welded a right- and left-hand flight together at the center, so when it's spinning all the mash is forced to the center of the tank, which circulates it pretty well. The agitator runs by chain drive with 15 to 1 reduction and spins at about 100 rpm. The first time Jim tried cooking grain, the steam jacket worked fine, but the settling grain cooked to the side of the tank "like oatmeal." Instead of using a faster agitation speed, he now augments agitation with a pump during the middle of the cooking, which is when the stuff tends to burn.

Jim's frustrations were not just with his boiler and sticking grains. These were things that a little work would fix. "The hardest part about building this damn thing was having nothing out here and trying to hunt everything down." This was a complaint Matt and I heard almost every place we visited that was far from cities and industrial centers (where information and access to industrial equipment is comparatively easy).

"The hardest thing for me to find was a reasonable explosion-proof motor for the reflux pump. I never did find anything reasonable. I use an old bulky $\frac{1}{3}$-horsepower motor." (The motor is a lot larger than he needed.)

"I had a difficult time finding thermometers—especially one for measuring my **bottoms**." (This is the temperature at the bottom of the rectifier column.) "I finally found one in a hydraulic parts catalog. Everything in here—electrical, boiler, and everything—cost about $25,000. I used the scrounge method. All the metal is secondary steel. The electrical work I didn't do. We had a guy come in real cheap because I traded some farming with him."

I would estimate that if Jim had tried to buy the plant he built from a distillery manufacturing company, he would have easily spent $125,000 or more.

Nothing stops Les Shook; he has more energy at the age when most men retire than most men half his age, and he's not afraid to try almost anything.

LES SHOOK

Les has lived on the same piece of land in the heart of apple country in Sonoma County, California, for his entire life, as did his father before him. Les took one of my earliest classes, and built a still that he's been refining ever since.

Nothing stops Les Shook; he has more energy at the age when most men retire than most men half his age, and he's not afraid to try almost anything. I know few other people who would even try to build a farm tractor from scratch, let alone

Fig. 27-12 Les Shook. With the author and Jerusalem artichoke tubers.

MATT FARRUGGIO

succeed. And he's designed an extremely efficient wood boiler for heating his house, but still maintains solar panels, on principle.

Les was one of the first farmers to grow Jerusalem artichokes in California, and a pioneer in experimenting with the fermentation of Jerusalem artichoke stalks. There was so little information on that feedstock back then that no one knew how to modify a potato planter for tubers. Having no seed cutter or planter, Les bought a ton of tubers and spent a few nights with his wife cutting them up by hand. He hired a couple of farm workers to plant the pieces in furrows he dug with his tractor. It took a day or so to plant two acres. Calculating

Fig. 27-13 Les Shook's personal still. *Wood-fired, pump-agitated, six-inch column distillery with a smallish 200-gallon tank. The brick firebox is very energy-efficient.*

MATT FARRUGGIO

his costs for the entire planting, it came to far less than renting a potato planter and seed cutter would have. I really appreciate people like Les, who remember that technology only helps make things a little easier, not necessarily cheaper.

Les cut his stalks down, as what literature he could find instructed, and discovered that the stalks don't fill back in as thick as was claimed. The next winter (1981–82) was a devastating one for chokes. An incredible amount of rain deluged California, and the muddy conditions didn't allow him to harvest his tubers. The following spring, the chokes came up sporadically, and in a losing battle with grass. "It was certain that cutting the stalks weakens the tubers a great deal, "and they aren't able to survive heavy rains and muddy soil."

He had held some tubers aside from his main field and planted them for a food crop for his wife and himself. That crop was so prolific he was able to save them for seed for a second try the year I visited him.

Les was disappointed in his experiments with stalk fermentation—he was able to produce just barely enough alcohol to run his still before going to low wines. His yields indicate he was getting over 90% efficiency in fermenting the 4% or so of sugar in the stalks.

He learned the importance of having access so he could clean his packing, and of baffling the entrance to the column when he distills his mash. "At first the alcohol came out nice and clear, and suddenly the pressure started rising in the still, and all kinds of brown color and actual stalk mash was coming out of the condenser. I guess I had what you'd call a severe foaming problem."

Les has salvaged three large tanks and uses a brake-cooling water tank from a logging truck as his still tank. The quality of his workmanship has been well tested: Once, when he forgot to open the valve to his fermentation lock and the carbon dioxide couldn't escape, his homemade hatch (see Chapter 10, Figure 10-3) withstood 15 psi, without any leaks.

His crop has shown no noticeable insect damage. When I asked if he had any trouble with gophers eating his Jerusalem artichokes, he nodded. "They must have gotten at least a couple of dozen plants." Leaving him with a little under 25,000 plants.

Les switched to fermenting tubers and values his stalks as an excellent source for compost. He uses little or no chemical fertilizer on his farm, and certainly none in his incredible garden. He credits the vigor of his 15-foot-tall Jerusalem artichokes to his organic soil. "Usually chokes don't have fertile seeds. Mine did. Two of my downwind neighbors had them sprout up in their yard. Luckily, they were tickled, and are encouraging them to grow."

Les has found several great feedstock resources in his area. Apple cider and vinegar production facilities a few minutes from his site have offered to provide sweet lees and vinegar lees (surpluses from the bottom of the tanks), both of which contain sugar. The vinegar lees also have alcohol in them.

Les is a talented craftsman; he built a fine wooden shed around his distillery from scratch, cutting the lumber from local trees. He's also a great scrounger, is self-sufficient, and has invested only a few thousand dollars in his plant.

IAN CRAWFORD

Ian Crawford, managing partner and founder of the NOPEC ("No OPEC") fuel cooperative in the San Francisco Bay Area, jumped into the alcohol movement with no science background, no mechanical skills (he didn't know a timing light from the northern lights), and no business background. What he had going for him was the willingness to learn, a strong idealistic environmental philosophy, a great deal of anger toward oil companies and other plunderers, persistence, and access to advice and lists of interested people from my company's cooperative service. After a period of intensive study, he began to build a micro-plant. Several friends began construction; others donated money and space.

Ian's enthusiasm for alcohol fuel goes beyond price and the desire to supply himself and a handful of friends with alcohol; it also serves his environmental convictions. He comments on the reasons many people start co-ops. "Let's be blunt. If you're like most people, you're only organizing your co-op because you've been dragged into it, kicking and screaming, by obscene gas prices. You don't want a lot of co-op members. You don't want a lot of responsibility. More than anything, you don't want this project to take up a lot of your time."

What Ian Crawford had going for him was the willingness to learn, a strong idealistic environmental philosophy, anger toward oil companies and other plunderers, persistence, and access to advice and lists of interested people from my company's cooperative service.

But it will take time. And there will be a lot of what seems like wasted effort and a lot of problems before things run smoothly. Ian's original co-op never made it. His membership was small, and when several members graduated from aircraft maintenance school and moved to Saudi Arabia, the co-op was dissolved. He began again, with a larger membership, thinking he'd be able to afford some pretty sophisticated equipment and cut down on the work involved in production. With no mechanical experience, his costs escalated. Because of some less than brilliant distillery plans, and his welder's failed workmanship, he was continuously laboring over parts of the firebox. To survive, the co-op had to bring in additional capital, which meant increasing membership.

Ian's site location difficulties are noteworthy. He had a hard time in an urban environment where, at the time, alcohol production was not widely understood by the public or the bureaucracies. The co-op had to relocate. They finally found an acre of inexpensive, open, county-owned, undeveloped land that cost only $50 a month (they provide their own electricity and water). So the co-op only had to deal with county officials for approval. Despite NOPEC's good fortune, Ian notes that the San Francisco Bay Area may be the most difficult area in the country in which to find industrial land; he says that, if he had to do it over again, he'd look for a sublet situation inside the building of another business.

Scrounging for cheap materials saved them. "Our best finds were the wood and our tanks. We paid about $1100 for the tanks—I picked them up at a soap factory. They'd be worth $15,000 if they were new."

Ian says NOPEC got the wood—200 cords of it—for free. In the city, the co-op had been renting from the University of California, Berkeley. According to Ian, "When the University evicted us, there was another victimized tenant who had to remove thousands of damaged pallets. The contractors ground them up with bulldozers and delivered them to our site rather than to the dump. That's enough to make about 150,000 gallons of alcohol, which is what we'll produce in a couple of years."

Feedstocks are no problem for NOPEC. "There are over 80 major food processing plants within 30 miles of our distillery. Any one of them throws out many times the amount of material we need to make 50,000+ gallons of alcohol per year."

In the past year and a half, Ian has learned how to repair and convert his vehicle, plan welding jobs and construction, locate and bid competitively for supplies and services, manage a business, do the necessary taxes and bookkeeping, and maintain a safe production environment. He's familiarized himself with government permit procedures, hired and trained employees, learned basic physics, engineering, and marketing, and how to swing a deal with the right person at the right time.

He's also learned how to make alcohol. "If I could learn all that in a year by just getting in, asking lots of questions, and doing it, almost any intelligent person can do it, too. Be ready for an incredible experience—in every way."

Ian shared a few tips: "I recommend you build your plant modularly. It's better to build two eight-inch stills than one giant ten-inch or 12-inch still, to have two medium-sized generators rather than one big one, to have two pumps, to have two part-time employees rather than one full-time, etc. If one of the two components goes down, you can still get by."

"Also, add various systems one at a time. While we debug the main plant, we're wasting our carbon dioxide. It would make me crazy to try to build and run a carbon dioxide recovery system at the same time. Once things get settled down, we intend to collect and market carbon dioxide, DDG, and then single-cell protein. Eventually, alcohol will be only half of NOPEC's revenues."

Taking this trip was one of the most satisfying parts of working on the first version of this book, back in the early 1980s. As a teacher, it's always gratifying to see your students go out and not only successfully do what they were taught, but take it further. In permaculture, we often say design's potential is theoretically unlimited. No matter how many yields you might have developed, how many functions you have stacked, or how ingeniously you have reused everything, another designer will look at what you've done and say, "Did you ever think to add in…?" Before you know it, you'll stand amazed once again.

Share. Organize. Win.

BOOK 6

A VISION FOR THE NATION

Every so often, someone within my earshot will say about me something like, "He's a visionary." I always try to respond by adding, "… but practical."

Let me give you an example of a vision that manifested itself in my home county of Santa Cruz, California. The county had been spraying herbicides along all the county roads for years. Thousands of gallons drained off the roadsides and accumulated in the Monterey Bay Sanctuary, killing all the toads in the creeks. Without the toads to keep garden pests in check, homeowners felt they had to resort to spraying many more pesticides.

People protested, signed petitions, showed up at county supervisor meetings, and attended endless pesticide use meetings. But it never budged the people in charge of the spraying. All the protesters accomplished was to get the county to agree that if people would cut the weeds down in front of their houses along the county roads, they could elect not to have their property sprayed. But the county would not spend a dime to notify people of this.

So, to accomplish this vision of unsprayed county roads, I spoke to the most serious activist and presented her with an official-looking form that I had drawn up on my computer. It appeared to be an official application to the county to exempt the applicant's property from spray and an agreement to cut their own weeds down. She would deliver this form to one home, skip a few houses and go to another, and so on. Residents had to send applications not to public works but to the elected supervisor of their district.

The public works department could have dealt with discrete blocks of houses exempting themselves. But the Swiss cheese made of their spray route, combined with the blizzard of applications descending on the supervisors, clearly made spraying an impractical, expensive, liability-ridden, and unpopular activity to continue. In order to put an end to the campaign, the supervisors ended roadside spraying, something years of polite protesting never accomplished. I like winning better than protesting.

What follows are some core strategies that I've presented to a great many people who, after hearing it for the first time, have almost always said something along the lines of, "Sign me up." So sign them up.

CHAPTER 28

FUELING A REVOLUTION: PROPOSED INCENTIVES AND REGULATORY CHANGES TO RAPIDLY MAKE THE U.S. A RENEWABLES-POWERED COUNTRY

How could alcohol make the U.S. a renewables-powered country in the shortest possible time? What kind of government subsidies would alcohol actually need to thrive and take over enough market share so that we no longer need oil products for transportation? Let's look at some incentives (carrots) and regulatory changes (the stick) that could have a major impact.

In drafting these provisions, we need to avoid the mistakes of the past where rewards were given for "alternative" fuels. MegaOilron has proven quite

HOW COULD ALCOHOL MAKE THE U.S. A RENEWABLES-

POWERED COUNTRY IN THE SHORTEST POSSIBLE TIME?

WHAT KIND OF GOVERNMENT SUBSIDIES WOULD

ALCOHOL ACTUALLY NEED TO THRIVE AND TAKE OVER

ENOUGH MARKET SHARE SO THAT WE NO LONGER NEED

OIL PRODUCTS FOR TRANSPORTATION?

Fig. 28-1

adept at perverting incentives for alternative fuel in order to fill their own coffers. So these proposed benefits and regulations are designed to make sure that our transportation fuels are renewable, in unlimited supplies, and reduce greenhouse gases. That pretty much limits the field to power sources that have their basis in solar energy.

A rudimentary analysis of the incentives below should show a net *gain* in taxes for the federal government. The dramatically increased economic activity would generate far more income tax and reduce the costs of social programs (unemployment, farm subsidy payments, medical effects of fossil fuel pollution, etc.), easily offsetting the tax incentives.

DISTRIBUTION INCENTIVES: THE MAIN CARROT

Increase the Federal Tax Credit

Break the bottleneck in distribution. Right now, there are 175,000 gas stations in the U.S., almost all of which are owned by the oil companies. The federal tax credit should increase to 75% (from the

current 30%) of the costs of establishing a dedicated renewable fuel station, up to a maximum of $120,000 per station. This would attract a large amount of capital to lease/finance above-ground stations on attractive terms to driver-owned co-ops or other businesses interested in setting up facilities to compete with oil.

This credit should be available only to new stations that sell only renewable fuel. The credit should not be available to any company that has more than 10% of its ownership in a fossil fuel production or distribution company. It's about establishing competition in the energy distribution field. This credit should continue until 75% of transportation fuel is from renewable sources.

Establish a Separate Alcohol Fuel Retail Credit

For alcohol retailed as E-100, add an additional 50-cent credit to the VEETC credit (see the Redefine Alcohol Fuel section later in this chapter). This credit would be extended only to new retail stations that sell E-100, to stimulate establishment of an alternative distribution system to the one we have now. This credit would drop as the percentage of renewable fuels in the national supply goes up (if the level of renewables goes up to 50%, the tax credit would be worth only 25 cents).

There should be a penalty of 1000% of the total tax credit collected by the entity operating the site, if the station elects to discontinue the sale of renewable fuel before 2025, or is ever sold to an entity that sells nonrenewable fuel at the site. There should also be a provision that heads off oil companies from establishing paper corporations to absorb the penalty and then go out of business. Proceeds of any penalty payments should go into a fund to establish new dedicated renewable fuel stations.

PATRIOTIC AMERICAN (?) OIL COMPANIES

While "American" oil companies rail against the minor subsidies extended to the alcohol movement, there is a serious question as to their moral right or even legal right to complain at all.

U.S. oil and gas companies have at least 882 subsidiaries located in oil-free tax havens such as the Cayman Islands, Bermuda, and even the tiny European principality of Liechtenstein, according to an investigation by the Center for Public Integrity.[1] The investigation further revealed that at least a half-dozen U.S. oil and gas companies have actually re-incorporated in tax haven countries, which make them truly un-American.

In 1989, Harken Energy, with CEO George W. Bush in charge, set up Harken Bahrain Oil Co. in the Cayman Islands to oversee a drilling contract with the government of Bahrain. Vice President Dick Cheney was also a big fan of locating subsidiaries in tax havens during his days as CEO of Halliburton Corp. An analysis of Halliburton's filings with the Securities and Exchange Commission, conducted by watchdog group Citizen Works, showed that while Cheney was CEO between 1995 and 2000, the number of subsidiaries the company operated in tax havens rose from nine to 44.[2]

Typically, moving to these havens reduced the corporations' tax liability by 90%, and some even got multimillion-dollar tax refunds.[3]

MEGAOILRON'S GIFT

When Linda Bilmes, a former assistant secretary at the Department of Commerce, finished calculating all the costs for a U.S. military presence in the Mideast until 2010, the figure comes to a staggering $1.3 trillion dollars.[4] That's $11,300 for every household in America. Think of it as MegaOilron's gift to your children.

Increase the Basis of the VEETC Production Credit

The basis (the rate used for calculation) of the VEETC production credit should be increased to $1 per gallon, putting it on a par with biodiesel; the credit should then be reduced over time to match the percentage of the transportation fuel supply that is renewable. Today, about 2% of the U.S. fuel supply is renewable, so the credit should be 98 cents. The basis should be held steady in 2007 dollars to account for inflation.

PRODUCTION INCENTIVES: CARROTS FOR FARMERS AND OTHER ALCOHOL MAKERS

The following credits should be enacted to stimulate fuel alcohol production. The producer can either take the credits, or pass them down to the distributors or retailers. Preferably, these credits should be financed by an elimination of the Oil Depletion Allowance (see Regulatory Initiatives—The Stick section below).

Implement a Federal Mini-Plant Tax Credit

In addition to the current "small plant" tax credit of ten cents for plants under 60 million gallons, the federal government should implement a mini-plant tax credit of 50 cents for plants producing fewer than five million gallons per year to be sold as E-100. An additional credit of 25 cents per gallon, for a total of 75 cents, should be made to plants that produce less than one million gallons per year, sold as E-100.

This credit will stimulate production from waste products and diversify feedstock sources in the name of national fuel security. Diversified feedstock sources are in the interests of national security, since we wouldn't be putting all our energy eggs in one basket (corn). By making sure we use a wide spectrum of crops, any weather or disease problem that affects one energy crop is unlikely to affect most of the others. Use of waste materials is also a diversifying strategy for energy production.

The goal is to disseminate the economic benefits of alcohol fuel to smaller farming co-ops, individual farmers, and others, while creating millions of jobs. These plants would then be likely sources for fueling community-supported-energy projects. This credit should continue until 75% of transportation fuel is renewable.

Implement a Loan Program for Alcohol Plants

The goal should be to establish several alcohol plants of various sizes going in each county. To help facilitate obtaining the capital for this fast growth, the Department of Energy should manage a 50-billion-dollar loan guarantee fund, so that banks can feel safe in loaning money to alcohol projects at any scale, knowing the project, once blessed by the DOE, will have its loan guaranteed against failure. Given the lack of risk, loan interest rates should be fixed at no higher than 5%.

This loan guarantee program should phase out for plants larger than 50 million gallons per year in 2015, 25–50 million gallons per year in 2020, and 5–25 million gallons per year by 2025. For smaller plants, the program would phase out once transportation fuel has been 95% replaced by renewable fuel.

How to fund this? If repealing the Oil Depletion Allowance isn't enough (see Regulatory Initiatives—The Stick below), a tax of five cents a gallon on gasoline would do the job in a few short years.

Fig. 28-2

Implement Feedstock Tax Credits

Provide a separate tax credit with a basis of $1 per gallon for alcohol fuel produced from cellulose, hemicellulose, alternative alcohol crops (crops other than corn), or combinations of alternative crops with higher yields than corn per acre per year. This credit should also apply to alcohol produced from marine algae.

This credit is needed to prod plant designers, entrepreneurs, and farmers to diversify feedstocks, especially on the small scale. Diverse feedstocks have many benefits to agriculture and help ensure the nation's fuel supply cannot be devastated if a monoculture crop is damaged by weather or disease. Diversifying feedstocks is also necessary to prevent market manipulation by MegaOilron. The credit should go down in proportion to the amount of renewable fuel, dropping as the percentage of renewable fuels in the national supply goes up, including adjustment for inflation. This credit should replace the current system of saleable "credits" for cellulose/hemicellulosic feedstocks.

"Power always thinks it has a great soul and vast views beyond the comprehension of the weak; and that it is doing God's service when it is violating all his laws."

—JOHN QUINCY ADAMS

Implement Renewable Process Energy Credits

There should be a tax credit of 50% on equipment used to generate a renewable source of energy to run an alcohol plant. This would stimulate the input of private capital for production of process energy from solar, biogas (methane), or biomass, instead of coal or natural gas. To obtain this tax credit, 90% of the plant's process energy would have to come from renewable sources.

Ideally, this credit should apply to all businesses, not just alcohol fuel plants, and should continue until 75% of transportation and electrical generation energy is renewable. This would jump-start a huge new industrial manufacturing sector in the U.S. that would benefit every business.

VEHICLE INCENTIVES: THE DRIVER'S CARROT

Implement Tax Credits for Vehicle Conversions

Provide up to a $1500 tax credit on conversion of an existing automobile or light truck to alcohol or flex fuel. Currently, no conversion is permitted unless it is certified by the EPA. That rule should be abolished as discriminatory, and conversions should be permitted as long as the vehicle can meet existing smog emission regulations. Smog testing of alcohol vehicles should be no more frequent than what's required for gasoline.

Provide a flex-fuel conversion credit of $2500 for trucks with a gross vehicle weight over 8000 pounds. For vehicles over 26,000 pounds, the credit should be $5,000.

To stimulate innovation and a new manufacturing industry, double the above credits when *replacing* a vehicle's engine with an E-100 alcohol-only engine that gets equal or better mileage than

---

FROM OXYGENATE STANDARD TO RENEWABLE FUEL STANDARD: A BAD IDEA

The oxygenate standard included in the Clean Air Act was abolished in favor of the Renewable Fuel Standard. This was a bad idea, since the oxygenate standard required oil companies to add chemicals that contained oxygen—such as alcohol—to their hydrocarbon fuel, which dramatically lowered carbon monoxide, especially in densely populated cities. No MegaOilron-loving president could have administratively dismissed the oxygenate standard, since its requirements were related to health issues under the Clean Air Act—which is not true with the RFS.

For many years, MegaOilron dealt with the issue by suing to prevent ethanol from being mandated to provide the oxygen. Instead, they insisted that they could use MTBE, an oil product, to reduce carbon monoxide emissions. But unlike alcohol, MTBE turned out to be carcinogenic. Although Tom DeLay (R-Texas) tried to get a liability waiver to exempt MegaOilron for liability for the groundwater pollution caused by MTBE, Democrats held their ground and refused to include it in the 2005 Energy Bill. Knowing that the days of MTBE were numbered, MegaOilron didn't rely on getting the liability waiver, but instead forced Democrats to abolish the oxygenate standard if they wanted to pass the RFS.

The oil companies' "compromise" to permit passage of the RFS was actually a masterful piece of politics. In theory, the RFS was supposed to spur the production of more alcohol. In reality, the levels of alcohol mandated by the RFS were less than the amount already being produced by the time the bill passed! So the oil companies pulled a fast one yet again, engineering the RFS to mean almost nothing to them, while abolishing the hated oxygenate standard.

---

the stock gasoline engine of the same displacement. This would reward those early pioneers who converted their vehicles to dedicated renewable fuel and provide incentives for auto companies to make the advanced engines.

Implement Tax Credits for Buyers of Dedicated Alcohol Cars

Provide a $2000 tax credit for buyers of dedicated alcohol cars or light trucks that can get at least 10% better mileage on alcohol than current gasoline engines of the same displacement. This allows the auto companies to charge more for newly designed high-compression engines and recoup their investment, while giving consumers a break on the higher costs. Exempt dedicated alcohol vehicles from smog testing after passing three consecutive tests. These credits should stay in place until 75% of transportation fuel is renewable.

Implement Incentive for Adding Cold-Start Devices to FFVs

Provide a $250 payment to drivers of FFVs for adding cold-start devices. This would help eliminate the E-85 issue discussed below in the Redefine Alcohol Fuel section.

REGULATORY INITIATIVES: THE STICK

Sign the Kyoto Accords

First of all, we need to elect politicians who will agree to sign the Kyoto Accords, to commit the U.S. to do its part along with the rest of the world in reducing greenhouse gases. This should be every citizen's nonnegotiable demand of our leaders. The U.S. should not be a rogue nation when it comes to climate change. As this book has shown, alcohol fuel can play a huge role in reversing the greenhouse effect.

Increase the Renewable Fuel Standard

To make the Renewable Fuel Standard (RFS) into a mandate that would make a difference, it needs to be increased with the dual goals of eliminating dependence on fossil fuels and urgently responding to the greenhouse effect. The government should bolster the RFS to reach 75% by 2025. It should be front-loaded to bring the U.S. into compliance with the Kyoto treaty.

This is totally reachable with the incentives in this chapter. Make "75% by 2025" the slogan every politician has to support in order to get elected. Pass language that makes it impossible to convert the Renewable Fuel Standard to some sort of Alternative Fuel Standard that rewards fossil

Fig. 28-3

fuel producers for coal to liquid fuel, tar sands, oil shale, nuclear power, or hydrogen produced from a nonrenewable source of energy.

Redefine Alcohol Fuel

Alcohol fuel should be redefined legally to mean E-100, as opposed to the current E-85, to eliminate the use of gasoline. E-100 should refer to an energy-conserving mixture of 96% alcohol/4% water, and a nonpetroleum denaturant, duplicating Brazil's long-term positive experience. In areas with bitter winter cold, a seasonal blend of 200-proof alcohol with non-petroleum denaturants added would constitute E-100.

Reform the Denaturing Law

Denaturing alcohol with gasoline has to be abolished in favor of using sustainable, renewable, biodegradable denaturants like 1% turpentine, a product of tree farms. (Turpentine is already permitted as a denaturant by the BATF/Treasury Department, but not when the alcohol is for fuel use.) This would revive the sustainable turpentine industry in the pine forests of the Southeast. A non-petroleum bittering agent should be approved to be used in conjunction with turpentine. Allyl mercaptan, an extract of garlic, would be effective.

This change will serve to simplify the design and manufacture of high-mileage flex-fuel and dedicated alcohol engines. It should be phased in by 2010 and should be implemented in conjunction with the incentive to install cold-start devices.

Implement Flexible Fuel Mandates

Starting in 2009, all new cars sold in the U.S. should be required to be E-100 flexible-fuel-capable, with cold-start devices, and with alcohol mileage no less than 85% of gasoline mileage in the same vehicle. Brazil's flex-fuel revolution took only three years; manufacturers have the experience to do this.

Manufacturers should also be required to raise the mileage level of new flex-fuel cars running on E-100 alcohol to equivalent to 2006 gasoline mileage or better by 2011.

Implement Federal Oversight of Alcohol Siting and Storage

Federal law should preempt state laws and local laws in regulating the siting of E-100 ethanol stations on any property locally zoned as commercial or agricultural. Federal law should prevent local fire marshals from disallowing storage of up to 500 gallons of alcohol fuel in a single-wall steel tank with exterior containment, in any zone in which fuel oil or propane storage is currently permitted. The alcohol can be for all uses—heating, lighting, cooking, generating electricity (in either ICE or fuel cells), and, of course, automobile fuel.

Implement Relief for Micro-Generators

Federal law should mandate that electricity micro-generated renewably has to be bought preferentially by the local utility at the rate prevalent at the time of day it is generated. No more discrimination against solar electricity or alcohol cogeneration. This would mean that there would be no more free rides for utilities that receive more electricity from a customer than the customer uses.

It also means that electricity generated midday would earn more than electricity generated at night. Home alcohol cogenerators would kick on at noon, heat water, and put electricity into the grid when it's needed most. Utility companies would set up a schedule with the cogenerator owners to stagger startup and shutdown times. This setup provides hot water (or air conditioning) for people by the time they come home from work. The electricity they draw at night will typically be about the same as or less than they put in to the grid during the day. If the homeowner used less than he fed into the grid, he would be paid by the utility company on the difference.

This setup, by the way, would sound the death knell for new nuclear power plants or fossil fuel midday peaker generators, which desperately depend on high midday utility rates to make a profit.

Abolish the Oil Depletion Allowance

I find it amazing and outrageous that oil companies get tax credit for using up the last of our oil. The theory is that they made capital investments to recover the oil, and, as the oil is depleted, they want credit to offset their expenditures in recovering a limited resource. The more they use up, the more they deplete the supply, the more tax credit they get. That's got to stop pronto.

Close Tax Haven Loopholes

Loopholes that let oil companies incorporate in tax havens in other countries have to be closed. MegaOilron has to be made to pay taxes just like the rest of us. *Just closing these giveaways could finance the entire transition to renewable fuel.* If the oil companies even think of trying to raise prices to make up for paying their share of taxes, then it's time to dust off the anti-trust regulations again and break the corporations up like we did in the early 1900s.

Prohibit Oil Company Gaming

Monitor futures trading contracts in feedstocks for alcohol production and in alcohol fuel. Mechanisms must be put in place to prevent oil companies from gaming the futures market. Artificially creating volatility in the futures market discourages investors and distorts pricing of alcohol. Without regulation, and given the immense capital the Oilygarchy controls, it can manipulate the market at will.

Fig. 28-4 Beauty Salon de la Congress. *"Dearie, we'll change the color, lift the eyebrows, and you'll be adorable."*

Start working on your legislators. Make sure their office owns a copy of this book, and check <www.permaculture.com> for up-to-date drafts of legislation for them to introduce.

Share. Organize. Win.

Endnotes

1. Bob Williams and Jonathan Werve, "Gimme Shelter (from Taxes)," *Center for Public Integrity*, July 16, 2004, 1.

2. Williams and Werve, 2.

3. Williams and Werve, 3.

4. Linda Bilmes, "The Trillion-Dollar War," *The New York Times*, August 20, 2005.

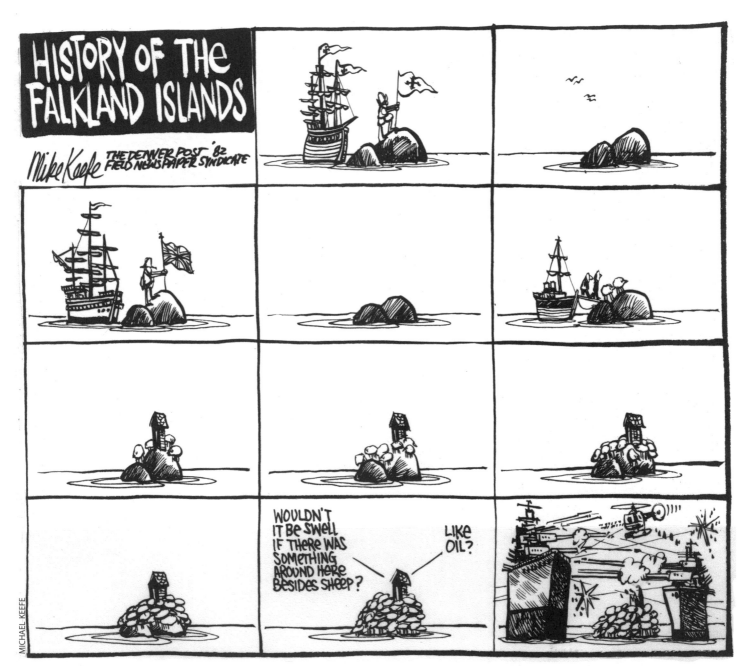

Fig. 28-5

If you've gotten this far, you realize that there are many ways to participate in our energy revolution. You could farm, you might be a mechanic, a book-keeper, an organizer of some sort, tax accountant, lawyer, welder, truck driver, contractor, or you might work at any number of practical occupations that make direct or indirect contributions to a collective effort to produce and distribute alcohol fuel. You may already have daydreamed about what your job would be like if you were somehow connected to a biofuels infrastructure.

THE COMMUNITY-SUPPORTED AGRICULTURE (CSA) MODEL

I'd like to tell you a bit about how I used to farm and how that suggests a structure that could form the hard-core nucleus of an energy revolution. When I farmed, I grew and sold my produce through the structure of community-supported agriculture (CSA). The idea is that a group of eaters contracts with a farmer to produce what eaters want. Of course, the eaters can be specific about how they want their farmer to grow crops (organically), raise animals (humanely), process certain foods after harvest, e.g., pickles or olives (nontoxically), and have input into what their food dollar does in the world.

The farmer gets certain benefits from this arrangement and gives up others. When the budget for the farm is developed for the year, all costs are listed, including a salary for the farmer. The number of people to be served by the CSA is divided into the total costs, and that, plus a little extra for unexpected expenses, is what the members pay. So people are getting food at the cost of production, including a fair salary for the farmer, which should be somewhere near the median income of the people eating the food.

So the farmer gives up the possibility of making a killing on her lettuce crop if a flood wipes out the bulk of the crop elsewhere and drives the price up.

LET'S LOOK AT HOW YOU COULD APPLY THE COMMUNITY-SUPPORTED AGRICULTURE MODEL TO CREATE A COMMUNITY-SUPPORTED ENERGY PROJECT. ONE WAY IS FOR CONSUMERS TO CONTRACT WITH A FARMER OR COOPERATIVE TO SUPPLY THEM WITH ALCOHOL FUEL.

PHOTO COURTESY OF DAVID BLUME

Fig. 29-1 The author at a 1981 workshop at the Fort Mason Center in San Francisco. I'm explaining the different parts of a six-inch packed column still to students.

But on the other hand, the farmer will not fall victim to the whims of Nature, because all the members agree to share the risk with her. If there's some crop failure, then everyone gets less food. Of course, the farmer grows a wide variety, both so that the risk of losing everything is small and because people like diversity in their food supply. So this cooperative arrangement generally satisfies everyone.

There are other differences and benefits for the farmer, as well. Normally, a farmer gets a loan from a bank (at high interest) to buy seed, chemicals, and fuel, and to make payments on her equipment, *putting up her farm as collateral.* Then the farmer plants a crop, waits three or more months, and then has all the expenses of harvesting her crop

and transporting it to a terminal. There, a terminal manager gives the farmer a receipt, not cash, and the corporation takes maybe 60–90 days to pay the farmer, if it can get away with it. All the while, the farmer is paying interest on her loan. With luck, there's enough left when she's finally paid to tide her over the winter, just in time to borrow against the farm again—unless something happens that is not covered by crop insurance, in which case the farmer loses her land. You've heard the saying, "I'd bet the farm on it"—this is literally what farmers must do each season.

In a CSA, the farmer gets the money from the eaters up front, which is when she needs the money, thus avoiding bank loans (and their interest). The

Fig. 29-2 Co-op fuel pump. *This is the site of the first co-op inspired by this book. This thousand-gallon, horizontal, single-wall fuel tank tucks in nicely behind the fence, out of view of potentially nosy neighbors. Note the fuel filter on the left, which leads to the meter and then to the pump hose.*

Fig. 29-3 Fill 'er up! The member can now pull the hose out to the street or driveway to fill the vehicle.

farmer grows the crop using this advance payment from the eaters, and they progressively get paid back in food over the season. This is much easier on the farmer and much less risky, too.

If done well, the urban-rural connection is strong. Members may get a newsletter that makes them feel part of the farm throughout the season. The city folks set up neighborhood drop-off points where the shares of food are delivered; they get to meet each other and form relationships.

The CSA subscriber benefits by getting fresher food, too. Most supermarket produce is already seven to 14 days old by the time you buy it. First, it is bought from the farmer by a broker, who then sells it to a wholesale distributor, who in turn sells it to a retailer. Moving the produce through this system takes time. The farmer may even have picked the produce up to a week before the broker could arrange pickup, storing it in her walk-in refrigerator. In a CSA, the produce is usually picked either the day before or the day of delivery to the consumer. All the middlemen are cut out of the process,

resulting in more money going to the producer and better-quality food going to the consumer.

CREATING COMMUNITY-SUPPORTED ENERGY (CSE)

Let's look at how you could apply this CSA model to create a community-supported energy (CSE) project. I want to emphasize this is not a recipe, but simply one configuration of many that might be employed. In this case, the consumers are organizing to contract with a farmer or cooperative to supply them with alcohol fuel.

Physical Site

Let's say the consumers form a driver-owned cooperative and find a location, usually commercially zoned, where they can receive their fuel and dispense it to the members, typically from an above-ground station. In Santa Cruz County in California, I was surprised to find that fuel could be dispensed at land zoned any of the four types of commercial zoning, even agricultural. I guess the

Fig. 29-4 Removing the access hatch. *To begin fueling, the member removes the access hatch covering the meter and pump nozzle.*

Fig. 29-5 Replacing the hatch. *The hatch, which looks just like the fence, goes back in place once filling is done.*

oil companies made sure a long time ago that there would be few restrictions on stations.

The generic term for this sort of station is the **cardlock station**; commercial fuel buyers commonly use these. Members have a membership card that they swipe through the reader on the pump like a credit card; they are billed once a month for their fuel.

An above-ground station might cost between $25,000–$75,000, depending on regional air pollution regulations and requirements for vapor recovery. It would have a card reader and one or, better yet, two pumps. It should hold somewhere between 4000 and 10,000 gallons, and have electricity and a phone line (or cell phone technology) for card transactions.

It's nice to have a roof over the fueling area for rainy days, plus a roof for shade on top of the tank—which could take the form of solar panels that power the pumps, card reader, and lights at

Fig. 29-6
500-gallon fuel dispenser. *This size tank should be legal wherever fuel oil or propane tanks are permitted.*

night. Not only would this reduce the amount of fuel lost to evaporation, since the sunlight is being turned into electricity rather than heating the tank, it would also reduce the amount of fossil fuel energy needed to pump the fuel by some 500+ kWh per year and would probably sell some energy into the grid.

Organizational Structure

The co-op would charge an initial membership fee during its formation to cover equipment, legal and permit costs, and the first batch of alcohol. Members would pay a small amount per gallon over the purchase price of the fuel for the manager's reasonable salary, equal to about the median hourly wage of the members, and enough to cover ongoing maintenance and to build up a bank account for a reserve to replace the station in 15 to 20 years.

The co-op manager would ideally be the person who organized the members together. He would be in charge of ordering fuel, be a source of information to members, do the bookkeeping and reporting requirements for taxes, recruit new individuals/fleets, and report to the members.

The management of the station can be done in many different ways; one option would be to have an active board, elected by the members, that meets quarterly.

You might be thinking that this sounds an awful lot like a nonprofit organization, but there are good reasons why this will most likely be a profit-making small corporation. As discussed in Chapter 26, a limited liability corporation, or LLC, is the commonly used structure for alcohol co-ops. All the alcohol users would be member-owners of the corporation.

The most important reason to form an LLC and for the members to be owners of the LLC is that it can pass along the alcohol fuel tax credits to its members. Drivers buying alcohol at a normal fuel station cannot get the current federal tax credit of 51 to 61 cents per gallon, but member-owners can.

Fig. 29-7 Small gravity tank. A tank this size could serve half a dozen or more people if it were filled every couple of months. Notice the inexpensive livestock-watering tank used for containment underneath and the clear sight tube to determine fuel level in the tank.

DAVID BLUME

So, if a driver-owned station purchases bulk E-98 at, say, $2.50 a gallon, it can pass along the 61-cent-per-gallon tax credit, bringing down the price of the fuel to $1.91 per gallon. The credit amount may go up in the future as current proposals make their way through Congress; an additional credit of 35 cents per gallon for businesses distributing E-85 is on the table.

The federal tax credit should apply whether the fuel is used for transportation or other purposes. State tax credits will vary on a state-by-state basis. If the alcohol sold is used as auto fuel, the co-op will be responsible for charging sales tax, if applicable. The member purchasing it will be responsible for paying federal and state (if any) road tax. Highway taxes do not apply if the purchaser does not burn the fuel on the highway, using it, for example, to cogenerate heat and electricity at home, or to power off-road vehicles or garden equipment.

The LLC owners can also divide up federal and, in some states, state tax credits for installing the station. This can amount to more than half of the cost of setting up the station, or 30% if only the federal tax credit applies. This makes investing in the station much less expensive for the member-owners.

Credits for establishing the station can go to the founding members, but all members should get the per-gallon tax credits. Depending on how you write your LLC by-laws, the CSE project does not have to allot the tax credit equally, though. For instance, if one of your members is a nonprofit organization (e.g., a van transport service for senior citizens), it doesn't pay income taxes, so it doesn't need any tax credits. In that case, you might give the nonprofit a cash discount and transfer the tax credits to others in the LLC who do need them. Currently, the per-gallon VEETC tax credit is paid back to the LLC as cash by the IRS.

This also means that tax credits could be used to help finance the LLC. If one or more people had put up all the original capital to get the station going, then they would be the ones to get the 30% or more tax credits. Those founders would get their investment paid back at say, 6% interest over a couple of years by adding a few cents per gallon to the fuel's sale price. So these tax credits become a benefit of financing the co-op with a low-interest-rate loan.

Perhaps if a financing group sets up the station, the investors could elect to take all repayment of their money as tax credit. For instance, instead of repaying the initial capital, the co-op could dedicate, say, 150% of the amount of the investment as tax credits (both setup and initial per-gallon credits) to the investors. The investors make a 50% return on their money in tax savings the first year, and the co-op owns the station free and clear.

By now, you are getting the picture of why the structure of an LLC gives you maximum flexibility to work the angles legally. Here's another reason: With all the alcohol being used within your LLC, you could force the BATF/Treasury Department to license you to legally receive **specially denatured alcohol (SDA)**. SDA requires a special permit because it's much easier to convert it to drinking alcohol, compared to **completely denatured alcohol (CDA)**, which can't be made drinkable. Companies that make things with alcohol are permitted to buy SDA in order to make whatever product they are in the business of producing. The denaturants are generally other parts of whatever the product's formula normally contains.

If you want to use a completely nonpetroleum denaturing additive with your alcohol, the difference between SDA and CDA might be crucial, since the approved complete denaturants seem to all be petroleum-based. A nonpetroleum fuel formula may relieve you of several additional regulatory burdens, as well as give you the great feeling of not handing the oil companies *any* of your money for fuel. But if no nonpetroleum CDA formula is approved by the BATF/Treasury Department, all of our co-ops can demand licensing under the SDA provisions—the BATF won't like it, but it doesn't really have the ultimate legal authority to stop us (and our lawyers).

Band Together

Starting a co-op is so much easier today than it was when I first did this back in the 1980s. At that time, there wasn't any affordable alcohol to buy, so co-ops had to go into both production and distribution. The learning curve was so steep that none of the more than 100 co-ops that started as a result of my classes survived after the price of gasoline plummeted in the late '80s.

In the '80s the motivation wasn't Peak Oil, the greenhouse effect, or displeasure at corporations

Fig. 29-8

running our government—it was anger at OPEC and the price of gas. When the price dropped, so did most people's motivation to make and distribute alcohol. Today we can start by buying alcohol for distribution and then learn how to make it to replace the alcohol that's being purchased, a much less steep learning curve.

Your organization should join together with other co-ops to form regional buying groups, like our Alcoholics Unanimous, that allow you to buy alcohol on moderately long-term, stable contracts with producers of your choice. Nowadays, we can choose whom we want to buy alcohol from, whether it is a corn farmer cooperative, Archer Daniels Midland Corporation (which doesn't buy GMO corn), a large corporate recycler of food waste, or a local guy recycling food waste in a small still. This will help stabilize the price of alcohol, make sure farmers get a fair price for their crops, and ensure that you will have alcohol if supplies get tight as oil runs out.

Banding together with other co-ops can also get you out of paying for outside companies to truck alcohol to your station. A group of co-ops can easily afford a couple of used fuel trucks to move the alcohol from the farm distillery, keeping all the stations topped up. Part-time drivers can come from the membership. You can also reduce the per-gallon expense of owning the trucks by delivering alcohol to personal fuel tanks in suburban and

Fig. 29-9 Fuel oil tank alcohol dispenser. These mass-produced 275 gallon oval fuel oil tanks are rated for flammable liquids. As you can see here, they make dandy private filling stations. Get a permit for it using fuel oil and add your electric pump and vent later, after approval. You can use the same tank to fuel your cogenerator.

DAVID BLUME

rural areas surrounding the city. This could create mini-co-ops that don't need to go to the expense of building a cardlock station.

I personally think that co-ops should go out of their way to buy the output of local mini-stills—which might put out only 10,000 to 50,000 gallons per year of 192-proof fuel. In fact, your co-op should even pay a little more for mini-still fuel, since the more you localize production and use, the more everyone in your community financially benefits. Exporting your fuel purchase capital out of state is better than exporting it out of the country to a transnational corporation, but local production, patronage, and marketing should be priorities. Besides, the local police ought to get to eat fresh donuts every day, with the mini-distilleries taking away the day-olds.

Farmers, Byproducts, and Integrated Production

Once your co-op is up and running and has a thousand members or so (and especially if you're urban), you should talk to local farmers about contracting with them to grow and process energy crops specifically for your co-op. This allows you, for instance, to get alcohol of 192 proof instead of water-free 200 proof, saving the extra expense and energy needed to remove the last bit of water. (The large ethanol plants supply the oil companies with octane-boosting additives, which their corporate customers demand be 200 proof, and denatured with gasoline.)

When you contract for feedstocks, if you insist that your fuel must be produced organically, you will directly reduce the use of pesticides and other chemicals. You will help farmers diversify cropping in your region, and, if the farmer chooses energy crops that don't share pests with the local monoculture, growing them will frequently give her nearly pest-free crops. Since many excellent rotation crops increase farmers' incomes over corn or soybeans, you may have less trouble than you think getting some of them to take on your contract.

If the farmer uses strategies outlined in this book, you may find that your group may be collectively buying organic produce, organic fish, organic chicken or eggs, or even organic beef and abalone from the farmer, in addition to the alcohol. In fact, you might fill up from the farmer's fuel truck when you pick up your veggies, or, alternatively, have

the farmer top off your home fuel tank when he delivers veggies to you.

The key to getting a farm to adopt an integrated production model (see Chapter 11), developing and marketing byproducts, is for some members of the co-op to take on one or more of the byproduct businesses at the back end of the alcohol process. The farmer is probably so busy, she might just look at the byproduct businesses as too much to handle.

If the farmer sees not only a contract for the energy crop, but also sees percentages of member-owned co-product businesses as passive income for her, then it becomes worth it growing a crop that Dad and Granddad never considered.

The co-op members could, for instance, buy the mash output from the farmer at relatively low cost—but one that reduces the farmer's risk and costs—or perhaps sharecrop with the farmer in production of high-value byproducts.

These byproduct businesses would be run on-site at the farm. Members could either pay the farmer some rent for space, or pay a share of the

take. If they were to grow fish and return the fish water and fish emulsion as fertilizer to the land (see Chapter 11), the farmer wouldn't have to buy fertilizer and could make organically grown fuel.

In permaculture, we recognize the value of an old system used decades ago on the Niels Bohr farm, as it applies to byproduct businesses. In this system, the farm owner/system designer licenses these subsidiary occupations to others, rather than having employees. A farmer producing the main crop and getting percentages (perhaps 10% of the gross) of the licensed co-product businesses gets the benefits of a diversified farm without having to manage employees.

Fig. 29-10

"In this, the farmers themselves are the stockholders. They bring their raw material to the still, have it made into alcohol, and take part of the alcohol home—sufficient for their own uses—as well as the byproduct. Then at the end of the year the farmer stockholders get a dividend on the business done by their still."

—*FRIENDS, FAMILIES & FORAYS: SCENES FROM THE LIFE AND TIMES OF HENRY FORD,* 1916[1]

The U.S. Department of Agriculture has several programs for value-added production on the farm; those of you who are urban distribution co-op members may want to research what funds are available for your farmer to capitalize these licensed occupations.

A minimum and maximum performance standard would be set for each operation, perhaps a net of $30,000 minimum to $150,000 maximum per year. If an operator can't make the minimum after a number of years, he has to give up his license for someone else to try. If the income is greater than $150,000, then it is time to issue another license and have two producers of that co-product. This system of capping income prevents what happened at the Bohr farm, when licensees made so much money that they retired and stopped full production.

Let's speculate wildly. If each of the approximately 650,000 remaining farmers in the U.S. produced 250,000 gallons per year of alcohol, we would replace all 160 billion gallons of transportation fuel in the U.S.

ALTERNATIVE FUEL VEHICLES

COPYRIGHT 1992–2002 ANDREW B. SINGER

SINGER

Fig. 29-11

A thousand-person distribution co-op using 500,000 gallons per year at a single station could easily contract for crops and processing done on as little as 250–1000 acres. That's about the size of the average Midwest farm today. A larger co-op could contract with three farmers, for instance, who would each grow one section (640 acres). This would minimize risk for both the farmers and the co-op, allowing the farmer to transition some acreage to contract growing for the co-op, while still doing what she currently knows well. The farmers could share crop-specific equipment as they rotate through different crops.

Use whatever crops work in your local climate: beets, sorghum, cattails, Jerusalem artichokes, chestnuts and hazelnuts, or buffalo gourds growing under mesquite. Soon you may be using hemp, willow, sudan grass, cattails, or coppiced tree crops for cellulosic alcohol.

If the distribution co-op is going to finance a farmer to take on producing alcohol, insist that at least three or four different crops are grown, so that all of you aren't gambling on a single crop. This means having an alcohol plant that can handle diverse feedstocks, and having a greater variety of farm equipment to work the multiple crops.

As soon as the farmers start to see the big improvements in their soil and profits, you may start to find plenty of self-organized farmers offering alcohol to distribution co-ops. After all, that's what happened in the CSA movement. It went from consumers being the initiators to farmers proactively organizing and looking for urban customers.

Where CSEs Can Lead

So let's speculate wildly. If each of the approximately 650,000 remaining farmers in the U.S. produced 250,000 gallons per year of alcohol, we would replace all the gasoline and half the diesel fuel used in the U.S. That's with each farmer producing only about 30 gallons per hour on a continuous basis for 325 days a year. An integrated plant this size would fit on less than an acre; the farmer could grow a few scores of acres in high-value food production, and the rest of her acreage in bulk fuel crop production. Each farmer, depending on climate and local markets, could supply all the transportation fuel needs of 250 to 500 people, and supply several times this number of customers with all the produce, fish, shrimp, dairy, flowers, fruit, nuts, and meat they need.

Each farm/plant would be the source of jobs/ licenses, directly and indirectly, for perhaps 25 to 40 people, depending on the product mix. That's nearly 26 million permanent jobs in production, and another few million in distribution. This sort of economic improvement has already begun to be felt in rural Brazil (see Chapter 5) and the U.S.[2]

Exporting these new small plants would be a huge boon to the farm equipment manufacturing sector for many years to come, since these small scales can work in niches all over the world. All over the planet, we might see a reverse migration, with people moving to the rural areas instead of cities, to take various jobs in the new, diffused, fuel industry. Instead of agriculture employing fewer than 2% of workers, integrated food and energy agriculture might directly and indirectly employ 20% of the workers in the U.S.

The predatory empires of agribusiness, oil, gas, and coal, and the interrelated transnational corporations, could be dismantled by regional cooperatives that decentralize production and manufacturing. How could a profit-driven oil-based transnational corporation compete with solid, rural-urban LLC cooperatives?

Massive military expenditures to defend the profits of oil, gas, and coal companies worldwide would dry up, freeing hundreds of billions of dollars a year of tax receipts which currently go to subsidize these industries. And who would join the military when the country has full employment due to the new fuel infrastructure? With no oil to fight for and no enemies on our borders, why maintain an expensive military?

In a perfect world, the military would reinvent itself into a new Civilian Conservation Corps, asking a year of national service from each 19-year-old to repair the awful mess the previous generations have left behind. We have lots of swales to build, forests to plant, roads and bridges to repair, railroads to improve, and skills to teach these young adults.

There are no huge technological breakthroughs necessary to start this revolution right now. Remember, making alcohol is humankind's second oldest profession. We don't need any trillion-dollar hydrogen pipeline systems, no complete reworking of the modern automobile, no trillions of dollars of capital. Most of all, no giant corporations are needed or even wanted. *I've always said*

"The vision of the nation's political leaders is limited to their meager two- to four-year terms of office; the desire for reelection influences all their decisions. I have found, sadly, that a global view of reality and a sense of moral responsibility for humanity's future are very rare among political figures. But if we can't trust our representatives, whom can we trust? The answer is simple: no one but ourselves. We must educate ourselves … and then move powerfully as individuals accepting full responsibility for preserving our planet for our descendants."

—DR. HELEN CALDICOTT

that if you don't like transnational capitalism, then stop giving transnational corporations your capital. The innovation that is continuing in all aspects of alcohol fuel is happening at breakneck speed all over the planet—even more so in developing countries than in the U.S. It can only get better, more efficient, and less expensive.

GET MOVING

Most of what you need to get started is in this book. The rest is up to you. It's up to you to educate your family, friends, and co-workers about alcohol fuel. You need to start putting alcohol in your unmodified car today. If you physically can't do that, if there isn't an E-85 pump in your area, then you need to start organizing people to help find a good site; all of you throw a few hundred dollars each into an escrow fund until you raise the full amount you need for a station and the first delivery of fuel. Don't let anything stop you. It is scarcely hyperbole to say that the future of the planet is at stake.

Fig. 29-12

Contact us at <www.permaculture.com> to see if we can send someone to help with your organizing/membership drive. Get the list of drivers with flexible-fuel vehicles from your state DMV, and send them a postcard about an organizing meeting for your co-op. Call around to find a nearby source to truck the alcohol to you. Put a flyer on the windshield of every contractor's truck you see.

Start talking to farmers. Get a copy of *Alcohol Can Be a Gas!* for your farm extension agent's lending library, and for the libraries of your local vocational agriculture program or Future Farmers of America chapter. Offer a $100 prize to the best alcohol-related farm project the FFA kids work on. Join the Farm Bureau, go to their mixers, look for someone to give a copy of the book to who has got the chutzpah to build a still on his farm and grow your fuel for you.

Learn to square dance; you'll meet lots of retired elders with land who want someone to farm it. Plunge in, grow cover crops to build the soil while building your distillery, greenhouses, etc.—and then farm!

If you are a lawyer, help the co-op customize the standard LLC for your local circumstances. Get your other lawyer buddies to put up the start-up money for the co-op in exchange for extra tax credit. Find nonprofits that need discounted fuel and don't need tax credit.

FUTURE TRIPPING

If we someday move to hydrogen production, it will almost certainly be done with on-board reformers using alcohol and cheap catalysts like nickel, or rhodium and cerium converters—leaving scarce monopoly platinum for other critical uses. No, we wouldn't plug our alcohol hybrid electric cars into the grid at night. Why bother with the grid? But we might plug our homes *into our cars,* using the vehicle's engine to provide electricity, charge house batteries, and heat and store hot water after we come home from work (see Chapter 24).

But with vaporized alcohol already a breaking reality for today's inexpensive ICE vehicles, the advantages of reformed hydrogen fuel dwindle to economic uselessness when it comes to running cars. Micro-turbines running on alcohol will be used to electrify and produce the hot water for individual homes and businesses. Alcohol delivered to simple tanks at home will become more common, replacing today's fuel oil or propane. Already, there are alcohol-powered fuel cells small enough to replace our toxic batteries in laptop computers and the zillion other battery-powered gizmos.

If you are a bookkeeper or accountant, get the co-op's books set up so they are in order right from the start so that everyone gets their proper tax credit.

Give your mechanic a copy of this book and a drum of fuel so he can play with alcohol first on his own car.

Spend a few bucks and go to truck driving school, so you can pass the suicide jockey (flammable liquids driver) test. That can be your contribution to the co-op, going twice a month to pick up your co-op's fuel—and hopefully also the fuel for the new co-op forming a mile down the road.

There's nothing about this we can't do if we work together. Once you are organized, you are unstoppable. And, remember, there's more to be organized about than fuel—there's food, there's childcare, there's taking care of each other when things are rough, and, most of all, there's dancing and parties, too. Patronize the food byproducts of your fuel farm's CSE, which can be picked up each week when you fill your tank.

The government can't take care of you and its corporate backers at the same time. The corporations don't care about you. They are busy selling toys to people who are like those who fiddled while Rome burned. It's up to us, each one of us, to take back our lives, take back our planet, and take care of ourselves.

Yes, I am talking about revolution here. It's a revolution to be accomplished without weapons, without harming people or the Earth, or declaring any person the enemy. It's a revolution based on withdrawing our support for, and therefore our permission to be governed by, entities that seem bent on planetary suicide—while fully taking on the responsibility for providing a healed planet and therefore a future worth living for all of Earth's life.

We can do it. Let the transnational corporations stall, sputter, and fall down on their own, without your money.

Share. Organize. Win.

Endnotes

1. Ford R. Bryan, "Friends, Families & Forays: Scenes from the Life and Times of Henry Ford," *Detroit News,* December 13, 1916.

2. Edward Epstein "Ethanol as Economic Elixir: Fuel Alternative Helps Revive a Struggling Illinois Farm Region," *The San Francisco Chronicle,* June 12, 2006, sec. A1.

IN CLOSING, I WOULD LIKE TO LEAVE YOU WITH THE WORDS OF HOWARD ZINN:

"[People] want change but feel powerless, alone, do not want to be the blade of grass that sticks up above the others and is cut down. They wait for a sign from someone else who will make the first move, or the second. And at certain times in history there are certain intrepid people who take the risk that if they make that first move others will follow quickly enough to prevent their being cut down. And if we understand this, we might make that first move.

"… And if we do act, in however small a way, we don't have to wait for some grand utopian future. The future is an infinite succession of presents, and to live now as we think human beings should live, in defiance of all that is bad around us, is itself a marvelous victory."

—HOWARD ZINN

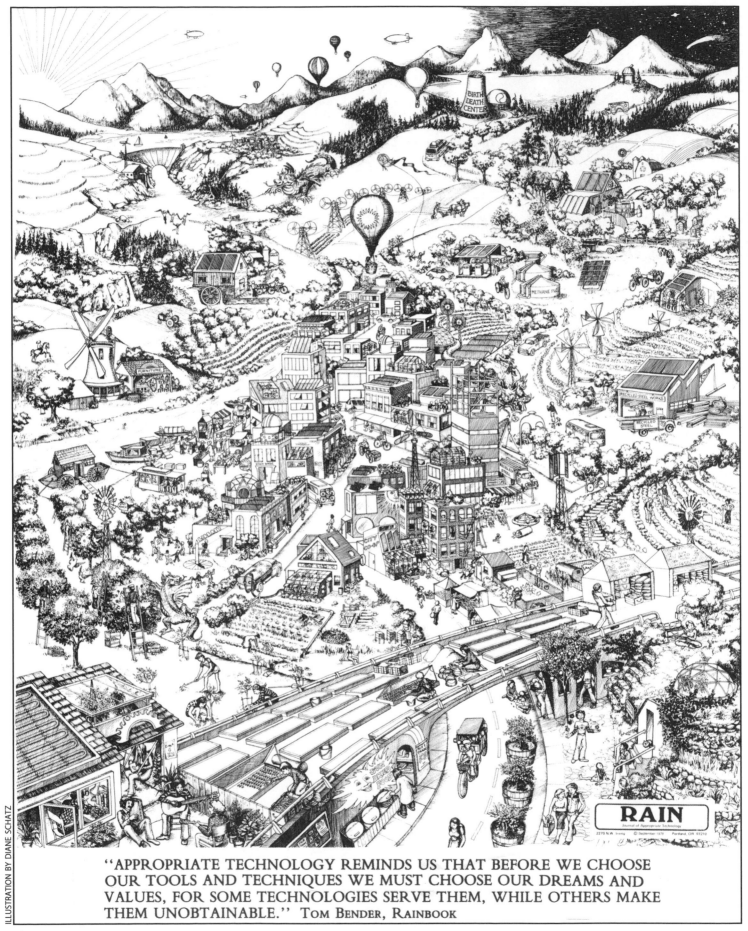

ILLUSTRATION BY DIANE SCHATZ

"APPROPRIATE TECHNOLOGY REMINDS US THAT BEFORE WE CHOOSE
OUR TOOLS AND TECHNIQUES WE MUST CHOOSE OUR DREAMS AND
VALUES, FOR SOME TECHNOLOGIES SERVE THEM, WHILE OTHERS MAKE
THEM UNOBTAINABLE." TOM BENDER, Rainbook

Fig. 29-13

MORE...

BACK MATTER

More?!? Yep. I just can't help myself.

The issues of entomologist David Pimentel's influence over the public discourse on alcohol fuel could have been dealt with throughout this book. But I instead gathered this subject up to form Appendix A.

Instructions for a couple of car conversions are back here in Appendix B. We plan to detail other specific conversions in future editions of the book, so look for our appendices to grow.

Since reference material in the middle of a book can interfere with the flow, the engineering tables that are useful in building small alcohol plants (Chapter 10) have been gathered together in Appendix C.

For those of you who have had your curiosity piqued about permaculture design, I've said a bit more about it, right after the appendices. And you can find out much more at our website, <www.permaculture.com>.

We've also included a couple of pages about our permaculture approach to our tree tax program.

Then come the glossary and index, which I hope make it easier to use this book. Any word that was in boldface in the text is defined for you in the glossary. And, since I know I always appreciate a very detailed index, I am glad to provide you with one here. If you come across terms you'd like added to either the glossary or index, please drop us a line at <info@permaculture.com>.

APPENDIX A

ETHANOL AND EROEI: HOW THE DEBATE HAS BEEN DOMINATED BY ONE VIEW

For 25 years, David Pimentel, Ph.D. at Cornell University, and, in recent years, Tad Patzek, Ph.D. at the University of California, Berkeley, have been responsible for the academic basis for most of the anti-ethanol sensibilities in the mainstream press, managing perceptions that have even leaked into Hollywood television (a 2005 episode of *The West*

FOR 25 YEARS, DAVID PIMENTEL AND, IN RECENT YEARS, TAD PATZEK HAVE BEEN RESPONSIBLE FOR THE ACADEMIC BASIS FOR MOST OF THE ANTI-ETHANOL SENSIBILITIES IN THE MAINSTREAM PRESS.

DAVID BLUME

Fig. A-1 Dr. Pimentel's tractor. *This enormous 7000 series John Deere is a close match to the seven-ton tractor in Dr. Pimentel's 2005 study. It can pull a 12-row corn planter that could plant the entire farm in under four hours (in air-conditioned comfort). Tractors of this size are used on farms up to 25 times the size of the farm described in Dr. Pimentel's study.*

Wing was an example). Although dismissed by academics in the field, their studies continue to receive extensive coverage in both business and environmental circles. Political realities today cannot reverse the damage done. Pimentel, now approaching 80 years of age, is a darling of the Peak Oil movement. He and Dr. Patzek have been essentially alone in publishing studies alleging that production of alcohol fuel, among other things:

– Has a negative energy balance;
– Is an unethical use of food;
– Pollutes the air;
– Costs the consumer money via subsidies;
– Takes 61% more fuel to go the same number of miles;
– Produces 13 gallons of sewage for every gallon of alcohol produced.

Dr. Pimentel is an entomologist, a studier of bugs, and Dr. Patzek is a physicist and engineer. Neither of them is trained in ecology. So they are straying far afield. This was amply borne out in their recent study[1] when both co-authors failed to catch their misuse of net primary productivity, a very basic concept in describing world photosynthesis.[2] In doing so, both also understate the photosynthetic efficiency of plants in general and corn in particular (so it can't be dismissed as a typo) by ten times, fully undermining their paper's first major conclusion that plants are 100 times less efficient than solar panels.

Pimentel's lack of expertise also explains his continuing choice to publish with the International Association for Mathematical Geology's *Nonrenewable Resources*, now renamed *Natural Resources Research*, (which handles "all aspects of *non-renewable* [author's emphasis] resources, both metallic and non-metallic..."),[3] not a journal known for peer-reviewing biological papers or those on renewable energy. His peer reviewers all missed the same glaring errors mentioned above.

In their most recent study,[4] Drs. Pimentel and Patzek cite a self-described "independent" DOE study by the Energy Research Advisory Board (1980)[5] as their "credible" source as to why we should believe their negative energy balance allegations. Far from being independent, the study in question was actually led by Pimentel himself, who was employed by Mobil Oil at that time. This was not disclosed to the DOE.[6] In light of this, the conclusion of the ERAB study was not surprising:

The U.S. should abandon attempts at producing ethanol and instead rely on the Mobil process for making synthetic gasoline from coal. Pimentel today still champions coal,[7] while his co-author Dr. Patzek stumps for nuclear power.[8]

The scandal that the study caused at the time resulted in South Dakota Senator George McGovern convening a Senate investigation to probe whether "scientists with ties to Mobil Oil ... would rob hundreds of thousands of American farmers of the opportunity to benefit from gasohol development."[9]

This dust-up should have ended any normal academic's career. Among statured, publishing, peer-reviewed scientists, no other study has come close to confirming Pimentel's allegations—and many are uncharacteristically candid in pointing out his repeated use of inappropriate or out-of-date data, or data so lacking in documentation as to be unable to be evaluated. This is the equivalent of coming to blows in academia.[10]

Pimentel publicly claims to have never taken money from oil companies, although he grants it's possible that oil companies have donated money to Cornell, his sponsoring university. Yet he admitted in a 2004 radio interview[11] that he took thousands of dollars, and that he was exposed in 1982 by investigative reporter Jack Anderson as being secretly on the payroll of Mobil Oil.[12] Following the 1982 exposé, Mobil Oil even took out a large ad to defend Pimentel, while admitting that it paid him.[13]

Pimentel has been a prolific publisher of roughly 475 studies and appears to almost never be at a loss for funding. That's rather unique in the world of organic agriculture. For comparison, the entire USDA got its first full-time funded position studying organic agriculture only a few years ago.

Dr. Pimentel's work in entomology and organic agriculture methods is rigorous and well documented. When one compares that work to his work on alcohol fuel, it would appear that two completely different people are publishing. In 2005, he and Dr. Patzek claimed that it takes 29% more fossil energy to make alcohol fuel than it contains. He almost simultaneously published a very solid piece of work showing that organically produced corn saved over 30% of the energy used to grow it chemically.[14] Yet in his alcohol study he did not cite his own work in positing his energy figures for ethanol production!

EROEI (energy returned on energy invested) is a way of evaluating how much energy is used in production of a fuel or energy source. If the energy in the fuel (measured as heating value) is greater than the energy used to produce it, then the fuel is considered positive. So a positive EROEI of 25% would mean that the fuel contains 25% more energy than was used to create it. If it takes more energy to make the fuel than is contained in it, then the EROEI is considered negative.

In making his most famous allegation on EROEI, Pimentel relies on several figures. Let's take his most controversial one, the energy it takes to build farm equipment. Pimentel has been claiming for 25 years that his inclusion of this embedded energy figure is what makes his study more accurate. This figure is higher than every other item he cites for growing corn, except for his hotly disputed figures for the energy embedded in nitrogen fertilizer.

The farm equipment figure was first published in a 1980 book, which Pimentel edited.[15] The first chapter was written by Otto Doering III and was an attempt to characterize the energy that went into farm equipment, starting with the metal being mined, smelted, and formed, and including the oil that went into the tires—a figure that was actually higher than the energy cost of the steel. Although Pimentel has never said so specifically, it is abundantly clear that this is where he obtained his embedded energy data. Although Doering himself said that it was impossible to accurately calculate the embedded energy in farm equipment[16] and cautioned against using his study as evidence for that, Pimentel pays no heed.

In his 2005 study, he finally specifies that he is talking about a six- to seven-ton tractor, an eight- to ten-ton harvester, and a smaller, unspecified amount of other equipment not deemed sufficiently worthy to assign a specific weight. Although

Fig. A-2
Comparative results of ethanol energy balance studies, 1995–2005.

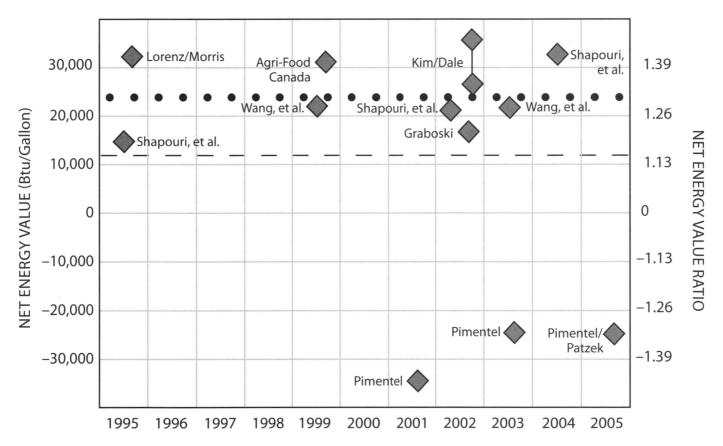

Comparative Results of Ethanol Energy Balance Studies
1995–2005

● ● ● ● ● ● ● Average NEV Minus Pimentel: 24,336 Btu (1.32 NEV Ratio)
— — — — — Average NEV With Pimentel: 11,739 Btu (1.15 NEV Ratio)
Calculations assume the low heat value energy content of ethanol is 76,330 Btu/gallon.

in earlier papers he and Doering assumed a 12-year life expectancy from or for the equipment, Pimentel reduced it to ten years in 2005, apparently in order to keep his weight of equipment per acre at his historically constant 55 kg/hectare.

At any rate, it is clear that this will be the last paper where he will be able to claim such a significant energy figure for farm equipment. Why? Now that he has finally committed to the size of the equipment in writing, it's possible to measure the degree to which the energy figure is overstated. Translated into U.S. measures, 55 kg/ha is 49 pounds per acre. Since tractors don't last forever, Pimentel alleges that this is the amount of the equipment that is used up in the farming of an acre of corn. If we generously assume a total of 20 tons of equipment with a ten-year life, as specified by Pimentel, we come to the mathematical conclusion that the average farm using this equipment is only 81.6 acres, about one-eighth of a square mile.[17] Farm equipment of this size and weight, among the largest tractors made, can work about 2000 acres of corn or sugarcane (more than three square miles). I could plant this entire 81-acre farm using this sized tractor before lunch! When you take into account that the real-world average life of farm equipment of this size, in both the United States and in Brazil,[18] is 25.7 years, not 10 years, the size of Pimentel's farm shrinks to less than 33 acres. His reported energy figure, therefore, is at least 61 times greater than reality.

But wait! Pimentel is counting the tractor energy starting from mining the ore. In reality, steel and tire rubber in the U.S. contains on average 76% recycled material[19,20] and is, of course, recycled again at the end of its life. So Pimentel's figure is actually a minimum of 196 times too high. Were we to pick nits (such as pointing out that large tractors like this are phasing out their eight huge tires in favor of two rubber bulldozer treads that use a fraction of the rubber), the figure is probably about 500 times too high. Of course, these rubber tracks can be made from alcohol-based synthetic rubber, further diminishing the petroleum energy figure to statistical insignificance.

The average-sized corn operation in the United States is approaching 2000 acres and heading for 4000 acres in major corn-growing states. (Even though the average-sized farm is less than 1000 acres, many farms are rented out to make the operation much larger.) In 2003, the one year in which

Pimentel cites Hulsbergen[21] instead of himself for the equipment energy figure, one finds that the study site, an organic research farm in Germany, was only about 40 acres, without a single stalk of corn grown anywhere on it! Pimentel has had to quietly retract that citation in his latest study and go back to citing himself. Even more embarrassing, Hulsbergen showed that sugar beets had a much higher EROEI than either of the grains in the study.

Isais Macedo of Brazil has analyzed, in excruciating detail, the energy involved in production of farm equipment, alcohol plant components, and all travel associated with every aspect of producing and distributing fuel. He is the recognized world authority on analysis of embedded energy in agriculture. His study goes into exceptionally rigorous, verifiable detail concerning both embedded energy and the greenhouse gases emitted at each step of the process. Yet, oddly, Pimentel never once cites this information—nor mentions it in any of his studies, in any context. In Macedo's study, alcohol from sugarcane garners a 9 to 1 positive energy return.[22] What's even more important is that virtually none of the energy used is fossil-fuel-based, so its ratio of *renewable* energy output per *fossil* energy input is much, much higher.

In case after case, Pimentel's figures are dramatically overstated, frequently by one to three orders of magnitude (tens to thousands of times too high). Major examples over the years have included:

- Assuming far lower yields of corn per acre than the USDA unambiguously states;
- Assuming energy figures for irrigation when almost all corn depends solely on rainfall; the figures he quotes are off by an order of magnitude for the small amount of acreage that is irrigated;[23]
- Assuming application rates for fertilizer like lime at ten times the normal rate;
- Assuming 90-mile transportation distances for grain, when the verifiable figure historically is less than half that, while modern plants are being based on less than 20 miles;[24]
- Assuming very high energy cost of nitrogen fertilizer (his largest energy cost) by citing international figures, which are double to triple the real energy cost in the U.S.;
- Assuming the energy to distill is very high by claiming fermentation yields only 8% alcohol, when the real-life historical figure has

been 15%, but currently is well over 20%;

- Assuming the yield of alcohol per ton of corn is much lower than is actually realized in modern alcohol plants;

- Assuming energy to run the plant is much higher that it is. For some time now, plant manufacturers have provided a money-back guarantee that the plant will consume less than 34,000 Btu/gallon, including the nearly 15,000 Btu required to dry the DDGS;

- Assuming that alcohol has to be dehydrated from 96% to 99+% purity (which takes a separate energy step) for use as an auto fuel. (Somehow the Brazilians missed this detail and blend 96% alcohol/4% water with their gas without mishap);

- Assuming the energy for dehydration of alcohol is based on old technology, compared to the modern, extremely low-energy-consumption, pressure swing, corn grit or molecular sieve adsorption methods than have been in use most of the last 20 years;

- Assuming the food eaten by the farmer as "gasoline equivalent" energy;

- Assuming high figures for the metal used in the alcohol plant by stating unrealistically short working life and without taking normal metal recycling into account;

- Assuming that all the energy used in the process should be attributed to the alcohol, when about half of it needs to be attributed to producing dried animal feed and carbon dioxide co-products;

- Assuming that liquid left over after fermentation is sewage to be disposed of at the energy cost of running an aerobic sewage treatment plant, when, in reality, none of the liquid waste is so treated. The liquid is evaporated, and the condensed solubles are combined with the dried byproduct grains;

- Assuming that the liquid left over with its load of solubles is sewage rather than stillage—a source of methane capable of producing more energy than is used in the entire plant. This is standard operating procedure in India and is now being adopted in the U.S;[25] The leftover liquid should be counted as an energy credit, not debit;

- Assuming that ethanol causes air pollution (from volatiles released in the drying of grain), when in reality all alcohol plants now

recover those volatiles, burn them, and generate a positive energy return from them;

- Assuming that heat energy must come from fossil sources, when the process energy of the majority of the alcohol fuel produced in the world is made from renewable biomass energy sources;

- Assuming that corn is representative of alcohol production when it is a minority crop in world alcohol production compared to sugarcane, beets etc.;[26]

- Assuming that the DDGS is comparable to soybean meal in feed/energy quality in the face of 100 years of experience and science demonstrating that it is far more valuable than soybeans or the original corn it came from;

- Assuming, incorrectly, that the heating value of ethanol equates to mileage, and then undervaluing ethanol by 39% in relation to gasoline. Current production flexible-fuel cars in Sweden get roughly equivalent mileage on both fuels, and dedicated alcohol engines get superior mileage to their gasoline counterparts.[27]

- Assuming that all Btu are the same, and converting coal or natural gas Btu to "gasoline equivalents" to inflate the apparent use of petroleum to six times its actual use. (Although the current process energy used in alcohol plants in the U.S. is either coal or natural gas, it certainly is not petroleum. New plants are now being designed to eliminate all fossil fuel use while self-producing their own natural gas);

- Assuming that the tax incentives provided to alcohol cost taxpayers money, when it has been clearly demonstrated that the return to the treasury is several times the cost of the incentives due to taxes collected on alcohol fuel's domestic economic activity;[28]

- Assuming that farmers derive no benefit from alcohol production and that it's all a plot by the big corporations to loot the treasury. The majority of alcohol today is produced by farmer-owned cooperatives, not transnational grain corporations.

To top all that off, when Pimentel changes his assumptions by as much as 1700% between 2003 and 2005, he diverts attention away from it by switching the units of measure so that the change would not be apparent to anyone, such as a reporter, who would try to compare it. In the chart comparing his 2003 and 2005 studies (see Figure A-3), I've converted the recent study to the units of measure (Btu) used in the prior study. You'll notice that his net change figure, the one that reporters should make the effort to compare, is barely altered (–4.95%), despite many assumptions changing.

Pimentel is all for accounting for energy costs relating to alcohol production, but does not consider energy gains from alcohol's use. He accounts for the embedded energy of the tractor as a cost, but does not include an energy credit for tripling the engine life of the thousands of car engines that would be running on the alcohol produced by that tractor. He is in favor of accounting for the transportation of all the materials involved in making alcohol, but does not account for displacing the energy required to power tankers 11,000 miles to deliver crude oil. He accounts for the petroleum energy it takes to make tires, but neglects to mention that the same tires can be made from alcohol (and were in World War II), or that the Archer Daniels Midland alcohol plant in Decatur, Illinois, uses one-third of the waste tires in that state to fuel its boilers, and therefore should garner an energy credit.

Perhaps Pimentel's most grievous misrepresentation has to do with the value of DDGS. It is incontestably proven that feeding DDGS instead of the original corn to animals increases meat or milk output (see Chapter 13). This alone would destroy his energy balance studies. So he assiduously avoids the energy credit for reducing the amount of feed to produce the same amount of meat or dairy.

These omissions also make mincemeat of his attempt at the moral high ground, of caring about feeding the world with our grain. If we fed a large part of our animals' diet as DDGS instead of corn, we would have more than enough surplus from this one act alone to feed every malnourished person in the world, from the newly freed surplus grain. And let's not forget that DDGS is a compact nutritious source of protein for *people*, without the unneeded starch.

In this appendix, I have hit only the high points of the misrepresentations contained in the studies by Drs. Pimentel and Patzek. It should come as no surprise that no other statured scientists in the world today concur with the conclusions made in the Pimentel studies.

But when the American Petroleum Institute cranks up its awesome press release mill and wallpapers the entire world media, a contrived paper can suddenly become the "truth" in the mind of the public. In an era when corporate funding of research is the norm, you cannot expect members of the scientific community to do the policing of their own that they once did, when to do so might endanger their own funding and survival.

Creative Concealment

Fig. A-3 Creative concealment.
When converted to common units, it's easy to see how the change of units of measure masked the inexplicably divergent figures of the two versions of the study. The technology involved has not measurably changed during this period.

ENERGY TO PRODUCE ONE HECTARE OF CORN	2003 BTU × 1000	2005 BTU × 1000	BTU CHANGE	YIELD ADJUSTED	% CHANGE
Labor	1000	1832	832.14	825.89	82.59
Machinery	5656	4037	–1618.95	–1606.79	–28.41
Diesel	3600	3978	377.56	374.73	10.41
Gasoline	2212	1606	–605.90	–601.35	–27.19
Nitrogen	10,952	9708	–1244.05	–1234.70	–11.27
Phosphorus	876	1071	194.73	193.27	22.06
Potassium	744	995	251.38	249.49	33.53
Lime	880	1249	369.19	366.41	41.64
Seeds	2080	2062	–17.85	–17.72	–0.85
Irrigation	3764	1269	–2494.99	–2476.25	–65.79
Herbicides	840	2459	1618.71	1606.56	191.26
Insecticides	60	1110	1050.39	1042.50	1737.50
Electricity	136	135	–1.17	–1.16	–0.85
Transport	1072	670	401.80	–398.78	–37.20
TOTAL	33,872	32,181	–1690.61	–1677.92	–4.95

DAVID BLUME

Endnotes

1. David Pimentel and Tad W. Patzek, "Ethanol Production Using Corn, Switchgrass, and Wood; Biodiesel Production Using Soybean and Sunflower," *Natural Resources Research* 14:1 (2005), 65–76.

2. David Morris, *The Carbohydrate Economy, Biofuels and the Net Energy Debate* (Minneapolis: Institute for Local Self Reliance, August 2005).

3. *Guidelines for Contributors,* Natural Resources Research, <http://207.176.140.93/index. php?option=com_content&task=view&id=113&Itemid =132> (April 2007).

4. Pimentel and Patzek.

5. Energy Research Advisory Board, *Gasohol* (Washington, DC: U.S. Dept. of Energy, 1980).

6. Jack Anderson, "Gasohol Program: Prey to Big Oil," Washington Post, May 24, 1980.

7. Stephen Thompson, "Running on Empty?" *Rural Cooperatives* Magazine, September 2005, www.rurdev. usda.gov/rbs/pub/sep05/running.htm.

8. David Pescovitz, "Ethanol Stirs Eco-Debate," *Lab Notes: Research from the College of Engineering. University of California, Berkeley* 5:3 (March 2005).

9. "Science and Politics Don't Mix," in Mobil Oil ad, *The New York Times,* June 19, 1980, Sec. A23.

10. Alexander E. Farrell, et al., "Ethanol Can Contribute to Energy and Environmental Goals," *Science* Magazine, 311:5760 (January 27, 2006), 506–08.

11. *Public Planet,* radio show, hosted by Jodi Selene, KVMR, Nevada City, California, 2004.

12. Anderson.

13. "Science and Politics Don't Mix."

14. David Pimentel, et al., "Environmental, Energetic, and Economic Comparisons of Organic and Conventional Farming Systems," *BioScience* 55:7 (July 2005), 573–82.

15. David Pimentel, *CRC Handbook of Energy Utilization in Agriculture* (Boca Raton, Florida, U.S.: CRC Press, 1980), 9–14.

16. Otto Doering III, in Pimentel, *CRC Handbook of Energy Utilization in Agriculture,* 9–14. ["There is no precise way to account for the energy used indirectly in agricultural production." ... "A tremendous amount of virtually unobtainable information would be required to make a precise accounting of the actual energy embodied in a specific stock of farm machinery for any given farming operation."]

17. 40,000 pounds divided by 49 pounds divided by 10 years.

18. Dr. Josmar Pagliuso, Universidade de São Paulo, Brasil, Unpublished Data from Actual Experience of Sugar Cane Farmers Producing Table Sugar and Alcohol Fuel Over 20 Years.

19. *Energy Savings from Recycling,* Consumer's Choice Council, May 2001, <www.ems.org/cgi-bin/ Gprint2002.pl?file-Energy_policy.recycling.rx> (July 8, 2005).

20. *Management of Scrap Tires,* U.S. Environmental Protection Agency, <www.epa.gov/epaoswer/non-hw/ muncpl/tires/ground.htm> (July 8, 2005). [Cites up to 80% recycling of tires, primarily into road asphalt.]

21. K.J. Hülsbergen, et al., "A Method of Energy Balancing in Crop Production and Its Application in a Long-Term Fertilizer Trial," *Agriculture Ecosystems and Environment* 86 (2001), 307.

22. Isaias Carvalho Macedo, *Energy Balance of the Sugar Cane and Ethanol Production in the Cooperated Sugar Mills,* CT Brasil, Ministério da Ciência e Tecnologia, The United Nations Framework Convention on Climate Change (1996).

23. Michael S. Graboski and John McClelland, *A Rebuttal to "Ethanol Fuels: Energy, Economics and Environmental Impacts" by D. Pimentel* (May 2002).

24. Hosein Shapouri, James A. Duffield, and Michael Wang, "The Energy Balance of Corn Ethanol: An Update," *Agricultural Economic Report* 813, U.S. Department of Agriculture (July 2002).

25. Nathan Leaf, "Big Farm Plant Is Planned," *Wisconsin State Journal,* December 8, 2005, Sec. E-1.

26. Christoph Berg, *World Fuel Ethanol Analysis and Outlook* (April 2004), 5.

27. Matthew Brusstar, et al., *High Efficiency and Low Emissions from a Port-Injected Engine with Neat Alcohol Fuels* (Washington, DC: U.S. Environmental Protection Agency, Society of Automotive Engineers, 2002).

28. John M. Urbanchuk, *Contribution of the Ethanol Industry to the Economy of the United States* (Renewable Fuels Association, January 2005), 1–4.

Fig. A-4

Although previous chapters contain all the information you need to convert vehicles, they don't convey the flavor of the process of conversion. I thought I'd write about a couple of the conversions we did so you can understand the process we went through in experimenting with conversion techniques and see how our thinking progressed during the conversions. As you will see, we had a pretty good time figuring this stuff out. I hope you have as much fun as we did, when you do your vehicles.

TOYOTA TUNDRA CONVERSION

There have been many conversion experiments with American cars over the years, resulting in today's flexible-fuel vehicles, but little has been published on the conversion of Asian vehicles—even though they are among the most popular vehicles in the U.S. and Canada.

We were certain that a full aftermarket computer, like the FAST or DFI systems, could be used to convert the pickup owned by Michael Bock. But we looked to see if a less expensive option might do the trick. For this, we turned to Fodge Engineering, legendary race car builders, to work with us on the conversion.

When we first met Dan Fodge, he asked why we were trying to convert to E-98 instead of sticking with E-85? I said, "E-85 was the lazy man's way to do it, allowing the automakers to avoid having to install a cold-start system. We will put in a cold-start system, so we'll use as little petroleum as possible." "Good answer," he said. "Let's get to work."

Dan is not a type A personality; he's a double A kind of guy. Our first step was to install a wide-band sensor in the exhaust pipe to give us data that a normal oxygen sensor lacks. The plan was to drop the exhaust pipe to make drilling the hole and welding on a fitting easy.

When the pipe wouldn't budge and sockets started breaking, another mechanic would have said, "Let's take it to a muffler shop to get it off."

ALTHOUGH PREVIOUS CHAPTERS CONTAIN ALL THE INFO YOU NEED TO CONVERT VEHICLES, THEY DON'T CONVEY THE FLAVOR OF THE PROCESS OF CONVERSION. I THOUGHT I'D WRITE ABOUT A COUPLE OF THE CONVERSIONS WE DID SO YOU CAN UNDERSTAND THE PROCESS WE WENT THROUGH IN EXPERIMENTING WITH CONVERSION TECHNIQUES AND SEE HOW OUR THINKING PROGRESSED DURING THE CONVERSIONS. AS YOU WILL SEE, WE HAD A PRETTY GOOD TIME FIGURING THIS STUFF OUT. I HOPE YOU HAVE AS MUCH FUN AS WE DID, WHEN YOU DO YOUR VEHICLES.

DAVID BLUME

Fig. B-1 Acura cold-start system. *The bottle of ether is at (1), with the valve at the bottom out of sight. The ether injector is at (2). Ether is injected into the air supply hose leading to the manifold. Note the fuel pressure gauge for calibrating for alcohol or for going back to gasoline at (3).*

But Dan said, "We'll just do it on the car." So under the car he went with a pneumatic drill and drilled an approximately 3/4-inch hole in the pipe and welded on a fitting he just made on the lathe from a chunk of roundstock! The fitting had the same threads on the inside as a spark plug (or oxygen sensor), and the outside fit in the hole, which he then arB-welded in place, leaving no air leaks. He then inserted the wide-band sensor into the hole and tightened it up.

The sensor was connected to a handheld air/fuel analyzer up in the Tundra's cab, which told us what the ratio of air to fuel was. We tested it, running on gas, and it showed reliably at 14.7 to 1 air to fuel ratio.

Since the metal ends were of a larger diameter, they would not fit into the sockets. Now, most mechanics would throw up their hands and say, "Sorry guys, it's the day before Thanksgiving; it'll be next week until we can get an order of proper injectors in." But this was no barrier to Dan.

We then installed a mass airflow tuning computer in the wiring of the airflow device on the car. Our little computer intercepted the signal from the mass airflow sensor and modified it before sending it on to the car's ECU. On the box of the little computer, we had four small knobs that allowed us to tune the air/fuel ratio at different engine conditions. The first knob controlled the mixture at idle, the second controlled early acceleration, the third controlled wide open throttle (WOT), and the fourth controlled cruising speed (see Figure 16-1).

On a carburetor, you have the idle circuit, the accelerator pump for early acceleration, the power valve for WOT, and the main jet for cruising. So, what this little box does is give us control of all the various conditions of fuel and air and make them adjustable. Each knob allows us to turn it to the right of center to make the mixture richer and to the left to make it leaner. This is similar to the kind of knob you have when setting the balance on your stereo speakers.

On the first test run, with all the knobs turned to the highest fuel setting, we were getting fuel to air ratios of around 16 to 1. This indicated to Dan that our fuel injectors were limiting the amount of fuel getting to the cylinders, and 16 to 1 was so lean that we had nowhere near enough power. Dang.

That meant that we had to tear down the top of the engine, and get the intake manifolds off to get to the fuel rails and injectors.

Fuel rails are the common pipeline bringing fuel to all the injectors and ending in a fuel pressure regulator which allows some of the fuel back to the tank. Rich (another mechanic), Dan, and I quickly pulled apart the manifold assemblies, giving the engine the appearance of a filleted fish.

Dan had anticipated the possibility that the injectors wouldn't be big enough, so he had ordered better-quality injectors, with a 30% higher flow rate. Wouldn't you know it, the new injectors weren't quite the right diameter to mount on this engine. Now, fuel injectors don't screw in like a spark plug. The end of the injector has an O-ring fitted around it, as does the top of the injector. The end of the injector is pushed down snugly into a socket in the engine, and the top pushes into a similar socket in the bottom of the fuel rail. The O-rings seal against leaks. When you bolt down the rail, the compression alone, with the O-rings in place, seals the injector in.

Since the metal ends were of a larger diameter, they would not fit into the sockets. Now, most mechanics would throw up their hands and say, "Sorry guys, it's the day before Thanksgiving; it'll be next week until we can get an order of proper injectors in." But this was no barrier to Dan. He just chucked them in his lathe and turned them down to the right size, and we bought some new O-rings. Once installed, we put the manifold back together in minutes, with all three of us wrenching nuts and bolts into place, as well as plugging lines and electrical connectors back together.

So, we fired up the air fuel meter and tried to start the Tundra. It ran at idle for about 20 seconds, and when Rich pressed down on the accelerator, it coughed and died. Then it wouldn't start at all. This was a bad moment.

We were all perplexed. Did we flood the engine somehow with the bigger injectors? Pulling out two spark plugs, we found them dry. We tried starting it again, and it would sort of cough and sputter. We all thought it sounded more like an ignition timing problem. Rich suggested that maybe we somehow switched the electrical connections to two of the fuel injectors? No, we could peek in and see they were right. Then Rich found a random electrical connector on a harness, that we had forgotten to plug back in from when we reassembled the manifolds. Rich got behind the wheel, and, vroom, the engine started right up.

It was time for another test drive. This time the truck ran okay, but our air/fuel mixture never got any richer than 13.6 to 1, and was generally a lot leaner. I was all for checking the exhaust gas temperature. If it was cool enough, i.e., less than gasoline's 1400°F, I was ready to call the job done, even though it was a little weak on acceleration.

But Dan insisted that the engine still wasn't getting enough fuel, and certainly not enough performance for him. So we went through the whole routine again with a slightly larger set of injectors that Dan had lying around loose in his shop (which, of course, needed to be machined to fit).

The truck changed, from feeling like a suburban soccer mom's anemic van into a powerhouse. We ran up and down country roads, roaring, and cruising, and everything in between.

The difference was immediate. It changed the truck from feeling like a suburban soccer mom's anemic van into a powerhouse. We ran up and down country roads, roaring, and cruising, and everything in between. Rich drove, giving Dan feedback on performance, while Dan tuned the truck while watching the air/fuel gauge installed in the exhaust pipe. Tom the cameraman filmed, and I was ecstatic.

Amazingly enough, we tuned it at different circuits between 12.5 to 1 up to about 14 to 1 at cruise. There was no stumbling, no hesitation, when downshifting the automatic transmission at lower rpms, and acceleration was really powerful. There was no odor to the exhaust whatsoever. The wide range of air/fuel mixtures that we were able to run was phenomenal, and we were using far less fuel than should have been possible. Spark advance seemed like it was handled automatically, with a knock sensor.

This increase in injector size probably means that we cannot tune back to gasoline, so this is an alcohol-only conversion at the present time. Using different injectors with a wider range, like those on FFVs, probably would allow us to go dual-fuel.

Dan and his son were more than excited. They had been talking for some years about ethanol. Both of them said, "It could solve so many of our

country's problems. We could even stop subsidizing tobacco growing and have those farmers switch to fuel crops."

Rich drove more than 100 miles home with no hitches. We planned emissions testing and more careful tuning with the computer's software; we needed to put it into a laptop, to refine the coarse settings we made with the knobs.

Our first emissions test, at a California dynamometer smog-testing center, running the Tundra on E-98, was quite a sight. The technician just kept shaking his head and finally said, "If we ran all the cars on this stuff, we wouldn't need smog tests."

Our first emissions test, at a California dynamometer smog-testing center, running the Tundra on E-98, was quite a sight. The technician just kept shaking his head and finally said, "If we ran all the cars on this stuff, we wouldn't need smog tests." Our readings were zero on carbon monoxide, zero on hydrocarbons, and 4 ppm on NOx. Normally, the technician told us, he sees hydrocarbons at 60 ppm with a 100-ppm maximum allowable. The normal reading for carbon monoxide is 0.2% with a 0.9% maximum allowable. With NOx readings, he usually sees 400–500 ppm with a maximum allowable of 600 ppm. California smog-testing is the most rigorous in the country.

ACURA INTEGRA CONVERSION

Paul Robbins, Tom Valens, and I all got together in Sacramento and worked on a different kind of conversion, using a full replacement computer for Paul's Integra, which allows us to program just about anything in his car down to the tiniest detail.

Paul's goal was to have a three-part conversion system: one version for gasoline, one for high performance (read: goes really fast), and one for a fuel efficiency mode. One of our goals was to do this using all stock fuel injectors, unlike the Tundra conversion, which required new fuel injectors.

Without getting into the gory details, it went amazingly well, meeting our first goal of high performance. The four of us tore up and down a quiet suburban street like teenagers with their new hot toy, making adjustments, while Paul got his jollies redlining his Japanese car like he was a NASCAR driver. This program also lets us use special settings during cold-start and warm-up. It ran really cool, as Paul confirmed by feeling the exhaust pipe, which would have usually taken a couple of layers of skin right off.

I went home early, so I missed the last part, since Paul and the mechanic worked together to adjust the vehicle timing at the high end. Paul told me that he had a more or less white-knuckled grip, roaring down a rural road at 115 mph, while the mechanic calmly tootled with this setting and that setting, to get even more power out of it on his laptop connected to the computer. Remember, this Integra only has four cylinders!

The conversion to alcohol was so powerful that, during Paul's tests, he couldn't tell any difference in acceleration between when the vehicle was empty, and when it was carrying all four of us guys full of Mexican food and margaritas.

Our conversion did use quite a bit more alcohol than gasoline (Paul almost didn't make it home). But as he pointed out, it was so powerful that, during his tests, he couldn't tell any difference in acceleration between when the vehicle was empty, and when it was carrying all four of us guys full of Mexican food and margaritas. He was also pleased with the torque that he got as soon as he touched the pedal.

Now that we know we can get enough alcohol through the original injectors for even high performance, we know it will do the job for normal high-mileage performance. We had boosted the fuel pressure from a stock 32 psi to a 60 psi reading on the pressure gauge we installed. In taking the fuel pressure up to 60 psi, the exhaust air/fuel gauge went from 18 to 1 (very lean) down to 12.5 to 1. Later, when Paul checked, he found that the pressure had actually gone up to 80 psi. I guess we didn't tighten the lock nut enough.

We did this with an aftermarket fuel pressure regulator, which allows you to determine how much fuel is returned to the tank. By restricting the back-flow, the pressure available to the injectors is raised. This gives us the effect of having larger fuel injectors, since each time they open they deliver more fuel in the same amount of time, as they would at lower fuel pressure. Using an adjustable regulator means that you can pretty easily switch back to gasoline, as Paul has done several times. You could do it by the number of turns of the

adjustment screw, but we installed a fuel pressure gauge in the fuel line, so we could be accurate during switching.

Part of the change was in spark timing. Once we had it running on a fuel mixture that read around 13 to 1 on our exhaust air/fuel analyzer (EA/F), we started working with the spark timing. Reverting to tactics we used to use on carburetors, we planned to advance the timing until we reached the first time we peaked in manifold vacuum pressure (MVP) (see Chapter 16). Normally, we'd start at about 15° before top dead center (BTDC) in setting up old-style distributors.

When we checked the Acura, it was already starting at 17.5° BTDC. Wow, do times ever change! So we started adding to the timing advance one-third of a degree at a time on the computer screen and watched the MVP on another computer window. We watched it go up and up and up. But we also noticed a change in the EA/F. While we were messing with the advance, the mixture seemed to become richer and went down as low as 10 to 1. So it was clear that we were getting different calculations from the computer for pulse width as we advanced the timing.

We finally reached a point where the MVP stopped going up. As we kept raising the advance, the MVP stayed on a plateau until we got into the low 30s of degrees BTDC, and then it started down again. This was such a pleasure compared to trying to do this with a 1980s analyzer. We now had a choice. In the old days, we went with the first time the MVP peaked, but now we could pick a point anywhere along the plateau, since we could now see the whole thing. We chose a point midway and set the timing at 25° BTDC at idle. It was particularly rewarding to watch how much less fuel we were consuming as we advanced the timing.

At this point, we took it out on the road with Tom the cameraman and me in the back seat, while Paul drove the car and Henry the mechanic operated the laptop connected to the computer. We did test drives up to 7000 rpm (redline), smoothing out acceleration curves and making the car perform at a high level. This is much harder driving than 99% of the public would expect from their four-cylinder engine. The power output was impressive, and at no time in any of the various acceleration phases did we find the fuel output of the injectors inadequate to the somewhat extreme requests that we were making electronically. Paul went home from this first round of working on his car with a high-performance fuel and spark map saved, but he was only getting about 26 mpg compared to his normal 32 mpg on gasoline. Ahh, but there were still more things to do.

On the second trip to work with Henry, the goal was to develop a higher mileage map. This meant that we would be departing the very safe, rich mixtures that cool the engine, and raising the air/fuel mixture quite a bit. The danger in doing this operation was that the leaner we went, the greater the chance we would have too high a combustion temperature, and literally burn up the engine—if we went beyond lean mixtures into abnormal combustion at the ultra-lean end of the range.

Normally, lean burning will produce somewhat cooler temperatures than stoichiometric.

So before meeting with Henry, Paul had a new opening welded on his exhaust pipe, where he could insert an exhaust gas temperature gauge. Paul first gathered data while running on gasoline and soon discovered that the peak temperature he measured in the exhaust pipe was close to 1400°F. So now we knew how hot we could go on alcohol without exceeding the limits of engine cooling.

Paul worked with Henry and slashed the amount of fuel that they were mixing with air, fooled around some more with the timing, and found that long before they could even get much over 1300°F, the vehicle would cough and buck from not getting enough fuel. So it turned out that it wasn't even physically possible to overheat his car by leaning!

Fig. B-2 Acura fuel heat. The plumbing in the foreground is the primary fuel heater (1) and a bypass line (2) to avoid heating the fuel when running on gasoline. On the left you can see the fuel pressure gauge (3). Hot water enters at (4) and exits at (5). Warm fuel enters the heater at (6) and exits at (7) to go to the fuel rail. Excess fuel from the fuel rail goes through the adjustable fuel pressure regulator at (8) and then has its surplus heat scavenged in the fuel preheater (9) which warms the cold fuel before going to (1).

> *By cranking up the pressure, we dramatically reduced the amount of return fuel, and our heat exchangers could then do their job efficiently with very little energy lost in the fuel going back to the tank.*

However, Paul—now being spoiled by the get up and go of his performance map—couldn't bring himself to totally castrate the power aspects of his car. So he and Henry left in some richer mixtures for acceleration phases, while keeping the cruising and idle very lean. This allows Paul to easily blast pass soccer moms talking on their cell phones while driving their SUVs. So it's not truly the highest mileage map. He had increased his mileage from 26 to about 30 mpg, still a little lower than his mileage on gasoline.

We worked together on the next step. Our plan was to heat the fuel quite a bit to improve atomization, and therefore mileage. This was new territory for me, since in the 1980s, using carburetors, fuel pressure was around five psi. So the boiling point of alcohol was pretty close to its normal 173°F. If you heated the fuel too much, you'd get either vapor lock or sudsing in the carb.

But this new fuel injection era theoretically opened up some interesting possibilities. With his fuel pressure running about 60 psi, the boiling point of the alcohol was far above the 205°F temperature of his engine coolant. So we figured that if we used a countercurrent heat exchanger, we should be able to pack in a lot of extra Btus to get his fuel up to 205°F. When it was sprayed into the vacuum of the engine, atomization should be much improved, and a small part of the fuel might even vaporize.

We found a nice little countercurrent heat exchanger and ran the hot water down the central tube and the fuel in a spiral path around the central tube. We were sadly disappointed when the fuel only got into the 150s. The problem was that, unlike old carb systems, a large part of the fuel passed though the pressure regulator and headed back to the tank. So our heat exchanger was trying to heat both the fuel that was injected and the fuel returning to the tank.

So we added another heat exchanger. This one was a fuel-to-fuel heat exchanger, where the heated fuel from the pressure regulator passed down the central tube, and the cold fuel from the tank spiraled around the outside. This scavenged much of the heat that was being wasted and preheated the fuel before it went into the main heat exchanger. This raised the temperature into the 170s, so we were still disappointed. We diddled around with insulation and a few other things, and, one night, it hit me what we had to do. When Paul called the next day, he was kind of frustrated with the heater project, and I said, "Paul I want you to try one last thing—crank up the fuel pressure another 20 pounds. I know it will shorten the pulse width, but I think it will solve our heating problems." He was doubtful, but he said he'd go and try it.

Half an hour later, he called me, practically jumping out of his skin. "It works, it works, my fuel is over 200°F!" By cranking up the pressure, we dramatically reduced the amount of return fuel, and our heat exchangers could then do their job efficiently with very little energy lost in the fuel going back to the tank. Within a few days, Paul left me another message on my answering machine. "Well Dave, I gotta say that I had the worst day of city driving and as far as I can measure it, I am getting 33 mpg. But highway driving on my way home, I'm sure it wasn't any less than 35 mpg. That's higher than I ever got on gasoline!"

Now it was my turn to be excited. This was truly new territory.

Temperature Conversion

°F	°C	°F	°C	°F	°C	°F	°C	°F	°C	°F	°C
32	0										
33	0.56	63	17.22	93	33.89	123	50.56	153	67.22	183	83.89
34	1.11	64	17.78	94	34.44	124	51.11	154	67.78	184	84.44
35	1.67	65	18.33	95	35	125	51.67	155	68.33	185	85
36	2.22	66	18.89	96	35.56	126	52.22	156	68.89	186	85.56
37	2.78	67	19.44	97	36.11	127	52.78	157	69.44	187	86.11
38	3.33	68	20	98	36.67	128	53.33	158	70	188	86.67
39	3.89	69	20.56	99	37.22	129	53.89	159	70.56	189	87.22
40	4.44	70	21.11	100	37.78	130	54.44	160	71.11	190	87.78
41	5	71	21.67	101	38.33	131	55	161	71.67	191	88.33
42	5.56	72	22.22	102	38.89	132	55.56	162	72.22	192	88.89
43	6.11	73	22.78	103	39.44	133	56.11	163	72.78	193	89.44
44	6.67	74	23.33	104	40	134	56.67	164	73.33	194	90
45	7.22	75	23.89	105	40.56	135	57.22	165	73.89	195	90.56
46	7.78	76	24.44	106	41.11	136	57.78	166	74.44	196	91.11
47	8.33	77	25	107	41.67	137	58.33	167	75	197	91.67
48	8.89	78	25.56	108	42.22	138	58.89	168	75.56	198	92.22
49	9.44	79	26.11	109	42.78	139	59.44	169	76.11	199	92.78
50	10	80	26.67	110	43.33	140	60	170	76.67	200	93.33
51	10.56	81	27.22	111	43.89	141	60.56	171	77.22	201	93.89
52	11.11	82	27.78	112	44.44	142	61.11	172	77.78	202	94.44
53	11.67	83	28.33	113	45	143	61.67	173	78.33	203	95
54	12.22	84	28.89	114	45.56	144	62.22	174	78.89	204	95.56
55	12.78	85	29.44	115	46.11	145	62.78	175	79.44	205	96.11
56	13.33	86	30	116	46.67	146	63.33	176	80	206	96.67
57	13.89	87	30.56	117	47.22	147	63.89	177	80.56	207	97.22
58	14.44	88	31.11	118	47.78	148	64.44	178	81.11	208	97.78
59	15	89	31.67	119	48.33	149	65	179	81.67	209	98.33
60	15.56	90	32.22	120	48.89	150	65.56	180	82.22	210	98.89
61	16.11	91	32.78	121	49.44	151	66.11	181	82.78	211	99.44
62	16.67	92	33.33	122	50	152	66.67	182	83.33	212	100

Fig. C-1 Temperature conversion between Fahrenheit and Celsius.

Common Conversions

TO CONVERT FROM	TO	MULTIPLY BY
Acres	Square Feet	43,560
	Square Kilometers	.00405
	Square Yards	4840
	Square Meters	4046.8564
Acre Feet	Cubic Feet	43,560
	Cubic Meters	1233.4818
	Cubic Yards	1613.333
Barrels (Petroleum)	Cubic Feet	5.614583
	Gallons	42
	Liters	158.98284
British Thermal Unit (Btu)	Btu (IST)	.999346
	Calories/Gram	251.99576
	Foot-Pounds	777.649
Btu/Hour	Horsepower (Boiler)	2.98563×10^{-5}
	Horsepower (Electric)	.000392594
	Horsepower (Metric)	.000398199
	Watts	.292875
	Calories, Kg/Hour	.251996
	Lbs. Ice Melted/Hour	.41828
Gallons	Cubic Feet	.13368
	Liters	3.7853
	Lbs. Water (4°C)	8.33585
Grams/Liter	Parts per Million (ppm)	1000
	Lbs. per Cubic Foot	.06242621
Hectares	Acres	2.47
	Square Meters	10,000
Horsepower (Boiler)	Kilowatts	9.8
	Lbs. of 212°F Steam	34.5
	Btu/Hour	33,445.7
Horsepower (Electric)	Btu/Hour	2547.16
	Calories, Gram/Second	178.298
	Foot-Pound/Second	550.221
	Horsepower (Boiler)	.0760487
	Horsepower (Metric)	1.0142777
	Horsepower (Mechanical)	1.00040
	Watts	746
Horsepower (Mechanical)	Tons of Refrigeration	.21204
	Horsepower (Electric)	.999598
Inches	Centimeters	2.54
Inches Mercury (Hg)	Feet of Air (60°F)	926.24

Fig. C-2 Common conversions.

Common Conversions (continued)

TO CONVERT FROM	TO	MULTIPLY BY
Inches Mercury at 32°F	Feet of Water (39.2°F)	1.132957
	Millimeters of Hg (60°F)	25.4
Inches Mercury at 60°F	Pounds Per Square Inch (psi)	70.5269
Inches of Water at 4°C	Inches of Hg (32°C)	.0735539
	Pounds per square inch	.03612628
Kilograms	Pounds (U.S.)	2.2046226
Kilowatts (Int.)	Btu/Hour	3414.43
	Horsepower (Boiler)	.101959
	Horsepower (Electric)	1.34070
	Horsepower (Metric)	1.35985
Liters	Gallons (U.S. Liquid)	.02641794
	Quarts	1.056718
Liters/Second	Gallons/Millimeter	15.85077
	Gallons/Second	.2641794
Ounces (Avoirdupois)	Grams	28.34952
Ounces (U.S. Fluid)	Cubic Centimeter (cc)	29.573730
	Cubic Inches	1.8046875
	Liters	.029572702
	Milliliters (ml)	29.572702
Pounds (Avdp.)	Kilograms	.45359237
Pounds/Cubic Foot	Kilograms/Cubic Meter	16.018463
Pounds of H_2O Evaporated from and at 212°F	Btu (IST)	970.2
Pounds per Square Inch (psi)	Centimeters of Mercury (0°C)	5.17149
	Centimeters of H_2O (4°C)	70.3089
	Grams/Square Centimeter	70.306958
	Inches of Mercury (32°F)	2.03602
	Inches of H_2O (39.2°F)	27.6807
Short Ton	Kilograms	907.18474
	Pounds	2000
	Metric Ton	.90718
Tons of Refrigeration (U.S. Standard)	Btu (IST)	288,000
	Cal., Kg. (Mean)	72,517.9
	Lbs. of Ice Melted	2009.1
Tons of Refrigeration (U.S. Commercial)	Btu (IST)	12,000

Fig. C-2 Common conversions (continued).

Pressure Loss from Pipe Friction

(New Schedule 40 Steel Pipe)
Loss in Pounds per Square Inch per Foot of Pipe*

GPM	PIPE SIZE	VISCOSITY, SSU																
		32 (water)	50	100	200	400	600	800	1000	2000	3000	4000	5000	6000	7000	8000	9000	10,000
1-1/2	3/8	.033	.050	.14	.28	.60	.87	1.2	1.5	3.3	4.5	6.0	7.5	8.8				
	1/2	.013	.020	.055	.11	.24	.35	.47	.60	1.3	1.8	2.4	3.0	3.5	4.2	5.0	5.4	6.0
	3/4	.0038	.0065	.018	.038	.080	.12	.16	.20	.40	.60	.80	1.0	1.2	1.4	1.6	1.8	2.0
	1	.0010	.0025	.0070	.015	.030	.045	.060	.075	.15	.23	.30	.36	.45	.52	.60	.67	.73
3-1/2	1/2	.060	.10	.13	.27	.56	.85	1.1	1.4	2.8	4.3	5.6	7.0	8.5	9.8			
	3/4	.014	.015	.044	.090	.18	.28	.36	.45	.90	1.4	1.9	2.3	2.8	3.2	3.7	4.1	4.6
	1	.0045	.0060	.016	.035	.070	.10	.13	.18	.35	.50	.70	.85	1.0	1.2	1.3	1.6	1.8
	1-1/4	.0011	.0020	.0055	.011	.023	.035	.046	.059	.12	.17	.24	.29	.34	.40	.46	.52	.59
5	3/4	.029	.045	.060	.13	.26	.40	.52	.65	1.3	2.0	2.6	3.2	4.0	4.5	5.2	6.0	6.5
	1	.0090	.0092	.018	.050	.10	.15	.20	.25	.50	.72	1.0	1.3	1.5	1.8	2.0	2.2	2.5
	1-1/4	.0022	.0028	.0079	.016	.033	.050	.066	.083	.17	.25	.33	.41	.50	.56	.66	.72	.83
	1-1/2	.0012	.0015	.0041	.0090	.018	.027	.036	.045	.090	.13	.18	.23	.27	.32	.36	.40	.45
7	3/4	.055	.075	.090	.18	.36	.55	.73	.90	1.8	2.8	3.6	4.5	5.5	6.2	7.3	8.1	9.0
	1	.016	.025	.032	.070	.14	.21	.28	.35	.70	1.1	1.4	1.8	2.1	2.5	2.8	3.1	3.5
	1-1/4	.0040	.009	.011	.023	.046	.070	.092	.11	.23	.35	.46	.60	.70	.80	.92	1.0	1.1
	1-1/2	.0019	.0021	.0060	.013	.025	.038	.050	.062	.13	.19	.25	.31	.37	.45	.50	.55	.62
10	3/4	.10	.14	.14	.26	.52	.80	1.1	1.3	2.6	4.0	5.2	6.4	8.0	9.0			
	1	.030	.045	.047	.10	.20	.30	.40	.50	1.0	1.5	2.0	2.5	3.0	3.5	4.0	4.5	5.0
	1-1/4	.0080	.013	.016	.033	.066	.10	.13	.17	.34	.50	.68	.85	1.0	1.2	1.3	1.5	1.7
	1-1/2	.0035	.0055	.0085	.018	.036	.053	.071	.090	.18	.27	.35	.45	.54	.62	.71	.81	.90
15	1	.064	.092	.14	.15	.30	.45	.60	.75	1.5	2.3	3.0	3.8	4.5	5.2	6.0	7.0	7.5
	1-1/4	.016	.025	.025	.050	.10	.15	.20	.25	.50	.75	1.0	1.3	1.5	1.8	2.0	2.3	2.5
	1-1/2	.0075	.011	.013	.026	.052	.080	.11	.13	.28	.40	.52	.66	.80	.92	1.1	1.2	1.3
	2	.0022	.0036	.0047	.010	.020	.030	.040	.050	.10	.15	.20	.25	.30	.35	.40	.45	.50
18	1	.090	.12	.17	.18	.36	.54	.70	.90	1.8	2.7	3.6	4.5	5.4	6.1	7.0	8.0	9.0
	1-1/4	.023	.030	.033	.060	.12	.18	.24	.30	.60	.90	1.2	1.5	1.8	2.1	2.4	2.8	3.0
	1-1/2	.011	.016	.016	.032	.064	.098	.13	.16	.32	.49	.64	.82	.98	1.1	1.3	1.5	1.6
	2	.0031	.0050	.0056	.012	.024	.036	.050	.060	.12	.18	.24	.30	.36	.42	.50	.55	.60
20	1	.11	.15	.20	.28	.40	.60	.80	1.0	2.0	3.0	4.0	5.0	6.0	7.0	8.0	9.0	10.0
	1-1/4	.028	.040	.060	.065	.13	.20	.26	.32	.65	1.0	1.3	1.6	2.0	2.3	2.6	3.0	3.2
	1-1/2	.013	.018	.019	.036	.071	.11	.15	.18	.36	.53	.70	.80	1.1	1.3	1.5	1.7	1.8
	2	.0039	.0058	.0061	.013	.026	.040	.054	.067	.13	.20	.27	.34	.40	.48	.54	.60	.67
25	1-1/4	.042	.060	.075	.080	.16	.25	.34	.42	.82	1.3	1.6	2.1	2.5	2.9	3.4	3.7	4.2
	1-1/2	.020	.029	.035	.045	.090	.13	.18	.23	.45	.67	.90	1.1	1.3	1.6	1.8	2.0	2.3
	2	.0058	.0083	.0085	.017	.033	.050	.069	.083	.17	.25	.33	.42	.50	.60	.69	.78	.83
	2-1/2	.0025	.0036	.0038	.0080	.016	.025	.032	.038	.080	.14	.16	.20	.25	.29	.32	.36	.38
30	1-1/4	.060	.083	.10	.10	.20	.30	.40	.50	1.0	1.5	2.0	2.5	3.0	3.5	4.0	4.5	5.0
	1-1/2	.027	.040	.045	.054	.11	.16	.21	.28	.52	.80	1.1	1.4	1.6	1.9	2.1	2.4	2.8
	2	.0080	.012	.016	.020	.040	.060	.080	.10	.20	.30	.40	.50	.60	.70	.80	.90	1.0
	2-1/2	.0034	.0047	.0048	.0095	.019	.030	.038	.047	.098	.15	.19	.24	.30	.35	.38	.44	.47
35	1-1/4	.080	.11	.13	.13	.23	.35	.46	.59	1.1	1.8	2.3	2.9	3.5	4.0	4.6	5.2	5.9
	1-1/2	.037	.052	.065	.065	.13	.19	.25	.32	.62	.94	1.3	1.6	1.9	2.3	2.5	2.8	3.2
	2	.011	.015	.020	.023	.046	.070	.094	.12	.23	.35	.46	.59	.70	.81	.94	1.1	1.2
	2-1/2	.0045	.0065	.009	.011	.023	.035	.045	.056	.11	.17	.22	.28	.35	.40	.45	.51	.56
40	1-1/2	.047	.066	.078	.080	.15	.22	.29	.36	.72	1.1	1.5	1.8	2.2	2.5	2.9	3.2	3.6
	2	.013	.020	.024	.026	.053	.080	.11	.13	.30	.40	.53	.68	.80	.92	1.1	1.2	1.3
	2-1/2	.0056	.0084	.011	.013	.025	.039	.050	.064	.13	.19	.25	.31	.39	.45	.50	.58	.64
	3	.0020	.0025	.0025	.0053	.011	.016	.022	.027	.055	.082	.11	.13	.16	.20	.22	.25	.27
50	1-1/2	.072	.097	.10	.10	.18	.28	.36	.46	.90	1.4	1.8	2.3	2.8	3.2	3.6	4.0	4.6
	2	.020	.029	.033	.033	.067	.10	.13	.17	.34	.50	.68	.83	1.0	1.1	1.3	1.5	1.7
	2-1/2	.0085	.012	.016	.016	.032	.050	.064	.080	.16	.24	.32	.40	.50	.59	.64	.72	.80
	3	.0030	.0045	.0060	.0068	.014	.020	.028	.035	.070	.10	.13	.17	.20	.24	.28	.31	.35
60	1-1/2	.10	.14	.16	.16	.22	.32	.43	.54	1.0	1.6	2.2	2.8	3.2	3.8	4.3	4.9	5.4
	2	.029	.040	.044	.044	.080	.12	.16	.20	.40	.60	.80	1.0	1.2	1.4	1.6	1.8	2.0
	2-1/2	.012	.017	.022	.019	.038	.059	.078	.097	.19	.29	.38	.49	.59	.70	.78	.88	.97
	3	.0040	.0060	.0080	.0080	.017	.025	.032	.040	.081	.13	.16	.20	.25	.28	.32	.37	.40
80	2	.050	.068	.086	.093	.10	.16	.22	.28	.52	.80	1.0	1.3	1.6	1.9	2.2	2.5	2.8
	2-1/2	.020	.028	.037	.045	.050	.079	.10	.13	.26	.39	.50	.65	.79	.90	1.0	1.1	1.3
	3	.0070	.010	.012	.012	.022	.032	.044	.054	.11	.17	.22	.28	.32	.37	.44	.50	.54
	4	.0018	.0027	.0030	.0035	.0072	.011	.015	.018	.036	.056	.074	.091	.11	.13	.15	.17	.18
90	2	.063	.082	.10	.11	.12	.18	.25	.30	.60	.90	1.2	1.5	1.8	2.2	2.5	2.8	3.0
	2-1/2	.025	.035	.045	.052	.058	.089	.11	.14	.29	.44	.58	.73	.89	1.0	1.1	1.3	1.4
	3	.0089	.013	.016	.022	.025	.037	.049	.060	.13	.19	.25	.30	.37	.42	.49	.55	.60
	4	.0022	.0034	.0040	.0040	.0081	.013	.016	.020	.040	.062	.081	.10	.13	.14	.16	.18	.20
100	2	.080	.10	.13	.13	.13	.20	.28	.34	.68	1.0	1.3	1.7	2.0	2.4	2.8	3.1	3.4
	2-1/2	.032	.043	.055	.060	.063	.099	.13	.16	.33	.50	.63	.80	.99	1.1	1.3	1.5	1.6
	3	.011	.015	.019	.024	.027	.040	.053	.068	.14	.21	.27	.35	.40	.47	.53	.61	.68
	4	.0028	.0040	.0046	.0046	.0091	.014	.018	.023	.045	.070	.092	.11	.14	.16	.18	.21	.23

*For liquids with a specific gravity other than 1.00, multiply the value from the above table by the specific gravity of the liquid. For old pipe, add 20% to the above values.

Figures to right of orange line are laminar flow. Figures to left of orange line are turbulent flow.

To convert the above values to kPa (kilopascals) per metre of pipe, multiply by 22.6.

To convert the above values to kg per cm2 per metre of pipe, multiply by 0.23.

Fig. C-3 Pressure loss from pipe friction.

Pressure Loss from Pipe Friction (continued)

GPM	PIPE SIZE	VISCOSITY, SSU												
		15,000	20,000	25,000	30,000	40,000	50,000	60,000	70,000	80,000	90,000	100,000	150,000	250,000
1-1/2	1-1/4	.37	.50	.62	.73	1.0	1.3	1.5	1.7	1.9	2.2	2.5	3.7	6.2
	1-1/2	.20	.27	.35	.40	.53	.69	.80	.92	1.1	1.2	1.3	2.0	3.5
	2	.075	.10	.13	.15	.20	.25	.30	.35	.40	.46	.50	.75	1.3
	2-1/2	.036	.050	.060	.072	.095	.12	.14	.17	.20	.23	.25	.36	.60
3-1/2	1-1/4	.88	1.2	1.5	1.7	2.4	2.9	3.5	4.0	4.5	5.1	5.9	8.8	
	1-1/2	.47	.60	.80	.92	1.2	1.6	1.8	2.3	2.5	2.8	3.1	4.7	8.0
	2	.18	.23	.29	.35	.46	.57	.70	.85	.93	1.1	1.2	1.8	2.9
	2-1/2	.085	.11	.14	.17	.22	.28	.34	.40	.45	.50	.55	.85	1.4
5	1-1/2	.66	.89	1.1	1.3	1.8	2.3	2.7	3.2	3.6	4.1	4.5	6.6	
	2	.25	.33	.41	.50	.67	.82	1.0	1.2	1.3	1.5	1.7	2.5	4.1
	2-1/2	.13	.16	.21	.25	.33	.41	.50	.59	.66	.75	.81	1.3	2.1
	3	.050	.070	.085	.10	.13	.17	.20	.24	.28	.30	.34	.50	.85
7	1-1/2	.92	1.3	1.6	1.9	2.5	3.1	3.8	4.5	5.0	5.5	6.1	9.2	
	2	.35	.46	.59	.70	.93	1.1	1.4	1.7	1.9	2.1	2.4	3.5	5.8
	2-1/2	.17	.23	.28	.34	.45	.55	.68	.80	.90	1.0	1.1	1.7	2.8
	3	.070	.095	.12	.15	.19	.24	.29	.34	.38	.43	.47	.70	1.2
10	1-1/2	1.3	1.8	2.3	2.7	3.5	4.5	5.4	6.3	7.1	8.0	8.9		
	2	.40	.65	.84	1.0	1.3	1.7	2.0	2.4	2.8	3.0	3.3	4.0	8.4
	2-1/2	.25	.33	.40	.49	.64	.80	.98	1.1	1.3	1.5	1.6	2.5	4.0
	3	.10	.14	.17	.20	.27	.35	.40	.48	.55	.61	.69	1.0	1.7
15	2	.75	1.0	1.3	1.5	2.0	2.5	3.0	3.6	4.1	4.6	5.0	7.5	
	2-1/2	.36	.50	.60	.72	.95	1.2	1.4	1.7	2.0	2.3	2.5	3.6	5.0
	3	.15	.20	.25	.30	.40	.50	.60	.70	.80	.90	1.0	1.5	2.5
	4	.050	.066	.085	.10	.13	.17	.21	.24	.28	.31	.34	.50	.85
18	2	.90	1.2	1.5	1.8	2.4	3.0	3.7	4.3	4.9	5.4	6.0	9.0	
	2-1/2	.44	.59	.72	.88	1.1	1.4	1.7	2.0	2.3	2.6	2.9	4.4	7.2
	3	.18	.25	.30	.36	.50	.60	.71	.85	.98	1.1	1.2	1.8	3.0
	4	.060	.080	.10	.13	.17	.20	.25	.28	.32	.37	.41	.60	1.0
20	2	1.0	1.3	1.7	2.0	2.7	3.4	4.1	4.8	5.4	6.1	6.8	10.0	
	2-1/2	.49	.65	.80	.96	1.3	1.6	1.9	2.3	2.6	2.9	3.2	4.9	8.0
	3	.20	.28	.34	.41	.54	.69	.80	.95	1.1	1.2	1.3	2.0	3.4
	4	.069	.090	.11	.14	.18	.23	.28	.31	.36	.41	.46	.69	1.1
25	2-1/2	.60	.80	1.0	1.2	1.6	2.0	2.4	2.9	3.2	3.7	4.0	6.0	10.0
	3	.25	.35	.42	.51	.70	.85	1.0	1.1	1.3	1.6	1.7	2.5	4.2
	4	.085	.11	.14	.18	.23	.28	.35	.40	.45	.52	.58	.85	1.4
	6	.016	.022	.028	.032	.043	.053	.064	.074	.085	.095	.11	.16	.28
30	2-1/2	.72	.99	1.2	1.4	1.9	2.4	2.8	3.4	4.0	4.5	4.9	7.2	
	3	.30	.40	.50	.61	.81	1.0	1.2	1.4	1.6	1.8	2.0	3.0	5.0
	4	.10	.13	.18	.21	.28	.34	.42	.49	.55	.64	.70	1.0	1.8
	6	.020	.026	.033	.040	.051	.065	.078	.092	.10	.12	.13	.20	.33
35	2-1/2	.85	1.1	1.4	1.7	2.3	2.8	3.4	4.0	4.5	5.0	5.5	8.5	
	3	.35	.48	.60	.72	.95	1.2	1.4	1.7	1.9	2.1	2.4	3.5	6.0
	4	.12	.16	.20	.25	.32	.40	.50	.55	.64	.73	.80	1.2	2.0
	6	.023	.030	.039	.046	.060	.076	.091	.10	.12	.13	.15	.23	.39
40	2-1/2	.97	1.3	1.6	2.0	2.5	3.2	3.8	4.5	5.0	5.8	6.3	9.7	
	3	.40	.55	.69	.82	1.1	1.3	1.6	1.9	2.2	2.5	2.7	4.0	6.9
	4	.14	.18	.23	.28	.37	.46	.57	.65	.73	.83	.90	1.4	2.3
	6	.027	.035	.045	.052	.070	.089	.10	.12	.14	.16	.18	.27	.45
50	2-1/2	1.2	1.6	2.0	2.4	3.2	4.0	4.8	5.5	6.4	7.3	8.0		
	3	.50	.70	.85	1.0	1.4	1.7	2.0	2.4	2.8	3.1	3.4	5.0	8.5
	4	.17	.23	.29	.35	.46	.60	.70	.81	.90	1.0	1.1	1.7	2.9
	6	.033	.044	.055	.065	.086	.11	.13	.15	.17	.19	.22	.33	.55
60	3	.60	.81	1.0	1.3	1.6	2.0	2.5	2.9	3.2	3.7	4.0	6.0	10.0
	4	.20	.27	.35	.41	.55	.70	.84	.99	1.1	1.3	1.4	2.0	3.5
	6	.040	.052	.065	.079	.10	.13	.15	.18	.20	.24	.26	.40	.65
	8	.014	.018	.023	.027	.036	.045	.054	.063	.072	.081	.090	.14	.23
80	3	.80	1.1	1.4	1.7	2.2	2.8	3.2	3.8	4.3	5.0	5.4	8.0	
	4	.27	.36	.46	.55	.74	.91	1.1	1.3	1.5	1.7	1.8	2.7	4.6
	6	.052	.070	.090	.10	.14	.18	.21	.25	.28	.31	.35	.52	.90
	8	.018	.024	.030	.036	.048	.060	.072	.085	.096	.11	.12	.18	.30
90	3	.91	1.2	1.6	1.9	2.5	3.0	3.7	4.3	4.9	5.5	6.1	9.1	
	4	.30	.40	.51	.62	.83	1.0	1.3	1.4	1.6	1.8	2.1	3.0	5.1
	6	.060	.079	.10	.12	.15	.20	.23	.27	.31	.36	.39	.60	.79
	8	.020	.027	.034	.040	.055	.067	.080	.095	.11	.12	.13	.20	.34
100	3	1.0	1.4	1.7	2.1	2.8	3.4	4.0	4.7	5.4	6.1	6.9	10.0	
	4	.35	.45	.60	.70	.91	1.1	1.4	1.6	1.8	2.1	2.3	3.5	6.0
	6	.065	.085	.11	.13	.18	.22	.26	.30	.35	.38	.44	.65	1.1
	8	.023	.030	.037	.045	.060	.073	.090	.10	.12	.13	.15	.23	.37

*For liquids with a specific gravity other than 1.00, multiply the value from the above table by the specific gravity of the liquid. For old pipe, add 20% to the above values.

All figures on this page are laminar flow.

To convert the above values to kPa (kilopascals) per metre of pipe, multiply by 22.6.

To convert the above values to kg per cm² per metre of pipe, multiply by 0.23.

Fig. C-3 Pressure loss from pipe friction (continued).

Pressure Loss from Pipe Friction (continued)

GPM	PIPE SIZE	VISCOSITY, SSU												
		15,000	20,000	25,000	30,000	40,000	50,000	60,000	70,000	80,000	90,000	100,000	150,000	250,000
120	3	1.2	1.6	2.0	2.5	3.2	4.0	4.9	5.8	6.5	7.5	8.0		
	4	.40	.53	.70	.84	1.1	1.4	1.7	2.0	2.2	2.5	2.8	4.0	7.0
	6	.080	.10	.13	.15	.21	.26	.31	.36	.41	.47	.52	.80	1.3
	8	.023	.035	.045	.055	.072	.090	.11	.13	.14	.16	.18	.23	.45
140	3	1.4	1.9	2.4	2.9	3.8	4.7	5.8	6.8	7.6	8.5	9.5		
	4	.47	.62	.81	.99	1.3	1.6	2.0	2.3	2.5	2.8	3.2	4.7	8.1
	6	.091	.12	.15	.18	.25	.30	.36	.42	.48	.55	.60	.81	1.5
	8	.031	.042	.052	.063	.085	.10	.13	.15	.17	.19	.21	.31	.52
150	3	1.5	2.0	2.5	3.1	4.0	5.1	6.1	7.1	8.1	9.1			
	4	.51	.68	.88	1.0	1.4	1.7	2.1	2.4	2.7	3.2	3.5	5.1	8.8
	6	.099	.13	.16	.19	.26	.32	.38	.46	.51	.57	.65	.99	1.6
	8	.033	.045	.055	.066	.090	.11	.13	.16	.18	.21	.23	.33	.55
160	4	.55	.71	.92	1.1	1.5	1.8	2.3	2.6	3.0	3.4	3.6	5.5	9.2
	6	.10	.14	.18	.21	.28	.35	.41	.48	.55	.62	.70	1.0	1.8
	8	.036	.048	.060	.072	.096	.12	.14	.17	.19	.21	.24	.36	.60
	10	.015	.020	.025	.030	.039	.049	.058	.070	.079	.090	.099	.15	.25
180	4	.61	.80	1.0	1.3	1.7	2.1	2.5	2.9	3.2	3.7	4.1	6.1	10.0
	6	.12	.16	.20	.23	.31	.40	.47	.55	.61	.70	.79	1.2	2.0
	8	.040	.052	.068	.080	.11	.13	.16	.19	.21	.24	.28	.40	.68
	10	.017	.022	.027	.033	.044	.055	.066	.077	.088	.099	.11	.17	.27
200	4	.70	.90	1.2	1.4	1.9	2.3	2.8	3.2	3.6	4.2	4.5	7.0	
	6	.13	.18	.22	.26	.35	.45	.51	.60	.70	.78	.85	1.3	2.2
	8	.045	.060	.075	.090	.12	.15	.18	.21	.24	.28	.30	.45	.75
	10	.018	.025	.030	.036	.048	.060	.071	.085	.098	.11	.12	.18	.30
250	4	.85	1.1	1.5	1.8	2.3	2.8	3.5	4.0	4.5	5.2	5.8	8.5	
	6	.17	.22	.28	.32	.44	.55	.64	.75	.86	1.0	1.1	1.7	2.8
	8	.056	.074	.092	.11	.15	.18	.22	.26	.30	.34	.37	.56	.92
	10	.023	.030	.038	.046	.060	.075	.090	.10	.12	.14	.15	.23	.38
300	4	1.0	1.3	1.8	2.1	2.8	3.5	4.2	4.7	5.4	6.2	7.0	10.0	
	6	.20	.26	.33	.40	.51	.65	.78	.90	1.0	1.2	1.3	2.0	3.3
	8	.068	.090	.11	.13	.18	.22	.27	.31	.35	.40	.45	.68	1.1
	10	.028	.036	.045	.055	.062	.090	.11	.13	.15	.17	.18	.28	.45
400	4	1.4	1.8	2.3	2.8	3.7	4.6	5.5	6.4	7.3	8.2	9.1		
	6	.26	.35	.45	.51	.70	.88	1.0	1.2	1.4	1.6	1.8	2.6	4.5
	8	.090	.12	.15	.18	.24	.30	.36	.41	.47	.54	.60	.90	1.5
	10	.037	.048	.060	.073	.096	.12	.15	.17	.19	.22	.25	.37	.60
450	4	1.5	2.0	2.6	3.1	4.2	5.0	6.0	7.0	8.0	9.0	10.0		
	6	.30	.40	.50	.60	.80	1.0	1.2	1.4	1.6	1.8	2.0	3.0	5.0
	8	.10	.14	.17	.20	.28	.34	.40	.46	.54	.61	.68	1.0	1.7
	10	.042	.055	.070	.082	.11	.14	.16	.19	.22	.25	.28	.42	.70
500	4	1.7	2.3	2.9	3.5	4.6	5.7	7.0	8.0	9.0	10.0			
	6	.33	.44	.55	.66	.87	1.0	1.3	1.5	1.8	2.0	2.2	3.3	5.5
	8	.11	.15	.19	.23	.30	.37	.45	.51	.60	.66	.74	1.1	1.9
	10	.046	.060	.075	.091	.12	.15	.18	.21	.25	.28	.30	.46	.75
600	4	2.0	2.8	3.5	4.2	5.5	6.9	8.3	9.5					
	6	.40	.51	.65	.80	1.0	1.3	1.5	1.8	2.1	2.4	2.6	4.0	6.5
	8	.13	.18	.23	.27	.36	.45	.54	.63	.72	.81	.90	1.3	2.3
	10	.055	.072	.090	.11	.15	.18	.22	.25	.29	.32	.37	.55	.90
750	6	.50	.65	.82	1.0	1.3	1.6	2.0	2.3	2.5	2.9	3.2	5.0	8.2
	8	.17	.22	.28	.34	.45	.55	.65	.79	.90	.98	1.1	1.7	2.8
	10	.070	.090	.11	.14	.18	.23	.27	.32	.37	.41	.46	.70	1.1
	12	.032	.043	.055	.066	.090	.11	.14	.16	.18	.20	.23	.32	.55
800	6	.52	.70	.89	1.0	1.4	1.6	2.1	2.3	2.7	3.1	3.5	5.2	8.9
	8	.18	.24	.30	.36	.48	.60	.71	.84	.95	1.0	1.2	1.8	3.0
	10	.072	.096	.12	.15	.19	.25	.29	.34	.40	.45	.50	.72	1.2
	12	.035	.046	.060	.070	.096	.12	.15	.17	.18	.21	.25	.35	.60
1000	6	.65	.86	1.1	1.3	1.7	2.2	2.6	3.0	3.5	3.9	4.5	6.5	
	8	.23	.30	.37	.45	.60	.74	.90	1.0	1.1	1.3	1.5	2.3	3.7
	10	.091	.12	.15	.18	.25	.30	.36	.42	.49	.55	.61	.91	1.5
	12	.045	.059	.075	.090	.12	.15	.18	.21	.24	.27	.30	.45	.75
1050	6	.70	.90	1.1	1.3	1.8	2.3	2.7	3.1	3.6	4.1	4.7	7.0	
	8	.24	.31	.40	.47	.62	.80	.94	1.0	1.2	1.3	1.5	2.4	4.0
	10	.098	.13	.16	.20	.26	.32	.39	.45	.51	.59	.65	.98	1.6
	12	.047	.061	.080	.095	.13	.16	.19	.22	.25	.29	.31	.47	.80

ᵇFor liquids with a specific gravity other than 1.00, multiply the value from the above table by the specific gravity of the liquid. For old pipe, add 20% to the above values.

All figures on this page are laminar flow.

To convert the above values to kPa (kilopascals) per metre of pipe, multiply by 22.6.

To convert the above values to kg per cm2 per metre of pipe, multiply by 0.23.

Fig. C-3 Pressure loss from pipe friction (continued).

Pressure Loss from Pipe Friction (continued)

GPM	PIPE SIZE	32 (water)	50	100	200	400	600	800	1000	2000	3000	4000	5000	6000	7000	8000	9000	10,000
								VISCOSITY, SSU										
120	2	.11	.14	.15	.18	.18	.24	.32	.40	.80	1.1	1.5	2.0	2.4	2.9	3.2	3.7	4.0
	2-1/2	.045	.060	.075	.078	.078	.12	.15	.19	.40	.60	.77	.99	1.2	1.3	1.5	1.8	1.9
	3	.015	.020	.026	.032	.032	.050	.065	.080	.16	.25	.32	.40	.50	.56	.65	.72	.80
	4	.0040	.0057	.0072	.010	.011	.017	.022	.028	.054	.083	.11	.14	.17	.19	.22	.24	.28
140	2-1/2	.060	.078	.10	.11	.11	.14	.18	.23	.45	.68	.90	1.1	1.3	1.6	1.8	2.0	2.3
	3	.020	.027	.034	.038	.038	.058	.076	.095	.19	.29	.38	.46	.58	.66	.76	.85	.95
	4	.0054	.0075	.0098	.011	.013	.020	.025	.031	.063	.10	.13	.16	.20	.23	.25	.29	.32
	6	.00067	.0010	.0013	.0013	.0024	.0037	.0050	.0060	.012	.018	.024	.030	.037	.042	.050	.055	.060
150	2-1/2	.065	.085	.11	.13	.14	.14	.19	.24	.50	.70	.95	1.2	1.4	1.6	1.9	2.2	2.4
	3	.022	.030	.038	.040	.040	.060	.080	.10	.20	.30	.40	.50	.60	.70	.80	.90	1.0
	4	.0060	.0085	.011	.013	.014	.021	.027	.035	.078	.10	.14	.17	.21	.24	.27	.32	.35
	6	.00075	.0011	.0013	.0013	.0026	.0040	.0052	.0065	.013	.020	.026	.032	.040	.047	.052	.058	.065
160	2-1/2	.0077	.10	.11	.11	.11	.15	.20	.25	.50	.75	1.0	1.3	1.5	1.8	2.0	2.3	2.5
	3	.025	.035	.044	.050	.050	.065	.087	.11	.22	.33	.44	.55	.65	.76	.87	.98	1.1
	4	.0070	.0095	.012	.014	.015	.022	.030	.037	.071	.11	.15	.18	.22	.26	.30	.33	.37
	6	.00086	.0012	.0015	.0015	.0028	.0042	.0055	.0070	.014	.021	.028	.035	.041	.049	.055	.064	.070
180	2-1/2	.10	.12	.15	.18	.18	.18	.23	.29	.58	.87	1.1	1.5	1.8	2.0	2.3	2.6	2.9
	3	.032	.042	.053	.065	.071	.074	.10	.12	.25	.37	.50	.62	.74	.85	1.0	1.1	1.2
	4	.0084	.012	.015	.016	.016	.025	.032	.041	.081	.13	.17	.21	.25	.30	.32	.37	.41
	6	.0011	.0016	.0020	.0027	.0031	.0047	.0063	.0080	.016	.023	.031	.040	.047	.055	.063	.070	.080
200	2-1/2	.12	.14	.18	.19	.20	.20	.25	.32	.63	.96	1.3	1.6	1.9	2.2	2.5	2.8	3.2
	3	.040	.052	.064	.075	.078	.081	.11	.13	.27	.42	.55	.70	.81	.95	1.1	1.2	1.3
	4	.010	.014	.018	.020	.020	.027	.036	.045	.090	.14	.18	.23	.28	.32	.36	.41	.45
	6	.0013	.0019	.0025	.0032	.0035	.0052	.0070	.0089	.018	.026	.035	.045	.052	.060	.070	.079	.089
250	3	.060	.075	.092	.10	.11	.11	.14	.17	.35	.50	.68	.84	1.0	1.2	1.4	1.5	1.7
	4	.016	.021	.026	.031	.033	.035	.045	.058	.11	.18	.23	.29	.35	.40	.45	.52	.58
	6	.0020	.0028	.0035	.0042	.0044	.0066	.0088	.011	.022	.033	.044	.055	.066	.077	.088	.099	.11
	8	.00051	.00079	.0010	.0013	.0015	.0022	.0027	.0037	.0075	.011	.015	.019	.023	.028	.030	.034	.037
300	3	.085	.10	.13	.15	.17	.18	.18	.20	.40	.60	.80	1.0	1.2	1.4	1.6	1.8	2.0
	4	.022	.030	.036	.042	.044	.045	.055	.070	.14	.21	.28	.35	.42	.48	.55	.62	.70
	6	.0028	.0040	.0050	.0058	.0060	.0080	.010	.013	.026	.040	.052	.065	.080	.090	.10	.11	.13
	8	.00070	.0011	.0014	.0017	.0018	.0027	.0033	.0045	.0090	.013	.018	.023	.027	.031	.035	.040	.045
400	3	.15	.18	.21	.25	.26	.26	.27	.28	.56	.84	1.1	1.4	1.7	1.8	2.1	2.4	2.8
	4	.040	.050	.060	.070	.073	.075	.078	.090	.18	.28	.37	.46	.55	.64	.72	.82	.90
	6	.0047	.0065	.0080	.0097	.010	.010	.014	.017	.035	.051	.070	.089	.10	.12	.14	.16	.17
	8	.0012	.0018	.0023	.0027	.0027	.0035	.0045	.0060	.012	.018	.024	.030	.035	.041	.047	.053	.060
450	4	.048	.060	.073	.088	.095	.098	.10	.10	.20	.30	.40	.50	.60	.70	.80	.90	1.0
	6	.0060	.0080	.010	.012	.013	.013	.016	.020	.040	.060	.080	.10	.12	.14	.16	.18	.20
	8	.0016	.0022	.0029	.0033	.0033	.0040	.0050	.0066	.013	.020	.027	.034	.040	.046	.053	.060	.068
	10	.00052	.00075	.00095	.0012	.0012	.0016	.0022	.0028	.0055	.0082	.011	.014	.016	.019	.022	.025	.028
500	4	.060	.071	.090	.11	.12	.13	.13	.13	.23	.35	.46	.57	.70	.80	.90	1.0	1.1
	6	.0074	.010	.012	.014	.016	.016	.018	.022	.044	.065	.086	.10	.13	.15	.18	.20	.22
	8	.0018	.0026	.0034	.0041	.0043	.0045	.0055	.0063	.015	.023	.030	.037	.045	.051	.060	.066	.075
	10	.00061	.00090	.0011	.0013	.0013	.0018	.0024	.0030	.0060	.0090	.012	.015	.018	.021	.025	.027	.030
600	4	.085	.10	.12	.14	.17	.20	.23	.25	.28	.42	.55	.70	.82	.93	1.0	1.2	1.4
	6	.010	.014	.016	.020	.022	.023	.024	.026	.051	.079	.10	.13	.16	.18	.21	.23	.26
	8	.0026	.0036	.0046	.0054	.0056	.0058	.0066	.0090	.018	.028	.036	.045	.054	.061	.071	.081	.090
	10	.00086	.0012	.0016	.0020	.0021	.0022	.0029	.0036	.0072	.011	.015	.018	.022	.025	.029	.033	.036
750	4	.13	.15	.18	.22	.27	.28	.29	.30	.34	.51	.70	.88	1.1	1.2	1.3	1.5	1.8
	6	.015	.020	.025	.028	.030	.031	.032	.032	.064	.10	.12	.16	.20	.22	.25	.29	.32
	8	.0040	.0055	.0065	.0081	.0090	.0095	.010	.011	.023	.034	.045	.055	.066	.080	.090	.10	.11
	10	.0013	.0018	.0022	.0027	.0028	.0028	.0036	.0045	.0090	.014	.018	.022	.027	.032	.036	.041	.045
800	6	.018	.024	.027	.032	.032	.033	.033	.035	.070	.10	.13	.17	.21	.25	.28	.31	.35
	8	.0046	.0062	.0080	.0095	.010	.011	.011	.012	.024	.036	.048	.060	.072	.084	.096	.10	.12
	10	.0014	.0020	.0026	.0032	.0033	.0033	.0038	.0050	.0098	.015	.020	.025	.029	.034	.040	.045	.050
	12	.00060	.00090	.0011	.0014	.0015	.0015	.0019	.0024	.0047	.0070	.0095	.012	.014	.017	.019	.022	.024
1000	6	.028	.035	.040	.050	.057	.065	.072	.079	.086	.13	.17	.21	.26	.30	.35	.39	.45
	8	.0070	.0093	.011	.014	.014	.015	.015	.015	.030	.045	.060	.075	.090	.10	.11	.12	.15
	10	.0022	.0030	.0038	.0047	.0047	.0048	.0049	.0060	.012	.018	.024	.030	.036	.042	.048	.055	.060
	12	.0095	.0013	.0017	.0020	.0022	.0022	.0024	.0030	.0060	.0090	.012	.015	.018	.021	.024	.027	.030
1050	6	.030	.037	.045	.054	.062	.070	.078	.085	.090	.13	.18	.23	.28	.31	.36	.40	.46
	8	.0080	.010	.012	.015	.015	.016	.016	.016	.031	.047	.063	.080	.094	.10	.12	.13	.16
	10	.0025	.0034	.0043	.0047	.0050	.0051	.0051	.0064	.013	.020	.026	.032	.039	.045	.051	.060	.065
	12	.0010	.0014	.0018	.0022	.0024	.0025	.0025	.0031	.0062	.0093	.013	.016	.019	.022	.026	.029	.032

*For liquids with a specific gravity other than 1.00, multiply the value from the above table by the specific gravity of the liquid. For old pipe, add 20% to the above values.

Figures to right of orange line are laminar flow. Figures to left of orange line are turbulent flow.

To convert the above values to kPa (kilopascals) per metre of pipe, multiply by 22.6.

To convert the above values to kg per cm2 per metre of pipe, multiply by 0.23.

Fig. C-3 Pressure loss from pipe friction (continued).

COURTESY OF VIKING PUMP, INC., A UNIT OF IDEX CORPORATION

Converting Pressure into Feet Head of Water

Pounds per Square Inch	Feet Head	Pounds per Square Inch	Feet Head	Pounds per Square Inch	Feet Head
1	2.31	40	92.36	170	392.52
2	4.62	50	115.45	180	415.61
3	6.93	60	138.54	190	438.90
4	9.24	70	161.63	200	461.78
5	11.54	80	184.72	225	519.51
6	13.85	90	207.81	250	577.24
7	16.16	100	230.90	275	643.03
8	18.47	110	253.98	300	692.69
9	20.78	120	277.07	325	750.41
10	23.09	125	288.62	350	808.13
15	34.63	130	300.16	375	865.89
20	46.18	140	323.25	400	922.58
25	57.72	150	346.34	500	1154.48
30	69.27	160	369.43	1,000	2308.

Fig. C-4 *Converting pressure into feet head of water.*

Converting Feet Head of Water into Pressure

Feet Head	Pounds per Square Inch	Feet Head	Pounds per Square Inch	Feet Head	Pounds per Square Inch
1	.43	60	25.99	200	86.62
2	.87	70	30.32	225	97.45
3	1.30	80	34.65	250	108.27
4	1.73	90	38.98	275	119.10
5	2.17	100	43.31	300	129.93
6	2.60	110	47.64	325	140.75
7	3.03	120	51.97	350	151.58
8	3.40	130	56.30	400	173.24
9	3.90	140	60.63	500	216.55
10	4.33	150	64.96	600	259.85
20	8.66	160	69.29	700	303.16
30	12.99	170	73.63	800	346.47
40	17.32	180	77.96	900	389.78
50	21.65	190	83.29	1,000	433.09

Fig. C-5 *Converting feet head of water into pressure.*

Equivalent Values of Pressure

Inches of Mercury	Feet of Water	Pounds per Square Inch	Inches of Mercury	Feet of Water	Pounds per Square Inch	Inches of Mercury	Feet of Water	Pounds per Square Inch
1	1.13	0.49	11	12.45	5.39	21	23.78	10.3
2	2.26	0.98	12	13.57	5.87	22	24.88	10.8
3	3.39	1.47	13	14.70	6.37	23	26.00	11.28
4	4.52	1.95	14	15.82	6.86	24	27.15	11.75
5	5.65	2.44	15	16.96	7.35	25	28.26	12.25
6	6.78	2.93	16	18.09	7.84	26	29.40	12.73
7	7.91	3.42	17	19.22	8.33	27	30.52	13.23
8	9.04	3.91	18	20.35	8.82	28	31.65	13.73
9	10.17	4.40	19	21.75	9.31	29	32.80	14.22
10	11.30	4.89	20	22.60	9.80	29.929	33.947	14.6969

Fig. C-6 *Equivalent values of pressure.*

Atmospheric Pressure, Barometer Reading, and Equivalent Head of Water at Different Altitudes

Altitude above Sea Level in Feet	Atmospheric Pressure Pounds per Square Inch	Barometer Reading Inches of Mercury	Equivalent Head of Water Feet
0	14.7	29.929	33.95
1000	14.2	28.8	32.7
2000	13.6	27.7	31.6
3000	13.1	26.7	30.2
4000	12.6	25.7	29.1
5000	12.1	24.7	27.9
6000	11.7	23.8	27.0
7000	11.2	22.9	25.9
8000	10.8	22.1	24.9
9000	10.4	21.2	24.0
10000	10.0	20.4	23.1

For feet head of liquid, Divide feet head of water by specific gravity of liquid pumped.

Fig. C-7 *Atmospheric pressure, barometer reading, and equivalent head of water at different altitudes.*

Approximate Comparison of Vacuum and Absolute Pressures at Sea Level

Vacuum in Inches Mercury	Vacuum in MM. Mercury	Absolute Pressure in Lbs. per Sq. In.	Absolute Pressure in Inches Mercury	Absolute Pressure in MM. Mercury	Absolute Pressure in Inches Water	Absolute Pressure in Feet Water	Feet Suction Lift	Atmospheres
0	0.0	14.7	29.9	759.5	407	33.9	0.00	1.00
2	50.8	13.7	27.9	709	380	31.6	2.27	0.93
4	101.6	12.7	25.9	658	352	29.4	4.53	0.86
6	152.4	11.7	23.8	605	324	27.1	6.80	0.79
8	203.2	10.8	22.0	559	299	24.9	9.07	0.73
10	254.0	9.78	19.9	505	271	22.6	11.34	0.66
12	304.8	8.79	17.9	455	243	20.3	13.61	0.60
14	355.6	7.81	15.9	404	216	18.1	15.88	0.53
16	406.4	6.83	13.9	353	189	15.8	18.14	0.46
18	457.2	5.84	11.9	302	162	13.5	20.41	0.40
20	508.0	4.86	9.9	251	135	11.2	22.68	0.33
22	558.8	3.88	7.9	201	107	8.95	24.95	0.26
24	609.6	2.89	5.9	150	80	6.69	27.22	0.197
26	660.4	1.91	3.9	99	53	4.42	29.48	0.13
28	711.2	0.92	1.9	48	26	2.15	31.75	0.063
29.9	759.5	0.00	0.0	00	00	0.00	33.91	0.00

Fig. C-8 *Approximate comparison of vacuum and absolute pressures at sea level.*

Comparative Equivalents of Liquid Measures and Weights

Measures and Weights for Comparison	MEASURE AND WEIGHT EQUIVALENTS OF ITEMS IN FIRST COLUMN						
	U.S. Gallon	Imperial Gallon	Cubic Inch	Cubic Foot	Cubic Meter	Liter	Pounds of Water
U.S. GALLON	1.	.833	231.	.1337	.00378	3.785	8.33
IMPERIAL GAL.	1.20	1.	277.27	.1604	.00454	4.542	10.
CUBIC INCH	.0043	.00358	1.	.00057	.000016	.0163	.0358
CUBIC FOOT	7.48	6.235	1728.	1.	.02827	28.312	62.355
CUBIC METER	264.17	220.05	61023.	35.319	1.	1000.	2200.54
LITER	.26417	.2200	61.023	.0353	.001	1.	2.2005
POUNDS OF WATER	.12	.1	27.72	.016	.00045	.454	1.

Fig. C-9 Comparative equivalents of liquid measures and weights.

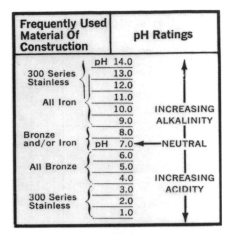

Fig. C-10 pH ratings of frequently used material of construction.

Number of Gallons in Round Vertical Tanks

Depth of Liquid in Feet	DIAMETER IN FEET OF ROUND TANKS OR CISTERNS																
	5	6	7	8	9	10	11	12	13	14	15	16	18	20	22	24	26
5	725	1060	1440	1875	2308	2925	3550	4237	4960	5765	6698	7520	9516	11750	14215	16918	18358
6	870	1270	1728	2250	2855	3510	4260	5084	5952	6918	8038	9024	11419	14100	17059	20302	22030
7	1015	1480	2016	2625	3330	4095	4970	5931	6944	8071	9378	10528	13322	16450	19902	23680	25701
8	1160	1690	2304	3000	3805	4680	5680	6778	7936	9224	10718	12032	15225	18800	22745	27070	29372
9	1305	1900	2592	3375	4280	5265	6390	7625	8928	10377	12058	13536	17128	21150	25588	30454	33043
10	1450	2110	2880	3750	4755	5850	7100	8472	9920	11530	13398	15040	19031	23500	28431	33838	36714
11	1595	2320	3168	4125	5230	6435	7810	9319	10912	12683	14738	16544	20934	25850	31274	37222	40385
12	1740	2530	3456	4500	5705	7020	8520	10166	11904	13836	16078	18048	22837	28200	34117	40606	44056
13	1885	2740	3744	4875	6180	7605	9230	11013	12896	14989	17418	19552	24740	30550	36960	43990	47727
14	2030	2950	4032	5250	6655	8190	9940	11860	13888	16142	18758	21056	26643	32900	39803	47374	51398
15	2175	3160	4320	5625	7130	8775	10650	12707	14880	17295	20098	22260	28546	35250	42646	50758	55069
16	2320	3370	4608	6000	7605	9360	11360	13554	15872	18448	21438	24064	30449	37600	45489	54142	58740
17	2465	3580	4896	6375	8080	9945	12070	14401	16864	19601	22778	25568	32352	39950	48332	57520	62411
18	2610	3790	5184	6750	8535	10530	12780	15248	17856	20754	24118	27072	34255	42300	51175	60910	66082
19	2755	4000	5472	7125	9010	11115	13490	16095	18848	21907	25458	28576	36158	44650	54018	64294	69753
20	2900	4210	5760	7500	9490	11700	14200	16942	19840	23060	26798	30080	38062	47000	56861	67678	73424

Fig. C-11 Number of gallons in round vertical tanks.

Loss in PSI Pressure per 100 Feet of Smooth Bore Rubber Hose
Data is for liquid having viscosity of 38 SSU.

U.S. GPM	ACTUAL INSIDE DIAMETER IN INCHES										
	1/2	5/8	3/4	1	1-1/4	1-1/2	2	2-1/2	3	4	5
1½	2.8	0.7	0.5								
2½	7.6	2.1	1.1								
5	28.5	9.6	4.0	1.1	0.4	0.2					
10	101.0	33.8	14.0	4.1	1.2	0.5	0.2				
15	....	70.0	30.0	8.9	2.5	1.1	0.4	0.1			
20		112.0	53.0	14.0	4.3	1.8	0.7	0.2			
25			79.0	22.0	6.5	2.9	1.0	0.3			
30			112.0	31.0	9.2	4.0	1.4	0.4	0.1		
35			147.0	41.0	12.0	5.3	1.8	0.5	0.2		
40				53.0	15.0	6.7	2.4	0.6	0.3		
45				66.0	19.0	8.4	3.0	0.8	0.4		
50				80.0	24.0	10.0	3.6	1.0	0.5		
60				101.0	35.0	14.0	5.1	1.4	0.6		
70					45.0	19.0	6.6	1.8	0.8		
80					58.0	24.0	8.6	2.3	1.1		
90					71.0	30.0	11.0	3.0	1.4	0.3	
100					88.0	37.0	12.5	3.5	1.7	0.4	0.1
125					132.0	55.0	20.0	5.3	2.5	0.6	0.2
150					183.0	78.0	27.0	7.5	3.5	0.7	0.3
175						100.0	37.0	10.0	4.6	1.1	0.4
200						133.0	46.0	13.0	5.9	1.4	0.5
250							70.0	19.0	9.1	2.1	0.7
300							95.0	27.0	12.0	2.9	1.0
350							126.0	36.0	17.0	4.0	1.3
400								46.0	21.0	5.1	1.7
450								57.0	26.0	6.3	2.1
500								70.0	32.0	7.4	2.6
1000									116.0	27.0	9.6

EXAMPLE: What pressure is required at intake end of a 150 ft. line of 1½-in. hose joined in 50 ft. lengths with shank coupling? A delivery of 50 gal. of No. 2 fuel oil per minute is desired. Consulting the table we find the hose required 10 PSI per 100 ft. or 15 PSI for the 150 ft. Adding 5% for each of three sets of couplings, we have a total of 17.25 PSI.

Fig. C-12 Loss of PSI pressure per 100 feet of smooth bore rubber hose.

Resistance of Valves and Fittings to Flow of Fluids

Example

The dotted line shows that the resistance of a 6-inch Standard Elbow is equivalent to approximately 16 feet of 6-inch Standard Pipe.

Note

For sudden enlargements or sudden contractions, use the smaller diameter, d, on the pipe size scale.

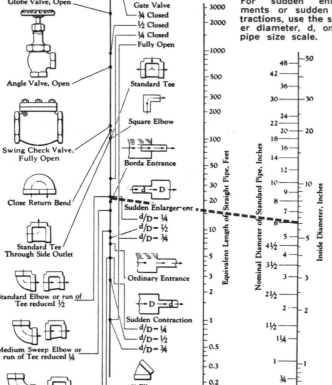

Fig. C-13 Resistance of valves and fittings to flow of fluids.

Metric-English Pressure Units

KILOGRAMS PER SQUARE CENTIMETER	POUNDS PER SQUARE INCH
0.1	1.42
0.2	2.85
0.3	4.27
0.4	5.69
0.5	7.11
0.6	8.54
0.7	9.96
0.8	11.38
0.9	12.81
1.0	14.2
1.5	21.3
2	28.5
3	42.7
4	56.9
5	71.1
6	85.4
7	99.6
8	114
9	128
10	142
15	213
20	285
30	427
40	569
50	712
100	1423

Fig. C-14 Metric-English pressure units.

Standard Pipe Data

All dimensions and weights are nominal.

Size	Diameters		Thick-ness	Length of Pipe per Sq. Ft. of		Length of Pipe Containing One Cubic Ft.	Weight per Ft., Plain Ends	Weight of Water per Ft.
	External	Internal		External Surface	Internal Surface			
Inches	Inches	Inches	Inches	Feet	Feet	Feet	Pounds	Pounds
1/8	.405	.269	.068	9.431	14.199	2533.775	.244	.025
1/4	.540	.364	.088	7.073	10.493	1383.789	.424	.045
3/8	.675	.493	.091	5.658	7.747	754.360	.567	.083
1/2	.840	.622	.109	4.547	6.141	473.906	.850	.132
3/4	1.050	.824	.113	3.637	4.635	270.034	1.130	.231
1	1.315	1.049	.133	2.904	3.641	166.618	1.678	.375
1 1/4	1.660	1.380	.140	2.301	2.767	96.275	2.272	.65
1 1/2	1.900	1.610	.145	2.010	2.372	70.733	2.717	.88
2	2.375	2.067	.154	1.608	1.847	42.913	3.652	1.45
2 1/2	2.875	2.469	.203	1.328	1.547	30.077	5.793	2.07
3	3.500	3.068	.216	1.091	1.245	19.479	7.575	3.20
4	4.500	4.026	.237	.848	.948	11.312	10.790	5.50
5	5.563	5.047	.258	.686	.756	7.198	14.617	8.67
6	6.625	6.065	.280	.576	.629	4.984	18.974	12.51
8	8.625	7.981	.322	.442	.478	2.878	28.554	21.70
10	10.750	10.020	.365	.355	.381	1.826	40.483	34.20

Fig. C-15 Standard pipe data.

Extra-Strong Pipe Data

All dimensions and weights are nominal.

Size	Diameters		Thick-ness	Length of Pipe per Sq. Ft. of		Length of Pipe Containing One Cubic Ft.	Weight per Ft., Plain Ends	Weight of Water per Ft.
	External	Internal		External Surface	Internal Surface			
Inches	Inches	Inches	Inches	Feet	Feet	Feet	Pounds	Pounds
1/8	.405	.215	.095	9.431	17.766	3966.392	.314	.016
1/4	.540	.302	.119	7.073	12.648	2010.290	.535	.031
3/8	.675	.423	.126	5.658	9.030	1024.689	.738	.061
1/2	.840	.546	.147	4.547	6.995	615.017	1.087	.102
3/4	1.050	.742	.154	3.637	5.147	333.016	1.473	.188
1	1.315	.957	.179	2.904	3.991	200.193	2.171	.312
1 1/4	1.660	1.278	.191	2.301	2.988	112.256	2.996	.56
1 1/2	1.900	1.500	.200	2.010	2.546	81.487	3.631	.77
2	2.375	1.939	.218	1.608	1.969	48.766	5.022	1.28
2 1/2	2.875	2.323	.276	1.328	1.644	33.976	7.661	1.87
3	3.500	2.900	.300	1.091	1.317	21.801	10.252	2.86
4	4.500	3.826	.337	.848	.998	12.525	14.983	4.98
5	5.563	4.813	.375	.686	.793	7.915	20.778	7.88
6	6.625	5.761	.432	.576	.663	5.524	28.573	11.29
8	8.625	7.625	.500	.442	.500	3.154	43.388	19.78
10	10.750	9.750	.500	.355	.391	1.929	54.735	32.35

Fig. C-16 Extra-strong pipe data.

Metric-English Capacity Units

Liters per Minute	Gallons per Minute	Cubic Meters per Hour	Gallons per Minute
1	0.264	0.1	0.44
2	0.528	0.2	0.88
3	0.792	0.3	1.32
4	1.056	0.4	1.76
5	1.32	0.5	2.20
6	1.58	0.6	2.64
7	1.85	0.7	3.08
8	2.11	0.8	3.52
9	2.38	0.9	3.96
10	2.64	1.0	4.4
25	6.6	1.5	6.6
50	13.2	2.0	8.8
75	19.8	4.0	17.6
100	26.4	6.0	26.4
200	52.8	8.0	35.2
300	79.2	10	44
400	106	20	88
500	132	30	132
600	158	40	176
700	185	50	220
800	211	60	264
900	238	70	308
1000	264	80	352
2000	528	90	396
3000	792	100	440
4000	1056	200	880
5000	1320	300	1320
7500	1980	400	1760
10,000	2640	500	2200

Fig. C-17 Metric-English capacity units.

Properties of Saturated Steam

Pressure—Pounds per Square Inch		Degrees F. Temperature	Specific Volume Cubic Feet per Pound
Absolute	Gauge		
14.696	0.0	212.00	26.80
50.0	35.3	281.01	8.515
55.0	40.3	287.07	7.787
60.0	45.3	292.71	7.175
65.0	50.3	297.97	6.655
70.0	55.3	302.92	6.206
75.0	60.3	307.60	5.816
80.0	65.3	312.03	5.472
85.0	70.3	316.25	5.168
90.0	75.3	320.27	4.896
95.0	80.3	324.12	4.652
100.0	85.3	327.81	4.432
105.0	90.3	331.36	4.232
110.0	95.3	334.77	4.049
115.0	100.3	338.07	3.882
120.0	105.3	341.25	3.728
125.0	110.3	344.33	3.587
130.0	115.3	347.32	3.455
135.0	120.3	350.21	3.333
140.0	125.3	353.02	3.220
150.0	135.3	358.42	3.015
160.0	145.3	363.53	2.834
170.0	155.3	368.41	2.675
180.0	165.3	373.06	2.532
190.0	175.3	377.51	2.404
200.0	185.3	381.79	2.288

Fig. C-18 Properties of saturated steam.

Friction Loss in Standard Valves and Fittings
Table gives equivalent lengths in feet of straight pipe.

Type of Fitting	NOMINAL PIPE DIAMETER, IN INCHES												
	1/2	3/4	1	1-1/4	1-1/2	2	2-1/2	3	4	5	6	8	10
Gate Valve (open)	.35	.50	.60	.80	1.2	1.2	1.4	1.7	2.3	2.8	3.5	4.5	5.7
Globe Valve (open)	17	22	27	38	44	53	68	80	120	140	170	220	280
Angle Valve (open)	8	12	14	18	22	28	33	42	53	70	84	120	140
Standard Elbow	1.5	2.2	2.7	3.6	4.5	5.2	6.5	8.0	11.0	14	16	21	26
Medium Sweep Elbow	1.3	1.8	2.3	3.0	3.6	4.6	5.5	7.0	9.0	12.0	14.0	18.0	22.0
Long Sweep Elbow	1.0	1.3	1.7	2.3	2.8	3.5	4.3	5.2	7.0	9.0	11.0	14.0	17.0
Tee (straight thru)	1.0	1.3	1.7	2.3	2.8	3.5	4.3	5.2	7.0	9.0	11.0	14.0	17.0
Tee (right angle flow)	3.2	4.5	5.7	7.5	9.0	12.0	14.0	16.0	22.0	27.0	33.0	43.0	53.0
Return Bend	3.5	5.0	6.0	8.5	10.0	13.0	15.0	18.0	24.0	30.0	37.0	50.0	63.0

Fig. C-19 Friction loss in standard valves and fittings.

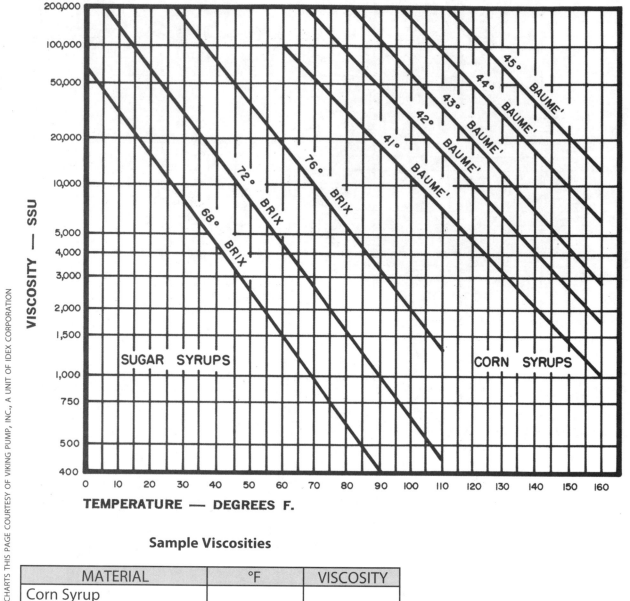

Fig. C-20 Effect of temperature on viscosity. Baume is a measure that indicates the potential alcohol. So a measure of 40 Baume should potentially yield 40% alcohol.

Viscosity as a Factor of Temperature

VISCOSITY — SSU

TEMPERATURE — DEGREES F.

CHARTS THIS PAGE COURTESY OF VIKING PUMP, INC., A UNIT OF IDEX CORPORATION

Sample Viscosities

MATERIAL	°F	VISCOSITY
Corn Syrup		
Specific Gravity 1.43	100	250,000
Mashed Potatoes		20,000
Molasses		
Specific Gravity 1.43	100	12,000
	130	4500
Specific Gravity 1.45	100	33,000
	130	9000
Specific Gravity 1.48	100	130,000
	130	30,000
Pear Juice	160	4000
Heavy Sugar Solutions	70	230
60%	100	90
70%	70	1700
	100	400
76%	70	10,000
	100	2000
Tomato Paste (33% solids)		32,500

Fig. C-21 Sample viscosities. Listed here are examples of the viscosities of a few materials, to give you some idea of what the SSU rating of your feedstock might be. Note that raising the temperature of your feedstock will reduce its viscosity.

PERMACULTURE

A FEW CLOSING WORDS

Permaculture is the art and science of designing human beings' place in the environment. Permaculture design teaches you to understand and mirror the patterns found in healthy natural environments. You can then build profitable, productive, sustainable, cultivated ecosystems, which include people, and which have the same diversity, stability, and resilience as natural ecosystems.

Permaculture designs can be applied in households, major agricultural enterprises, and even entire bioregions. Permaculture integrates disciplines relating to food, shelter, energy, water, trees and plants, wildlife, livestock, weather, waste management, economics. and social sciences. These integrated designs create systems capable of yielding far more benefits than conventional systems. Permaculture can reclaim devastated lands, roll back deserts, build just social/economic systems, and design planet-based livelihoods.

Permaculture departs from any other design system in that it is guided by a common-sense ethical system. This system forms the criteria for design decisions. These base criteria are:

- Care of the Earth,

- Care of the People of the Earth, and

- System surpluses distributed in accordance with the first two ethics.

Briefly, when a design component isn't ecologically sound, community-building, and careful in its use of resources, then it's pretty unlikely that it will work out in the long run. Most design systems are defined by a "market-driven" ethic. Such designs are subservient to the conclusions of a short-term cost/benefit analysis, discounting or ignoring such factors as environmental degradation or destruction of human community.

PERMACULTURE TEACHES YOU TO UNDERSTAND AND MIRROR THE PATTERNS FOUND IN HEALTHY NATURAL ENVIRONMENTS. YOU CAN THEN BUILD PROFITABLE, PRODUCTIVE, SUSTAINABLE ECOSYSTEMS.

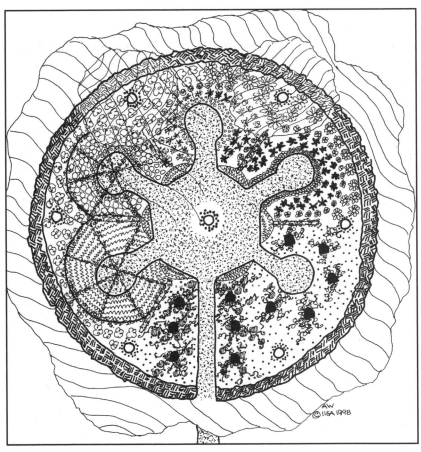

Fig. p-1. A mandala garden. *Mandala gardens are minimum path gardens. This example is a 30-foot-diameter layout. Things visited or harvested often are planted close to the path. Fruit trees and other seasonally harvested crops are planted furthest from the path. A nitrogen-fixing Farmer's tree is planted in the center, lightly shading the whole mandala. Six fruit trees are planted near the periphery. The paths are filled with organic matter, which is turned into compost by worms. The crosshatched outer ring is a hedge that produces fruits and nuts and acts as an animal barrier and windbreak. This garden would feed at least four people.*

The permaculture ethic is the basis of sustainability and also makes excellent, long-term business sense. Systems designed with these principles are ecologically sound, economically stable, community-building, and don't leave future generations with a cleanup bill for today's enterprise.

The author has certified over 500 people as permaculture designers and teaches a certification course once a year, most years. He also consults with landowners on the permacultural potential of their property.

To learn more, visit <www.permaculture.com>.

OUR TREE TAX

Although a portion of the paper in this book is recycled, it still originally came from a tree at some time. One "average" tree only provides about a three-foot-tall pile of newsprint. Some authors assess themselves a tree tax, often about 25 cents per book, to plant replacements for the biomass that was cut down to produce their books. So they cheerfully send off a check to some organization that plants trees.

I take this responsibility personally, however; and, in the fashion of permaculture, I believe each designed action should serve at least three functions. Function one is to replace the biomass used for printing this book, while absorbing carbon dioxide from the atmosphere. So I plant. I prepare the land, fence it if necessary, plant the seedlings, transplant them to permanent locations on swales, and stay responsible for them until maturity.

Function two, I make sure I plant high-value hardwood trees. Why? Because when tropical rainforests are cut down, the loggers only use one or two trees per acre for timber, and the rest get bulldozed into a pile and burned. By planting 400 high-value trees per acre, I am effectively replacing what would be taken from 200–400 acres of rainforest.

The third function has to do with an old saying from India. "Trees are better than sons." Which means that, in your old age, your sons might take care of you, but your trees definitely will. So I am planting special fast-growing trees that will mature in time to fund my retirement. When I cut them down, I will do so carefully, in order to make sure a new tree sprouts from the stump, that will live on long after I do.

I am available to plant trees for your retirement.

For more information, you may write me at <info@permaculture.com>.

SOME AUTHORS ASSESS THEMSELVES A TREE TAX, OFTEN ABOUT 25 CENTS PER BOOK, TO PLANT REPLACEMENTS FOR THE BIOMASS THAT WAS CUT DOWN TO PRODUCE THEIR BOOKS. I TAKE THIS RESPONSIBILITY PERSONALLY, HOWEVER; AND, IN THE FASHION OF PERMACULTURE, I BELIEVE EACH DESIGNED ACTION SHOULD SERVE AT LEAST THREE FUNCTIONS. FUNCTION ONE IS TO REPLACE THE BIOMASS USED FOR PRINTING THIS BOOK, WHILE ABSORBING CARBON DIOXIDE FROM THE ATMOSPHERE. SO I PLANT....

ERIC LANG

Fig. t-1 Planting for the future. *Here I am planting a dormant fast growing black walnut seedling in a bed enriched with compost.*

Fig. t-2 Tree nursery. *Anticipating the first printing of the book, I planted 500 walnut trees in April 2007, in this fenced nursery with drip irrigation. In the winter, I will transplant them to their permanent home in a polyculture with other trees.*

Fig. t-3 Budding out. *Befitting the occasion, as the book was finalized in June, the little buds have quickened and begun to open into new leaves.*

Fig. t-4 Growing up. *Not long after the first printing of the book went to press in September, the trees were eight feet tall!*

FREE MEMBERSHIP IN ALCOHOLICS UNANIMOUS! | 8627

Welcome to the *Alcohol Can Be A Gas!* community. Thank you for your foresight and support of our project in purchasing this book. As a reward, we would like to offer you a free six-month membership in Alcoholics Unanimous (David Blume's nickname for those who support the book and alcohol fuel). As a member, you will receive a newsletter by email from time to time, and you'll be notified when an alcohol fuel workshop is coming to town, and about legislation you might want to track and respond to. You will also be eligible for exclusive informational tools and forums, and you'll be kept up to date as we begin to offer such useful products as car conversion kits, distillery building supplies, prefabricated alcohol fueling stations, special pricing on alcohol-fueled vehicles, and whatever else we discover the community needs. Just return this card (or a photocopy of it) to the IIIEA.

NAME _____ STREET _____

EMAIL _____ CITY _____ STATE _____ ZIP _____

Where did you purchase this book? _____ How did you hear about it? _____

❏ Please notify me of workshops in my area.

❏ I can help get the word out about workshops in my area.

❏ I am interested in organizing an alcohol fuel station in my neighborhood.

❏ I am a farmer who wants to grow energy crops and make alcohol for an alcohol station.

iiEA
The International Institute
for Ecological Agriculture
309 Cedar St. #127
Santa Cruz, CA 95060

GLOSSARY

FROM ABE TO ZEOLITE

ABE   A mixture of approximately 30% acetone, 60% butanol, and 10% ethanol (with some water), obtained in the process of distilling butanol.

absorption   A process in which one substance permeates another. Contrast with adsorb.

accelerator pump   A device that gives an extra squirt of fuel to maintain a combustible air/fuel ratio during acceleration. The pump sprays raw fuel into the venturi in an effort to make up for a drop in manifold vacuum.

acetaldehyde   A colorless volatile liquid aldehyde obtained by oxidizing ethanol; a possibly carcinogenic pollutant produced in very small quantities by alcohol-burning engines; disappears completely with a modern three-way catalytic converter.

acetic acid   The acid that gives vinegar its characteristic taste; what bacteria can turn your mash into. High acetic acid in low-proof alcohol is corrosive.

acetone   A colorless, flammable liquid byproduct of butanol distillation, easily soluble in water, ethanol, ether, etc. It works as a wide-ranging solvent; when used instead of water in weak acid hydrolysis, lignin and cellulose are quickly broken down.

acid hydrolysis   The chemical reaction of breaking apart of a polymer (long chains of molecules) into component parts by breaking the bonds between the components with water, acid, and energy.

acidity   The measure of how many hydrogen ions a solution contains; any solution with a pH of less than 7 is acidic. Contrast with alkalinity.

adsorption   The adhesion, in an extremely thin layer, of the molecules of gases, dissolved substances, or liquids, to the outer or inner surfaces of solid bodies. Contrast with absorption.

adventitious   Describes a shoot or root produced on an unusual place, such as a root produced on a branch.

agitator   A device, such as a stirrer, that provides complete mixing and uniform dispersion of all components in a mixture.

alcohol   A group of organic chemical compounds composed of carbon, hydrogen, and oxygen; a series of molecules that vary in chain length and are composed of a hydrocarbon plus a hydroxyl group; includes methanol, ethanol, isopropyl alcohol, and others.

alcohol dryer   Device used to remove remaining water from alcohol after distillation.

aldehyde   Any of a class of highly reactive organic chemical compounds obtained by oxidation of primary alcohols, used in the manufacture of resins, dyes, and organic acids.

alkaline   Describes a substance, such as lime or soda, capable of reacting with an acid to form a salt and water, or capable of accepting or neutralizing hydrogen ions. Alkalis release hydroxide ions when dissolved in water. An alkaline solution has a pH greater than 7. Also called base. Contrast with acid.

alpha amylase   An enzyme that converts starch into medium-length sugar molecules such as dextrins.

American Type Culture Collection (ATTC)   A nonprofit biological resource center that acquires, authenticates, produces, preserves, develops, and distributes standard reference microorganisms.

American Petroleum Institute (API)   A national trade association that advocates for the oil and natural gas industry.

amino acids   The building blocks of protein, which appear naturally in plants and animals.

anaerobic   Without oxygen.

anhydrous   Describes a compound that does not contain water.

anhydrous alcohol   Refers to purified ethanol, containing no more than 1% water; literally means no-water alcohol, but in reality is low-water alcohol. National Formulary grade is 99+% ethyl alcohol by weight. Also called absolute alcohol.

anneal   To treat with heat in order to alter the microstructure of a material, causing changes in such properties as strength and hardness.

annual   Describes a plant that lives a year or less.

annual gallon of production capacity   The cost of an alcohol plant amortized over its rated capacity per year. If a plant costs $1 per annual gallon of capaciity and it is able to produce 100 million gallons per year it will cost $100 million to build.

aquifer   A layer or layers of permeable rock that can contain or transmit groundwater.

GLOSSARY From ABE to Zeolite   **549**

aromatic An organic compound, e.g., benzene, containing a planar unsaturated ring of atoms stabilized by an interaction of the bonds forming the ring; used as octane enhancer, see BTX; called "aromatic" because many of these compounds have a sweet scent.

aromatic hydrocarbon (AH) A hydrocarbon (hydrogen/carbon compound) with a molecular structure that incorporates one or more planar sets of six carbon atoms that are connected by electrons. Also called an arene.

atmospheric distillery A still that uses ambient air pressure. Contrast with vacuum distillation.

atmospheric pressure Pressure exerted by the atmosphere; it has a mean value of about 14.7 pounds per square inch at sea level. Also called barometric pressure.

auger dewatering press A machine with a rotating helical shaft, used to compress wet material in a screened barrel against a spring-loaded opening, causing most of the free liquid to be discharged through screen openings.

auto-ignition point The lowest temperature at which a heated liquid's vapors will self-ignite and burn, without exposure to any flame or spark source.

azeotropic distillation Distillation that uses any technique to break an azeotrope (a mixture of two or more liquids that are attracted to each other too much to be separated by simple heating); e.g., adding benzene to water and ethanol, permitting distillation of anhydrous ethanol, with the water staying combined with the benzene.

backslop The liquid left over after distillation; spent mash. Contains a mixture of unfermentable materials and dead yeast that can provide most of the nutrients that a batch of new living yeast need.

bagasse The fibrous byproduct left over after sugarcane is crushed to extract the juice used in making sugar or alcohol.

balling hydrometer See Brix hydrometer.

ball valve A valve that opens by turning a handle attached to a ball inside the valve. The ball has a hole through the middle; when the hole is in line with both ends of the valve, flow will occur; when the valve is closed, the hole is perpendicular to the valve openings, and flow is blocked.

barley malt Barley that is subjected to a malting process in which the seeds are germinated and then quickly dried before the plant develops. Malted barley has enzymes helpful in turning starches into sugars during fermentation.

barrel A liquid measure equal to 42 American gallons or about 306 pounds; one barrel equals 5.6 cubic feet or 0.159 cubic meters; for crude oil, one barrel is about 0.122 metric tons, 0.134 long tons, and 0.150 short tons.

barstock Raw material obtained from metal manufacturers in the form of long bars; may be round, square, or hexagonal.

batch distillation A process in which a quantity (a batch) of feedstock is loaded, distilled to the desired point, and removed; the still is then reloaded with fresh mixture to distill. Contrast with continuous distillation.

batch hydrolysis The cooking of a feedstock with water as part of series of steps to convert a fixed quantity of feedstock to fermentable sugars.

batch method A process where a feedstock is run through all the steps of a process before emptying the process vessel and starting over with new feedstock. Contrast with continuous processes in which feedstock is steadily added and reacted.

batch plant A facility where raw materials are loaded, and products removed after the reactions complete. Contrast with a plant where raw materials are fed and products removed on a continuous basis.

bearing A part of a machine that takes on friction, as between a rotating part and its housing.

beer The product of fermentation by microorganisms in fermented mash; usually refers to the alcoholic solution remaining after yeast fermentation of sugars.

before top dead center (BTDC) Ignition system timing is specified as degrees before or after top dead center. Top dead center for cylinder one is often marked on the crankshaft pulley, flywheel, or dynamic balancer and indicates the piston at the full extent of its travel, compressing the air/fuel mixture into its smallest volume.

benzene A volatile liquid hydrocarbon, present in coal tar and petroleum, used in chemical synthesis; highly carcinogenic; combined with toluene and xylene (BTX) to raise the octane level of gasoline. Also called benzol.

berm An artificial embankment or ridge.

beta amylase An enzyme that converts starch into sugars; found in higher plants such as soybean, sweet potato, and barley, and in some microorganisms.

biocide A chemical substance added to other materials (typically liquids) to kill microorganisms, algae, mosquitoes, plants, insects, and living things in general.

biofilter Living material used to filter or chemically process pollutants, e.g., for processing wastewater..

biological oxygen demand (BOD) The amount of dissolved oxygen being consumed as microbes break down organic matter, useful in determining water pollution levels. A high BOD means that oxygen levels are falling, because of, e.g., poorly treated wastewater or high nitrate runoff from farmland.

biomass Any material, excluding fossil fuels, that was or is a living organism, and can be potentially used as fuel. Includes forest and mill residues, agricultural crops and wastes, wood and wood wastes, animal wastes, livestock operation residues, aquatic plants, fast-growing trees and plants, and municipal and industrial wastes. Also means the total mass of organisms in a given area or volume.

biorefine   To refine biomass into multiple products. A biorefinery is a place that takes raw materials, e.g., a crop, and makes, e.g., alcohol, animal feed, and/or other products

blending octane value   May be greater than, equal to, or less than the value calculated from the volumetric average of the octane numbers of the blend components.When blending equal amounts of two fuels based on their blending octane rating, the resulting mixture will have a final octane rating precisely between the two.

blow-by   When a car's rings are worn or sooty, the engine's combustion causes fuel, air, and moisture to be forced past the rings into the crankcase, lowering compression, horsepower, and mileage, and contaminating the oil in the crankcase.

blower   A mechanical device that creates a current of air used to move, dry, or heat something. Also, a slang term for turbo- or superchargers.

boiling point   The temperature, at a fixed pressure, at which a liquid bubbles and turns to vapor.

boost pressure   The increase above atmospheric pressure that is produced inside the intake manifold by a supercharger.

boroscope   A flexible fiber optic device that lets you look at the inside of anything that its cable can thread through.

bottoms   What remains at the bottom of a column or reboiler following distillation of the lighter components.

breakaway fitting   A fitting installed in the filler hose of a fuel dispenser; designed to separate, shutting off fuel from the dispenser, if the driver leaves the filling nozzle in the vehicle and drives away, pulling on the filler hose.

breaker points   See points.

brewer's yeast   The dried, pulverized cells of *Saccharomyces cerevisiae*, a type of fungus; a rich source of B-complex vitamins, protein (providing all essential amino acids), and minerals; usually a byproduct of brewing; not to be confused with nutritional yeast.

British thermal unit (Btu or BTU)   The standard unit for measuring quantity of heat energy; the amount of heat required to raise the temperature of one pound of water one degree Fahrenheit under stated conditions of pressure and temperature (equal to 252 calories, 788 foot-pounds, 1,055 joules, and 0.293 watt-hours);

Brix (°Bx)   A measurement of the mass ratio of dissolved sucrose to water in a solution. A 25°Bx solution has 25 grams of sucrose per 100 grams of solution—there are 25 grams of sucrose sugar and 75 grams of water.

Brix hydrometer   A triple-scale wine hydrometer used to record the specific gravity of a solution containing sugar. Also called a balling hydrometer.

BTX   A compound of benzene, toluene, and xylene; highly carcinogenic; added to gasoline to increase octane.

bubble cap column still   A distillery that uses an upright cylinder filled with a series of donut-shaped plates stacked one on top of another, spaced a few inches apart. During distillation, vapors rise from the bottom plate into the liquid atop the next plate, through the donut hole which is fitted with an open two-piece cap. The vapor is then forced to bubble into the liquid on the plate, transferring heat. This raises the temperature of the liquid to boiling, so that a higher-grade vapor is produced. By the time the vapors reach the top of a column with eight or more plates, the proof is very high.

buffer   A solution that resists changes in pH when an acid or base is added to it, or upon dilution.

bung wrench   Tool used for tightening and loosening barrel plugs (bung caps).

Bureau of Alcohol, Tobacco, and Firearms (BATF)   The U.S. federal government agency responsible for the issuance of permits for the production of alcohol.

bushing   A bearing used in a round hole to constrain, guide, or reduce friction, usually for a revolving shaft.

butanol   Either of two butyl alcohols derived from butane; used as solvents and in organic synthesis; butyl alcohol.

butyl rubber   A synthetic rubber produced by copolymerization of a butylene with isoprene, nearly impermeable to air and used in tires, inner tubes, and insulation.

butyric acid   Either of two colorless isomeric acids, used in disinfectants, emulsifying agents, and pharmaceuticals. Also called butanoic acid.

calorie   The amount of heat required to raise the temperature of 1 kilogram of water by 1°C at 1 atmospheric pressure. Also called a large calorie.

cam   In machinery, a projection on a rotating part; it slides to make contact with another part, imparting motion to it; in cross-section, it is often egg-shaped so that when it turns, the lobed portion of the "egg" can raise and lower a part in contact with it.

camlock fitting   A quick-connect coupling where a cam is turned to engage in, e.g., a slot.Also known as a cam and groove coupling.

camshaft   A rod with a cam or cams attached to it, which operates the valves in an internal combustion engine.

canopy   The uppermost level of a mature forest.

carbohydrate   A chemical term describing compounds made up of carbon, hydrogen, and oxygen; includes sugars and coumounds made from sugars such as starches, cellulose, inulin, and other oligosaccharides. Contrast with hydrocarbons, which do not contain oxygen.

carbon dioxide (CO_2)   A gas produced by breathing or by burning carbon and organic compounds; a by-product of fermentation. Under normal conditions, it is stable, inert, and nontoxic. It is naturally present in fresh air and is absorbed by plants in photosynthesis.

carbon dioxide compressor A device that uses an internal screw mechanism and cooling to compress CO_2 gas.

carbon monoxide (CO) A colorless, odorless, flammable, highly poisonous gas formed by incomplete combustion of carbon or a carbonaceous material, such as gasoline.

carburetor A device in an internal combustion engine that mixes air with a spray of fuel; it changes the air/fuel ratio according to what the engine needs when starting, idling, or with added load or altitude changes.

carburetor float A buoyant part inside the carburetor that controls a needle valve to open and close the fuel inlet.

cardlock station A type of fuel station, usually unattended, where vehicles refuel, often using a plastic card for access.

castable refractory Ceramic material that can handle very high temperatures; usually mixed with a bonding agent, so that when water is added, it becomes strong and set in a mold.

catalytic combustor A ceramic honeycomb device containing catalytic (involving a catalyst) metals, used to force smoke particles to burn at a much lower temperature than they ordinarily would.

catalytic converter A part of a vehicle's exhaust system that uses a catalyst to cause unburned CO, HC, and NOx to combust at lower termperatures turning them into less toxic materials, such as carbon dioxide, heat, and water, before they get to the tailpipe.

cellulase A generic term for an enzyme capable of splitting cellulose; different cellulases attack different chemical bonds in the cellulose chain.

cellulose The insoluble, main constituent of plant cell walls and of vegetable fibers; a polysaccharide consisting of chains of glucose monomers.

cellulose digestion factors Enzymes, co-enzymes, and precursors that assist in metabolism of cellulose.

cellulytic Able to reduce cellulose to component sugars.

Celsius (°C) (Degrees Celsius) A temperature scale measured in degrees; water freezes at 0°C and boils at 100°C.

centrifugal pump A device that moves a fluid by rotation.

centrifuge A device where a rapidly rotating container applies centrifugal force to its contents (centrifugal means moving away from a center); used to separate fluids of different densities (e.g., cream from milk) or liquids from solids.

ceramic saddle A specially shaped hardened clay ring used as packing for distillation columns.

cetane A measure for fuel that indicates the autoignition characteristics of a fuel in a compression-ignition engine.

cetane improver A hydrocarbon added to fuel to improve starting and running when cold, in a compression-ignition (diesel) engine; it shortens the time from when the fuel enters the ignition chamber until it ignites. White smoke, poor mileage, and less power in cold weather may indicate low cetane.

CGM See corn gluten meal.

channel stock Steel formed into a flat-bottomed U shape during manufacture.

chelate A compound containing a molecule or ion bound to a metal atom at two or more points to form a stable ring structure; chelation is the process of binding to the metal. Chelating agents are used to remove toxic metals (e.g., lead, cadmium), or used to extract metals.

chemurgist A person who develops industrial and chemical products from organic raw materials, especially from farm products; the word is a combination of *chemistry* and *metallurgy*; it was coined to describe the work of George Washington Carver.

chisel plow A tool used to loosen and aerate soil, while leaving crop residue at the top of the soil; it reduces the effects of compaction and helps break up hardpan; it does not turn the soil, maximizing erosion prevention by keeping organic matter on the soil surface.

chitin [pronounced ki-tin] The major component of the exoskeleton of arthropods (e.g., insects and crustaceans) and in the cell walls of fungi; a fibrous substance consisting of polysaccharides.

chill days A measurement that describes the number of degrees below a fixed temperature in the cool season; similar to degree days.

chlorinated hydrocarbons Hydrocarbons in which one or more hydrogen atoms have been replaced with chlorine, e.g, dioxin, DDT, and chlordane.

choke A valve in the carburetor that reduces the amount of air in the fuel mixture when the engine is started. Also, a knob that controls the valve.

cleaning-in-place (CIP) system A system to automatically clean and disinfect (e.g., an alcohol plant) without major disassembly.

closed loop Describes a vehicle operating within the abilities of its oxygen sensor to monitor and adjust the amount of fuel sent to the engine.

cob A material made of sand, clay, and straw; used to build houses.

cogeneration The process of simultaneously producing electrical energy and another form of useful thermal energy (such as heat or steam) from the same fuel source; e.g., when water is being boiled to generate electricity, the leftover steam can be used for industrial processes or for heating.

cogenerator A facility that produces electricity and another form of useful thermal energy (such as heat or steam).

coil See immersion coil heat exchanger.

column Vertical vessel used to increase the degree of separation of liquid mixtures by distillation or extraction; usually cylindrical or spiral in shape.

collar In a machine, a band or pipe that restrains or connects. In fuel injection, it is the housing that contains the throttle plate and fuel injectors.

combined heat and power See cogeneration.

combustion The process of burning something; e.g., fossil fuels. A substance is combined with oxygen, producing heat and light.

combustion chamber An enclosed space in which combustion takes place, e.g., in an engine or furnace.

community-supported agriculture (CSA) A way of food production, sales, and distribution that reduces farmers' financial risks, while increasing the quality of food and improving the care of the land. A number of people (members) pay a farmer so as to be included in a regular delivery or pick-up of the vegetables and fruits he grows (and perhaps also flowers, herbs, milk or meat). The farmer thus gets his salary paid and his working capital upfront; his customers share the risks of poor harvests, but reap the benefits of very fresh good grown in ways they feel good about.

community-supported energy (CSE) Similar to community-supported agriculture; instead of investing in potatoes, carrots, or cucumbers, members invest locally in energy projects that provide energy security and many other benefits.

completely denatured alcohol (CDA) Alcohol denatured so it cannot be made drinkable; currently all CDAs approved by the BATF are petroleum-based. See specially denatured alcohol (SDA) for alternatives.

complex sugars Sugars composed of three or more units of a single sugar molecule (monosaccharides); their complicated structure is why they are called "complex" carbohydrates. The chemical name for the largest type of complex carbohydrate is "polysaccharide," meaning "many sugars." Generally not fermentable with yeast; contrast with fermentable sugars.

compost Partially decayed organic material that is applied to the soil to fertilize it and increase its humus content, as well as to inoculate it with a large population of beneficial microorganisms.

compost tea A solution made by soaking compost in aerated water; the "tea" is used as a liquid fertilizer/soil inoculant.

compound A combination of two or more distinct elements.

compression ratio Degree to which the fuel mixture in an internal combustion engine is compressed before ignition; the volume of the combustion chamber with the piston farthest out divided by the volume with the piston in the full-compression position. A compression ratio of six means that the piston compresses the mixture to one-sixth its original volume.

compression-ignition engine An internal combustion engine that has no ignition system; air is compressed until it's hot enough to autoignite fuel injected into the cylinder; a diesel engine.

compressor A device that uses compression to increase the pressure of a gas, e.g. to power a turbine.

concentration The measure of how much of a substance is mixed with another substance.

condensed distiller's solubles (CDS) The syrup created by evaporating thin stillage into a syrup; can be used as animal feed, or combined with distiller's dried grains (DDG) to create distiller's dried grains with solubles (DDGS) to make a storable feed.

condenser A device that removes heat from a gas or vapor in order to reduce it to a liquid.

conservation vent A device that lets air, gas, or liquid pass out of or into a confined space, while minimizing evaporation losses or contamination.

constant injection system (CIS) A fuel injection system that delivers a constant supply of fuel to the engine instead of injecting fuel at specific times.

constructed wetlands An artificial marsh created to serve as a biofilter, removing pollutants, e.g., heavy metals, from wastewater, stormwater runoff, or sewage treatment, and as habitat for wildlife.

continuous distillation Instead of processing one batch at a time; this system continuously pumps mash into the still from fermentation tanks.

control coil A heat exchanger used to control temperature at a specific area in a distilling column.

control pressure In a constant injection system (CIS), a fuel pressure that responds to changes in the air inflow and controls the fuel distributor. When fuel pressure is too high, a needle valve will restrict fuel; when pressure is too low, a pump sends fuel to the pressure control chamber and the fuel chamber.

conventional oil Oil produced using the traditional oil well method.

conversion The fermentation step in which beta or glucoamylase enzymes convert complex sugars to fermentable sugars. Also called saccharification.

cooker A tank or vessel that usually contains a source of heat and is fitted with an agitator. Also called a mash cooker.

cooling coil A coiled arrangement of pipe or tubing for the transfer of heat between two fluids. See immersion coil heat exchanger.

cooling jacket A device containing a heat exchange medium; it surrounds the outside of a tank to lower the temperature of the tank contents.

cooling tower A device that evaporates a portion of the water to be cooled, to reduce the temperature of the majority of the water.

coppicing A traditional method of woodland management. Trees are radically pruned each year to stimulate new growth, also providing a biomass yield from the prunings.

co-products The resulting substances, materials, and energy that accompany the production of ethanol by fermentation process. Also includes products made onsite from these surpluses, with or without addition of other materials/energy.

corn gluten meal (CGM) A byproduct of commercial corn milling that contains the protein part of the corn. CGM is used as feed for cattle, poultry, fish, and dogs, and will not harm people.

corporate average fuel economy (CAFE) Standards for fuel economy regulated by the U.S. National Highway Traffic Safety Administration and the Environmental Protection Agency. These describe the fuel economy of a manufacturer's fleet of passenger vehicles or light trucks with a gross vehicle weight rating of 8500 pounds or less.

countercurrent stream A basic principle of distillation that alcohol rises as a vapor and water drops as a liquid.

counterflow heat exchanger A shell and tube heat exchanger in which two fluids enter from opposite ends; transfers more heat than a parallel-flow heat exchanger, where the fluids enter at the same end.

coupling A device normally used to connect two pieces of pipe; may join the same or dissimilar types or sizes of pipes, depending on its construction.

CPVC (chlorinated polyvinyl chloride) Behaves similarly to PVC, but at much higher temperatures; used, e.g., for hot corrosive liquids and hot water distribution.

Crabtree effect An inhibition of cellular respiration by high concentrations of glucose.

cross-linked polyethylene (PEX) A thermoset material, able to become permanently rigid when heated or cured; the "X" stands for cross-linking, which chemically changes the polyethylene to make it flexible and chemical-resistant.

cubic feet per minute (cfm) A measurement used to express the rate of movement of gases.

cultivar A plant variety that has been produced by selective breeding, e.g., Sweet 100 is a specific cultivar of tomato.

cyanobacteria Blue-green algae; microorganisms related to bacteria but capable of photosynthesis.

cylinder The space in which a piston travels in an internal combustion engine.

DDG See distiller's dried grains

DDGS See distiller's dried grains with solubles

DDS See distiller's dried solubles

dead zones Areas of water with very elevated nitrogen levels and low levels of oxygen; aquatic life have a hard time surviving. The nitrogen causes a population boom of microscopic algae; when the algae die, their decomposition uses up much of the oxygen in the water.

degree day A measurement of the amount of heat necessary for a plant to reach a maturity benchmark; e.g., a variety of corn may require 1800 degree days to fully set and develop its ears. A degree day is the number of degrees above the criteria set for that plant. In the case of corn, that level is 50°F, so one day that reaches 85 degrees counts as 35 degree days. Contrast with chill days. (Some plants, e.g., sugarcane, require both degree days and chill days to be satisfied.)

dehydration The removal of 95% or more of the water from a substance by exposing to high temperature; useful in preserving for storage. Also, the process of removing the last few percent of water from distilled alcohol to produce anhydrous alchohol.

denaturant A substance added to alcohol to make it unfit for human consumption.

denatured alcohol Alcohol that has had a substance added to it in order to make it undrinkable; the denaturing agent may be gasoline or other substances specified by the Bureau of Alcohol, Tobacco, and Firearms.

Department of Energy (DOE) A U.S. federal agency charged with overseeing energy. It collects data on energy production and consumption, operates the petroleum reserve, and oversees numerous research programs.

desiccant A substance having an affinity for water; used for drying purposes.

destructive distillation A process where organic substances, e.g., wood, coal, and oil shale, are decomposed by heat in the absence of air; they are then distilled to produce, e.g., coke, charcoal, oils, and gases.

dewater To remove the free water from a solid substance.

dextrin A gummy substance made by hydrolizing starch using heat or acids; used as a thickening agent or in adhesives. Generally, dextrins are made of multiple sugar molecules, and are too large to be consumed by yeast without further breakdown to simpler sugars.

dextrose The predominant naturally occuring form of glucose; specifically, the right-handed version of glucose generally referred to as D-glucose.

diaphragm pump A positive displacement pump that uses a flexible diaphragm instead of a piston. It can handle slurries containing a moderate amount of grit and solids. Also called a membrane pump.

diesters A compound containing two ester groupings; an ester is an organic compound where the hydrogen of an acid has been replaced with an alkyl or other group. Naturally occurring fats and essential oils are often esters of fatty acids.

diethyl ether A colorless liquid, slightly soluble in water; used as a reagent and solvent; volatile and highly flammable. Also called ether, ethyl ether, ethyl oxide, and ethylic ether.

diffusion The movement of concentrated dissolved solids in a solution into areas of lower solids concentration.

dimethyl ether (DME) A flammable, colorless liquid, soluble in water and alcohol; used as a solvent, extractant, reaction medium, and refrigerant; proposed as a liquid fuel by the coal industry. Also called methyl ether or wood ether.

direct refluxing Using liquid alcohol spraying instead of a heat exchanger to control column temperature and purity of output.

displacement The total volume of air/fuel mixture in all of an engine's cylinders, including the cylinder head in one cycle; used as a measure of engine size; expressed in cubic inches or liters. At different times, various manufacturers have offered several engine displacements (sizes) by using the same block but varying the volume of the heads, e.g., Chevy 235-, 250-, and 292-cubic-inch six-cylinder engines.

distillate The portion of a liquid that is removed as a vapor and condensed during distillation.

distillation The process of separating a mixture's components by their differences in boiling point; the liquid forms a vapor when heated; the vapors are collected and condensed into liquids.

distiller's dried grains (DDG) The solid, granular portion left over from the mash after distillation; usually screened or centrifuged out of the liquid part of the mash (stillage), and then dried.

distiller's dried grains with solubles (DDGS) The substance that remains when, in making alcohol from grain, all of the starch is removed, but all of the protein and fat, some of the cellulose, and a wide array of vitamins and minerals remain, along with yeast from fermentation; useful as animal feed. In the case of DDGS, after the DDG is separated from the spent mash, the liquid portion is evaporated to concentrate the soluble components to a syrup, which is then blended with the DDG.

distiller's dried solubles (DDS) Condensed distiller's solubles (CDS) that have been dried to be changed from a syrup to essentially a solid.

distiller's solubles (DS) After distillation, the liquid part of the mash that remains following removal of dried distiller's grains (DDG); it contains about 10% solids that are too fine to be separated, and also many nutrients dissolved in solution. Also called thin stillage.

distillery A device for separating liquids based on their boiling points and their latent heat of vaporization; synonym of still. Also, a plant or works where alcohol is made by distillation.

distributor A device that applies electric current in proper sequence to the spark plugs of an engine.

doubler still A distillery that uses intermediate tanks to condense and revaporize between the main tank and condenser.

downcomer A pipe for the net downward transport of water from one plate to the next lower plate in a distillery.

downdraft Inlet located to produce a strong downward air current, e.g., into a cob stove or the carburetor of an internal combustion engine.

downspeeding Changing the gearing of the transmission so it matches the higher power output of alcohol at the right point in the torque curve.

drum dryer A machine that tumbles the material to be dried in a cylinder, through which heated air is blown.

DS See distiller's solubles.

dump Liquid falling back through the holes in a perforated plate column, due to pressure loss.

dry-milling The process by which whole dry seed is ground and fermented in its entirety. This is changing, as alcohol plants are now creating co-products from the grain before fermentation, rather than after. Some very large plants do wet-milling, where corn is first soaked in acidic water, and separated into many components, and only the corn starch is fermented.

duty cycle meter A device that measures an intermittently operating machine's cycle of operation; it expresses the "on" or "off" period as a percentage of total period; it divides the pulse by the period between pulses and expresses this ratio in percent.

dwell meter Device used to measure and adjust the dwell angle and length of contact (the dwell) of the points.

E-85 A fuel mixture of 85% ethanol and 15% gasoline, by volume on average; legally includes mixtures of as little as 70 to 98% alcohol.

E-98 200-proof ethyl alcohol with 2% denaturants added.

ECU See electronic control unit.

elastomer See fluorocarbon elastomer.

electronic control unit (ECU) The engine's computer "brain;" an electronic system that controls various aspects of an internal combustion engine's operation, e.g., how much fuel is injected into each cylinder during each engine cycle, ignition timing, cam timing, boost level (in turbocharged cars).

electronic fuel injection (EFI) A system that meters fuel very precisely, then atomizes the fuel by forcibly pumping it through a small nozzle under high pressure.

embodied energy The quantity of energy that is needed to manufacture and get a material, product, or service to the point of use—e.g., energy for the raw material extraction, transportation, manufacturing, assembly, and installation—and finally to dispose of it at the end of its useful life.

energy balance See energy returned on energy invested.

energy returned on energy invested (EROEI) The ratio of the amount of energy expended to obtain a resource, compared with the amount of energy obtained from that resource.

enrichment The process of increasing the percentage of alcohol (raising the proof) of a distillation by condensing and revaporizing the vapors.

ensiling Anaerobically fermenting an animal feed crop for preservation.

enzymatic hydrolysis A process that uses enzymes to break down starch or cellulose into sugar.

enzyme A substance produced by microorganisms that acts as a catalyst to promote chemical processes without itself being altered or destroyed.

EROEI See energy returned on energy invested.

ester An organic compound made by replacing the hydrogen of an acid by an alkyl or other organic group. Many naturally occurring fats and essential oils are esters of fatty acids.

ethanol The alcohol product of fermentation that is used in alcohol beverages and for industrial purposes. Also called ethyl alcohol.

evaporation Conversion of liquid to the vapor state by the addition of latent heat of vaporization.

evaporative emissions Vapors that escape when gasoline evaporates, generally from the fuel tank or other parts of the fuel delivery system. See permeation emissions.

exhaust gas analyzer A device that analyzes a vehicle's gaseous products to see how efficient the combustion process is.

exhaust gas recirculation (EGR) The process of recirulating some of an engine's exhaust gas back to the engine cylinders, where it dilutes the incoming air with inert gas, thus lowering peak combustion temperatures and (in diesel engines) reducing the amount of excess oxygen. Because of the lower temperatures, NOx formation is much reduced.

exhaust gas temperature (EGT) gauge A device that measures the temperature of the spent gas leaving an internal combustion engine.

expansion valve A valve through which a fluid is expanded from one pressure to a lower pressure at a controlled rate; e.g., an air conditioner has an expansion valve to allow liquefied freon to flow into the heat absorption coils, where it expands into a gas.

Fahrenheit (°F) (Degrees Fahrenheit) A temperature scale where the boiling point of water is 212° and the freezing point is 32°. To convert Fahrenheit to Celsius, subtract 32, multiply by five, and divide by nine (at sea level).

Farmer's tree A broad-canopied tree that provides light shade, and that often has nitrogen-fixing bacteria on its roots, exuding large quantities of sugars through it roots to the rhizosphere, providing an excellent microenvironment for production of human-oriented food or products.

feedback system A system that regulates itself by feeding back part of its output to itself; e.g., a heating system uses a thermostat to monitor and adjust its output.

feedstock The base raw material that is the source of sugar for fermentation.

fip (female internal pipe thread) Pipe threads that appear on the pipe's inner diameter. Contrast with mip (male internal pipe thread).

fermentable sugars Sugars (usually glucose), derived from starch and cellulose, that can be converted to ethanol. Also called reducing sugars or monosaccharides.

fermentables Liquids, or powders dissolved into liquids, that are the source of energy for yeast, e.g., malt extract, sugar, honey, or molasses.

fermentation The chemical breakdown of a substance, using yeast, bacteria, or other microorganisms, that usually includes effervescence and heat output.

fermentation lock A water trap that releases the yeast's exhaled carbon dioxide but keeps oxygen and bacteria from entering the tank, thus preventing contamination.

fermentation tank The container where the biological magic of brewing takes place, where yeast changes the sugars in the wort into alcohol over a period of hours or days.

fermenter See fermentation tank.

fertigation The process of applying fertilizers, soil amendments, or other water-soluble substances through an irrigation system.

field capacity The amount of water retained in the soil after the excess has drained away (which usually takes 2 to 3 days after a rain); sand, for example, has a very low field capacity.

film strength The measurement of a lubricant's ability to keep an unbroken film over surfaces. Synthetic oil can have a higher film strength than regular oil.

fin tubing A length of tubing or pipe with a fin (flattened projection) around it; used to increase heat transfer, as in heat exchangers.

firebox The chamber of a solid fuel appliance where the fuel is located and where primary combustion occurs.

fire-tubes The portion of a heater where hot, gaseous combustion products pass through tubes that are surrounded by heat exchange fluid.

five-carbon (C5) sugar Any of a class of monosaccharides having five carbon atoms per molecule; includes xylose.

flame front The flame that develops when any flammable mixture of vapor or gas contacts an ignition source.

flame propagation The spread of a flame in a combustible environment outward from the point at which the combustion started.

flash heater A device that very rapidly heats a liquid by exposing small amounts of it to high temperature and using high flow rates, e.g., a high-temperature portable pressure washer.

flashpoint The minimum temperature at which a substance gives off enough vapor to ignite in air.

flatstock Raw material purchased from metal manufacturers in the form of metal plates with square edges and no detail.

flexible-fuel vehicle (FFV) A vehicle that can alternate between sources of fuel, e.g., alcohol and gasoline, and can run on any ratio of the two fuels.

flighting A continuous helix formed from bar or cut plate into the desired diameter and pitch to fit conveyor screw pipes or shaftless screw applications; also called an auger or feedscrew.

flinty Describes a starch with chains of glucose molecules bunched together in semi-crystalline masses of 1,000 or more; the starch granules are surrounded by protein that is hard to digest; a flinty starch, e.g., corn, requires premalting before fermentation.

float bowl A chamber in a carburetor that holds a small amount of liquid fuel; it serves as a constant-level reservoir of fuel that is metered into the passing flow of air.

float switch A device used to sense the level of liquid within a tank; it may trigger, e.g., a pump or an indicator.

float valve A device that regulates fluid level by using a float to control an inlet valve; a higher fluid level will force the valve closed, a lower level will permit it to open; e.g., the float valve in a toilet's water tank.

flow-control valve A valve where the opening is controlled by the rate of flow of the fluid through it. Also called rate-of-flow control valve.

fluorocarbon elastomer A rubber-like plastic that contains fluorine; has high resistance chemicals, heat, oils, oxidation, and solvents.

fluoroelastomer See fluorocarbon elastomer.

fogger A device that spreads a cloud of vaporized liquid.

fooler technology A technique that causes a sensor to change information or to adapt to changed information about the system it is monitoring; e.g., it can fool an engine into thinking it's running gasoline, when it's really running gas.

foot-candles (ft-c) A unit of measure of the intensity of light falling on a surface; one lumen is equal to the amount of light put forth by one candle one foot away.

foreshot The first third of a distillation run; vapors rising from the mash during the foreshot are very high-proof.

formic acid A colorless volatile liquid made from carbon monoxide and steam, using catalysts; a component of methanol exhaust; used in dyeing textiles, tanning leather, and creating latex rubber; also occurs in the bodies of red ants and in the stingers of bees. Also called methanoic acid.

fossil fuel Any fuel formed from the remains of living organisms, e.g., crude oil, natural gas, or coal.

fractional distillation The separation of a liquid mixture into components with different boiling points.

fractional still A distillery that has outlets at intervals up the column, which allow for the removal of substances having different boiling points. The ("lightest") products with the lowest boiling points exit from the top of the column, and the ("heaviest") products with the highest boiling points exit from the bottom.

freon The trade name for a chlorofluorocarbon (CFC) used as a refrigerant, aerosol propellant, and solvent.

frequency valve A valve used to control the flow of a fluid. In an engine, it is a solenoid valve similar to that on a fuel injector but used instead to detemine how much fuel to allow to return to the gas tank, which in turn determines the pressure in the fuel rail that supplies the fuel injectors.

fructose A very sweet monosaccharide (simple) sugar occurring in many fruits and honey. Also called fruit sugar.

fuel cell   A device that contains electrodes immersed in an electroylyte, that produces electric current directly from solar or chemical energy, using external supplies of fuel and an oxidant that react to the electrolyte; a hydrogen cell, e.g., uses hydrogen as fuel and oxygen as oxidant.

fuel/feed plant   An organized operation for loading feedstock, processing and producing fuel from that feedstock, and offloading the fuel and byproducts for use or sale.

fuel injector   The nozzle and valve through which fuel is sprayed into the combustion chamber.

fumigation   The addition of ethanol to the intake air manifold; has been used as a method of displacing diesel fuel with ethanol in compression-ignition engines.

furfural   Or furfuraldehyde, is a oily, colorless liquid that has a pleasant almond odor; upon exposure to air it quickly turns dark brown, yellow, or black. It is soluble in ethanol and ether and somewhat soluble in water, and commonly used as a solvent and in the manufacture of pesticides, resins, and nylon. A derivative of furan, it is prepared commercially by dehydration of pentose sugars obtained from cornstalks and corncobs, husks of oat and peanut, and other waste products.

fusel oil   A clear, colorless, poisonous liquid mixture of alcohols obtained as a byproduct of grain fermentation; generally amyl, isoamyl, propyl, isopropyl, butyl, isobutyl alcohols and acetic and lactic acids. A mixture of several alcohols (chiefly amyl alcohol) produced as a byproduct of alcoholic fermentation, fusel oils, also sometimes called fusel alcohols, or potato oil in Europe, are higher order (more than two carbons) alcohols formed by fermentation and present in cider, mead, beer, wine, and spirits to varying degrees. The term fusel is German for "bad liquor". During distillation, fusel alcohols are concentrated in the "tails" at the end of the distillation run. They have an oily consistency, which is noticeable to the distiller, hence the name fusel oil. These heavier alcohols can be almost completely separated in a reflux still.

gasket   A shaped piece or ring of rubber or other material sealing the junction between two surfaces in an engine or other device.

gasohol   A mixture of gasoline and ethyl alcohol used as fuel in internal combustion engines.

gasoline   A volatile, flammable liquid obtained from petroleum, used as fuel for spark-ignition internal combustion engines; it has no fixed formula and has a boiling range from about 85°F to 450°F.

gate valve   A valve that has a sliding part to control the size of the opening through which fluid can travel; often designed to be fully open or fully closed, and thus may not be suitable for fluid regulation; when open, usually results in very low loss to friction. Also called a sluice valve.

gelatinization   An effect that can occur during the process of making starches soluble in water, using heat or a combination of heat and enzymes. The starch grains swell and rupture; the outer layers absorb water and lose their crystalline structure; the outer layers become adhesive and form a sticky gel.

glow plug   A small electric heating element used to heat the combustion chamber of an engine (usually diesel) to help ignition during a cold-start.

glucoamylase   An enzyme of microbial origin that breaks down carbohydrates into fermentable sugars.

glucose   The most common (monosaccharide) simple sugar; an important energy source in living organisms; a component of many carbohydrates.

GMO (genetically modified organism)   An organism or crop whose genetic material has been changed to produce desired characteristics, using recombinant DNA technology (combining DNA molecules from different sources, such as a tomato and a fish). GMO is not the same as natural selective breeding, which occurs between individuals of the same species, e.g., between a beefsteak tomato and a cherry tomato, by combining the pollen from one donor with the embryo of the other.

grain fermentation factors (GFF)   In distiller's feeds, the combination of cellulose digestion factors, unknown growth stimulants, and urea proteins that substantially stimulate growth and proliferation of rumen microflora.

greenhouse effect   The process where solar radiation is trapped in the Earth's atmosphere, as greenhouse gases let incoming sunlight (visible radiation) through, but will not then absorb the heat (infrared radiation) as it radiates back off the Earth.

greenhouse gas   Any of the atmospheric gases that contribute to the greenhouse effect by absorbing infrared radiation, e.g. carbon dioxide.

grid   Describes the interconnecting electricity distribution systems among utility companies.

gristmill   A machine for grinding grain.

hammermill   A tool used to smash, crush, and tear large raw materials into smaller pieces.

hard plumbing   Pipe that has been made more permanent, either because it is threaded or permanently glued with slip fittings (solvent-welded connections).

hardened munitions   Depleted uranium weapons that penetrate tanks, bunkers, and caves. Upon impact, much of the depleted uranium becomes a fine dust of uranium oxide; this is easily carried on the wind, blanketing the impact area. According to medical literature, when these particles are ingested or inhaled, they enter the lungs and blood, spread throughout the body, and can cause cancers and kidney ailments, as well as deformities and cancers in babies whose parents were exposed. Can be used in a "dirty bomb" designed to spread radioactive material with the intent of making an area uninhabitable. Use of hardened munitions should be considered nuclear war.

hardpan　A hard, impervious soil layer, caused by cementation of soil particles, that impairs drainage and which many plant's roots cannot penetrate, resulting in poor growth; caused by tractors or animals compacting the soil to several inches below the surface.

harmonic balancer　A device connected to the crankshaft of an engine to reduce torsional (twisting) vibration. Also called crank pulley damper, torsional damper, vibration damper.

hatch　An opening or door of relatively small size that allows passage from one area to another.

heat exchanger　A device for transferring heat from one medium to another; parallel metal surfaces keep the fluids separated while heat transfers from the hot fluid/gas to the cooler fluid/gas; used in refrigeration, air conditioning,space heating, the generating of electricity, and the processing of chemicals; e.g., the radiator in a car.

heating value　A measure used to describe the amount of heat released during combustion of food or fuel, based on the perfect combustion of the material with an ideal amount of oxygen.

heavy metal　A high-density metal or high-atomic-weight element; it cannot be degraded or destroyed, so persists in the environment; e.g., arsenic, lead, mercury. .

hectare (ha)　A metric unit of area, equal to 10,000 square meters, commonly used for measuring land area; one hectare equals 2.471 acres; one acre equals 0.4047 hectare.

hemicellulose　Polysaccharides with a simpler structure than cellulose; they help make up the cell walls of trees and plants, along with cellulose and lignin. While cellulose is resistant to hydrolysis, hemicellulose has a random structure with little strength, and so is easily hydrolyzed.

high wine　The high-proof alcohol, at least 140-proof, distilled from mash.

histological　Having to do with microscopic structures of tissues; histological damage would be, e.g., the failure of cell walls or the impairment of liver cells' detoxification functions.

homogeneous charge compression ignition (HCCI)　A form of internal combustion in which ignition occurs at several places at the same time, so that nearly all the fuel and air mixture burns at the same time.

hopper　A wide, bin-like entry for feeding large pieces of solid matter into a machine. It is open on top, and becomes thinner at the bottom where it feeds into the machine.

hose barb　A connector stem with raised ridges inserted into the end of a hose to help it stay in place; the barb is usually permanently attached to a threaded adapter for connecting to another fitting. Also called hose tail or hose end.

humus　[pronounced hew-muss] Fibrous, partially decayed plant matter, necessary for keeping soil healthy; the organic component of soil, created by decomposition of plant material by soil microorganisms; holds soluble soil nutrients and makes them available to roots; prevents nutrients from being lost by washing out of the topsoil.

hydraulic motor　A motor activated by water or other liquid under pressure.

hydrocarbon (HC)　A compound of hydrogen and carbon, e.g., the chief components of petroleum and natural gas. Contrast with carbohydrate.

hydrolysis　The cooking of mash; the chemical breakdown of organic materials through the use of water.

hydrometer　A tool used to determine the specific gravity of liquids. The lower the density of the substance, the lower the hydrometer will sink.

hydrophilic　Having an affinity for water; readily absorbing, or dissolving in, water.

idle　(of an engine) To run slowly while out of gear or disconnected from a load; running only to sustain its running instead of doing any useful work.

idle circuit　A device, inside the carburetor, that lets fuel enter through an idle jet, while more air mixes with this fuel through an air bleed to keep the engine running when the throttle is closed; a simple bypass system that allows the engine to get a richer mixture when idling. The air/fuel mixture at idle is adjusted by a needle-type jet, which restricts how much fuel the engine can suck through the idle jet opening in the venturi wall.

ignition timing　The process of setting the time that a spark will occur in the combustion chamber; greatly affects performance of an internal combustion engine.

immersion coil heat exchanger　A coil of metal tubing immersed in the liquid that needs heated or cooled. One end of the tube connects to a reservoir of heat exchange fluid, and the other end drains the coil after exchanging the heat. Using a reservoir of cold water with an immersion chiller coil offers a simple, effective way to quickly cool hot wort.

inches of mercury　A unit used when measuring atmospheric pressure; used in measuring vacuum; the pressure exerted by a column of mercury of one inch high at 32°F at the standard acceleration of gravity; called inches of mercury because some barometers use the height of mercury in a sealed tube as a measuring device. Also called inHg.

intercooler　A small radiator heat exchanger in the exhaust gas duct of the compressor to absorb heat; a device for cooling gas between successive compressions. Also used in some vehicles to cool air from a turbocharger on its way to the engine.

internal combustion engine (ICE)　A device in which the combustion of a mixture of fuel and air produces hot gases that push against turbine blades or pistons or rotors.

inulin   A sweet, nondigestible carbohydrate found in the roots of various plants; used to replace starch in foods for diabetics.

isopentane   An extremely volatile and flammable liquid at room temperature and pressure; the normal boiling point is about 80°F, so it will boil and evaporate away on a warm day; derived from petroleum and used mainly as a solvent; sometimes added to gasoline to help with cold-starting, or added to alcohol as a denaturant. Also called 2-methylbutane.

joule   A measure of the amount of work done; a unit of electrical, mechanical, and thermal energy equal to the work done when a current of one ampere is passed through a resistance of one ohm for one second; about the amount of energy it takes to toss an apple vertically a foot.

kerogen   A complex fossilized organic material found in oil shale that yields petroleum products when distilled.

ketone   An organic compound made by oxidizing secondary alcohols, e.g., acetone; may be used in perfumes and as a solvent.

keyline   A contour line passing through a piece of land's keypoint (where the lower and flatter portion of a valley suddenly steepens); contour plowing parallel to the keyline will cause rainfall to soak into the plowlines instead of running off downhill.

kilojoule (kj)   A unit of energy or work equal to 1000 joules.

kilowatt-hour (kWh)   A unit of electric energy equal to the work done by 1000 watts in one hour.

knock   See pre-ignition.

knock sensor   A microphone that picks up the frequency of the abnormal combustion of preignition, also known as pinging or knocking; the sensor sends an electrical signal to the ECU to be used in calculating ignition timing or turbocharge boost.

krill   A small, shrimplike crustacean of the open seas that eats plankton and in turn becomes an important source of food for whales, various fishes, and birds.

lactic acid   The acid formed as a result of fermentation of carbohydrates by bacteria called *Lactobacillus;* found in sour milk, cheese, yogurt, molasses, various fruits, and wines.

lactose   The least sweet of all natural sugars; occurs naturally in milk. Also called "milk sugar."

ladder tank   An above-ground tank with a ladder for accessing the top.

laminar flow   Flow in which the fluid or gas travels smoothly or in regular paths, e.g., a cool thin layer moving on the outside, with hot gas moving in the core.

laminarin   A polysaccharide carbohydrate that serves as an energy storage compound for Laminaria and other brown algae-like kelp; produced by photosynthesis; feeds hundreds of sea animals; an important part of the Japanese diet; roughly equivalent to starch in land plants.

latent heat of vaporization   The amount of energy required to bring a liquid to a vapor state; the heat absorbed when a substance changes phase from liquid to gas, e.g. the heat lost by air when liquid water changes into vapor; water's high latent heat of vaporization represents the amount of extra heat that can be stored in water, at its boiling point, without turning it into steam.

lathe chips   The unwanted metal pieces removed from a workpiece against a cutting tool; a lathe holds and turns the workpiece.

leaf protein concentrate (LPC)   A concentrated form of the proteins found in the leaves of plants; as a food source for people and animals, potentially the cheapest and most abundant source of available protein.

leaned out   Slang for the opposite of a rich mixture of fuel. See rich mixture.

lees   The sediment surpluses from the bottom of the wine tank or barrel.

leguminous   Describes plants of the pea family (*Leguminosae*); includes peas, beans, lentils, clover, and alfalfa; seeds are in pods, and root nodules usually contain symbiotic bacteria that can fix nitrogen.

lifting auger   A spiral screw inside an outer tube that pushes grain from the lower end to the top end; used to transfer grains from storage to the grinding site to the cooker; usually powered by an electric motor.

lignified   Describes a plant that has changed into wood or become woody through the deposit of lignin in cell walls.

lignin   A complex polymer found in the cell walls of many plants that makes them woody and rigid; after cellulose, the most abundant organic material on Earth; it is removed from wood pulp in paper-making and used as a binder in particleboard, as a soil conditioner, in adhesives, etc.

lignocellulose   A complex of lignin and cellulose found in the cell walls of woody plants.

limited liability corporation (LLC)   A legal form for a business that is a hybrid of a partnership and a corporation; a popular choice of business owners who want to limit their liability exposure to their percentage of ownership or equity interest in the company; a pass-through tax entity.

linoleic acid   An polyunsaturated fatty acid essential in the human diet; an important component of drying oils, such as linseed oil.

line pressure   The pressure under which an air or hydraulic system operates.

linkage   A system of interconnected machine elements, e.g., rods, used to transmit power or motion; an example is transmitting motion from the accelerator pedal to operate a carburetor.

liquefaction   The change in the phase of a substance to the liquid state; in fermentation, the conversion of water-insoluble carbohydrate (starch) to water-soluble carbohydrate (dextrins).

liquefied natural gas (LNG)   Natural gas in its liquid form; when cooled to –259° Fahrenheit, natural gas becomes a clear, odorless liquid.

livingry   A term coined by R. Buckminster Fuller; he says, "The essence of livingry is human-life advantaging and environment-controlling. With the highest aeronautical and engineering facilities of the world redirected from weaponry to livingry production, all humanity would have the option of becoming enduringly successful."

long-term fuel trim   The variation from an ideal air/fuel mixture that is allowed by a vehicle's ECU on an ongoing basis.

low-grade steam   The condensed byproduct of the use of high-pressure steam in various industrial operations; its temperature can be 250–300°F, or even higher with inefficient operations. Also called wet steam .

low wine   A traditional term for the end of a batch distillation run in which the last 10–20% of the alcohol is recovered without refluxing; this remaining alcohol is generally recovered at low proof and contains high-boiling-point components such as the fusel oils.

lysine   An amino acid essential to the growth of protein molecules; usually lacking in grain or green forage; usually needs to be added to animal feed; available in DDGS and some mushroom substrate.

main metering jet   A needle-valve-regulated opening in the carburetor that keeps the float bowl full, while providing the engine with sufficient fuel for normal loads.

malt   See barley malt.

maltose   A sugar produced by the breakdown of starch, e.g., by malt enzymes; contains two glucose molecules. Also called malt sugar.

manifold   The part of an internal combustion engine that delivers air and fuel to the cylinders, or that leads from the cylinders to the exhaust pipe, e.g., the exhaust manifold.

manifold absolute pressure (MAP) sensor   A device that measures the mass of air entering the engine and sends that information to the vehicle's ECU, so it can regulate fuel and spark.

mannitol   A widely occurring sugar alcohol compound; found in bacteria, fungi, algae, and plants; similar to xylitol or sorbitol; added to foods to thicken, stabilize, and sweeten; used in foods for diabetics.

mantle   A mesh of threads impregnated with the rare earths cerium and thorium that gives off light when heated by a flame.

mariculture   The cultivation of marine organisms for food or for other products, e.g., oysters and algae.

mash   In alcohol-making, the grain and water mixture once you have added yeast and fermentation begins; before yeast is added, the mixture is called a wort.

mash cooker   See cooker.

mash pot   The tank portion of a batch distillery that holds the fermented liquid to be distilled; mash is boiled in this tank to distill the alcohol.

mash pump   A device that uses suction or pressure to move mash to or from fermentation tanks.

mass airflow sensor   A device used to measure the mass of air entering an engine, calculate its density (which varies with temperature and pressure), and deliver the information to the engine.

medium density polyethylene (MDPE)   A type of thermoplastic (become moldable when heated and become hardened when cooled, repeatedly) containing cross-link bonds, with good resistance to cracking; used for water tanks, gas pipes and fittings, sacks and bags, etc.

MegaOilron   A relatively small clan of government, military, and energy industry elites.

membrane   A sheet polymer capable of separating liquid solutions.

mercury vacuum scale   A device that measures atmospheric pressure. See inches of mercury.

metering rod   In an internal combustion engine, the long, tapered metallic pin fitted to the main nozzle of a carburetor to measure or meter the amount of gasoline flowed by it at various speeds; in fuel injection systems, it is the tapered rod that regulates the fuel pressure that reaches the injectors in a constant injection system. Also called a metering pin.

methane   A colorless, odorless, flammable gas created by anaerobic fermentation of organic compounds by bacteria; the major component of natural gas.

methane digester   A simple biological system that limits access to oxygen, causing microbes in waste to generate methane and carbon dioxide. Also called an anaerobic digester (AD).

methane hydrates   A form of water ice that contains a large amount of methane within its crystal structure; once thought to occur only in the outer regions of the solar system, but very large deposits have been found under sediments on the Earth's ocean floor. Also called methane clathrate or methane ice.

methanol   A toxic, colorless, volatile, flammable liquid alcohol; formerly made by destructive distillation of wood; now primarily made by oxidizing methane. Also called methyl alcohol or wood alcohol.

methanobacters Methane-making, anaerobic bacteria widely regarded as the oldest form of life on Earth; can be technologically exploited with processes that rely on the exclusion of oxygen.

methionine An essential amino acid found in most proteins; an essential nutrient in the diet of vertebrates.

methyl tertiary butyl ether (MTBE) A volatile, flammable, colorless liquid that was originally used in gasoline as an octane enhancer and lead substitute; more recently used to reduce engine exhaust emissions.

microflora Microscopic plants; the plants of a microhabitat, including in the digestive tracts of ruminent animals.

micro-plant An easy-to-run alcohol production system that uses one tank for cooking, fermentation, and distillation; only capable of distilling alcohol every third or fourth day, since the tank is tied up as a fermentation vessel for several days.

micro-still A vessel serving as a distillery, both cooker and fermenter.

middle cut The second cut of the output from the still, containing mainly pure alcohol; the best part of the distillation, obtained after the foreshot. Also known as the heart of the distillation.

mild steel Common, plain steel that rusts, containing 0.05%-0.25% carbon, making it strong but low in resilience and elasticity.

milling A mechanical grinding of grain or feedstock into pieces small enough to make the starch accessible to the chemicals and enzymes used in cooking, conversion, and fermentation.

mip (male internal pipe thread) Fittings with the threads on the outside surface. Contrast with fip (female internal pipe thread).

MMT Methylcyclopentadienyl manganese tricarbonyl; a manganese-based chemical compound used as a gasoline additive in marine engines and in gasoline in some states, to control low-melting-point ash deposits.

molecular sieve A zeolite containing tiny pores, used as a filter that lets some molecules pass through, but prevents others; offers a very high surface area for adsorption of molecules; can remove the last bits of water from alcohol.

molecule A group of atoms bonded together into the smallest particle of matter that is the same chemically as the whole mass.

monoculture The cultivation of a single crop on a farm, or in a region or country.

motor-speed controller An electronic device that lets an electric motor operate outside its normal fixed revolutions per minute without damage.

mulch A protective covering of substances, e.g., leaves, clippings, plastic sheeting, straw, sawdust, or bark, that is spread on the surface of the soil around plants to protect the soil and the roots of the plants from moisture evaporation and freezing, and to control weeds.

multi-port fuel injection A process where fuel is injected into the intake port upstream of the cylinder's intake valve, instead of at a central point within an intake manifold. Also called multi-point fuel injection.

mutagen An agent that can induce or increase the frequency of mutation in an organism, e.g., a chemical, ultraviolet light, or radiation.

mycelium [plural mycelia] The part of a fungus that consists of a network of threadlike filaments that absorb nutrients; exists below ground, or within another substrate.

mycorrhizae Fungi that form mutually beneficial relationships with the roots of higher plants. Plants support fungi by providing sugar and a hospitable environment; fungi support plants by providing increased surface area for water uptake and by selectively absorbing essential minerals.

mycorrhizal symbiosis Interdependent relationship in which the host plant receives mineral nutrients while the fungus obtains photosynthetically derived carbon compounds.

natural gas condensate (NGC) A low-density mixture of hydrocarbon liquids present in raw natural gas.

neat ethanol Ethanol that is not diluted or mixed with anything else.

needle valve A valve closed by a thin tapering part fitting into a conical seat, used to regulate the flow of a liquid or gas.

neoprene A synthetic polymer resembling rubber, resistant to oil and heat, used in fuel systems.

net metering A system where renewable energy generators, e.g., solar panels, are connected to a public-utility power grid so as to transfer surplus power to the grid; lets customers offset the cost of power from the utility.

net primary productivity (NPP) The production of organic compounds from carbon dioxide, mostly through photosynthesis, excluding the energy used for respiration; the rate at which a biosystem accumulates energy or biomass. Worldwide, NPP represents the photosynthesis of land plants, waterborne algae, and plankton.

nipple A short section of pipe with a screw thread at each end for coupling.

nitrile elastomer A synthetic rubber copolymer with a high resistance to oil, fuel, and chemicals; used for gloves, sealants, footwear, and extreme automotive applications.

nitrogen-fixing The chemical processes by which atmospheric nitrogen is assimilated into organic compounds, esp. by certain microorganisms as part of the nitrogen cycle.

nitrogen oxides (NOx) Highly reactive gases, containing nitrogen and oxygen, which form when fuel is burned at high temperatures; primary sources are motor vehicles, electric utilities, and other industrial, commercial, and residential sources that burn fuels.

nonphytate phosphorus Phosphorus with improved bioavailability, as in DDGS. Contrast with phosphorus stored in plant tissues in the form of phytic acid, which is not digestible by non-ruminant animals since they lack the digestive enzyme phytase; the unabsorbed phytate passes through the gastrointestinal tract, elevating the amount of phosphorus in the manure, leading to environmental problems.

octane A chemical compound with 18 isomers, one of which is used as the 100 point on the octane rating scale; a colorless flammable hydrocarbon of the alkane series obtained in petroleum refining.

octane number A rating which indicates the tendency to knock when a fuel is used in a standard internal combustion engine under standard conditions as compared to the knocking characteristics of pure chemical octane. Also called octane rating.

oil shale Any fine-grained sedimentary rock that contains enough organic matter to yield significant quantities of petroleum when heated.

on-board diagnostics (OBDII) A vehicle's self-diagnostic system; reports state of health information for the engine and various sub-systems; allows control of functions and diagnosis of problems.

open compressor A refrigerant compressor with a shaft or other moving part extending through its casing to be driven by an outside source of power, thus, requiring a shaft seal rubbing contact between fixed and moving part.

open loop Having no means of feedback, no way to compare the output with the input. In the case of automotive control systems, the ECU relies on stored default data rather than responding to real-time data input from sensors. Contrast with closed loop.

organophosphate A petroleum-based compound containing phosphorous, used in fertilizers, insecticides, and nerve gases.

osmotic pressure The pressure that would have to be exerted on a pure solvent to keep it from passing into a given solution via osmosis; a measurement used to express the concentration of a solution.

overpressure vent An opening that prevents excess air, gas, or liquid buildup in a confined space.

overstory The uppermost layer of foliage that forms a forest canopy.

oxygen sensor In an internal combustion engine, a device that measures the concentration of oxygen that remains in exhaust gas, and sends the information to the electronic control unit so it can adjust to maintain efficient combustion.

oxygenate standard The Clean Air Act's requirement that hydrocarbon fuel contain the addition of oxygen-containing fuel to reduce the carbon monoxide ouput of engines. At one time, MTBE was used to meet this standard, as well as ethanol. The oxygenate standard was lifted in 2005 in return for the establishment of a Renewable Fuel Standard.

oyster mushroom A widely distributed edible fungus, *Pleurotus ostreatus*, with a good-tasting, grayish-brown, fan-shaped cap and a very short (or no) stem.

packed column A distillery column filled with material, e.g., bronze wool, pall rings, or lathe chips, that gives alcohol/water vapor places to condense on and revaporize; also used in cleaning gases of contaminants in conjunction with a downward flow of water or other solution.

packed column still A distillery that uses a column which contains materials that help dissipate heat, allowing a gradual, continuous rise in proof and lowering of boiling point as you go higher; vapor is condensed and revaporized thousands of times on its way to the top of the column.

packing gland The metal part that compresses and holds packing in place in a stuffing box, commonly used for waterproofing a propeller shaft.

pall ring A specially shaped steel ring used as packing for distillation columns.

particulate (mash) Containing distinct particles.

particulate emissions Tiny particles of solid (a smoke) or liquid (an aerosol) suspended in a gas, varying in size from a few molecules up to the size where particles can no longer be carried by the gas. Also called particulate matter.

Peak Oil A theory that states that after the high point of Earth's oil production has been reached, the rate of oil production enters a terminal decline. Also called the Hubbert Peak theory.

peaker A power plant that runs only when there is a high demand (peak demand) for electricity, e.g., during afternoons in the summer when air conditioning use is high.

pectin The jelly-like carbohydrate similar to starch that helps cement plant cells together and in which cell wall components are embedded.

pectinase Enzymes that break down pectin into simple carbohydrates, used to speed extraction of fruit juices.

perennial A plant living three or more years.

perforated plate column still A distillery using a column containing plates with holes in them; water and alcohol vapors rise through holes in the plates and are cooled by liquid flowing across the plates; the alcohol remains a vapor and rises, while the water is cooled to a liquid and sinks through downcomers.

permaculture (from "permanent agriculture" and "permanent culture") Agricultural ecosystems that are sustainable; also, the design of systems that create sustainable human habitats (e.g., food production, housing, community development) by learning from Nature's patterns.

permeation emissions The amount of evaporative emissions that gets through flexible fuel lines.

pesticide A substance used to destroy insects or other creatures considered harmful to plants or animals.

pH (from "potential of hydrogen") A measure of the acidity or alkalinity of a solution, where 7 is neutral; lower numbers are acid; higher numbers are alkaline.

phase-change energy The energy used to cause transitions between solid, liquid, and gaseous phases of matter, due to temperature or pressure. See latent heat of vaporization.

phase separation A condition in which a mixture separates out, such as alcohol and water separating from gasoline in a fuel tank, in distinct layers.

phosphoric acid A chemical compound made up of phosphorus, oxygen, and hydrogen; used in making fertilizers and soaps and in food processing.

photochemical smog Air pollution produced by the action of sunlight on hydrocarbons, nitrogen oxides, and other pollutants in the atmosphere.

photosaturation The point at which a plant's photosynthetic process shuts down because of excess sunlight. Also called solar saturation.

pin-lug fitting A quick-release type of coupling, usually with two pins arranged so that when one connector is inserted and twisted, its pins lock into the other connector; typically used for fire hoses.

pinging See pre-ignition.

piston A disk or short cylinder that fits closely within a tube in which it moves up and down against a liquid or gas; used in an internal combustion engine or in a pump to create motion.

pitching solution A solution of wort and yeast used to inoculate a batch of mash, which is started hours before being added to the main batch; contains rapidly reproducing yeast at a high concentration.

pith The spongy central cylinder of the stems of higher plants, e.g., the soft core near the center of a tree trunk.

points Mechanical switches for opening and closing electrical circuits; each of a set of electrical contacts in the distributor of a motor vehicle. The purpose of the contact breaker is to interrupt the current flowing in the primary circuit of the ignition coil. When this occurs, the collapsing current induces high voltage in the secondary winding of the coil, which has very many more turns. This causes a very large voltage to appear at the coil output for a short period—enough to arc across the electrodes of a spark plug. Also called breaker points.

polar liquid A liquid with an unequal charge distribution (dipole moment), e.g., water.

polar solvent A substance, generally liquid, in which other materials dissolve to form a solution, which substance has an unequal charge distribution (dipole moment), e.g., water.

polyculture Growing several crops in the same space, in imitation of natural ecosystems, avoiding large stands of single crops.

polycyclic organic materials (POM) A broad class of compounds formed mainly from combustion; present in the air as particulates; emitted from cigarette smoke, vehicle exhaust, wood burning, etc.; probable human carcinogens.

polyethylene (in the UK, polythene) A tough, light, flexible resin made by polymerizing ethylene; used for plastic bags, food containers, and other packaging.

polyol ester The product of a high-temperature reaction of an organic fatty acid with a polyhydric alcohol; added to oils; fortified with antioxidants, corrosion inhibitors, metal deactivators, and viscosity improvers to produce high-pressure, high-performance hydraulic fluid, synthetic oil, and lubricants. Also see ester.

polyolefin plastics The largest group of thermoplastics (become moldable when heated and become hardened when cooled, repeatedly), e.g., polyethylene and polypropylene; characterized by flexibility, toughness, and chemical resistance; have low cost and a wide range of uses.

porting Enlarging and reshaping an engine's intake and exhaust systems (ports) for enhanced aerodynamic flow, so that more air/fuel mixture can smoothly enter the compression chamber.

positive crankcase ventilation (PCV) A system to remove harmful vapors from the engine and to keep those vapors from being expelled into the atmosphere; uses manifold vacuum to draw harmful gases from the crankcase into the intake manifold, so they are carried with incoming air/fuel back into the combustion chamber, and burned.

positive displacement pump A pump that causes a fluid to move by trapping a volume of the liquid, raising its pressure, and then forcing (displacing) that trapped volume into a receiving pipe.

potentiometer A device for measuring an electromotive (producing or tending to produce an electric current) force.

pound of steam A pound of water, in the vapor phase; not to be confused with steam pressure, expressed in pounds per square inch (psi).

pounds per square inch (psi) A measure referring to the pressure exerted on one square inch of an object's surface.

power valve A carburetor device that opens when the engine is under load (e.g., accelerating), making the air/fuel mixture richer.

preignition   Premature combustion of the air/fuel mixture in an internal combustion engine; occurs when the air/fuel mixture in the cylinder has been ignited by the spark plug, but the unburned mixture in the combustion chamber explodes before the flame front can reach it, causing combustion to occur before the optimum moment. The resulting shockwave reverberates in the combustion chamber, creating a metallic "pinging" sound, and pressures increase; commonly caused by hot carbon deposits. Similar to pinging and knocking.

premalting   Adding the enzyme alpha amylase to keep the wort from becoming too thick to be stirred during hydrolysis (cooking with water), by breaking down starch into maltose and dextrin.

pressure drop   The difference in pressure between two points in a system, caused by resistance to flow, e.g., the decrease in pressure from one point in a pipe or tube to another point downstream, usually the result of friction of the water against the tube.

pressure loss   See pressure drop.

pressure regulator   A device used to maintain system pressure within a specified range.

pressure relief valve   A device used to limit pressure in a system by letting the pressurized fluid flow out of the pressure vessel through an auxiliary passage, e.g., a radiator cap.

pressure swing vacuum regeneration   A process used to separate alcohol and water with molecular sieves. The water-saturated sieve material is regenerated for continued use by rapidly evacuating the sieve chamber, lowering the boiling point, and using the residual heat to evaporate the water from the adsorbant without addition of heat.

prime   When first using a pump, to remove the air from the pump, displacing it with fluid.

prion   A protein particle similar to a virus but lacking nucleic acid; believed to be the cause of brain diseases; highly resistant to destruction.

process energy   Energy needed in any form for manufacturing a product.

progressive cavity pump   A kind of pump that moves fluid using a sequence of small cavities, which move as the pump rotor rotates.

proline   An amino acid that is a constituent of most proteins.

proof   A measure of distilled alcohol content; 1% equals 2 proof.

proof gallon   One gallon of 100-proof alcohol (50% alcohol by volume); the U.S. government taxes alcohol by the proof gallon.

propanol   A highly flammable liquid alcohol used as a solvent and antiseptic. Also called propyl alcohol.

protease   An enzyme that breaks peptide bonds between amino acids of proteins.

protein   An organic compound that consists of large molecules made of one or more long chains of amino acids; an essential part of all living organisms, as a structural component of muscle, hair, collagen, etc., and as enzymes and antibodies.

puddling   The condensing of fuel out of the airstream onto the cool manifold walls.

pulse width   The time interval between the leading edge and trailing edge of a pulse at the point where the amplitude is a particular fraction (e.g., 50%) of the peak value. In fuel injection, the amount of time an injector remains open, often expressed as a percentage of open full time (100%).

pump chamber   The area inside the pump housing where material to be pumped gathers before being pumped outside of the housing.

rail   The common fuel supply line that feeds each of the individual fuel lines leading to each fuel injector.

ratoon   A shoot sprouting from the base of a plant, e.g., banana, pineapple, or sugar cane; to propagate from ratoons in a system used probably since humans first noticed that regrowth of new shoots followed the cuttings of certain crops at harvest, producing a new crop without replanting.

reboiler   A heat exchanger that boils the liquid from the bottom of a distillation column to create vapors, which are then returned to the column, to drive the distillation.

rectifier   The upper portion of a distillation column, where rising vapor is enriched by interaction with a countercurrent falling stream of condensed liquid, which leaves the bottom of the rectifier for the stripper or alternatively the rectifier reboiler.

refine   A process by which a raw material is taken through additional steps or processes to become one or more finished products.

reflux   The process of boiling a liquid so that any vapor produced is liquefied and returned. In distillation, where a portion of the finished distillate is returned to the column as part of the distillation control system and to subject the distillate to additional distillation.

reflux ratio   The ratio of the amount of alcohol condensate being refluxed to the amount being withdrawn as product.

reformer   A high-temperature catalytic device for reducing a fuel to hydrogen and other components, for use in an ICE or fuel cell.

refractometer   A device used to determine a substance's refractive index, or a property related to its refractive index, e.g., used to measure a solution's sugar content by measuring the bend of light through a droplet of filtered mash.

refractory   Resistant to heat. See castable refractory.

Reid Vapor Pressure (RVP)   A measure, expressed in pounds per square inch, of the vapor pressure of gasoline, voltaile crude oil, and other volatile petroleum products.

Renewable Fuel Standard (RFS) A U.S. Environmental Protection Agency program designed to significantly increase the volume of renewable fuel that is blended into gasoline.

renewable resources Renewable energy; resources that are not depleted by their use, e.g., solar or wind energy, energy from growing plants.

revenuer A derogatory term used by producers of unregistered beverage alcohol to refer to a U.S. government agent who collects the excise tax on beverages; dates back to shortly after the Whiskey Rebellion of the 1790's.

rhizome A horizontal underground plant stem that can produce both the upward shoot and the downward root system of a new plant.

rhizosphere The soil region around plant roots where chemistry and microbiology are influenced by the roots' growth, respiration, and nutrient exchange.

rich mixture A mixture that contains too much fuel in relation to air to combust stoichiometrically.

riser A pipe leading upwards.

road base A mixture of gravel or recycled concrete with specific proportions of different-sized rock, designed to compact densely when used a a road surface.

rocket stove A cooking device that burns sticks; typically made from large cans, heavy pipe, or concrete; easy to construct, using very low-cost materials; about twice as efficient, and much cleaner, than open-fire cooking methods.

roundstock Raw material purchased from metal manufacturers in the form of rods without detail.

rpm gauge See tachometer.

rumen The first stomach of a ruminant, where it partially digests food or cud with the help of bacteria, before eventually passing it on to the second stomach.

ruminant A hoofed mammal with either two or four toes on each foot, that chews the cud regurgitated from its rumen, e.g., cattle, sheep, antelope, deer, and giraffes.

run The length of time it takes to complete one distillation; sometimes used to mean a single distillation.

running octane value A phrase coined by the author synonymous with the formula of Research Octane plus Motor Octane divided by 2. Contrast with blending octane value.

RV cams Camshafts that were originally designed for use in heavy vehicle and towing applications; they give more torque earlier in the stroke, producing a lot of torque at low rpm, such as when you're just starting off pulling an RV.

saccharification The process of hydrolyzing a complex carbohydrate into a simpler soluble fermentable sugar, such as glucose or maltose. Also called conversion.

saccharometer A type of hydrometer that measures the amount of sugar in a solution.

saccharomyces A class of single-cell yeasts that selectively consume simple sugars.

Saccharomyces cerevisiae Common distiller's yeast, an important ingredient used in making bread, wine, and beer.

sandwich adapter A device installed between the oil filter landing on the engine block and the filter to route the oil to an oil cooler, oil pressure sensor, fuel heater, or oil temperature gauge.

Saybolt Seconds Universal (SSU) A measure of viscosity for a wide variety of substances, including used for thick, sticky oils; the time in seconds for 60 milliliters of oil to flow through a standard capillary tube at a given temperature.

screen cylinder A filter or grating made from wire cloth or perforated metal, with parallel sides and a circular section.

screw press A device used to squeeze out moisture from plant pulp, with pressure exerted by means of a screw mechanism.

scrubber A filter that removes particles and contaminants from the air or other gases.

scrubbing pads A grocery item used for cleaning pots and pans which can be used as packing on small-diameter distillery columns.

separator A device that uses centrifugal force to separate the solid from the liquid portion of the mash, or to remove solids before fermentation if you are performing an all-liquid fermentation.

sequester To form a stable compound with an ion, atom, or molecule, so that it is no longer available for reactions. In the case of greenhouse gases, to convert them to solid form from their gaseous form.

severe reforming The conversion of natural gas to carbon dioxde and hydrogen by the use of energy-intensive, very high-pressure/high-temperature steam.

shaker table A device that creates a mechanical vibration to separate materials, and to move materials to a collection area. Also called a shake table.

sheave A wheel with a grooved rim to guide a rope, belt, or cable, as in a pulley block.

shell and tube heat exchanger A type of heat exchanger common in oil refineries and other large chemical processes consisting of a shell (a large vessel) with a bundle of tubes inside it.

shiitake mushroom An edible fungus that is an excellent source of amino acids, vegetable proteins, iron, thiamine, riboflavin, niacin, and vitamins B6 and B12; is considered a "white rot" fungus, capable of feeding on both cellulose and lignin.

side-draft Describes a carburetor that sucks air in from the side instead of from the top.

sight tube A small window or tube that allows observation or measuring of material moving through a system. Also called a sight gauge.

silage Grass or other green fodder that is compacted and stored in airtight conditions, usually in a silo, without first being dried, then used as animal feed in the winter.

simple sugar See fermentable sugar.

simultaneous saccharification and fermentation (SSF) A process that eliminates saccharification as a separate step of fermentation; because of new yeasts and enzymes, processing can move directly from liquefaction to fermentation while simultaneously saccharifying the mash, saving a step and helping eliminate contamination problems.

single-cell protein (SCP) A protein extracted from algae, fungi, yeasts, or bacteria that are grown in a nutrient solution; used as a substitute for protein-rich foods, especially in animal feeds.

single-phase The kind of power that comes out of a normal home electrical outlet, having an electric circuit that consists of one alternating current (typically 120V AC or 230V AC, depending on the country) that is carried between two wires (live and neutral), and sometimes a third ground wire for safety.

six-carbon (C6) sugar A monosaccharide (simple sugar) with six carbons in its backbone, e.g., glucose or fructose.

slag box Used to prevent particulate matter from plugging moonshine distillery condensers.

slurry A liquid containing suspended solids, e.g., milled corn in water.

slurrying Mixing dry feedstock with water prior to cooking.

small plant An alcohol production facility that produces 15,000–75,000 gallons per year, the next larger size from a micro-plant. Sometimes called a mini-plant.

small producer tax credit A U.S. Department of Energy credit that can be set against tax liability to benefit small producers of alcohol; consists of a tax credit of 10 cents per gallon for up to 60 million gallons per year.

solar-hydrogen-fuel-cell-engine A combination of several different technologies, whereby hydrogen is produced by some solar-originating source of energy, such as photovoltaic panel electricity. The electricity is then used in electrolysis to split hydrogen from water. The hydrogen is then used to produce electricity and surplus heat in a fuel cell. Finally, this electricity is used to drive an electric motor in a vehicle.

solar income The amount of solar energy that falls on an area as sunlight.

solenoid valve A valve that typically controls the flow of air, fuel, or water by opening and closing by running or stopping an electrical current through a coil of wire.

soybean meal (SBM) An excellent source of protein, produced by cracking, heating, and flaking soybeans; highly digestible, with high energy content; used extensively in feed for swine, cattle, poultry, and aquaculture.

sparge To agitate by introducing air or compressed gas into a liquid.

spark-ignition engine An internal combustion engine where the air/fuel mixture is ignited with a spark. Contrast with compression-ignition engine.

spark plug A device inserted in the head of an internal-combustion engine cylinder that ignites the fuel mixture by using an electric spark.

specially denatured alcohol (SDA) Alcohol that has been treated, usually with methanol, so that it can be used in, e.g., mouthwash or food flavorings, but that will make a person sick if they consume it for the alcohol content.

specific gravity The ratio of the density of a substance to the density of an equal volume of a standard: for liquids or solids, distilled water at 4°C.

spent grains The nonfermentable solids remaining after fermentation of a grain mash.

starch An odorless, tasteless white polysaccharide that occurs widely in plant tissue as storage for carbohydrates; an important constituent of the human diet; obtained mainly from cereals and potatoes.

stave A narrow strip of wood that helps to form a barrel.

still Short name for the distillery, the apparatus used in distillation.

stillage The residual mash after distillation.

still pot See mash pot.

stoichiometric ratio A mixture occurring when all of the fuel in a vehicle's combustion chamber is combined with all the free oxygen; it represents a chemically balanced air-fuel ratio (AFR) mass ratio of air to fuel present during combustion, called the air-fuel ratio (AFR). When all the fuel combines with all the free oxygen, typically within the combustion chamber, the mixture is chemically balanced. AFR is an important measure for anti-pollution and performance tuning reasons.

stover   The leaves and stalks of corn, sorghum, or soybean plants that are left in a field after harvest; similar to straw, the residue left after a cereal grain or grass has been harvested at maturity for its seed.

stripper   The lower portion of a plate distillation column, where descending liquid is progressively depleted of its volatile components, by the introduction of heat at the base. Enriched alcohol vapor leaves the stripper to enter the rectifier.

strong acid hydrolysis   A method of breaking down cellulose, involving exposing it to a strong acid, generally at ambient pressure.

substrate   The material on which an organism grows or from which it obtains its nourishment.

sucrose   A crystalline disaccahride (two glucose molecules) carbohydrate found in many plants, e.g., sugarcane, sugar beets, and maple trees.

sugar alcohol   A hydrogenated class of carbohydrates, e.g., sorbitol, mannitol, and xylitol, that can replace table sugar for most uses, but which is more slowly or incompletely absorbed by the digestive system than sugars.

sulfuric acid   The most commonly used acid for correcting pH for enzyme and cooking purposes, as well as for neutralizing mash at the end of fermentation, before distillation; reacts violently when added to hot water. Also called oil of vitriol.

sump   A pit or depression into which liquid collects.

sump pump   A device placed inside a sump to remove the excess liquid that has collected there.

supercharger   A compressor, driven by belts or by an electric motor, that compresses the air entering an internal combustion engine. Also called a blower. Contrast with turbocharger.

supercritical steam   Water/steam at such a pressure that there is no clear distinction between its fluid and gaseous phases, i.e., it is a homogeneous liquid. Thus, in energy generation, steam and water do not have to be separated out; they can be processed together.

swale   A ditch on the contour, which holds water, allowing it to infiltrate the downslope soil; also catches soil, becoming a fertile area.

tachometer   A device that measures the working speed of an engine, typically in revolutions per minute of the main shaft. Also called an rpm gauge.

tang   A corkscrewed insert used to create turbulence and slow down flow, to increase contact time for more effective heat exchange. Also called a turbulator.

tank head   An opening for access to the interior of a tank, which can be sealed tightly while the tank is in use; may contain venting, or vacuum-relieving apparatus.

tar sands   A stratum of sand or sandstone containing petroleum precursors; to avoid public disclosure of Peak Oil, governments have improperly redefined "conventional oil" to include both tar sands and natural gas concentrates.

tax credit   An amount of money that can be offset against a tax liability dollar for dollar.

tax deduction   A percentage amount allowed to reduce taxable income.

terneplate   Sheet iron or steel plated with an alloy of three or four parts of lead to one part of tin, used as a gas tank coating in old vehicles.

tetraethyl lead   A colorless, poisonous, oily liquid, used as an antiknock agent in gasoline.

tetrahydrofurfural (tetrahydrofurfuryl) nitrate   A product of preparing tetrahydrofurfuryl alcohol (THFA) with a nitrate; used as a cetane improver so that ethanol will combust properly in a diesel engine.

thermal efficiency   A measure of productive use of energy of a thermal device, e.g., an internal-combustion engine, a boiler, or a heat pump, expressed as a percentage. The input to the device is heat; the output is mechanical work, or heat, or both. Sometimes called energy efficiency.

thermal mass   A large amount of material that absorbs and stores heat during periods when surplus heat is available and releases the heat during periods when it is needed, e.g., sunlight falling on an interior masonry wall during the day, which radiates heat at night.

thermocouple probe   A temperature-sensing device that takes advantage of the properties of certain metals that produce a predictable voltage when heated. This volatage output can be used by various devices to operate equipment based on temperature.

thermophilic   Describes a thermophile: a bacterium or other microorganism capable of growing and surviving at higher-than-normal temperatures.

thin stillage   The water-soluble portion of a fermented mash plus the mashing water; stillage with the solids removed. Also called distiller's solubles.

three-phase   An electric circuit that consists of three separate currents delivered at one-third cycle intervals by means of a three-wire circuit; typically used to power motors that operate at 200 volts or higher.

throttle   A valve that directly controls how much air enters the engine; often a butterfly valve; located, on fuel-injection engines, in the throttle body; or located in the carburetor. Also used to refer to any device that regulates the power or speed of an engine.

throttle body fuel injection (TBI)   A type of electronic fuel injection system that resembles a carburetor, except that there is no fuel bowl, float, or metering jets; one or a pair of injector(s) spray fuel directly into the throttle bore(s).

throttle linkage   See linkage.

throttle plate A butterfly valve inside the throttle body of a carburetor that regulates the amount of air/fuel entering the engine.

tilapia A spiny-finned freshwater fish of the family Cichlidae, native to Africa; economically important as food, in native regions and where introduced or grown on fish farms; increasingly important in aquaculture worldwife. Tilapias have a pleasant, mild-tasting flesh.

timing light A stroboscope used to set the ignition timing of an internal combustion engine.

timing tape A metallic or paper strip with calibrated markings, attached with adhesive to the balancer or pulley on an engine to read the tape against the zero mark on the timing plate.

tipping fee A charge to take a quantity of waste off your hands, usually assessed by a waste facility.

toluene A colorless, liquid, carcinogenic hydrocarbon present in coal tar and petroleum and used as an octane booster in gasoline.

top dead center The position of an engine piston and its crankshaft arm when at the top or outer end of its stroke.

torque A turning or twisting force that tends to cause rotation; the measure of a force's tendency to produce torsion and rotation about an axis.

torque curves A mathematic model of acceleration; a car will accelerate hardest at its torque curve peak in any particular gear, and will not accelerate as hard below or above that peak.

transmission An assembly of gears and related parts which transmits power from the engine to a driving axle. Also called a gearbox.

triethylene glycol dinitrate (TEGDN) A nitrated ester chemically similar to nitroglycerin, but much more stable; used as a cetane improver for straight alcohol being used as a diesel fuel.

turbocharger A compressor driven by exhaust gases, used to increase the power output of an internal combustion engine by compressing air that is entering the engine.

turkey tails A common woodland fungi; its name comes from the banding pattern on mushroom bodies that look like the tail of a turkey.

understory The plants that grow beneath a forest canopy.

urea proteins Soluble nitrogenous organic compounds denatured by urea (carbamide).

urethane A synthetic crystalline compound used in making pesticides and fungicides; also used as the base of polymeric plastics, e.g., polyurethane.

vacuum advance A mechanism that allows the distributor to supply a more optimum spark timing proportional to the load and speed output.

vacuum breaker A device used to relieve a vacuum formed in a supply line or tank to prevent backflow or tank collapse.

vacuum distillation The separation of liquids under reduced vapor pressure with or without heating; reduces the boiling points of the liquids being separated; used when liquids to be distilled have high atmospheric boiling points. Sometimes referred to as low temperature distillation.

vacuum distillery A device utilizing reduced pressure above a solution to be distilled to evaporate the volatile liquid.

vacuum gauge A gauge for measuring negative atmospheric pressure.

vacuum vent An opening that lets air, gas, or liquid pass into or out of a confined space in order to avoid pressure-change damage to a tank.

valve A device that regulates the flow of substances by opening, closing, or partially blocking various passageways.

vane A thin, rigid, flat surface that is attached to a rotator, and pushes against air or fluid.

vapor lock An interruption in flow through a fuel line caused by vaporization; fuel boiling to bubbles blocks the flow before it gets to the fuel pump.

vaporize To change from a liquid or a solid to a vapor, as in heating water to steam.

vapor pressure A measure of the evaporative nature of a fuel; the pressure of a gaseous substance in equilibrium with its liquid or solid form.

vaporized alcohol engine An internal combustion engine that heats and releases liquid alcohol into an expansion chamber, causing it to vaporize, then mixes it with air for proper combustion.

venturi A narrowed passage in the carburetor's air inlet that causes airflow to speed up and air pressure to drop, resulting in atomized fuel being drawn out of the carburetor bowl.

vinasse The residue left in a still after distilling sugarcane mash.

viscosity The state of a liquid or gas being thick, sticky, and semifluid in consistency, because of internal friction between the molecules; the measure of a fluid's resistance to flowing.

volatile Describes a substance that evaporates easily at normal temperatures.

Volumetric Ethanol Excise Tax Credit (VEETC) A U.S. tax credit for every gallon of pure ethanol blended into gasoline, refundable quarterly; an E-10 blend will have a credit of 51 cents per gallon, and E-85 will have a credit of 43.35 cents per gallon.

vortex A mass of whirling fluid or air.

w/w Means "by weight"; describes the concentration of a substance in a mixture or solution; e.g., 1% w/w means that the mass of the substance is 1% of the total mass of the mixture or solution. Also called g/g (metric).

wastegate A valve in a turbocharger that regulates the pressure at which exhaust gases pass to the turbine by opening or closing a vent to the exterior. The primary function of the wastegate is to regulate boost pressure in turbocharger systems and to protect the engine and turbocharger.

water column A measure of gas pressure. Most gas appliances operate at 11 inches water column while propane appliances operate at 7 inches of water column.

watt A unit of power, equivalent to one joule per second; the power in an electric circuit where the potential difference is one volt and the current is one ampere.

weak acid hydrolysis A process to break down cellulose using a reactor, high temperature, high pressure, and weak sulfuric acid.

weir A low dam; in distillation, a low barrier on a continuous distillation plate that lets thick mash overflow without clogging perforation or bubble caps.

wet distiller's grains Solids collected after distillation and filtering of a grain-based mash; generally contains approximately 75% water.

whey The liquid portion of milk that remains when protein, or casein, coagulates to become curd; the leftover liquid from the making of cheese. Contains protein and approximately 5% lactose.

whole stillage The undried "bottoms" from the beer well made up of nonfermentable solids, distiller's solubles, and the mashing water.

wide-band sensor A device that measures an engine's air/fuel ratio directly, instead of switching back and forth from rich to lean like conventional exhaust gas oxygen sensors.

wide-open throttle (WOT) An internal combustion engine's maximum intake of air and fuel; occurs when the accelerator is pressed down all the way so that the throttle plates inside the carburetor or throttle body are "wide open," providing the least resistance to the incoming air.

windrow A long narrow heap of compost, originally used to describe hay drying.

wine gallon Aproximately 3.79 liters, the same size as a U.S. gallon.

wiring harness An array of insulated conductors bound together in a specific arrangement suitable for use only in equipment for which the harness was designed; it may include electrical connectors.

wood alcohol See methanol.

wort In alcohol-making, the grain and water mixture created by mashing and boiling the feedstock, after yeast is added, it's called a mash.

xerophyte A plant that needs very little water.

xylene A volatile liquid hydrocarbon obtained by distilling wood, coal tar, or petroleum; carcinogenic; used as an octane booster in gasoline.

xylitol A five-carbon sugar alcohol derived from xylose; used as a sweetener in regular and diabetic foods, and in dyeing and tanning. Also called wood sugar.

xylose A simple five-carbon sugar extracted from cellulosic feedstock, e.g., wood or straw. Hemicellulose is converted into soluble sugars such as xylose, which are fermented into ethanol or methane.

yeast A microscopic fungus consisting of single oval cells that reproduce by budding, that can change sugar into alcohol and carbon dioxide by fermentation.

yeast breeder A fermenter specifically designed to rapidly multiply yeast and maintain them in their aerobic reproductive form before addition to wort.

Yeoman plow A soil tillage device designed to gently lift and loosen beneath the soil's surface, without disturbing the surface much..

zeolite A mineral with a lattice structure arranged to form a honeycomb of interconnecting channels and pores, used to attract and trap odors and gases in molecular sieves.

431

THE *ALCOHOL CAN BE A GAS!* BOOK AND DVD

Have you borrowed this book so many times that your friends or the librarian are about to cut you off? Maybe your friends are using your DVD or book more than you are? Can't get your copy back from your mechanic? Is one of your bonehead legislators in need of some clear thinking about ethanol? How about the person writing editorials in your local paper? Is your extension agent calling you all the time to look something up for him in your book? Want to make a powerful gift to your Future Farmers of America chapter? The answer to all these questions is simple—it's time to buy a book or DVD!

THE *ALCOHOL CAN BE A GAS!* T-SHIRT

This classy shirt is what all the stylish revolutionaries wear. Guaranteed to start conversations at parties, on the street, and with your friends. Printed on undyed organic cotton with water-based ink to ensure that your shirt purchase doesn't end up in MegaOilron coffers. No polyester, no pesticides, no herbicides, and no chemical fertilizers are used in production. Manufacturer guarantees no sweatshop labor is used to make the shirts. Proceeds go to fund alcohol fuel education. Both women's and men's styles available.

ORDER ONLINE AT

<WWW.PERMACULTURE.COM>,

OR USE THE ORDER FORM ON

THE REVERSE OF THIS PAGE.

T-Shirt Front

T-Shirt Back

QUICK ORDER FORM

(Feel free to photocopy.)

iiEA

**The International Institute
for Ecological Agriculture**

You can use this quick order form to get Alcohol Can Be a Gas! books, DVDs, and T-shirts, if you don't want to bother with going online to our electronic shopping cart. But online ordering will give you faster options for shipping if that's important to you and you can use PayPal, if you prefer, instead of the credit cards listed below. Online orders typically go out within 48 hours. For online orders, go to <www.permaculture.com>.

But if you would like to order without using our online shopping cart, here are your choices:

Email orders: Send your own hand-typed order containing the information below to <ourstore@permaculture.com>.

Telephone orders: Call 1-888-PERMACUlture (888-737-6228).

Postal orders: Send to 309 Cedar St. #127, Santa Cruz, CA 95060. In addition to credit cards, we accept checks, money orders, or well-concealed cash (at your own risk) on mailed-in orders.

International orders by mail: Email or write us for the postage & handling amount for your order.

PLEASE SEND THE FOLLOWING ITEMS:

Be sure to specify quantity. For T-shirts, circle the size and Men's or Women's style.

QTY	ITEM	SIZE/STYLE	PRICE	P & H
	Alcohol Can Be a Gas! — Hardcover		59.00	8.00
	Alcohol Can Be a Gas! — Softcover		47.00	7.00
	Alcohol Can Be a Gas! — DVD (2 hrs. 40 min.)		20.00	5.00
	Alcohol Can Be a Gas! — T-Shirt	Men S M L X L Women S M L	20.00	5.00
	Alcohol Can Be a Gas! — T-Shirt	Men S M L X L Women S M L	20.00	5.00
	Alcohol Can Be a Gas! — T-Shirt	Men S M L X L Women S M L	20.00	5.00
		TOTAL		

Orders shipped to California addresses must have sales tax added (7.75% on total price, not including P&H).

SHIP TO:

SHIPPING LOCATION PHONE (with area code) _____

NAME _____ STREET _____

CITY _____ STATE _____ ZIP _____ COUNTRY _____

PAYMENT:

EMAIL ADDRESS **(necessary to get your free newsletter; helpful for any order problems)** _____

❏ CHECK ❏ MONEY ORDER ❏ CASH ❏ VISA ❏ MASTERCARD ❏ AMEX ❏ DISCOVER

CARD # _____ EXPIR. DATE _____ SECURITY CODE _____
(from back of card)

NAME ON CARD _____

BILLING ADDRESS:

(where you receive your credit card statement) ❏ **SAME AS SHIPPING ADDRESS**

NAME _____ STREET _____

CITY _____ STATE _____ ZIP _____ COUNTRY _____

GET INVOLVED

HOW YOU CAN HELP SPREAD THE WORD

By now, you realize that the design and implementation of an energy revolution using small-scale ethanol production is simple common sense—and at the same time a radical departure from the way corporations and government currently do things. You know that MegaOilron historically stops at nothing to make sure that the public perception of ethanol is tightly managed. But you now know the truth and can't be swayed by their propaganda. It is truly up to us citizens to make the change.

In the 1970s and '80s, I had little trouble making myself heard in the media. I made over 750 radio television and print appearances on the subject of ethanol—and that was without a book published! At the time, there were still vestiges of a diverse free press.

Well, times have changed, and with the corporatization and concentration of the media, getting the message of an ethanol revolution into the public sphere has become next to impossible. While promoting a workshop in Fall 2007 in Chicago, I was unable to get even one radio, television, or newspaper appearance despite literally hundreds of person-hours of work by the media team sponsoring the workshop. A complete and total wall has been erected against any positive stories about ethanol, built brick by brick in a months-long relentless campaign by the American Petroleum Institute.

We need to depend on people like yourself to get the word out on a grassroots level, person to person, by email, blogs, letters to the editor, or by lobbying your local librarian or hosting a gathering at your home to get out the word.

This time around, the revolution will not be televised. This time, the revolution starts with a whisper. It's time we share what we know, organize to bring it about—and then we win. On the next page are some ways to circumvent the wall of silence, to pull the public discourse on ethanol back, away from MegaOilron, and bring about the solar-based future we need to survive as a species.

WITH THE CORPORATIZATION AND CONCENTRATION OF THE MEDIA, GETTING THE MESSAGE OF AN ETHANOL REVOLUTION INTO THE PUBLIC SPHERE HAS BECOME NEXT TO IMPOSSIBLE. A COMPLETE AND TOTAL WALL HAS BEEN ERECTED AGAINST ANY POSITIVE STORIES ABOUT ETHANOL, BUILT BRICK BY BRICK IN A MONTHS-LONG RELENTLESS CAMPAIGN BY THE AMERICAN PETROLEUM INSTITUTE. WE NEED TO DEPEND ON PEOPLE LIKE YOURSELF TO GET THE WORD OUT ON A GRASSROOTS LEVEL, PERSON TO PERSON. IT'S TIME WE SHARE WHAT WE KNOW, ORGANIZE TO BRING IT ABOUT— AND THEN WE WIN.

BECOME AN AFFILIATE

- Make money while helping to fuel the ethanol revolution, by becoming an *Alcohol Can Be a Gas!* internet affiliate.
- You get an <alcoholcanbeagas.com> link, which has a code that identifies you. When people click on the link—which you provide them in emails, blogs, on your site, etc.—and they buy something on our site, you get 10%.
- Affiliates get early notices of any new interviews or articles that Dave Blume writes, to send out to their network.

TELL PEOPLE ABOUT US

- Download and distribute flyers and fact sheets from our website: for your bookstore, library, bulletin boards, friends, and associates.
- Email your friends to tell them about the book and our website (and you'll earn affiliate $$ on their purchases from us).
- Link from your website to our website (earn affiliate $$).
- Say nice things about us on your blog (earn affiliate $$).
- Check out the press room on the *Alcohol Can Be a Gas!* website. Share the video and audio clips there (earn affiliate $$).
- Write letters to the editors of newspapers and magazines.
- Post reviews and rate us on websites for Amazon, Barnes and Noble, Powell's, etc.
- Post reviews and rate our videos on YouTube and Google Video. Send emails to the sites asking that the videos be featured.

HOST A HOME THEATER PARTY

- Show the *Alcohol Can Be a Gas!* DVD to a gathering of your friends or other interested folks.
- Share collateral materials we'll provide.
- Arrange for a live call-in question and answer session with Dave Blume after the presentation.

LET US HEAR FROM YOU

- Let us know what you think of *Alcohol Can Be a Gas!* We want your feedback.
- We want to hear your alcohol-fuel-related ideas, plans, and stories and share them.
- We want to share your conversion, still-building, and byproduct business experiences for our newsletters and website. Each of these areas will get their own forum on our website.
- Let us know of people, organizations, reporters, publications, TV shows, anyone or anything you think might help get the word out.

HELP MAKE THE ETHANOL DISCUSSION HONEST

- Bust the myths! (Download the flyer from our website.)
- Correct misinformation (disinformation) whenever and wherever you see it, especially on talk radio—and make sure you mention <alcoholcanbeagas.com>.
- Dedicated letter writers: Help us flood editors when they permit bad, inaccurate stories about alcohol fuel to run in their publications. We can provide you with text to use in your letters, so we'll have a rapid, intelligent response team to fight back against the American Petroleum Institute propaganda machine.

VOLUNTEER

- When you have one to five hours per week or a chunk of time that you'd like to volunteer, please contact us. We'll give you interesting, educational, and entertaining volunteer work that really will make a difference.

BUY FROM OUR WEB STORE

- Send books and DVDs to your friends.
- Wear our T-shirt.

INFLUENCE INFLUENTIAL PEOPLE

- Send copies of *Alcohol Can Be a Gas!* to influential people for half-price.

ATTEND OR HOST ONE OF DAVE'S WORKSHOPS

- Enjoy all the benefits of Dave's life-changing, full-day, alcohol-fuel-related presentations.
- Tailored to the needs of the audience.

DONATE

- In "cash or kind." Tax-deductible. Financial contributions are always needed, as are materials and human energy.

JOIN US

- Join Alcoholics Unanimous.
- Subscribe to our newsletter.
- Get on our mailing list.

PLUS, OF COURSE

- Start a community-supported energy co-op.
- Ask for ethanol at the pump.
- Convert your car.

SEE THE "GET INVOLVED" MENU AT <WWW.ALCOHOLCANBEAGAS.COM>

- Learn more about ways you can help spread the alcohol fuel revolution.